# Lecture Notes
# in Business Information Processing 464

## Series Editors

Wil van der Aalst ⓘ, *RWTH Aachen University, Aachen, Germany*

Sudha Ram ⓘ, *University of Arizona, Tucson, AZ, USA*

Michael Rosemann ⓘ, *Queensland University of Technology, Brisbane, QLD, Australia*

Clemens Szyperski, *Microsoft Research, Redmond, WA, USA*

Giancarlo Guizzardi ⓘ, *University of Twente, Enschede, The Netherlands*

LNBIP reports state-of-the-art results in areas related to business information systems and industrial application software development – timely, at a high level, and in both printed and electronic form.

The type of material published includes

- Proceedings (published in time for the respective event)
- Postproceedings (consisting of thoroughly revised and/or extended final papers)
- Other edited monographs (such as, for example, project reports or invited volumes)
- Tutorials (coherently integrated collections of lectures given at advanced courses, seminars, schools, etc.)
- Award-winning or exceptional theses

LNBIP is abstracted/indexed in DBLP, EI and Scopus. LNBIP volumes are also submitted for the inclusion in ISI Proceedings.

Maria Papadaki · Paulo Rupino da Cunha ·
Marinos Themistocleous · Klitos Christodoulou
Editors

# Information Systems

19th European, Mediterranean,
and Middle Eastern Conference, EMCIS 2022
Virtual Event, December 21–22, 2022
Proceedings

*Editors*
Maria Papadaki
British University in Dubai
Dubai, United Arab Emirates

Marinos Themistocleous ⓘ
University of Nicosia
Nicosia, Cyprus

Paulo Rupino da Cunha ⓘ
University of Coimbra
Coimbra, Portugal

Klitos Christodoulou
University of Nicosia
Nicosia, Cyprus

ISSN 1865-1348        ISSN 1865-1356 (electronic)
Lecture Notes in Business Information Processing
ISBN 978-3-031-30693-8        ISBN 978-3-031-30694-5 (eBook)
https://doi.org/10.1007/978-3-031-30694-5

This Springer imprint is published by the registered company Springer Nature Switzerland AG
The registered company address is: Gewerbestrasse 11, 6330 Cham, Switzerland

# Preface

The 19th edition of the European, Mediterranean, and Middle Eastern Conference on Information Systems (EMCIS) took place 21–22 December 2022. Although we had originally planned to return to an in-person event, by the time we had to make final decisions, concerning news about air travel disruptions and a potential Covid wave in the Winter led us to maintain EMCIS online for, we hope, one last time. We look forward to celebrating EMCIS's 20th anniversary face-to-face with our community in 2023.

The distance did not prevent an exciting exchange of ideas, though. We have accepted 47 papers from 136 submissions, corresponding to an acceptance rate of 35%, across the following 14 tracks:

- Big Data and Analytics
- Blockchain Technology and Applications
- Cloud Computing
- Digital Services and Social Media
- Digital Governance
- Emerging Computing Technologies and Trends for Business Process Management
- Enterprise Systems
- Information Systems Security and Information Privacy Protection
- Healthcare Information Systems
- Management and Organizational Issues in Information Systems
- IT Governance and Alignment
- Innovative Research Projects
- Metaverse
- Artificial Intelligence

As usual, papers were double-blind reviewed by at least two reviewers. In addition, any submissions made by track chairs were reviewed by a member of the EMCIS Executive Committee and a member of the international committee. Finally, any submissions made by conference chairs were reviewed by two senior external reviewers.

EMCIS proved once more to be truly international, with authors coming from 24 countries and participants from 26 countries as summarized below and sorted by the number of participants:

- Germany
- Greece
- UAE
- Cyprus
- Portugal
- Poland
- France
- USA
- UK

- Finland
- Tunisia
- Sweden
- South Africa
- Austria
- Romania
- Norway
- Italy
- Croatia
- Turkey
- Sudan
- Slovakia
- Kosovo
- Holland
- Estonia
- Bulgaria
- Afghanistan

The papers were accepted for their theoretical and practical excellence and promising results. We hope the readers will find them interesting and consider joining us in the next edition of the conference.

December 2022                                  Maria Papadaki
                                        Paulo Rupino da Cunha
                                       Marinos Themistocleous
                                         Klitos Christodoulou

# Organization

## Conference Chairs

Maria Papadaki      Columbia University, USA and British University in Dubai, UAE

Paulo Rupino da Cunha      University of Coimbra, Portugal

## Conference Executive Committee

Marinos Themistocleous      University of Nicosia, Cyprus
(Program Chair)

Muhammad Kamal      Coventry University, UK
(Publications Chair)

Nikolay Mehandjiev      University of Manchester, UK
(Local Organizing Chair)

Gianluigi Viscusi      Imperial College Business School, London, UK
(Public Relations Chair)

## International Committee

| | |
|---|---|
| Aggeliki Tsohou | Ionian University, Greece |
| Alan Serrano | Brunel University London, UK |
| Andriana Prentza | University of Piraeus, Greece |
| Angeliki Kokkinaki | University of Nicosia, Cyprus |
| António Trigo | Coimbra Business School, Portugal |
| Besart Hajrizi | University of Mitrovica, Kosovo |
| Catarina Ferreira da Silva | University Institute of Lisbon, Portugal |
| Celina M. Olszak | University of Economics in Katowice, Poland |
| Charalampos Alexopoulos | University of the Aegean, Greece |
| Chinello Francesco | Aarhus University, Denmark |
| Elias Iosif | University of Nicosia, Cyprus |
| Ella Kolkowska | Örebro University, Sweden |
| Euripidis N. Loukis | University of the Aegean, Greece |
| Federico Pigni | Grenoble Ecole de Management, France |
| Gianluigi Viscusi | Imperial College Business School, UK |
| Grażyna Paliwoda-Pękosz | Cracow University of Economics, Poland |

| | |
|---|---|
| Paweł Wołoszyn | Cracow University of Economics, Poland |
| Heidi Gautschi | IMD Lausanne, Switzerland |
| Heinz Roland Weistroffer | Virginia Commonwealth University, USA |
| Hemin Jiang | University of Science and Technology of China, China |
| Horst Treiblmaier | Modul University, Vienna, Austria |
| Ibrahim Osman | American University of Beirut, Lebanon |
| Janusz Stal | Cracow University of Economics, Poland |
| Khalid Al Marri | British University in Dubai, UAE |
| Klitos Christodoulou | University of Nicosia, Cyprus |
| Lasse Berntzen | University of South-Eastern Norway, Norway |
| Leonidas Katelaris | University of Nicosia, Cyprus |
| Luning Liu | Harbin Institute of Technology, China |
| Małgorzata Pańkowska | University of Economics in Katowice, Poland |
| Manar Abu Talib | Zayed University, UAE |
| Marijn Janssen | Delft University of Technology, The Netherlands |
| Mariusz Grabowski | Cracow University of Economics, Poland |
| Miguel Mira da Silva | University of Lisbon, Portugal |
| Milena Krumova | Technical University of Sofia, Bulgaria |
| Mohamed Sellami | Télécom SudParis, France |
| Claudio Pacchierotti | University of Rennes, France |
| Paulo Henrique de Souza Bermejo | Universidade Federal de Lavras, Brazil |
| Paulo Melo | University of Coimbra, Portugal |
| Peter Love | Curtin University, Australia |
| Piotr Soja | Cracow University of Economics, Poland |
| Przemysław Lech | University of Gdańsk, Poland |
| Richard Kirkham | University of Manchester, UK |
| Ricardo Jimenes Peris | Universidad Politécnica de Madrid (UPM), Spain |
| Sevgi Özkan | Middle East Technical University, Turkey |
| Slim Kallel | University of Sfax, Tunisia |
| Sofiane Tebboune | Manchester Metropolitan University, UK |
| Soulla Louca | University of Nicosia, Cyprus |
| Tillal Eldabi | Ahlia University, Bahrain |
| Vishanth Weerakkody | University of Bradford, UK |
| Vincenzo Morabito | Bocconi University, Italy |
| Wafi Al-Karaghouli | Brunel University London, UK |
| Walid Gaaloul | Télécom SudParis, France |
| Yannis Charalabidis | University of the Aegean, Greece |

# Contents

## Emerging Computing Technologies and Trends for Business Process Management

## Enterprise Systems

## Information System Security and Information Privacy Protection

**Innovative Research Projects**

**IT Governance and Alignment**

**Management and Organizational Issues in Information Systems**

**Metaverse**

# Artificial Intelligence

# How Blockchain and Artificial Intelligence influence Digital Sovereignty

Martha Klare[✉][iD], Lisa Verlande[iD], Maximilian Greiner[iD],
and Ulrike Lechner[iD]

Universität der Bundeswehr München, 85579 Neubiberg, Germany
{martha.klare,lisa.verlande,maximilian.greiner,ulrike.lechner}@unibw.de

**Abstract.** Digital Sovereignty is an ascending field that is viewed from
a society, politics and enterprise perspective. When considering the situation from an economic standpoint, it is still unclear how to understand the issue of what businesses can do to strengthen their Digital
Sovereignty. Artificial Intelligence and Blockchain are disruptive technologies that have significant impact on Digital Transformation. This
article studies the relationship between Digital Sovereignty and novel
technologies, such as Artificial Intelligence and Blockchain. A quantitative survey with 163 respondents is the empirical basis of the analysis.
We propose seven measures to strengthen Digital Sovereignty: preserve
Data Sovereignty, address concerns and create awareness, define responsibilities, co-create transformation, expand expertise, promote freedom
of choice and measurement criteria. The proposed measures support formulating a Digital Sovereignty strategy to ensure the vision, goals and
requirements for balancing heteronomy and autarky in a self-determined
manner.

**Keywords:** Digital Transformation · Digital Sovereignty ·
Blockchain · Artificial Intelligence

## 1 Introduction

In the context of Digital Transformation, Digital Sovereignty becomes a strategic goal. Considering the critical dependency on disruptive technologies such
as the Internet of Things, Big Data, Virtual Reality, 5G, Artificial Intelligence
(AI), or Blockchain, geopolitical issues and aspects such as independence, self-
determination, trust and credibility are becoming more meaningful. The Digital
Transformation leads to a change in Information Technology (IT) which makes
the consideration of Digital Sovereignty indispensable [16]. An initial approach
by researchers to break down Digital Sovereignty into dimensions leads to the
distinction between the perspectives of state, economics and individuals as well
as the relationships between each other [17]. Glasze et al. [19] and Ciriumaru
[10] discuss challenges for central societal areas such as law, technology and
ethics to express Digital Sovereignty. Researchers such as Fries et al. [17] or
Pohle [30] make the first attempt to transfer the factors and influences of Digital

M. Papadaki et al. (Eds.): EMCIS 2022, LNBIP 464, pp. 3–16, 2023.
https://doi.org/10.1007/978-3-031-30694-5_1

Sovereignty, as well as the linkages between the state, the economy and individuals into logical modeling. From an economic perspective, one way could be the exploration of growing technologies. Two growing technologies in the literature are AI and Blockchain [8,22]. However, considering how to build or strengthen Digital Sovereignty with the selected technologies, it remains unclear how to interpret the answer. While there are many policy proposals for dealing with the promotion of Digital Sovereignty, concrete requirements for key technologies are scarce. Therefore, we decided to explore the impact of Blockchain and AI on Digital Sovereignty. Blockchain seems to contribute meaningfully to Digital Sovereignty due to its decentralization, multiple and distributed authorities, and tamper-resistance [3]. The link between AI research and Digital Sovereignty is obvious when constructing and designing trustworthy, credible and autonomous AI systems [33]. Furthermore, researchers adopting the Information Systems perspective are highly concerned with the issue of Data Sovereignty. For example, open networks [11] or remote evaluation [4] are proposed to ensure an appropriate level of Data Sovereignty. Our approach is to critically examine the impact of AI and Blockchain on Digital Sovereignty in organizations and their relationships. Furthermore, linking the concepts can help identify new potentials, research areas and directions in all three areas (AI, Blockchain and Digital Sovereignty) [20]. Finally, we create a list of measures to develop concrete recommendations for corporate management. Three research questions guide the analysis:

RQ1: What is the companies' state of Digital Sovereignty?
RQ2: How do AI and Blockchain support companies' sense of Digital Sovereignty?
RQ3: How to strengthen Digital Sovereignty?

Guided by these research questions, we investigate to what extent AI or Blockchain supports Digital Sovereignty and how Digital Sovereignty impacts the development and use of these technologies in companies. Therefore, this article contributes to the body of knowledge by outlining the impact of new technologies on enhancing corporate Digital Sovereignty based on empirical data.

This article is structured as follows. First, the research methodology is presented (Sect. 2), followed by the theoretical background (Sect. 3) that establishes a correlation between Blockchain, AI and Digital Sovereignty in terms of Digital Transformation. Section 4 presents the survey results, including the impact of AI and Blockchain on Digital Sovereignty. Finally, the conclusions and an outlook for further research are highlighted (Sect. 5).

## 2    Research Design

For our analysis, we employ a survey method guided by Lehmann et al. [23]. The following steps were taken within the study: determination of target group, creation of questionnaires, 19 conducted pretests, data collection and analysis. The survey uses semi-open and closed questions, one-dimensional scales and

Multiple-Choice-Questions. In addition, some of the questions were filter questions, ensuring that participants completed the questions they were most familiar with but that each question was answered multiple times. First, we carried out a statistical analysis following Muller et al. [28]. Network diagrams are used to depict and communicate our results. Finally, we create clusters to find similarities and proceed to the detailed analysis [28].

Data was collected over two months between July and August of 2022. The survey was conducted in Germany and targeted businesses and decision makers in Germany. We collected 163 person data sets (107 completed data sets) considering an error rate of 10%. The person data sets include 57 from small and medium-sized enterprises and 77 from large companies.

## 3 Theoretical Background

The following section provides definitional delimitation of the terms Digital Sovereignty, Blockchain and AI.

### 3.1 Digital Sovereignty

Digital Sovereignty is a politically motivated goal and is often understood as a strategic balancing act between heteronomy and autarky [13]. States, companies and individuals want to be able to act in the digital world without cutting themselves off from others or becoming too dependent [31]. Thus, factors such as freedom of choice, independence in selecting IT products and the security of one's own country, citizen, or company play an important role in becoming more digitally sovereign [27]. Digital Sovereignty is not a fundamental value on its own. It is part of the values of liberal democracy, open competition policy and efforts toward greater digital sustainability, especially in German research [18]. Digital Sovereignty can help drive Digital Transformation by clarifying as state, business or individual how they want to use hardware or software and capabilities of others [27, 29]. In the digital sense, this means that politicians, researchers and decision makers in companies should increasingly ask themselves what technical, organizational and conceptual requirements for key technologies are needed to become truly more independent [12]. In the following section, the current state of research on Digital Sovereignty is expanded by subjecting two key technologies (Blockchain and AI) to the particular analysis. To broaden the understanding, information is provided on the extent company members believe that the implementation of AI or Blockchain can help strengthen Digital Sovereignty.

### 3.2 Blockchain in the Digital Transformation

Blockchain is a kind of Distributed Ledger Technology and can be seen as the standard for digital transactions as B2B and B2C companies increasingly shift to the digital market [25]. Blockchain is seen as an accelerator of business processes, enabling reliable and secure transactions in various business areas to be fully automated [1]. Because of increasing networking and connectivity, the technology allows agreements to be set up largely forgery-proof and in a transparent manner [5]. For us, Blockchain is a technical solution to manage data in a distributed infrastructure without a central instance in a traceable and tamper-proof way by consensus [36]. Moreover, Blockchain allows verifying transactions without a central instance in a transparent and trustworthy process [7]. Christidis and Devetsikiotis [9] proposed four main benefits of Blockchain technology: tolerance of node failures, a single view of events, data ownership without a central authority and transparent, verifiable, predictable as well as audible activities. Given these elements, we can assume a solid connection to Digital Sovereignty, as issues of ownership, transparency, trust, ethics and authority play a particular role. As outlined, the field of Blockchain and Digital Sovereignty holds significant research potential. To understand the current situation and derive assumptions, it is critical to explore recent adoptions, the state of the technology and the potential impact on Digital Sovereignty.

### 3.3   Artificial Intelligence in the Digital Transformation

As a result of digitalization, AI is experiencing a resurgence within research and various industries. This renewed upswing is due to the technological drivers within the Digital Transformation: improved computing capacities, faster connections and larger connectivity areas enable the usability of AI systems on a larger scale [6]. In this study, we define AI as a system that can make predictions and decisions based on algorithms and existing data sources through learning. AI algorithms focus on mapping three essential human cognitive abilities - learning, reasoning and self-reflection [32]. However, AI in the context of Digital Transformation also has much potential for debate. Particularly around ethical issues, risks, credibility, trust and even fairness, the question arises regarding how to achieve these. At these points, we see the connection to Digital Sovereignty, which becomes even more apparent when we look at the guiding principles of the European Union's AI Expert Group [15]: respect for human autonomy, harm avoidance, fairness and explainability. A deeper insight into the current use, purpose, obstacles and concerns of AI systems in enterprises is therefore crucial to set up AI systems for the future in a more digitally sovereign way. So, it is critical to uncover and profitably leverage the interfaces and goals between Digital Sovereignty and AI. One way to achieve more Digital Sovereignty in AI is through the recent machine learning approach called Federated Learning (FL), which we will discuss in our paper in Sect. 4.3.

## 4   Results

As the debate on Digital Sovereignty is still young, it is important to determine the level of awareness of the topic of Digital Sovereignty in companies and among decision makers. Therefore, we want to use the survey data to determine what measures have already been taken in connection with Digital Sovereignty.

## 4.1 Towards Digital Sovereignty in Enterprises

This section presents, analyzes and contextualizes the findings on the topic of Digital Sovereignty.

**Current State of Digital Sovereignty.** Employees and decision makers in companies are grappling with the idea of Digital Sovereignty. 79% of respondents state that they are aware of the topic and the interest in Sovereignty is strong to very strong at 67% (Fig. 1).

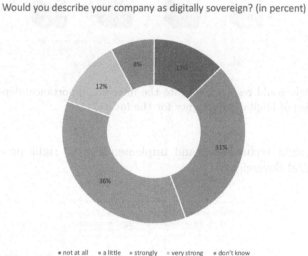

Would you describe your company as digitally sovereign? (in percent)

■ not at all ■ a little ■ strongly ■ very strong ■ don't know

**Fig. 1.** Would you describe your company as digitally sovereign?

The importance of the topic to the companies is slightly lower. However, 60% of people rate the importance high to very high (Fig. 2).

We asked how decision makers compare their state of Digital Sovereignty towards the competition and the result showed that the majority see themselves in the midfield compared with other companies. When asked which areas in companies are seen as already digitally sovereign, it does not seem clear to the respondents where Digital Sovereignty plays a central role in the company and why.

**Low Sovereignty vs. High Sovereignty in Companies.** The respondents were divided into two clusters based on their self-assessment regarding Digital Sovereignty in the company (group 1: low Digital Sovereignty, group 2: high Digital Sovereignty). It is striking that the proportion of companies from group 1 see low Data Sovereignty as the main reason for this (84%) and those who rate Digital Sovereignty in their own company as high (group 2), in turn, see accountability (31%), their systems (31%) and transparency (14%) as the reasons (Fig. 3). The results are consistent with Blossfeld et al. [2], who see selecting the

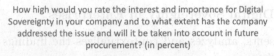

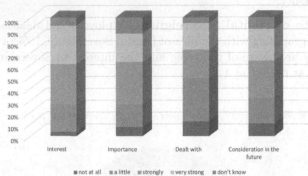

**Fig. 2.** How high would respondents rate the interest, importance, dependencies and the consideration of Digital Sovereignty for the future?

right people, right technologies and implementing the right processes as key drivers of Digital Sovereignty.

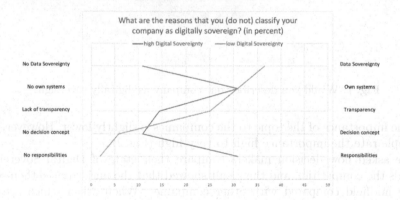

**Fig. 3.** What are the reasons that you (do not) classify your company as digitally sovereign?

**Lessons from the Past for Recommendations for Future Action.** We asked the respondents who considered their company to be digitally sovereign what measures their company has used in the past to strengthen its Digital Sovereignty. The results show that 30% of respondents have strengthened their Digital Sovereignty in the past by using vendor-independent modules. In addition, co-designing IT solutions (26%) and creating a strategy (13%) have also helped companies. Still, 63% of respondents said that their company had not paid

Is there a possibility of strengthening Digital Sovereignty with the following means?
(in percent)

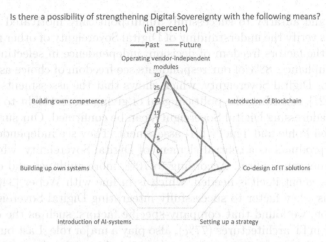

**Fig. 4.** Means of strengthening Digital Sovereignty

attention to ensuring that new technologies were digitally sovereign in their procurement in the past (Fig. 4). Now all participants were again asked to give their assessment. We asked what means they believe they can use to strengthen Digital Sovereignty in the future. According to the respondents, the goal of strengthening Digital Sovereignty in the future should be to gain more control over one's data, i.e., Data Sovereignty. Thus, 89% of respondents see the topic of Data Sovereignty as (very) important for strengthening Digital Sovereignty. The surveyed persons state that promoting research and development (34%) can significantly help companies to enhance Digital Sovereignty. In the future, employees and decision makers would use vendor-independent modules to strengthen their company's Digital Sovereignty (27%). Building and using Blockchain technology also appears to be a way to increase Digital Sovereignty in the future (up 6% points from the past). The importance of AI in enhancing Digital Sovereignty has increased by 5% points in the future compared to the past. Respondents believe that using strategic guidelines will help them better understand the complex subject (13%). As an alternative to the classic AI system, FL is seen by most respondents as a positive approach to strengthening Digital Sovereignty because only 10% denied that FL would help.

Obstacles to strengthening Digital Sovereignty, according to the respondents, are the main challenges in adapting the company's internal IT infrastructure (27%), information deficits (19%), lack of responsibilities (19%) and financing additional costs. The state is expected to support research and development (34%), more interoperability in national products (22%) and the development of a national data strategy (13%) to guide companies. Some respondents also mention expanding greater collaboration with EU countries (12%), more training opportunities (10%) and an open source strategy (5%).

**Expanding the Understanding of Digital Sovereignty.** As mentioned before, we understand Digital Sovereignty as a strategic balancing act between heteronomy and autarky to successfully fulfill politically motivated goals. Our survey results verify the understanding of Digital Sovereignty of other researchers in regard to the factors freedom of decision, independence in selecting IT products, and compliance. 88% of our respondents see freedom of choice as an important factor for Digital Sovereignty, which shows that the assessments by Kagermann et al. [21], regarding the political goal of giving more weight to freedom of choice when addressing Digital Sovereignty, can be confirmed. Our survey results also supported Pohle and Thiel's [31] assessment. They see independence in the choice of IT products as a central element of Digital Sovereignty, which 77% of the respondents can confirm. According to 70%, more guidance and compliance by the management itself is needed, which is in line with Weber [34], who sees compliance as a key factor to successfully integrating Digital Sovereignty.

In addition, we found that company-specific factors, such as the creation of transparency in IT architectures (77%), also play a major role. Last but not least, the decision makers see collaboration with stakeholders who share similar values (58%) and accessibility to the companies' source codes (57%) as prerequisites for strengthening Digital Sovereignty.

### 4.2 Towards Artificial Intelligence and Blockchain in Enterprises

Regarding Blockchain and AI, we wanted to discover the awareness for the two technologies, their use cases, properties and purposes. In addition, respondents were asked whether their company works with the technologies and whether they would say Blockchain and AI would support Digital Sovereignty.

**Blockchain.** According to our survey, it can be assumed that 74% is not aware of Blockchain. In addition, we find out that many respondents are not sure whether their company invests in Blockchain projects. Thus, 30% stated that they do not know the status of Blockchain implementation in their company, while 33% noted that there are implemented and planned projects. This data can be directly related to Blockchain familiarity. The following three potential applications remain the most widely perceived: real-time certificate management (21%), cryptocurrencies (19%) and sensor data collection (16%). Blockchain, then, is best known for its tamper resistance (25%), the immutability of data (28%) due to the hash function, resilience (15%) and transparency due to the decentralized storage of information (15%).

**Artificial Intelligence.** Although AI is already used in many areas of daily life, more than 3/4 of respondents still say they have little or no knowledge about AI. As shown by the results, the best-known applications are intelligent systems (25%), voice control (15%) and text recognition (13%). However, it can be assumed here that while AI is known as a term, there is a significant lack of understanding of how AI works. In addition, 39% of respondents indicated that their company had implemented projects involving AI and another 21% of companies were already in the project planning phase. So, more than half of the respondents have already been confronted with AI in their daily work.

The cause for failure of Blockchain and AI projects has also been investigated. Although both technologies serve different purposes and are used in different areas, the picture is almost identical with regard to the question of what could be causes for failure in technology development (Fig. 5). Participants indicated that lack of expertise is the most common reason Blockchain or AI adoption fails. While interest in adopting AI and Blockchain is considered high, additional reasons include additional cost, legal concerns, implementation difficulty, lack of infrastructure and lack of standards (Fig. 5). Note that when asked about ways to strengthen Digital Sovereignty, Blockchain and AI are below 20% for both past and future (Fig. 4). Still, a later question revealed an approval rate of 54% (Blockchain) and 69% (AI) regarding positive support for Digital Sovereignty. The findings show a low understanding of how to strength strategic goals and underscore the importance of awareness and clear measurement criteria.

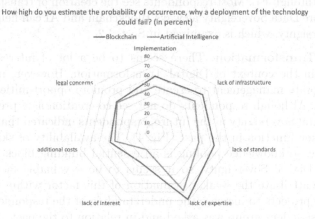

**Fig. 5.** Criteria of Blockchain and AI

## 4.3 Implications and Discussions of Artificial Intelligence and Blockchain for Digital Sovereignty

Digital Sovereignty is a strategic goal. To work towards this strategic goal, based on our data, we see the need to address the adoption of Blockchain and Federated Learning (Sect. 4.1). In addition, we have developed seven measures for consideration in a Digital Sovereignty strategy that pay attention to understanding, acknowledging and best implementing the vision, goals and requirements for a digitally sovereign enterprise. We use clusters and network diagrams to restructure the results, create codes and compare them to data from literature. We then further explored the data in two rounds of workshops within the research group, the first round consisting of the elementary collection of unstructured measures and the second for refinement and alignment. In doing so, our data analysis revealed that the following measures are relevant for future approaches.

**Preserving Data Sovereignty.** Preserving Data Sovereignty is a fundamental goal for respondents' approach to Digital Sovereignty (Fig. 3). FL is a new, decentralized approach first named by McMahan et al. [26], which shares models rather than data to protect user privacy. Over 90% of respondents can envision using FL instead of current machine learning methods to maintain Data Sovereignty (Sect. 4.1). Blockchain can also support Data Sovereignty due to the tamper resistance and immutability of the data. At the same time, one advantage of implementing Blockchain is the visibility that can be created within a supply chain, for example. While respondents cannot imagine sharing information with competing companies and thus losing a competitive advantage, sharing with suppliers and customers is entirely conceivable (Sect. 4.1). Blockchain can thus help create a secure and sovereign enterprise while maintaining Data Sovereignty. As Lu et al. [24] suggest, coupling the two technologies can have a game-changing effect by securing communications through Blockchain while maintaining Data Sovereignty through FL. Most respondents see the creation of transparency, fairness, and more Data Sovereignty as ways Blockchain and AI can help strengthen Digital Sovereignty, which is examined in Sect. 4.2.

**Co-create Transformation.** There seems to be a lot of interest in Digital Sovereignty in the context of Digital Transformation. However, it is also the task of corporate management to create and promote opportunities for further development. Although respondents do not see co-creation as a prerequisite for creating Digital Sovereignty in the future, respondents indicated that co-creation had been a key function in the past (Fig. 4). The availability of skilled labor is limited today, so knowledge pooling is an essential building block for transformation and Digital Sovereignty, so it seems to be a valuable factor for later projects. We attribute the weaker evaluation of this factor within the question about future projects to increase the understanding of the respondents, as here, the entire respondent group was asked and in relation to the past, only the persons who previously stated that projects have already been implemented (Fig. 4). Technologies like FL and Blockchain can support this journey by making collaborations secure and transparent while preserving an organization's expertise. In addition, realigning processes, adapting roles, breaking down old structures and forging new paths are critical governance milestones that help transform companies into digitally sovereign enterprises. This contrasts with an increased self-sufficiency idea that dominates the discussion in Dreo et al. [14].

**Expand Expertise.** The lack of expertise is a significant obstacle to successfully implementing a digitally sovereign strategy and incorporating Blockchain and AI (Fig. 5). Here, with company-wide sensitization, it is necessary to prepare specialists for the changes. This includes additional training and education measures, direct involvement in the ongoing structural change and creating a sovereignty strategy. In addition, this may involve encouraging skilled personnel to engage in their research and development as resources allow. Designing an open communication architecture, not only internally but also to external partners, research institutions and communities, as well as being open and flexible to new ideas and willing to implement them. The emphasis is on active listening, especially for management.

**Encourage Freedom of Choice.** An important aspect of Digital Sovereignty is freedom of choice. When selecting services, products and systems, freedom of choice must continue to be promoted in the market. Monopolies or limited choice are barriers to leveraging Digital Sovereignty, as the respondents indicated in last part of Sect. 4.1. Integrating FL and Blockchain seems essential in supporting Digital Sovereignty by increasing the flexibility and interconnectivity of services as they imply a certain degree of autonomy and data protection.

**Define Responsibilities.** According to the statements of the respondents who already consider their company to be digitally sovereign, creating clear responsibilities is the key to a more digitally sovereign company (Fig. 1). The creation of responsibilities, roles and decision-making concepts is already discussed in governance approaches like Weill and Ross [35] and is seen as critical to the success of companies. In this way, information deficits, such as those that have been identified as barriers to the successful incorporation of Blockchain and AI, can be reduced (Fig. 5).

**Eliminate Concerns and Create Awareness.** To successfully implement a sovereignty strategy, concerns and less knowledge must be dispelled by a clearer picture of what Digital Sovereignty is and how it can be integrated. Digital Sovereignty, Blockchain and AI are all highly rated in terms of interest and significance by the respondents. In this context, more than half of the respondents even stated that they would describe their company as digitally sovereign, yet most respondents rate their familiarity with Digital Sovereignty as low to moderate (Fig. 2). At this point, it is clear that respondents perceive the term as value-creating and relevant but are unaware of what they associate with it and what the focus is in adoption and implementation. The gap between the respondents' importance and sense of Digital Sovereignty shows that creating and training employees can help to expand awareness and strengthen Digital Sovereignty in companies. Furthermore, raising awareness is crucial for successful integration when introducing Blockchain and AI, as the survey also showed knowledge gaps in this area (Sect. 4.2). Here, it is up to the research community to raise awareness of the technology by highlighting its added value.

**Measurement Criteria.** Within the previous sections, we could detect a certain skepticism and ignorance among the respondents. To mitigate this, we proposed awareness as one measure to create trust. Another measure could be the creation of measurement criteria. Political goals currently overshadow the concept of Digital Sovereignty. More business measurement approaches are needed to quantify the achievement of goals regarding this topic. Measurable and controllable criteria can help to make Digital Sovereignty more tangible and thus facilitate implementation for companies.

# 5    Conclusion and Outlook

Our findings show that the surveyed companies feel that Blockchain and AI - especially Federated Learning - can help strengthen Digital Sovereignty. Our paper extends the existing body of knowledge in that we conducted preliminary research on the impact of Blockchain and AI in the context of Digital Sovereignty. We expanded our understanding of Digital Sovereignty by identifying requirements for raising Digital Sovereignty from companies' perspective, making technological recommendations and proposing seven measures based on our data. From the survey and comparison with existing literature, we were able to formulate key requirements for a strategy to strengthen Digital Sovereignty. First, it is critical to develop awareness of data ownership and derive how to maintain control over data. In addition, companies need to make their employees more aware of Digital Sovereignty and Digital Transformation. This includes assigning roles and responsibilities within corporate governance to set a company up for Sovereignty. To act with digital confidence, it is also necessary to rethink and restructure outdated processes, roles and structures. Another measure to consider is the development and expansion of know-how. Additionally, selecting services, products, or systems is crucial to maintaining autonomy and protecting sensitive data. To establish tangible Digital Sovereignty for companies, there is a need for measurement criteria that can provide orientation for design. Finally, the main goal should be to formulate a strategy to ensure the comprehension of visions, goals and requirements for the specific company environment.

Based on the results, developing targeted measurement criteria for Digital Sovereignty and creating an initial strategy paper for companies are approaches that require further description, explanation and validation based on concrete design studies within further research projects.

**Acknowledgments.** The Project LIONS is funded by dtec.bw - Digitalization and Technology Research Center of the Bundeswehr which we gratefully acknowledge. We would also like to thank Bayern Innovativ for funding the project "Federated Learning Enhancing IT Security (FLEIS)" (DIK 0241_2104_0080). We thank our partners and fellow researchers and all study participants.

*Contributions* can be attributed to the coauthors as described below:

– Martha Klare addressed the volunteers with the survey and wrote the results section on Digital Sovereignty in Enterprises (4.1).

– Lisa Verlande wrote the section Towards Artificial Intelligence and Blockchain in Enterprises (4.2) and the Implications and Discussions of Artificial Intelligence and Blockchain for Digital Sovereignty (4.3).

– Maximilian Greiner added implications regarding the Discussions on Blockchain for Digital Sovereignty (4.3).

– The authors jointly designed the study and co-authored the introduction (Sect. 1), methodology (Sect. 2), theoretical background (Sect. 3) and conclusion (Sect. 5).

– Ulrike Lechner contributed to the research design, data collection, interpretation and revisions of the article.

# References

1. Akter, S., Michael, K., Uddin, M.R., McCarthy, G., Rahman, M.: Transforming business using digital innovations: the application of AI, blockchain, cloud and data analytics. Ann. Oper. Res. **308**(1), 7–39 (2020). https://doi.org/10.1007/s10479-020-03620-w
2. Blossfeld, H.P., et al.: Digitale Souveränität und Bildung. Waxmann, Gutachten. Münster (2018)
3. Bons, R.W.H., Versendaal, J., Zavolokina, L., Shi, W.L.: Potential and limits of Blockchain technology for networked businesses. Electron. Mark. **30**(2), 189–194 (2020). https://doi.org/10.1007/s12525-020-00421-8
4. Bruckner, F., Howar, F.: Utilizing remote evaluation for providing data sovereignty in data-sharing ecosystems. In: 54th Hawaii International Conference on System Sciences (2021). https://doi.org/10.24251/HICSS.2021.842
5. Buhl, H.U., Schweizer, A., Urbach, N.: Blockchain-Technologie als Schlüssel für die Zukunft. Zeitschrift für das gesamte Kreditwesen **12** (2017)
6. Calp, M.H.: The role of artificial intelligence within the scope of digital transformation in enterprises. In: Advanced MIS and Digital Transformation for Increased Creativity and Innovation in Business. IGI Global (2020)
7. Carayannis, E.G., Christodoulou, K., Christodoulou, P., Chatzichristofis, S.A., Zinonos, Z.: Known unknowns in an era of technological and viral disruptions-implications for theory, policy, and practice. J. Knowl. Econ. **13**(1), 587–610 (2022)
8. Chatterjee, R., Chatterjee, R.: An overview of the emerging technology: blockchain. In: 3rd International Conference on Computational Intelligence and Networks (CINE). IEEE (2017)
9. Christidis, K., Devetsikiotis, M.: Blockchains and smart contracts for the Internet of Things. IEEE Access **4** (2016). https://doi.org/10.1109/ACCESS.2016.2566339
10. Circiumaru, A.: The EU's digital sovereignty-the role of artificial intelligence and competition policy. SSRN: 3831815 (2021)
11. Dalmolen, S., Bastiaansen, H., Kollenstart, M., Punter, M.: Infrastructural sovereignty over agreement and transaction data ('metadata') in an open network-model for multilateral sharing of sensitive data. In: 40th International Conference on Information Systems, ICIS 2019. Association for Information Systems (2020)
12. Bundesregierung, D.: Datenstrategie der Bundesregierung: Eine Innovationsstrategie für gesellschaftlichen Fortschritt und nachhaltiges Wachstum. Bundeskanzleramt, Berlin (2021)
13. Diekmann, G.: Digitale Souveränität: Positionsbestimmung und erste Handlungsempfehlungen für Deutschland und Europa. Bitkom, Berlin (2015)
14. Dreo, G., Eiseler, V., Gehrke, W., Helmbrecht, U., Hommel, W., Zahn, J., et al.: Europäische Digitale Souveränität: Weg zum Erfolg?-Ein Bericht zur Jahrestagung CODE 2020. Zeitschrift für Außen-und Sicherheitspolitik **13**(4), 399–404 (2020)
15. Europäische Kommission, Generaldirektion Kommunikationsnetze, Inhalte und Technologien Kommission: Ethikleitlinien für eine vertrauenswürdige KI. Publications Office (2019). https://doi.org/data.europa.eu/doi/10.2759/22710
16. Friedrichsen, M., Bisa, P.J. (eds.): Digitale Souveränität. Springer Fachmedien Wiesbaden, Wiesbaden (2016). https://doi.org/10.1007/978-3-658-07349-7, http://link.springer.com/10.1007/978-3-658-07349-7
17. Fries, I., Greiner, M., Hofmeier, M., Hrestic, R., Lechner, U., Wendeborn, T.: Towards a layer model for digital sovereignty: a holistic approach. In: Proceedings of the 17th International Conference on Critical Information Infrastructures Security (CRITIS 2022). Springer (2022). in preparation

18. Fritzsche, K., Pohle, J., Bauer, S., Haenel, F., Eichbaum, F.: Digitalisierung nachhaltig und souverän gestalten. DINA, Positionspapier, CO (2022)
19. Glasze, G., Odzuck, E., Staples, R. (eds.): Was heißt digitale Souveränität?: Diskurse, Praktiken und Voraussetzungen »individueller« und »staatlicher Souveränität« im digitalen Zeitalter. Transcript Verlag (2022)
20. Hartmann, E.A.: Digitale Souveränität: Soziotechnische Bewertung und Gestaltung von Anwendungen algorithmischer Systeme. In: Digitalisierung souverän gestalten II. Springer Vieweg, Berlin, Heidelberg (2022)
21. Kagermann, H., Streibich, K.H.: Digitale Souveränität - Status quo und Handlungsfelder. acatech IMPULS (2021)
22. Lee, J., Suh, T., Roy, D., Baucus, M.: Emerging technology and business model innovation: the case of artificial intelligence. J. Open Innov. Technol. Market Complex. **5**(3), 44 (2019)
23. Lehmann, D.R., Gupta, S., Steckel, J.H., Gupta, S.: Marketing Research. Addison-Wesley Reading, Boston (1998)
24. Lu, Y., Huang, X., Dai, Y., Maharjan, S., Zhang, Y.: Blockchain and federated learning for privacy-preserved data sharing in industrial IoT. IEEE Trans. Ind. Inf. **16**(6), 4177–4186 (2019)
25. Massaro, M.: Digital transformation in the healthcare sector through blockchain technology. Insights Acad. Res. Bus. Dev. Technovation, 102386 (2021)
26. McMahan, B., Moore, E., Ramage, D., Hampson, S., Arcas, B.A.: Communication-efficient learning of deep networks from decentralized data. In: Artificial Intelligence and Statistics. PMLR (2017)
27. Merkel, A., Frederiksen, M., Marin, S., Kallas, K.: Letter to the President-Digital Sovereignty. European Commission, Berlin (2021)
28. Müller-Böling, D., Klandt, H.: Methoden empirischer Wirtschafts-und Sozialforschung. Eine Einführung mit wirtschaftswissenschaftlichem Schwerpunkt **3** (1996)
29. Philipp, U.: Schriftliche Frage an die Bundesregierung (2022). https://mdb.anke.domscheit-berg.de/wp-content/uploads/2022/03/1-76-Domscheit-BergGeschwaerzt.pdf
30. Pohle, J.: Digital sovereignty. a new key concept of digital policy in Germany and Europe. Konrad-Adenauer-Stiftung, Berlin (2020)
31. Pohle, J., Thiel, T.: Digitale Souveränität – Von der Karriere eines einenden und doch problematischen Konzepts. In: Der Wert der Digitalisierung: Gemeinwohl in der digitalen Welt. Bielefeld: Transcript Verlag (2021)
32. Shinde, P.P., Shah, S.: A review of machine learning and deep learning applications. In: 4th International Conference on Computing Communication Control and Automation (ICCUBEA). IEEE (2018)
33. Vogt, R.: Digitale Souveränität und Künstliche Intelligenz für den Menschen. In: Digitalisierung souverän gestalten. Springer Vieweg, Berlin (2021)
34. Weber, H.: Digitale Souveränität. Informatik Spektrum 1–12 (2022)
35. Weill, P., Ross, J.W.: IT Governance: How Top Performers Manage IT Decision Rights for Superior Results. Harvard Business Press, Boston (2004)
36. Xu, M., Chen, X., Kou, G.: A systematic review of blockchain. Fin. Innov. **5**(1), 1–14 (2019). https://doi.org/10.1186/s40854-019-0147-z

# The Concept of a New Neural Map for Clustering, Data Visualization and Prediction with Probability Distribution Approximation

Janusz Morajda(⊠)

Krakow University of Economics, Krakow, Poland
morajdaj@uek.krakow.pl

**Abstract.** The paper submits a proposition of a new data analysis tool (named PDCM – *Probability Distribution generating and Clustering Maps*) constructed in the form of an extended neural map and dedicated to a variety of tasks, such as specific clustering, visualization, prediction (together with its possible visual analysis and justification) and generation of approximated Bayesian *a posteriori* probability density distribution for dependent variable. Basic theoretical aspects concerning the structure and training process of the proposed model have been presented. Also, research involving application of the PDCM method for real estate market data (Boston dataset) has been shown together with promising research results and conclusions.

**Keywords:** neural maps · neural networks · SOM · Self-Organizing Prediction Maps · data analysis · data exploration · clustering · prediction · data visualization · probability distribution · management · real estate

## 1 Introduction

In data analysis and data mining processes, various models can be constructed depending on the goal of data processing. Frequently considered tasks concern:

a) data clustering – aimed to explain group structure of data patterns,
b) data visualization – in order to better understanding of (multidimensional) data sets,
c) prediction of values or classes of certain (dependent) variable for new patterns,
d) justification/explanation of such predictions.

These tasks (when performed on the same dataset for a given problem) are usually executed by different and unrelated tools (dedicated to specific, separate kinds of issues), what can make interpretation difficulties for a researcher or practitioner during data analyses. For example, such diverse techniques as:

- k-means classifiers, hierarchical clustering methods, or Kohonen neural networks (Self Organizing Maps - SOM) [8] are intended exclusively to clustering problems

© The Author(s), under exclusive license to Springer Nature Switzerland AG 2023
M. Papadaki et al. (Eds.): EMCIS 2022, LNBIP 464, pp. 17–32, 2023.
https://doi.org/10.1007/978-3-031-30694-5_2

[28]; some of them are also equipped with visualisation capabilities (i.e. dendrograms in hierarchical methods, or 2-dimensional groups presentation on a plane (map) in case of Kohonen SOMs).

- all linear and nonlinear regression tools, and all machine learning regression techniques (e.g. perceptrons, GRNN networks and others) are aimed to solve prediction problems concerning forecasting/evaluation of real values of a dependent variable,
- all pattern classifiers (like decision trees, k-nearest neighbour method, naive Bayes classifier, specific types of neural networks, etc.) are dedicated to assigning proper class (from a previously approved finite set of classes) to a new (multidimensional) pattern (vector of features).

In paper [16] Morajda and Paliwoda-Pękosz presented a concept of Self-Organizing Prediction Map (SOPM) that is a kind of a neural map based on standard Kohonen SOM network (see Sect. 2), and links (within one tool) all tasks a) b) c) d) listed above. The SOPM model is described here in Sect. 3.

The main goal of present article is introducing a concept of another neural map – an extension of SOPM, which apart from tasks a) b) c) d) generates the probability distributions for predictions of dependent variables for new patterns. This new model is named PDCM (from *Probability Distribution generating and Clustering Maps*). The idea of PDCM (as certain enhancement of SOPM) is presented in Sect. 4. Section 5 submits a research concerning application of proposed PDCM model in real estate data analysis.

## 2   Neural Maps – Literature Background

Kohonen neural network, particularly Self-Organizing Map (SOM) was proposed by T. Kohonen (see e.g. [8]) and has been accepted as a basic type of a neural map. In general SOMs are devoted to solve clustering problems (i.e. identification of groups of similar objects treated as multivariate vectors of features) with additional possibility of visualization of recognized groups in the plane (2-dimensional map). They perform the process of cluster analysis (patterns grouping) with the mapping of groups existing in a multidimensional feature space onto a two-dimensional map (rectangular structure of neurons in a plane), with maintaining topology of group distribution. Signals delivered by neurons placed in the rectangular map can then be analysed numerically and can be used for construction of graphs that visualize clusters arrangement.

Many publications report usefulness of these tools in various domains, e.g. in genetics (gene data clustering [20]); chemistry (antioxidants classification in biodiesel [7]); computer systems security (network intrusion detection [12]) and others.

A great many research works show usefulness of SOMs in management (e.g. in decision-support processes) and in business/economic data analysis. Such applications may particularly concern: business failure prediction [27], city infrastructure management [10], waste management [21], company performance analysis and clustering of companies [3], technological processes designing [25], analysis of social media with utilization in tourism businesses [9], generating of transaction strategies in financial markets [15], identification of bank risk profiles and failure prediction [24], detection of tax evasion [1], protected area management [5], water resources management [2, 22], corporate behaviours analysis [6], and others.

Numerous modifications of original SOM networks have been proposed in literature, let us show here only a small sample of various published postulates: a structure composed of many hierarchically (layered) SOM maps, named HSOM, was proposed in [19], the SOM model, in which the coordinates of neurons on the map are not constant, but are subject to dynamic changes during the learning process, was proposed in [14]; multilayer, hierarchical neural architectures based on Kohonen networks implementing clustering, used to recognize certain types of images were proposed in [11]. A good review of various variants of SOM has been presented by Moshou in monography [18].

## 3 Self-Organizing Prediction Maps (SOPM)

Original SOM model and its derivative tools, dedicated to clustering and its visualization, are usually not applicable in prediction problems. In turn, prediction neural networks are not equipped with explanation and/or visualization capabilities. A certain solution to these problems is the concept of Self-Organizing Prediction Map (SOPM) – as a modification of SOMs – proposed by Morajda and Paliwoda-Pękosz in paper [16] (certain modifications of SOPM, called FLOPM, has also been submitted by the same authors in [17]). SOPMs enable:

- clustering of available (used for model training) patterns (features vectors) with special respect to a selected feature of special meaning (denoted here by $\chi$)
- visualization of clustering results in the special map,
- making predictions of the special feature $\chi$ for new patterns, together with numerical and visual analysis of these predictions.

The basic assumption for SOPM is that a selected variable $\chi$ (one of features $x_i$ describing each pattern included into a dataset undergoing analysis) has a special (key) meaning in a data mining process or is accepted as a dependent variable in prediction task. In clustering of patterns from a given dataset, the research inquiry can involve recognition and visualization of clusters' arrangement with separate, particular consideration of variable $\chi$. Consequently in SOPM models, the idea of modification (in relation to SOM) of the projection between the multidimensional feature space (the analysed patterns are positioned in) and set of neurons placed in a rectangular $XY$ map, relies on following rules (see Fig. 1):

a)  the key variable $\chi$ is projected only on the coordinate $Y$,
b)  other variables (features) $x_j$ $(j = 1, 2, ..., n; x_j \neq \chi)$ are mapped only on the coordinate $X$.

The projection is realized by a special training algorithm using analysed dataset.

If the variable $\chi$ is qualitative and expressed on the ordinal scale with finite set of ordered values, then the projection $\chi$ onto $Y$ according to the rule a) is simple: subsequent rows in the SOPM map represent subsequent ordered values of $\chi$ (number of rows is equal to number of $\chi$ values).

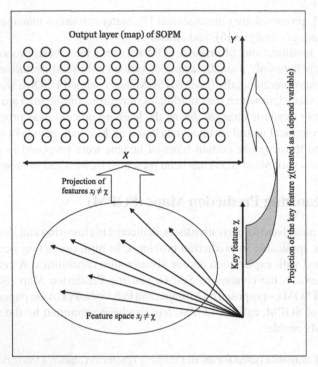

**Fig. 1.** Scheme of mapping of a multidimensional feature space onto the rectangular layer of neurons in SOPM model (small circles represent positions of neurons in the map). *Source:* [16]

Let us now assume that variable $\chi$ is continuous, i.e. takes real values from certain range $D \subset \mathfrak{R}$. Let $N$ denotes the number of rows of neurons in SOPM. The projection $\chi$ onto $Y$ is then executed as follows:

- all values of $\chi$ from training dataset are sorted into an ascending sequence, which then is cut into $N$ equally numerous subsequences,
- each subsequence determines certain range $R_i$ ($i = 1, 2,..., N$) of $\chi$ values, so that for any $i$ each value from $R_i$ is not greater than any value from $R_{i+1}$, and $R_1 \cup R_2 \cup ... \cup R_N = D$,
- ranges $R_i$ ($i = 1, 2,..., N$) are assigned to subsequent rows of SOPM, in turn these rows have numerical coordinates $y_i$ ($i = 1, 2,..., N$) on $Y$ axis, where $y_i$ is the centre of $R_i$,
- consequently each value of variable $\chi$ can be assigned to a certain row (range $R_i$) and finally projected to respective value $y_i$.

The proposed training algorithm of SOPM is a simple modification of well-known SOM training procedure (see e.g. [8]), adjusted to the above mentioned concept of SOPM data mapping as follows:

**Step 1.** Consider a set $\{(\mathbf{x}_p, \chi_p), p = 1, 2, ..., J\}$ as a training dataset (representing analysed phenomenon), where $\mathbf{x}_p$ is a vector of features (variables) $\neq \chi_p$

**Step 2.** $p \leftarrow 1$

**Step 3.** Deliver $\mathbf{x}_p$ to the SOPM input; find the row of neurons corresponding to $\chi_p$

**Step 4.** In the selected row find a neuron generating the lowest signal (as a distance between $\mathbf{x}_i$ and neuron's weight vector $\mathbf{w}$) and approve it as a *winning* neuron $n_p$

**Step 5.** In the SOPM output map determine the neighbourhood for (around) $n_p$

**Step 6.** Adjust weights $\mathbf{w}$ (by adding $\Delta\mathbf{w}$ to $\mathbf{w}$) for all neurons from this neighbourhood according to the rule:

$$\Delta\mathbf{w} = \eta \cdot s(n_m) \cdot (\mathbf{x}_p - \mathbf{w}) \tag{1}$$

where $\eta$ is a learning coefficient $(0 < \eta < 1)$, and $s(n_m)$, where $0 < s(n_m) \leq 1$ and $s(n_p) = 1$, is the value of neighbourhood function for a being trained neuron $n_m$ belonging to the determined neighbourhood of $n_p$

**Step 7.** $p \leftarrow p + 1$; if $p \leq J$ go to **step 3**, otherwise go to **step 8**

**Step 8.** If the end-of-training condition is not fulfilled go to **step 2**, otherwise **stop**.

It should be noted that main and crucial difference between SOPM training algorithm and the classical training procedure for Kohonen's SOM relies on constraint of selection the winning neuron (and then its neighbourhood) from strictly selected row of map neurons, i.e. the row corresponding to the current value of $\chi_p$. It should also be noted that a classic SOM training algorithm is fully unsupervised, but in SOPM this procedure is mixed: supervised as regards the key variable $\chi$, and unsupervised with respect to all other features (see [16] for details).

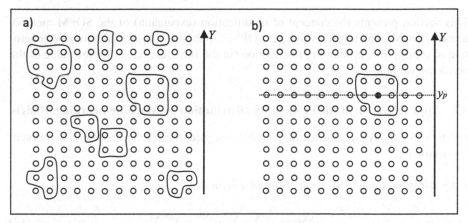

**Fig. 2.** Interpretation of SOPM maps (small circles represent particular neurons distributed in a rectangular map). a) Hypothetical effect of clustering process (indicated groups of neurons represent multidimensional clusters of patterns). b) Utilisation of a trained model in prediction of $\chi$ variable for a new pattern (explanation in text).

After completing the training algorithm, each pattern from the analysed dataset is assigned to a certain neuron in the output map of SOPM – it is the finally winning neuron (see Step 4) for the given pattern. Consequently, particular neurons "collect" assigned patterns, then groups of such neighbouring neurons in the map, having large "collections" of patterns, represent corresponding clusters of objects placed in multi-dimensional feature space (see hypothetical example in Fig. 2a). However, in case of SOPMs, such a clustering is executed also with respect to special variable $\chi$ projected onto a separate (vertical) axis in the map of neurons.

Apart from such special clustering and its visualization, a trained SOPM can also be utilised for prediction of $\chi$ variable for new patterns (see example in Fig. 2b)). After delivering input vector **x** of a new pattern to the model input, a winning neuron (i.e. generating the lowest signal) out of the whole map is being found (black circle in Fig. 2b)). The coordinate $y_p$ of the row it belongs to (dotted line in Fig. 2b)) is a predicted value of $\chi$ for this pattern. Moreover, if a winning neuron belongs to a certain previously identified group, the considered pattern can be assigned to the corresponding cluster of objects. It delivers additional information that better explains the prediction result. Also, further fine-tuned prediction process applied only to the identified cluster (with use of other methods) is possible.

Additionally, numerical (or graphical) analysis of signals from a neighbourhood of winning neuron can (informally) show uncertainty of the prediction: if the winning neuron is distinctly identified then the uncertainty is lower, however if there are many neighbouring neurons (belonging to many rows) generating similar signals – the uncertainty of the prediction is higher.

## 4   Proposition of PDCM Neural Map as an Extended Version of SOPM, Enabling Probabilistic Prediction

This section presents the concept of modification (expansion) of the SOPM method, named PDCM, which (preserving all SOPM capabilities) allows additionally genera-tion of *a posteriori* probability distribution (in the Bayesian sense) for the value of the predicted variable $\chi$.

### 4.1   Approximation of the Probability Distribution by Machine Learning Models

Let us consider any machine learning model designed to solve the classification problem and trained:

- by minimisation of SSE (*sum of squared errors*):

$$SSE = \frac{1}{2} \sum_p \sum_i (\theta_i^{(p)} - y_i^{(p)})^2 \qquad (2)$$

where: $y_i^{(p)}$ – signal of $i$-th output neuron for $p$-th training pattern,

$\theta_i^{(p)}$ – desired training value of $i$-th output neuron (related to $i$-th class) for $p$-th learning pattern.

- using training output (desirable) values as 1 an 0 as follows:

$$\theta_i^{(p)} = \begin{cases} 1 & \text{if } i \text{ indicates correct class for } p\text{th pattern} \\ 0 & \text{otherwise} \end{cases} \tag{3}$$

Ruck et al. in [26] showed that a multilayer perceptron (or other supervised machine learning model - there are no formal limitations to its structure) designed to solve the classification problem and trained according to postulates (2) and (3), approximates the Bayesian optimal discriminant function. Moreover it was shown that the output signals $y_i$ of such a model approximate (in the sense of minimisation of SSE) the Bayesian *a posteriori* probabilities of belonging of the input vector $\mathbf{x}$ to particular classes, i.e.:

$$y_i(\mathbf{x}) \approx P(\omega_i|\mathbf{x}) \quad \text{for outputs (classes)} \quad i = 1, 2, \ldots, N \tag{4}$$

where $\omega_i$ denotes $i$-th class.

If a predicted output variable $\chi$ is continuous, i.e. takes real values, a given class $\omega_i$ is related to a specific range $R_i$ of the variable $\chi$ (see Sect. 3). Then the set of all outputs $y_i$ ($i = 1, 2, \ldots, N$) can be used to approximate the entire conditional Bayesian probability density distribution (*a posteriori*) for the predicted variable $\chi$ (under the condition that the vector $\mathbf{x}$ has appeared). However, in order to approximate the probability distribution for $\chi$, it is necessary to scale the $y_i$ signals linearly. The scaling factor depends on the length of the range $D$ ($D$ is the set of all $\chi$ values, $D = R_1 \cup R_2 \cup \ldots \cup R_N$) and on the value $N$; this scaling should ensure that the area under the probability distribution graph (i.e. total probability) is equal to 1. This condition can be written as:

$$\lim_{N \to \infty} \sum_{i=1}^{N} (\lambda \cdot y_i) \frac{\chi_{\max} - \chi_{\min}}{N} = 1, \tag{5}$$

where: $\lambda$ – scaling factor (multiplier of $y_i$ signals),

$\chi_{\min}, \chi_{\max}$ – limits of range $D$.

As $\sum_{i=1}^{N} P(\omega_i|\mathbf{x}) = 1$, and consequently $\sum_{i=1}^{N} y_i = 1$ (with exact approximation (4)), condition (5) becomes fulfilled (independently from value $N$) for scaling factor:

$$\lambda = \frac{N}{\chi_{\max} - \chi_{\min}}. \tag{6}$$

It should be noted that despite theoretical considerations, in machine learning practice many elements can influence accuracy of approximation (4). For example model architecture, training parameters, selection of patterns in training set or (here) accepted number $N$ of ranges $R_i$ may have significant impact on this accuracy.

### 4.2 Main Assumptions of the PDCM Model as a Modified SOPM Network

In order to adopt the SOPM model to additional task of creating probability distribution for undergoing prediction variable $\chi$, the structure of the model should be expanded with a new layer of nodes that generate output signals $y_i$ ($i = 1, 2, \ldots, N$) approximating probabilities according to (4), see Fig. 3. During the training stage, desirable signals for these nodes are 0 or 1 according to (3). Each node aggregates signals from map neurons belonging to a certain map row representing given range $R_i$ of the variable $\chi$. There are no weights assigned to connections between middle layer and output layer (such connections are shown – for clarity reasons – only for the first and last rows in Fig. 3).

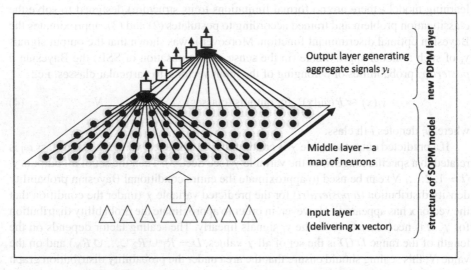

**Fig. 3.** Postulated structure of PDCM network.

Firstly, let us consider signals generating by neurons of middle layer (map). So far, in SOPM, this signals represented (for a given input vector $\mathbf{x}$) values of $d = \|\mathbf{x} - \mathbf{w}\|$, i.e. distance between $\mathbf{x}$ and neuron weight vector $\mathbf{w}$. Now let us introduce an exponential activation function for these neurons, so that they generate signals:

$$y_{ik} = e^{-d} = e^{-\|\mathbf{x}-\mathbf{w}\|} \tag{7}$$

where $i$, $k$ are coordinates of the neuron in the map.

In PDCM model, signals (7) from all neurons belonging to a given $i$-th map row (corresponding to a range $R_i$, $i = 1, 2,..., N$) are aggregated by an output node (see Fig. 3) to an output signal $y_i$ according to the postulated formula:

$$y_i = \sqrt[K]{y_{i1} \cdot y_{i2} \cdot ... \cdot y_{iK}} = (y_{i1} \cdot y_{i2} \cdot ... \cdot y_{iK})^{\frac{1}{K}} \tag{8}$$

where $K$ is the number of neurons in the row (horizontal size of the map).

### 4.3 Proposed Training Procedure of PDCM

As in the SOPM model, the training of PDCM network is dual: supervised due to the key variable $\chi$ (which is dependent variable in estimation and forecasting problems), and unsupervised due to other variables. This approach enables (as in SOPMs) the implementation of the clustering process (cluster analysis) in the feature map along with its visualization, and solving the problem of estimating (predicting) the variable $\chi$. However, while in the case of SOPM, the supervision relies only on indicating the appropriate row of the map (corresponding to the value of $\chi$ for a considered pattern **x**), in the PDCM model, similarly to classic supervised neural networks, desired output signals $\theta$ (according to rule (3)) must be given for all nodes of output layer.

Let us note that due to specific character of connections between middle layer and output layer of PDCM (Fig. 3), and assuming (at first) no relations between rows (no neighbourhood at all) in the map, it is possible to consider separately $N$ parts of the whole model (each containing one $- i$-th $-$ row of the map and one corresponding output node); let us name such $i$-th part of the model by PDCM$i$ ($i = 1, 2,..., N$), (the neighbourhood aspect during the training stage of whole PDCM model will be considered later in Subsect. 4.4).

Below, a single step of PDCM$i$ training for $p$-th training pattern $\mathbf{x}_p$ is presented (indexes $i$ and $p$ are then omitted due to better clarity). The minimized error function (2), denoted here by $E$, is now given by formula:

$$E(\mathbf{W}) = \frac{1}{2}(\theta - y)^2 = \frac{1}{2}\delta^2 \qquad (9)$$

where: $\delta = \theta - y$ denotes output error value for a given ($p$-th) pattern
   **W** denotes the vector of all weights of PDCM$i$.

The training procedure (based on classic backpropagation algorithm) aim to minimize function (9) using desired output signals $\theta$ determined according to rule (3).

Let $\mathbf{w}_k$ ($k = 1, 2, ..., K$) denotes weight vector of $k$-th neuron of PDCM$i$ $-$ on the whole PDCM map it is the neuron having coordinates ($i, k$). A formula describing one step of adjusting weights $\mathbf{w}_k$ (by a correction vector $\Delta \mathbf{w}_k$), based on steepest gradient method (used also in classic backpropagation algorithm), is given by equation:

$$\Delta \mathbf{w}_k = \eta \cdot (-\nabla E(\mathbf{w}_k)) \quad (k = 1, 2, \ldots, K) \qquad (10)$$

where: $\nabla E(\mathbf{w}_k)$ is a part (relating to $\mathbf{w}_k$) of gradient of error function (2) in point $\mathbf{w}_k$,
   $\eta$ is a value of training coefficient, $0 < \eta < 1$.

Now the problem of specifying the training algorithm for the PDCM network (as a modification of the error backpropagation algorithm in the version with independent weight correction for each training pattern) relies on finding the vectors $\nabla E(\mathbf{w}_k)$. As:

$$\nabla E(\mathbf{w}_k) = \frac{\partial E(\mathbf{W})}{\partial \mathbf{w}_k} = \frac{\partial E(\mathbf{W})}{\partial y} \cdot \frac{\partial y}{\partial y_k} \cdot \frac{\partial y_k}{\partial \mathbf{w}_k} \qquad (11)$$

then, considering dependencies (7), (8) and (9) we obtain (note that for clarity reasons the index $i$ has been omitted everywhere, particularly for $y$, $y_k$ and $\mathbf{w}_k$):

$$\nabla E(w_k) = -\delta \frac{1}{K}(y_1 \cdot y_2 \cdot \ldots \cdot y_K)^{\frac{1}{K}-1} \cdot \frac{y_1 \cdot y_2 \cdot \ldots \cdot y_K}{y_k} \cdot e^{-d} \cdot \left(-\frac{\partial d}{\partial \mathbf{w}_k}\right) \qquad (12)$$

and, after simplification, taking again into consideration (7) and (8):

$$\nabla E(w_k) = \frac{1}{K}\delta \cdot y \cdot \left(\frac{\partial d}{\partial \mathbf{w}_k}\right). \qquad (13)$$

Assuming the Euclidean metric to determine the distance in the weights space, the distance $d = \|\mathbf{x} - \mathbf{w}_k\|$ is expressed by:

$$d = \sqrt{\sum_j (x_j - w_{jk})^2} \tag{14}$$

where $j$ is the index for all subsequent elements of vectors $\mathbf{x}$ and $\mathbf{w}_k$.

Now, assuming that $d \neq 0$, we obtain

$$\frac{\partial d}{\partial w_{jk}} = \frac{1}{2d} 2 \cdot (x_j - w_{jk}) \cdot (-1) \tag{15}$$

and then, after applying it in (13), the gradient is determined as

$$\nabla E(w_k) = -\frac{1}{K} \delta \cdot y \cdot \frac{\mathbf{x} - \mathbf{w}_k}{\|\mathbf{x} - \mathbf{w}_k\|} \tag{16}$$

Finally, considering (10) and (16), the one-step weight correction vector $\Delta \mathbf{w}_k$, is determined as:

$$\Delta \mathbf{w}_k = \frac{1}{K} \eta \cdot \delta \cdot y \cdot \frac{\mathbf{x} - \mathbf{w}_k}{\|\mathbf{x} - \mathbf{w}_k\|} \quad (k = 1, 2, \ldots, K). \tag{17}$$

It should be noted that the last factor in Eq. (17) represents a unit-length vector directed from the point $\mathbf{w}_k$ towards the point $\mathbf{x}$. The direction determined in this way is the direction of the entire weight correction vector $\Delta \mathbf{w}_k$ (anyway its orientation and length may vary and depend on other factors in (17)).

Such training steps are repeated for all training patterns from a considered dataset.

### 4.4 Generalized Training Procedure Taking into Account Neighbourhood Aspects

Following the methodology of training SOM and SOPM networks, let us now consider – for the PDCM network – the possibility of introducing the idea of neighbourhood and the related principle of similar method of training for topologically adjacent map neurons in middle layer (mapping adjacent areas of the feature space).

For a single PDCM$i$ network ($i = 1, 2, \ldots, N$) the neighbourhood concept involves the requirement to differentiate the lengths of weight correction vectors $\Delta \mathbf{w}_k$ for $k = 1, 2, \ldots, K$ (both when the considered PDCM$i$ contains a winning neuron and when it does not). Then the training rule (17) should be modified as follows:

$$\Delta \mathbf{w}_k = a_k \frac{1}{K} \eta \cdot \delta \cdot y \cdot \frac{\mathbf{x} - \mathbf{w}_k}{\|\mathbf{x} - \mathbf{w}_k\|} \quad (k = 1, 2, \ldots, K), \tag{18}$$

where $a_k$ are neighbourhood coefficients (values of a neighbourhood function $s$) responsible for differentiating lengths of vectors $\Delta \mathbf{w}_k$.

Certain theoretical analyses executed by author has led to the conclusion that for a single PDCM$i$ network the dependency:

$$\sum_{k=1}^{K} a_k = K \tag{19}$$

should be ensured.

Considering now the neighbourhood aspect for the whole PDCM during the training stage (i.e. the essential aspect for ensuring a proper organization of the map – PDCM middle layer – in order to perform clustering process), there is a need for introducing a certain neighbourhood function $s$, like in models SOM and SOPM. Here, the function $s$ should be responsible for determining the neighbourhood coefficients $a_k$ for all neurons in the map, during a given training step. The „centre" of function $s$ is always the winning neuron, selected separately for each training pattern exactly according to rules accepted in SOPM (see Sect. 3). However in PDCM, the neighbourhood function $s$ should additionally take into account the rule (3) and Eq. (19). Optimal selection of function $s$ is the matter of on-going experiments, current results of such exploratory analyses were implemented by author in researches shown in next section.

The definition of learning rule (formula (18)) for PDCM network, supplemented by approving a method of determining the neighbourhood coefficients ensuring the implementation of the pattern grouping process, allows creation the network training algorithm. The algorithm has been implemented in form a computer program written by the author in C++, which is the basis for the research discussed in the next section.

## 5  Application of the PDCM Model in Real Estate Market Analysis

Below, results of application of the PDCM model in the issue of real estate value estimation are presented. A Boston housing dataset, available in the UCI ML Repository (https://archive.ics.uci.edu/ml/machine-learning-databases/housing/), has been used in presented research. This set has been often exploited in many researches concerning clustering or regression problems (see e.g. [4, 13, 23]).

The Boston dataset contains 506 patterns described by 14 numeric variables. The dependent feature $\chi$ represents the median of the estate prices in the given census area (MEDV). The remaining 13 variables describe certain features influencing real estate values and, after standardisation, constitute the input vector to PDCM. Six observations out of 506 (their numbers in the original Boston dataset are: 48, 137, 197, 314, 435, 457) were selected randomly for ultimate testing (creating the test set); the remaining 500 patterns create the training set.

The following PDCM model parameters were adopted:

- number of training epochs (presentations of whole training set): 300,
- PDCM map dimensions: $X$ axis – 10 ($K = 10$), $Y$ axis – 20 ($N = 20$),
- each range $R_1, R_2, ..., R_N$ contains 25 values of $\chi$, taken from the training set,
- the rule (18) has been adopted for model training,
- training coefficient $\eta$ has decreased linearly during the training from 0.7 to 0.07

- initial weights for map neurons were selected randomly from range $[-1.5, 1.5]$.

The effect of clustering, expressed by the numbers of training patterns **x** assigned to particular neurons of the map (middle layer) of the PDCM network, is presented in tabular and graphical form in Fig. 4 (note also a relation to Fig. 2a)).

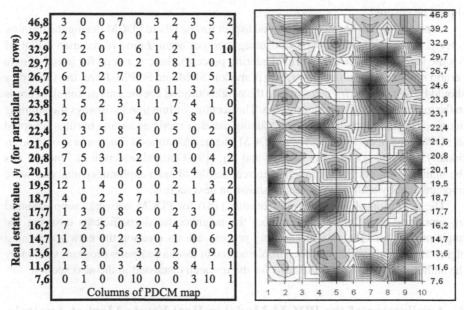

| Real estate value $y_i$ (for particular map rows) | 1 | 2 | 3 | 4 | 5 | 6 | 7 | 8 | 9 | 10 |
|---|---|---|---|---|---|---|---|---|---|---|
| 46,8 | 3 | 0 | 0 | 7 | 0 | 3 | 2 | 3 | 5 | 2 |
| 39,2 | 2 | 5 | 6 | 0 | 0 | 1 | 4 | 0 | 5 | 2 |
| 32,9 | 1 | 2 | 0 | 1 | 6 | 1 | 2 | 1 | 1 | 10 |
| 29,7 | 0 | 0 | 3 | 0 | 2 | 0 | 8 | 11 | 0 | 1 |
| 26,7 | 6 | 1 | 2 | 7 | 0 | 1 | 2 | 0 | 5 | 1 |
| 24,6 | 1 | 2 | 0 | 1 | 0 | 0 | 11 | 3 | 2 | 5 |
| 23,8 | 1 | 5 | 2 | 3 | 1 | 1 | 7 | 4 | 1 | 0 |
| 23,1 | 2 | 0 | 1 | 0 | 4 | 0 | 5 | 8 | 0 | 5 |
| 22,4 | 1 | 3 | 5 | 8 | 1 | 0 | 5 | 0 | 2 | 0 |
| 21,6 | 9 | 0 | 0 | 0 | 6 | 1 | 0 | 0 | 0 | 9 |
| 20,8 | 7 | 5 | 3 | 1 | 2 | 0 | 1 | 0 | 4 | 2 |
| 20,1 | 1 | 0 | 1 | 0 | 6 | 0 | 3 | 4 | 0 | 10 |
| 19,5 | 12 | 1 | 4 | 0 | 0 | 0 | 2 | 1 | 3 | 2 |
| 18,7 | 4 | 0 | 2 | 5 | 7 | 1 | 1 | 1 | 4 | 0 |
| 17,7 | 1 | 3 | 0 | 8 | 6 | 0 | 2 | 3 | 0 | 2 |
| 16,2 | 7 | 2 | 5 | 0 | 2 | 0 | 4 | 0 | 0 | 5 |
| 14,7 | 11 | 0 | 0 | 2 | 3 | 0 | 1 | 0 | 6 | 2 |
| 13,6 | 2 | 2 | 0 | 5 | 3 | 2 | 2 | 0 | 9 | 0 |
| 11,6 | 1 | 3 | 0 | 3 | 4 | 0 | 8 | 4 | 1 | 1 |
| 7,6 | 0 | 1 | 0 | 0 | 10 | 0 | 0 | 3 | 10 | 1 |

Columns of PDCM map

**Fig. 4.** Clustering results of the Boston real estate dataset

Analysis of clustering results (Fig. 4) shows the tendency to create clusters of training patterns assigned to the topologically adjacent PDCM map neurons. This method also allows (like SOPM) the identification of clusters with respect to values of a particular feature (variable $\chi$ – here median of the estate prices MEDV).

The PDCM method also (like SOPM) allows, for a given new pattern, determining a prediction value of variable $\chi$ (along with a visual or numerical informal assessment of prediction uncertainty), based on identification of winning neuron (and signal analysis of adjacent neurons). However in PDCM (contrary to SOPM) it is additionally possible to generate (approximate) formal *a posteriori* probability density distribution for the predicted (estimated) variable $\chi$ – this property is a key functional feature of this model. The results of testing the PDCM network in the real estate valuation process for six test cases are presented below.

Figure 5 shows graphically output signals generated according to formula (7) by neurons of the PDCM map (middle layer) in response to selected (exemplary) test patterns 1 and 3. The darker area in the graph, the stronger neuron's signal. The winning neuron (generating the highest signal – see formula (7), and indicating the point prediction of $\chi$ for a given pattern) is placed in the black area. For the test pattern 1 the prediction value is 18.7, for the test pattern 3 the prediction is 32.9.

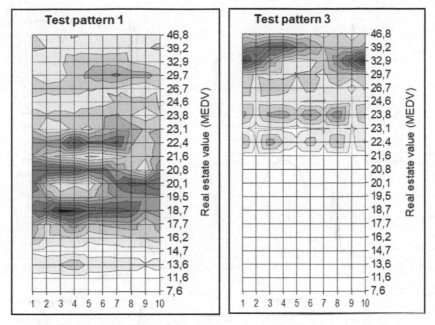

**Fig. 5.** Output signals of map neurons for two selected test pattern 1 and 3

The testing effects (based on point predictions generated by winning – i.e. generating the strongest signal – neuron) for all six testing patterns are presented in Table 1. The mean absolute error in estimating the MEDV value for the test set is **1.55**, what (for this specific dataset) should be appointed as a good result.

**Table 1.** Testing results (based on point predictions) for all six testing patterns

| Test pattern | Actual values | | Predicted values | | Absolute error | |
|---|---|---|---|---|---|---|
| | MEDV | Range $R_i$ | MEDV | Range $R_i$ | MEDV | Range $R_i$ |
| 1 | 16.6 | 5 | 18.7 | 7 | 2.1 | 2 |
| 2 | 17.4 | 6 | 17.7 | 6 | 0.3 | 0 |
| 3 | 33.3 | 18 | 32.9 | 18 | 0.4 | 0 |
| 4 | 21.6 | 11 | 23.1 | 13 | 1.5 | 2 |
| 5 | 11.7 | 2 | 14.7 | 4 | 3.0 | 2 |
| 6 | 12.7 | 2 | 14.7 | 4 | 2.0 | 2 |

Figure 6 show the graphs of estimated probability density distributions for MEDV for six test patterns. These distributions have been approximated on the basis of signals generated by output nodes (output layer) of the PDCM model, according to formula (4). Black triangular mark on the horizontal axis shows the actual value of the MEDV.

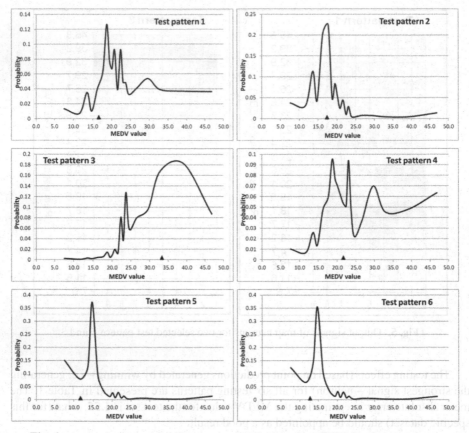

**Fig. 6.** Approximated probability density distributions for MEDV for six test patterns

It should be noted that apart from above mentioned possibilities of data analyses performed by proposed PDCM model, next data exploration options, based on further investigation in signals of all map neurons, are delivered by this method. For example, for the test pattern 3, a fairly large area of strong signals can be identified at the top of PDCM map (see Fig. 5); this fact is also confirmed by the shape of the probability density function (Fig. 6, pattern 3). The neuron generating the strongest signal in this area (the winning neuron) with coordinates: $X = 10$, $Y = 18$ (column = 10, MEDV = 32.9) collects 10 training patterns assigned to it (see table in Fig. 4, bold number **10**). These 10 cases may constitute (for a real estate market analyst) a comparative base helpful in additional justification of the estate price estimation.

# 6    Conclusions

The paper submits a concept of new neural tool PDCM, dedicated to relatively wide range of data analysis and data exploration tasks, i.e. special clustering, clusters' visualization, dependent variable prediction (together with its possible visual analysis and justification)

and probabilistic prediction on the basis of approximations of *a posteriori* probability density distribution. Basic theoretical considerations concerning the proposed model have been shown. Also, the presented analyses of application of the PDCM model for the real estate market data indicate the significant effectiveness of this tool and quite rich possibilities of this method in data mining.

However, much future research should yet be done; especially desirable analyses should concern selection of: PDCM parameters (e.g. map dimensions), coefficients used in training algorithm and shape of neighbourhood function (neighbourhood coefficients). Also, testing the model with use of other datasets will be beneficial.

**Acknowledgements.** The publication is financed by the Krakow University of Economics under the program "Support for Conference Activity - WAK-2022".

# References

1. Assylbekov, Z., Melnykov, I., Bekishev, R., Baltabayeva, A., Bissengaliyeva, D., Mamlin, E.: Detecting value-added tax evasion by business entities of Kazakhstan. In: International Conference on Intelligent Decision Technologies, pp. 37–49. Springer, Cham (2016)
2. Chang, F.J., Huang, C.W., Cheng, S.T., Chang, L.C.: Conservation of groundwater from over-exploitation – scientific analyses for groundwater resources management. Sci. Total Environ. **598**, 828–838 (2017)
3. Dameri, R.P., Garelli, R., Resta, M.: Neural networks in accounting: clustering firm performance using financial reporting data. J. Inf. Syst. **34**(2), 149 166 (2020)
4. Dembczyński, K., Kotłowski, W., Słowiński, R.: Solving regression by learning an ensemble of decision rules. In: Artificial Intelligence and Soft Computing ICAISC 2008, pp. 533–544. Springer, Berlin, Heidelberg (2008)
5. Hossu, C.A., Ioja, I.C., Nita, M.R., Hartel, T., Badiu, D.L., Hersperger, A.M.: Need for a cross-sector approach in protected area management. Land Use Policy **69**, 586–597 (2017)
6. Iamandi, I.E., Constantin, L.G., Munteanu, S.M., Cernat-Gruici, B.: Mapping the ESG behavior of European companies. A holistic Kohonen approach. Sustainability **11**, 3276 (2019)
7. Kimura, M., et al.: Application of the self-organizing map in the classification of natural antioxidants in commercial biodiesel. Biofuels **12**, 673–678 (2018)
8. Kohonen, T.: Self-Organizing Maps, Series in Information Sciences (31). Springer-Verlag, Heidelberg (1995)
9. Le, T., Pardo, P., Claster, W.: Application of Artificial Neural Network in Social Media Data Analysis: A Case of Lodging Business in Philadelphia, in Artificial Neural Network Modelling, pp. 369–376. Springer, Cham (2016)
10. Liang, Y.M., Yin, X.F., Chang, D.O.U., Yang, L.I.U.: Application of SOM Neural Network in the Construction of Urban Ramp Driving Cycle, DEStech Transactions on Computer Science and Engineering, International Conference on Artificial Intelligence and Computing Science (ICAICS 2019), pp. 240–244 (2019)
11. Lis, B., Szczepaniak. P.S., Tomczyk .A.: Multi-layer Kohonen network and texture recognition. In: Soft Computing Tools, Techniques and Applications. AOW EXIT, Warsaw (2004)
12. Liu, J., Xu, L.: Improvement of som classification algorithm and application effect analysis in intrusion detection. In: Patnaik, S., Jain, V. (eds.) Recent Developments in Intelligent Computing, Communication and Devices. AISC, vol. 752, pp. 559–565. Springer, Singapore (2019). https://doi.org/10.1007/978-981-10-8944-2_65

13. Lydia, E.L., Bindu, G.H., Sirisham, A., Kiran, P.P.: Electronic governance of housing price using Boston dataset implementing through deep learning mechanism. Int. J. Recent Technol. Eng. 7(6S2), 560–563 (2019)
14. Merkl, D., Rauber, A.: Alternative ways for cluster visualization in self-organizing maps. In: Proceedings of the Workshop on Self-Organizing Maps (WSOM'97), Helsinki (1997)
15. Morajda, J., Domaradzki, R.: Application of cluster analysis performed by SOM neural network to the creation of financial transaction strategies. J. Appl. Comput. Sci, 13(1), 87–98 (2005)
16. Morajda, J., Paliwoda-Pękosz, G.: An enhancement of Kohonen neural networks for predictive analytics: Self-Organizing Prediction Maps. In: AMCIS 2020 Proceedings, vol. 6 (2020)
17. Morajda, J., Paliwoda-Pękosz, G.: A concept of FLOPM: neural maps with floating nodes for classification and prediction. In: AMCIS 2021 Proceedings, vol. 16 (2021)
18. Moshou, D.: Artificial Neural Maps. Applications. VDM Verlag Dr. Müller, Saarbrücken, Concepts, Architectures (2009)
19. Mükkulainen, R.: Script recognition with hierarchical feature maps. Connection Sci. 2, 83–101 (1990)
20. Nan, F., Li, Y., Jia, X., Dong, L., Chen, Y.: Application of improved SOM network in gene data cluster analysis. Measurement 145, 370–378 (2019)
21. Niska, H., Serkkola, A.: Data analytics approach to create waste generation profiles for waste management and collection. Waste Manage. 77, 477–485 (2018)
22. Padulano, R., Del Giudice, G.: Pattern detection and scaling laws of daily water demand by SOM: an application to the WDN of Naples, Italy. Water Resour. Manag. 33(2), 739–755 (2019)
23. Peng, J., Xia, Y.: A cutting algorithm for the minimum sum-of-squared error clustering. In: Proceedings of the Fifth SIAM International Conference on Data Mining. Newport Beach, California, pp. 150–160 (2005)
24. Rashkovan, V., Pokidin, D.: Ukrainian banks' business models clustering: application of Kohonen neural networks. Visnyk of the National Bank of Ukraine 2016(238), 13–38 (2016)
25. Rojek, I.: Technological process planning by the use of neural networks. AI EDAM 31(1), 1–15 (2017)
26. Ruck, D.W., Rogers, S.K., Kabrisky, M., Oxley, M.E., Suter, B.W.: The multilayered perceptron as an approximation to a Bayes optimal discriminant function. IEEE Trans. Neural Networks 1, 296–298 (1990)
27. Wang, L., Wu, C.: Business failure prediction based on two-stage selective ensemble with manifold learning algorithm and kernel-based fuzzy self-organizing map. Knowl.-Based Syst. 121, 99–110 (2017)
28. Wierzchoń, T., Kłopotek, M.: Algorithms of Cluster Analysis. Institute of Computer Science, Polish Academy of Sciences, Warsaw (2015)

# Significance of Eco-Cement Constituents to Its Mechanical Properties, by Machine Learning Algorithms

Natia R. Anastasi[✉] [iD] and Nikolaos P. Bakas [iD]

Civil Engineering Department, School of Architecture, Engineering, Land and Environmental Sciences, Neapolis University Pafos, 2 Danais Avenue, 8042 Paphos, Cyprus
natia.anastasi@nup.ac.cy

**Abstract.** A novel proposed material for the potential replacement of cement in some of its applications was evaluated. This new material labeled Eco-Cement comprised of Biomass from the dairy and poultry industries, Urea, Cement Kiln Dust, Rice Husk Ash, Sand, and Water and was manufactured in a range of weight ratios. In this work, a comprehensive analysis of the ingredients in varying weight percentages of the novel material was manufactured and the corresponding strength and strains of the material were studied. A variety of concrete pastes using amounts of sought ratios were produced, molded into blocks, and allowed to cure under laboratory conditions. Unconfined compression tests were performed using a deformation control compressive strength machine. The strength and strains were evaluated from the initial zero load step incrementally, until the failure of each specimen. The resulting database was analyzed by utilizing Linear Regression, Random Forests, and the Gradient Boosting machine learning methods. Extensive sensitivity analysis with the machine learning algorithms, reveal certain patterns, which were established with three different methods. Furthermore, we present the analysis of the corresponding literature with Bibliometric techniques.

**Keywords:** Cement · Biomass · Compressive Strength · Machine Learning

## 1 Introduction

The need to use unwanted side products or trash in the preparation of useful materials is a general drive of modern industrial research. This is elaborated in such processes as the use of recycled tires in road construction [7, 10, 12], recycled glass in the production of heat insulation products [4] or more intriguing applications, like the use of recycled bottles in the manufacturing of concrete [11, 15]. Concrete is one of the most abundant materials on earth. It is composed of cement, water, and aggregates. The manufacturing of cement is amongst the processes that are most severely detrimental to the environment. On the contrary, the material studied herein, Eco-Cement [1], is a composite material of Bacteria, Urea, Calcium Carbonate, Cement Kiln Dust (CKD), the hydraulic agent Rice Husk Ash (RHA), and Sand. It utilizes waste products of numerous industries such as the cement manufacturing industry, the dairy industry, the poultry growing industry making it an overall environmentally desired product (Fig. 1).

M. Papadaki et al. (Eds.): EMCIS 2022, LNBIP 464, pp. 33–39, 2023.
https://doi.org/10.1007/978-3-031-30694-5_3

## 2 Experimental Design

CKD, Cement Kiln Dust is a solid waste of the cement manufacturing industry. It is a super-fine grained solid that accumulates during the manufacturing of cement on the air pollution control filters. The physical and chemical characteristics of this material vary from plant to plant and depend heavily on the raw materials used in the manufacturing of Portland cement. It is highly alkaline and it is considered to be a mixture of primarily unreactive starting materials. Studies have been performed in order to investigate the use of CKD as a cementitious material in concrete and mortar. It was found that whilst substitution of certain percentages of cement with CKD did not have advert effects on the mechanical characteristics of the resulting material, with substitution above certain limits the effects on said properties were detrimental. The diminished mechanical abilities of CKD-containing concrete are due to the low levels of calcite, the calcium carbonate polymorph speculated to have the highest strength tolerances.

Precipitation of calcium carbonate is facilitated though a plethora of processes, amongst them biomineralization. Bacteria have shown to induce calcium carbonate precipitation through urea hydrolysis. Urease is a Nickel-containing enzyme that catalyzes the hydrolysis of urea into ammonia and carbon dioxide. At specifically tuned pH the hydrolysis products are ammonium and carbonate.

$$CO(NH_2)_2 + H_2O \rightarrow 2NH_4^+ + CO_3^{-2} \tag{1}$$

The produced carbonate ions precipitate in the presence of calcium ions as calcium carbonate crystals:

$$Ca^{+2} + CO_3^{-2} \rightarrow CaCO_3 \tag{2}$$

## 3 Linear Regression

We applied basic statistical analysis to the obtained dataset, and afterwards a machine learning investigation. We normalize each one of the independent and dependent variables $v_i$ by subtracting its mean value $\mu_i$ and dividing by its standard deviation $\sigma_i$, and we get each normalized ones

$$v_i^n = (v_i - \mu_i)/\sigma_i \tag{3}$$

Initially, we checked the multiple correlations among the independent variables, and obtained a high correlation among Water and CKD (0.9646), and Water with Biomass (0.8532), as depicted in Fig. 2. This was due to the mixing procedure, with mixtures of high CKD, and Biomass needed more water to get to a solid-state. Accordingly, as the aim of the regression analysis is to assess the impact of each independent variable on the dependent, and not predictive modeling, we exclude water from the model. The new model has five variables, which are Urea, Biomass, CKD, RHA, and Sand. In order to check the multicollinearity, we perform linear regression, with each independent variable as response and the remaining independent variables as predictors, and compute the Variance Inflation Factor (VIF)

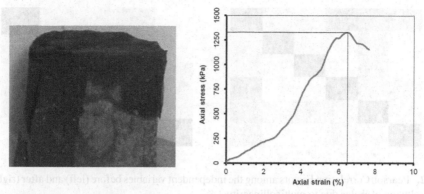

**Fig. 1.** Example strength test: (left) Photograph of the specimen after failure, (right) Stress-strain curve.

$$VIF_i = /(1 - R_i^2) \tag{4}$$

for each independent variable i. In the new model, we get the VIF values for each variable: Urea (2.261), Biomass (5.551), CKD (2.791), RHA (2.858), Sand (2.255). It is suggested that the independent variable has a VIF < 10 [13], so we keep these five variables.

We use compressive strength as the dependent variable, and the regression analysis results are presented in Table 1, and Fig. 5. Interestingly, we obtained a low p-value for Biomass (0.0148), with high and positive weight (0.747) indicating Biomass as a significant factor for the compressive strength. Furthermore, the CKD exhibited a p-value of 0.0267, with a negative weight (−0.4792), indicating that the CKD has a negative effect on the compressive strength. This is because of the highly exothermic reaction of the hydration of CKD raises the temperature to levels above the tollerances of the enzyme, resulting in denaturing and less concentration of active enzyme participating in biocementation [6]. These numerical results are verified by the Machine Learning investigation in the following section.

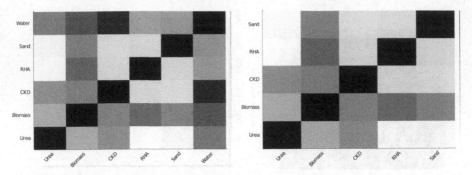

**Fig. 2.** Pearson Correlation Factors among the independent variables before (left) and after (right) the exclusion of Water due to multicollinearity.

## 4 Machine Learning Models

In order to quantify the significance of each ingredient to Eco-Cement's mechanical properties, a variety of regression analyses were performed. The analyses were accomplished utilizing MIT's Julia programming language [2] and custom codes written by the authors. The dependent variable was the compressive strength (Fig. 3) and the independent, the ingredients of Eco-Cement, divided with each specimen's volume. The dataset contained limited observations (N = 55), hence three different regression methods were utilized and compared: Multiple Linear Regression (MLR), Random Forests [3, 16], and Gradient Boosting [5, 8, 18]. Accordingly, a modified version of the Profile method [9, 14] is utilized, in order to investigate each variable's contribution to the dependent variable. In particular, each input variable varies within its given (raw) range while all the other input variables are kept constant in a certain value. This constant takes three discrete values: 25% Percentile, Median, and 75% Percentile. In Table 2, the accuracy metrics for each regression method are presented: the Pearson Correlation Coefficient (COR), the Mean Absolute Error (MAE), the Root Mean Squared Error (RMSE), the Mean Absolute Percentage Error (MAPE), the Maximum Absolute Percentage Error (MAXAPE), and the Alpha metric (Table 1).

**Table 1.** Regression analysis metrics for compressive strength

| METHODS | COR | MAE | RMSE | MAPE | MAXAPE | Alpha |
|---|---|---|---|---|---|---|
| RandomForest | 0.661 | 172.553 | 220.357 | 0.177 | 1.387 | 0.217 |
| GradientBoosting | 0.901 | 101.121 | 131.363 | 0.110 | 0.936 | 0.617 |
| LinearRegression | 0.471 | 177.965 | 236.693 | 0.181 | 1.516 | 0.222 |

The metrics for the three methods indicate low predictive performance. The reasons for this performance may be attributed to the relatively small sample database and a poor transcription of solution chemistry [17] to solid state.

Through detailed sensitivity analysis, however we were able to reveal some patterns on how each independent variable influences the dependent. In Fig. 3, the sensitivity curves for CKD are plotted. Linear Regression Exhibits a clear decreasing pattern, while Random Forests and Gradient Boosting a decreasing trend for low values of CKD and almost constant for the rest. This can be speculatively attributed to a more exothermic reaction at higher concentrations of CKD resulting in the possible denaturing on the enzyme and affecting the cementation process. Furthermore, in Fig. 4, for higher values of Urea we obtain lower values of compressive strength, however with a lower variation. Again, we can speculate that higher concentrations of urea, result in an increase of the pH hence causing denaturing of the enzyme and affecting biocementation.

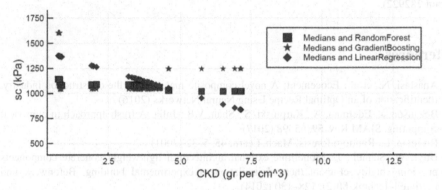

**Fig. 3.** Sensitivity curves for CKD.

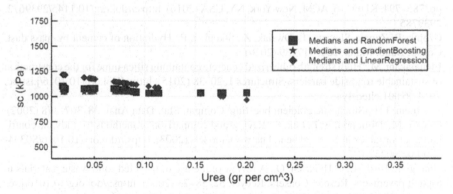

**Fig. 4.** Sensitivity curves for urea.

## 5   Conclusions

The composition of a new material using waste byproducts of various industries, achieves compressive strengths that fall within the range of a material used as mortar. An optimal recipe is not easily achieved as there is a non-straightforward behavior that relates to the compressive strength and the amounts of the components. Statistical analysis and machine learning algorithms extracted specific patterns of the importance of Eco-Cement ingredients, with respect to its compressive strength, especially for the Urea and CKD. The difficulty in obtaining a clear dependency of the compressive strength, based on its constituents, is a well-known problem for a variety of materials, as even if we repeat the same experiment with the same composition of a material, the resulting strength might diverge. Hence this study serves as an initial point of investigation for an ecofriendly material, which could be used as a mortar, with significant economic and environmental advantages.

**Acknowledgements.** This work was made under the frame of the ECOCE-MENT project (FP7 - Grant 282922).

# References

1. Anastasi, N., et al.: Ecocement: A novel composite material for the construction industry. Identification of an Optimal Recipe Using Neural Networks (2016)
2. Bezanson, J., Edelman, A., Karpinski, S., Shah, V.B.: Julia: A fresh approach to numerical computing. SIAM Rev. **59**, 65–98 (2017)
3. Breiman, L.: Random forests. Mach. Learn. **45**, 5–32 (2001)
4. Breit, W., Schnell, J.: Monolithic external architectural lightweight concrete components providing highly efficient thermal insulation-an experimental building. Betonwerk und Fertigteil-Technik **80**(2), 128–130 (2014)
5. Chen, T., Guestrin, C.: XGBoost: A scalable tree boosting system. In: Proceedings of the 22nd ACM SIGKDD International Conference on Knowledge Discovery and Data Mining. pp. 785–794. KDD '16, ACM, New York, NY, USA (2016). https://doi.org/10.1145/2939672. 2939785
6. Czapik, P., Zapała-Sławeta, J., Owsiak, Z., Stkepień, P.: Hydration of cement by- pass dust. Constr. Build. Mater. **231**, 117139 (2020)
7. Elchalakani, M.: High strength rubberized concrete containing silica fume for the construction of sustainable road side barriers. Structures **1**, 20–38 (2015). https://doi.org/10.1016/j.istruc. 2014.06.001,citedBy43
8. Friedman, J.H.: Stochastic gradient boosting. Comput. Stat. Data Anal. **38**, 367–378 (2002)
9. Gevrey, M., Dimopoulos, I., Lek, S.: Review and comparison of methods to study the contribution of variables in artificial neural network models (2003). https://doi.org/10.1016/S0304-3800(02)00257-0
10. Huang, Y., Bird, R.N., Heidrich, O.: A review of the use of recycled solid waste materials in asphalt pavements. Resour. Conserv. Recycl. **52**, 58–73 (2007). https://doi.org/10.1016/j.res conrec.2007.02.002,citedBy284
11. Irwan, J.M., et al.: The mechanical properties of pet fiber reinforced concrete from recycled bottle wastes. Adv. Mat. Res. **795**, 347–351 (2013). https://doi.org/10.4028/www.scient ific.net/AMR.795.347, cited By 14; Conference of 2nd International Conference on Sustainable Materials, ICoSM 2013; Conference Date: 26 March 2013 Through 27 March 2013; Conference Code:100411
12. Myhre, M., Mackillop, D.A.: Rubber recycling. Rubber Chem. Technol. **75**, 429–474 (2002). https://doi.org/10.5254/1.3547678,citedBy106
13. Neter, J., Kutner, M.H., Nachtsheim, C.J., Wasserman, W.: Applied linear statistical models (1996)

14. Olden, J.D., Jackson, D.A.: Illuminating the "black box": a randomization approach for understanding variable contributions in artificial neural networks. Ecol. Model. (2002). https://doi.org/10.1016/S0304-3800(02)00064-9
15. Pelisser, F., Montedo, O.R.K., Gleize, P.J.P., Roman, H.R.: Mechanical properties of recycled pet fibers in concrete. Mater. Res. **15**, 679–686 (2012). https://doi.org/10.1590/S1516-143 92012005000088
16. Sadeghi, B.: Decisiontree.jl (2013)
17. Whiffin, V.S.: Microbial caco3 precipitation for the production of biocement (2004)
18. Xu, B., Chen, T.: Xgboost.jl (2014)

# Next Generation Process Material Tracking and Analytics for the Process Industries Using Machine Learning Algorithms

Symon Doe[1], Christoforos Kassianides[1], Symeon Kassianides[1(✉)], Nikos Bakas[2] (iD), and Christos Christodoulou[2] (iD)

[1] Hyperion Systems Engineering Ltd., Athalassis Str. 36, 2201 Nicosia, Cyprus
s.kassianides@hyperionsystems.net

[2] Computation-Based Science and Technology Research Center, The Cyprus Institute, Konstantinou Kavafi Str. 20, 2021 Nicosia, Cyprus

**Abstract.** Demand for plastics is the key driver for petrochemicals and they are predicted to account for more than a third of the growth in world oil demand by 2030. While Plastics provide substantial benefits to society their production presents detrimental environmental challenges that need immediate attention and intervention. Typical Polymer production is a combination of Continuous and Batch processes which makes high level traceability of raw materials, products, while maintaining the quality of the end-product a big challenge. To address this, we developed HYPPOS software that is based on a mathematical material tracking algorithm that discretises the continuous manufacturing processes into identifiable slices of material and tracks them as they move through the manufacturing stages. HYPPOS continuously collects data in real time from the plant instrumentation and feeds this data to its tracking model. HYPPOS algorithm can associate any parameter with each individual slice of material and uses this to infer the quality of the material providing visibility to the operator at an early stage of the process. Many plants rely on off-line laboratory measurements for the critical quality parameters, rather than having online automated measurement instruments. To address this limitation, Machine learning (ML) algorithms were built to implement soft sensors for the quality prediction using the existing plant instrumentation. The ML algorithms were trained, validated, and tested using historic measurements from the plant using a variety of ML models, comprising Linear and Polynomial Regression (LR, PR), XGBoost (XGB) and Random Forests (RF), as well as Artificial Neural Networks (ANNs). The errors of the predictions have thoroughly been analysed, to identify specific patterns such as heteroscedasticity and bias in the residual errors. Finally, the importance of each feature to the target variable has been assessed, utilizing the p-values of the linear model, XGBoost importance, as well as sensitivity analysis of all models, using stochastic perturbation of the input variables. The first results are very encouraging, and we are in the process of integrating them with HYPPOS.

**Keywords:** Decision Support Tool · Machine Learning · Petrochemicals

M. Papadaki et al. (Eds.): EMCIS 2022, LNBIP 464, pp. 40–46, 2023.
https://doi.org/10.1007/978-3-031-30694-5_4

# 1 HYPPOS Software

## 1.1 Description

HYPPOS, Hyperion Predictive Production Online Software, is an innovative all-in-one deep tech decision-support tool that integrates with production level and business management level IT components of polymer production plants.

HYPPOS provides seamless integration and collects data from various external systems, such as Enterprise Resource Planning (ERP) systems, and Laboratory Information Management Systems (LIMS), as well as Automatic/Distributed Control Systems (DCS) on the production floor. HYPPOS transforms data collected into information and applies different inferencing algorithms, to provide real-time decision support and insight to operations personnel. HYPPOS addresses specific industry challenges related to quality traceability, real-time analysis, rapid identification, accurate reporting, process visualization, and quality consistency.

With real-time process data and effective communication between plant operators and managers on a single platform, corrective actions can be taken promptly to improve product and batch quality. HYPPOS detects off-spec/off-grade material making its exact location visible to operators in real-time allowing separation from the rest of the batch, leading to the production of first-time right material.

HYPPOS polymer plants can save up to 40% of production losses and therefore have a significant positive impact on production efficiency, profitability, and with consequential contribution in lowering the overall CO2 carbon emissions generated by the polymer industry.

The new EU Climate law targets to curb climate change by the Bloc, cutting carbon emissions by at least 55% by 2030 (compared to 1990 levels). By 2030 GHG emissions from plastics production will reach 1.26 Gt CO2e per year. The polymer industry is a major contributor with an industry with a CAGR of 5.1% (2020–2030), and its share will only increase. Actions need to be taken now.

## 1.2 Tracking Model

HYPPOS features an integrated tracking system based on a process flow model, which is driven by measurements from the plant. The tracking system can be thought of as a conveyor belt carrying material through the process equipment advancing at the same speed as the production stream. The model drives the "tracking conveyor", which tracks discrete sections of the production stream called "Quanta". The Quanta represents a mass of material entering the production stream over a defined time interval (for example, one minute). The Quanta act as a container for information to be tracked by collecting information at each point during production as they advance through the process equipment. Now, any information relating to materials, quality, or process measurements can be associated with the Quanta on the belt, and the information can easily be retrieved at a later stage (Fig. 1).

In HYPPOS, Hyperion has combined the material tracking technology and technologies for on-line polymer analysis and quality measurements (OLPA) into a single product, enabling tracking to be performed in real-time and providing the ability to

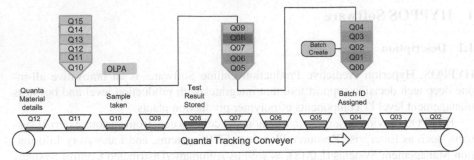

**Fig. 1.** Illustration of Quanta Tracking Concept

display the location of material with specific OLPA results as it passes through the production process.

In essence, HYPPOS displays the quality of material in real-time and in each process equipment where OLPA measurements are available. This real-time visibility on the quality of material enables the operator to take corrective actions as and when needed with the ultimate benefit of reducing the production of off-spec material and product waste. In the following figure, the slices of material are presented in the form of colored quality bands. For example, green means in-spec, yellow means suspect, red means off-spec (Fig. 2).

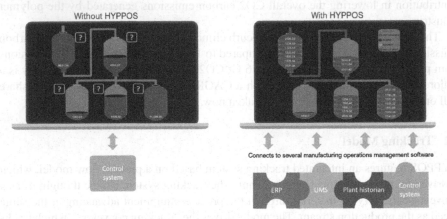

**Fig. 2.** Real-time Discretization, Visibility, Traceability and Quality Tracking

### 1.3   Results

HYPPOS Material Tracking model has been implemented to a Polymers plant as a main action to its digital transformation. The following benefits were reported:

- All continuous production history (reactor and extruder areas) were associated with the final batch product in real-time and available online.

- It offered a single consolidated source of information within a specific product batch, spanning across the whole production cycle (continuous & batch).
- Through the provided material tracking it enabled the immediate troubleshooting of production issues and the investigation of any customer complaints/claims.
- Online monitoring and real-time calculation of raw material consumption (e.g., monomer, catalyst, extruder additives) enabled the end-user to maintain good product quality (on-spec) as well to monitor the associated production cost.
- It allowed the real-time and online monitoring of production, by a range of users and managers of the organization. Facilitating the "single version of the truth" eliminated the need for the manual preparation of reports by the process engineers for their managers which could be prone to errors. Therefore, HYPPOS increased and made collaboration more effective.

The unavailability of on-line polymer analysers prevented the implementation of the OLPA algorithm/part of HYPPOS for the real-time tracking of quality. The unavailability of such quality analysers in most polymer plants primarily due to their high cost led to the idea of extending HYPPOS functionality to implement soft sensors using Artificial Intelligence/Machine Learning to provide quality predictions in real-time. The machine learning algorithms built and tested are described in the following section.

## 2 Machine Learning Algorithms

The aim of each applied ML algorithm is to predict the material quality (MFR – Melt Flow Rate) at polymer production plants, based on historical time series data of a variety of associated features, potentially driving the evolution of the material over time. The topic is significant in industrial, as well as research settings [1, 2]. The features are the time series' values of specific physical parameters, such as temperature, pressure, etc., measured on or before the reactors. Accordingly, the given dataset is comprised of approximately three thousand signals measured in different timestamps among them, as well as with the target variable.

The key steps of the data analysis process are the following:

- Collect production plant data and make a structured data engineering process
- Clean and transform the production plant data ready to be used by the machine learning model
- Produce descriptive statistics, plots and metrics
- Develop a machine learning model that predicts the material quality based on the selected and processed features, along with accuracy metrics and error analysis

The project started with an investigation of the statistical properties of the studied variables, followed by predictive modelling with Machine Learning Algorithms. In any predictive modelling computation, the predicted values always differ from the given ones in the raw dataset [3]. Hence, and despite the extended effort made to create a robust model, the predictions deviate from the actual value of MFR, especially in certain timestamps.

We train the ML models, after the data processing as automated by the HYPPOS program, enriched by new data. PR [4], XGB [5, 6] and RF [7, 8] and ANN [9] models, have been tuned with cross-validation, to identify their optimal hyperparameter, by training multiple instances of each model, and selecting the one with the best possible generalization accuracy. Despite the high amount of statistical noise existing in the studied dataset, we attained to train robust models in terms of their prediction capabilities on unforeseen data, and the Mean Absolute Percentage Errors (MAPE) in the majority of the test-set observations were less than 10% (Table 1 and Fig. 3).

**Table 1.** Performance Metrics for all ML models.

| Method | Dataset | Pearson | M.A.P.E | M.A.M.P.E | M.A.E | R.M.S.E | alpha | beta |
|---|---|---|---|---|---|---|---|---|
| Linear | Train | 0.9804 | 0.2099 | 0.1490 | 0.9894 | 1.4307 | 0.9612 | 0.2580 |
| Linear | Test | 0.9735 | 0.2323 | 0.1581 | 1.2888 | 1.9349 | 0.9505 | 0.5958 |
| Polynomial | Train | 0.9835 | 0.1875 | 0.1307 | 0.8679 | 1.2395 | 0.9708 | 0.1936 |
| Polynomial | Test | 0.9735 | 0.2104 | 0.1453 | 1.1849 | 1.8655 | 0.9540 | 0.5364 |
| XGBoost | Train | 0.9999 | 0.0122 | 0.0080 | 0.0594 | 0.0777 | 0.9981 | 0.0096 |
| XGBoost | Test | 0.9816 | 0.0657 | 0.0588 | 0.4795 | 1.6844 | 1.0192 | 0.0243 |
| ANNMBN | Train | 0.9935 | 0.0882 | 0.0750 | 0.4979 | 0.8286 | 0.9906 | 0.0813 |
| ANNMBN | Test | 0.9742 | 0.1530 | 0.1124 | 0.9167 | 1.9725 | 1.0102 | -0.1627 |
| Random Forests | Train | 0.9969 | 0.0441 | 0.0393 | 0.2607 | 0.5770 | 0.9853 | 0.0826 |
| Random Forests | Test | 0.9819 | 0.0688 | 0.0635 | 0.5177 | 1.6996 | 1.0275 | 0.0155 |

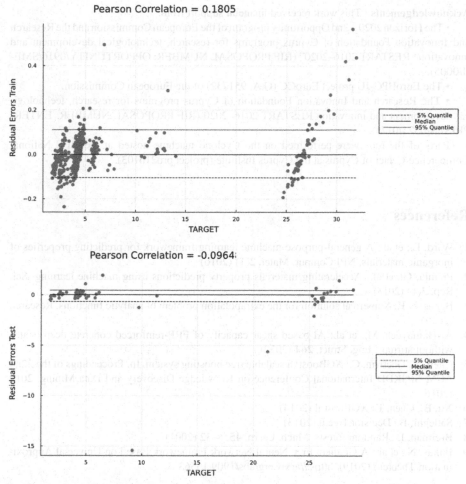

**Fig. 3.** Residual Errors of XGBoost model in Train and Test Sets. We see a good performance in the Test set, with certain points comprising high errors.

## 3 Conclusions

The given dataset comprises a rich variety of potential features for the prediction of MFR, with specific features exhibiting a high Pearson's Correlation with MFR. However, these correlations change with time vastly, along with the corresponding outliers in the Features. These Signal Outliers disorientate the predictions made by any of the utilized models. To confront this issue, we applied five ML models, with an exhaustive search for their hyperparameters. The best-performing model for the current dataset is the XGBoost one, followed by Random Forests with similar accuracy. We believe that this is a prototype algorithm capable of predicting the levels of MFR, in actual industrial plant conditions.

**Acknowledgements.** This work received financial support from:

• The Horizon 2020 – 2nd Opportunity program of the European Commission and the Research and Innovation Foundation of Cyprus programs for research, technological development and innovation "RESTART 2016–2020" (RIF PROPOSAL NUMBER: OPPORTUNITY/0916/SME-II/0005).

• The EuroHPC-JU project EuroCC (G.A. 951732) of the European Commission.

• The Research and Innovation Foundation of Cyprus programs for research, technological development and innovation RESTART 2016–2020 (RIF PROPOSAL NUMBER: ENTER-PRISES/0521/0175).

Parts of the runs were performed on the Cyclone machine hosted at the HPC National Competence Center of Cyprus at the Cyprus Institute (project pro21b103).

# References

1. Ward, L., et al.: A general-purpose machine learning framework for predicting properties of inorganic materials. NPJ Comput. Mater. **2**(1) (2016)
2. Pilania, G., et al.: Accelerating materials property predictions using machine learning. Sci. Rep. **3**(1) (2013)
3. Bakas, N.P.: Numerical solution for the extrapolation problem of analytic functions. Research (2019)
4. Al-Hamaydeh, M., et al.: AI-based shear capacity of FRP-reinforced concrete deep beams without stirrups. Eng. Struct. **264** (2022)
5. Chen, T., Guestrin, C.: XGBoost: a scalable tree boosting system. In: Proceedings of the 22nd ACM SIGKDD International Conference on Knowledge Discovery and Data Mining, 2016 (2016)
6. Xu, B., Chen, T.: XGBoost.jl (2014)
7. Sadeghi, B.: DecisionTree.jl (2013)
8. Breiman, L.: Random forests. Mach. Learn. **45**, 5–32 (2001)
9. Bakas, N., et al.: A Gradient Free Neural Network Framework Based on Universal Approximation Theorem (2019). https://arxiv.org/abs/1909.13563

# Identification of Multiple Sclerosis Signals' Dependence on Patients' Medical Conditions Through Stochastic Perturbation of Features in Five Machine Learning Models

Spyros Lavdas[1,2(✉)], Dimitrios Sklavounos[2], Panagiotis Gkonis[3], Panagiotis Siaperas[4], and Nikolaos Bakas[2,5]

[1] Department of Computer Science, Neapolis University Pafos, 8042 Paphos, Cyprus
s.lavdas@nup.ac.cy
[2] Department of Computer Science, Athens Metropolitan College, Marousi, Greece
{dsklavounos,nbakas}@mitropolitiko.edu.gr
[3] Department of Digital Industry Technologies, National and Kapodistrian
University of Athens, Dirfies Messapies, Greece
pgkonis@uoa.gr
[4] Institute of Occupational Science and Rehabilitation Metropolitan College, Athens
Metropolitan College, Marousi, Greece
psiaperas@mitropolitiko.edu.gr
[5] National Infrastructures for Research and Technology – GRNET, Athens, Greece

**Abstract.** Multiple sclerosis (MS) is a disease that deteriorates the central human nervous system, which can potentially cause significant brain, spinal cord and visual problems. Based on recent studies, MS has affected 3 million people with a prevalence rate of 3.9%. To this end, a wealth of information about MS has been produced, which makes MS the ideal candidate for applying artificial intelligence (AI) techniques for early diagnosis through a machine learning (ML) exploration framework. Accordingly, the current work studies to what extent the nervous system has been degenerated by analyzing data from a recently published dataset. Such data has been derived by motor evoked potential (MEP) measurements conducted in each patient hospital visit. Therefore, five machine learning models have been trained with cross-validation, in order to obtain the best one with good generalization properties. We compare the accuracy of all models utilizing various metrics (maximum obtained accuracy is ∼96% with XGBoost model). Furthermore, we use sensitivity analysis in order to explain the dependence of the target variable on the input parameters statistically.

**Keywords:** multiple sclerosis · EDSS · artificial intelligence · gradient boosting · polynomial regression · hyperparameter tuning · random forests

© The Author(s), under exclusive license to Springer Nature Switzerland AG 2023
M. Papadaki et al. (Eds.): EMCIS 2022, LNBIP 464, pp. 47–59, 2023.
https://doi.org/10.1007/978-3-031-30694-5_5

# 1   Introduction

Multiple sclerosis (MS) is a well-known autoimmune chronic disease which causes visuals, sensor and motor problems having as a direct consequence the deterioration of the functional status of the central nervous system (CNS). Besides, it is difficult enough to timely detect MS disease as there are no certain symptoms and physical findings that dictate its diagnosis. To this end, a multitude of medical approaches have been adopted in order to accurately assess and diagnose MS. To date, the most applicable medical treatment is magnetic resonance imaging (MRI) which is a non-invasive imaging technology that creates anatomical images.

## 1.1   Related Work

Most of the MRI research works for MS diagnosis are based on the implementation of machine learning (ML) or deep learning (DL) techniques in MRI scans in order to extract critical conclusions from brain images, [25, 26]. Although such type of processes is accurate and robust, it has turned out to be time-consuming and susceptible to manual errors, [21]. Instead, ML techniques have been rapidly evolved as the most promising player in the arena of MS decision support systems during the last decade. Such type of techniques does not require any prior knowledge or experience related to MS from clinicians facilitating the most accurate and objective diagnosis.

In particular, the most widely ML and DL-employed techniques have incorporated multiple data sources as input parameters such as clinical data, MRI scans, optical coherence tomography (OCT) data and motor evoked potential (MEP) measurements, [2, 18]. Some representative works will be presented in order to clarify the importance of ML models for MS decision support. To start with, [9]. In this paper the authors proposed a machine learning pipeline for clinical questionnaires analysis which aimed at detecting MS disease course. In particular, patient-reported outcomes (PRO) questionnaires were used in order to capture the self-perception of the MS disease. Besides, in a recent work [16], serum and CSF levels of forty-five cytokines were analyzed to identify MS diagnostic markers. Thus, cytokines were analyzed using multiplex immunoassay. Analysis of variance-based parameters and Pearson correlation coefficient scores were employed in order to utilize the appropriate input parameters for classification purposes. In the same context, [1], text mining methods were introduced in transcriptomic data analysis of multiple sclerosis disease for the first time. A complete predictive model was developed by taking into consideration consecutive transcriptomic data preprocessing procedures. Besides, the KmerFIDF method was utilized as a feature extraction method and linear discriminant analysis for dimensionality reduction. Additionally, in [17], a support vector machine (SVM) method with tenfold validation performed on specific properties of patients' blood, such as zinc, adiponectin, total radical-trapping antioxidant parameter and, sulfhydryl, in order to predict MS with high sensitivity, specificity, and accuracy.

There are also published works which are not strongly dependent on medical data, but their analysis has been built from raw data such as gait disturbances [17] or exhaled breath analysis [8]. In both of the former works, four classification algorithms were employed in total: i) Logistic Regression (LR), ii) XGBoost (XGB) iii) SVM and iv) artificial neural network (ANN) model in order to analyze the imported raw data for MS prediction and classification purposes. Noteworthy, classifications and predictions have been enhanced by including the parameter of expanded disability status scale (EDSS), [14], which is a method of quantifying disability in multiple sclerosis and tacking down the evolution of the disability. Namely, it holds values from 0 (healthy person) to 10 (death). In this context, all of the following indicative published works, [12,13,27] have adopted the EDSS parameter as the target of their classification ML techniques. To this end, a multitude of ML techniques has been utilized, such as Bayesian, random decision trees as well as simple logistic-linear regression.

Apart from the EDSS parameter, there is a specific additional type of data that improved the accuracy and sensitivity of ML models. Such type of data is derived from MEP measurements, namely conducted measurements which quantify the conductivity of the CNS. In [5,23], MEP measurements were carried out and were further analyzed by random forests and linear regression classifiers. To this end, the current study is based on MEP measurements and utilizes the EDSS as a target parameter for the employed ML algorithms. The required dataset regarding the MEP measurements has been derived from a recently published paper, [24].

It is worth noting that the evaluation of a patient's disability is a multi-parameter medical process which is prone to EDSS miscalculation due to manual errors, and it is time-consuming as well. Thus, the estimation of EDSS through an automated analysis of a nervous system signal pulse, as suggested by the current work, could accelerate all the procedures in terms of MS prognosis and medical treatment.

The rest of the current research work is organized as follows: In Sect. 2, the structure of the used dataset and the metadata regarding patients is presented. The key features of the five employed machine learning models are described in the next Sect. 3 while the analysis of the derived results is developed in Sect. 4. Finally, the main conclusions and the proposal for future work are given in the last section.

## 2    Description of Dataset

The dataset derived from the work of Yperman et al., [24], contains data regarding electrical signals which have propagated through the NS and detected from either hand or foot. In particular, the brain of each patient has been stimulated through a magnetic stimulator and an external trigger system leading to the creation of a signal pulse which propagates along the nervous system. Samples of the resulting signal are detected, are stored and exported into a file. To be more specific, 2000 time points in a time window of 100 ms determine the signal shape. In total, the dataset includes information about:

– metadata of patients (963 records) such as age, gender, time of hospital visit, type of machine that conducted the measurements, teams that carried out the MEP measurements, etc. For more information, see [24].
– MEP measurements (96290 records).
– EDSS values (7414 records) for specific patients.

As it has already been mentioned, the target of the ML techniques is the EDSS value for each patient at the specific time of MEP measurement. To achieve this, critical properties of the resulting signals should be taken into consideration in order to be used as input parameters to the machine learning models. The following table describes all the input parameters used for the development and analysis of the suggested prediction model.

**Table 1.** Description of the utilized input parameters.

| Input parameters | Description |
| --- | --- |
| Age | Age of the patient |
| AnatomyAH | AnatomyAH = 1 corresponds to the MEP hand measurement, while AnatomyAH=0 corresponds to the MEP feet measurement |
| R | R=1 corresponds to right(hand or foot), while R=0 corresponds to left (hand or foot) |
| maximum peak | Maximum peak voltage of signal (mV) |
| FilterTrue | FilterTrue=1 corresponds that the machine has applied frequency filter to the raw MEP measurements, while FilterTrue=0 means that there was not any filter applied to the MEP measurement |
| Time minimum | The timepoint when the minimum peak occurs |
| half power max(min) | The width of the pulse when the maximum(minimum) peak occurs |
| Time maximum | The timepoint when the maximum peak occcured |
| time of first local-0.25(0.5,0.75) min | The timepoint when a local peak obtains voltage $\geq 0.25(0.5, 0.75)$ of the minimum peak |
| mean | Voltage average of the pulse |
| minimum peak | Minimum peak voltage of signal (mV) |
| delta time of first local-0.25(0.5,0.75) min | Time difference between time of first local-0.25(0.5,0.75) and time minimum |
| Male | Male=1 corresponds to male, while Male=0 corresponds to female |
| delta time max min | Time difference between the times that maximum and minimum occur |
| time of first local-0.25(0.5,0.75) max | The timepoint when a local peak obtains voltage $\leq 0.25(0.5, 0.75)$ of the maximum peak |
| Energy | The energy of the signal pulse calculated by $\int_0^{100\ ms} |U(t)|^2 dt$ |
| std deviation | Standard deviation |
| TeamA | TeamA=1 means that the group A conducted the MEP measurement, while TeamA=0 corresponds to team B |
| MachineA | MachineA=1 corresponds to the machine A that used to carry out the MEP measurement, while MachineA=0 corresponds to machine B |
| N min(max) local 0.25(0.5,0.75) | Number of minimum(maximum) locals when the peak voltage corresponds to voltages of $\geq(\leq)0.25(0.5, 0.75)$ of minimum(maximum) peaks |

# 3   Machine Learning Models

The following models for approximating Features-Target relationship have been employed.

1. Linear Regression as a baseline model for the next ones.
2. Polynomial Regression. As it is necessary to select from a vast pool of potential nonlinear features along with their number, the ITSO [4] as well as PROS [19] Optimization Algorithms for Feature Selection have been adopted, which have also been found experimentally vastly efficient.
3. Gradient Boosting [7,10,22] with hyperparameter tuning. Particularly, the grid search method with cross-validation has been utilized.
4. Random Forests [6], as implemented in [20]
5. Artificial Neural Networks [3].

For each model, the following computations are carried out:

- The accuracy among the Prediction and Target-variable for the Train and Test Sets.
- Error analysis: Residual Errors vs Target diagrams, Probability Density Functions, Cumulative Density Functions, for the Train and Test Sets as well.

The former approach is beneficial for detecting specific patterns occurring in the prediction and, hence, enhancing the generalization capability and reliability of the model.

# 4   Results

The application of the aforementioned ML techniques produced a vast amount of constructive results which shed light on the MS disease. It should be highlighted, that input parameter with high discrepancies or abrupt deviations of the signal voltage magnitude in the time domain has been removed from the calculations. In the same, context, all the spikes occurred at the beginning of the detected pulse have been deleted. Figure 1 shows the pairwise correlations among the input parameters. Particularly, we reform the correlation matrix, such that the non-zero values are concentrated around the diagonal of the matrix. This is performed with the CuthillMcKee Method [11,15]. This way, the clusters of associated features are computationally identified.

## 4.1   Performance of ML Models

After an exhaustive search for the hyperparameters of the studied ML models, we identified that RF and XGB exhibited the best performance (Figs. 2, 3). The error metrics are depicted in Fig. 4. Although the final models have adequate accuracy, some errors occur, especially for lower values of EDSS. Hence we run a data adequacy check (Fig. 5), where it is demonstrated that the accuracy is increased with the number of training samples.

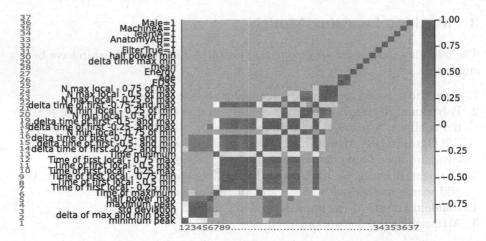

**Fig. 1.** CuthillMcKee representation of the input parameters, see Table 1. Note that the x-axis label is exactly the same as that of y-axis. For the sake of readability, the x-axis labels are described with numbers which correspond to specific input parameters, as shown on the left side of the graph.

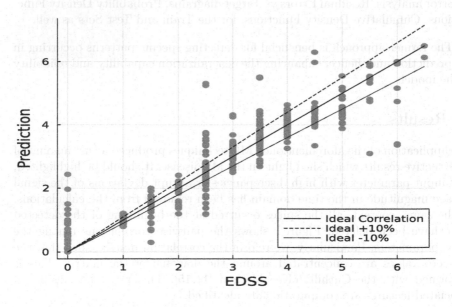

**Fig. 2.** Target vs predicted EDSS derived from linear model.

## 4.2 Sensitivity Analysis

Finally, we run a sensitivity analysis of the input parameters with respect to the target, by keeping all features constant to a particular value (median, 25% and 75% quantiles), and change the studied feature within its given values in the initial dataset. Accordingly, we predict using each one of the models. Figures 6

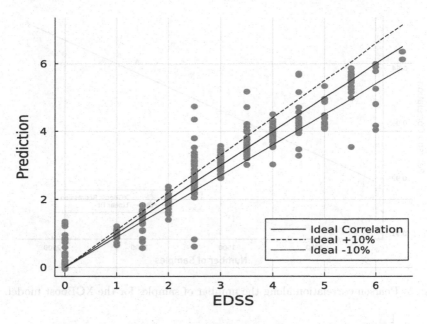

**Fig. 3.** Target vs predicted EDSS derived from XGBoost.

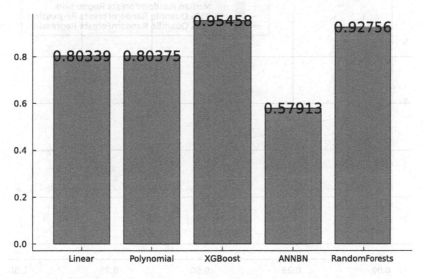

**Fig. 4.** Error metrics for 5 machine learning methods implemented in the current analysis.

and 7 illustrate that the Anatomy AH, when it is equal to the unit, the EDSS results in lower values. Furthermore, Figs. 8 and 10 depict an increasing pattern of EDSS with respect to age.

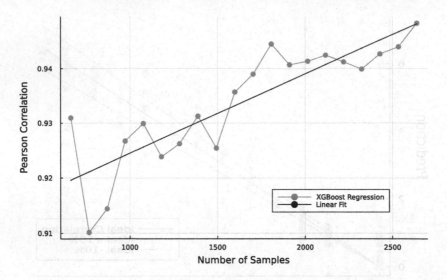

**Fig. 5.** Pearson correlation along the number of samples for the XGBoost model.

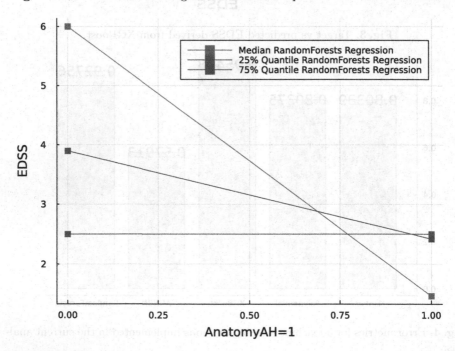

**Fig. 6.** Sensitivity analysis for AnatomyAH=1 based on the RandomForests model.

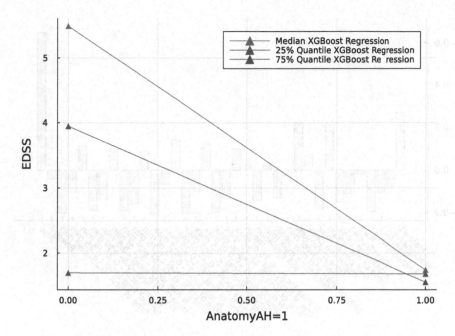

**Fig. 7.** Sensitivity analysis for AnatomyAH=1 based on the XGBoost model.

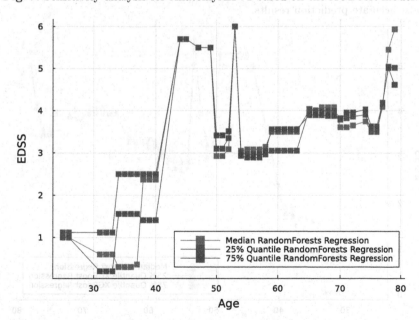

**Fig. 8.** Presentation of increasing pattern between the input parameter of age and EDSS emerged from the RandomForests model.

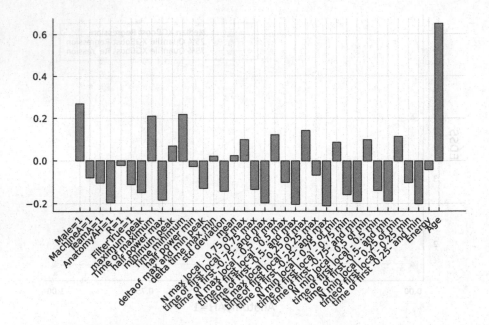

**Fig. 9.** Features importance derived by XGBoost model, which was proved to provide the most accurate prediction results.

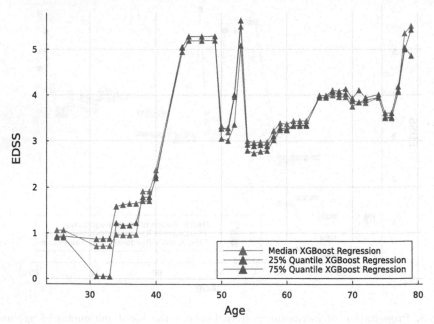

**Fig. 10.** Resentation of increasing pattern between the input parameter of age and EDSS emerged from the XGBoost model.

# 5 Conclusions

The present research analysis in concert with former descriptive Figs. 1-3 manifest that the highest Pearson correlation was achieved through the implementation of XGBoost, reaching approximately 96% accuracy. On the contrary, the lowest accuracy of 58% was obtained by the artificial neural network. Thus, the most appropriate ML model to predict the value of EDSS from a propagating signal on the nervous system is XGBoost.

One of the most critical conclusions drawn in the current MS analysis is depicted in Fig. 9 wherein the most effective input parameters of the prediction model are presented. In particular, the age, gender, anatomyAH and maximum peak are the prominent ones. Among the former significant input parameters are also the energy, time difference of local peaks as well as the time points at either local maximums or minimums occur.

Besides, the Pearson correlation follows an increasing linear trend along with the increase of input training data, Fig. 5. Hence, the suggested prediction model can be clarified as efficient and robust.

The early results of the current research work dictate its extension to the prediction of the forthcoming evolution of EDSS for a certain MS patient. This is of great importance as it will determine the most appropriate long-term medical treatment to enhance the quality of patient's life.

# References

1. Ali, N.M., Shaheen, M., Mabrouk, M.S., Aborizka, M.A.: A novel approach of transcriptomic microrna analysis using text mining methods: an early detection of multiple sclerosis disease. IEEE Access **9**, 120024–120033 (2021). https://doi.org/10.1109/access.2021.3109069
2. Aslam, N., et al.: Multiple sclerosis diagnosis using machine learning and deep learning: Challenges and opportunities. Sensors **22**(20) (2022). https://doi.org/10.3390/s22207856, https://www.mdpi.com/1424-8220/22/20/7856
3. Bakas, N.P., Langousis, A., Nicolaou, M., Chatzichristofis, S.A.: A gradient free neural network framework based on universal approximation theorem. arXiv preprint arXiv:1909.13563 (2019)
4. Bakas, N.P., Plevris, V., Langousis, A., Chatzichristofis, S.A.: ITSO: a novel inverse transform sampling-based optimization algorithm for stochastic search. Stoch. Env. Res. Risk Assess. **36**(1), 67–76 (2021). https://doi.org/10.1007/s00477-021-02025-w
5. Bejarano, B., et al.: Computational classifiers for predicting the short-term course of multiple sclerosis. BMC Neurol. **11**(1) (2011). https://doi.org/10.1186/1471-2377-11-67
6. Breiman, L.: Random Forest. Machine Learning **45**(1), 5–32 (2001)
7. Chen, T., Guestrin, C.: Xgboost. In: Proceedings of the 22nd ACM SIGKDD International Conference on Knowledge Discovery and Data Mining (2016). https://doi.org/10.1145/2939672.2939785
8. Ettema, A.R., Lenders, M.W., Vliegen, J., Slettenaar, A., Tjepkema-Cloostermans, M.C., de Vos, C.C.: Detecting multiple sclerosis via breath analysis using an eNose,

a pilot study. J. Breath Res. **15**(2), 027101 (2021). https://doi.org/10.1088/1752-7163/abd080

9. Fiorini, S., Verri, A., Tacchino, A., Ponzio, M., Brichetto, G., Barla, A.: A machine learning pipeline for multiple sclerosis course detection from clinical scales and patient reported outcomes. In: 2015 37th Annual International Conference of the IEEE Engineering in Medicine and Biology Society (EMBC), pp. 4443–4446 (2015). https://doi.org/10.1109/EMBC.2015.7319381

10. Friedman, J.H.: Stochastic gradient boosting. Comput. Stat. Data Anal. **38**(4), 367–378 (2002)

11. Gates, R.L.: Cuthillmckee.jl (2022). https://github.com/rleegates/CuthillMcKee.jl

12. Ion-Mărgineanu, A., et al.: Machine learning approach for classifying multiple sclerosis courses by combining clinical data with lesion loads and magnetic resonance metabolic features. Front. Neurosci. **11**, 398 (2017). https://doi.org/10.3389/fnins.2017.00398

13. Kocevar, G., et al.: Graph theory-based brain connectivity for automatic classification of multiple sclerosis clinical courses. Front. Neurosci. **10**, 478 (2016). https://doi.org/10.3389/fnins.2016.00478

14. Kurtzke, J.F.: Rating neurologic impairment in multiple sclerosis. Neurology **33**(11), 1444–1444 (1983). https://doi.org/10.1212/WNL.33.11.1444, https://n.neurology.org/content/33/11/1444

15. Liu, W.H., Sherman, A.H.: Comparative analysis of the Cuthill-Mckee and the reverse Cuthill-Mckee ordering algorithms for sparse matrices. SIAM J. Numer. Anal. **13**(2), 198–213 (1976)

16. Martynova, E., et al.: Serum and cerebrospinal fluid cytokine biomarkers for diagnosis of multiple sclerosis. Mediat. Inflamm. **2020**, 1–10 (2020). https://doi.org/10.1155/2020/2727042

17. Mezzaroba, L., et al.: Antioxidant and anti-inflammatory diagnostic biomarkers in multiple sclerosis: a machine learning study. Mol. Neurobiol. **57**(5), 2167–2178 (2020). https://doi.org/10.1007/s12035-019-01856-7

18. Nabizadeh, F., et al.: Artificial intelligence in the diagnosis of multiple sclerosis: a systematic review. Multiple Sclerosis Related Disord. **59**, 103673 (2022). https://doi.org/10.1016/j.msard.2022.103673, https://www.sciencedirect.com/science/article/pii/S2211034822001882

19. Plevris, V., Bakas, N.P., Solorzano, G.: Pure random orthogonal search (pros): a plain and elegant parameterless algorithm for global optimization. Appl. Sci. **11**(11), 5053 (2021). https://doi.org/10.3390/app11115053

20. Sadeghi, B.: Decisiontree.jl (2013)

21. Stafford, I.S., Kellermann, M., Mossotto, E., Beattie, R.M., MacArthur, B.D., Ennis, S.: A systematic review of the applications of artificial intelligence and machine learning in autoimmune diseases. NPJ Digit. Med. **3**(1) (2020). https://doi.org/10.1038/s41746-020-0229-3

22. Xu, B., Chen, T.: Xgboost.jl (2014)

23. Yperman, J., et al.: Machine learning analysis of motor evoked potential time series to predict disability progression in multiple sclerosis. BMC Neurol. **20**(1), 1–15 (2020). https://doi.org/10.1186/s12883-020-01672-w

24. Yperman, J., Popescu, V., Wijmeersch, B.V., Becker, T., Peeters, L.: Motor evoked potentials for multiple sclerosis, a multiyear follow-up dataset. Sci. Data **9**(1), 207 (2022). https://doi.org/10.1038/s41597-022-01335-0

25. Zeng, C., Gu, L., Liu, Z., Zhao, S.: Review of deep learning approaches for the segmentation of multiple sclerosis lesions on brain MRI. Front. Neuroinformatics **14**, 610967 (2020). https://doi.org/10.3389/fninf.2020.610967

26. Zhang, Y., et al.: Comparison of machine learning methods for stationary wavelet entropy-based multiple sclerosis detection: decision tree, k-nearest neighbors, and support vector machine. Simulation **92**(9), 861–871 (2016). https://doi.org/10.1177/0037549716666962

27. Zhao, Y., et al.: Exploration of machine learning techniques in predicting multiple sclerosis disease course. PLOS ONE **12**(4), e0174866 (2017). https://doi.org/10.1371/journal.pone.0174866

# Big Data and Analytics

# Applying Predictive Analytics Algorithms to Support Sales Volume Forecasting

Jörg H. Mayer[1]([✉]), Milena Meinecke[1], Reiner Quick[1], Frank Kusterer[2], and Patrick Kessler[3]

[1] Darmstadt University of Technology, Hochschulstrasse 1, 64289 Darmstadt, Germany
jhmayer@t-online.de
[2] UE University of Europe for Applied Sciences, Reiterweg 26B, 58636 Iserlohn, Germany
[3] WILO SE, Wilopark 1, 44263 Dortmund, Germany

**Abstract.** Caompanies struggle with predictive analytics (PA), which aims to be a "modern" crystal ball. But how does one choose the "right" algorithms? Based on the findings from a sales volume forecasting case study, this article presents six design guidelines on how to apply PA algorithms properly: (1) When fixing the objective of your forecast, start with reflecting the available data. (2) Considering the available data and forecast horizon, develop a strategy for the training phase, ultimately the model's deployment. (3) Choose algorithms first that act as an orientation as well as a benchmark for more elaborated models. (4) Continue with time series algorithms such as (S)ARIMA and Holt-Winters. Take automated parameter setting into consideration. (5) Integrate additional input by applying ML-based algorithms such as LASSO Regression. (6) Besides accuracy, process efficiency and transparency determine the most suitable approaches.

**Keywords:** Forecasting · Predictive Analytics · Artificial Intelligence · Machine Learning · Design Science Research in Information Systems · Design Guidelines · Case Study: Manufacturing Industry

## 1 Introduction

Companies that rely solely on human judgment or simply extrapolate historical values neglect predictive analytics (PA) to improve their forecasting. Although studies have revealed that machine-learning-(ML)-based algorithms can increase the value of big data, practice falls short on applying such algorithms [1]. For example, we found only few case studies such as Wang [2], Gerritsen et al. [3], Sagaert et al. [4], and Concalves et al. [5] addressing the integration of external variables such as leading indicators into PA sales forecasts. We argue that practitioners often lack guidance when applying predictive analytics on big data.

Accordingly, the objective of this article is to present both findings from a sales volume forecasting for a reference company and – derived from these findings – associated design guidelines on how to apply PA algorithms properly. We take an international pump system provider as our reference aiming to increase forecast accuracy, whilst

M. Papadaki et al. (Eds.): EMCIS 2022, LNBIP 464, pp. 63–76, 2023.
https://doi.org/10.1007/978-3-031-30694-5_6

also improving process efficiency and transparency compared to its current approach. Efficiency covers the reduction of manual actions whilst generating comparable outcomes. Transparency refers to the reproducibility and traceability of actions. We pose two research questions (RQ):

1. What constitutes design guidelines on how to apply PA algorithms?
2. Do the proposed design guidelines fulfill what are they meant to do (validity) and are they useful beyond the environment in which they were developed (utility and generalizability)?

To create things that serve human purposes [6], ultimately to build a better world [7], we follow Design Science Research (DSR) in IS [8, 9] for which the publication scheme from Gregor and Hevner [10] gave us direction. The lack of experience in applying PA algorithms for sales (volume) forecasting motivates this article *(introduction)*. Highlighting several research gaps, we contextualize our research questions *(literature review)*. Addressing these gaps, we adopt a single case study *(method)*. Emphasizing a staged research process with iterative "build" and "evaluate" activities [10], we come up with design guidelines *(artifact description,* related to RQ 1). Testing their validity and utility [11], we perform expert interviews within and beyond the reference company *(evaluation,* related to RQ 2). Comparing our results with prior work and examining how they relate to the article's objective and research questions, we close with a summary, limitations of our work, and avenues for future research *(discussion and conclusion)*[12].

## 2 Literature Review

Following Webster and Watson [13] as well as vom Brocke [14, 15], we identified relevant articles in a four-stage process. (1) We focused on leading IS journals [16] as well as selected business, management, and accounting journals [17], complemented by proceedings from major IS conferences [18] *(outlet search)*. For a practitioners' perspective, we considered journals such as MIS Quarterly Executive and Harvard Business Review. (2) Accessing the identified outlets, we used ScienceDirect, EBSCOhost, Springer Link, and AIS eLibrary *(database search)*. (3) Then, we searched for articles through their titles, abstracts, and keywords *(keyword search)* – limiting the results to the last ten years.

Integrating this strategy for our research, we combined the keywords "sales (volume)" or "demand" with "planning," "forecasting," or "predicting." To limit our findings to data-driven approaches, we then added the keywords "analytics" or "model." This search resulted in *31 relevant papers* that focus on sales or demand forecasting within a business setting applying data-driven methods. Finally, (4) we conducted a *backward and forward search* identifying another sixteen publications such as Verstraete et al. [19].

For our gap analysis, we structured the publications into three clusters: (1) The *application area* describes the context in which the existing PA models have been developed. Furthermore, we examined (2) *forecasting techniques*. (3) Lastly, we differentiated between *research methods*, addressing the way in which researchers collect and understand data [20]. Table 1 summarizes the take-aways of our literature review.

**Table 1.** Publications Shaping our Research Approach

| Author | Research Gap | Take away for proposed model |
|---|---|---|
| (1) **Application area:** [1–5, 21–23] | Regarding RQ #1: Papers focusing on PA for manufacturing companies and their production planning are underrepresented | We contribute a use case focusing on forecasting for sales volumes at the product group level |
| (2) **Forecasting techniques:** [4, 19] | Regarding RQ #1: There is a lack of research on advanced forecasting approaches beyond judgmental and statistical forecasts | We propose to test a series of different forecast approaches including *ML-based algorithms,* to increase accuracy whilst improving process efficiency and transparency |
| (3) **Research method**: [22, 24] | Regarding RQ #2: We examined a lack of qualitative research within a company setting | With a *single case study*, we aim to expand the current body of know-ledge, focusing on a company setting, and present design guidelines for future discussions |

**Application Area.** Most of the reviewed publications focused on PA models in the *retail industry*. Predicting online sales, we found articles about leveraging product reviews [25], social media posts, and other online activities via search engines [26]. Authors such as Abolghasemi et al. [27] considered the impact of promotions. To gain customer feedback, *sentiment analyses* were frequently conducted [28].

Focusing on production planning, we found few papers such as Fildes et al. [21] and Chen and Chien [22]. The examined forecasting horizon differed between short-term [23] and strategic planning [4]. With varying aggregations level, Wijnhoven and Plant [29] focus on car sales, Ma and Fildes [30] examined *forecasting at the product level*.

**Forecasting Techniques.** A growing number of articles examined *ML-based algorithms* such as artificial neural networks and support vector machines. They can model more complex relationships for leveraging big data [31]. *Big data* refers to the growing amount of diverse data being generated by different enterprise resource-planning processes, business intelligence systems, sensors such as smart meters, social media listening, and other sources [32]. However, research lacks deploying ML techniques, especially when adding leading indicators to potentially increase forecast accuracy. This is often done manually so that outcomes are susceptible to bias and lack transparency [33]. Based on this insight, we will apply different statistical and ML models to increase accuracy, whilst improving process efficiency and transparency compared to the judgmental forecast in place (Table 1).

**Research Method.** Although there are articles on sales volume prediction with a focus on quantitative research [21, 34], they are barely employed in practice [19, 22, 24, 35].

Research neglects qualitative research examining the integration of forecast solutions into day-to-day business. Thus, with a single *case study*, we aim to expand the current body of knowledge.

## 3 Method

Case studies bridge the gap between practice and academia when few activities on this research topic exist and practical insights are proposed as significant [36, 37]. Compared to surveys, they provide more substantial in-depth information and enable researchers to study their artifacts in a natural setting [38].

Compared to multiple case studies, which yield a multitude of results [39], *single case studies* are more suitable when the research topic is complex, and thus, relevant starting points for research are not easy to obtain [40]. Guided by our research questions (Sect. 1) and the findings from our literature review (Sect. 2), we decided to conduct a *single case study*. We took an international pump system provider (sales: 1.5 billion EUR; 8,000 employees) as our reference company. It produces pumps for various applications such as industrial production and building services and their sales are influenced by seasonal fluctuations and affected by continuous technological progress.

We focused on the sales volume forecasting for two product groups within Germany with a forecast horizon of twelve months. The pumps are installed in heating devices for residential and commercial buildings. Our project team consisted of the authors of this paper and members of the reference company's controlling department. Modeling our PA solution, we followed CRISP-DM [41] and gathered learnings throughout its different phases. We finally deducted design guidelines in an iterative manner by discussing these learnings within our project team and how they might be helpful for implementing PA forecasting solutions in similar use cases.

In order to gather feedback about their utility and validity, we present them in semi-structured expert itnerviews (RQ 2, Sect. 1) which took on average 45 min. These interviews combine both a comparable structure within a series of interviews whilst being flexible when interviewees want to share their individual way of thinking and any hidden facets [42]. We performed these interviews with the Head of Finance and three employees from Controlling and Supply Chain Planning of the reference company. Including an external perspective, we conducted interviews with two data science experts from the working group "Digital Finance" [43].

## 4 Artifact Design

Emphasizing a staged research process with iterative "build" and "evaluate" activities, we started with the results from our literature review (Sect. 2) and followed CRISP-DM. We gained **(1) a business understanding** by examining the forecasting process of sales volumes currently in place at our reference company. The twelve-month rolling forecast is performed at the sub-group level and later aggregated to the group level. It is conducted as follows: (a) After cleansing the historical sales volume data from exceptional influences, (b) the reference company currently applies a Holt-Winters algorithm within SAP Advanced Planning and Optimization (SAP APO, R. 4.1). (c) The results are then manually adjusted by local planners considering short-term influences such as sales promotions as well as general economic climate and technological trends.

The cleansing step is skipped, as data relating to exceptional events are often missing. The Holt-Winters parameters had not been altered in recent years. With the different sales regions in focus, we observed differences in the local planner's manual forecasting adjustments. Due to a series of unstructured adjustments, the current procedure is time-consuming and lacks transparency resulting in forecasts that are prone to bias with no monitoring or documentation of achieved accuracies in place.

We gathered possible influences on the sales volumes in seven expert interviews with local demand planners who have been involved in the forecast process and members of the sales department who are knowledgeable about the specific market conditions. We interviewed all available employees until saturation. Based on the qualitative content analysis of our interviews, we compiled a list of possible long- and short-term influences. Their effect on the sales volume was perceived as high by the interviewees, which motivated us to include them in the PA model.

Differentiating between influences and indicators, the latter being specific data that can be processed by the model, we devoted considerable effort to identifying influences on sales volumes and finding leading indicators representing them. We recommend limiting the effort of collecting and analyzing indicators, as their benefit for the forecast accuracy is often questionable as measurable data often do not accurately represent underlying constructs [44]. In our experience, starting the modeling with historical time series data only is a more efficient approach since those data are often readily available. Considering the overall objective of the forecast, budget, expertise, and available data, we recommend to iteratively specifying the scope of the forecast and then gradually collecting additional data when necessary.

*Design guideline #1: When fixing the objective of your forecast, start with reflecting the available data.*

Within the **(2) data understanding** process, we examined the historical sales volume data in order to apprehend their trend and seasonality. We took nine years of monthly sales volume data without missing data points, but with structural changes caused by product changes. Both product groups are highly seasonal with a decreasing trend (Fig. 1). By decomposing the time series into its seasonal trend and residual component, this becomes apparent (Fig. 1). Furthermore, we analyzed indicators representing the identified influences from the expert interviews (Table 2). Public statistics databases acted as our source for indicators such as economic sentiment indices [45] and building permits [46]. We were unable to gather indicators reflecting some aspects mentioned by the interviewees such as regulatory requirements and technological shifts. Our search resulted in six indicators (Table 2).

During **(3) data preparation**, indicators and historical data were prepared according to the requirements of different forecasting algorithms. Handling seasonality was crucial to the model's success. Removing seasonality helped to detect underlying patterns within time series. Whether seasonality should be removed before modeling depends on various factors like the size of available datasets, the stability of seasonal patterns, and the

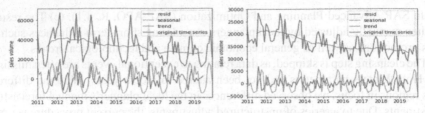

**Fig. 1.** Decomposition of Product Group 1(left) and 2 (right).

algorithm applied. We prepared two datasets, one with unadjusted data and the other one using seasonal differencing. The latter is subtracting the previous year's observation to remove seasonal fluctuations. In the case of our sentiment indicators, we used the seasonally adjusted data already available.

**Table 2.** Supposed Influences and Indicators of Sales Volume.

| Influences | Leading indicators (not all influences represented) |
|---|---|
| General economic development | Economic sentiment index (Eurostats) |
| Sector-specific development | Confidence indicators for sectors industrial, construction, |
| Regulatory requirements | and retail (Eurostats) |
| Internal and external competition | Consumer confidence indicator (Eurostats) |
| | Building permits issued (Statistisches Bundesamt) |

Additionally, we standardized the indicators by subtracting their mean and dividing the result by their standard deviation to remove differences in scale, which is necessary for multivariate models. Preparing the data is as important as choosing the right algorithm and having different test datasets is necessary for evaluating the performance of your algorithm accordingly.

Comparing the accuracy of various approaches, the best error measure depends on the individual use case [47]. We used the mean absolute percentage error (*MAPE*, Table 3) because it is easily comprehensible for business users. We performed stepwise *cross-validation*, splitting our dataset into overlapping time periods with three years of training data, one year representing the forecast horizon, and using the consecutive month as the test point. This approach validates the model efficiently while always considering the time restrictions of the data to generate valid results.

After the deployment of the model, the accuracy must be continuously monitored and documented, as this was a major hurdle for the use case, having only data from the past three years from the current forecast.

*Design guideline #2: Considering the available data and forecast horizon, develop a strategy for the training phase, ultimately the model's deployment.*

For **(4) modeling**, we adopted seasonal naïve, taking the unaltered observation of the previous year as our prediction. Understanding the time series' composition (step 2) helped to choose a simple algorithm as a starting point and reference for other modeling approaches. Depending on the use case, even simple algorithms such as naïve methods can help companies improve their existing forecasting processes [48]. These models do not require extensive data preparation and therefore are easy to implement.

*Design guideline #3: Choose algorithms first that act as both an orientation as well as a benchmark for more elaborated models.*

The seasonal naïve approach could significantly lower the MAPE compared to the automated SAP APO forecast and even outperformed the manually adjusted forecast, in doing so, setting the seasonal naïve as the benchmark to beat (Table 3). Moving on to time series algorithms, we applied Seasonal Autoregressive Integrated Moving Average (SARIMA) and Holt-Winters. Besides dealing with potential outliers and missing data, the predefined algorithms do not require additional data preparation. Performing a stepwise forecast, we used the Python packages *pmdarima* and *statsmodels* instead of manual parameter setting, which is often challenging. We recommend opting for popular methods, as the available packages are usually well-maintained, and resources on their application are widely available.

**Table 3.** Results of several Forecast Algorithms

| Method | MAPE (Product group 1) | MAPE (Product group 2) |
|---|---|---|
| SAP APO (Holt-Winters) | 38.76% | 23.61% |
| Local planner | 22.18% | 16.67% |
| Seasonal naïve | 21.5% | 14.68% |
| SARIMA | 37.26% | 27.75% |
| Holt-Winters (dynamic parameters) | 22.05% | 16.44% |
| LASSO linear regression | 17.49% | 14.96% |

Although we could not achieve an improvement compared to seasonal naïve, we improved the accuracy of the automated SAP APO forecast using the Holt-Winters algorithm (Table 3). Using automated parameter setting allows to continuously update the algorithm's parameters, leading to better results.

*Design guideline #4: Continue with time series algorithms such as (S)ARIMA and Holt-Winters. Take automated parameter setting into consideration.*

Due to the assumption that external factors significantly affect sales volumes, we tried to further improve the forecasting accuracy by integrating *leading indicators*. Performing a correlation analysis, we tried to identify influential variables and their associated time lag. Exogenous variables can easily be added to the SARIMA or Holt-Winters algorithms used in our case study or to regression models. However, the procedure of manually selecting and integrating indicators is complex, especially when dealing with multiple indicators and different time lags.

In our case, we could not determine any unambiguous relation between sales volume and our indicators. Thus, instead of selecting the indicators manually, we used the LASSO algorithm with a linear regression model to select the indicators automatically. The algorithm consists of an additional penalization term that prevents overfitting and selects the best fitting variables. The predefined Python function we used can be found in the scikit-learn library. By adding the six indicators as well as the historical time series data and their respective 12-to-24-month lags, the LASSO algorithm could automatically select the best input variables with the optimal time shift.

For our two product groups, we could only slightly improve the forecast accuracy of the benchmark by including additional indicators with LASSO regression. The indicators chosen by the LASSO algorithm were mostly historical sales volume data, and other indicators were not consistent over the years. This could be because our time series are not significantly influenced by external indicators, or because our selection of indicators might not represent the actual influences on sales volume. Due to the nature and restricted size of our available datasets, we did not include other ML algorithms such as neural networks and vector autoregression.

*Design guideline #5: Integrate additional input automatically by applying ML-based algorithms such as LASSO Regression.*

(5) **Evaluating** the results of our models (Table 3), for our time series, we could not significantly improve our long-term forecasts (twelve months) over our simple benchmark (seasonal naïve, PG1: 21.5%, PG2: 14.68%) by adding additional data or using more complex methods. Compared to the automatic forecast currently in place (SAP APO, PG1: 38.76%, PG2: 23.61%), we could achieve a significant improvement in this use case. Even the manually adjusted forecast currently in place did not yield better results for twelve-month forecasts (PG1: 22.18%, PG2: 16.67%). This indicates that the current method is not adequate for our examined products and aggregation level.

From the algorithms that we considered, seasonal naïve yielded the highest accuracy except for PG1, which had better accuracy with the LASSO regression (17.49%). Considering the inconsistent coefficients generated by the algorithm, the improvement does not seem significant. When including other evaluation criteria transparency and process efficiency, the seasonal naïve has the advantage of entailing minimum effort for collecting data and generating the forecast, while being transparent and easy to understand for the recipients of the forecast. Process efficiency can encompass a variety of factors such as computational resources as well as human resources for developing and maintaining the forecast. Especially, when manual reviews and adjustments are included in the process, transparency is an important criterion when selecting a forecasting approach. The acceptance and abilities of the involved employees must be considered.

Therefore, we propose forecasting the sales volumes of our examined product groups by applying seasonal naïve for twelve-month forecasts. By reducing the need for manual adjustments for those product groups, the planners' expertise could be targeted toward more challenging forecasting tasks. Identifying adequate methods by systematically testing different approaches and continuously monitoring them can lead to more accurate forecasts, even when using simple data-driven approaches. Robust and tangible results are to be preferred over complex and costly approaches, as accuracy is not the only criterion that should be considered.

*Design guideline #6: Besides accuracy, process efficiency and transparency determine the most suitable approaches.*

For the final CRISP-DM step **(6) deployment** of our prototype, we consider the insights from our use case. The forecast of the product groups can be improved by adapting the current SAP APO forecast to a seasonal naïve approach. Alternatively, the dynamic adaption of Holt-Winters parameters could improve the forecast in comparison to the current static approach. Concerning the manual adjustment by local planners, the consideration of long-term influences such as economic indicators might not improve the forecast and the effort of manually adjusting forecasts could be decreased challenging some of the assumptions about exogenous influences.

In addition to the new forecasting approach, the forecast process must be monitored continuously, updating applied methods when necessary and setting the basis for more advanced algorithms in the future. The technical deployment of such a forecasting and monitoring tool at the reference company is still in an initial stage.

## 5  Evaluation

Starting with RQ 1, we began our evaluation interviews with a question about the *understandability* of our design guidelines at hand, followed by their potential for implementation. All participants agreed that the design guidelines are easy to understand. The participants were constituted in a *mixed team* of business users and IS experts.

Another aspect that we reflected on intensively is the fact that extensive projects typically aim at implementing advanced forecasting solutions right from the beginning of the project (Sect. 2). We argue for an *iterative approach* including starting with straightforward methods and continuing with more advanced PA models as needed. We learned in our case study that this is especially true when companies lack the necessary data and data science expertise. Simpler solutions can already be beneficial often being more effective than complex ML approaches (Sect. 4).

Following RQ 2 and incorporating the criteria proposed by Gregor and Hevner [10], we structured our evaluation interviews as follows. **Validity:** When asked if the design guidelines would help in applying PA methods, the participants from Controlling and Supply Chain Planning stated: "The systematic approach helps to structure the selection of an appropriate forecasting method with the presented algorithms being a helpful guidance." One of the external experts added that such design guidelines would support a *situational artifact design*, managing "pure" standardization vs "pure" individualization of future forecasting applications.

Compared to the current forecasting of the reference company, we asked the respondents whether the design guidelines could improve the forecast *accuracy*. The answers varied. Half of the interviewees agreed, whereas others were skeptical as the "expectations are very high" and only use cases that are "inadequately maintained" can achieve significant improvement. The training of existing staff and additional hiring is necessary but difficult, as these profiles are barely available, especially for medium sized companies, stated the Head of Finance.

Regarding process *efficiency*, all participants agreed that the design guidelines can help to reduce manual effort, especially for the *operational controllers*, however their implementation requires adequate infrastructure that allows the convenient testing and monitoring of data models.

Regarding *transparency* the Head of Finance stated that he is not really interested in the "technical" details of different algorithms, but rather in the results of the process. Nonetheless, he was interested in the kind of data influencing the forecast, especially when it comes to *leading indicators*. One of the experts agreed with this answer, in that knowing what data influences is valuable for a greater acceptance of algorithms, whereas understanding the individual algorithm in detail was not deemed necessary.

Summarizing the *validity* of our design guidelines, the Head of Finance said: "For a wider application of PA, statistical knowledge among employees and the variety of products, aggregation levels, and influences within the reference company, all pose a challenge." Therefore, a partially automated application that allows the easy comparison of different algorithms for specific use cases is key to enabling more fact-driven decision-making, whilst reducing human bias and improving efficiency.

**Utility and Generalizability:** Moving on to the utility and generalizability of our design guidelines, we asked the participants how they could be useful for other product groups and applications beyond the sales-volume domain at our reference company. All interviewees stated that the design guidelines would *support more fact-driven decisions* within the reference company – probably in other companies as well. One of the employees from the Controlling and Supply Chain Planning department explained that the design guidelines are applicable in sales volume forecasting beyond the examined product groups as they provide a general starting point for challenging existing forecast routines.

When it comes to more granular or high-frequency data – for example for detailed warehouse planning – our approach may not fit. When *scaling our model* to establish a cohesive approach across the company, one employee from the Controlling and Supply Chain Planning department emphasized the issue of data availability and inconsistencies between different departments. This impedes the usage of more advanced methods such as ML algorithms. Consequently, providing a solid data architecture is a crucial step for effectively implementing PA.

When asked about the value of our design guidelines beyond *improving forecasting performance*, the Head of Finance mentioned that dealing with PA with a more hands-on approach has helped him to gain a better understanding.

## 6  Discussion and Conclusion

As their current forecast process requires substantial manual effort and lacks transparency, this article laid out an improved sales volume forecasting for an international pump system provider. The associated design guidelines – derived from these findings – should help to apply PA algorithms properly.

Based on an examined lack of statistical knowledge among the employees of the reference company (Sect. 5) and a large variety of products, aggregation levels, influences, and potential leading indicators, easy-to-handle PA applications were fundamental to enabling business users to drive their forecasting. Thus, for *practice*, our approach should help to get started more effectively with PA – without having extensive data science knowledge.

For *research purposes*, we updated current PA research in a new *industry setting*. Starting with a literature review, we laid out the role of PA in business forecasting. Besides forecast accuracy, we have argued that additional evaluation criteria such as process efficiency and transparency should determine the most suitable approaches as well. Considering the perspective of business users that are hesitant to apply modern forecasting techniques, we detailed the process of applying PA algorithms for practice.

Although business users showed great interest in including leading indicators and thereby making the model "more explainable," the benefit of leading indicators as input for predictive models is questionable. In fact, leading indicators might be better suited to modeling *causal relations* in explanatory models.

Even though ML-based algorithms can outperform simpler algorithms, their limiting factor is often the availability of data. However, our use case showed that even simple models with limited data can improve the forecasting performance.

Newer studies often present specialized solutions neglecting quick win use cases [49]. Our research aims at closing this gap by providing comprehensive design guidelines with a focus on *business users*.

However, our research inevitably reveals certain *limitations*. Although single case studies offer a broad range of advantages, their limited generalizability requires complementary use cases. Furthermore, the lack of data availability prevented us from using more complex methods. Focusing on the most common algorithms kept the project tractable. For future research, we will examine how the software providers keep responding to the PA trend, offering *Auto ML solutions* that aim at enabling business users. The feedback from our interviewees has shown that the interest in understanding the influences on a forecast is high, but the understanding of algorithms itself is not deemed necessary (Sect. 5). This shows that a *black-box approach* does not conform to the expectations of a "modern" PA model. Responding to that, *explainable artificial intelligence* (XAI) is certainly worth being considered in greater detail [50]. Accordingly, a final topic of research could be how the algorithm understandability correlates with its *acceptance* by business users.

# References

1. Gilliland, M.: The value added by machine learning approaches in forecasting. Int. J. Forecast. **1**(36), 161–166 (2020)
2. Wang, Ch.-H.: Considering economic indicators and dynamic channel interactions to conduct sales forecasting for retail sectors. Comput. Ind. Eng. **165**, 107965 (2022)
3. Gerritsen, D., Reshadat, V.: Identifying leading indicators for tactical truck parts' sales predictions using LASSO. In: Arai, K. (ed.) IntelliSys 2021. LNNS, vol. 295, pp. 518–535. Springer, Cham (2022). https://doi.org/10.1007/978-3-030-82196-8_38

4. Sagaert, Y.R., Aghezzaf, E.-H., Kourentzes, N., Desmet, B.: Temporal big data for tactical sales forecasting in the tire industry. Interfaces **2**(48), 121–129 (2018)
5. Gonçalves, J.N.C., Cortez, P., Sameiro Carvalho, M., Frazão, N.M.: A multivariate approach for multi-step demand forecasting in assembly industries: empirical evidence from an automotive supply chain. Dec. Support Syst. **142**, 113452 (2021)
6. Simon, H.A.: The Science of the Artificial. MIT Press, Cambridge, Massachusetts (1996)
7. Walls, J.G., Widmeyer, G.R., El Sawy, O.A.: Building an information system design theory for Vigilant EIS. Inf. Syst. Res. **3**(1), 36–59 (1992)
8. Hevner, A.R., March, S.T., Park, J., Ram, S.: Design science in information systems research. MIS Q. **28**(1), 75–105 (2004)
9. Vom Brocke, J., Winter, R., Hevner, A., Maedche, A.: Accumulation and evolution of design knowledge in design science research - a journey through time and space. J. Assoc. Inf. Syst. **21**(3), 520–544 (2020)
10. Gregor, S., Hevner, A.R.: Positioning and presenting design science research for maximum impact. MIS Q. **37**(2), 337–355 (2013)
11. Peffers, K., Tuunanen, T., Rothenberger, M.A., Chatterjee, S.: A design science research methodology for information systems research. J. Manag. Inf. Syst. **24**(3), 45–77 (2007)
12. Esswein, M., Mayer, J.H., Stoffel, S., Quick, R.: Predictive analytics – a modern crystal ball? answers from a cash flow case study. In: Proceedings of the 27th European Conference on Information Systems, pp. 1–16 (2019)
13. Webster, J., Watson, R.T.: Analyzing the past to prepare for the future: writing a literature review. MIS Q. **2**(26), 13–23 (2002)
14. Vom Brocke, J., Simons, A., Niehaves, B., Riemer, K., Plattfaut, R., Cleven, A.: Reconstructing the giant: on the importance of rigor in documenting the literature search process. In: Newell, S., Whitley, E.A., Pouloudi, N., Wareham, J., Mathiassen, L. (eds.) Proceedings of the 17th European Conference on Information Systems (2009)
15. vom Brocke, J., Simons, A., Riemer, K., Niehaves, B., Plattfaut, R., Cleven, A.: Standing on the shoulders of giants: challenges and recommendations of literature search in information systems research. Commun. Assoc. Inform. Syst. **37**, 205–224 (2015). https://doi.org/10.17705/1CAIS.03709
16. AIS Senior Scholar's Basket of Journals: https://aisnet.org/page/SeniorScholarBasket/. Last accessed 30 Nov 2022
17. Scimago Journal & Country Rank, Business, Management, and Accounting: https://www.scimagojr.com/journalrank.php?area=1400. Last accessed 30 Nov 2022
18. AIS Conferences: https://aisnet.org/page/Conferences/. Last accessed 30 Nov 2022
19. Verstraete, G., Aghezzaf, E.-H., Desmet, B.: A leading macroeconomic indicators' based framework to automatically generate tactical sales forecasts. Comput. Ind. Eng. **139**(1), 1–10 (2020)
20. Myers, M.D.: Qualitative research in information systems. MIS Q. **21**(2), 241 (1997)
21. Fildes, R., Goodwin, P., Lawrence, M., Nikolopoulos, K.: Effective forecasting and judgmental adjustments: an empirical evaluation and strategies for improvement in supply-chain planning. Int. J. Forecast. **25**(1), 3–23 (2009)
22. Chen, Y.-J., Chien, C.-F.: An empirical study of demand forecasting of non-volatile memory for smart production of semiconductor manufacturing. Int. J. Prod. Res. **56**(13), 4629–4643 (2018)
23. Wang, C.-H., Yun, Y.: Demand planning and sales forecasting for motherboard manufacturers considering dynamic interactions of computer products. Comput. Ind. Eng. **149**, 1–8 (2020)
24. Wu, S.D., Kempf, K.G., Atan, M.O., Aytac, B., Shirodkar, S.A., Mishra, A.: Improving new-product forecasting at intel corporation. Interfaces **40**(5), 385–396 (2010)
25. Liu, Y., Feng, J., Liao, X.: When online reviews meet sales volume information: is more or accurate information always better? Inf. Syst. Res. **28**(4), 723–743 (2017)

26. Geva, T., Oestreicher-Singer, G., Efron, N., Shimshoni, Y.: Using forum and search data for sales predictions of high-involvement products. MIS Q. **41**(1), 65–82 (2017)
27. Abolghasemi, M., Hurley, J., Eshragh, A., Fahimnia, B.: Demand forecasting in the presence of systematic events: cases in capturing sales promotions. Int. J. Prod. Econ. **230**, 1–28 (2020)
28. Qiu, J.: A predictive model for customer purchase behavior in e-commerce context. In: Proceeding of the 19th Pacific Asia Conference on Information Systems, p. 369. Chengdu, China (2014)
29. Wijnhoven, F., Plant, O.: Sentiment analysis and google trends data for predicting car sales. In: Proceedings of the 38th International Conference on Information Systems, pp. 1–16 (2017)
30. Ma, S., Fildes, R.: Retail sales forecasting with meta-learning. Eur. J. Oper. Res. **288**(1), 111–128 (2021)
31. Tsoumakas, G.: A survey of machine learning techniques for food sales prediction. Artif. Intell. Rev. **52**(1), 441–447 (2018)
32. Grover, V., Chiang, R.H., Liang, T.-P., Zhang, D.: Creating strategic business value from big data analytics: a research framework. J. Manag. Inf. Syst. **35**(2), 388–423 (2018)
33. Benthaus, J., Skodda, C.: Investigating consumer information search behavior and consumer emotions to improve sales forecasting. In: Proceedings of the 21st Americas Conference on Information Systems, pp. 1–12 (2015)
34. Chong, A.Y.L., Li, B., Ngai, E.W., Ch'ng, E., Lee, F.: Predicting online product sales via online reviews, sentiments, and promotion strategies. Int. J. Oper. Prod. Manag. **36**(4), 358–383 (2016)
35. Blackburn, R., Lurz, K., Priese, B., Göb, R., Darkow, I.-L.: A predictive analytics approach for demand forecasting in the process industry. Intl. Trans. in Op. Res. **22**(3), 407–428 (2015)
36. Flyvbjerg, B.: Case study. In: Denzin, N.K., Lincoln, Y.S. (eds.) The SAGE Handbook of Qualitative Research, pp. 301–316. SAGE, Los Angeles, London, New Delhi, Singapore, Washington DC, Melbourne (2018)
37. Benbasat, I., Goldstein, D.K., Mead, M.: The Case Research Strategy in Studies of Information Systems. MIS Q. **11**(3), 369–386 (1987)
38. Dul, J., Hak, T.: Case study Methodology in Business Research. Butterworth-Heinemann, Amsterdam (2007)
39. Yin, R.K.: The case study crisis: some answers. Adm. Sci. Q. **26**(1), 58–65 (1981)
40. Gustafsson, J.: Single case studies vs. multiple case studies: A comparative study (2017)
41. Chapman, P., et al.: CRISP-DM 1.0: Step-by-step data mining guide (2000)
42. Qu, S.Q., Dumay, J.: The qualitative research interview. Qual. Res. Account. Manag. **8**(3), 238–264 (2011)
43. Working Group "Digital Finance" Schmalenbach-Gesellschaft: https://www.schmalenbach.org/index.php/arbeitskreise/finanz-und-rechnungswesen-steuern/digital-finance. Last accessed 29 Nov 2022
44. Shmueli, G.: To explain or to predict? Stat. Sci. **25**(3), 289–310 (2010)
45. Eurostat: Confidence Indicators: https://ec.europa.eu/eurostat/databrowser/view/teibs020/default/table?lang=en (2021). Last accessed 20 Apr 2021
46. Statistisches Bundesamt: Monthly issued building permits for Germany. https://www-genesis.destatis.de/genesis//online?operation=table&code=31111-0002&bypass=true&levelindex=0&levelid=1620400463162#abreadcrumb (2021). Last accessed 20 Apr 2021
47. Kumar, A., Shankar, R., Aljohani, N.R.: A big data driven framework for demand-driven forecasting with effects of marketing-mix variables. Ind. Mark. Manage. **90**, 493–507 (2020)
48. Tibshirani, R.: Regression shrinkage and selection via the lasso. J. Roy. Stat. Soc.: Ser. B (Methodol.) **58**(1), 267–288 (1996)

49. Schröer, C., Kruse, F., Gómez, J.M.: A Systematic literature review on applying CRISP-DM process model. Procedia Comput. Sci. **181**, 526–534 (2021)
50. Barredo Arrieta, A., et al.: Explainable artificial intelligence (XAI): concepts, taxonomies, opportunities and challenges toward responsible AI. Inform. Fusion **58**, 82–115 (2020)

# AIOps in Higher Education Institutions

Thabo Sakasa and Tendani Mawela(✉) (iD)

University of Pretoria, Hatfield, South Africa

u21249212@tuks.co.za, tendani.mawela@up.ac.za

**Abstract.** The study explored the opportunities, barriers and enablers for Artificial intelligence for IT operations (AIOps) in the higher education sector. It adopted the case study strategy supported by an interpretive and qualitative approach. Primary data was collected through semi-structured interviews with IT management and staff at two institutions of higher learning. The Technology Organisation Environment (TOE) framework was the theoretical lens used to understand the perceptions of the respondents regarding the barriers and enablers towards implementing AIOps. The results highlighted various themes related to the implementation of AIOps including: IT infrastructure, skills challenges, executive support, budget constraints, IT strategy and governance.

**Keywords:** Artificial Intelligence · Data Analytics · Machine Learning · IT Operations · Higher Education

## 1 Introduction and Background

Information Technology (IT) operations involves the activities of identifying, integrating and managing different products and processes to provide a stable, responsive and robust IT environment [1]. A healthy IT organisation provides a competitive advantage for organisations in a fast-paced market [2]. IT operations deal with the management of software and hardware, IT support, network administration, device management and their related functions in an organisation. Developments around artificial intelligence (AI) and big data analytics offer an opportunity for IT operations to improve how IT operations are managed. Artificial intelligence offers the opportunity to overcome IT operations challenges such as monitoring, alerts, redundancies, downtime and slow response [3]. Artificial intelligence for IT operations (AIOps) is a combination of big data and machine learning which optimises the related IT operations processes and tasks. The link between big data and AI has facilitated the development of AIOps. Big data and machine learning are central to AIOps to understand real-time insights into problems and incidents that affect the IT operations environment. Such insights are predictive and are accompanied by automated recommendations to address IT related problems.

IT underpins all of a higher education institution's principal activities. In many areas, IT contributes directly to the efficiency of the organisation's operations and in others, it is vital for the existence of the activity. IT has direct and indirect roles in supporting and enabling higher education institutions' vision and strategic objectives. The IT

© The Author(s), under exclusive license to Springer Nature Switzerland AG 2023
M. Papadaki et al. (Eds.): EMCIS 2022, LNBIP 464, pp. 77–90, 2023.
https://doi.org/10.1007/978-3-031-30694-5_7

function needs to be operated efficiently and effectively to enable the organisation to perform optimally. IT architectures are complex environments with many points of failures. Identifying these points of failures and resolving them on time presents a challenge to various organisations. Traditional IT operations, based on domain centric monitoring are unable to cope with the volumes of data produced that provide the required insight into the IT environment [4]. IT operations that are not optimally managed will negatively affect the performance of the organisation [5]. Scholars point out that the complexity of the IT environment requires a diverse set of skills [6]. In such circumstances, the IT operations team requires additional resources and approaches to manage the IT environment optimally to avoid downtime that can have negative consequences for the organisation. The impact and consequences of IT incidents that are not resolved can be severe. IT incidents can affect the adoption of useful technologies as shown by the reluctance of some faculties and lecturers [7, 8]. There is an opportunity for universities to consider implementing AIOps. The extant literature has limited studies focusing on AIOps and in particular in the higher education context. The study contributes to research about IT operations, big data, artificial intelligence and machine learning. This paper reports on a study that investigated the enablers and barriers to the implementation of AIOps in the higher education sector. The research question which underpinned the study was " *"What are the barriers and enablers to the implementation of AIOps?"*.

## 2   Literature Overview

### 2.1   Artificial Intelligence for IT Operations

The authors in [6] refer to AIOps as, "a set of diagnostic and predictive tools, automation and humans-in-the-loop capabilities that will enable operation teams to embrace change". AIOps is thus a combination of capabilities that guide IT operations teams in handling the challenges of complexity and high-performance demands that characterise IT operations, utilizing new and powerful technologies that provide the potential to transform IT operations. AIOps provides the ability to conduct real-time analysis of the entire IT operations environment, detecting and addressing issues and incidents at the same time. Artificial Intelligence for IT Operations is about the application of algorithms in IT operations. It is about the use of AI and machine learning to automate tasks and processes that are undertaken to plan and manage IT operations. AIOPs can add value to IT Operations in the areas of: log analysis, capacity planning, infrastructure scaling and cost management. Also, with the wide acceptance and usage of cloud computing, the scale and complexity of services has increased, leading to challenges for the IT operations team, challenges that can be addressed by AIOps. AIOPs may also be considered for server management, application support, and service desk management [9].

### 2.2   IT Operations Challenges

The challenges faced by IT operations are attributed to developments and innovations in IT. IT is becoming more complicated and sophisticated. It is highlighted by [6] that, "IT operations personnel spend a significant amount of time and energy trying to keep everything up and running and are always under mounting pressure to predict when things will go wrong and be prepared with fixes and lengthy root cause analysis documents". Some of the challenges for IT Operations include: scalability, shifting technology landscape, availability and reliability of technology, multiple and diverse platforms, tools and devices [1, 10–12]. Additionally, the IT environment contends with short release cycles, increasing negative and costly IT incidents [6, 13].

## 2.3 IT Operations in Higher Education

The focus areas for IT Operations in higher education, include student admissions, student data, learning systems, staff administration, payroll and financial accounting, inventory systems and library systems. A study on ICT investment in higher education [14] concluded that while quality and quantity were important factors, managers needed to prioritise the avoidance or elimination of downtime. IT is central to the effective provision of teaching and learning in higher education. It is against this background that IT operations are required to perform optimally for universities to achieve their objectives. Prior research [9] calls for a coordinated collaboration between institutions of higher learning and the software engineering industry.

## 2.4 Enablers and Barriers for AIOPs

The following barriers towards innovations such as AIOps are highlighted in the literature: Uncertainty in organisations on whether AIOps is a hype or a true innovation, since it is a major shift from traditional IT operations [4]. Due to the transformative nature of AIOps to IT operations, which requires realignment of people, processes and technology, it may discourage organisations to commit to AIOps [2, 15]. The IT architecture required for AIOps in terms of hardware and software may require large investments [16, 17]. Also, a lack of skilled personnel may inhibit the successful implementation of AIOps [15]. A lack of understanding of big data and low confidence in managing big data is a potential barrier to AIOps [18, 19].

The following outlines the enablers that may support a move towards AIOps as noted in the literature: Cloud computing and Network and Data Center Operations with AI can support the implementation of AIOps [6, 9]. Additionally, coordinated collaboration between institutions of higher learning and the software engineering industry [9, 11]. The ability to correlate millions of data, analysis of the data, determination of patterns and presentation of data in meaningful formats is an essential part of driving AIOps [4, 18]. Trends indicate that training in AIOps related fields by companies and some tertiary institutions is also supporting upskilling towards AIOps initiatives [20].

## 3 Theoretical Framework

The study investigated AIOps and the barriers and enablers that affect its implementation in the higher education sector and relied on the technology-organisation-environment (TOE) framework as a theoretical lens. The TOE framework [21] takes the approach that an organisation's ability to adopt technological innovation is affected by three elements, which are technology, organisation and environment. This framework was used as a basis of understanding the barriers and enablers of the implementation of AIOps as well as the analysis of the data that was collected. The framework points to a convergence of organisational factors supported by technological factors and the macro environmental context leading to a better opportunity for the successful adoption of innovations like AIOps. The TOE framework as articulated by [21] and [22] was useful for understanding how organisations may consider AIOps as well as the barriers and enablers towards adopting it. Research indicates that the TOE framework is capable of being applied broadly and has the potential to explain a number of technological, industrial and national or cultural contexts about the enablers and barriers that affect innovations such as AIOps [23, 24].

## 4   Research Method

### 4.1   Overview

The research study adopted the interpretive philosophical paradigm. The exploratory and qualitative study aimed to explore and illuminate the barriers and enablers to the implementation of AIOps in the higher education setting which has limited research studies. The research adopted a case study strategy [25] and included respondents from IT operations teams of universities. The research was based on two organisations in the higher education sector. The case study organisations are herewith referred to as Organisation A and Organisation B. Organisation A is a traditional university which offers mainly contact learning and a few distance learning classes while Organisation B is technical university offering mainly distance learning classes and some contact learning. Organisation A has 39 953 students enrolled across undergraduate and postgraduate degree programs. It has staff complement of 5818 including academic, administrative and management staff. The IT department of Organisation A has 110 staff members. Organisation B has 65 920 students enrolled with a total staff complement of 3199. The IT department of Organisation B has 115 members of staff.

### 4.2   Population and Sampling

Purposeful sampling [26] was applied to identify professionals who are working in the IT operations environment at higher education institutions. The choice of this population was based on the understanding that they would be knowledgeable and best placed to be able to answer the research questions pertaining to the implementation of AIOps. A total of 18 participants across the two organisations participated in the interviews. 56% of respondents were male and 44% female. 61% were in operational roles in IT while the balance held supervisory, management or strategic positions.

### 4.3   Data Collection and Analysis

An interview guide aligned to the TOE framework was compiled and ethics clearance obtained for the research study. Semi-structured individual interviews were conducted with the members of the IT operations teams at both institutions. All interviewees signed consent forms and the interviews were recorded. The recordings were transcribed. We relied on the thematic analysis approach as recommended by [27] to analyze the data. The TOE framework informed the coding and interpretation of the data.

## 5    Findings and Discussion

### 5.1    Technology

The technology theme deals with the IT architecture, infrastructure, automated diagnostic and predictive tools, automated mechanism of tracking, monitoring and resolving incidents, automated correlation between the infrastructure and applications teams as well as the analysis of logs. As highlighted by [4] current IT architectures are organised in silos, which presents a challenge for the IT operations team. With these silos, the IT operations team is unable to combine, analyze and correlate data from the different domains to provide insights that the IT operation's teams require to manage the environment proactively.

**IT Architecture and Infrastructure**
With regards to the IT architecture and infrastructure of Organisation B's readiness for AIOps, the Microsoft Enterprise Specialist said he *"cannot see that happening"*, while the Deputy Director ICT Infrastructure said, *"currently we are not ready for that, but we are working towards that"*. At Organisation A, the ICT Support Manager said, *"I'd say yes. I do think that the systems we have now are able to integrate with artificial intelligence and the latest technologies. The reason why I am saying so is at the moment from an audio-visual perspective, we have implemented a smart class room...It's a device whereby it can tell you that a certain device has gone down. You know, and obviously that pulls in from other systems. So, we already have that type of integration to put Artificial Intelligence on top of it obviously if possible... So yes, I do believe we have, we can do that. There is that capability"*.

This is an interesting remark considering that the readiness of an IT architecture to implement AIOps is a complex endeavor and incorporates more than one tool or audio-visual integration. However, it is noted that the remark does acknowledge the significance of integration of solutions in order to build a foundation for AIOps. The IT operations environment at Organisation B is generally at an entry level with regards to the implementation of AIOps. The IT architecture and infrastructure is not ready for AI, data analytics, big data and machine learning. A few of the participants expressed that they were unsure of the readiness of their organisation to implement AIOps. At Organisation A, the IT architecture / infrastructure was described as capable of integrating AI, machine learning and data analytics, especially from an audio-visual perspective. Another participant confirmed the same position and added that further investment would be required. From the data it appears this university is partially ready. According to Organisation's A annual report, the infrastructure and new technologies required to support the new digital strategy have been implemented, the infrastructure has been improved, classroom technologies have been upgraded as well as laboratories and the training of the academic staff. This indicates a plan being implemented for a solid foundation once the implementation of AIOps is considered. The overall feedback though is that the IT infrastructure is not fully ready for AIOps implementation, which points to this aspect as being a barrier. It reflects the silo approach, which [4] indicated as the reason for IT operations teams being unable to correlate data from different domains to provide the required insights. The authors in [28] emphasise the need for infrastructure and operations leaders to prioritise architecture if they want to succeed in transforming the service desk to incorporate machine learning, big data and other AI technologies. They argue that those who do not take this approach will become increasingly irrelevant. Authors [6] also purport that an AIOps IT infrastructure should break the silo-based tools and integrate data related to events, metrics, logs, job data, tickets and monitoring.

**Automated Diagnostic Tools**

When asked about automated diagnostic tools a participant highlighted that: *"So, in the IT ops environment at the moment, I'm reluctant to say there is none. We have just installed or implemented the core Advanced Technology Systems Laboratory (ATSL). Therefore, our service desk is equipped. In addition, obviously the tool is new but we would like to use the tool in a way whereby we could start automating ... so it is early days. But the tool is able to do so."* - ICT Support Manager, Organisation A.

This finding is consistent with literature in terms of the location of diagnostic tools at the service desk. Such a location enables the tools to interact with the service desk agents. The study by [29] concludes that 40% of all large companies will combine data and machine learning functions to replace the service desk partially. According to [28], the service desk needs to be transformed into a proactive function through AI. The authors in [20] reported on a service desk that was equipped with diagnostic tools, which enabled it to learn from previous logged problems, and the applicable solutions to provide. Capacity planning illustrates the significance of diagnostic tools when it comes to AIOps. Currently it is indicated that IT operations teams perform capacity planning manually. They rely on existing specifications and analyze performance related shortcomings based on these specifications. This becomes an ineffective approach for large institutional environments such as Organisations A and B, characterised by large, complex, multi-tier applications and different service providers. Without diagnostic tools, the IT operations team is estimating the number of CPU cores required, the amount of RAM storage required and the network bandwidth required.

On the automated diagnostic tools, the situation from both Organisation A and B is generally not optimal. There were a few silo tools, for example in the network environment. There are no predictive tools used across their IT operations. As highlighted by [6] AIOps includes, "a set of diagnostic and predictive tools, automation and humans-in-the-loop capabilities that will enable operation teams to embrace change". Organisation A and B may need to consider the implementation of automated diagnostic tools as a foundation for the implementation of AIOps. There is a change required for the successful implementation of the diagnostic tools. The IT operations teams need could implement a change management process to revisit some of their past approaches and may consider the potential of AIOps.

**Tracking and Monitoring**

*"The only thing that we use is a manual system. That is how we track every movement*

*but we don't have something that is predictive."-* Learning Technologist, Organisation A.

From the data collected, there was not much implemented with regards to tracking and monitoring. There are elementary disparate tools utilised, for example, at Organisation B, the network team has a particular application for Wi-Fi monitoring. According to [2], distributed environments are characterised by different applications running on different platforms, which presents a challenge for tracking problems, resulting in downtime, which affects organisations negatively. The organisations interviewed in this study face this challenge. AI offers the opportunity to overcome IT operations challenges relating to monitoring, alerts, redundancies, downtimes and slow response [3]. An interesting aspect is that the interviews at Organisation A were delayed due to a downtime incident that affected the entire institution. There were teams running around trying to diagnose and fix the problem. There was pressure on the IT operations team since there were assessments also scheduled for that morning. It became apparent that tracking and monitoring was not at the required levels as it took some time to resolve the incident. The literature indicates that, where tracking and monitoring is not optimal, there will be endless downtimes and wastage of resources, which will eventually frustrate the end users. [8] stated that some faculties and lecturers in the higher education environment have expressed reluctance in adopting online learning management systems due to the impact of IT performance. In a complex environment with numerous points of failures and many access points the authors in [6] advise that organisations need an automated way of tracking, monitoring and resolving issues before they result in downtime.

## Correlation

*"Yes. Yes. There is, there is. A little bit. Yes, we need some improvement. You see there is a process especially on applications that is managed by us. But because there are now many other pieces managed by another company. So, we rely on one another. But on our applications, there is a process which we should upgrade." -* Deputy Director: ICT Infrastructure Organisation B.

Correlation in the AIOps environment is the collection of data from different sources (servers, network, operating systems, databases, etc.) and consolidating this data to determine relationships between infrastructure and applications [4]. There is an indication of a lack of understanding of the type of correlation required for the implementation of AIOps from the participants that were interviewed. Correlation between the infrastructure and applications teams is deemed to be at a basic level. Instead of the notion of correlation, the participants referred to teamwork. Correlation requires an improvement in both organisations. According to [7] correlation is critical. It is an interesting observation that the concept of correlation from both institutions seems to be very low in both understanding and execution. Since the environments are complex at both institutions, it would be expected that correlation would be central in the IT operations management approach. As outlined by [4] correlation covers the data about the application performance, events logs, transactions and more, including the network and storage resources dealing with the delivery of the application. Without this information, the IT operations team is operating with insufficient information and has limited understanding of what is happening with the applications. These may have a negative impact on the ability of IT operations to deliver a good service.

**Log Analysis**
*"It's all manual it's not automated. But I believe all the new products will have an application like that and then we will start with some of that."*- Information Security Specialist, Organisation B.

The analysis of logs from an IT operations team's perspective is linked to data analysis. Where there is no culture of analysing data, there will be very little analysis of logs for the purposes of managing the environment optimally. Research [3] maintains that log analysis is the obvious place to introduce AI. Literature points out that every component of the IT operations e.g.: hardware, software, operating system, servers, applications and databases generate logs that can be analyzed by machine learning algorithms. This can assist the IT operations in identifying problems proactively. Log analysis at both organisations was found to be at an elementary stage. They indicated plans to improve however it was not applicable in all the IT operations areas. Log analysis focuses on the proactive analysis and visualisation of the data generated by the logs for problem identification. The IT operations teams rely on these logs in order to gain an understanding of the performance and health status of the different IT components. This process often requires significant human intervention, with the teams analyzing these logs and determining the root cause of the problem or potential problem as well as determining the intervention approach. AIOps introduces machine-learning algorithms, which can proactively find problems and potential problems before they happen [19].

### 5.2 Organisation

This section discusses the organisational theme. From the data, the organisational aspect consists of management support for AIOps, skills set of the IT operations team, the awareness levels of the team, the utilization of data analytics and the budget allocation for IT operations. "The ability to adopt AIOps depends not only on the availability of monitoring data and automation systems, but also the alignment of people and processes" [2].

**Management Support**
*"I think management actually care and that they want to go in a certain direction on IT. But I think they do have a clear plan on how to get there. So, I think I've got a feeling that support is there in principle, but we don't have a plan that is in place. You know the plan is to be co-created between management and operations."*- Research Computing, Organisation A.

IT operations, is essentially a management function that is driven by various processes to ensure availability and performance of IT systems. For such a function to succeed, the support of management is required. IT operations play a crucial and supportive role in organisations. A healthy IT function provides key competitive advantages for organisations in a fast-paced market [2]. Management support incorporates the allocation of the required resources, human and financial as well as ensuring that stakeholders cooperate to achieve efficient IT operations. In some cases, management is also required to undergo changes and training to understand and support the implementation of AIOps. Management support at both educational institutions was perceived as being generally positive by the respondents. There is room for improvement, as it was highlighted that sometimes management does not understand some of the IT concepts. Management support is often a prerequisite to the successful implementation of strategic projects such as AIOps. There is evidence that Organisation A's management is supportive of the development of the infrastructure and new technologies required to support the digital strategy, which will enable better IT support, access to broadband and faster Wi-Fi connections, the development of smart classrooms, simulation laboratories and the training of academics to participate in the digital era. They did note some challenges regarding the network related to the upgrade project, which has taken longer to complete along with performance issues. These challenges are being addressed by the organisation. Respondents indicated that an internal oversight and management committee was established to address the challenges, and to ensure that the project remains on track until it is completed. This was noted as an enabler towards implementing AIOps.

**IT Skills**
*"We have no knowledge of that. Not there yet right now, there are no data analysts in the environment. The IT environment people who deal with the analyzing data are not really qualified for that. I think that is one of the reasons we are speaking of this. We can always do with more than what we are getting from the government, in terms of funding. We do the best we can. But we are not there yet."* -Microsoft Enterprise Specialist, Organisation B.

The skills required for AIOps are not at the required level at both institutions. Both organisations report that there is a need for up skilling. AIOps skills represent an enhancement of the current IT operations teams' skills set. Such skills will evolve over time to include knowledge of machine learning, programming and security. It is argued by [17] that the skills required for digital transformation should be developed within the organisation to enable the growth of a digital mind set. The issue of data scientist or data analysts is an interesting one. Participants from both institutions acknowledge the importance of such skills in the IT operations environment. However, both organisations reported that they have no data analysts. The IT operations team needs training and education in AIOps and big data analysis. There will be a need to consider involving data scientists to augment the team with the necessary big data skills. The literature highlights that the lack of appropriate skills is one of the reasons that organisations are not utilizing big data to their advantage [16]. The overall analysis of the AIOps, big data and machine learning skills of the IT teams at both institutions is that they do not exist. The skills aspect of the two institutions thus represents a barrier to the implementation of AIOPs.

**IT Operations' Team Awareness**
*"Yeah, so, there is the awareness and I think it is based on us coming to the strategy session years ago. When we sat down, we said where we are going to. Where is the world now and how do we get there? And we sat down we considered and know what things are coming up....So, we are definitely aware and we keep on you know researching. You know and always look at the technology curve. Just to see what is coming up so we can*

*plan and be ready for what's coming up next. So, we are definitely with it.*"- ICT Support Manager, Organisation A.

On the IT operations team's awareness, both institutions recorded positive feedback. Literature emphasises that there are no easy steps to implement AIOps [6]. There is a need for the IT operations team to keep up with developments, focus on supporting the business and also transform as a team as well. It has been indicated that a high level of awareness creates a receptive climate for change [12]. The authors argue that in organisations where there is overall lack of awareness and understanding of the value of big data and AIOps, the staff will see little value in executing these initiatives. The implementation of AIOps would require a receptive environment as well as a strong change management programme. Although both institutions recorded positive results on awareness, there is an indication that there is some way to go before these teams can be confirmed as being aware of the holistic transformation they need to undertake to implement the requirements of AIOps.

**Budget**

*"...And the money that is required we are still saying, can we have that? And the answer is always: there is no budget...And it is more even last year if I'm not mistaken. We had a couple of posts that we were looking for. I don't even think we got the money with budget cuts."*- Senior Manager Business Solutions, Organisation A.

University budgets are under pressure due to a variety of factors such as an increase in the number of students enrolling for higher education, declining grants and the fast developments in technology. In such circumstances, IT operations may be considered a lower priority when it comes to budget allocations for projects such as implementing AIOps. Universities may be satisfied with maintaining the status quo when it comes to IT operations and not necessarily willing to allocate budget towards innovations such as AIOps in the context of budget constraints. However, this may affect IT operations negatively in the future. The budget is generally viewed as inadequate for a move towards AIOps. There is a strong view that the budget partially addresses what is required for a university in the digital transformation era. Thus, budgetary constraints are deemed as a barrier towards AIOps.

### 5.3 Environment

This theme covers the policy provisions that govern the university's IT operations, the IT strategy as well as governance. Students at universities are exposed to the technological developments of the digital transformation era, which includes mobility and exposure to a variety of applications. They tend to be more proactive and demanding of IT resources such as bandwidth, availability and support. It is against this background that universities have to formulate policies that meet the student and staff demands in a manner that is cost effective while not hindering developments and advancements. Universities require appropriate policies to manage risks, and should review these policies regularly to ensure compliance.

**Policies**

*"They exist, but then I've forgotten the terms that was used then but essentially, it needs to*

*be said that they exist but they need constant updates. They need to be a living document but it needs to be relevant to the university's circumstances. Relevant to big data, machine learning. We have a very strong digitization drive."* - Senior Manager Business Solutions (ERP), Organisation A.

Organisation A reports that the information security and the cyber security policies that were approved in the last two years have been implemented. It further reports that the university council has paid more attention to the revision of policies, with a focus on IT and compliance with a view to improve the governance framework. This has been reflected from the data collected from the participants. What is perhaps missing is an understanding that organizations may also need technological tools in order to implement policies successfully, especially in a large and complex environment. It is one thing to have a revised policy and another to have a successfully implemented policy. The institutions have a gap in as far as the tools that are required for a successful implementation of IT policies. From the data, the policy element at both institutions is overall positive due to the existence of an applicable policy framework.

**Strategy**
*"We definitely use them to build that team... So, one of the things I think we try to stay better ahead in that sort of focus point. If I look at it some of the universities have the resources. It's a question of how do you utilise them."* - Information Security Specialist, Organisation B.

The IT strategy is also an important indicator of the university's readiness or willingness to undertake the implementation of AIOps. AI has the potential to assist organisations, however, it is complex and requires the development of a comprehensive strategy in order to realise its benefits. Literature [20] pointed out that there is a need for organisations to take incremental steps with the currently available technology while planning for transformational change in the not-too-distant future. This reflects the importance of considering AIOps as part of the strategy for the provision of IT services. The feedback on the IT strategy was positive for both institutions. There was an acknowledgment of its existence, in some interviews partially, in some cases hundred percent with the digital drive being prominent. Organisation B reported that the Office of CIO and Executive Director has delivered on The Digital Transformation Strategy, which aims to unify technology plans through a redesigned enterprise architecture framework, and redeveloped business processes to support a digitized environment. This is an indication of IT strategic objectives being supported at Executive level, which is a positive development.

**Governance**
*"I think it is very good in governance... With good governance you get happy customers."*
– Research Computing, Organisation A.

Governance deals with the planning arrangements that ensure IT resources and services meet the needs of the students and staff in an efficient and economic manner, while complying with legal and regulatory requirements. To implement governance, universities investigate best practice models and frameworks for IT, especially those that are adopted by higher education institutions [30]. As universities have many unique characteristics, governance normally reflects such characteristics at particular times. There is no one model that fits all universities. Feedback from participants reflected a good position for governance from both institutions. There is an indication of having IT governance, and planning as well as defined governance structures. In some cases, governance is viewed as delaying decision making but necessary. At Organisation A, a new operational model which involves the conclusion of service level agreements (SLA's) between IT and various other units within the institution has been established with the objective of managing IT's performance in this regard. Several structures were created to manage risk, security, compliance and efficiencies around the use of ICT infrastructure. This is another enabler towards an environment that may exploit AIOps.

## 6   Conclusion

### 6.1   Recommendations

It is suggested that to move towards AIOps and the benefits it promises, management and IT operations practitioners consider planning for the replacement of the traditional domain-based IT architectures with the consolidated unified architectures that include cloud computing, virtualization and agile development. Such a plan should be accompanied by the required diagnostic predictive tools, tracking and monitoring as well as how the correlation of data will be achieved. It is recommended that management and IT practitioners consider implementing effective and optimal log analysis as a foundation for the implementation of AIOps. Also, it is suggested that management and IT operations teams establish a project to raise awareness on AIOps across the organisation. There is a need to allocate a sustainable budget to prepare for the implementation of AIOps that will include training, workshops, tools and pilot initiatives for AIOps. Senior management, the Executive, the academics and students should be made aware of the potential value and benefits that AIOps will provide to the teaching and learning processes. It is recommended that management and IT practitioners should also establish formal initiatives to understand, manage and leverage on the data that is being generated from the IT operations environment as a preparation for the implementation of AIOps. Furthermore, it is recommended that higher education management and IT operations practitioners consider collaborating with the IT industry to develop an understanding of the context, trends and tools that relate to AIOps for higher education institutions. Management may consider implementing a pilot project that will automate the enforcement and execution of policies as a precursor to the implementation of AIOps.

### 6.2   Contribution and Concluding Remarks

The study contributes to the literature on the subject of AIOps which authors such as [9, 11] and [29] have highlighted is required. Additionally, research on AIOps in the higher education sector is limited. The study highlighted the challenges, complexity and effort required to implement AIOps in higher education. This study brings forth the perspectives of IT operations practitioners on the barriers and enablers affecting the implementation of AIOps in institutions of higher learning. From a theoretical stance the study is an example of how the TOE framework may inform the understanding of the complex issues surrounding the adoption and implementation of AIOps in higher education. On a practical level, the study offered several recommendations that IT management and executives may consider when implementing AIOPs. This study was limited to two institutions of higher learning. Future studies should consider conducting a study at multiple organisations across a variety of sectors. This would assist in better understanding the dynamics surrounding the adoption and implementation of AIOps. Future research should involve defining the baseline technological readiness for AIOps and how this could assist organisations in transitioning from the traditional IT operations models.

# References

1. Schiesser, R.: IT Systems Management, 4th edn. Pearson Education Inc., Boston (2010)
2. Tarun, G.: AIOps combines machine learning and automation to transform IT operations. USA (2017). Retrieved from https://www.cio.com/article/3198505/cloud-computing/aiops-combines-machine-learning-and-automation-to-transform-it-operations.html
3. Janakiram, M.S.V.: Artificial Intelligence is set to change the face of IT (2017). Retrieved from: https://www.forbes.com/sites/janakirammsv/2017/07/16/artificial-intelligence-is-set-to-change-the-face-of-it-operations/#44a99d771d21
4. Padhye, S., Nayak, B., Signore, E.: AIOPs. John Wiley & Sons Inc., New Jersey (2018)
5. Oats, M.: The high cost of "unpredictable" IT outages and disruptions. Netherlands (2017). Retrieved from https://www.intellimagic.com/resources/blog/high-cost-upredictable-outages-disruptions/
6. Mohanty, S., Vyas, S.: How to Compete in the Age of Artificial Intelligence: Implementing a Collaborative Human-Machine Strategy for your Business. Apress Media, New York (2018)
7. Loom Systems: The quantified benefits of Loom Systems AIOps solution. San Francisco, USA (2019). Retrieved from https://www.loomsystems.com/whitepaper-quantified-benefits-of-loom-systems-aiops-solution
8. Smale, M.A., Regalado, M.: Digital Technology as Affordance and Barrier Higher Education. Palgrave Macmillan, New York, USA (2017)
9. Dang, Y., Lin, Q., Huang, P.: AIOps: real-world challenges and research innovations. In: 2019 IEEE/ACM 41st International Conference on Software Engineering: Companion Proceedings (ICSE-Companion), pp. 4–5. IEEE (2019)
10. Apronti, A.T., Elbanna, A.: Re-imagining technology adoption research beyond development and implementation: ITOps as the New Frontier of IS research. In: Sharma, S.K., Dwivedi, Y.K., Metri, B., Rana, N.P. (eds.) TDIT 2020. IAICT, vol. 617, pp. 307–319. Springer, Cham (2020). https://doi.org/10.1007/978-3-030-64849-7_28
11. Qi, J., Wu, F., Li, L., Shu, H.: Artificial intelligence applications in the telecommunications industry. Expert. Syst. 24(4), 271–291 (2007)
12. Chaffey, D.: E-business and E-Commerce Management: Strategy, Implementation and Practice, 4th edn. Pearson Education Ltd., England (2009)
13. Ahmed, M.: Five Corporate IT Failures that Caused Huge Disruption. Financial Times, London (2017)
14. Ssempembwa, J., Canene, A.P., Mugabe, M.: ICT investment in Rwandan higher education: highlighting the cost of downtime and end-users' operations. Kampala Int. Univ. Res. Digest 1(1), 19–29 (2007)
15. Kalema, B.M., Mkgadi, M.: Developing countries organisations' readiness for Big Data analytics. Probl. Perspect. Manag. 15(1–1), 260–270 (2017)
16. Alharthi, K., Krotov, V., Bowman, M.: Addressing barriers to big data. Bus. Horiz. 60, 285–292 (2017)
17. Hess, T., Matt, C., Benlian, A., Wiesbock, F.: Options for formulating a digital transformation strategy. MIS Q. Exec. 15(2), 123–139 (2016)

18. Kibria, M.G., Nguyen, K., Villardi, G.P., Ishizu, O.Z.: Big data analytics, machine learning and artificial intelligence in next-generation wireless networks. IEEE Access **6**, 2169–3536 (2018)
19. Thankachan, K.: Data driven decision making for application support. In: Proceedings of the International Conference on Inventive Computing and Informatics (ICICI 2017), pp. 716–720. IEEE, Bangalore (2017)
20. Davenport, T.H., Ronanki, R.: Artificial intelligence for the real world. Harvard Business Rev. 1–10 (2018)
21. Tornatzky, L.G., Fleischer, M., Chakrabarti, A.K.: Processes of Technological Innovation. Lexington books (1990)
22. Sun, S., Cegielski, C.G., Jia, L., Hall, D.J.: Understanding the factors affecting the organisational adoption of big data. J. Comput. Inform. Syst. **58**(3), 193–203 (2018)
23. Baker, J.: The technology-organisation-environment framework. In: Dwivedi, Y.K. (eds.) Information systems theory: explaining and predicting our digital society, vol. 1. Springer Science Business Media (2011)
24. Hoti, E.: The Technological, Organisational and Environmental framework of IS innovation adaption in small and medium enterprises: evidence from research over the last 10 years. Int. J. Bus. Manag. **3**(4), 1–14 (2015)
25. Yin, R.K.: Case Study Research: Design and Methods, 3rd edn. Sage Publications, California (2003)
26. Palinkas, L.A., Horwitz, M.A., Green, C.A., Wisdom, J.P., Duan, N., Hoagwood, K.: Purposeful sampling for qualitative data collection and analysis in mixed method implementation research. Admin. Policy Mental Health and Mental Health Serv. Res. **42**(5), 533–544 (2015)
27. Nowell, L., Norris, J.M., White, D.E., Moules, N.J.: Thematic analysis: Striving to meet the trustworthiness criteria. Int. J. Qual. Methods **16**, 1–13 (2017)
28. Elliot, B., Andrews, W.: A framework for Applying AI in the Enterprise. Gartner, Stamford (2017)
29. Andenmatten, M.: AIOPs - Artificial Intelligence for IT operations. HMD Pract. Bus. Inform. Syst. **56**(2), 332–344 (2019)
30. Bianchi, I.S., Sousa, R.D.: IT governance for public universities: Proposal for a framework using Design Science Research (2015)

# Digital Content Profiling Based on User Engagement Features

Pawel Misiorek[1]([⊠]) [ID], Michal Ciesielczyk[2] [ID], and Bartosz Rzycki[2]

[1] Faculty of Computing and Telecommunications, Institute of Computing Science, Poznan University of Technology, ul. Piotrowo 2, 61-138 Poznan, Poland
pawel.misiorek@put.poznan.pl
[2] Deep.BI Poland, Warszawa, Poland
{michal.ciesielczyk,bartosz.rzycki}@deep.bi
https://deep.bi

**Abstract.** Exploring audience engagement with digital media content may lead to many various benefits. In this paper, we study how adding engagement-based features to the article description can influence the efficiency of algorithms aimed at detecting digital media readers' propensity to buy a subscription. Based on the propensity score, the publishers can optimize a decision to display a paywall. Moreover, it is observed that more and more page views are of new or anonymous users. Consequently, the decision concerning the paywall application has to rely only on digital content features. In order to address this application scenario, we propose a novel digital content enrichment framework based on the engagement statistics of users reading a given article. We experimentally evaluate the performance of machine learning algorithms for predicting the propensity to subscribe using the dataset based on events describing the behavior of users exploring the digital news site of the large media publisher. The results of experiments demonstrate that enrichment of article profiles with user engagement features significantly improves prediction models' efficiency.

**Keywords:** Digital content scoring · Behavioural profiles for articles · Subscription for digital content · Streaming data processing · Big Data

## 1 Introduction

The digitization in the media industry forces the vast majority of enterprises to rethink and reorganize the revenue model of their organizations. That is why those companies transformed into the digital space to stay profitable and embrace contemporary trends, mainly focusing on the online part of the business. Nevertheless, low barriers to entry into the industry, widespread access to content on similar topics, and reduced attention span among readers make the competition even more fierce. Two primary ways for gaining a competitive advantage emerge for digital media businesses focus on customers or content.

M. Papadaki et al. (Eds.): EMCIS 2022, LNBIP 464, pp. 91–104, 2023.
https://doi.org/10.1007/978-3-031-30694-5_8

In the first case, the organization personalizes the content and strategy based on the user behavior using such tools as cookie tracking, data analysis, dynamic paywall, and dynamic pricing [2,11]. In the second case, attention is paid to analyzing the characteristics of the article and recognizing whether there are any repeating patterns among successful articles in order to provide adequate feedback to content producers - journalists, editors, and the editorial board. The importance of such content analysis methods is increasing in the highly competitive digital media business [1,7]. That is especially relevant considering that companies are striving for more data about users in order to personalize the offer and increase their chances of subscription. However, only a small percentage of readers register on the website and leave contact data. Moreover, it is observed that more and more page views are of new or anonymous users. That is why organizations turn to other ways of increasing the pool of subscribers. Firstly, getting to know the specifics of the created articles enables to optimize the paywall strategy of the organization, including the decision of which articles should be available for the user before displaying the paywall. Secondly, analysis of articles can be a feedback for both the editorial board and authors themselves when it comes to preferences and tastes of the users as well as successful publishing strategy.

The next big step for content analysis is the implementation of artificial intelligence to enhance the business processes of the publishers [1,3,7]. As discussed in this paper, one of the outcomes of such implementation is data-based scoring for the content to better represent its chance of success. Success may be defined differently for every enterprise, data team, and editorial board. In this paper, we focus on success expressed as the situation where an article increases the chances of the user to subscribe. Therefore, adequate machine learning models for Propensity to Subscribe (P2S) are applied to provide a score of the article – how likely it is that the user will subscribe having read that article. The models include variables mentioned above, such as when users read the article, the daily/weekly patterns of interest, behavioral features of the article, number of referred visits (using a link), time of reading and attention, age of the article, and the interaction with a paywall. On top of that, custom variables that help assess the article are added and calculated.

Inspired by the research on modeling user engagement profiles for detection of reader's propensity to subscribe presented in [11], we introduce the content scoring solution based on articles' engagement profiles and aimed to be applied to enhance the dynamic paywall policy. However, unlike in the cases of the analysis of the behavior of readers [11], P2S models for articles are a highly dynamic issue and so the method and approach cannot be copied. That is why, in this paper, we propose the new architecture for building scalable and efficient content scoring solution.

The paper contribution is as follows. We propose the novel content profiling framework being a part of the Deep Glue System [11] responsible for managing and optimizing the access for digital media users. In particular, we describe the article profiles based on comprehensive engagement statistics of users reading this article. Furthermore, we demonstrate how such profiles can be enriched and dynamically updated in real-time and then applied to propensity to subscribe modeling and paywall control. Finally, we experimentally evaluate the performance of machine learning algorithms which utilize the proposed digital content profiles for the application scenario of predicting the propensity to subscribe based on article features.

## 2    Related Work

The digital content profiling for detecting users' propensity to subscribe is an underexplored research problem [1,2,7]. Many studies concerning the general content scoring problem have already been published, i.e., [10,12], but none of them is focused on digital article scoring aimed at optimizing subscription sales. The most relevant research results on modeling and measuring user engagement with digital articles are presented in [1,7]. Unlike our approach, Carlton et al. [1] study the problem of engagement prediction. Furthermore, they use short video content as their application scenario. On the other hand, the author of [7] analyses user engagement patterns with page views of news articles. Specifically, he investigates the relationship between engagement levels and information gained in the articles' text. In contrast to his research, we are not limited to news articles. Moreover, we are focused on user features closely related to the subscription process, e.g., describing user interactions with a paywall. In [3], Davoudi et al. propose the subscription prediction model using user engagement measures. Additionally, in more recent research presented in [2] the authors propose engagement-based paywall control policies. However, their research is not focused on modeling article profiles and does not investigate the influence of engagement-based profiling on the efficiency of machine learning models.

In the case of research on user profile enrichment techniques, many solutions focus on social media applications [5,9]. Unlike in the case of our method, they are based on processing textual data [9] extracted from social services such as Twitter [5]. Our approach is closer to the research of Tang et al. [13] and Li et al. [8], which propose to build time-agnostic temporal features based on aggregations in a specific time window as some time-forgetting mechanism. However, their studies apply to real-time recommendation systems [13], and streaming service churn prediction [8]. Our research is strictly connected to studies presented in [11], in which the framework for building digital media user profiles using their engagement features has been presented. However, in this paper we introduce the propensity-to-subscribe scoring solution based on articles' engagement profiles, which is aimed to optimize the dynamic paywall mechanism for the case of new or anonymous users.

## 3    Content Profiling Based on Behavioral Features

In this paper, we introduce the content profiling framework to enrich the page view events with additional engagement-based features of articles. These new features may be seen as the current article profile based on statistics concerning article readers. Specifically, for a given article $a$ and a given timestamp $t$, the article profile $p(a, t)$ is formally modeled as a sequence of features:

$$p(a, t) = (f_1, \ldots, f_m),$$

where $m$ is the total number of profile features. The new features are generated using events collected in various periods before the time $t$, which usually corresponds to the timestamp of the enriched event. The details about profile feature types and the description of specific features applied in tests presented in this paper are presented in Tables 1 and 2, respectively.

**Table 1.** Article profile features.

| Feature type | Description |
| --- | --- |
| Counters | Features counting the number of specific events in a given time window (e.g., today, yesterday, lastN days), e.g., the number of page views, the number of conversions just after reading the article, the number of paywall displays, the number of paywall clicks |
| Total sums | Sums in a given time window, e.g., total attention time for users reading the article |
| Averages | Averages of a given feature, e.g., average attention time per user in a given time window |
| Ratios | Ratios of some features, e.g., ratios to paywall clicks to paywall displays in a given time window, ratio subscribers to users reading the article |
| Unique values counters | Counters of unique values of a given raw feature, e.g., the number of unique user locations, the number of unique users, the number of unique subscribers |
| Segment-based features | Features defined by user engagement segments, e.g., numbers of unique users from a given engagement segments (e.g. engaged, perspective, new, fly-by, won-back, etc.) |
| Percentage features | Percentages of occurrences for some raw feature values, e.g., percentages of visits from a given source (e.g., home, search, social media), percentages of users from a given segment. |
| Dynamics features | Dynamics of change of a given feature in time, e.g. dynamics of the change in the number of page views between today and yesterday, or today and last week |

The proposed article profiles contain the information intended to be helpful when predicting if reading the given article may increase the user's propensity to subscribe for content. In particular, it includes the most recent historical data on article page views, readers' attention time, types of users reading the article, user engagement segments, traffic sources, statistics of paywall displays and clicks, and the number of subscription purchases. Most of the features are aggregation features, including counters of specific events (e.g., the number of subscriptions sold just after reading the article) or total sums of a given original numeric feature (e.g., the number of seconds spent in the system) in the given time window (e.g., today, yesterday or during last 7 days). Additionally, profiles include features based on simple statistics such as the average or percentage of occurrence of some feature values, including segment-based features corresponding to readers from different user groups. Finally, we defined custom features based on predefined formulas involving the current values of original or enriched profile features. Some of them are just simple ratios, and others describe dynamics of given feature change in time, e.g., modeling differences between today and yesterday or between today and last week.

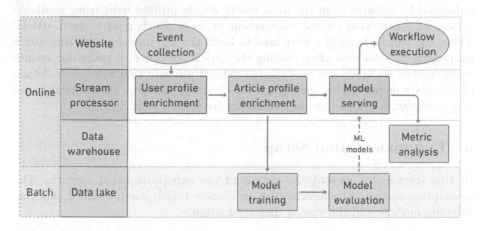

**Fig. 1.** Deep Glue content scoring architecture diagram.

The overview of the content scoring system architecture is depicted in Fig. 1. Stream of events (describing all user-article interactions) is collected in real-time and stored on a distributed messaging system (Apache Kafka[1]), which is one of the Stream processor components. Each event is enriched on Apache Flink[2] with current engagement features. First, the corresponding user profile, updated on every interaction (based on the solution presented in [11]), is added. Subsequently, the events are enhanced with article profile features (as described in Table 1). Events enriched with engagement features are used to generate predictions controlling the workflow execution for every article on the website. The performance can be monitored online using metrics emitted to a Data warehouse solution. Machine learning model used for serving is trained and evaluated offline, periodically, in batch manner. Models that passed the evaluation are serialized and pushed to the Stream processor environment.

---

[1] https://kafka.apache.org.
[2] https://flink.apache.org.

## 4   Experimentation Dataset

We use the unique dataset containing the real data collected based on the traffic on a digital media webpage. It consists of events describing the article views of users exploring the content of a digital site of a large media publisher. Articles published on the website are news, reports or reviews on politics, technology, environment, business, and economics. They have various characteristics including both short news with timely content which are popular for a limited time and then become irrelevant as well as reports or reviews with content which continues to be relevant long time after its publication date.

In this paper we use the data collected during the second half of 2021. The raw data contains around 100M events corresponding to page views of 200K unique articles viewed by 50M unique users. Each event has been dynamically enhanced by features from the most recent article profiles built using available historical information on the engagement of users which read a given article. Then, the enriched samples were used to build the ML models predicting user's propensity to subscribe after reading the article. In order to make our results reproducible, we made our anonymized dataset publicly available[3]. The details of dataset's preprocessing including data cleaning, filtering, engagement-based enhancement, and final dataset's statistics are presented in Sect. 5.

## 5   Experimentation Setup

In this section, we present the details of the experimentation scenario. The description contains the information on dataset preparation, building machine learning models, and the way of their evaluation.

**Dataset Prepartion.** The original dataset – introduced in Sect. 4 – consists of 100M events corresponding to article views. Just after its collection, each event was enriched by the most recent article profile available in the profile store. The profiles contain engagement features from the Deep Glue Content Profiling System described in Sect. 3. The outline of article profile features used in experiments presented in this paper is presented in Table 2. The article views are labeled based on the information on the subscription purchase just after reading a given article. Specifically, we selected 2098 new subscription purchases (i.e. non-renewal purchases from newly acquired users) with registered information about the last seen article. Moreover, we excluded all the articles views of users with active subscription from the datasets. Then, due to the fact that the labeled dataset was highly imbalanced, we decided to randomly downsample the events with negative label with the downsampling ratio set experimentally to 0.02. The final data samples were generated synthetically based on the characteristics of data collected. The basic statistics of the datasets are summarized in Table 3.

---

[3] https://www.kaggle.com/datasets/pawelwmm/contentscoringdataset.

**Table 2.** Features used in experiments.

| Feature Group | Description |
|---|---|
| Raw features (non-profile) | article author, topic, number of days from publication, weekday, day, hour of a view |
| Attention time/page views (article profile) | attention time today, yesterday, last $N$ days ($N = 7, 30, 60$); average attention time today, yesterday, last $N$ days ($N = 7, 30, 60$); # of page views today, yesterday, last $N$ days ($N = 7, 30, 60$); dynamics of the change in # of page views between today and yesterday, today and last week, and yesterday and last week; |
| Paywall/conversion (article profile) | # of conversions to subscribed users # of paywall clicks today, yesterday, last $N$ days ($N = 7, 30, 60$); # of paywall displays today, yesterday, last $N$ days ($N = 7, 30, 60$); ratio of paywall clicks to paywall displays today, yesterday, last $N$ days ($N = 7, 30, 60$); |
| User types (article profile) | # of unique users (which read the article); # of unique subscribers; # unique user locations (city, region, country) of users which read the article; # of unique readers (reader: a user with # of recent views above a threshold); # of unique subscriber readers; ratio of subscribers to users, ratio of readers to users, ratio of subscriber readers to subscribers # of users from a given engagement segments (engaged, perspective, new, fly-by, won-back, etc.) percentage of users from a given engagement segment |
| Traffic source (article profile) | # of user reading sessions # percentage of visits from a given source: home, search, social media, other |

**Table 3.** Basic statistics of a preprocessed dataset used in experiments.

| | |
|---|---|
| Total number of samples | 106,998 |
| Number of positive samples | 2,098 |
| Number of negative samples | 104,900 |
| Number of unique articles | 19,602 |
| Number of unique article authors | 1,868 |

**Experimental Scenarios.** We tested the effectiveness of our approach using two experimentation scenarios: (i) a basic off-line scenario assuming 10 repetitions of the experiment based on different random splits to train and testing data, and (ii) an additional real-world scenario assuming efficiency evaluation of models built using historical data. Both scenarios are based on the dataset presented in Sect. 5. The purpose of off-line tests was to obtain the reliable results provided as averages of 10 individual results. Each repetition is based on different random splits on training and test data for a training ratio equal to 0.75. For the real-world scenario, the model was built using the data collected during 20 weeks, and then evaluated during the next 5 week period. By choosing time as a factor for partitioning the data, we could mimic the real-time nature of the target infrastructure. The goal of the real-world experiment was to demonstrate the impact of real-time profile enrichment with time-agnostic behavioral features on the efficiency of propensity-to-subscribe modeling.

## 5.1 Approaches Under Comparison

To demonstrate the efficiency improvement caused by engagement-based enrichment of article profiles, we compared the following approaches:

- the *baseline* prediction algorithm based on the ML model utilizing the raw features describing the article (see Table 2), i.e., article author, topic, number of days from publication, as well as weekday, day, hour of a page view,
- the *basic profile* prediction algorithm utilizing the content profile enriched by basic engagement-based features based on general counters, i.e., total number of distinct users which read the article, total sum of readers' attention time, total numbers of page views, paywall displays, paywall clicks, and conversions,
- the *full profile* prediction algorithm utilizing each feature of digital content profile introduced in Table 2.

Furthermore, in order to provide some more detailed and insightful discussion, we delivered the additional efficiency comparison for models build using different thematically-grouped parts of engagement-based content profiles. We followed the group definition introduced in Table 2. Specifically, we compared the impact of features related to (i) attention time and page views statistics, (ii) paywall and conversion statistics, (iii) types of users which read the article, and (iv) the traffic source.

We used the CatBoost classifier [4] (implemented using its official library [14]) to build machine learning models compared in this paper. We applied CatBoost with default parameters and predefined *random_state* for our experiments. We indicate all the raw features presented in Table 2 as categorical features. The classifier choice was driven by technological constraints and business needs. Firstly, we were limited to algorithms that did not cause high execution latency, which was crucial to ensure the high-quality real-time performance of the infrastructure. Secondly, since most crucial article basic features, such as the author's name, and the article topic, are categorical, we chose the solution known to handle this kind of data effectively. The efficiency of algorithms was evaluated using the test set by means of standard ML measures [6], i.e., the Area Under the ROC curve (AUC), the average precision (AP), accuracy, balanced accuracy - included due to the fact of dealing with highly imbalanced data, precision, recall, and F1. The evaluation results are presented in Sect. 6.

# 6    Results

In this section, the results of experiments introduced in Sect. 5 are presented.

## 6.1    Results of Off-Line Experiments

The results of off-line experiments are presented in Figs. 2, 3, 4 and 5, and then summarized using Tables 4 and 5.

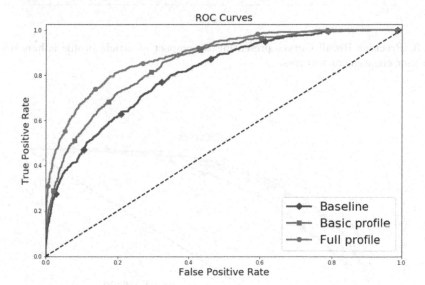

**Fig. 2.** ROC Curves presenting the impact of article profile enhancement with user engagement features.

Comparing AUC curves (see Fig. 2) proves that the models exploiting enriched data have achieved better efficiency than the models based on raw features describing the articles. We can also observe the quality progress implied by applying full profile features when looking at prediction efficiency through precision-recall curves (see Fig. 3). This observation confirms the importance of adding more specific features, such as counters and averages within the shorter time window concerning events from today or yesterday, and custom features modeling ratios, percentages, or change dynamics. The values of most popular machine learning measures [6] collected in Tables 4 confirm the crucial role

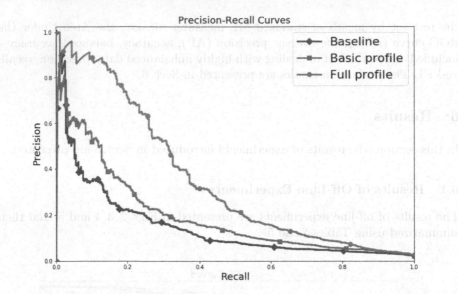

**Fig. 3.** Precision Recall Curves presenting the impact of article profile enhancement with user engagement features.

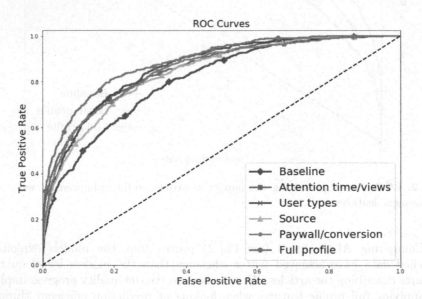

**Fig. 4.** ROC Curves presenting the impact of various groups of profile features.

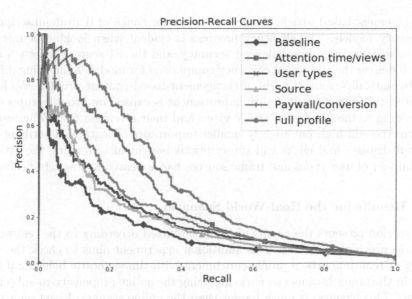

**Fig. 5.** Precision Recall Curves presenting the impact of various groups of profile features.

**Table 4.** Results of experiments presented as means and standard deviations for series of 10 experiment iterations (baseline vs user engagement profiles).

| Measure | Baseline | Basic Profile | Full Profile |
|---|---|---|---|
| AP | 0.1650(±0.0088) | 0.2023(±0.0092) | 0.3465(±0.0156) |
| AUC | 0.8084(±0.0051) | 0.8476(±0.0031) | 0.8838(±0.0058) |
| Accuracy | 0.9805(±0.0009) | 0.9807(±0.0007) | 0.9826(±0.0007) |
| Balanced acc. | 0.5287(±0.0061) | 0.5399(±0.0048) | 0.5939(±0.0071) |
| Precision | 0.6167(±0.0465) | 0.6255(±0.0608) | 0.7511(±0.0383) |
| Recall | 0.0581(±0.0123) | 0.0808(±0.0099) | 0.1890(±0.0144) |
| F1 | 0.1059(±0.0207) | 0.1426(±0.0146) | 0.3013(±0.0159) |

**Table 5.** Results of experiments presented as means and standard deviations for series of 10 experiment iterations (impact of different groups of features).

| Measure | Attention time/views | Paywall/ Conversion | User types | Source |
|---|---|---|---|---|
| AP | 0.3031(±0.0138) | 0.2316(±0.0125) | 0.2047(±0.0099) | 0.1943(±0.0104) |
| AUC | 0.8523(±0.0065) | 0.8474(±0.0050) | 0.8517(±0.0041) | 0.8402(±0.0044) |
| Accuracy | 0.9821(±0.0009) | 0.9811(±0.0008) | 0.9808(±0.0008) | 0.9807(±0.0008) |
| Balanced acc. | 0.5755(±0.0096) | 0.5545(±0.0051) | 0.5418(±0.0058) | 0.5379(±0.0046) |
| Precision | 0.7486(±0.0309) | 0.6575(±0.0430) | 0.6346(±0.0524) | 0.6273(±0.0462) |
| Recall | 0.1521(±0.0193) | 0.1102(±0.0103) | 0.0846(±0.0117) | 0.0768(±0.0093) |
| F1 | 0.2522(±0.0265) | 0.1884(±0.0148) | 0.1490(±0.0188) | 0.1365(±0.0144) |

of engagement-based article profiles in the performance of digital subscription propensity models. The efficiency progress is evident when looking at metrics such as average precision, balanced accuracy, and the F1 score. Figure 4–5 and Table 5 deliver the additional efficiency comparison for models build using different thematically-grouped parts of engagement-based content profiles. We have observed that the biggest quality improvement is caused by using features corresponding to the number of article views and their attention time. The results confirm the still high but slightly smaller importance of features describing the paywall displays and clicks and subscriptions bought after reading the article. The impact of user types and traffic sources has appeared as less significant.

## 6.2   Results for the Real-World Scenario

This section presents the results of tests conducted according to the real-world scenario provided in Sect. 5. This additional experiment aims to check the efficiency of real-time article profile enrichment with time-agnostic behavioral features in the target business scenario involving the online propensity-to-subscribe scoring. This scenario is much harder than the offline scenario based on a random split of train and testing sets since it requires testing the performance using new data, which usually is different from the one used for training. The results are collected in Tables 6 and 7. The biggest performance decrease is observed for baseline models when comparing the metric values with the corresponding results of the offline scenario. In the case of models involving the use of article profiles, this deterioration is much smaller, especially when we limit to basic most-general features. The results confirm that the proposed profile features are more general, time-agnostic, and, therefore, more useful when applying the models in the target online environment. In addition, similar results to those presented were obtained in online tests using a real system, which further reinforces this hypothesis.

**Table 6.** Results of the experiment for the real-world scenario (baseline vs user engagement profiles).

| Measure | Baseline | Basic Profile | Full Profile |
|---|---|---|---|
| AP | 0.0769 | 0.1358 | 0.2715 |
| AUC | 0.7401 | 0.8543 | 0.8788 |
| Accuracy | 0.9872 | 0.9871 | 0.9882 |
| Balanced accuracy | 0.5128 | 0.5184 | 0.5769 |
| Precision | 0.6000 | 0.4815 | 0.6750 |
| Recall | 0.0258 | 0.0372 | 0.1547 |
| F1 | 0.0495 | 0.0691 | 0.2517 |

**Table 7.** Results of the experiment for the real-world scenario (impact of different groups of features).

| Measure | Attention time/views | Paywall/ Conversion | User types | Source |
|---|---|---|---|---|
| AP | 0.2677 | 0.1731 | 0.1208 | 0.1094 |
| AUC | 0.8603 | 0.8363 | 0.8607 | 0.8447 |
| Accuracy | 0.9884 | 0.9873 | 0.9870 | 0.9871 |
| Balanced accuracy | 0.5685 | 0.5439 | 0.5141 | 0.5184 |
| Precision | 0.7742 | 0.5536 | 0.4545 | 0.4815 |
| Recall | 0.1375 | 0.0889 | 0.0287 | 0.0372 |
| F1 | 0.2336 | 0.1531 | 0.0539 | 0.0691 |

## 7 Conclusions

In this paper, we present that real-time article profile enrichment with time-agnostic features based on users' engagement leads to significantly improved machine learning models to detect the user's propensity to buy. We demonstrate that AI-based accurate targeting of the users interested enough in the offer to pay for access, is ready to become a standard in the digital media industry.

Our findings indicate that to attract more subscribers, the media companies should invest into modeling data-driven features describing article performance. While metrics such as the number of unique users or total number of views are useful in many ways, by themselves are not enough to infer a user's propensity to buy. A fit for purpose ML model can properly take into account multiple factors to predict attractiveness of any particular article in real-time, outperforming any attribution model heuristic that is currently widely used in the industry.

As article attractiveness is usually a highly dynamic concept – especially in the case of news – modeling additional data-driven features describing the most recent readers' engagement has turned out to be an efficient tool for propensity-to-subscribe model improvement. Our profiling method delivers user engagement features that are more context-independent and time-agnostic. Consequently, they are more generalizable and applicable to many scenarios within the industry, especially when the ML models are usually served in an environment that differs from the one where the models are built. Finally, our research opens future work directions, including extending the framework with additional features modeling the content type, such as video, text, or rich-media articles.

**Acknowledgments.** This work was supported by the Polish National Centre for Research and Development, grant POIR.01.01.01-00-1352/17-00 and by Poznan University of Technology, grant 0311/SBAD/0727.

# References

1. Carlton, J., Brown, A., Jay, C., Keane, J.: Using Interaction Data to Predict Engagement with Interactive Media, pp. 1258–1266. Association for Computing Machinery, New York, NY, USA (2021). https://doi.org/10.1145/3474085.3475631
2. Davoudi, H., Rashidi, Z., An, A., Zihayat, M., Edall, G.: Paywall policy learning in digital news media. IEEE Trans. Knowl. Data Eng. **33**(10), 3394–3409 (2021). https://doi.org/10.1109/TKDE.2020.2969419
3. Davoudi, H., Zihayat, M., An, A.: Time-aware subscription prediction model for user acquisition in digital news media. In: Proceedings of the 2017 SIAM International Conference on Data Mining (SDM), pp. 135–143. SIAM (2017). https://doi.org/10.1137/1.9781611974973.16
4. Dorogush, A.V., Gulin, A., Gusev, G., Kazeev, N., Prokhorenkova, L.O., Vorobev, A.: Fighting biases with dynamic boosting. CoRR abs/1706.09516 (2017). http://arxiv.org/abs/1706.09516
5. Fei, Y., Lv, C., Feng, Y., Zhao, D.: Real-time filtering on interest profiles in twitter stream. In: Proceedings of the 16th ACM/IEEE-CS on Joint Conference on Digital Libraries, pp. 263–264. JCDL 2016, ACM, New York, NY, USA (2016). https://doi.org/10.1145/2910896.2925462
6. Flach, P.: Machine Learning: The Art and Science of Algorithms That Make Sense of Data. Cambridge University Press, New York, NY, USA (2012)
7. Grinberg, N.: Identifying modes of user engagement with online news and their relationship to information gain in text. In: Proceedings of the 2018 World Wide Web Conference, pp. 1745–1754. WWW '18 (2018). https://doi.org/10.1145/3178876.3186180
8. Li, H., Vu, Q.H., Pham, T.L., Nguyen, T.T., Chen, S., Lee, S.: An ensemble approach to streaming service churn prediction. In: WSDM Cup 2018 Workshop, The 11th ACM International Conference on Web Search and Data Mining, Los Angeles, California, USA, pp. 1–8 (2018). https://wsdm-cup-2018.kkbox.events/
9. Liang, S.: Collaborative, dynamic and diversified user profiling. In: Proceedings of the AAAI Conference on Artificial Intelligence, vol. 33, no. 01, pp. 4269–4276 (2019). https://doi.org/10.1609/aaai.v33i01.33014269
10. Madnani, N., Loukina, A., Cahill, A.: A large scale quantitative exploration of modeling strategies for content scoring. In: Proceedings of the 12th Workshop on Innovative Use of NLP for Building Educational Applications, pp. 457–467 (2017). https://doi.org/10.18653/v1/W17-5052
11. Misiorek, P., Warmuz, J., Kaczmarek, D., Ciesielczyk, M.: Modeling user engagement profiles for detection of digital subscription propensity. In: Themistocleous, M., Papadaki, M. (eds.) Information Systems. EMCIS 2021. LNBIP, vol. 437, pp. 55–68. Springer International Publishing, Cham (2022). https://doi.org/10.1007/978-3-030-95947-0_5
12. Riordan, B., Flor, M., Pugh, R.: How to account for mispellings: quantifying the benefit of character representations in neural content scoring models. In: Proceedings of the Fourteenth Workshop on Innovative Use of NLP for Building Educational Applications, pp. 116–126 (2019). https://doi.org/10.18653/v1/W19-4411
13. Tang, X., Xu, Y., Geva, S.: Integrating time forgetting mechanisms into topic-based user interest profiling. In: 2013 IEEE/WIC/ACM International Joint Conferences on Web Intelligence (WI) and Intelligent Agent Technologies (IAT), vol. 3, pp. 1–4 (2013). https://doi.org/10.1109/WI-IAT.2013.132
14. Yandex: Catboost - open-source gradient boosting library. https://catboost.ai/. Accessed October 2022

# A Sensor-Based Unit for Diagnostics and Optimization of Solar Panel Installations

Lasse Berntzen[1]($\boxtimes$), Saeed Teimourzadeh[2], Paula Anghelita[3], Qian Ming[1], and Viorel Ursu[3]

[1] University of South-Eastern Norway, Kongsberg, Norway
lasse.berntzen@usn.no
[2] EPRA Electric Energy Co, Ankara, Turkey
[3] ICPE, Bucharest, Romania

**Abstract.** To solve the problem of relatively low efficiency in most of nowadays solar panels, this paper proposes a new solution by diagnosing and optimizing solar panel installations. Dr. Solar, a sensor-based device, acquires information on solar radiation, geographical position, inclination, roll, panel temperature, and ambient temperature and humidity. The sensor-based device is mounted to the solar panel for a few days while collecting data. The cloud-based service gathers information from the inverter while the device is moved to other installations. Combined with the data of power generation from the inverter, a cloud service is offered for processing and analyzing the data to maximize power production by adjusting the panel installation. During several short-term experiments in different places, Dr. Solar provided reliable data for analysis on the cloud platform, and a web interface was developed to calculate power production. The solution has increased photovoltaic panels' efficiency by optimizing adjustments to the panel's physical position (tilt and angle). The paper describes the sensor-based device, the cloud service, and the algorithms used.

**Keywords:** photovoltaic energy · solar panels · optimization · diagnostics · big data · analytics · sensors · Dr. Solar

## 1 Introduction

Photovoltaic energy has just reached a new dimension. According to the research of SolarPower Europe, in May 2022, the total global installed solar capacity has passed the 1 TW threshold [1]. That is 500 times increase compared to 2002, when the capacity was only 2 GW. It took 16 years, until 2018, to reach the 500 GW level, and only three years later, the amount has doubled to 1,000 GW or 1 TW. The forecast for the rest of 2022 shows an increase of 200 GW in one year. Solar power has become the third most significant renewable energy source by installed capacity, after hydro and wind power [1].

At the same time, however, solar still meets only a small share of around 4% of the global electricity demand, while non-renewable sources provide over 70%. In 2022, global solar additional installation capacity expects to increase by 36% to 228.5 GW.

M. Papadaki et al. (Eds.): EMCIS 2022, LNBIP 464, pp. 105–121, 2023.
https://doi.org/10.1007/978-3-031-30694-5_9

The world will see solid demand for solar in the next four years, growing from 255.8 GW of additional capacity in 2023 to 347 GW in 2026. It will likely add 314.2 GW in 2025 [1].

Most commercial solar panels have efficiencies from 15% to 20% [2]. Therefore there is a demand to improve the efficiency of energy conversion. Although emerging perovskite solar cells have a high conversion efficiency (over 25%), there is still a long way for perovskite solar cells to get commercialized since perovskite crystals are easily decomposed in a humid environment. At the moment, the lack of stability is a significant disadvantage.

This paper reports on a project to diagnose and optimize the use of solar panel installations by developing a prototype sensor platform fixed to the panels and delivering data to a cloud-based service for analytics. The project was a collaborative effort between ICPE (Romania), EPRA (Turkey), and the University of South-Eastern Norway in 2021 and 2022. Photovoltaic panel output depends on several factors. The idea of Dr. Solar is to optimize the placement of the solar panels based on an analysis of such elements.

The essential characteristics of solar panels include the following:

- Structural characteristics (roof-top mounted, ground mounted, wall mounted)
- Type (monocrystalline, polycrystalline, thin film)
- Geographical position (latitude, longitude, altitude)
- Magnetic orientation, inclination, and base slop (roll)

  Operational characteristics.

- Solar radiation level
- Solar panel back side temperature
- Ambient temperature
- Ambient relative humidity

The installation of the photovoltaic panels should consider some factors to make the most use of solar radiation: the material types, the geographical position, the elevation, inclination/angle, and solar radiation conditions. The correct measurement and adjustment of these factors ensure that photovoltaic panels produce maximum energy by being exposed to the greatest intensity of solar radiation during operation. Photovoltaic cells employ solar radiation in their operation; hence the cells are affected by the weather or condition of the atmosphere to which they are exposed. Here, the temperature is an essential factor which can be evaluated from two standpoints, i.e., ambient temperature and solar cell surface temperature. The electrical efficiency of solar cells depends on the ambient temperature. Solar cells are made of semiconductor materials like the most used crystalline silicon, and semiconductors are sensitive to temperature. Solar panels represent a negative temperature coefficient, and their performance declines when the temperature increases. The main reason is that an increment in the temperature decreases the band gap of a semiconductor which decreases the open circuit voltage of solar cells and the output power [3].

Regarding the cell surface temperature, the surface temperature can also be affected by the ambient temperature. However, the challenging point is the occurrence of hotspot areas. Hot spots are areas of high temperature that affect a solar cell by consuming energy instead of generating it [4]. As the solar cells are connected in series, a weak cell or group of cells will affect the energy production of all the cells on the same string.

The other factor is humidity. Solar radiation is electromagnetic waves, and when the light consisting of photon strikes the water layer, which is denser, refraction appears and decreases the light's intensity. Because the efficiency depends upon the value of the maximum power point of the solar cell, the effect of humidity deviates from the maximum power point, decreasing solar cell efficiency [5].

Humidity readily affects the efficiency of the solar cells and creates a minimal layer of water on their surface. It also decreases the total power output produced. Panjwani et al. [6] conducted more than four experiments with water absorbents to improve efficiency and make the system more efficient.

The next section provides a conceptual description of the Dr. Solar system. The following sections describe data collection, data analysis, and results. Finally, the last section contains the conclusion and ideas for further work.

## 2   Conceptual Description

Dr. Solar is a system designed to assess photovoltaic systems by collecting relevant measurement parameters followed by an analysis in the cloud using dedicated algorithms that give indications for optimizing energy production. The system consists of the following main parts:

- Dr. Solar box containing various sensors
- The inverter that collects and provides the data concerning energy production
- A cloud-based platform that provides analytics service based on data obtained from the Dr. Solar box sensors and the inverter

The results from the cloud-based service are presented to the user as a report which contains, along with the presentation of the results regarding the configuration of the installed photovoltaic system, indications related to possible adjustments of the system to increase energy production. The inverter production-related problems are also included in the report. Figure 1 shows the main parts that compose the Dr. Solar solution.

The inverter transforms the DC low voltage gained from the solar panels into 220–240 V AC used in households. The prototype of the Dr. Solar device was used with a Huawei inverter SUN2000L-3KTL model to demonstrate the functionality.

The inverter collects and submits data about its input and output parameters, such as input and grid currents, input and grid voltages, active and reactive power, and energy production.

The Dr. Solar box also collects and uploads data into the same cloud-based platform, where analytics service is implemented using a GSM-based mobile router. A router is necessary since the solar panels are usually not placed in areas covered by Wi-Fi connection. The Dr. Solar box is set to collect data every 5 min for three consecutive days, ideally when radiation exceeds 600 W/m$^2$. Both the Dr. Solar device and inverter provide data at five minutes intervals that are polled by the cloud-based analytics service for the time interval between 10:00 am to 2:00 pm.

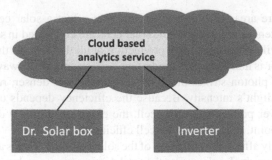

**Fig. 1.** The main components of Dr. Solar

## 3  Collecting Data

The Dr. Solar box is a sensor-based module designed to collect data from sensors and send them to the cloud platform to be processed. The Dr. Solar box is attached to the solar panel and stays there for a few days. The device can then be moved to a new location to collect and submit new data to be processed by the cloud service. The cost of the box does not justify a permanent mounting in only one location, except for large installations. During a few days, the Dr. Solar box will collect the data used for analytics. The sharing of the Dr. Solar box between many photovoltaic systems users justifies the costs of producing the sensors units/boxes.

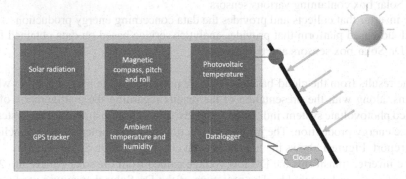

**Fig. 2.** Dr. Solar Box

Figure 2 shows the different building blocks of the Dr. Solar box. The Dr. Solar box includes several sensors relevant to photovoltaic systems investigation, as follows:

- Solar radiation
- Magnetic compass, inclination, and roll (base slope)
- Photovoltaic panel temperature (measures the back-side surface temperature of the solar panel)
- GPS tracker
- Ambient temperature and relative humidity

Figure 3 shows the prototype of the Dr. Solar device. The radiation sensor is on the top front view right side. The wire on the left side is for the surface temperature sensor.

**Fig. 3.** The Dr. Solar box

Figure 4 shows how the Dr. Solar device prototype mounted on the solar panel.

**Fig. 4.** The Dr. Solar box fixed to the solar panel

Table 1 briefly describes the different components of the Dr. Solar box. The components are also shown in Fig. 5

The controller is a data logger that also has a protocol stack. The data logger can be used as a web client and server. It is the datalogger that submits information to the cloud-based service.

**Table 1.** Dr. Solar box sensors

| Sensor | Model | Description |
|---|---|---|
| Solar radiation sensor (pyranometer) | SR05-D1A3-PV | Measures solar radiation received by the plane surface |
| Temperature sensor | NTC 10k | Measures the surface temperature of the photovoltaic panel |
| Magnetic compass and inclinometer | DCS456M | This sensor is used to find the photovoltaic panel orientation and pitch. Together with the GPS, it provides positioning data for the prosumer-installed photovoltaic panels, useful for knowing the incident angle of the solar radiation |
| Temperature and humidity sensor | Lumel18D | A sensor that measures ambient temperature and relative humidity |
| GPS tracker | iUni GT02 | GPS tracker has a built-in GSM module for mobile communication. The device is waterproof and mainly used for theft protection of cars and motorbikes. The GPS tracker is used to obtain the exact position of the photovoltaic panel |

**Fig. 5.** Dr. Solar sensors: (a) Solar radiation sensor, (b) Ambient temperature and relative humidity sensor, (c) Photovoltaic panel temperature sensor, (d) Magnetic compass and inclinometer, (e) GPS tracker

## 4    Analyzing the Data

The analytics process is visualized in Fig. 6. The left side shows the Dr. Solar device and the inverter. The cloud-based service is in the middle, and the web-based user-interface is on the right.

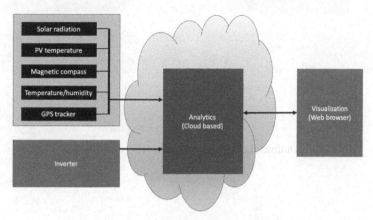

**Fig. 6.** The analytics process visualized

Dr. Solar is the implementation of an idea of a cloud-based analysis and diagnosis platform for photovoltaic prosumers, shortened as (PVADIP-C). The proposed platform provides a cost-effective solution for small on-grid photovoltaic systems. The relevant onsite data are collected and sent to a dedicated cloud platform to be processed, analyzed, and diagnosed. Hence, there is no need to implement expensive control and optimization engines at small-scale photovoltaic locations. Two main functionalities are envisioned for the PVADIP-C. The outline of the proposed method is depicted in Fig. 7. As can be seen, the first functionality is an operational decision-making engine that optimizes a photovoltaic-based prosumer's operation. The second engine is the asset management engine which identifies the possible failures in the photovoltaic-based prosumer.

### 4.1    Operational Decision Support

This section deals with a small-scale photovoltaic system proposed operational decision support mechanism. Figure 8 depicts a schematic of a load point which includes a connection to the upstream grid, small-scale photovoltaic system, fixed loads, controllable loads (washer & dryer, dishwasher, and pump), storage device, and plug-in electric vehicle.

For optimal operation, the objective is minimum cost while satisfying technical constraints pertaining to nodal power balance, power transactions with the upstream network, and permissible operation range of elements.

The objective function can be formulated as follows:

$$Minimize\ OF = \sum_{t \in I_T} \left( \lambda_t^{Buy} P_t^{Buy} - \lambda_t^{Sell} P_t^{Sell} \right) \tag{1}$$

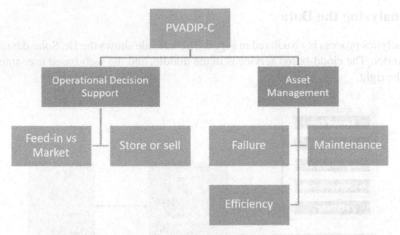

**Fig. 7.** Outline of the proposed methodology for PVADIP-C

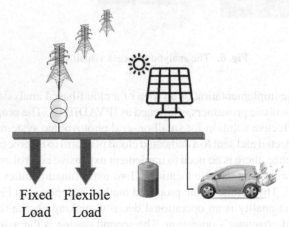

Fixed    Flexible
Load      Load

**Fig. 8.** Schematic of a load point with a small-scale photovoltaic system, flexible loads, and storage devices

$t, I_T$ I ndex and set of time
*Buy, Sell* Superscript for buying from and selling to the market
$P$ Active power
$\lambda$ Price per kilowatt

The primary constraint is to maintain nodal power balance, which is formulated as follows:

$$P_t^{Gen} - P_t^{Load} = 0 \quad \forall t \in I_T \tag{2}$$

$$P_t^{Gen} = P_t^{Buy} + P_t^{PV} + P_t^{Bat-} + P_t^{EV-} \quad \forall t \in I_T \tag{3}$$

$$P_t^{Load} = P_t^{Sell} + P_t^{Fix} + P_t^{Flex} + P_t^{Bat+} + P_t^{EV+} \quad \forall t \in I_T \tag{4}$$

*PV, Bat, EV, Fix, and Flex* are superscripts for solar generation, plug-in electrical vehicles, fixed load at home, and flexible load at home, respectively. Here, $+$ and $-$ superscripts show the charging\discharging stage of battery storage and plug-in electric vehicle, respectively.

The technical constraints for small-scale photovoltaic operation are represented by (5)–(23). The power transacted between the load point and upstream network should be within a pre-defined range modeled by (5) and (6).

$$0 \le P_t^{Buy} \le \alpha_t^M P_t^{Buy,Max} \tag{5}$$

$$0 \le P_t^{Sell} \le \left(1 - \alpha_t^M\right) P_t^{Sell,Max} \tag{6}$$

*M* is the symbol for market-related quantities, *Min* and *Max* are symbols for lower and upper limits, respectively. In (5) and (6), the binary variable is used to avoid enabling both selling and buying options simultaneously. The upper and lower limits of photovoltaic generation are modeled by (7)

$$P_t^{PV,Min} \le P_t^{PV} \le P_t^{PV,Max} \tag{7}$$

The state-of-charge (SOC) of the storage device at each period is calculated by (8) [7]. Constraint (10) is limitations on SOC, and (11)–(12) is limitations on charge/discharge power associated with storage.

$$SoC_t^{Bat} = SoC_{(t-1)}^{Bat} + \frac{\eta^{Bat}\Delta t}{E^{Bat,Max}}\left(P_{(t-1)}^{Bat+} - (\eta^{Bat})^{-2}P_{(t-1)}^{Bat-}\right) \tag{8}$$

$$SoC^{Bat,Min} \le SoC_t^{Bat} \le SoC^{Bat,Max} \tag{9}$$

$$0 \le P_t^{Bat+} \le \alpha_t^{Bat} P^{Bat+,Max} \tag{10}$$

$$0 \le P_t^{Bat-} \le \eta^{Bat}\left(1 - \alpha_t^{Bat}\right) P^{Bat-,Max} \tag{11}$$

$\eta$ is the conversion efficiency coefficient, $\alpha$ is a binary decision variable, and E is the energy capacity for the storage unit. The EV charge/discharge constraints are modeled by (12)-(13) associated with storage. In (14), the SOC of parking lots in various time intervals is computed. Constraints on the target SOC of parking lots are stated by (15)-(17), respectively.

$$0 \le P_t^{EV+} \le \alpha_t^{EV} P^{EV+,Max} \tag{12}$$

$$0 \le P_t^{EV-} \le \eta^{EV}\left(1 - \alpha_t^{EV}\right) P^{EV-,Max} \tag{13}$$

$$SoC_t^{EV} = SoC_{(t-1)}^{EV} + \frac{\eta^{EV}\Delta t}{E^{EV,Max}}\left(P_{(t-1)}^{EV+} - (\eta^{EV})^{-2}P_{(t-1)}^{EV-}\right) \tag{14}$$

$$SoC^{EV,Min} \le SoC_t^{EV} \le SoC^{EV,Max} \tag{15}$$

$$0 \le P_t^{EV+} \le \alpha_t^{EV} P^{EV+,Max} \left( 1 - SoC_t^{EV} / 1 - SoC_t^{Sat-EV} \right) \tag{16}$$

$$0 \le P_t^{EV-} \le \left( 1 - \alpha_t^{EV} \right) P^{EV-,Max} \left( 1 - SoC_t^{EV} / 1 - SoC_t^{Sat-EV} \right) \tag{17}$$

Equations (18)–(23) model controllable loads as follows:

$$P_t^{Flex} = \alpha_t^{Flex} P^{Flex,Dep} \tag{18}$$

$$\sum_{t \in I_T} \alpha_t^{Flex} = T^{Flex} \tag{19}$$

$$\alpha_t^{Flex} - \alpha_{t-1}^{Flex} = \beta_t^{Flex} - \zeta_t^{Flex} \quad \forall t \in [2, 24] \tag{20}$$

$$\sum_{i=t-T^{Flex}+1}^{t} \beta_i^{Flex} \le \alpha_t^{Flex} \quad \forall t \in \left[ T^{Flex} + 1, 24 \right] \tag{21}$$

$$\sum_{i=t-DT^{Flex}+1}^{t} \zeta_i^{Flex} \le 1 - \alpha_t^{Flex} \quad \forall t \in \left[ DT^{Flex} + 1, 24 \right] \tag{22}$$

$$DT^{Flex} = 24 - T^{Flex} \tag{23}$$

*Dep* is superscript for deployment, $T$ is the time required for the complete operation of the equipment and are auxiliary binary variables. Equation (18) expresses that the power consumed by the flexible load at time t, i.e., is equal to its nominal power consumption while deployment if it is in operation, i.e., otherwise, is 0. In addition, the operation time of a flexible load should be equal to the required time for its complete operation, modeled by (19). Here, continuous operation of flexible load should also be considered. In other words, when the flexible load is committed, it should be continuously in operation until the time required for its complete operation (interruption is not allowed). To this end, the suit of constraints represented by (20)–(23) is added.

The outputs of the optimization engine are:

- Optimal amount of photovoltaic-based generation to be consumed at each hour.
- Optimal amount of power to be traded with the upstream network.
- Optimal charging\discharging pattern for battery storage and electric vehicle.
- Optimal commitment of controllable loads.
- Hourly cost of community operation.

The required inputs for the proposed model are:

- Hourly energy price ($/kWh).
- Hourly fix load (kW).
- Hourly generation of the photovoltaic generation (kW).
- Installed capacity for renewable generation, battery storage, and electric vehicle.

- Permissible operating range of devices.

From the inputs mentioned above, the hourly fix load and the photovoltaic generation can be acquired from the Dr. Solar box. The user or the operator can enter additional information. With this data in place, the devised model offers the optimal operation for a load point equipped with a small-scale photovoltaic system. Note that Dr. Solar's optimal decision support service can be offered when Dr. Solar is installed for a considerable period at the customer's place, for instance, a month.

## 4.2 Asset Management

This section deals with devising an asset management approach for small-scale photovoltaic systems. The main objective is to detect any failure or deficiency that results in the efficient operation of a small-scale photovoltaic system. The outline of the proposed approach for asset management of small-scale photovoltaic systems is depicted in Fig. 9.

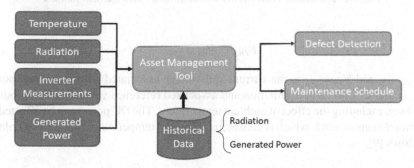

**Fig. 9.** Outline of the proposed approach for asset management

The main objective is to estimate the small-scale photovoltaic system's output power and ensure that all system elements are intact. The estimated output power is then compared with the measured power, and a failure occurrence is reported in case of substantial difference. Here, the proposed methodology calculates the DC side parameters for accurate decision-making. The detailed calculation within the asset management toolbox in Fig. 9 is depicted in Fig. 10.

In Fig. 10, three types of inputs are required. The first category is the measurement which is solar irradiation and ambient temperature. The second and third groups are the data on solar cells and the inverter, accessible through the data sheets. Once the input data are acquired, the next step is calculating the power at the DC side of the inverter. The DC power is calculated as [8]:

$$I_{mpp} = I_{mpp,ref} \frac{G}{G_{ref}} \tag{24}$$

$$V_{mpp} = V_{mpp,ref} \ln\left(e + c\left(G - G_{ref}\right)\right) \tag{25}$$

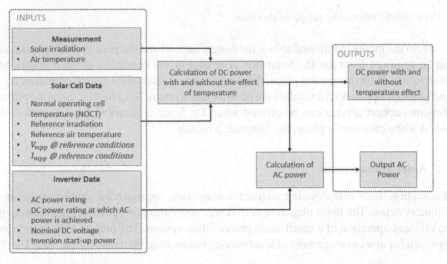

**Fig. 10.** The detailed calculation within the asset management toolbox

$$P_{DC} = V_{mpp}I_{mpp} \qquad (26)$$

$I_{mpp, ref}$, and $V_{mmp, ref}$ is the current and voltage associated with maximum power point. G and $G_{ref}$ are solar irradiation and associated reference value. $P_{DC}$ is the output DC power, excluding the effect of ambient temperature. The DC power can be affected by the ambient temperature, which is considered. The cell temperature should be calculated as follows [9]:

$$T_c = T_{air} + \left(T_{NOCT} - T_{ref}\right)\frac{G}{G_{ref}} \qquad (27)$$

$T_{air}$ is the ambient temperature, $T_{NOCT}$ is the normal operating cell temperature, and $T_{ref}$ is the reference temperature. Once the cell temperature is calculated, the DC power, including the ambient temperature effect, can be calculated as follows:

$$I_{mpp}^{Temp} = I_{mpp,ref}\left(1 + a\left(T_c - T_{ref}\right)\right)\frac{G}{G_{ref}} \qquad (28)$$

$$V_{mpp}^{Temp} = V_{mpp, ref}\left(1 + b\left(T_c - T_{ref}\right)\right) \\ \times \ln\left(e + c\left(G - G_{ref}\right)\right) \qquad (29)$$

$$P_{DC} = V_{mpp}I_{mpp} \qquad (30)$$

The superscript *Temp* stands for ambient temperature-affected quantities.

As can be seen, the DC power in Fig. 10 is calculated for both with and without the effect of temperature. The main reason to calculate these two parameters is to identify an occurrence of overheating. The next step is calculating the AC power, which is realized using the DC power and inverter data outputs. The AC power is calculated as follows [10]:

$$P_{ac} = \left(\frac{P_{aco}}{A-B} - C(A-B)\right)\left(P_{DC}^{Temp} - B\right)$$
$$+ C\left(P_{DC}^{Temp} - B\right)^2 \tag{31}$$

$$A = P_{dco}(1 + C_1(V_{DC} - V_{dco})) \tag{32}$$

$$B = P_{so}(1 + C_2(V_{DC} - V_{dco})) \tag{33}$$

$$C = C_o(1 + C_3(V_{DC} - V_{dco})) \tag{34}$$

$P_{aco}$ is the maximum AC-power rating for an inverter at reference or nominal operating condition, $P_{dco}$ is the DC-power level at which the ac-power rating is achieved at the reference operating condition, $V_{dco}$ DC-voltage level at which the ac-power rating is achieved at the reference operating condition, $P_{so}$ is DC-power required to start the inversion process, or self-consumption by inverter, strongly influences inverter efficiency at low power levels, and $C_0$, $C_1$, $C_2$, and $C_3$ are constants.

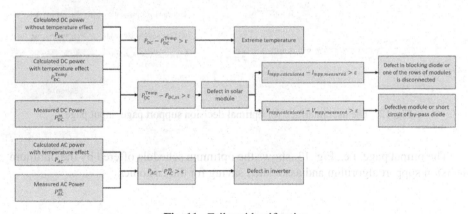

**Fig. 11.** Failure identification

Once the powers at AC and DC sides are calculated, they will be evaluated using the approach depicted in Fig. 11 to identify the failure.

## 5 Results

Dr. Solar employs an efficient web platform as the user interface where the user or the operator can check the measured values and implement desired settings. This section summarizes the input and output pages that the user/operator would be faced while working with Dr. Solar's solution.

## 5.1   Optimum Decision Support

Figure 12 depicts the input and output pages for the optimum decision support service of the Dr. Solar box. The input page acquires the daily fixed load and photovoltaic generation from Dr. Solar's measurements. The electricity prices are based on historical data. For the selected date, the user\operator is asked to choose the preference on including the battery storage system (if available) and electric vehicle (EV) (If available) in the optimization process. In addition, the availability of any other flexible load is also asked for and considered in the optimization process.

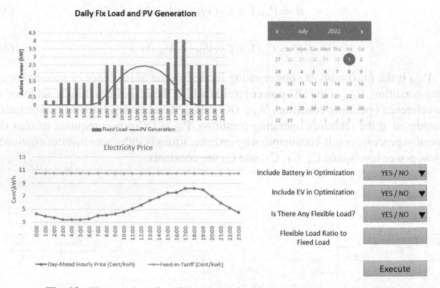

**Fig. 12.** The user interface for optimal decision support page: input page

The output page, i.e., Fig. 13, shows the optimum schedule offered by the optimum decision support algorithm and associated saving for the customer.

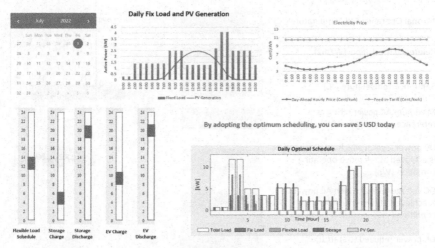

**Fig. 13.** The user interface for optimal decision support page: output page

## 5.2 Asset Management

Figure 14 depicts the inputs required for performing the asset management algorithms. The required inputs are based on the datasheet of the photovoltaic power plant, which the operator can easily access. Then, the asset management evaluation period is asked. At least three days are required for decision-making with an acceptable level of accuracy. When the evaluation is finished, the output page, as depicted in Fig. 15, will be shown to the operator/user. Here, the main output is the message which shows the potential deficiency in the solar photovoltaic system. The list of possible output messages is as follows:

- No failure is detected
- Defect in solar module

  - Defective module or short circuit of by-pass diode
  - A defect in the blocking diode or one of the rows of modules is disconnected

- Defect in inverter

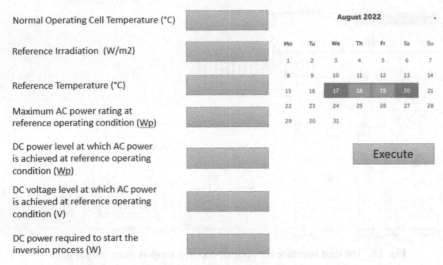

Fig. 14. The user interface for the asset management page: input page

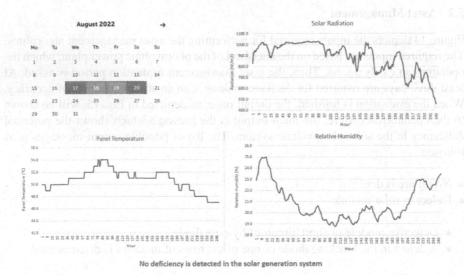

No deficiency is detected in the solar generation system

Fig. 15. The user interface for the asset management page: output page

## 6 Conclusion

This paper discusses the development of Dr. Solar, a sensor-based device for solar panel optimization and diagnostics. The device uses sensors to measure solar radiation, photovoltaic panel surface temperature, temperature, relative humidity, and position. The data collected by the Dr. Solar box is combined with data collected from the inverter and sent to a cloud service for processing. Using artificial intelligence algorithms, the data is analyzed, and the results are presented to the user through a web-based interface.

The results will provide the user with information on faults or degradations, as well as advice on optimizing adjustments of the physical position (tilt and angle).

## 6.1 Future Work

A companion paper (to be submitted) describes business model considerations. Another work in progress discusses neighborhood sentiments toward solar panel adoption based on aesthetics. For this research, we use a 3D household model with the possibility of adjusting tilt and angle.

**Acknowledgments.** This work was part of the project "Cloud-based analysis and diagnosis platform for photovoltaic (PV) prosumers" supported by the Manu Net scheme Grant number MNET20/NMCS-3779 and funded through the Research Council of Norway Grant number 322500, UEFISCDI – Executive Agency for Higher Education (Romania): Research, Development and Innovation Fund, contract no. 215/2020, and TÜBİTAK ARDEB 1071 (Turkey): Support Program for Increasing Capacity to Benefit from International Research Funds and Participation in International R&D Cooperation (Project No: 120N838).

# References

1. SolarPower Europe: Global Market Outlook for Solar Power 05.2022. https://api.solarpowe reurope.org/uploads/Solar_Power_Europe_Global_Market_Outlook_report_2022_2022_ V2_07aa98200a.pdf
2. University of Michigan: Photovoltaic Energy Factsheet. https://css.umich.edu/publications/ factsheets/energy/photovoltaic-energy-factsheet. Last accessed 19 Oct 2022
3. Dash, P.K., Gupta, N.C.: Effect of temperature on power output from different commercially available photovoltaic modules. Int. J. Eng. Res. Appl. **5**(1), 148–151 (2015)
4. Köntges, M., Kurtz, S., Packard, C., Jahn, U., Berger, K.A., Kato, K.: Performance and reliability of photovoltaic systems subtask 3.2: Review of failures of photovoltaic modules: IEA PVPS task 13: external final report IEA-PVPS. IEA. ISBN 978-3-906042-16-9 (2014)
5. Amajama, J., Oku, D.E.: Effect of relative humidity photovoltaic panels' output and solar illuminance/intensity. J. Sci. Eng. Res. **3**(4), 126–130 (2016)
6. Panjwani, M.K., Panjwani, S.K., Mangi, F.H., Khan, D., Meicheng, L.: Humid free efficient solar panel. In: 2nd International Conference on Energy Engineering and Smart Materials: ICEESM, vol. 1884, no. 1 (2017)
7. Teimourzadeh, S., Tor, O.B., Cebeci, M.E., Bara, A., Oprea, S.V.: A three-stage approach for resilience-constrained scheduling of networked microgrids. J. Mod. Power Syst. Clean Energy **7**(4), 705–715 (2019)
8. Cui, C., Zou, Y., Wei, L., Wang, Y.: Evaluating combination models of solar irradiance on inclined surfaces and forecasting photovoltaic power generation. IET Smart Grid **2**, 123–130 (2019)
9. Ciulla, G., Lo Brano, V., Moreci, E.: Forecasting the cell temperature of PV modules with an adaptive system. Int. J. Photoenergy **2013**, 1–10 (2013). https://doi.org/10.1155/2013/192854
10. Boyson, W.E., Galbraith, G.M., King, D.L., Gonz, S.: Performance Model for Grid-Connected Photovoltaic Inverters. Sandia National Laboratories (SNL), Albuquerque, NM, and Livermore, CA (2007)

# Big Data in the Innovation Process – A Bibliometric Analysis and Discussion

Zornitsa Yordanova[✉] [iD]

University of National and World Economy, Sofia, Bulgaria
zornitsayordanova@unwe.bg

**Abstract.** Companies are already using big data to develop innovations and improve efficiency and productivity in many business processes. The innovation process is one of the most important processes at the company level and beyond, as it impacts the company's overall performance and potential. Innovation is a data-intensive process, so the development of new technologies, especially big data, has a significant impact on innovation. The purpose of this research is to examine how big data is currently influencing the innovation process. The current scope of big data in innovation processes is defined by a literature analysis of 110 publications indexed in the Web of Science, Social Science Citation Index. This represents any article that uses both "big data" and "innovation process" in the title, abstract, and author words. In addition to state-of-the-art analysis, we identify topics that have not yet been explored to advance future research in this area and in some clusters, allowing further depth and analysis of the topic.

**Keywords:** Innovation Process · Big Data · Innovation Management · Technology Management · Bibliometric Analysis

## 1 Introduction

Big data as a tool for innovation continues to grow in popularity [1], but it is imperative to examine how the nature and steps of the innovation process change as a result of its inclusion in its development and application, as they become components of the process and the knowledge management within enterprises [2]. Research on innovation's curves and transformation is urgently needed in light of its significant influence on overall innovations it brings as a main systematic means of innovating across teams, organizations, and industries, as well as the growing research into technologies involved in its digital transformation [3, 4]. Recently, many studies have provided evidence on the scope and level of big data (BD) use for improving firm performance [5] but still, an overall perspective and mapping of the whole literature for BD use in innovation management is missing in regard to the innovation process as a separate theoretical sub-filed of the innovation management science. Thousands of comprehensive research has already revealed the role of BD in co-innovation, open and collaborative innovation [6], service innovation [7], and new product development [8] as well as in some

M. Papadaki et al. (Eds.): EMCIS 2022, LNBIP 464, pp. 122–133, 2023.
https://doi.org/10.1007/978-3-031-30694-5_10

particular economic fields such as agriculture [9], hospitality sector [7], banking [10], etc. The majority of the literature indicates that BD has a positive impact on innovation performance [11], but it has yet to be explored how innovation patterns are changing. For addressing firms' goals and challenges, data has become increasingly critical [12]. The concept of business data management as a superior level of data management can open up the littoral capabilities of business data management towards external sources of data that firms could incorporate into their internal processes in order to gain operational efficiency, boost the efficiency of processes, support decision making [9] and accelerate go-to-market strategies as well as increase customer satisfaction [13]. Since BD has been hailed as the next big thing in innovation [14], it has not only been used for product development but has also been embedded into some companies' innovation processes to increase the sustainability and performance of any kind of innovative development.

The innovation process is among the essential business process in organizations that directly influence a firm's profits [15]. Innovation for firms is often assessed as the most important outcome of business processes and critical for a firm's performance [16]. All of these factors place the internal innovation process at the center of a firm's innovation capabilities and the entire company's ability to operate successfully.

As part of the firm's innovation process, big data incorporation and utilization are inevitable, especially when research indicates that BD can positively influence employee creativity [17] as well as company performance and decision-making [18]. Nevertheless, the literature does not provide a clear picture of the current state of the art or evidence regarding the successful implementation of BD. As a promising research area, proven by the increasing number of studies on the topic and generally on the topic of the use of BD in the business process of firms for improving the quality and results of them, this study also aims at identifying the areas for further research for the innovation processes in particular. Furthermore, to gain a clear understanding of the development of this research area, it is critical to examine the current literature in its entirety. According to Sheng, Amankwah-Amoah, and Wang [19], big data in business processes and big data in innovation management represent emerging research areas in which multidisciplinary expertise is required. Bibliometric analysis is a useful method to analyse cross-discipline, emerging, or rapidly changing research areas impacting diverse areas [20, 21]. In our case, we examine a firm's innovation process and how big data technology is influencing it.

The results from the study give insights into the current state of the art on the problem of BD usage in the innovation process and give directions for further research.

## 2   Theoretical Background

### 2.1   Firm's Innovation Process

The firm's innovation process or internal innovation process (FIP) has attracted the interest of researchers since the 1950s and it is still amongst the hottest sub-topics in management science. In general, the innovation process is a sequence of activities and strategic vision related to the development and commercialization of innovative outcomes from organizations to the audience/market or for internal needs. Its evolution has gone through different concepts and theories starting from more general and industry-based therefore internal-to-firm models. Rothwell [22] provided a systematization into five categories: The Technology Push Theory – in the 1950s, The Market Pull Theory – in the 1960s, The Coupling Innovation Process Theory – in the 1970s–1980s, The Functional Integrated Innovation Process Theory [23] – 1980s, The Systems Integration and Networking Innovation Process Theory - 1990s.

Later on, more FIP have been introduced and followed as principles by firms such as Open Innovation [24], User Innovation [25], Lean Startup framework [26], chain-linked models [27], the Information assurance of innovation process [22], Innovation process based on continuous improvement [28], etc. with diverse kinds of extensions and customizations. Opening the innovation process not only towards different stakeholders but also to different concepts [29] incl. Including emerging technologies such as artificial intelligence [30], technology acceptance models, or BD [31] have clearly defined benefits in several case studies within the scientific literature. Nowadays, FIP remains under-researched as modern techniques, technologies, and design tools for adjusting the innovation processes are frequently applied [32].

However, a common understanding of the innovation process is missing in the literature, as it is quite complex and specific to the firm's sector and approach [33]. Nevertheless, among the most important aspects of the innovation process is identified to be the need for gathering information from and transmitting information to several external information areas within and outside of the organization, called also innovation systems [34]. BD provides a huge amount of data, incl. Internal and external to the firm that may benefit the organizational innovation process [35]. BD use in the innovation process as a managerial approach influencing the process for developing not only technological innovations directly related to BD, but any kind of products or processes is still under-researched, which motivates this article.

## 2.2  Big Data in Innovation Management and Innovation Process

Innovation management is considered a comprehensive and general scientific field for innovation management in organizations. The innovation management and specifically firm's innovation process theories contain numerous sub-streams that are generally clustered into different themes and directions for research and practice. Here in this section, we are revealing some of these sub-streams of theories, which are already discussed and researched in the context of big data as part of the innovation process, incl. The use of big data, data analytics, decision-making through big data, big data for generating users' ideas, big data to support innovation commercialization, etc. The digital transformation of firms and even of entire industrial sectors calls for the introduction of new forms of human-machine collaboration through the increased use of big data and cognitive systems require completely new approaches [36]. As Trabucchi and Buganza scholars [37] suggested, scholars are constantly providing different strategies and methods to help firms grasp and understand the added value immersed in their data to foster innovation and improve the efficiency of existing processes. To enhance organizational learning in the innovation process, improved use of existing and new information and its better assimilation become imperative [2]. Furthermore, big data benefits not on the innovation process but also firm productivity in general [38].

## 3   Research Design

The research design is organized following the main principles of bibliometric analysis as this method can extract relevant information from a large number of publications and elicit the information for answers to the research questions in this study.

### 3.1  Data Selection

The scope of this research was defined by conducting a Boolean search in the Web of Science (WoS) database for extracting high-quality publications regarding the innovation process and big data that are mutually discussed and analyzed. The formula used was the following:

ALL: BIG DATA (Topic) and ALL: INNOVATION PROCESS (Topic), refined by: document types: article; language: English; Social Sciences Citation Index (SSCI).

After conducting the search, 110 publications met the search criteria. The details about the scope are provided in Table 1 and presented visually in Table 1.

**Table 1.** Data selection and scope for bibliometric analysis

| Timespan | 1998:2022 |
|---|---|
| Sources (Journals, Books, etc.) | 77 |
| Documents | 110 |
| Annual growth rate % | 12.25 |
| Average citations per document | 26.63 |
| Authors | 352 |
| Co-Authors per Doc | 3.3 |
| International co-authorships % | 34.55% |

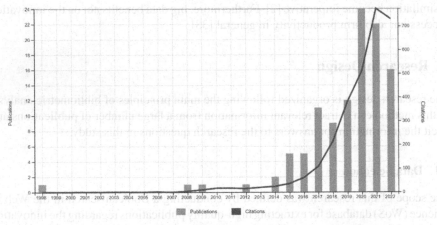

**Fig. 1.** Annual scientific production on BD in the innovation process

Figure 1 shows the exponential development of research fields that drive research. Even if the research does not focus on a specific research question, its value and contribution can be attributed to mapping the current state of the art in the interdisciplinary scientific field of big data in the innovation process and to one of the identified areas.

## 3.2  Bibliometric Analysis

Bibliometric analysis was applied as a systematically proven type of research by many scholars and is currently considered one of the most effective scientific methods for understanding the research field from a historical, holistic, and interdisciplinary perspective. Bibliometric analysis facilitates the mapping of the current research as well as identifies knowledge gaps, streams of research already done, and authors' information, and recognizes further research agenda. Bibliometric analysis is an effectively used method to explore the emergence of the domain of digitalization and innovation [39] and has thacanthushe research and forecast future research trends [25].

In this study, we applied the following bibliometric analysis to address the core topic of digitalization of the firm's innovation process in the particular field of using BD:

- Co-word analysis
- Top-tier Journals
- Themati

Thematicose of bibliometric analysis, we used R Studio and its package Biblioshiny.

## 4 Results and Discussion: Big Data in the Innovation Process

The first results show the most commonly met keywords used by authors, the most influential authors, and the respective journals publishing such research. Amongst the journals, three out of four have a more sustainable focus. Another interesting insight is the combination of artificial intelligence along with BD (Fig. 2).

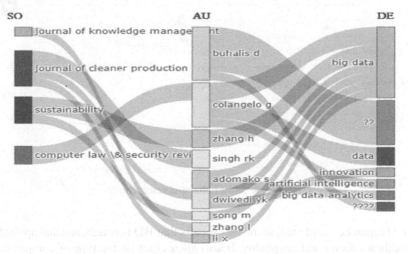

**Fig. 2.** Three plot analyses on sources, authors, and author keywords

Figure 3 reveals the most relevant sources for such research: Sustainability and Journal of cleaner production, which both are eco-oriented.

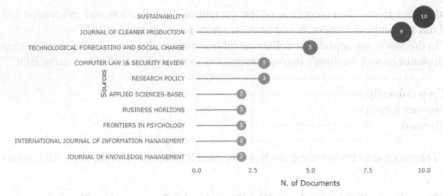

**Fig. 3.** Most relevant sources (journals)

Figure 4 presents the most influential studies on the matter. The one on first place has already been cited by 480 other studies.

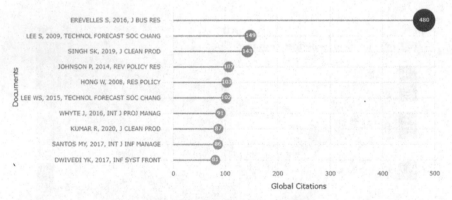

**Fig. 4.** Most influential research

The bigram co-word analysis reveals sectors that BD is researched and applied in – the healthcare sector and hospitality. It also gives clues on the type of companies that usually apply BD in the innovation process, which are the corporates. When it comes to business functions, BD is used in the supply chain and social media performance. Other technologies that are also researched in combination with BD are AI and could computing. The results are presented in Fig. 5.

The cumulative co-word analysis demonstrates the huge increase in the number of studies discussing BD and the innovation process after 2017. Mostly, it has been in the context of analytics, firm performance, and management (Fig. 6).

Trending topics are presented in Fig. 7 using bigrams word analysis. The trending words are arranged in time periods interesting observation brings the insight that in the last two years, BD has been mainly discussed in the context of digital transformation,

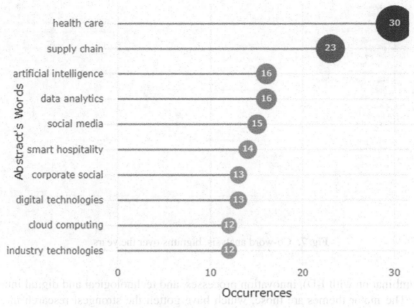

**Fig. 5.** Co-word analysis on abstracts (bigrams)

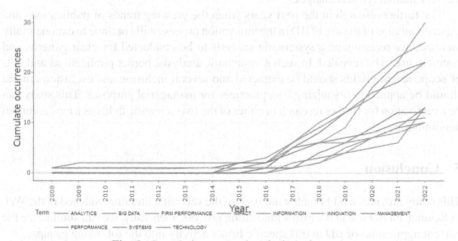

**Fig. 6.** Cumulative co-word analysis (unigrams)

digital innovation and, supply-chain management. The topic emerged as more a sustainable related application back in 2012 associated with social responsibility and innovation climate. In the last years, it was already incorporated into the innovation process along with other technologies bringing automation efficiency.

The thematic mapping shows different aspects of developing the topic in the literature. The basic topics are well researched and these are related to machine learning

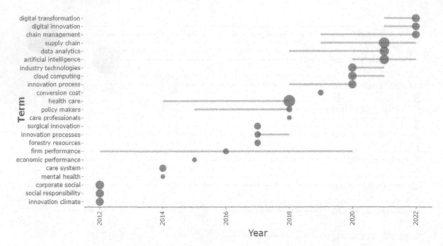

**Fig. 7.** Co-word analysis, bigrams over the years

(in combination with BD), innovation processes, and technological and digital innovations. The motor themes are those, which have gotten the strongest research interest recently. These are innovation systems, cloud computing, digital technologies, digital transformation, AI, and analytics.

For further research in the next years when the growing trends of publications and experimentation of the use of BD in the innovation process will continue to exponentially increase, we recommend a systematic analysis to be conducted for clear patterns and concrete use to be revealed. In such a systematic analysis, border publications and out-of-scope science fields should be extracted and several inclusion and exclusion criteria should be applied to organizing best practices for managerial purposes. This study can be a motivation for further research in either of the two scientific fields as a new common denominator.

## 5 Conclusion

This study provides a bibliometric analysis of the scientific literature indexed in the Web of Science on BD use in the firm's innovation process. Generally, we can summarize the current application of BD in this specific firm's activity into the following groups:

- Big data for decision-making in innovation
- Big data in innovation management and innovation process
- Big data to encourage innovation
- Big data to facilitate innovation
- Big data to scale up innovation (incl. Innovation performance)
- Big data as a methodological approach in innovation studies

Further research should focus on motor themes and still under-researched combinatory applications of BD along with other technologies such as cloud computing and

AI. From the innovation management domain, the main theoretical knowledge areas on which the scoped studies were focused on aro-innovation, service innovation, innovation process, product innovation, and process innovation.

**Acknowledgment.** This work was supported by the UNWE Research Programme (Research Grant No 9/2021).

# References

1. Ghasemaghaei, M., Calic, G.: Assessing the impact of big data on firm innovation performance: big data is not always better data. J. Bus. Res. **108**, 147–162 (2020). https://doi.org/10.1016/j.jbusres.2019.09.062
2. Johannessen, J.A., Olsen, B., Olaisen, J.: Aspects of innovation theory based on knowledge-management. Int. J. Inf. Manage. **19**(2), 121–139 (1999). https://doi.org/10.1016/s0268-4012(99)00004-3
3. Marion, T.J., Fixson, S.K.: The transformation of the innovation process: how digital tools are changing work, collaboration, and organizations in new product development. J. Prod. Innov. Manag. **38**(1), 192–215 (2020). https://doi.org/10.1111/jpim.12547
4. Appio, F.P., Frattini, F., Petruzzelli, A.M., Neirotti, P.: Digital transformation and innovation management: a synthesis of existing research and an agenda for future studies. J. Prod. Innov. Manag. **38**(1), 4–20 (2021). https://doi.org/10.1111/jpim.12562
5. Cappa, F., Oriani, R., Peruffo, E., McCarthy, I.: Big data for creating and capturing value in the digitalized environment: unpacking the effects of volume, variety, and veracity on firm performance. J. Prod. Innov. Manag. **38**(1), 49–67 (2020). https://doi.org/10.1111/jpim.12545
6. Bresciani, S., Ciampi, F., Meli, F., Ferraris, A.: Using big data for co-innovation processes: mapping the field of data-driven innovation, proposing theoretical developments and, providing a research agenda. Int. J. Inf. Manage. **60**, 102347 (2021). https://doi.org/10.1016/j.ijinfomgt.2021.102347
7. Shamim, S., Yang, Y., Zia, N.U., Shah, M.H.: Big data management capabilities in the hospitality sector: Service innovation and customer generated online quality ratings. Comput. Hum. Behav. **121**, 106777 (2021). https://doi.org/10.1016/j.chb.2021.106777
8. Duan, Y., Cao, G., Edwards, J.S.: Understanding the impact of business analytics on innovation. Eur. J. Oper. Res. **281**(3), 673–686 (2020). https://doi.org/10.1016/j.ejor.2018.06.021
9. Su, Y., Wang, X.: Innovation of agricultural economic management in the process of constructing smart agriculture by big data. Sustain. Comput.: Inform. Syst. **31**, 100579 (2021). https://doi.org/10.1016/j.suscom.2021.100579
10. Hassani, H., Huang, X., Silva, E.: Digitalization big data mining in banking. Big Data Cognitive Comput. **2**(3), 18 (2018). https://doi.org/10.3390/bdcc2030018
11. Lin, R., Xie, Z., Hao, Y., Wang, J.: Improving high-tech enterprise innovation in big data environment: a combinative view of internal and external governance. Int. J. Inf. Manage. **50**, 575–585 (2020). https://doi.org/10.1016/j.ijinfomgt.2018.11.009
12. Chen, H., Chiang, R.H.L., Storey, V.C.: Business intelligence and analytics: from big data to big impact. MIS Q. **36**(4), 1165 (2012). https://doi.org/10.2307/41703503
13. Akter, S., Wamba, S.F., Gunasekaran, A., Dubey, R., Childe, S.J.: How to improve firm performance using big data analytics capability and business strategy alignment? Int. J. Prod. Econ. **182**, 113–131 (2016). https://doi.org/10.1016/j.ijpe.2016.08.018

14. Gobble, M.M.: Big data: the next big thing in innovation. Res. Technol. Manag. **56**(1), 64–67 (2013). https://doi.org/10.5437/08956308x5601005
15. Galanakis, K.: Innovation process. Make sense using systems thinking. Technovation **26**(11), 1222–1232 (2006). https://doi.org/10.1016/j.technovation.2005.07.002
16. Hitt, M., Hoskisson, R., Johnson, R.A., Moesel, D.D.: The market for corporate control and firm innovation. Acad. Manag. J. **39**(5), 1084–1119 (1996). https://doi.org/10.2307/256993
17. Dammak, H., Dkhil, A., Cherifi, A., Gardoni, M.: Enterprise content management systems: a graphical approach to improve the creativity during ideation sessions—case study of an innovation competition "24 h of innovation." Int. J. Interact. Des. Manuf. (IJIDeM) **14**(3), 939–953 (2020). https://doi.org/10.1007/s12008-020-00691-8
18. Delen, D., Zolbanin, H.M.: The analytics paradigm in business research. J. Bus. Res. **90**, 186–195 (2018). https://doi.org/10.1016/j.jbusres.2018.05.013
19. Sheng, J., Amankwah-Amoah, J., Wang, X.: A multidisciplinary perspective of big data in management research. Int. J. Prod. Econ. **191**, 97–112 (2017). https://doi.org/10.1016/j.ijpe.2017.06.006
20. Dhir, S., Dhir, S.: Diversification: literature review and issues. Strateg. Chang. **24**(6), 569–588 (2015). https://doi.org/10.1002/jsc.2042
21. Dhir, S., Ongsakul, V., Ahmed, Z.U., Rajan, R.: Integration of knowledge and enhancing competitiveness: a case of acquisition of Zain by Bharti Airtel. J. Bus. Res. **119**, 674–684 (2020). https://doi.org/10.1016/j.jbusres.2019.02.056
22. Rothwell, R.: Industrial Innovation: Success, Strategy, Trends. The Handbook of Industrial Innovation. Edward Elgar, Aldershot (1994)
23. Imai, K., Nonaka, I., Fakeuchi, H.: Managing the new product development. In: Clark, K., Hayes, F. (eds.) The Uneasy Alliance. Harvard Business School Press (1985)
24. Chesbrough, H.W.: Open Innovation: The New Imperative for Creating and Profiting from Technology (First Trade Paper ed.). Harvard Business Review Press (2003).
25. von Hippel, E.: The dominant role of users in the scientific instrument innovation process. Res. Policy **5**(3), 212–239 (1976). https://doi.org/10.1016/0048-7333(76)90028-7
26. Ries, E.: The Lean Startup: How Today's Entrepreneurs Use Continuous Innovation to Create Radically Successful Businesses (Illustrated ed.). Currency (2011)
27. Kline, S.J., Rosenberg, N.: An overview of innovation. In: Rosenberg, N. (ed.) Studies on Science and The Innovation Process Selected Works of Nathan Rosenberg, Chapter 9, pp. 173–203, World Scientific Publishing Co. Pte. Ltd. (2009)
28. Tidd, J., Bessant, J.R.: Managing Innovation: Integrating Technological, Market and Organizational Change (7th edn.). Wiley (2020)
29. Gassmann, O.: Opening up the innovation process: towards an agenda. R D Manag. **36**(3), 223–228 (2006). https://doi.org/10.1111/j.1467-9310.2006.00437.x
30. Ohlberg, K.H., Salmeron, J.L.: Proposal-based innovation: a new approach to opening up the innovation process. In: Glauner, P., Plugmann, P. (eds.) Innovative Technologies for Market Leadership. FBF, pp. 201–231. Springer, Cham (2020). https://doi.org/10.1007/978-3-030-41309-5_14
31. Brous, P., Janssen, M., Herder, P.: The dual effects of the Internet of Things (IoT): a systematic review of the benefits and risks of IoT adoption by organizations. Int. J. Inf. Manage. **51**, 101952 (2020). https://doi.org/10.1016/j.ijinfomgt.2019.05.008
32. Lendel, V., Hittmár, T., Siantová, E.: Management of innovation processes in company. Procedia Econ. Finan. **23**, 861–866 (2015). https://doi.org/10.1016/s2212-5671(15)00382-2
33. Becheikh, N., Landry, R., Amara, N.: Lessons from innovation empirical studies in the manufacturing sector: a systematic review of the literature from 1993–2003. Technovation **26**(5–6), 644–664 (2006). https://doi.org/10.1016/j.technovation.2005.06.016
34. Tushman, M.L.: Special boundary roles in the innovation process. Adm. Sci. Q. **22**(4), 587 (1977). https://doi.org/10.2307/2392402

35. Lozada, N., Arias-Pérez, J., Perdomo-Charry, G.: Big data analytics capability and co-innovation: an empirical study. Heliyon **5**(10), e02541 (2019). https://doi.org/10.1016/j.hel iyon.2019.e02541
36. Vocke, C., Constantinescu, C., Popescu, D.: Status quo and quo vadis: creativity techniques and innovative methods for generating extended innovation processes. Procedia CIRP **91**, 39–42 (2020). https://doi.org/10.1016/j.procir.2020.02.148
37. Trabucchi, D., Buganza, T.: Data-driven innovation: switching the perspective on Big Data. Eur. J. Innov. Manag. **22**(1), 23–40 (2019). https://doi.org/10.1108/ejim-01-2018-0017
38. Wu, L., Lou, B., Hitt, L.M.: Data analytics supports decentralized innovation. SSRN Electron. J. (2019) https://doi.org/10.2139/ssrn.3351982
39. Zhang, X., et al.: A bibliometric analysis of digital innovation from 1998 to 2016. J. Manag. Scie. Eng. **2**(2), 95–115 (2017). https://doi.org/10.3724/SP.J.1383.202005

# Blockchain Technology and Applications

# EduCert – Blockchain-Based Management Information System for Issuing and Validating Academic Certificates

Diogo Melim[1] 🆔 and António Trigo[1,2](✉) 🆔

[1] Polytechnic of Coimbra, Coimbra Business School Research Centre, Instituto Superior de Contabilidade e Administração de Coimbra (ISCAC), 3045-601 Coimbra, Portugal
antonio.trigo@gmail.com, a2016042046@alumni.iscac.pt
[2] Centro ALGORITMI, University of Minho, 4804-533 Guimarães, Portugal

**Abstract.** The forgery of academic certificates is a global concern, being one of the best-known examples the case of fake doctors. This counterfeiting is facilitated because the validation of these certificates by the employers is, in most cases, through visual inspection of the certificates which, by itself, does not guarantee their veracity. In this sense, this work proposes a new certification approach based on blockchain technology that allows anyone to validate an academic certificate by placing the hash or the pdf of the academic certificate in a web application that performs this validation in a public blockchain network, without the need to authenticate.

**Keywords:** Blockchain · Authentication · Certification · Academic Degree · Diplomas · Higher Education · Forgery

## 1 Introduction

The validation of the authenticity of records of academic degrees and diplomas is still an archaic process that resorts in most cases to visual validation of the same by the clerks of the various organizations whether they are issued on paper or in electronic format, which even having digital signatures, are easy to forge putting in question their veracity. If the organization's clerks responsible for validating the academic degrees doubt the authenticity of the documents, they will have to contact the Higher Education Institution (HEI), either by e-mail or by phone, which takes time and resources both for the organization that wants to hire a new professional and for the HEI that must validate the document.

In addition to the issue of the practicality of validating documents, there is also the issue of forged certificates of academic degrees and diplomas, which has reached considerable volumes in recent years, believed to reach billions of dollars [1], and it is easy to obtain online forged certificates from prestigious universities.

Considering the above, a decentralized, blockchain-based degree and diploma certification prototype (source code available at https://github.com/dmelim/EduCert) is proposed to solve this problem, which will allow any user anywhere in the world to validate

© The Author(s), under exclusive license to Springer Nature Switzerland AG 2023
M. Papadaki et al. (Eds.): EMCIS 2022, LNBIP 464, pp. 137–149, 2023.
https://doi.org/10.1007/978-3-031-30694-5_11

whether the pdf document he or she holds, relating to a degree or diploma certificate issued by a HEI is valid.

The rest of the paper is structured as follows. In the next section a literature revision on the concepts and similar systems is presented. Next a section regarding the development of the system is presented. A section with the presentation of the application and some use cases is also provided. Finally, the paper ends with the conclusions section, where the limitations of the work and proposals for future work are presented, not only technical but also for the dissemination and promotion of the system developed.

## 2  Background

In this section introductory concepts about blockchains are presented as well as some examples of its use in the validation of academic degrees and diplomas found in the literature review.

### 2.1  Blockchain Concepts

Blockchain started as a concept rather than a term. It was introduced by Satoshi Nakamoto [2], an alias of the original creator of the concept, in October of 2008. This technology was developed to guarantee secure transactions between two parties without the need for a third party, thus creating the concept of trustless networks. The evolution of this technology has taken it to different domains beyond financial transactions, being considered today as one of the most important technologies of the present century [3].

### Blockchain

A blockchain consists of data sets that are composed of a chain of blocks, where each block contains a timestamp, the hash value of the previous block and a nonce, which is a random number for hash verification, thus ensuring the integrity of the entire blockchain up to the first block [4]. As Satoshi Nakamoto has put in the original bitcoin paper, "Each timestamp includes the previous timestamp in its hash, forming a chain, with each additional timestamp reinforcing the ones before it". [2].

The blockchain is distributed across several computers designated as nodes responsible for transactions and the addition of new blocks. Although there are different types of organization, this is originally a decentralized process where each node owns a copy of the blockchain and is rewarded for ensuring the functioning of the network [5]. To prevent network fraud, the original network of this concept, bitcoin uses a mechanism called proof-of-work [2], a decentralized consensus mechanism that requires network members to spend effort solving complex mathematical problems. This factor is a deterrent to actors who wish to threaten the network, as the computational power required to change a single block is enormous. Nakamoto came up with a principle that explains this concept well, "Proof-of-work is essentially one-CPU-one-vote" [2].

Given the high energy cost of maintaining a consensus mechanism using proof-of-work, other blockchain-based networks try to find other consensus mechanisms, being one of the most famous the proof-of-stake. In this mechanism members of the network instead of having to contribute to the network with their computational effort they contribute with the storage of network tokens in question. This mechanism assumes that if members of the network are required to store network tokens, they will not engage in fraudulent transactions that jeopardize their capital. As an example, the Ethereum network is migrating to this new mechanism by requiring members of the network to have a minimum of 32 ETH to participate in the network, which at current values is about $50,000. This consensus mechanism is being widely adopted in the recent years because, contrary to proof-of-work, it doesn't need computer power, and spends less electricity and so pollutes less [5].

**Smart Contracts**
Later Vitalik Buterin, decided to improve on the concepts created by Nakamoto, to create the Ethereum Blockchain. Buterin saw in Bitcoin more potential and along the years, the works of Nick Szabo, Namecoin, Colored coins and Metacoins inspired him to implement some ideas and developing them further [6].

One of these ideas was the implementation of smart contracts in the Ethereum network through a Turing complete programming language, which would make possible to build decentralized applications on Ethereum. Smart contracts are like normal contracts, an agreement between two or more parties, but instead of an entity checking if the conditions are completed, a computer automatically executes the contract when the conditions are met. This concept was introduced by Nick Szabo in the 1990's. [7].

To complete this objective Vitalik Buterin created Solidity, a smart contract programming language. Solidity makes it possible to work with the backend of Ethereum applications, also called Decentralized Applications, or DAPPS for short. This is a simple workflow in the Ethereum network [8].

**Permission vs. Permissionless**
There are two types of blockchains, private or permissionless and public or permissioned. Private blockchains restrict the users who can participate in transactions or the validation process. So only authorized users can participate in it. However private blockchains can have their history seen by external people to the network [9]. In contrast, public blockchains do not restrict anyone who wants to participate in the network or the validation process. Even so, the network can still have rules that need to be obeyed to participate in it [9].

## 2.2 Blockchain Certificate Authentication in Education

In the literature there are several experiences and studies about the application of BlockChain in the education area, in particular in the area of academic certificate management [10–16] Below are presented two projects whose characteristics were considered the most important for the development of the project presented in this paper.

Marella and Vijayan [11] developed a solution like the one proposed in this paper but based on a permissioned or private blockchain. Their problem was related to verifying the authenticity of the information stated on CV's. The authors chose to use Hyperledger Fabric [17], an open source blockchain technology that belongs to the permissioned or private blockchain subtype. Along with this they decided to use a consortium blockchain platform, which is a private blockchain that has a wider user base, for example instead of just one company using the blockchain, a group of organizations share this blockchain platform. To make it user friendly, a frontend, using, HTML and JavaScript was built. The information that is stored in the blockchain is the hash of the document, that can be calculated using an algorithm. For this they used the SHA-256 algorithm. Then this hash and the hash of the identification is stored in the blockchain as a message. To make the process easier 3 entities were created, the "peers", "Administrators" and the "Certification Authority". The peers take care of the process of submitting the hash to the blockchain. The administrator verifies the authenticity of the documents submitted and can approve or refuse their submission. The Certification Authority has the capability to give the certificates to the peers and administrators. The authors conclude this research presenting a comparison with a more traditional way of achieving this, using a central database application. The greatest reason they see to use blockchain instead is the immutability nature blockchain has. Giving it a layer of security and providing data integrity. Another reason is the scalability, in their case, it can have global impact, since hiring managers across the globe can check the information of candidates from various geographic locations [11].

Rama et al. [12] developed a software solution to identify fake or forged university credentials, using a decentralized application with Ethereum blockchain, using JavaScript and MetaMask to develop it. For the development phase they use Truffle and Ganache, a very useful set of tools made for the development of Decentralized Applications on Ethereum. To create the smart contract, Solidity was used. In their case, they propose that the blockchain is managed by a consortium of colleges and universities. If a student wants to use this to store their information, he or she needs to approach this consortium for approval. Certificates are given an individual identification, which is used as a verification mechanism.

## 3  Prototype

This section presents the development of the authentication system for academic degree and diploma certificates, from the enumeration of the requirements to be implemented to the actual coding, highlighting some of the codified components, presenting some of the screens of the front-end application for interaction with the blockchain where the hashes of the certificates are.

### 3.1  Requirements

The EduCert system is composed of two main packages. One that manages the entire cycle and life of degree and diploma certificates, which includes publishing their hash on the blockchain and registering them in the database. The other allows the management of users who have access to the platform, allowing the normal actions, such as their authentication in the system, registration/deletion of users and management of the respective permissions.

Figure 1 presents the use case diagram of the certificate management package. It was decided not to present the use case diagram relative to the user management package because the use cases are those commonly present in user management systems.

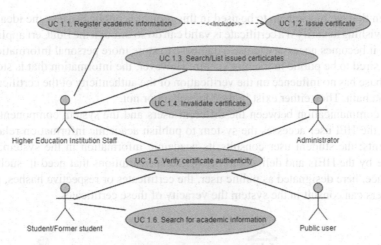

**Fig. 1.** Certificate management system use cases

The application as shown in Fig. 1 and Fig. 2 has four types of users: the HEI user, who manages the certificates in the system, having available the functionalities for registering the academic information in the system, which includes the issuing of certificates in the blockchain, the possibility to search the issued certificates and also invalidate the issued certificates (certificate invalidation can only be carried out in the database, not in the blockchain); the student user, who can consult his/her academic information including all the certificates issued by the different institutions of higher education; the public user who accesses the system to validate the certificates they have in their possession; and, finally, the administrator user who can perform in the system mainly operations related to the management of the users.

## 3.2 Architecture

As can be seen in Fig. 2 the EduCert system consists of an application server, where the frontend and backend are located, a database server and the blockchain network. The frontend allows the different users to interact with the system to perform different operations and communicate with the backend. The backend receives requests from the frontend through an Application Programmable Interface (API) to store information about students and their academic certificates both in the database and in the blockchain. The database stores all the information relative to the users and, in the case of the students, also the academic information including the certificates. Finally, the public blockchain where the hashes relative to the certificates are stored.

Given that the blockchain to be used in this project is public because the idea is that anyone wishing to verify if a certificate is valid can do so through the EduCert application or other, it becomes necessary to use a database to store more personal information that is not desired to be public in the blockchain. However, the information that is stored in the database has no influence on the verification of the authenticity of the certificates on the blockchain. They either exist in the blockchain or not.

The communication between the different users and the system components is as follows: the HEI user accesses the system to publish academic information relative to its students; the student user consults its academic information in the system, made available by the HEIs and delivers to the different institutions that need it, such as its work place, here designated as public user, the certificates or respective hashes, so that these users can consult in the system the veracity of these certificates.

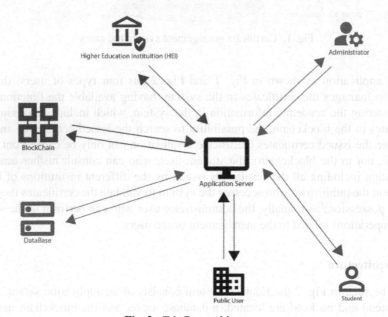

**Fig. 2.** EduCert architecture.

## 3.3 Technologies Used

The following technologies were used to create the EduCert system:

- React/Express for building the frontend;
- NodeJS for building the back-end system, which communicates with the database and the blockchain and provides an API, to be used by the frontend, with a set of functionalities to perform the different operations both with the database and the blockchain;
- MongoDB as the database engine, for storing the information;

- Fantom blockchain, for storing certificate hashes. This blockchain was chosen because it has a lot of similarities with Ethereum and has smaller costs. The similarities like Ethereum allow the use of the tools available for Ethereum like writing the contracts in the Solidity language and using tools from the Ethereum ecosystem like Remix for compiling the contracts or Ganache for testing them.
- MetaMask, to store Fantom tokens and connect the browser to the blockchain to pay for the publishing fees of the hash certificates.
- Javascript language, used to program some of the above-mentioned technologies.

### 3.4 Implementation

In this section some of the most important aspects of the coding of the EduCert system are presented, namely, the coding of the contract that sustains the publication of the certificates on the Fantom blockchain (Fig. 3), the example of the function of hash creation from the PDF file for validation on the blockchain (Fig. 4) and the backend functions for the publication (Fig. 5) and search/verification (Fig. 6) of the certificate validity, which are made available through the backend API. In the experimentation section of this document (Sect. 5) it is possible to see the visual part of the system made available to users through the front-end.

```solidity
// SPDX-License-Identifier: MIT
pragma solidity ^0.5.0;

contract Notary {
    struct Record {
        uint mineTime;
        uint blockNumber;
        address messageSender;
    }

    mapping (bytes32 => Record) private docHashes;

    function addDocHash (bytes32 hash) public {
        Record memory newRecord = Record(now, block.number, msg.sender);
        docHashes[hash] = newRecord;
    }

    function findDocHash (bytes32 hash) public view returns(uint, uint, address) {
        return (docHashes[hash].mineTime, docHashes[hash].blockNumber, docHashes[hash].messageSender);
    }
}
```

**Fig. 3.** Smart contract (solidity).

In Fig. 3 the smart contract code written for EduCert is presented, where it is possible to see the types of data involved (stamp, block number and the sender) as well as the functions that manipulate them (addDocHash and findDocHash). This contract was originally found in "Ether doc cert" GitHub repository [18]. To store a hash the user needs to pay gas fees, but to search it the user does not.

```
$(document).ready(function() {
    notary_init();
});

function hashForFile(callback) {
    input = document.getElementById("cert");
    if (!input.files[0]) {
        alert("Please select a file first");
    }
    else {
        file = input.files[0];
        fr = new FileReader();
        fr.onload = function (e) {
            content = e.target.result;
            var shaObj = new jsSHA("SHA-256", "ARRAYBUFFER");
            shaObj.update(content);
            var hash = "0x" + shaObj.getHash("HEX");
            callback(null, hash);
        };
        fr.readAsArrayBuffer(file);
    }
};
```

**Fig. 4.** Hashing algorithm.

Figure 4 presents the hashing algorithm that calculates the hash of the file (PDF certificate of the academic degree), using a JavaScript library that uses an algorithm, in this case SHA-256 to calculate the hash of a document, which will be the hash stored on the blockchain (see Fig. 5).

```
function send () {
    hashForFile(function (err, hash) {
        notary_send(hash, function(err, tx) {
            $("#responseText").html("<p>Hash Value Submited with Sucess to Fantom Network.</p>"
            + "<p>Hash Value: " + hash +"</p>"
            + "<p>Transaction ID: " + tx +"</p>"
            + "<p>The Contract has the following adress: " + address +"</p>"
            + "<p><b>The information can take some time to be avaliable</b></p>");
        });
    });
};
```

**Fig. 5.** Send hash to blockchain.

Figure 5 shows the method used to send the hash to the Fantom blockchain. This is done using the function "addDocHash" defined in the smart contract (Fig. 3) and a third-party module called Web3 [19]. This module makes it able to create functions that can interact with function inside a smart contract. In this case the "notary_send" function. This is linked to Metamask [20] which will signal to the user that a transfer is being attempted, which will be carried out, if the user agrees to it. This is also done through the web3.js Ethereum API library [19].

```
function find () {
    hashForFile(function (err, hash) {
        notary_find(hash, function(err, resultObj) {
            if (resultObj.blockNumber != 0) {
                var ipc = "";
                if (resultObj.messageSender == "0xACd7A7dCaA6A5d786d7b88aD00F00F4371219b66") {
                    ipc = "IPC";
                }
                else {
                    ipc = "Not IPC";
                }
                $("#responseText").html("<p>We found your document hash in the fantom Network!</p>"
                    + "<p>Hash Value: " + hash + "</p>"
                    + "<p>Block No.: " + resultObj.blockNumber + "</p>"
                    + "<p>Signing Date: " + resultObj.mineTime + "</p>"
                    + "<p>Signed by: " + ipc + "(" + resultObj.messageSender + ")" + "</p>"
                );
            } else {
                $("#responseText").html("<p>We didn't found your document hash in the fantom Network!</p>"
                    + "<p>Hash Value: " + hash + "</p>"
                );
            }
        });
    });
};
```

**Fig. 6.** Searching function.

The function presented in Fig. 6 searches the Fantom blockchain for the corresponding hash. It starts by verifying if the message sender, or the account that uploaded the hash originally, is a valid account. The value is hard coded, as this is a prototype. If it is an authorized account, it gives the user who searched for the hash more information about it. If it isn´t it notifies the user of it. If the hash is not on the network, the user will be notified with a different message.

## 4   Exemplification of Prototype Use

In this section two use cases of the EduCert system are presented, that of issuing an academic certificate by the HEI and that of validating the certificate by a public user (e.g., a user from a company where the former student/student is applying for a job).

### 4.1   HEI Issue Certificate

An administrative employee from HEI, in this example, Coimbra Business School from the Polytechnic of Coimbra, Portugal, accesses the system to issue an academic certificate relative to a student. To do so, he or she fills in a form with the necessary information to issue the certificate, which, after validated, will be used to generate the certificate. There is also the option of importing from Excel the information relative to several students for the issuing of the certificate.

After sending this information, the certificate in PDF is generated by the EduCert system (see Fig. 7) and the respective hash is calculated for storage in the Fantom blockchain.

iscac Politécnico de Coimbra COIMBRA BUSINESS SCHOOL 100

Diploma de Pós-Graduação

**Lorem ipsum dolor sit amet**, Presidente da Coimbra Business School | ISCAC, do Instituto Politécnico de Coimbra, conforme requerimento que fica arquivado nesta Escola, faz saber que,

O(a) aluno(a) **Lorem ipsum dolor sit amet**

filho(a) de Lorem ipsum dolor sit amet

e de Lorem ipsum dolor sit amet,

portador(a) do Cartão de Cidadão nº Lorem ipsum dolor sit amet,

concluiu nesta Escola, no dia 21 **de julho de 2021**, o **Lorem ipsum dolor sit amet**, com a classificação final de **catorze (14) valores**, a que corresponde a informação final de **Bom**.

Esta informação consta dos documentos arquivados na Coimbra Business School | ISCAC.

O presente documento vai autenticado com o selo branco em uso nesta Escola.

Coimbra Business School | ISCAC, **6 de janeiro de 2022**

O Presidente da Coimbra Business School | ISCAC

*Lorem ipsum dolor sit amet*

Lorem ipsum dolor sit amet

SPECIMEN

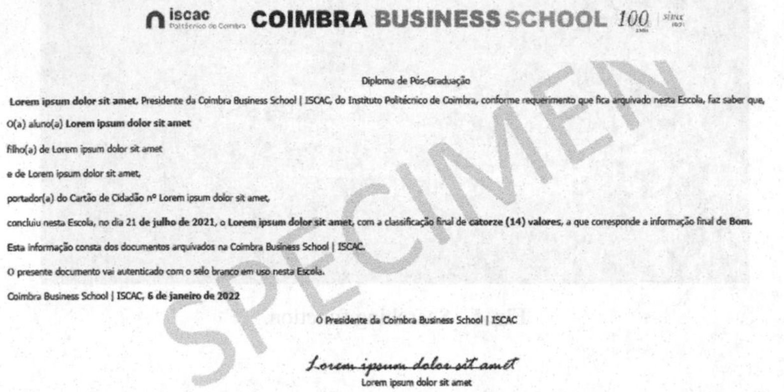

**Fig. 7.** PDF certificate example

After generating the certificate, the system asks the user if he/she wants to open the MetaMask to complete the transaction. If the user authorizes the hash of the original certificate is stored in the Fantom blockchain and a new PDF certificate (see Fig. 8) similar to the previous one, but with a QR Code that contains the web address, with the hash of the original certificate, for validation, is generated.

iscac Politécnico de Coimbra COIMBRA BUSINESS SCHOOL 100

Diploma de Pós-Graduação

**Lorem ipsum dolor sit amet**, Presidente da Coimbra Business School | ISCAC, do Instituto Politécnico de Coimbra, conforme requerimento que fica arquivado nesta Escola, faz saber que,

O(a) aluno(a) **Lorem ipsum dolor sit amet**

filho(a) de Lorem ipsum dolor sit amet

e de Lorem ipsum dolor sit amet,

portador(a) do Cartão de Cidadão nº Lorem ipsum dolor sit amet,

concluiu nesta Escola, no dia 21 **de julho de 2021**, o **Lorem ipsum dolor sit amet**, com a classificação final de **catorze (14) valores**, a que corresponde a informação final de **Bom**.

Esta informação consta dos documentos arquivados na Coimbra Business School | ISCAC.

O presente documento vai autenticado com o selo branco em uso nesta Escola.

Coimbra Business School | ISCAC, **6 de janeiro de 2022**

O Presidente da Coimbra Business School | ISCAC

*Lorem ipsum dolor sit amet*

Lorem ipsum dolor sit amet

SPECIMEN

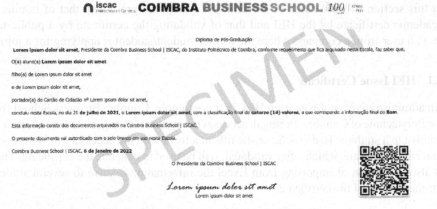

**Fig. 8.** PDF receipt certificate with QR Code

## 4.2 Verification of the CERTIFICate's Authenticity by the Public User

To verify the authenticity of a certificate issued by an HEI, the public user (e.g., a user from a company where the former student/student is applying for a job) takes a picture

of the QR Code (see Fig. 8) that will take him to a web page of the EduCert system and
tell him if the certificate in question is valid or not (see Fig. 9).

**Fig. 9.** EduCert system certificate search result

# 5  Conclusion

This paper presents the system developed for the validation of academic degree certificates, the EduCert system. This application has several components, with emphasis on
the frontend where applications were created for interaction with different users (HEI,
students and public or external users (who do not require authentication)) and the backend component that records in the database the different operations performed and in
the Fantom blockchain the certificate hashes.

In the testing phase, the system proved to be functional, being now necessary to test
it in a real context. Nevertheless, some limitations have already been detected for which
it will be necessary to find solutions, such as the immutability of the blockchain, which
does not allow deleting records. In other words, although records are marked as invalid
in the database, the same cannot be done at the blockchain level. These will always be
occasional situations, because as in the case of physical certificates, once issued they are
valid.

As future work there are, as already mentioned, tests at the level of the teaching
institution where this solution was developed, the Polytechnic of Coimbra and, in the
future, try to extend the use of this solution to other institutions.

As a conclusion would like to leave a summary of the costs per certificate to have
such a system. In the case of Fantom blockchain, chosen for this stage of the system,
the costs are about 0.026 FTM ($\approx$ 0.001 euros), while if chosen the Ethereum network

are about 0.12 euros. The Ethereum network may be the option for the final system because it gives more guarantees of future support, i.e., that it will not disappear as a platform. Added to this is the fact that the Ethereum network is looking for a new consensus mechanism (proof-of-stake) that will lower transaction prices (gas fees costs) which will make the system even cheaper.

# References

1. Grolleau, G., Lakhal, T., Mzoughi, N.: An introduction to the economics of fake degrees. J. Econ. Issues **42**(3), 673–693 (2008). https://doi.org/10.1080/00213624.2008.11507173
2. Nakamoto, S.: Bitcoin: A Peer-to-Peer Electronic Cash System. www.bitcoin.org (2008). Accessed: 29 Mar 2022
3. el Haddouti, S., Ech-Cherif El Kettani, M.D.: Analysis of identity management systems using blockchain technology. In: Proceedings - 2019 International Conference on Advanced Communication Technologies and Networking, CommNet 2019 (2019). https://doi.org/10.1109/COMMNET.2019.8742375
4. Nofer, M., Gomber, P., Hinz, O., Schiereck, D.: Blockchain. Bus. Inf. Syst. Eng. **59**(3), 183–187 (2017). https://doi.org/10.1007/S12599-017-0467-3/TABLES/1
5. Kim, C.: Ethereum 2.0: How it works and why it matters (2020)
6. Buterin, V.: A next generation smart contract & decentralized application platform. https://translatewhitepaper.com/wp-content/uploads/2021/04/EthereumOrijinal-ETH-English.pdf (2013). Accessed 12 Apr 2022
7. Zheng, Z., et al.: An overview on smart contracts: challenges, advances and platforms. Futur. Gener. Comput. Syst. **105**, 475–491 (2020). https://doi.org/10.1016/J.FUTURE.2019.12.019
8. Dannen, C.: Introducing Ethereum and Solidity: Foundations of Cryptocurrency and Blockchain Programming for Beginners. Apress, Berkeley, CA (2017). https://doi.org/10.1007/978-1-4842-2535-6
9. Solat, S., Calvez, P., Naït-Abdesselam, F.: Permissioned vs. Permissionless Blockchain: how and why there is only one right choice. J. Softw. **16**, 95–106 (2021). https://doi.org/10.17706/jsw.16.3.95-106
10. Nguyen, B.M., Dao, T.C., Do, B.L.: Towards a blockchain-based certificate authentication system in Vietnam. Peer J. Comput. Sci. **6**, 266 (2020). https://doi.org/10.7717/PEERJ-CS.266
11. Marella, V., Vijayan, A.: Document verification using blockchain for trusted CV information. In: AMCIS 2020 Proceedings. https://aisel.aisnet.org/amcis2020/adv_info_systems_research/adv_info_systems_research/12 (2020). Accessed 27 Mar 2022
12. Rama Reddy, T., Prasad Reddy, P.V.G.D., Srinivas, R., Raghavendran, C.V., Lalitha, R.V.S., Annapurna, B.: Proposing a reliable method of securing and verifying the credentials of graduates through blockchain. EURASIP J. Inf. Secur. **2021**(1), 1–9 (2021). https://doi.org/10.1186/s13635-021-00122-5
13. Alshahrani, M., Beloff, N., White, M.: Towards a blockchain-based smart certification system for higher education: an empirical study. Int. J. Comput. Dig. Syst. **11**(1), 553–571 (2022). https://doi.org/10.12785/ijcds/110145
14. Serranito, D., Vasconcelos, A., Guerreiro, S., Correia, M.: Blockchain ecosystem for verifiable qualifications. In: 2020 2nd Conference on Blockchain Research & Applications for Innovative Networks and Services (BRAINS), pp. 192–199 (2020). https://doi.org/10.1109/BRAINS49436.2020.9223305
15. Bahrami, M., Movahedian, A., Deldari, A.: A comprehensive blockchain-based solution for academic certificates management using smart contracts. In: 2020 10th International Conference on Computer and Knowledge Engineering (ICCKE), pp. 573–578 (2020). https://doi.org/10.1109/ICCKE50421.2020.9303656

16. University of Nicosia: Blockchain Certificates (Academic & Others). https://www.unic.ac.cy/iff/blockchain-certificates/ (2014). Accessed 10 Sep 2022
17. Hyperledger Foundation, "Hyperledger Fabric". https://www.hyperledger.org/use/fabric (2022). Accessed 12 Apr 2022
18. Dat Tran, "Ether doc cert," GitHub. https://github.com/datts68/ether-doc-cert (2022). Accessed 19 Jul 2022
19. ChainSafe: "Web3.js," Git Hub. https://github.com/ChainSafe/web3.js (2014). Accessed 20 Jul 2022
20. MetaMask: Metamask-extension. https://github.com/MetaMask/metamask-extension (2019). Accessed 20 Jul 2022

# A Distributed Multi-key Generation Protocol with a New Complaint Management Strategy

Rym Kalai[1]([⊠])[iD], Wafa Neji[2][iD], and Narjes Ben Rajeb[3][iD]

[1] LIPSIC Laboratory, Faculty of Sciences of Tunis, University of Tunis El Manar,
Tunis, Tunisia
rym.kalai@gmail.com
[2] LIPSIC Laboratory, The General Directorate of Technological Studies,
Higher Institute of Technological Studies of Beja, Beja, Tunisia
[3] LIPSIC Laboratory, National Institute of Applied Sciences and Technology,
University of Carthage, Tunis, Tunisia

**Abstract.** Privacy protection is a main goal in the majority of the Blockchain studies. However some dishonest users may abuse from the benefits of this property and the fact of not being identified to do illegal crimes. That is why several researches focus on implementing identity tracing to avoid the flaws related to privacy protection in Blockchain applications.

In this paper, we propose a Distributed Multi-Key Generation (DMKG) protocol without private channels built on the DMKG protocol of the Blockchain Traceable Scheme with Oversight Function (BTSOF) presented in [8].

Our protocol introduces a new strategy to manage complaints between participants that avoids them to publicly reveal the values of their shares of secrets. This new management of complaints and the use of public channels allow a precise identification of malicious participants. We prove that our solution satisfies the security requirements of the Verifiable Multi-Secret Sharing (VMSS) schemes and DMKG protocols.

**Keywords:** Distributed Multi-Key Generation (DMKG) · Complaints' management strategy · Multi-Secret Sharing (MSS)

## 1 Introduction

In 1979, Shamir [17] and Blakley [2] introduced the concept of Secret Sharing (SS). It is an important research content of cryptography especially to solve the problem of secret decentralization and to protect secret information from being lost, destructed, inaccessible or falling into the hands of unauthorized third parties [16]. A trusted party called dealer splits the secret into different pieces called shares and distributes them to other parties. Moreover $(t, n)$ threshold cryptosystems allow the dealer to share a secret information between a set of

M. Papadaki et al. (Eds.): EMCIS 2022, LNBIP 464, pp. 150–164, 2023.
https://doi.org/10.1007/978-3-031-30694-5_12

$n$ participants such that only a subgroup of no less than $t$ members among the $n$ participants $(2 \leq t < n)$ can cooperate together to recover the secret [20]. The shares of any smaller subset are unable to reveal any information about the secret. The security depends on this unique trusted party known as the dealer who is the only participant that holds the secret key and distributes secret shares to participants. This is the weak point of the system aiming to protect the secret information and guarantee that no party owns it [1]. Distributed Key Generation (DKG) protocols offer the solution. In 1991, Pedersen [14] presented the first DKG protocol with no trusted party. It involves many participants to cooperate together to generate and share the secret key. All honest participants have a valid share of a unique secret key. There is $n$ secret sharing processes: Each participant plays the role of a dealer on a single sharing process. Yet, DKG protocols permit to share only one secret among the participants. Multi-Secret Sharing (MSS) scheme was the key to solve this problem. In a MSS scheme, the dealer distributes several secrets among the participants at the same time. In addition, in $(t, n, l)$ threshold multi-secret sharing schemes, at least $t$ or more participants can easily pool their shares and reconstruct $l$ secrets at the same time [19]. Once again, it ends up with a single party holding all the secrets. Thus, this is the point of failure of the whole scheme. To overcome this problem, the idea is to do as in the DKG protocols except that the participants handle $t$ several secrets at the same time. It is the equivalent of having $n$ parallel executions of the MSS scheme. This is called Distributed Multi-Key Generation (DMKG) protocol introduced by Tianjun Ma et al. in [8] in 2020. However, not all participants are honest. Some may cheat and distribute invalid shares. The protocol must detect the cheating either by using an expensive public verification which affects the scalability of the DMKG protocol, or by using a complaint management strategy that allows participants to complain if they receive invalid shares [3,6,12,14]. The issue is that this strategy forces participants to reveal their shares in order to prove their honesty. In addition, a malicious participant may cheat and complain about receiving an invalid share to eliminate an honest participant. The challenge is to detect dishonest participants and to prove their honesty without revealing their shares [11,15].

In this contribution, we present our DMKG protocol which relies on the *Disputation* strategy proposed in [11] to manage complaints between participants. This strategy avoids participants to reveal their shares and disqualifies the malicious ones. However, it is limited to a unique secret sharing. We extend it in our DMKG protocol to suit the multi secret sharing context.

Note that we use only public channels and that all the shares of secrets are encrypted. The construction of our DMKG protocol is built on the DMKG protocol presented in [8]. The multi-secret sharing scheme is based on the verifiable Pedersen's secret sharing scheme [13] and leverage the Franklin and Yung multi-secret sharing scheme [5].

This paper is organized as follows: We start by presenting the related work of our research namely the multi-secret sharing scheme, the most important DMKG protocols and the Blockchain Traceable Scheme with Oversight Function

(BTSOF) defined in [8]. Then, we detail our new DMKG protocol with a new complaint management strategy based on BTSOF's DMKG protocol. Next, we prove the security of our protocol. Finally, we present a short comparison between our protocol and BTSOF's DMKG protocol.

## 2    State of the Art

The term Distributed Multi-Key Generation (DMKG) protocol was introduced by Tianjun Ma *et al.* in [8] in 2020 to describe a distributed and parallel executions of a MSS scheme. As mentioned in the previous section, the DMKG protocol is a main part of distributed encryption system. It solves the problem of the single trusted entity in MSS schemes. Each participant will play the role of the dealer. It involves the participants to cooperate together and share multiple secrets to generate a private and public keys. At the end, only an authorized subset of them can reconstruct the private key. The DMKG protocol proposed in [8] was used to design a Blockchain Traceable Scheme with Oversight Function (BTSOF). In 2021, Tianjun Ma *et al.* [10] improved the scheme and proposed a traceable scheme that is suitable for consortium Blockchain with leverages of the proposed DMKG protocol. On the other hand, to guarantee uniform distribution of the generated keys, the DMKG proposed by Tianjun Ma *et al.* in [8] is built on the Gennaro's DKG protocol [6]. This last, has proved that the presence of malicious participants disrupts the distribution of the generated keys and that all protocols based on Pedersen's DKG do not assure a uniform distribution. The proposed protocol uses private channels between participants and requires extra communications to deal with the complaints between participants. This increases the complexity of the protocol and makes it difficult to put into practice. Also, the private channels make it hard to determine the malicious participants.

In this paper, we propose a new DMKG protocol with a $(t, n)$ threshold multi secret sharing (TMSS) scheme based on the Franklin and Yung MSS scheme with no private channels. The MSS scheme is simultaneous: all the secrets are reconstructed at the same time in only one stage [7]. Moreover, it is a verifiable scheme based on Pedersen's Verifiable Secret Sharing (VSS). In addition, our DMKG protocol offers a new complaints management strategy that avoids participants' shares revealing.

## 3    DMKG Protocol of BTSOF

This section presents the DMKG protocol of Blockchain Traceable Scheme with Oversight Function (BTSOF) introduced in [8].

### 3.1    Notation

Let $p$ and $q$ denote two large prime numbers, such that $q|(p-1)$. $\mathbb{Z}_p$ denotes a group of order $p$ and $\mathbb{Z}_q$ denotes a group of order $q$. By default the exponential operation performs modulo $p$ operation. For example, $g^x$ denotes $g^x mod p$, where $g \in \mathbb{Z}_p$ and $x \in \mathbb{Z}_q$. $(pk_{tra}, sk_{tra})$ denotes the traceable public-private key pair.

## 3.2   Context

Some dishonest users can commit crimes by taking advantage of the fact of not being identified. Several researches focus on implementing identity tracing to avoid the flaws related to privacy protection in Blockchain applications. In 2020, Tianjun Ma et al. presented a Blockchain traceable scheme denoted SkyEye [9]. It allows the regulator to trace the users' identities of the Blockchain data. However, there are no restrictions and oversight measures. The regulator can arbitrarily trace the data. In [8], Tianjun Ma et al. proposed an improvement of SkyEye in BTSOF. The regulator must obtain the consent of the committee to enable tracing. He can trace one data, multiple data or data in multiple period. In this paper, we focus only on tracing all data of the $T$ period denoted $data^T$. The committee has $n$ participants $P_1, ..., P_n$. The traceable public-private key pair $(pk_{tra}^T, sk_{tra}^T)$ is generated by the committee in a $T$ period. The private key $sk_{tra}^T$ is sent by the committee to the regulator. The tracing process is as follows: Let $u$ denotes a user, for each user's data noted $data_u$ in $data^T$, the regulator gets the chameleon hash public key set $PK_C$ by decrypting each ciphertext of chameleon hash public key in $data_u$ using the private key $sk_{tra}^T$. He can obtain the users' true identity set $ID$ corresponding to $data_u$ by searching the database according to $PK_C$. The traceable public-private key pair is generated by the committee through the DMKG protocol. The protocol is based on the DKG protocol proposed by Gennaro et al. [6]. It uses a non-interactive VMSS scheme for the Cramer-Shoup public key encryption scheme [4]. The VMSS is formed from the combination of the Franklin and Yung MSS scheme and Pedersen's VSS.

## 3.3   Communication and Adversary Model

They assume that there is a set of $n$ probabilistic polynomial time participants $P_1, P_2, \ldots, P_n$ in a fully synchronous network. All participants have a common broadcast channel, and there are private channels between participants.

The adversary $A$ is static. At the beginning of the protocol, he chooses the corrupted participants. From the $n$ participants, he can corrupt at most $(t-1)$, such that $(t-1) < \frac{n}{2}$.

## 3.4   The DMKG Protocol

The traceable public-private key pair is generated by the committee through the DMKG protocol. In this paper, we focus specially on the generation of the traceable private key. We simplify the notations and note $sk$ the traceable private key and $pk$ the traceable public key. The protocol consists on $n$ parallel executions of the VMSS scheme defined in [8]. The generation of $sk$ is done in three phases as follows:

**Initialization Phase.** A trusted authority has chosen $g_1, h_1, g_2, h_2 \in \mathbb{Z}_p$, where $h_1 = g_1^{\gamma_1}$ and $h_2 = g_2^{\gamma_2}$ for $\gamma_1, \gamma_2 \in \mathbb{Z}_q$. The protocol aims to generate the private key $sk = (x_1, x_2, y_1, y_2, z)$.

**Distribution and Verification Phase.** For $i = 1, ..., n$, the committee member $P_i$ is the dealer and performs the following steps:

*Step 2.a* $P_i$ chooses at random $x_{1i}, x_{2i}, y_{1i}, y_{2i}, z_i \in \mathbb{Z}_q$ and $\beta_{0i}, \beta_{1i}, \beta_{2i}, \beta_{3i}, \beta_{4i} \in \mathbb{Z}_q$.

*Step 2.b* $P_i$ broadcasts $E_{1i} = g_1^{x_{1i}}.h_1^{\beta_{1i}}$, $E_{2i} = g_1^{y_{1i}}.h_1^{\beta_{2i}}$, $E_{3i} = g_2^{x_{2i}}.h_2^{\beta_{3i}}$, $E_{4i} = g_2^{y_{2i}}.h_2^{\beta_{4i}}$, $E_{5i} = g_1^{z_i}$

*Step 2.c* $P_i$ randomly chooses 6 polynomials $f(x), f'(x), g(x), g'(x), h(x)$ et $h'(x)$ of degree $t$ such that $f_i(x) = a_{i0} + a_{i1}x + ... + a_{it}x^t$, $f_i'(x) = b_{i0} + b_{i1}x + ... + b_{it}x^t$, $g_i(x) = a_{i0}' + a_{i1}'x + ... + a_{it}'x^t$, $g_i'(x) = b_{i0}' + b_{i1}'x + ... + b_{it}'x^t$, $h_i(x) = a_{i0}'' + a_{i1}''x + ... + a_{it}''x^t$, $h_i'(x) = b_{i0}'' + b_{i1}''x + ... + b_{it}''x^t$ where $f_i(-1) = x_{1i}$, $f_i(-2) = y_{1i}$, $f_i'(-1) = \beta_{1i}$, $f_i'(-2) = \beta_{2i}$, $g_i(-1) = x_{2i}$, $g_i(-2) = y_{2i}$, $g_i'(-1) = \beta_{3i}$, $g_i'(-2) = \beta_{4i}$, $h_i(0) = z_i = a_{i0}''$, and $h_i'(0) = \beta_{0i} = b_{i0}''$. The distribution process of $z_i$ uses the Pedersen-VSS scheme [13].

*Step 2.d* $P_i$ broadcasts : for $k = 0, ..., t$ $CM_{ik} = g_1^{a_{ik}}.h_1^{b_{ik}}.g_2^{a_{ik}'}.h_2^{b_{ik}'}$ and $cm_{ik} = g_1^{a_{ik}''}.h_1^{b_{ik}''}$ where $cm_{i0} = g_1^{a_{i0}''}.h_1^{b_{i0}''} = g_1^{z_i}.h_1^{\beta_{0i}}$.

*Step 2.e* For each $i = 1, ..., n$, for $\tau = 1, 2$, for each $j = 1, ..., n$ $P_j$ checks:

$$E_{\tau i}.E_{(\tau+2,i)} \stackrel{?}{=} \prod_{k=0}^{t}(CM_{ik}^{(-\tau)^k}) \tag{1}$$

If it fails for an index $i$, then $P_i$ is a dishonest participant and is disqualified, so $i \notin Q_{temp}$ where $Q_{temp}$ denotes the temporarily set of qualified participants.

*Step 2.f* For $i = 1, ..., n$, $P_i$ computes the shares $sf_{ij} = f_i(j)$, $sf_{ij}' = f_i'(j)$, $sg_{ij} = g_i(j)$, $sg_{ij}' = g_i'(j)$, $sh_{ij} = h_i(j)$ and $sh_{ij}' = h_i'(j)$. Then, $P_i$ sends secretly the set of shares $(sf_{ij}, sf_{ij}', sg_{ij}, sg_{ij}', sh_{ij}, sh_{ij}')$ to each participant $P_j$.

*Step 2.g* Each participant $P_j$ checks for each $i \in Q_{tem}$:

$$\begin{cases} (g_1^{sf_{ij}}.h_1^{sf_{ij}'}.g_2^{sg_{ij}}.h_2^{sg_{ij}'}) \stackrel{?}{=} \prod_{k=0}^{t}(CM_{ik})^{j^k} \\ g_1^{sh_{ij}}.h_1^{sh_{ij}'} \stackrel{?}{=} \prod_{k=0}^{t}(cm_{ik})^{j^k} \end{cases} \tag{2}$$

If it fails, then the participant $P_j$ complaints against $P_i$. Otherwise $P_j$ accepts the set of shares.

*Step 2.h* If the participant $P_i$, for $i \in Q_{tem}$, received a complaint from $P_j$ then $P_i$ reveals and broadcasts the set of shares that satisfy Eq. 2.

*Step 2.i* A participant $P_i$, $(i = 1, ..., n)$, is marked as disqualified if the number of complaints against $P_i$ is superior than $(t - 1)$ in the *Step 2.g* or if the values broadcasted by $P_i$ in *Step 2.h* do not satisfy Eq. 2.

*Step 2.j* Each participant $P_i$, $i \in Q_{tem}$, constructs a set of qualified participants $Q_{final}$. It contains the participants that are not disqualified in the *Step 2.e* and *Step 2.i*.

**Recovery Phase.** Each participant $P_i$, ($i = 1, ..., n$) computes the shares : $sf_i = \sum_{j \in Q_{final}} sf_{ji} modq$, $sf'_i = \sum_{j \in Q_{final}} sf'_{ji} modq$, $sg_i = \sum_{j \in Q_{final}} sg_{ji} modq$, $sg'_i = \sum_{j \in Q_{final}} sg'_{ji} modq$, $sh_i = \sum_{j \in Q_{final}} sh_{ji} modq$, $sh'_i = \sum_{j \in Q_{final}} sh'_{ji} modq$. The private key $sk$ is not computed. Each committee member $P_i$ with $i = 1...n$, has the traceable private key component $(x_{1i}, x_{2i}, y_{1i}, y_{2i}, z_i)$ so the private key $sk = (x_1 = \sum_{i \in Q_{final}} x_{1i} \ modq, \ x_2 = \sum_{i \in Q_{final}} x_{2i} \ modq, \ y_1 = \sum_{i \in Q_{final}} y_{1i} \ modq, \ y_2 = \sum_{i \in Q_{final}} y_{2i} \ modq, \ z = \sum_{i \in Q_{final}} z_i \ modq)$.

## 4  Proposed DMKG Protocol

After presenting the BTSOF's DMKG protocol, an attention should be drawn to security flaws encountered. First, the set of shares is publicly revealed when there is a complaint about a participant: when the participant $P_j$ finds that the set of shares sent by $P_i$ did not verify Eq. 1, then he complaints about $P_i$. This forces $P_i$ to reveal publicly the set of shares that satisfy the equation. In addition, the use of private point-to-point channels for secretly distributing the shares, hides dishonest participants: is it the dealer who distributed an invalid share or the participant who lies and claims to have received an invalid share? The protocol fails to identify who.

In this section, we present a DMKG protocol with only public channels. There is no private communication between participants. All the shares are encrypted. For the complaint handling between the participants, our protocol is characterized by a phase named *Complaint management* like [11] protocol. It solves participants' complaints and correctly identifies malicious participants without revealing any information on honest participants' shares. Our protocol suits the MSS context unlike [11].

### 4.1  Communication and Adversary Model

We consider the same adversary and communication model as [8]. However, in our contribution, there are no private channels between participants. All information including the set of shares are distributed through the public channel.

### 4.2  The Protocol

In our DMKG protocol, we separate the verification phase from the distribution phase. The generation of the private key $sk = (x_1, x_2, y_1, y_2, z)$ is done in five phases (Initialization, Distribution, Verification, Complaint management, Recovery).

**Initialization Phase.** During the initialization phase, we keep the same assumptions as the DMKG protocol in [8]. In addition, each participant $P_i$ selects a private key $sk_{P_i}$ and publishes a public key $pk_{P_i} = g^{sk_{P_i}}$ where $sk_{P_i} \in \mathbb{Z}_q$ and $g$ is a generator of prime order group. The public key $pk_{P_i}$ will be used to encrypt and send the shares.

**Distribution Phase.** We keep the same first steps (from *Step 2.a* to *Step 2.d*) in BTSOF' DMKG protocol except for the *Step 2.b*. In our protocol, all verifications are done after the end of the distribution phase.

*Step 2.e* Like BTSOF, for $j = 1, ..., n$, $P_i$ computes the shares $sf_{ij} = f_i(j)$, $sf'_{ij} = f'_i(j)$, $sg_{ij} = g_i(j)$, $sg'_{ij} = g'_i(j)$, $sh_{ij} = h_i(j)$, $sh'_{ij} = h'_i(j)$. For $j = 1, ..., n$, $P_i$ actually has the set of shares $(sf_{ij}, sf'_{ij}, sg_{ij}, sg'_{ij}, sh_{ij}, sh'_{ij})$ to be sent to $P_j$.

*Step 2.f* Unlike [8], to avoid using private channels between participants, our proposed idea is to encrypt the shares with a simple encryption function and then broadcast them instead of securely sending them to the recipient. In [11], Neji *et al.* encrypt the shares of a secret as follow: $E_{ij} = s_{ij}(pk_{P_j})^{s_{ij}} = s_{ij}(g^{sk_{P_j}})^{s_{ij}}$, where $s_{ij}$ is the share sent from participant $P_i$ to participant $P_j$ and $g$ is a generator of prime order group. As a reminder, each committee member $P_i$ has a public-private key pair $(pk_{P_i}, sk_{P_i})$. These keys will be used to encrypt the set of shares computed by $P_i$ following the encryption function in [11]. For $j = 1, ..., n$, $P_i$ computes: $Enc_{sf_{ij}} = sf_{ij}(pk_{P_j})^{sf_{ij}} = sf_{ij}(g^{sk_{P_j}})^{sf_{ij}}$, $Enc_{sf'_{ij}} = sf'_{ij}(pk_{P_j})^{sf'_{ij}} = sf'_{ij}(g^{sk_{P_j}})^{sf'_{ij}}$, $Enc_{sg_{ij}} = sg_{ij}(pk_{P_j})^{sg_{ij}} = sg_{ij}(g^{sk_{P_j}})^{sg_{ij}}$, $Enc_{sg'_{ij}} = sg'_{ij}(pk_{P_j})^{sg'_{ij}} = sg'_{ij}(g^{sk_{P_j}})^{sg'_{ij}}$, $Enc_{sh_{ij}} = sh_{ij}(pk_{P_j})^{sh_{ij}} = sh_{ij}(g^{sk_{P_j}})^{sh_{ij}}$, $Enc_{sh'_{ij}} = sh'_{ij}(pk_{P_j})^{sh'_{ij}} = sh'_{ij}(g^{sk_{P_j}})^{sh'_{ij}}$. The participant $P_i$ sends the set of encrypted shares $(Enc_{sf_{ij}}, Enc_{sf'_{ij}}, Enc_{sg_{ij}}, Enc_{sg'_{ij}}, Enc_{sh_{ij}}, Enc_{sh'_{ij}})$ **publicly** to $P_j$ for $j = 1, ..., n$.

## Verification Phase

*Step 3.a* Each participant $P_j$ decrypts the set of the shares received from the participant $P_i$ by using the private key $sk_{P_j}$. $sf_{ij} = Enc_{sf_{ij}}[(g^{sf_{ij}})^{sk_{P_j}}]^{-1}$, $sf'_{ij} = Enc_{sf'_{ij}}[(g^{sf'_{ij}})^{sk_{P_j}}]^{-1}$, $sg_{ij} = Enc_{sg_{ij}}[(g^{sg_{ij}})^{sk_{P_j}}]^{-1}$, $sg'_{ij} = Enc_{sg'_{ij}}[(g^{sg'_{ij}})^{sk_{P_j}}]^{-1}$, $sh_{ij} = Enc_{sh_{ij}}[(g^{sh_{ij}})^{sk_{P_j}}]^{-1}$, $sh'_{ij} = Enc_{sh'_{ij}}[(g^{sh'_{ij}})^{sk_{P_j}}]^{-1}$. At the end of this step, for each $i \in Q_{tem}$, each $P_j$ has a set of shares $(sf_{ij}, sf'_{ij}, sg_{ij}, sg'_{ij}, sh_{ij}, sh'_{ij})$ that needs to be verified.

*Step 3.b* From the moment that the participant $P_j$ has the set of decrypted shares, we could now verify Eq. 2 as in the *Step 2.g* of BTSOF. If it fails, then the participant $P_j$ complaints against $P_i$. This complaint will be examined in the *Complaint management* phase. Otherwise, $P_j$ accepts the set of shares.

**Complaint Management Phase.** This phase is executed whenever there is a complaint from a participant $P_j$ against a participant $P_i$. The complaint management strategy is inspired by the *Disputation* idea presented in [11,18] and adapted to MSS context. First of all, a participant $P_i$ is marked as disqualified ($i = 1, ..., n$) if the number of complaints against $P_i$ is superior than $(t-1)$ in the *Step 3.b*. Second, all participants that are not disqualified in the *Step 3.b* will be considered as temporarily qualified and $\in Q_{temp}$. Finally, if a participant $P_i$ is already disqualified, then all the complaints against $P_i$ are accepted. Else, both participants $P_i$ and $P_j$ have to prove their honesty to all the other participants. Let's note $Q$ the set of participants where $Q = \{Q_{temp} \setminus \{P_i, P_j\}\}$.

*Step 4.a* $P_j$ chooses at random $S'_j = \{a_j, b_j, a'_j, b'_j, a''_j, b''_j\}$ and publishes $S''_j = \{g_1{}^{a_j}, g_2{}^{a'_j}, g_1{}^{a''_j}, h_1{}^{b_j}, h_2{}^{b'_j}, h_1{}^{b''_j}\}$

*Step 4.b* Each $P_k$ with $k \in Q$ chooses at random $S'_k = \{a_k, b_k, a'_k, b'_k, a''_k, b''_k\}$ and publishes $S''_k = \{g_1{}^{a_k}, g_2{}^{a'_k}, g_1{}^{a''_k}, h_1{}^{b_k}, h_2{}^{b'_k}, h_1{}^{b''_k}\}$

*Step 4.c* $P_i$ computes and publishes: $\lambda_1 = sf_{ij}[g_1{}^{a_j} \cdot \prod_{k=1;P_k \in Q}^{n} g_1{}^{a_k}]^{sf_{ij}}$, $\lambda'_1 = sg_{ij}[g_2{}^{a'_j} \cdot \prod_{k=1;P_k \in Q}^{n} g_2{}^{a'_k}]^{sg_{ij}}$, $\lambda''_1 = sh_{ij}[g_1{}^{a''_j} \cdot \prod_{k=1;P_k \in Q}^{n} g_1{}^{a''_k}]^{sh_{ij}}$, $\lambda_2 = sf_{ij}(g_1{}^{a_j})^{sf_{ij}}$, $\lambda'_2 = sg_{ij}(g_2{}^{a'_j})^{sg_{ij}}$, $\lambda''_2 = sh_{ij}(g_1{}^{a''_j})^{sh_{ij}}$, $\gamma 1 = sf'_{ij}[h_1{}^{b_j} \cdot \prod_{k=1;P_k \in Q}^{n} h_1{}^{b_k}]^{sf'_{ij}}$, $\gamma'1 = sg'_{ij}[h_2{}^{b'_j} \cdot \prod_{k=1;P_k \in Q}^{n} h_2{}^{b'_k}]^{sg'_{ij}}$, $\gamma''1 = sh'_{ij}[h_1{}^{b''_j} \cdot \prod_{k=1;P_k \in Q}^{n} h_1{}^{b''_k}]^{sh'_{ij}}$, $\gamma 2 = sf'_{ij}(h_1{}^{b_j})^{sf'_{ij}}$, $\gamma'2 = sg'_{ij}(h_2{}^{b'_j})^{sg'_{ij}}$, $\gamma''2 = sh'_{ij}(h_1{}^{b''_j})^{sh'_{ij}}$.

*Step 4.d* Each participant $P_k$ with $k \in Q$ publishes his set $S'_k$. If any participant cheats by publishing an invalid set which does not correspond to $S''_k$, then he will be disqualified and removed from $Q$, and the process resumes at *Step 4.b*.

*Step 4.e* Each participant $P_k$ with $k \in Q$ computes : $\alpha = \frac{\lambda_1}{\lambda_2}$, $\alpha' = \frac{\lambda'_1}{\lambda'_2}$, $\alpha'' = \frac{\lambda''_1}{\lambda''_2}$, $\beta = \frac{\gamma_1}{\gamma_2}$, $\beta' = \frac{\gamma'_1}{\gamma'_2}$, $\beta'' = \frac{\gamma''_1}{\gamma''_2}$ and $r = \sum_{k=1}^{n} a_k$, $r' = \sum_{k=1}^{n} a'_k$, $r'' = \sum_{k=1}^{n} a''_k$, $t = \sum_{k=1}^{n} b_k$, $t' = \sum_{k=1}^{n} b'_k$, $t'' = \sum_{k=1}^{n} b''_k$. Then, $P_k$ computes: $\alpha^{\frac{1}{r}} = g_1{}^{sf_{ij}}$, $\alpha'^{\frac{1}{r'}} = g_2{}^{sg_{ij}}$, $\alpha''^{\frac{1}{r''}} = g_1{}^{sh_{ij}}$ and $\beta^{\frac{1}{t}} = h_1{}^{sf'_{ij}}$, $\beta'^{\frac{1}{t'}} = h_2{}^{sg'_{ij}}$, $\beta''^{\frac{1}{t''}} = h_1{}^{sh'_{ij}}$. $P_k$ verifies

$$\begin{cases} \alpha^{\frac{1}{r}}.\beta^{\frac{1}{t}}.\alpha'^{\frac{1}{r'}}.\beta'^{\frac{1}{t'}} \stackrel{?}{=} \prod_{k=0}^{t} (CM_{ik})^{j^k} \\ \alpha''^{\frac{1}{r''}}.\beta''^{\frac{1}{t''}} \stackrel{?}{=} \prod_{k=0}^{t} (cm_{ik})^{j^k} \end{cases} \tag{3}$$

If it fails, then the secret set $(sf_{ij}, sf'_{ij}, sg_{ij}, sg'_{ij}, sh_{ij}, sh'_{ij})$ is not correct. $P_k$ concludes that $P_i$ is dishonest and the protocols ends. Otherwise the protocol continues.

*Step 4.f* Each participant $P_j \notin Q$ computes: $sf_{ij} = \frac{\lambda_2}{(g_1{}^{sf_{ij}})^{a_j}}$, $sg_{ij} = \frac{\lambda'_2}{(g_2{}^{sg_{ij}})^{a'_j}}$, $sh_{ij} = \frac{\lambda''_2}{(g_1{}^{sh_{ij}})^{a''_j}}$ and $sf'_{ij} = \frac{\gamma_2}{(h_1{}^{sf'_{ij}})^{b_j}}$, $sg'_{ij} = \frac{\gamma'_2}{(h_2{}^{sg'_{ij}})^{b'_j}}$, $sh'_{ij} = \frac{\gamma''_2}{(h_1{}^{sh'_{ij}})^{b''_j}}$. $P_j$ verifies if these values check Eq. 2. If it checks, the participant $P_j$ has a valid set of shares. In this case, $P_j$ sends a confirmation to all participants in Q and the protocol ends. Else, $P_j$ sends the set $S'_j$ to all participants in $Q$.

*Step 4.g* Each participant $P_k \in Q$ checks if the set $S'_j$ sent by $P_j$ is valid. This means that $P_k$ verifies if the values $(g_1{}^{a_j}, g_2{}^{a'_j}, g_1{}^{a''_j}, h_1{}^{b_j}, h_2{}^{b'_j}, h_1{}^{b''_j})$ published by $P_j$ in *step 4.a* correspond to the values in the set $S'_j$ published by $P_j$ in *step 4.f*. If it fails, $P_k$ concludes that $P_j$ lied and that he is dishonest and the protocol ends. Otherwise, the protocol continues.

*Step 4.h* Each $P_k \in Q$ computes $(g_1{}^{sf_{ij}})^{a_j}$, $(g_2{}^{sg_{ij}})^{a'_j}$, $(g_1{}^{sh_{ij}})^{a''_j}$, $(h_1{}^{sf'_{ij}})^{b_j}$, $(h_2{}^{sg'_{ij}})^{b'_j}$, $(h_1{}^{sh'_{ij}})^{b''_j}$ and $e_1 = \frac{\lambda_2}{(g_1{}^{sf_{ij}})^{a_j}}$, $e'_1 = \frac{\lambda'_2}{(g_2{}^{sg_{ij}})^{a'_j}}$, $e''_1 = \frac{\lambda''_2}{(g_1{}^{sh_{ij}})^{a''_j}}$, $e_2 = \frac{\gamma_2}{(h_1{}^{sf'_{ij}})^{b_j}}$, $e'_2 = \frac{\gamma'_2}{(h_2{}^{sg'_{ij}})^{b'_j}}$, $e''_2 = \frac{\gamma''_2}{(h_1{}^{sh'_{ij}})^{b''_j}}$. $P_k$ verifies if the equations in *Step 4.c* are correct with : $e_1 = sf_{ij}$, $e'_1 = sg_{ij}$, $e''_1 = sh_{ij}$, $e_2 = sf'_{ij}$, $e'_2 = sg'_{ij}$, $e''_2 = sh'_{ij}$. If there is equality, then $P_k$ concludes that $P_j$ lied, else, $P_i$ lied.

*Step 4.i* Each $P_k \in Q$ broadcasts the result of the previous steps by indicating whether the dishonest participant is $P_i$ or $P_j$.

*Step 4.j* The dishonest participant is identified by counting the number of votes for $P_i$ or $P_j$ and selecting the maximum. Thus, thanks to the *Complaint management* phase, dishonest participants are identified and disqualified and we can construct $QUAL$ the set of non disqualified participants.

**Recovery Phase.** Each participant $P_i \in QUAL$ computes the shares: $sf_i = \sum_{j \in QUAL} sf_{ji} \bmod q$, $sf'_i = \sum_{j \in QUAL} sf'_{ji} \bmod q$, $sg_i = \sum_{j \in QUAL} sg_{ji} \bmod q$, $sg'_i = \sum_{j \in QUAL} sg'_{ji} \bmod q$, $sh_i = \sum_{j \in QUAL} sh_{ji} \bmod q$, $sh'_i = \sum_{j \in QUAL} sh'_{ji} \bmod q$.

As a result of this approach, $P_i$ obtains a set of qualified participants $QUAL$ and holds for $j \in QUAL$ the values $f_j(i)$, $g_j(i)$, and $h_j(i)$.

## 5   Security

Our DMKG protocol respects the same security requirements in [8] which are secrecy and correctness for distributed multi-key generation in the Cramer-Shoup encryption scheme [4] in the presence of an adversary $A$ that corrupts at most $(t-1)$ participants for any $(t-1) < \frac{n}{2}$.

## 5.1  Security Requirements

**Correctness.** The correctness of the protocol includes the following elements:

*C1* Any set of $(t + 1)$ shares provided by honest participants defines the same unique secret key $sk = (x_1, x_2, y_1, y_2, z)$.

*C2* There is an efficient algorithm that has as input $n$ shares submitted by participants and outputs the unique secret key $sk$ even if at most $(t - 1)$ invalid shares are submitted by dishonest participants.

*C3* All honest participants have the same public key $pk = (c_1, c_2, c_3) = (g_1^{x_1}, g_2^{x_2}, g_1^{y_1}, g_2^{y_2}, g_1^{z})$, where $sk = (x_1, x_2, y_1, y_2, z)$ is the unique secret key guaranteed by (C1).

*C4* The secret key $sk = (x_1, x_2, y_1, y_2, z)$ is uniformly distributed in $\mathbb{Z}_q$.

**Secrecy.** Since the shared private key $sk$ is a confidential information that should not be found by a party who is unauthorized to know it, the DMKG protocol verifies the secrecy requirement if the adversary gets nothing about $sk$, or any information on $sk$ except for the pubic key $pk$. We use the same simulation in [8] to formally express the secrecy of the DMKG and to prove that an adversary $A$ cannot compute the value of a secret key $sk$ even if he can corrupt at most $(t - 1)$ participants and can use their secret information. For each probabilistic polynomial-time adversary, a simulator takes as input the public key $pk$ and produces as an output a distribution that is indistinguishable from the adversary's view in the real run of the DMKG protocol that outputs the public key $pk$. This dishonest participant cannot compute the value of the secret key $sk$.

## 5.2  Security Proofs

In this section, we prove that our DMKG protocol satisfies the security requirements presented previously in 5. First of all, we need to present the following lemma about the Pedersen VSS scheme [6,13] and the VMSS scheme in [8].

**Lemma 1.** *In the presence of an adversary that corrupts at most $t$ participants, any $(t + 1)$ shares of the honest participants can reconstruct the secret $s$. If the dealer is honest, then all shares held by honest participants can interpolate to a unique polynomial of degree $t$.*

**Lemma 2.** *In the presence of an adversary that corrupts at most $(t - l + 1)$ participants, any $(t + 1)$ shares of the honest participants can reconstruct the secret set $S = s_1, \ldots, s_l$. If the dealer is honest, then all shares held by honest participants can interpolate to a unique polynomial of degree $t$.*

**Correctness Proof.**

**Theorem 1.** *Our DMKG protocol satisfies the security requirement (C1). All shares submitted by a set of honest participants define the same private key sk.*

*Proof.* At the end of the *Complaint Management* phase, if the participant $P_i$ is not disqualified, which means that $P_i \in QUAL$ and that as a dealer, he has correctly performed the Pedersen-VSS and VMSS scheme. We rely on Lemma 1 and Lemma 2. Then, all honest participants have valid shares of $P_i$ that they use to compute the polynomials $f_i(x)$, $g_i(x)$ and $h_i(x)$ where $f_i(-1) = x_{1i}$, $f_i(-2) = y_{1i}$, $g_i(-1) = x_{2i}$, $g_i(-2) = y_{2i}$, and $h_i(0) = z_i$.

For this purpose, for a set $Q$ of $(t+1)$ valid shares, a unique value $x_{1i}$ (respectively $x_{2i}$, $y_{1i}$, $y_{2i}$, $z_i$) is computed with the Lagrange interpolation as $x_{1i} = \sum_{j\in Q} \lambda_{1j} sf_{ij}$, $x_{2i} = \sum_{j\in Q} \lambda_{1j} sg_{ij}$, $y_{1i} = \sum_{j\in Q} \lambda_{2j} sf_{ij}$, $y_{2i} = \sum_{j\in Q} \lambda_{2j} sg_{ij}$, $z_i = \sum_{j\in Q} \lambda_{0j} sh_{ij}$ where $\lambda_{0j} = \prod_{k\in Q, k\neq j} \frac{-k}{j-k}$, $\lambda_{1j} = \prod_{k\in Q, k\neq j} \frac{-1-k}{j-k}$, $\lambda_{2j} = \prod_{k\in Q, k\neq j} \frac{-2-k}{j-k}$ are the coefficient of Lagrange. Thus, from these shares $x_1$ (respectively $x_2$, $y_1$, $y_2$, $z$) can be generated as $x_1 = \sum_{i\in QUAL} x_{1i} = \sum_{i\in QUAL} \sum_{j\in Q} \lambda_{1j} sf_{ij} = \sum_{j\in Q} \lambda_{1j} \sum_{i\in QUAL} sf_{ij} = \sum_{j\in Q} \lambda_{1j} sf_j$ (respectively $x_2 = \sum_{j\in Q} \lambda_{1j} sg_j$, $y_1 = \sum_{j\in Q} \lambda_{2j} sf_j$, $y_2 = \sum_{j\in Q} \lambda_{2j} sg_j$, $z = \sum_{j\in Q} \lambda_{0j} sh_j$). For any authorized set $Q$ of participants, the private key $sk = (x_1, x_2, y_1, y_2, z)$ is unique.

**Theorem 2.** *Our DMKG protocol satisfies the security requirement (C2). Only valid shares of honest participants who check successfully the shares verification are used to reconstruct the private key sk.*

*Proof.* The validity of the shares $(sf_i, sg_i, sh_i)$ submitted by the participant $P_i$ with $i \in QUAL$ can be checked as follows:

$$\begin{cases} g_1^{sf_i} g_2^{sg_i} = g_1^{\sum_{i\in QUAL} sf_{ij}} g_2^{\sum_{i\in QUAL} sg_{ij}} = \prod_{i\in QUAL} g_1^{sf_{ij}} g_2^{sg_{ij}} = \prod_{i\in QUAL} \prod_{k=0}^{t} (A_{ik})^{j^k} \\ g_1^{sh_i} = g_1^{\sum_{i\in QUAL} sh_{ji}} = \prod_{i\in QUAL} g_1^{sh_{ij}} = \prod_{i\in QUAL} \prod_{k=0}^{t} (A'_{ik})^{j^k} \end{cases}$$

Therefore, only valid shares $(sf_i, sg_i, sh_i)$ submitted by honest participants who passed successfully the check are used.

**Theorem 3.** *Our DMKG protocol satisfies the security requirement (C3). After the Complaint Management phase, all qualified participants compute the same value of the public key $pk = (c_1, c_2, c_3) = g_1^{x_1} g_2^{x_2}, g_1^{y_1} g_2^{y_2}, g_1^{z})$, where $sk = (x_1, x_2, y_1, y_2, z)$ is the unique secret key guaranteed by Theorem 1.*

*Proof.* Since we did not modify the generation of the public key used in [8], we use the same proof of *Theorem 3* to prove that all qualified participants from the *Complaint Management* phase compute the same value of the public key pk. More details in [8].

**Theorem 4.** *Our DMKG protocol satisfies the security requirement (C4). The values $x_1$, $x_2$, $y_1$, $y_2$ and $z$ of the secret key sk are uniformly distributed in $\mathbb{Z}_q$.*

*Proof.* At the end of the *Complaint Management* phase, only honest participants are in the set $QUAL$. Each participant $P_i$, $i \in QUAL$ has already selected a secret value $x_{1i}$ (respectively $x_{2i}$, $y_{1i}$, $y_{2i}$, $z_i$) randomly in $\mathbb{Z}_q$. As $x_1$ (respectively $x_2$, $y_1$, $y_2$, $z$) is defined as the sum of all $x_{1i}$ (respectively $x_{2i}$, $y_{1i}$, $y_{2i}$, $z_i$) then it can be guaranteed that $x_1$ (respectively $x_2$, $y_1$, $y_2$, $z$) is randomly chosen in $\mathbb{Z}_q$. At the end of the *Distribution* phase of our DMKG protocol, these values are already chosen and uniformly distributed via VMSS protocol. Neither the values $x_{1i}$ (respectively $x_{2i}$, $y_{1i}$, $y_{2i}$, $z_i$) nor the set $QUAL$ change later. Thereby, as $x_1 = \sum_{i \in QUAL} x_{1i}$, then $x_1$ is uniformly distributed in $\mathbb{Z}_q$. In the same way, $x_2$, $y_1$, $y_2$ and $z$ are also uniformly distributed.

**Secrecy Proof.** The secrecy of the DMKG is formally expressed by a simulator. Since we have kept the same simulator used in [8], we also keep the same proof of secrecy. We hence refer the reader to the aforementioned publication for further details.

# 6    Discussion and Comparison

The comparison of our DMKG protocol with BTSOF's DMKG protocol is summarized in Table 1. It takes into consideration (1) the communication model: private or public channels, (2) the computational complexity (3) the security of the DMKG protocol in the sense that the secrecy and correctness requirements are satisfied, (4) the efficiency of the complaint management strategy in the sense that malicious participants are identified and honest participants are not forced to reveal their shares.

The DMKG protocol used in [8] uses private channels to send the shares between participants. Also, it uses a complaint management strategy that is not efficient and does not clearly identify the dishonest participants. The strategy can be briefly described as follow: At the end of the *Distribution and verification* phase and specifically in the *Step 2.h* and *Step 2.i*, if the number of complaints against a participant $P_i$ is superior than $(t-1)$, then $P_i$ is disqualified. Otherwise, for each complaint from $P_j$ against $P_i$, $P_i$ reveals the shares sent to the participant $P_j$. In this case, if the revealed values do not check Eq. 2, then the participant $P_i$ is disqualified. This strategy is not effective because in some cases it may not eliminate dishonest participants. Indeed, $P_i$ can send valid shares to $P_j$ but $P_j$ pretends that these shares are invalid to force $P_i$ to reveal his valid shares. In this case, the dishonest participant $P_j$ will not be disqualified. Another case, $P_i$ can cheat and send invalid shares to $P_j$. Afterwards, during the management of $P_j$'s complaint against $P_i$, he reveals valid shares. He will not be disqualified. Otherwise, in the *Step 2.e*, all the participants must verify all broadcasted public information $E_{\tau i}$ for $\tau = 1, 2, 3, 4$ of each participant $P_i$. This verification is costly. Its complexity is $O(kn^2)$ where $k$ is a security parameter and $n$ is the number of participants.

In our DMKG protocol, there is no private channel between participants. It uses only public channels. All the shares are encrypted. Also, the protocol uses a complaint management strategy that does not depend of the number of complaints reported against a participant to disqualify him. If there is a complaint from $P_j$ against $P_i$ then both participants have to prove their honesty without revealing the values of their shares. Moreover, as previously described, this strategy is more efficient. It clearly identifies the malicious participants. In fact, thanks to the *Complaint Management* phase, the 2 cases described previously are identified and treated. (1) a participant who falsely pretends receiving invalid shares is disqualified and (2) a participant that sends invalid shares is disqualified too. In another hand, in our protocol there is no need to verify all broadcasted information of all participants. The verification is done in the *Complaint management* phase. It is executed only if there is at least one complaint between participants. Thus, the computational complexity is $O(kn)$ and it depends on the number of complaints reported by participants.

**Table 1.** Comparison of our DMKG protocol with BTSOF's DMKG protocol.

| Criteria | BTSOF's DMKG protocol | Our DMKG protocol |
|---|---|---|
| Communication Model | Public and private | Public |
| Computational complexity | $O(kn^2)$ | $O(kn)$ |
| Full security | Yes | Yes |
| Complaint management Efficiency | No | Yes |

## 7    Conclusion

In this paper, we come up with a Distributed Multi-Key Generation (DMKG) protocol based on the DMKG protocol of BTSOF with a new complaint management strategy. The main idea is to avoid honest participants publicly revealing their shares and to efficiently identify dishonest participants. This strategy relies on the *Disputation* strategy proposed in [11] and has been extended to suit the multi secret sharing context. We note that we use only public channels and that all the shares of secrets are encrypted. The proposed DMKG protocol is not limited to Blockchain traceability. It can be used as a building block in several domains such as electronic voting and any identity-based cryptography.

In this work, we focused on the generation of the traceable private key $sk$. In future works, we intend to use this contribution for the generation of the traceable public key and managing all the complaints between participants. We are working on implementing our protocol in order to validate our contribution with experimental results.

# References

1. Biswas, A.K., Dasgupta, M., Ray, S., Khan, M.K.: A probable cheating-free (t, n) threshold secret sharing scheme with enhanced blockchain. Comput. Electr. Eng. **100**, 107925 (2022)
2. Blakley, G.R.: Safeguarding cryptographic keys. In: Managing Requirements Knowledge, International Workshop on, pp. 313–313. IEEE Computer Society (1979)
3. Canetti, R., Gennaro, R., Jarecki, S., Krawczyk, H., Rabin, T.: Adaptive security for threshold cryptosystems. In: Wiener, M. (ed.) CRYPTO 1999. LNCS, vol. 1666, pp. 98–116. Springer, Heidelberg (1999). https://doi.org/10.1007/3-540-48405-1_7
4. Cramer, R., Shoup, V.: A practical public key cryptosystem provably secure against adaptive chosen ciphertext attack. In: Krawczyk, H. (ed.) CRYPTO 1998. LNCS, vol. 1462, pp. 13–25. Springer, Heidelberg (1998). https://doi.org/10.1007/BFb0055717
5. Franklin, M., Yung, M.: Communication complexity of secure computation. In: Proceedings of the 24th Annual ACM Symposium on Theory of Computing, pp. 699–710 (1992)
6. Gennaro, R., Jarecki, S., Krawczyk, H., Rabin, T.: Secure distributed key generation for discrete-log based cryptosystems. In: Stern, J. (ed.) EUROCRYPT 1999. LNCS, vol. 1592, pp. 295–310. Springer, Heidelberg (1999). https://doi.org/10.1007/3-540-48910-X_21
7. Kiamari, N., Hadian, M., Mashhadi, S.: Non-interactive verifiable LWE-based multi secret sharing scheme. Multimedia Tools Appl. pp. 1–13 (2022). https://doi.org/10.1007/s11042-022-13347-4
8. Ma, T., Xu, H., Li, P.: A blockchain traceable scheme with oversight function. In: Meng, W., Gollmann, D., Jensen, C.D., Zhou, J. (eds.) ICICS 2020. LNCS, vol. 12282, pp. 164–182. Springer, Cham (2020). https://doi.org/10.1007/978-3-030-61078-4_10
9. Ma, T., Xu, H., Li, P.: Skyeye: a traceable scheme for blockchain. Cryptology ePrint Archive (2020)
10. Ma, T., Xu, H., Li, P.: A traceable scheme for consortium blockchain. In: 2021 IEEE 9th International Conference on Smart City and Informatization (ISCI), pp. 39–46. IEEE (2021)
11. Neji, W., Blibech, K., Ben Rajeb, N.: Distributed key generation protocol with a new complaint management strategy. Secur. Commun. Netw. **9**(17), 4585–4595 (2016)
12. Pakniat, N., Noroozi, M., Eslami, Z.: Distributed key generation protocol with hierarchical threshold access structure. IET Inf. Secur. **9**(4), 248–255 (2015)
13. Pedersen, T.P.: Non-interactive and information-theoretic secure verifiable secret sharing. In: Feigenbaum, J. (ed.) CRYPTO 1991. LNCS, vol. 576, pp. 129–140. Springer, Heidelberg (1992). https://doi.org/10.1007/3-540-46766-1_9
14. Pedersen, T.P.: A threshold cryptosystem without a trusted party. In: Davies, D.W. (ed.) EUROCRYPT 1991. LNCS, vol. 547, pp. 522–526. Springer, Heidelberg (1991). https://doi.org/10.1007/3-540-46416-6_47
15. Schindler, P., Judmayer, A., Stifter, N., Weippl, E.: Distributed key generation with ethereum smart contracts. In: CIW'19: Cryptocurrency Implementers' Workshop (2019)

16. Shalini, I., Sathyanarayana, S., et al.: A comparative analysis of secret sharing schemes with special reference to e-commerce applications. In: 2015 International Conference on Emerging Research in Electronics, Computer Science and Technology (ICERECT), pp. 17–22. IEEE (2015)
17. Shamir, A.: How to share a secret. Commun. ACM **22**(11), 612–613 (1979)
18. Shil, A.B., Blibech, K., Robbana, R., Neji, W.: A new pvss scheme with a simple encryption function. arXiv preprint arXiv:1307.8209 (2013)
19. Yang, C.C., Chang, T.Y., Hwang, M.S.: A (t, n) multi-secret sharing scheme. Appl. Math. Comput. **151**(2), 483–490 (2004)
20. Zhou, X.: Threshold cryptosystem based fair off-line e-cash. In: 2008 2nd International Symposium on Intelligent Information Technology Application, vol. 3, pp. 692–696. IEEE (2008)

# The Importance of Blockchain for Ecomobility in Smart Cities: A Systematic Literature Review

Irénée Dondjio[✉]

University of Nicosia, Nicosia, Cyprus
dondjio.i@unic.ac.cy

**Abstract.** By 2050, cities will house at least 70% of the world's population. Already, metropolitan areas are responsible for 70% of all greenhouse gas emissions caused by energy usage. The expanding population of city dwellers has presented new challenges to contemporary urban areas. Congestion and other kinds of inefficient urban movement irritate city people and reduce their quality of life. Cities are reducing carbon dioxide emissions in an attempt to prevent global warming, which necessitates innovative new modes of transportation. Smart cities are innovative problem-solving projects for difficulties that have evolved as a consequence of increasing urbanization and changing environmental circumstances. Today's cities depend largely on people's capacity to travel freely and easily in order to minimize traffic congestion, provide a high quality of life for residents and tourists, and reduce carbon emissions and environmental deterioration. Everything is dependent on an effective and sustainable transportation infrastructure. Blockchain technology is being researched for smart city applications and is used in smart mobility for ridesharing, electric charging, platoon member interactions, and vehicle communication. Although exploratory research on this issue exists, it is spread over different use-cases and applications. This research conducts a thorough analysis of the literature on the implications of blockchain for smart city mobility and transportation, as well as its potential to boost efficiency in these sectors. The goal of this article is to synthesize and summarize existing information on the issue and contribute to closing the knowledge gap by researching how blockchain technology might improve ecomobility by reducing pollution, travel time, and congestion. The results of the literature review are divided into six use-case categories.

**Keywords:** Ecomobility · Smart Mobility · Blockchain · Smart Cities · Smart Contracts

## 1 Introduction and Research Problem

Rising pollution levels, the severe repercussions of climate change, and global warming are pressuring large cities to lessen the environmental impact of their means of transportation. According to an United Nation (UN) report, urban areas have a significant influence on the environment as they account for 67–76% of CO2 emissions and the amount of

M. Papadaki et al. (Eds.): EMCIS 2022, LNBIP 464, pp. 165–184, 2023.
https://doi.org/10.1007/978-3-031-30694-5_13

energy used. At the same time, metropolitan regions housed 55% of the world's population in 2018. It is anticipated that by 2050, this proportion would have increased to 70% (UN 2020). There are millions of vehicles on the road at any one moment, with devastating consequences for both the people who live there and the environment. In reality, transportation in all of its forms is responsible for the vast majority of the dangerous gas emissions generated in cities. In fact, transportation is responsible for around 26% of greenhouse gas emissions, with personal automobile trips contributing significantly (WHO 2019). Consequently, it appears that an ecologically friendly and intelligent transportation system is the promising way to address these issues Morris (2015). There is evidence that, although traffic is required for flexible movement within cities, it is harmful to people's wellbeing Morris (2015). Morris' study also discovered a substantial positive relationship between city life satisfaction and traffic time. Consequently, travelling time is important in cities and whole nations, and traffic has a definite detrimental impact on inhabitants' quality of life (Morris 2015). Urban areas are at the edge of smart and environmentally friendly transportation, embracing technology and policies to reduce carbon emissions from transportation, ease traffic congestion, and enhance air quality (Karger et al. 2021). Walking, public transportation, and cycling reduce pollution and promote sustainable mobility. Ecomobility supports practical, low-pollution, eco-friendly mobility and living conditions in local areas and municipalities. For a sustainable transportation, people should be prioritized and provided with more affordable, accessible, healthier, and cleaner alternatives to their present mobility behaviors. Automated and networked multimodal transportation, as well as smart traffic control technologies enabled by digitalization will play a growing role (Williams et al. 2012). Furthermore, the price of transportation should reflect its environmental and health effect. Transportation solutions are often tied to three primary concerns: urban sprawl, climate change, and equitable access to services and employment. Transportation, as previously noted, contributes considerably to CO2 emissions, and solutions may range from technological to behavioral to financial to infrastructural improvements. The primary cause of urban sprawl is the dependence on the private vehicle as the primary source of transportation (Woodcock et al. 2007). As result, the concept of smart cities emerged to solve economic and ecological concerns of the 21st century. A smart city improves people's quality of life and services by merging breakthrough technologies and new urban infrastructures (Batty et al. 2012). In fact, breakthrough technologies such as blockchain are developing at a tremendous speed. Blockchain technology excels in the efficient management of financial transactions. Its administration is completely decentralized, therefore there is no need for a central authority. According to Lopez and Farooq (2018), the great potential that blockchain has in the context of smart mobility has already been shown by research on smart cities and blockchain. For example, the technology may secure people's personally identifiable information on their mobility and defend their privacy (Lopez and Farooq 2018). Nevertheless, in spite of the preliminary work that has been done in this field, the studies are still disjointed and only seldom properly integrated.

The central aim of this paper is to help bridge the above mentioned gap by investigating how blockchain technology might enhance ecomobility by lowering pollution, travel time and congestion in Smart Cities. As a result the structure of this paper will be as follows: literature review, systematic literature revue, key findings, conclusion and discussion.

## 2  Literature Review

### 2.1  The Blockchain Technology

Blockchains are databases that record and preserve the transactional history of the data they were created to hold. It consists of interconnected "blocks" of data that cannot be altered after they have been added to the chain (Dondjio and Themistocleous 2022). The blockchain technology is sometimes referred to as digital ledger technology (DLT). This is due to the fact that it stores vast volumes of transaction information in a digital format on its ledger (CB Insights 2021). The introduction of cryptocurrency markets has contributed to the growing popularity of blockchain technology, illustrating how this innovation has the potential to shake up the financial industry.The characteristics of being decentralized, consensus-based, digital, immutable, chronological, and time-stamped have brought blockchain technology a great deal of notoriety (Deloitte (2017). As a result of these characteristics, blockchain has the potential to enhance a variety of aspects of the business environment, including as the reduction of risk, the visibility of supply chain activities, the eradication of fraudulent transactions, and general transparency (Teoh 2022).

### 2.2  Key Blockchain Characteristics

The most important features of blockchain technology are described in the subsequent paragraphs.

*Traceability* lets users trace blockchain transactions. Users may learn crucial transaction information by checking a data block. To monitor data, each system block is closely linked to the next one. *Transparency* allows system members to see and manage transactions. Members may publish transactions they enter into the system.It also detects and rejects questionable transactions. By letting stakeholders choose what data to send via a network, the system improves openness and security. *Security* is ensured by preventing outsiders from changing network data without consent. Blockchain transactions are guaranteed by *immutability.* The system stops users from removing or modifying validated transactions. Blockchain technology avoids system failure and builds stakeholder *trust.* (Dondjio and Themistocleous 2022).

The *blockchain trilemma* states that it is impossible to simultaneously maintain decentralization, security, and scalability in a distributed ledger system. One of these three properties is sacrificed whenever one of the elements is improved. Decentralization refers to the fact that a blockchain network does not depend on centralized points of control; security refers to the fact that a blockchain network can withstand and repel DDoS attacks; and scalability refers to the fact that it can manage enormous numbers of transactions (Teoh 2022).

## 2.3 Blockchain and Smart Contract

A smart contract is a self-executing contract in which the conditions of the buyer-seller agreement are directly encoded into lines of code. The code and the contract's agreements live on a decentralized and distributed blockchain network. Smart contracts enable anonymous parties to enter into trustworthy agreements and transactions without the requirement for a governance, legal system, central authority, or external enforcement mechanism (Antonopoulos and Wood 2018).

The creation of a smart contract may be accomplished on a number of different platforms, but one of the more prominent ones is Ethereum (Antonopoulos and Wood 2018). On the Ethereum platform, smart contract developers are free to create any decentralized application (DApp) of their choice. The decentralized programs execute precisely in accordance with the criteria specified in the code and do not run the danger of being censored, tricked, or experiencing downtime. Figure 1 depicts the execution process of a smart contract on the Ethereum network. Two parties achieve an agreement, which is then written in solidity code by a developer. The code is subsequently compiled to bytecode for processing by the Ethereum Virtual Machine (EVM). The participation of miners is essential for the contract to be processed on the blockchain. Once included, the contract is processed on the event date specified by the supplied code. The contract's execution releases payment to the relevant party, which may subsequently be confirmed by everyone (Sayeed et al. 2020).

## 2.4 Ecomobility and Smart Mobility

Ecomobility is a strategy for establishing and managing local regions and towns that promotes practical, low-pollution, environmentally friendly mobility and living conditions (Portillo 2018). Walking, public transit, and cycling are often confused with ecomobility. These three actions reduce pollution and promote environmentally friendly transportation. Ecomobility, also known as "smart mobility" or "sustainable mobility" is an example of an eco-innovation that involves responding to the needs of consumers and gaining their support (also known as "demand-pull innovation"); at the same time, however, it also involves asking which business ecosystem is most likely to provide efficient solutions to these needs (also known as "technology-push innovation") within the context of an institutional setting that is constantly changing (Boennec 2018). The concept "sustainable transportation and street mobility" refers to the wide issue of sustainable transportation in regards to its social, environmental, and economic implications (Beaume and Midler 2009). Sustainable transportation systems benefit the communities they serve in terms of environmental, social, and economic well-being (Dias et al. 2013). Many factors are considered when determining sustainability, such as the vehicles used, the source of energy, and the existing infrastructure supporting transportation, transportation operations and logistics, and transit-oriented development (Litman 2017). Eco-mobility allows cities to access goods and services in a sustainable way. It improves urban quality of life, expands sustainable travel options, promotes social cohesion, reduces greenhouse gas emissions and congestion, improves air quality, provides equitable transit options, boosts local green economies, and improves cyclist safety (Portillo 2018). To respond to rising pollution levels, the severe repercussions of climate change, the environmental and global warming effect of cities requires eco-friendly and sustainable action (Lauwers et al. 2025). To preserve economic development and quality of life, smart cities invest in human and social capital, transportation and ICT infrastructure, and democratic government. Smart towns reduce traffic congestion. Self-driving automobiles and sensors for urban infrastructure are options (Dameri 2017). Smart mobility, a component of smart cities, employs technology and approaches to improve transportation sustainability and efficiency by reducing pollutants, travel times, traffic, accidents, and urban footprint (Giffinger et al. 2007).

## 3 Systematic Literature Review

The purpose of this research is to map the state of the art in published papers on the use of blockchain for fostering ecomobility in smart cities by using elements such as novelty, drivers, models, and methods. Moreover, the research is based on an understanding of a variety of topics, including people and technology and challenges around the blockchain adoption. The Systematic Literature Review (SLR) technique will be divided into three (3) components, as illustrated in Fig. 1. At the first step (planning phase), the researcher realizes the need for an SLR and develops and tests a review approach. In the second step (conducting phase), the study sought for and collects primary studies, and data is collected and analyzed. The last stage (report phase) focuses on synthesizing the information obtained in earlier phases in order to communicate the results.

### 3.1 Planning Phase

● **Identifying the Research Question and goals for the review**

Given the significance of blockchain's potential usage in many industries, a systematic literature analysis was conducted to identify current research as well as possible use of blockchain technology for ecomobilty in smart city. To accomplish this goal, the following research question has been identified: **"How could blockchain technology be used to foster ecomobility in smart cities?"**

● **Identifying Keywords and Search Queries**

The keywords used to guide the search process were *"Blockchain"*, *"Smart Mobility"*, *"Ecomobility"* and *"Smart City"* (Table 1)

**Table 1.** Keywords and Search Queries

| Key Words | Search Queries |
|---|---|
| Blockchain | Query-1: "Blockchain" AND" Smart Mobility" |
| Smart Mobility | Query-2: "Blockchain" AND "Ecomobility" |
| Ecomobility | Query-3: "Ecomobility "AND "Smart Mobility" |
| Smart City | Query-4: "Ecomobility "AND "Smart Smart City" |

## 3.2 The Conducting Phase

• **Literature Selection**

A thorough search of the databases such as: JSTOR, EBSCOhost, IEEE Xplore Digital Library, ScienceDirect, and SpringerLink yielded a total of 1250 papers, with the "Preferred Reporting Items for systematic reviews and Meta-Analyses (PRISMA)" directing the researchers to identify, filter, and choose the most relevant material. The articles were imported into "Endnote," where they were combined, verified, reassessed, and corrected, and duplicates were removed all at once. The inclusion and exclusion

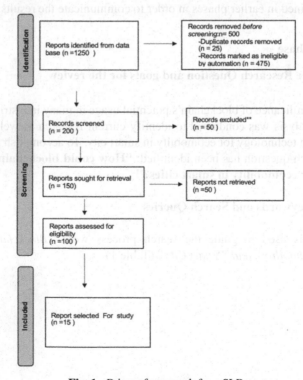

**Fig. 1.** Prisma framework for a SLR

criteria were applied to further filter and choose the most relevant articles while excluding those that were not (Table 2).

- **Inclusion and Exclusion Criteria**

**Table 2.** Inclusion and Exclusion Criteria

| Inclusion Criteria | Exclusion criteria |
|---|---|
| Peer review studies only | Grey literature |
| Academic | White papers and Non-academic papers |
| Full text only | Full text not available |
| English language | Non-English |
| Published on 2016 onwards | Published before 2016 |
| Within the study scope | Diverged from the study scope |

- **Study Screening, Selection and Data Extraction**

A manual scan of the references lists from the chosen papers was undertaken to ensure a thorough selection, and relevant additional articles were rechecked and included. Many abstracts and titles were obtained among numerous publications to determine whether or not the content was related to this investigation. In all, 150 studies were selected, and their full-texts were reviewed, downloaded, and re-evaluated in light of the inclusion and exclusion criteria. This approach further decreased the findings, and 100 papers that met the inclusion criteria were referred to the "Critical Appraisal Skills Programme (CASP)" for quality assessment. The majority of the papers were rejected because they did not match the inclusion criteria. After final rechecking, 15 studies were chosen for this evaluation.

### 3.3 Reporting Phase

- **Selected articles for the study**

Table 3 summarizes the articles chosen for this analysis based on the approach of Brereton et al (Brereton et al. 2007). The first column includes the research paper's authors, the second column lists the research piece's title, and the third column briefly summarizes the study work's emphasis.

**Table 3.** Articles selected for the study

| Authors | Titles | Subject's Goals and Summary |
|---|---|---|
| Chang and Chang 2018 | Application of Blockchain Technology to Smart City Service: A Case of Ridesharing | This article creates a decentralized blockchain using SmaRi. The technology offers users automatic execution, an immutable record, and decentralized decision-making. Peer-to-peer trades may support the sharing economy. Authors connect SmaRi to management systems. SmaRi may build functionalities by connecting to the original system. This study may advance smart cities |
| Huang et al. 2019 | NSC: A Security Model for Electric Vehicle and Charging Pile Management Based on Blockchain Ecosystem | This study presents a decentralized blockchain security paradigm based on the lightning network and smart contract (LNSC). Registration, scheduling, authentication, and billing are included. The new security model can be simply incorporated with existing scheduling techniques to improve EV-charging pile trading. This study presents actual experimental findings. These data show that our strategy improves vehicle security |
| Baza et al. 2019 | B-Ride: Ride Sharing With Privacy-Preservation, Trust and Fair Payment Atop Public Blockchain | B-Ride introduces a time-locked deposit scheme for ride-sharing using smart contracts and zero-knowledge set membership evidence. Drivers and riders must establish good faith by paying a deposit to the blockchain. Later, a motorist must confirm to the blockchain that he/she arrived on time. To protect rider/driver privacy, zero-knowledge set membership proof is used. To guarantee equitable remuneration, pay-as-you-drive is based on driver and rider distance. A reputation model rates drivers based on their prior behavior |
| KIM, 2019 | Impacts of Mobility on Performance of Blockchain in VANET | This paper presents a paradigm for analyzing the influence of mobility on a blockchain system's performance in a VANET based on three critical metrics: the chance of a successful block addition, the stability of a rendezvous, and the number of blocks exchanged during a rendezvous. Closed-form expressions and numerical data show blockchain performance in VANET situations |
| Nguyen et al. 2019 | Blockchain-Based Mobility-as-a-Service | Since public and private transportation companies must connect to this layer for MaaS to work, we suggest a blockchain-based MaaS to remove it. The approach promotes confidence and transparency for all parties and removes the need for separate MaaS agents. Computing power and resources are given to transportation providers at the network's edge, creating decentralized trust. The blockchain-based MaaS might become the key component of smart city mobility, lowering CO2 emissions |

*(continued)*

**Table 3.** (*continued*)

| Authors | Titles | Subject's Goals and Summary |
|---|---|---|
| Zhang and Wang, 2019 | Data-Driven Intelligent Transportation Systems: A Survey | The availability of a huge quantity of data may lead to a revolution in ITS development, converting a traditional technology-driven system into a more powerful multifunctional data-driven intelligent transportation system (D 2 ITS). D 2 ITS is becoming a more intelligent, privacy-aware system |
| Chen et al. 2020 | Smart-Contract-Based Economical Platooning in Blockchain-Enabled Urban Internet of Vehicles | This article proposes platoon-driving for autonomous cars in free-flow traffic to enhance urban traffic and prevent accidents. This concept groups path-matched vehicles into a platoon directed by the platoon leader (PH). A PH selection method is established to encourage cars to be PHs and sustain platoon dynamics. Next, a smart contract enables blockchain-based payments between the PH and PMs, eliminating harmful and misleading payments. The platoon model consumes less gasoline than the solo model |
| Yuan and Wang 2016 | Towards blockchain-based intelligent transportation systems | Blockchain might change centralized ICT systems (ITS). Blockchain can provide a safe, trustworthy, decentralized autonomous ITS environment, strengthening legacy infrastructure and crowdsourcing. Blockchain ITS (B 2 ITS). The paper proposes a seven-layer ITS-oriented blockchain architecture to solve B2 ITS research concerns. The paper examines blockchain's use in parallel transportation management systems (PtMS) Ride-sharing blockchain case study concludes the investigation. This study aims to encourage more B2 ITS research |
| Aujla et al.2020 | BloCkEd: Blockchain-Based Secure Data Processing Framework in Edge Envisioned V2X Environment | This proposes a blockchain-based safe data processing platform for edge V2X. (hereafter referred to as BloCkEd). A multi-layered edge-enabled V2X system model for BloCkEd is provided with a multi-objective optimization problem. BloCkEd uses efficient container-based data processing and blockchain-based data integrity management to reduce connection breakage and latency. We execute and analyze the suggested technique in Chandigarh, India, for latency, energy usage, and SLA compliance |

(*continued*)

**Table 3.** (*continued*)

| Authors | Titles | Subject's Goals and Summary |
|---|---|---|
| Ying et al. 2020 | BEHT: Blockchain-Based Efficient Highway Toll Paradigm for Opportunistic Autonomous Vehicle Platoon | Blockchain-based highway tolls are proposed for the opportunistic platoon. The blockchain records and verifies every registered vehicle's driving history. A roadside unit (RSU) distinguishes a single vehicle from a platoon and allocates lanes. An aggregate signature speeds up RSU authentication. This scenario's security risks are analyzed. The experiment shows that our strategy works well |
| Lei et al. 2017 | Blockchain-Based Dynamic Key Management for Heterogeneous Intelligent Transportation Systems | This paper introduces heterogeneous network key management. Security managers (SMs) record vehicle departure information, encapsulate block-to-transport keys, and rekey automobiles within the same security domain. First, a blockchain-based decentralized network. Blockchain may simplify VCS distributed key management. Second, dynamic transaction collection reduces vehicle handover time. Extensive simulations and analysis show the framework's effectiveness. The blockchain structure transfers keys faster than the structure with a central manager, and the dynamic approach allows SMs to meet shifting traffic volumes |
| Hawlitschek et al. 2018 | The limits of trust-free systems: A literature review on blockchain technology and trust in the sharing economy | This paper analyses how blockchain technology might tackle the trust problem in the sharing economy. By conducting a dual literature review, the authors found that 1) the conceptualization of trust differs between blockchain and the sharing economy, 2) blockchain technology is suitable to replace trust in platform providers, and 3) trust-free systems are hardly transferable to sharing economy interactions and will depend on the development of trusted interfaces for blockchain-based sharing economy |
| Kim et al. 2021 | Design of Secure Decentralized Car sharing System Using Blockchain | Traditional car-sharing is centralized, so there's a single point of failure. Decentralized car-sharing is key to overcoming the centralized dilemma. This research created a blockchain-based car-sharing system. Blockchain was utilized to decentralize car-sharing and assure data integrity. Anonymous authentication is possible for system participants. The suggested car-sharing system may be protected against assaults utilizing informal analysis, AVISPA simulation, and BAN logic analysis. The suggested scheme's computing and communication costs were studied |

(*continued*)

**Table 3.** (*continued*)

| Authors | Titles | Subject's Goals and Summary |
|---|---|---|
| Dungan and Pop 2022 | Blockchain-based solutions for smart mobility sustainability assurance | Blockchain, recognized for its relation to bitcoin and other cryptocurrencies, may suit ITS safety and security needs (ITS). This technology can store information securely, protecting it from security concerns. This streamlines and secures TMCs' sensor network data processing (Traffic Monitoring Centers). Cyberattacks may create erroneous traffic signal phases and green-interval settings at junctions. The results will be based on a comparison with existing IoT techniques for securing sensor traffic data |
| Orecchini, et al. 2018 | Blockchain Technology in Smart City: A New Opportunity for Smart Environment and Smart Mobility | Smart Cities use people-centered digital initiatives that lead to high-tech breakthroughs to develop capacity and opportunity. This paper investigates the possibility of integrating the innovative and multi-purpose Blockchain Technology in the smart city evolutionary process, and in particular in the Smart Environment and Smart Mobility by allowing renewable energy source traceability and by providing information about the kind of energy used to refuel, for example, the selected vehicle. More user knowledge of environmental and energy sustainability is projected |

## 4  Key Findings

The majority of the articles studied pointed the great potential that blockchain has in the context of Ecomobility and smart mobility. According to the findings, the blockchain technology is suitable for the implementation of a decentralized trust management system in automotive networks because of its decentralized nature, its transparency, and its inalterability. In addition, payments for parking and public transportation may be processed faster using blockchain-based transactions. The key findings of this study are presented below.

**Finding-1: Sustainable Mobility System Model**
Dungan and Pop (2022) provide a conceptual model for a sustainable smart transportation system that incorporates blockchain technology. Their model takes into account the Smart Sustainable City-Blockchain (SSCB) integration architecture and emphasizes its advantages in the transportation area. In addition, transportation energy management has been incorporated as an indication. The model (see Fig. 2 below) also reveals a particular interest in green transportation systems integrating intelligent energy consumption control and $CO_2$ emission reduction. Blockchain technology is being developed with the intention of delivering appropriate solutions for car-sharing services in terms of both privacy and security. In addition, payments for parking and public transportation may

be processed faster using blockchain-based transactions, with the only limiting factor being the amount of traffic on the network (Dungan and Pop 2022).

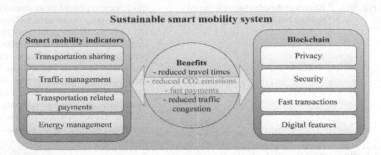

**Fig. 2.** Sustainable Mobility System Model by (Dungan and Pop 2022)

### Finding-2: Intelligent Transportation Systems (ITS)

Yuan and Wang (2016) highlight that Intelligent Transportation Systems (ITS) have emerged with the goal to apply several advantages to mobility, like improving travel security and increasing the performance of transportation systems. In addition, ITS enables smart vehicles to communicate with each other and access the internet. The authors take a broader perspective and propose a blockchain-based ITS model. As they believe that the advantages of blockchain models like decentralization, security, and trust make it highly suitable for such a purpose (Yuan and Wang 2016). Proposes a blockchain platform architecture (Fig. 3 illustrates ITS architectural layers).

*The physical layer* contains physical objects like cars, traffic lights, or devices. *The network layer* ensures data forwarding, verification, and distributed network establishment. *The incentive layer* describes the production and distribution of economic incentives in blockchains to drive network members to continue mining and verifying transactions. *The contract layer* is the basis for smart contracts, which ITS uses to manage and program physical and digital assets. *The application layer* shows ridesharing, logistics, and asset management.

### Finding-3: Vehicular Communication Systems (VCS'S)

Aside from the underlying architecture of ITSs, another component of smart mobility where blockchain might aid is vehicular communication. Vehicles may interact with each other (Vehicle-to-Vehicle, V2V) and with the infrastructure (Vehicle-to-Infrastructure, V2I), according to Cooperative-Intelligent Transportation Systems (C-ITS), as illustrated in Fig. 4. This broadens each vehicle's awareness of its surroundings and has fostered the creation of new productivity, safety, and entertainment applications, which are the foundation of the future C-ITS ecosystem (Javed and Hamida 2017). But these advances have been focussed on traditional cars, overlooking personal mobility options that are changing urban settings. Many people use bicycles, scooters, and electric motorcycles since they're eco-friendly and healthful. Lei et al. (2017) suggest that, despite recent advancements, VCSs still face vulnerabilities and concerns related to security and key

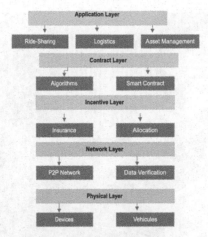

**Fig. 3.** An ITS-Oriented blockchain model (adapted)

management. As a result, the authors suggest a blockchain-based key-management strategy for securely sharing keys for communication across multiple devices or vehicles. Network members may use the blockchain-based technique to collect the trust values of other network participants and, based on this, assess the trustworthiness of received messages.

### Finding-4: Carsharing

Consider that you and your colleague are neighbours with automobiles posted on a carsharing platform. He is out of town, and your children need a vehicle. Instead of driving your vehicle home from the workplace, you might reserve his vehicle at his residence. While the underlying technology is unclear, well-known entrepreneur Elon Musk has discussed developing a comparable idea, which he calls the "Uber-killer." Nowadays cab services are accessed using a smartphone app. Due to government anti-pollution programs, increasing taxi charges, and the decreased demand for personal automobiles owning to remote work, the worldwide car sharing industry is expected to grow. Europe carsharing market size is expected to cross $4 billion valuation by 2024 (Graphical research 2021). However, car sharing may encounter trust issues between customers and car owners along with some security concerns.

The confidence that individuals place in one another as well as the faith that individuals place in the platform itself are two aspects of trust that are of special significance in the context of the sharing economy. First, having faith in one's contemporaries involves having faith in one's own customers and producers (Hawlitschek et al. 2018). Therefore, in the context of mobility, what this is referring to is the trust that exists between the riders and the drivers. Hawlitschek et al. (2018) went on to suggest that there is a connection between trust in the platform and trust in the platform for the sharing economy. To the best of our knowledge however, trust in the context of ridesharing based on blockchain technology has not yet been investigated at this point in time. Because faith in the platform as well as trust in one's fellow users is a vital component of the sharing

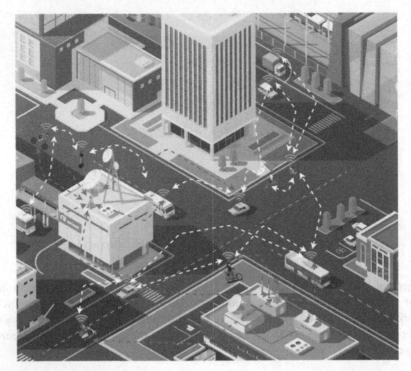

**Fig. 4.** Urban fully-connected vehicular scenario.: V2V/V2I links (Javed and Hamida 2017).

economy, it seems that this is one of the most significant research gaps that Blockchain technology might assist in addressing.

**Finding-5: Smart Contracts Connecting Stakeholders**

To ensure that clients are not overcharged and car owners are always compensated for rentals, (Kotik and Serhii (2022) developed a reliable blockchain model for car sharing (see Fig. 5). In this basic smart contract example, they demonstrated that payment may be flexible: the user simply pays for the time spent using the automobile, and income is sent straight to the car's balance. Car owners are always rewarded when their vehicles are utilized since the customer's deposit payment is locked in before the rental begins. On the event of an accident, the car owner will know who was driving since information saved in the blockchain cannot be modified and may be shared with other parties such as insurance companies. According to Kim et al. (2021) people may now simply utilize a shared automobile by performing basic operations on their mobile devices thanks to the growth of the Internet of Things. The car-sharing program, however, has security issues. Sensitive data is communicated for car-sharing on a public channel, including the user's identity, location data, and access code (Kim et al. 2021).The suggested car-sharing authentication approach proposed by Kim et al. (2021) uses blockchain with five entities: trust authority, stations, owner, vehicle, and user. A trust authority sets up the system and gives user credentials and pseudo-identities. Stations arrange the consortium blockchain with data storage and compute. The user requests car-sharing via the station. After authentication, the user gets the vehicle access code.

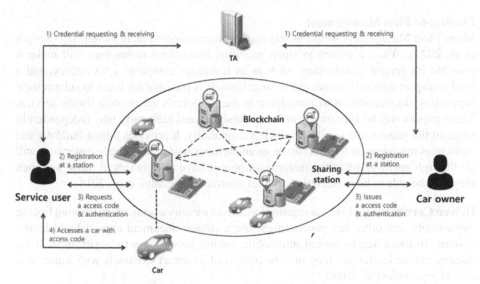

**Fig. 5.** The proposed system model by (Kim et al. 2021)

- **Trust Authority (TA)** - A trust authority is responsible for the initial setup of the system as well as the distribution of user credentials and pseudo-identities. Data storage and processing are handled by the stations that make up the consortium blockchain. The station receives a request from the user for carsharing. The user is given the vehicle access code after they have been validated.
- **Stations** - Stations arbitrate between car-sharing users and vehicle owners. The station gets the user's and owner's car-sharing credentials. The station validates and keeps credentials on blockchain. When a user requests car-sharing, the station authenticates them using blockchain data. It offers car-sharing by communicating owner information. The station saves service information in blockchain, which the trusted authority may utilize to arbitrate disputes.
- **User** - smartphone users may access the car-sharing service. User sends request and authentication messages to station to verify driving authorization. The station authenticates users using blockchain data. After authentication and acquiring the vehicle access code, the user may use their mobile device.
- **Owners** - By registering their car at the station, the owner turns it into a shared vehicle. When the station delivers a user's request to share a car, the owner produces an access code and sends it to the station to distribute to the user and vehicle.
- **Cars** - At the station, authorized users may share parked cars. Automotive modules include communication and tamper-proofing.

The vehicle's communication modules send it an access code to verify whether the user is permitted. Vehicle parameters are kept in a tamper-proof module for confidentiality.

**Finding-6: Fleet Management**
Many Fleet Management solutions rely on real-time remote vehicle monitoring (Karger et al. 2021). When it comes to smart mobility, blockchain technology will make it possible for several stakeholders, such as an insurance company, a tax agency, and a road transport authority, to act at the same time. It is possible for users to submit their supporting documentation to participate in the platform's automobile listing service. These papers may be kept on a distributed database, and each party may independently validate the request and certify a vehicle's availability. It provides a trust-building and consensus management mechanism on an anonymous basis. The vehicle paperwork will go through approval when the majority of parties have acknowledged them, and then they will be able to be preserved as a smart contract (Orecchini, et al. 2018).

**Driver Certification.** Like car registration, driver identity management, driving license authenticity, and other key papers might use a similar document approval and listing system. To take a nearby parked automobile, submit your license for verification. If the documents are legitimate, they may be preserved as Smart Contracts with immutable data (Orecchini, et al. 2018).

**Security.** Decentralized databases are safe and unchangeable. Imagine a motorist taking someone else's automobile. Every automobile owner whose car has been rented keeps the same database. Every transaction is confirmed and acknowledged (Orecchini, et al. 2018).

# 5   Conclusion

In terms of transportation, the previous several years have seen some progress towards future smart cities. Population growth affects urban mobility since cities were built decades or centuries ago with a much lower population and distinct infrastructural demands. The evolution towards cities and smart mobility will not involve making more space for vehicles or more roads; rather, it will undergo a cultural paradigm shift in which people will stop using their private cars and switch to shared transport, which may reduce the number of vehicles in circulation and the ecological footprint. In addition to the technology challenge, we also have a very obvious cultural shift since using our own automobiles provides comfort and convenience that won't be simple to give up. Citizens must comprehend the severe environmental repercussions of preserving the existing situation and the difficulties of managing so many automobiles in high-density cities. We will all benefit from this cultural and paradigm change.

People need inexpensive, accessible, healthier, and cleaner alternatives to their current mobility practices for sustainable transportation. Digitalization will boost automated and networked multimodal transportation and smart traffic management technology (Williams et al. 2012). Transport prices should reflect environmental and health impacts. For the completion of this paper, the author conducted a systematic literature review on blockchain's usage in smart mobility and transportation. 15 selected papers were examined to address the research question, and the use of blockchain technology in the ecomobility and smart mobility was identified. An analysis of the existing research literature indicates that there are five major primary categories of blockchain applications that may be used for the benefit of smart mobility, and these categories are as follows: Sustainable Mobility System (SMS), Intelligent Transportation Systems (ITS), Vehicular Communication System(VCSS), Carsharing, Fleet Management.

# 6 Discussion

As mentioned in this paper, Blockchain will be the next revolutionary smart transportation technology in coming years. However the privacy of the data related to their brief trips and the overall GDPR compliance of the Smart Mobility Solution will be the main concerns of the public. The stability, performance, and usability of the side chain's elements, such as payments and real-time location data services, are still ambiguous at this point. A few use-cases has already been addressed by researchers.

The environmental benefits of the proposed eco-mobility networks, such as energy efficiency and climate change mitigation and adaptation, are important, but community-based participatory urban renewal and landscape planning that involves local residents (commuters and users) in all decision-making is more crucial. Despite the initial work in this area, the studies are still fragmented and sometimes barely sufficiently integrated. This review aimed to closing this gap and providing an overview of the current state of the art in blockchain for ecomobility and smart mobility.

Despite the fact that the research addressed the primary issue, there are certain constraints to consider. To begin, the papers examined were restricted to those found in JSTOR, EBSCOhost, IEEE Xplore Digital Library, ScienceDirect, and SpringerLink. Second, despite the fact that this study utilized a wide list of keywords in major research databases, there is a risk of losing some essential work due to the dispersion in function of the keywords used in this area. Furthermore, the scalability and continuously evolving aspect of the blockchain technology remain significant issues, and there is a lack of work (for example, Framework) that combines the improvements offered by all research efforts. For instance it is expected that the number of nodes will expand fast, causing blockchain scalability concerns, and smart cars' bandwidth will be too low to handle the broadcasting overhead. Another issue could be that the algorithms sustaining the blockchain need more processing power than smart devices, thus there would be a large gap when using it in automotive businesses. Also smart contract adaptability may be another issue as many industries find it challenging to keep smart contract structure and standards because it contains so many hidden aspects that cannot be seen by rivals. An adaptive smart contract framework would be extremely beneficial. Furthermore, the application of blockchain in ecomobility and smart mobility, including testing, is still in its infancy. All of these results hint to a research gap and unanswered research problems that must be addressed further. As a result, there is a need for a more complete conceptual model that integrates, synthesizes, and orchestrates all key research works in an effective manner to aid people and/or organizations in the widespread adoption of blockchain.

# References

Antonopoulos, A.M., Wood, G.: Mastering Ethereum: Building Smart Contracts and DApps. O'reilly Media, Sebastopol (2018)

Aujla, G.S., Singh, A., Singh, M., Sharma, S., Kumar, N., Choo, K.K.R.: BloCkEd: blockchain-based secure data processing framework in edge envisioned V2X environment. IEEE Trans. Veh. Technol. 69(6), 5850–5863 (2020)

Batty, M., et al.: Smart cities of the future. Eur. Phys. J. Special Topics 214(1), 481–518 (2012). https://doi.org/10.1140/epjst/e2012-01703-3

Beaume, R., Midler, C.: From technology competition to reinventing individual ecomobility: new design strategies for electric vehicles. Int. J. Automot. Technol. Manage. 9(2), 174 (2009)

Baza, M., Lasla, N., Mahmoud, M.M., Srivastava, G., Abdallah, M.: B-ride: ride sharing with privacy-preservation, trust and fair payment atop public blockchain. IEEE Trans. Netw. Sci. Eng. 8(2), 1214–1229 (2019)

CB Insights Banking is only the beginning; 58 big industries blockchain could transform. CB Insights (2021). https://www.cbinsights.com/research/industries-disrupted-blockchain/. Accessed 15 June 2021, Return to ref 6 in article

Chang, S.E., Chang, C.Y.: Application of blockchain technology to smart city service: a case of ridesharing. In: 2018 IEEE International Conference on Internet of Things (iThings) and IEEE Green Computing and Communications (GreenCom) and IEEE Cyber, Physical and Social Computing (CPSCom) and IEEE Smart Data (SmartData), pp. 664–671. IEEE (2018)

Chen, C., Xiao, T., Qiu, T., Lv, N., Pei, Q.: Smart-contract-based economical platooning in blockchain-enabled urban internet of vehicles. IEEE Trans. Industr. Inf. 16(6), 4122–4133 (2019)

Dameri, R.P., Ricciardi, F.: Leveraging smart city projects for benefitting citizens: the role of ICTs. In: Rassia, S.T., Pardalos, P.M. (eds.) Smart city networks. SOIA, vol. 125, pp. 111–128. Springer, Cham (2017). https://doi.org/10.1007/978-3-319-61313-0_7

Deloitte Key characteristics of blockchain. Deloitte (2017). https://www2.deloitte.com/content/dam/Deloitte/in/Documents/industries/in-convergence-blockchain-key-characteristics-noexp.pdf. Accessed 15 June 2021

Dias, E., Linde, M., Rafiee, A., Koomen, E., Scholten, H.: Beauty and brains: integrating easy spatial design and advanced urban sustainability models Lecture Notes in Geoinformation and Cartography, pp. 469–484 (2013).https://doi.org/10.1007/978-3-642-37533-0_27

Dondjio, I., Themistocleous, M. (2022). Blockchain technology and waste management: a systematic literature review. In: Themistocleous, M., Papadaki, M. (eds.) Information Systems. EMCIS 2021. Lecture Notes in Business Information Processing, vol. 437, pp. 194–212. Springer, Cham. https://doi.org/10.1007/978-3-030-95947-0_14

Dungan, L., Pop, M.D.: Blockchain-based solutions for smart mobility sustainability assurance. In: IOP Conference Series: Materials Science and Engineering, vol. 1220, no. 1, p. 012057 (2022). https://doi.org/10.1088/1757-899x/1220/1/012057

Giffinger, R., Fertner, C., Kramar, H., Meijers, E.: City-ranking of European medium-sized cities. Cent. Reg. Sci. Vienna UT 9(1), 1–12 (2007)

Graphical Research: Car Sharing Market Trends 2021 - Regional Statistics and Forecasts 2024 | Europe, North America & APAC: Graphical Research. GlobeNewswire News Room (2021). https://www.globenewswire.com/news-release/2021/02/03/2168780/0/en/Car-Sharing-Market-Trends-2021-Regional-Statistics-and-Forecasts-2024-Europe-North-America-APAC-Graphical-Research.html. Accessed 5 Oct 2022

Hawlitschek, F., Notheisen, B., Teubner, T.: The limits of trust-free systems: a literature review on blockchain technology and trust in the sharing economy. Electron. Commer. Res. Appl. 29, 50–63 (2018). https://doi.org/10.1016/j.elerap.2018.03.005

Huang, X., Xu, C., Wang, P., Liu, H.: LNSC: a security model for electric vehicle and charging pile management based on blockchain ecosystem. IEEE Access **6**, 13565–13574 (2018)

Javed, M.A., Hamida, E.B.: On the interrelation of security, QoS, and safety in cooperative ITS. IEEE Trans. Intell. Transp. Syst. **18**(7), 1943–1957 (2017). https://doi.org/10.1109/tits.2016. 2614580

Karger, E., Jagals, M., Ahlemann, F.: Blockchain for smart mobility—literature review and future research agenda. Sustainability **13**(23), 13268 (2021). https://doi.org/10.3390/su132313268

Kim, M., Lee, J., Park, K., Park, Y., Park, K.H., Park, Y.: Design of secure decentralized car-sharing system using blockchain. IEEE Access **9**, 54796–54810 (2021). https://doi.org/10.1109/access. 2021.3071499

Kim, S.: Impacts of mobility on performance of blockchain in VANET. IEEE Access **7**, 68646–68655 (2019)

Kotik, A., Serhii, R.: How to implement a blockchain in a car sharing service using the cosmos network. Apriorit (2022). https://www.apriorit.com/dev-blog/733-blockchain-implement-blo ckchain-in-car-sharing-service-using-the-cosmos-network. Accessed 6 Oct 2022.

Lopez, D., Farooq, B.: A blockchain framework for smart mobility. In: 2018 IEEE International Smart Cities Conference (ISC2), pp. 1–7. IEEE (2018)

Lei, A., Cruickshank, H., Cao, Y., Asuquo, P., Ogah, C.P.A., Sun, Z.: Blockchain-based dynamic key management for heterogeneous intelligent transportation systems. IEEE Internet Things J. **4**(6), 1832–1843 (2017). https://doi.org/10.1109/jiot.2017.2740569

Litman, T.: Evaluating transportation equity. Victoria Transport Policy Institute (2017)

Morris, E.A.: Should we all just stay home? Travel, out-of-home activities, and life satisfaction. Transp. Res. Part A: Policy Pract. **78**, 519–536 (2015). https://doi.org/10.1016/j.tra.2015. 06.009

Nguyen, T.H., Partala, J., Pirttikangas, S.: Blockchain-based mobility-as-a-service. In: 2019 28th International Conference on Computer Communication and Networks (ICCCN), pp. 1–6. IEEE (2019)

Orecchini, F., Santiangeli, A., Zuccari, F., Pieroni, A., Suppa, T.: Blockchain technology in smart city: a new opportunity for smart environment and smart mobility. In: Vasant, P., Zelinka, I., Weber, G.-W. (eds.) ICO 2018. AISC, vol. 866, pp. 346–354. Springer, Cham (2019). https:// doi.org/10.1007/978-3-030-00979-3_36

Portillo, G.: What is and how important is ecomobility. Renovables Verdes (2018b). https://www. renovablesverdes.com/en/ecomovilidad/. Accessed 28 Sept 2022

Sayeed, S., Marco-Gisbert, H., Caira, T.: Smart contract: attacks and protections. IEEE Access **8**, 24416–24427 (2020). https://doi.org/10.1109/access.2020.2970495

World Health Organization (WHO) (2019). https://www.who.int/health-topics/air-pollution

United Nations. Urban Environment Related Mitigation Benefits and Co-benefits of Policies, Practices and Actions for Enhancing Mitigation Ambition and Options for Supporting Their Implementation. https://unfccc.int/sites/default/files/resource/docs/2017/tp/02.pdf. Accessed 1 Dec 2020

Williams, J.H., et al.: The technology path to deep greenhouse gas emissions cuts by 2050: the pivotal role of electricity. Science **335**(6064), 53–59 (2012). https://doi.org/10.1126/science. 1208365

Teoh, B.P.C.: Navigating the blockchain trilemma: a supply chain dilemma. In: Ismail, A., Dahalan, W.M., Öchsner, A. (eds.) Advanced Maritime Technologies and Applications. ASM, vol. 166, pp. 291–300. Springer, Cham (2022). https://doi.org/10.1007/978-3-030-89992-9_25

Woodcock, J., Banister, D., Edwards, P., Prentice, A.M., Roberts, I.: Energy and transport. Lancet **370**(9592), 1078–1088 (2007). https://doi.org/10.1016/s0140-6736(07)61254-9

Ying, Z., Yi, L., Ma, M.: BEHT: blockchain-based efficient highway toll paradigm for opportunistic autonomous vehicle platoon. Wireless Commun. Mob. Comput. **2020**, 1–13 (2020)

Yuan, Y., Wang, F.Y.: Towards blockchain-based intelligent transportation systems. In: 2016 IEEE 19th International Conference on Intelligent Transportation Systems (ITSC) (2016). https://doi.org/10.1109/itsc.2016.7795984

# Blockchain Governance – A Systematic Literature Review

Johannes Werner[✉]

Chair of Information and Communication Management, Technische Universität Berlin, Berlin, Germany
johannes.werner@tu-berlin.de

**Abstract.** The blockchain technology offers the possibility to decentralize platforms besides a variety of other possibilities. This may even result in the absence of a central authority that defines its governance. Consequently, a special kind of governance can emerge that takes into account the specifics of this technology. While governance in various research areas, such as corporate or IT governance, has been studied for a long time, the study of blockchain governance has only recently begun. Nevertheless, research on blockchain governance is increasing in recent years and an appropriate way to structure it is needed. Hence, the goal of this study is to identify the current state of research on blockchain governance. For this purpose, a systematic literature review is conducted and as a result, 25 studies were identified which address blockchain governance. Moreover, an adapted version of the IT governance cube is introduced, which allowed to structure the concepts of blockchain governance. Furthermore, it allowed to identify research areas that have obtained little attention yet.

**Keywords:** Blockchain · Governance · Literature Review

## 1 Introduction

The blockchain technology has received much attention from practitioners and researchers since its first practical implementation by the cryptocurrency Bitcoin [1]. Besides technological features this technology promises decentralization, data integrity, transparency, auditability, and automation as key features [2]. Thereby, especially the possibility of increasing efficiency by avoiding a middleman is a reason for the paid attention [3]. In the case of platforms, this leads to the possibility that a platform operator is no longer needed. While the number of use cases of blockchain technology seems to be unlimited [4], the adoption of this technology in industries still faces many challenges [5]. Additionally to the practical application of the blockchain technology it continues to gain attention from research and the number of related publications is increasing [6]. However, in information systems there is still little research on blockchain [7] and, in particular, research on the governance of blockchain [3, 8]. Due to the technological enabled features of blockchain, governance mechanisms can directly be implemented in the technology [9]. In addition, the need for a central authority, which defines the governance can be avoided [10]. Nevertheless, since this was determined some time has passed and thus, it can be assumed the research on blockchain governance has increased.

Thus, this paper covers the current state of the research on blockchain governance, and addresses the following research questions:

*What is the current state of research on blockchain governance?*
and
*How can the current state of research on blockchain governance be structured?*

The remainder of this paper is structured as followed: First, the used methodology of a systematic literature review is described and then, the results are outlined. Afterwards, in the discussion research gaps are identified and a conclusion is given.

## 2 Methodology

For the structured literature review the guidelines of Kitchenham [11] and Tranfield et al. [12] were applied. According to these guidelines, first of all, the need for a review was identified and a review protocol was developed, which includes the review questions, the literature search and the selection process as well as inclusion and exclusion criteria. In the remainder of this section, the procedure of conducting the review is described, more specifically the process of literature search and selection as well as the data extraction.

### 2.1 Literature Search

For the search of literature the recommendations of vom Brocke et al. [13] were applied. Thus, the search scope was defined based on the research question. In particular, a sequential process was applied, which should cover most of the relevant literature and thus should be comprehensive. As sources the Association for Information Systems eLibrary and the Web of Science were selected. The first source covers the leading conferences and journals in information systems research and the second one indexes multiple other bibliographic databases such as IEEE Xplore for computer science. For the literature search a keyword search was applied on title, abstract and keywords. The following search string was used:

*(Blockchain OR "Distributed Ledger" OR DLT) AND (governance OR govern\* OR "Decision rights")*

In the first part of this string, the term blockchain, distributed ledger and its abbreviation was used. The second part covers the search term governance. Additionally, a term using an asterisk (govern\*) was included to cover also articles using verbs, like governing, etc. Moreover, "decision rights" was added to cover also articles that do not use the term governance explicitly. The latter term was used because it is one of the most important dimensions of IT governance [14].

A pre-test was conducted to check the suitability of the search string. Thereby, the search was conducted and checked, whether the results contain two known articles in each database. This check was passed, so no further changes were made to the search string. The final literature search was conducted in May/June 2022.

## 2.2 Study Selection

For the selection of studies, the inclusion and exclusion criteria defined in the review protocol were applied. The first in- and exclusion criteria was the focus of the article. It was included, if it focuses on governance of blockchain regardless the concrete field of investigation. On the other hand, it was excluded, if it investigates the blockchain for governance [15] or on the application of blockchain for corporate governance or similar. No methodological restriction was made, so that empirical as well as conceptual articles or reviews were taken into account. In addition, no restrictions in terms of the research outcome were made, so case studies as well as frameworks were taken into account. As a further exclusion criteria the publication language was used and articles in other languages than English were excluded. The publication outlet was used as exclusion criteria and for the assessment of the quality. Thus, not peer-reviewed publications were excluded, which includes publications that are not published in journals or conference proceedings, as well as books or book chapters. An additional exclusion criteria was the date of publication, so only publications since 2008 were considered, which was the publication date of Bitcoin as the first practical implementation of blockchain technology [1].

The applied study selection process, which is outlined in Fig. 1, results in a final dataset of 25 publications. The selection process started with 1788 studies, which were found using the search string in the sources. First, in- and exclusion criteria were applied,

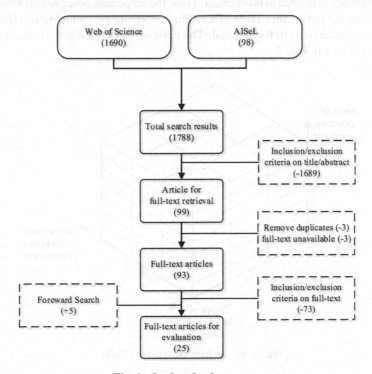

**Fig. 1.** Study selection process

e.g., publications in other languages than English were excluded as well as studies published as pre-prints. In the same step, the other in- and exclusion criteria based on the content were applied on title and abstract of the publications. This results in a set of articles for which full texts were retrieved. In this step, duplicates were also removed. Afterwards, the in- and exclusion criteria were applied on the full-text. In addition, a forward search was conducted to identify literature, which was not found by the initial keyword search [13]. As the field of study is very recent, no seminal papers were assumed to be found and thus, a backward search would only offer little added value [13].

## 2.3   Data Extraction

For the data extraction a concept matrix was created while reading the articles, according to Webster & Watson [16]. First, the IT Governance Cube [17] was used as a multidimensional method to synthesize the reviewed literature and to identify research gaps [16, 18]. In the IT Governance Cube [17] governance mechanisms are represented by the dimension *How is it governed?* (e.g., decision rights, architecture, etc.). However, two additional dimensions are necessary to describe this dimension. For instance, the dimension *What is governed?* (e.g., IT-Artifact, stakeholders) refers to the focus of the mechanisms and *Who is governed?* (e.g., ecosystem, firm) refers to its scope, i.e. the unit of analysis.

In this paper, an adapted form of the IT Governance Cube is used because the scope of the governance is defined as blockchain. Thus, the scope now describes on which layer of blockchain the governance takes place. The remaining two dimensions (*How?* and *Who?*) are considered as in the original. The cube of blockchain governance resulting from this is shown in Fig. 2.

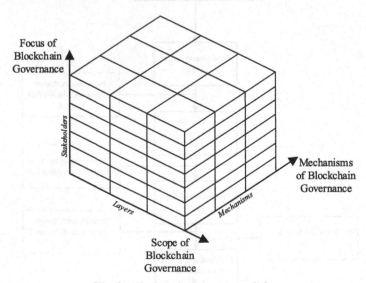

**Fig. 2.** Blockchain Governance Cube

# 3   Results

## 3.1   Focus of Blockchain Governance

For the dimension of focus of blockchain governance, its stakeholders are used as concepts. Thereby, the number of stakeholders involved depends on the specific use case. For example, four groups of stakeholders can be identified for permissionless blockchains [19], eight stakeholders for the Bitcoin blockchain [20] and twelve stakeholders for governance in the blockchain ecosystem [21]. Another study includes also stakeholders from the environment, like governments [22]. Since the reviewed literature refers to different scenarios of blockchain technology, not all stakeholders necessarily appear in it. For example, exchanges are a specific example of a stakeholder in the field of cryptocurrencies or miners in blockchains using Proof-of-Work.

In order to obtain a generic overview about the stakeholders in blockchain governance, a summarized form of the stakeholders for governance in the blockchain ecosystem [21] were used, namely: *nodes*, *developers*, *communities*, *users*, *token holders*, and *organizations*. In this process, not all possible ones were considered, some of them were aggregated, and one was discarded. As *nodes* all kind of nodes (miner nodes, full nodes and masternodes) were grouped together and as *organizations* foundations and consortia or federations were seen. *Token holders* includes token and coin holders, stakers, delegates and arbiters. The stakeholders *developers*, *communities* and *users* were taken over one to one and "Projects and DApps as stakeholders" were discarded. Additional stakeholders like technical suppliers or special community members, e.g., curators, are listed as *others*.

Nevertheless, a clear distinction between the stakeholders is not always possible and depends also on the analyzed case. Thus, overlapping stakeholders can occur, which belong to multiple categories [22]. For example, developers and users are often part of the community or token holders are also users, which might be caused by a necessary ownership of tokens for using the platform. Another example can be a member of the core developer team, who also holds tokens and is active in the community.

An overview of the stakeholders in the focus of blockchain governance in the reviewed literature is shown in Table 1.

**Table 1.** Focus of blockchain governance.

| Stakeholder | Sources |
| --- | --- |
| Nodes | [3, 8–10, 19, 21, 23–35] |
| Developers | [3, 8–10, 19, 21–24, 26–38] |
| Communities | [3, 9, 10, 19, 21–28, 30–35, 37–39] |
| Users | [8–10, 19, 21–23, 25–30, 32–39] |
| Token holders | [3, 8, 21–23, 25–27, 31, 34–39] |
| Organisations | [3, 9, 19, 21, 23, 30, 31, 34–36, 40] |
| Others | [3, 9, 21–24, 26, 28, 34, 40] |

## 3.2 Scope of Blockchain Governance

The scope of blockchain governance describes on which layer the governance takes place. For the purpose of this study, two classifications are merged. One of these differentiates in on-chain and off-chain governance and the other one describes where exactly the governance takes place on-chain. Figure 3 gives an overview about the used scope of blockchain governance. Beside the scope it also shows some of the stakeholders, which occur in off-chain governance.

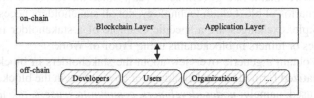

**Fig. 3.** Scope of Blockchain Governance

In this regard, on-chain governance describes rules and processes, which are directly implemented as code in a blockchain-based system, in which interactions between the participants take place. Hence, the technology is enforcing the governance. In contrast, off-chain governance refers to all other governance, which is not included in on-chain governance and which is not directly implemented as code. For instance, this includes governance of developers and communities belonging to the system. Furthermore, in contrast to on-chain governance, it is not automatically enforced by the system [31, 41]. The rules and processes of blockchain governance can be multi-layered, so that one layer has influence on another layer. As the case of "The DAO" and the resulting hard fork has shown, off-chain governance has an impact on on-chain governance [41]. With on-chain governance a further subdivision can be made on the basis of the blockchain stack. These include the blockchain layer, which consists of blockchain networks like Ethereum, and the application layer, which includes dApps and DAOs [7, 41]. There are other descriptions of the technological blockchain stake besides this one (see e.g., 42, 43), but due to their focus they are less suitable for the current study. The same applies to partitions of off-chain governance in community and development [3], which, however, does not take into account all of the previously identified stakeholders in the scope of blockchain governance.

In Table 2 the considered layers of the reviewed literature are shown. In most cases one of the two on-chain layer is combined with the off-chain layer, followed by a combination of both on-chain layers with the off-chain layer.

The inclusion of on- and off-chain layer often depends on the analyzed field, e.g., a voting process of a DAO is on the application layer on-chain due to its technological implementation, but also covers participants in the process, which were off-chain, e.g., users or token holders.

In addition to this distinction, the mutual influence of the two layers should be considered, as they are partly based on each other and influence each other. This can be described as reciprocal governance. In this concept blockchain governance includes governance by

**Table 2.** Scope of blockchain governance.

| Layer | Sources |
|---|---|
| On-chain blockchain layer | [3, 8–10, 19, 21–24, 26–29, 31, 33, 34, 40] |
| On-chain application layer | [3, 8, 10, 19, 21, 25–28, 30, 32, 34–39] |
| Off-chain | All |

blockchain, e.g., consensus, and governance of blockchain, e.g., decision making processes. The governance of blockchain influences the governance by blockchain through technological changes and vice versa by the application of the implemented governance [28]. An example for influences between off-chain and the blockchain layer is the specific governance mechanism of Ethereum Improvement Proposal (EIP) or Bitcoin Improvement Proposal (BIP). In this mechanism any community member can submit a change proposal for the underlying code of the blockchain [44]. Then, the community and the core developer team review the proposal and if necessary, improve it. At the final stage, the BIP and EIP differs. On EIP, the core developer team decides on including it in a new release of the protocol, based on their assessment. The same applies to BIPs, but with these mining nodes vote on the change by applying the update [24, 45]. Hence, the implementation of these changes on the on-chain layer has effects on the users, which can be located off-chain.

In case of the application layer, nearly the same process occurs in DAO, but it differs in terms of participants and in terms of its effects. A change proposal can be created by members of it and is discussed in the community. Then, a voting process by smart contracts in which token holders can participate takes place and afterwards, the proposed smart contracts will be deployed. In addition to voting on operational decisions, these may also include strategic decisions, such as system upgrades [38], which can lead to updates of the implemented rules [39]. Hence, the voting process for updates can also be updated, so the on-chain and off-chain layers affect each other [25].

## 3.3 Mechanisms of Blockchain Governance

The reviewed literature examines mechanisms of blockchain governance from various perspectives. At this point, the focus is on the governance mechanisms decision rights and incentives, as they are specific to blockchain governance and takes place on-chain and off-chain. Moreover, mechanisms of coordination and control are considered as blockchain specific mechanisms. Their coverage in the literature reviewed is shown in Table 3. Other governance mechanisms, which are also crucial for blockchain-based platforms, are for example accessibility or perceived trust [30, 34]. These are not considered here due to the reason they are not specific blockchain governance mechanisms and are mainly adapted from the research on platform governance.

The mechanisms of decision rights were structured based on the scope of blockchain governance, more specifically by the level on which the decisions takes place and on which level they have impact. Additionally, the stakeholders, who are involved into the decision process and have the power to decide, were taken into account.

**Table 3.** Mechanisms of blockchain governance.

| Mechanism | Sources |
|---|---|
| Decision rights | [3, 8, 10, 21, 23–26, 28, 33, 36, 38, 39, 44, 45] |
| Incentives | [3, 8, 19, 26, 30, 32, 34–38] |
| Coordination & control | [32, 35, 40] |

The consensus mechanisms are a special form of decision rights in blockchains. In this procedure mining nodes decide which copy of the transactional database is used by consensus between the participating nodes [10]. In addition, nodes do also have the final decision rights about adoption of software changes and forking the system [21]. Thus, the direct decision takes place on-chain and has impact on this level. The stakeholders directly involved in this process are nodes. In case of Proof-of-Stake, the same holds true but instead of nodes the token holders are involved [23]. Examples of changes to the software can be the BIP or the EIP described above. Another mechanism taking place on the same layer is forking, by which conflicts about decisions can be solved [8, 21]. More precisely, this mechanism can occur, if existing other procedures cannot lead to a decision, e.g., in the case of the DAO attack [31].

In case of the adoption of software changes or forking, the decision itself occurs on-chain, but the process thereof is located off-chain. For off-chain decision process the influence of individual community members on the community was shown exemplarity by BIPs. Thereby, single influencers, which were part of the community, use different tactics to enact their desired rules, which are to be technological implemented [33].

A special form of decision rights is voting, which occurs using smart contracts on the application layer on-chain. Although the decision on software changes or the state can also be seen as a kind of voting process [23, 24], the term voting for voting processes based on smart contract. In case of DAOs, holders of governance tokens are able to decide on proposals for the platform to change its processes [39]. In some cases it is also possible that users without governance tokens can submit proposals [21]. The most common voting system is based on the amount of owned tokens [25, 38], but also other voting models exist, like one vote per individual [21] or liquid democracy [25]. Regardless of the low participation in voting within DAOs [39], strategic voting tasks can improve its operational performance, whereas operational voting tasks can worsen it [38]. Moreover, by the transfer of decision rights from the former platform owners to token holders decentralization can be archived [36].

The next mechanism of blockchain governance described in the literature are incentives. It describes, who is awarded for what and how. The awarded stakeholders include nodes, developers, users and token holders [8]. Thus, the mechanism of incentives can occur on all layers of blockchain governance. According to their type incentives can be divided into monetary incentives, like mining fees for consensus, or non-monetary, like reputation [8, 35]. Whereas on the blockchain layer mining fees or similar monetary incentives occur, on the application layer both types of incentives exist [35]. Moreover, participation in decision rights can also be seen as an incentive [35, 38], as well as participation in the further development of a platform [8] or its usage [8, 36]. Mainly through this mechanism stakeholders should act in a desired behavior [8]. For example it can be used to motivate them to participate, to facilitate a specific output [32], or to act honestly [37]. Some of the incentive mechanisms like the consensus on a blockchain as the technical foundation of a blockchain are necessary and thus can be seen as crucial [8]. Besides the positive mechanism of incentives, which are awarded for desirable behavior, sanctions can be seen disincentives, which punish undesired behavior. By this, nodes can be suspended from the mining process for malicious behavior [21, 35] or token holders can be punished by burning their tokens [34].

The governance mechanisms of coordination and control describe who is how coordinated or controlled and by whom. On the technical layer, nodes are self-coordinated and -controlled by the used technical protocol. In case of an underlying blockchain technology, like Ethereum, changes in this mechanism depend on it. If an own blockchain is used, proposals can be used to alter the coordination mechanisms. Experienced users, developers or the organization evaluate these and decide on them. The developers themselves are in most cases coordinated and controlled by an organization, e.g., by off-chain contracts. Although an blockchain-based form may be desired, it is not yet implemented [35]. Beside these coordination and control mechanisms, also incentives, sanctions and decision rights, which were mentioned above, can serve as mechanisms for it. Moreover, smart contracts can serve as control mechanisms, because rules are implemented and enforced automatically [32]. The mechanisms of coordination and control were also analyzed in the field of enterprise blockchains. Thereby, the focus is on intra-organizational coordination and control. For control mechanisms, the allocation of decision rights is a main issue between organizations [40].

## 4 Discussion

The objective of this study is to review the current state of research of blockchain governance and to propose a way of structuring it. Therefore, a structured literature review was conducted, by which 25 studies on blockchain governance have been identified for analysis. For structuring the research on blockchain governance, an adapted form of the IT governance cube [17] was used, which offers the possibility to synthesize existing research and to identify research gaps [16, 18]. In the following the results of the literature review along the presented dimensions of blockchain governance are being discussed.

In the dimension of focus of blockchain governance several classifications for stakeholders of blockchain governance have been identified [19–22]. They differ in their range, e.g., they include environmental stakeholders, as well as in their levels of detail, e.g., they distinguish between types of nodes. To structure the identified literature, a smaller scope without environmental stakeholders was used. Furthermore, an adjusted level of detail was used to be as general as possible and, at the same time, as detailed as possible. This made it possible to obtain an overview of the considered stakeholders in the literature. Nevertheless, the classification of stakeholders is not mutually exclusive, since one actor can take on different stakeholder roles, e.g., token holder and developer. It can be seen that most of the literature investigate stakeholders involved in the governance mechanisms, like token holders, developers, communities and nodes. In contrast, less attention paid to stakeholders, like users, who may be only affected by blockchain governance. Consequently, a lack of deep understanding of the stakeholders can be seen [22].

In contrast to the original IT governance cube, the scope of blockchain governance describes on which layer the governance is located. For this purpose, first the classification of on- and off-chain governance [31, 41] was used. Further classifications of off-chain governance, like community and developers [3], were not used. They do not cover all stakeholders from the focus of blockchain governance that can be located off-chain, like users or organizations. To take into account the different concepts of blockchain, which are used in the literature, the on-chain layer was divided into application and blockchain. By using this classification it was possible to determine that the majority of the reviewed literature examined one of the two on-chain layers in combination with the off-chain layer. Only a few studies cover both of the on-chain layers. Moreover, the mutual influence of on- and off-chain layers could be presented by this classification.

The mechanisms of blockchain governance were narrowed down to mechanisms, which are blockchain specific and are located on-chain at least partly. These were decision rights and incentives as well as coordination and control mechanisms.

For the decision rights the processes of finally making the decision takes place on-chain, regardless on which layer exactly. Thus, stakeholders like nodes or token holders (e.g., [23, 24]) are involved. Nevertheless, other stakeholders are involved in the upstream decision-making process, which can takes place on the off-chain layer. Examples for this are the influence of individuals on the decision-making process [33] or in the decision-making process by developers [35].

In the case of incentives, it was possible to distinguish between who receives which incentives from whom and what type of incentives these are. It was possible to observe that incentives can occur on all layers of blockchain governance and that sometimes they have only an indirect relation to the on-chain layer. The participation in the usage of a blockchain-based platform as an incentive is an example for this [8, 36]. In addition to incentives, which award a desired behavior it is also possible to punish undesired behaviors by disincentives, so called sanctions. However, these ones are only marginally considered in terms of focus and scope.

The governance mechanisms of control is closely related to decision rights and incentives. Both can be used as control mechanisms, to control the behavior of stakeholders by motivating them to behave in a desired way. The emerging coordination mechanisms can be differentiated by the on- and off-chain. Thereby, coordination mechanisms affecting stakeholders like nodes are implemented in the code, are self-executing and are located on-chain. In contrast, the coordination of other stakeholders, like developers or communities, are not implemented in the code and are thus located off-chain.

## 5   Conclusion and Outlook

This study provides an overview of the current state of research on blockchain governance. For this purpose, a structured literature review was conducted. As a result of the literature search and study selection 25 studies have been identified that address blockchain governance. An adapted form of the IT governance cube with the dimensions scope, focus and mechanisms was used to structure the identified literature. This allows to provide an overview of the identified literature and to categorize its concepts. Future research can use this as a tool to categorize research on blockchain governance and thus as a structure to identify research gaps. Moreover, in this study it was possible to identify research areas that had received little attention so far and thus should be investigated more closely by future research. For the focus of blockchain governance these are stakeholders, who do not directly participate in it, but are only affected by it. For example, a user may be affected by governance decisions of voting processes, in which he cannot participate. Moreover, the appearance of stakeholders in multiple roles has been underrepresented so far, e.g., developers who also hold tokens. For the scope of blockchain governance the mutual influence between on- and off-chain layers should be studied in more detail, in particular the influence of off-chain decisions on the on-chain decision-making process and vice versa. In the field of governance mechanisms, it was identified that the upstream decision-making process should be examined more closely, which is related to the mutual influence between layers. Moreover, the focus and scope of disincentives or sanctions should be examined. To overcome these shortcomings, future research is suggested to include qualitative methods such as case studies, which will allow investigating them in detail.

A natural limitation of this study finds its cause in the used search term and in the used databases by which the search result was limited. In addition, subjective interpretations may have affected the selection of literature and the assignment of concepts. In addition to the methodical limitations, narrowing down the governance mechanisms to blockchain specific ones is a limitation on its own, which necessities further research. To overcome this limitation and to extend this study also other governance mechanisms should be taken into account, e.g., mechanisms of platform governance.

# References

1. Nakamoto, S.: Bitcoin: a peer-to-peer electronic cash system (2008)
2. Fridgen, G., Radszuwill, S., Urbach, N., Utz, L.: Cross-organizational workflow management using blockchain technology-towards applicability, auditability, and automation. In: Proceedings of the 51st Hawaii International Conference on System Sciences (2018)
3. van Pelt, R., Jansen, S., Baars, D., Overbeek, S.: Defining blockchain governance: a framework for analysis and comparison. Inf. Syst. Manage. **38**, 21–41 (2021). https://doi.org/10.1080/10580530.2020.1720046
4. Swan, M.: Blockchain: Blueprint for a New Economy. O'Reilly Media, Inc. (2015)
5. Al-Jaroodi, J., Mohamed, N.: Blockchain in industries: a survey. IEEE Access **7**, 36500–36515 (2019). https://doi.org/10.1109/ACCESS.2019.2903554
6. Yli-Huumo, J., Ko, D., Choi, S., Park, S., Smolander, K.: Where is current research on blockchain technology? A systematic review. PLoS One **11**, e0163477 (2016). https://doi.org/10.1371/journal.pone.0163477
7. Rossi, M., Mueller-Bloch, C., Thatcher, J.B., Beck, R.: Blockchain research in information systems: current trends and an inclusive future research agenda. JAIS **20**, 1388–1403 (2019). https://doi.org/10.17705/1jais.00571
8. Beck, R., Muller-Bloch, C., King, J.L.: Governance in the blockchain economy: a framework and research agenda. J. Assoc. Inf. Syst. **19**, 1020–1034 (2018). https://doi.org/10.17705/1jais.00518

9. Rikken, O., Janssen, M., Kwee, Z.: Governance challenges of blockchain and decentralized autonomous organizations. Inf. Polity **24**, 397–417 (2019). https://doi.org/10.3233/IP-190154
10. Zachariadis, M., Hileman, G., Scott, S.V.: Governance and control in distributed ledgers: understanding the challenges facing blockchain technology in financial services. Inf. Org. **29**, 105–117 (2019). https://doi.org/10.1016/j.infoandorg.2019.03.001
11. Kitchenham, B.: Procedures for Performing Systematic Reviews. Keele University, Keele, UK (2004)
12. Tranfield, D., Denyer, D., Smart, P.: Towards a methodology for developing evidence-informed management knowledge by means of systematic review. Br. J. Manage. **14**, 207–222 (2003)
13. vom Brocke, J., Simons, A., Riemer, K., Niehaves, B., Plattfaut, R., Cleven, A.: Standing on the shoulders of giants: challenges and recommendations of literature search in information systems research. CAIS **37** (2015). https://doi.org/10.17705/1CAIS.03709
14. Weill, P.: Don't just lead, govern: how top-performing firms govern IT. MIS Q. Exec. **3**, 1–17 (2004)
15. Ølnes, S., Ubacht, J., Janssen, M.: Blockchain in government: benefits and implications of distributed ledger technology for information sharing. Gov. Inf. Q. **34**, 355–364 (2017). https://doi.org/10.1016/j.giq.2017.09.007
16. Webster, J., Watson, R.T.: Analyzing the past to prepare for the future: writing a literature review. MIS Q. **26**, 13–23 (2002)
17. Tiwana, A., Konsynski, B., Venkatraman, N.: Special issue: information technology and organizational governance: the IT governance cube. J. Manage. Inf. Syst. **30**(3), 7–12 (2013). https://doi.org/10.2753/MIS0742-1222300301
18. Brocke, J.V., Simons, A., Niehaves, B., Niehaves, B., Reimer, K., Plattfaut, R., Cleven, A.: Reconstructing the giant: on the importance of rigour in documenting the literature search process. (2009). ECIS 2009 Proceedings. 161. https://aisel.aisnet.org/ecis2009/161
19. Anthony, B.: Toward a collaborative governance model for distributed ledger technology adoption in organizations. Environ. Syst. Decis. **42**, 1–19 (2022). https://doi.org/10.1007/s10669-022-09852-4
20. Islam, N., Mäntymäki, M., Turunen, M.: Understanding the role of actor heterogeneity in blockchain splits: an actor-network perspective of bitcoin forks. In: Proceedings of the 52nd Hawaii International Conference on System Sciences (2019)
21. Honkanen, P., Nylund, M., Westerlund, M.: Organizational building blocks for blockchain governance: a survey of 241 blockchain white papers. Front. Blockchain **4** (2021). https://doi.org/10.3389/fbloc.2021.613115
22. Schmid, R., Ziolkowski, R., Schwabe, G.: Together or not? Exploring stakeholders in public and permissionless blockchains. In: Bui, T.X. (ed.) Proceedings of the 55th Annual Hawaii International Conference on System Sciences, pp. 6093–6102. Department of IT Management Shidler College of Business University of Hawaii at Manoa, Honolulu, HI (2022)
23. Allen, D.W.E., Berg, C.: Blockchain governance: what we can learn from the economics of corporate governance. J. Br. Blockchain Assoc. **3**, 46–52 (2020). https://doi.org/10.31585/jbba-3-1-(8)2020
24. Hsieh, Y.-Y., Vergne, J.-P.: Bitcoin and the rise of decentralized autonomous organizations. J. Org. Des. **7**(1), 1–16 (2018). https://doi.org/10.1186/s41469-018-0038-1
25. Kaal, W.A.: Decentralized corporate governance via blockchain technology. Ann. Corp. Govern. **5**, 101–147 (2020). https://doi.org/10.1561/109.00000025
26. Laatikainen, G., Li, M., Abrahamsson, P.: Blockchain governance: a dynamic view. In: Wang, X., Martini, A., Nguyen-Duc, A., Stray, V. (eds.) ICSOB 2021. LNBIP, vol. 434, pp. 66–80. Springer, Cham (2021). https://doi.org/10.1007/978-3-030-91983-2_6

27. Leewis, S., Smit, K., van Meerten, J.: An explorative dive into decision rights and governance of blockchain: a literature review and empirical study. Pacific Asia J. Assoc. Inf. Syst. **13**, 25–56 (2021). https://doi.org/10.17705/1pais.13302

28. Li, Y., Zhou, Y.: Research on the reciprocal mechanism of hybrid governance in blockchain. J. Econ. Manage. Res. 1–5 (2021). https://doi.org/10.47363/JESMR/2021(2)121

29. Miscione, G., Ziolkowski, R., Zavolokina, L., Schwabe, G.: Tribal Governance: The Business of Blockchain Authentication. In: Proceedings of the 51st Hawaii International Conference on System Sciences (2018)

30. Perscheid, G., Ostern, N.K., Moormann, J.: Determining Platform Governance: Framework for Classifying Governance Types (2020). ICIS 2020 Proceedings. 8. https://aisel.aisnet.org/icis2020/governance_is/governance_is/8

31. Reijers, W., et al.: Now the code runs itself: on-chain and off-chain governance of blockchain technologies. Topoi **40**(4), 821–831 (2018). https://doi.org/10.1007/s11245-018-9626-5

32. Schmeiss, J., Hoelzle, K., Tech, R.P.G.: Designing governance mechanisms in platform ecosystems: addressing the paradox of openness through blockchain technology. Calif. Manage. Rev. **62**, 121–143 (2019). https://doi.org/10.1177/0008125619883618

33. Thapa, R., Sharma, P., Hüllmann, J.A., Savarimuthu, B.T.R.: Identifying Influence Mechanisms in Permissionless Blockchain Communities: The Bitcoin Case (2021). ICIS 2021 Proceedings. 8. https://aisel.aisnet.org/icis2021/fintech/fintech/8

34. Werner, J., Frost, S., Zarnekow, R.: towards a taxonomy for governance mechanisms of blockchain-based platforms. In: ECIS 2020 Research Papers (2020)

35. Ziolkowski, R., Miscione, G., Schwabe, G.: Exploring Decentralized Autonomous Organizations: Towards Shared Interests and 'Code is Constitution' (2020). ICIS 2020 Proceedings. 12. https://aisel.aisnet.org/icis2020/blockchain_fintech/blockchain_fintech/12

36. Burda, M.C., Locca, M.P., Staykova, K.: Decision rights decentralization in de-fi platforms. In: ECIS 2022 Research Papers (2022)

37. Mini, T., Ellinger, E.W., Gregory, R.W., Widjaja, T.: An Exploration of Governing via IT in Decentralized Autonomous Organizations (2021). ICIS 2021 Proceedings. 1. https://aisel.aisnet.org/icis2021/gen_topics/gen_topics/1

38. Zhao, X., Ai, P., Lai, F., Luo, X., Benitez, J.: Task management in decentralized autonomous organization. J. Oper. Manage. **68**(6–7), 649–674 (2022). https://doi.org/10.1002/joom.1179

39. Faqir-Rhazoui, Y., Arroyo, J., Hassan, S.: A comparative analysis of the platforms for decentralized autonomous organizations in the Ethereum blockchain. J. Internet Serv. Appl. **12**(1), 1–20 (2021). https://doi.org/10.1186/s13174-021-00139-6

40. Goldsby, C., Hanisch, M.: The boon and bane of blockchain: getting the governance right. Calif. Manage. Rev. **64**, 141–168 (2022). https://doi.org/10.1177/00081256221080747

41. de Filippi, P., McMullen, G.: Governance of Blockchain Systems: Governance of and by Distributed Infrastructure (2018)

42. Glaser, F.: Pervasive decentralisation of digital infrastructures: a framework for blockchain enabled system and use case analysis. In: 50th Hawaii International Conference on System Sciences (HICSS-50), 1543–1552. Waikoloa Village, Hawaii, January 4–7 (2017)

43. Gao, W., Hatcher, W.G., Yu, W.: A survey of blockchain: techniques, applications, and challenges. In: 2018 27th International Conference on Computer Communication and Networks (ICCCN), pp. 1–11 (2018)

44. Reyes, C.: (Un) corporate crypto-governance. Fordham Law Rev. **88**, 1875–1922 (2020)

45. Parkin, J.: The senatorial governance of Bitcoin: making (de)centralized money. Econ. Soc. **48**, 463–487 (2019). https://doi.org/10.1080/03085147.2019.1678262

# Cloud Computing

# Cybersecurity and Data Quality in Cloud Computing: A Research Framework

Hongjiang Xu(✉)

School of Business, Butler University, 4600 Sunset Avenue, Indianapolis, IN 46208, USA
hxu@butler.edu

**Abstract.** In the cloud computing environment, cybersecurity and data quality capacities and vulnerabilities can have impact on could computing performance. Cloud computing resiliency addresses risks and capabilities in order to make cloud computing stronger and less susceptible to disruptions. Cloud computing provides many opportunities, however, it also presents some cybersecurity and data quality issues. In this paper, we propose a research framework for cybersecurity and data quality in the cloud computing environment.

**Keywords:** Cybersecurity · Cloud Computing · Data Quality · Risks · Capabilities

## 1 Introduction

More and more systems are operating in the cloud computing environment. Cloud computing brings many benefits as the same time presents many challenges and issues. Particularly in regards to cybersecurity. Organizations' cybersecurity and data quality capacity and vulnerability will likely to have impact on how their systems can operate in the cloud computing environment.

Cybersecurity has some unique features compare to traditional IT security. The effectiveness and efficiency of traditional IT security protection mechanisms are being reconsidered, as the characteristics of cloud computing deployment model differs widely from the traditional architectures (Ramachandra, Iftikhar, & Khan, 2017; Zissis & Lekkas, 2012). Cybersecurity have impact on cloud computing performance (Xu & Mahenthiran, 2021).

Ensure data quality (DQ) is one of the primary job for information systems. The output and decision making will be impacted by poor quality information from the system. There are many measurements for information quality (Strong, Lee, & Wang, 1997; Xu, Horn Nord, Brown, & Daryl Nord, 2002). Systems operate in cloud computing cannot achieve overall effectiveness without good quality of data (Almutiry, Wills, Alwabel, Crowder, & Walters, 2013), such as cloud-based health information systems, and cloud ERP.

Taking all of this into consideration, this paper aims to answer the following research question: What is the influence of cybersecurity and data quality on cloud computing resiliency? To do that, in the next sections we provide a theoretical background of from the fields and propose a research framework.

M. Papadaki et al. (Eds.): EMCIS 2022, LNBIP 464, pp. 201–208, 2023.
https://doi.org/10.1007/978-3-031-30694-5_15

## 2 Theoretical Background

### 2.1 Cybersecurity

Cybersecurity is a critical aspect for cloud computing and supply network. When more and more computing and supply chain network, data storage, and transactions moved from the traditional local hosted hardware and software to the cloud, the cyber space, there are potentially many cybersecurity threats and vulnerabilities.

One the cybersecurity concerns is that the large amount of data stored in the cloud, including critical information, which would attract highly skilled hackers who would want to steal the information for unauthorized users for financial gains (Srinivasamurthy, Liu, Vasilakos, & Xiong, 2013). The cybersecurity is even more critical when a business has sensitive information such as intellectual property, trade secrets, and personally identifiable information about their customers, employees, and suppliers that make security breaches a significant cost to the firms (Kamara & Lauter, 2010). Cybersecurity concerns are one of the major barriers to the adoption of cloud computing (Chen & Zhao, 2012). Therefore, to manage costs, organizations must learn to manage the cybersecurity and privacy risks (Kamara & Lauter, 2010), and learn how to deal with cybersecurity threats and try to manage and reduce the cybersecurity vulnerabilities.

Cybersecurity in the cyber space works differently than the normal IT security due to the potential threats coming from the cyber space, which makes it harder to *prevent, detect and respond to the cyber-attacks*. However, many of the general IT security theories still apply. Such as one of the basic and major security concerns is data security, it is also true in cybersecurity. Cybersecurity requires high level of protection of information.

There are many reasons for the vulnerability of cybersecurity, such as unauthorized access or breach into the system, capacity to store data in comparatively small space, complexity of code, negligence (Ani, 2011). There are also many technologies organizations can implement to help ensure cybersecurity. Such as: 1. Vulnerability scanners. 2. Intrusion prevention system. 3. Intrusion detection system. 4. Network and application firewall (Razzaq, Hur, Ahmad, & Masood, 2013).

### 2.2 Data Quality

The quality of information in any type of the systems is as important as the security for the information. As the data quality control theory of Garbage-in garbage-out (GIGO) is true for all information systems. Cloud computing is not an exception. There are many factors that impact data quality of the system. Those factors are in few categories: information systems characteristics, data quality characteristics, organizational factors, stakeholders' related factors and external factors (Xu, 2013).

To ensure high quality information, the measurements of quality of data need to be understood and used. Information quality problem pattern concept has been used to measure date quality in different type of systems (Xu et al., 2002). DQ problem patterns include:

- Intrinsic DQ pattern: multiple sources of same data, questionable believability, judgement involved in data production, questionable objectivity, poor reputation, and little added value, leading to data not used.

- Accessibility DQ pattern: lack of computing resources, poor accessibility, access security, interpretability and understandability, concise and consistent representation, amount of data, and timeliness, leading to barriers to data accessibility.
- Contextual DQ pattern: operational data production problems, changing data consumer needs, incomplete data, poor relevancy, distributed computing: inconsistent representation, and little value added, leading to data utilisation difficulty (Strong et al., 1997).

## 2.3 Cloud Computing

Cloud computing is one of the trends in the recent years, and many more businesses are joining it, but the theory for data quality and cybersecurity management in cloud computing is still weak. Many businesses are moving toward cloud by force or try to follow the new development, or catch up with their peers and competitors, without fully understand the implications of such action. Many risks associated with cloud computing, especially the data quality and cybersecurity issues need to be understood, and studied. The capacity of the cybersecurity might have impact on the cloud computing's performance (Xu & Mahenthiran, 2021).

The effectiveness and efficiency of traditional IT protection mechanisms are being reconsidered, as the characteristics of cloud deployment model differs widely from the traditional architectures (Ramachandra, Iftikhar, & Khan, 2017; Zissis & Lekkas, 2012). Cloud computing has three deliver models, which are infrastructure as a service (IaaS) that is multi-tenant cloud layers are provided by the service provider and shared with contracted clients, cloud platform as a service (PaaS) where the cloud provider provisions not just the operating system but also provides a development stack (e.g., a database), and cloud software as a service (SaaS) model that provides the complete application stack (e.g., cloud based accounting system) (Ramachandra et al., 2017). And these three delivery models can be deployed either as a private cloud, public cloud, or hybrid cloud. Currently, it is generally believed that small and medium size firm users (SMEs) require services more in the area of offering infrastructure and SaaS, because they do not have the necessary skills, time or resources to setup an application ecosystem and manage it (Khan & Al-Yasiri, 2016). There needs to be research that systematically examining the system development life cycle (SDLC) for cloud consumers to incorporate various technological advancements to improve security at a very fundamental level. Additionally, research need be done in regards to what are the impact of cybersecurity and information quality on cloud computing resiliency.

Cloud computing resiliency addresses risks and capabilities in order to make cloud computing stronger and less susceptible to disruptions. Cloud computing provides many opportunities, however, it also presents some cybersecurity and data quality issues.

## 3    The Research Framework

In order to prevent and reduce cybersecurity threats and vulnerabilities, there are a few areas of cybersecurity that we are interested in investigating: cyber data security in cloud computing, budget and investment for cybersecurity, security policy, and the human aspect of information security. Cloud computing is defined as applications delivered as services over the Internet and data centers provide those services (Armbrust et al., 2010). Cybersecurity is one of the major concerns, as cloud computing enables the migration of system processing and data storage to the cloud, which increased the number of potential cyberattacks (Drew, August 2012). Therefore, it is important to understand the security and privacy risks in cloud computing and develop appropriate solutions for it to be successful (Takabi, Joshi, & Ahn, 2010). Security is implicit within the capabilities of cloud computing. There are many issues and concerns regarding cybersecurity. For example, one of the cybersecurity concerns for cloud computing is who is responsible for the security: is it solely the storage provider's responsibility, or it is also on the entity that leases the storage for its applications and data? (Kaufman, 2009).

## 3.1  Research Questions

In this paper, we propose a research framework on cybersecurity and data quality's impact on the resiliency of cloud computing. In particular, the objectives of the study is try to answer the following research questions:

(1)  Does a higher level of awareness of the cybersecurity issues in firms lead to better cybersecurity risk management policies for cloud computing,
(2)  Does better cybersecurity policies lead to higher level of cloud computing resiliency,
(3)  What are the impact of cybersecurity and data quality vulnerabilities to cloud computing resiliency?

We develop two hypotheses for this question as following:

H1: Cybersecurity vulnerabilities have negative influence on cloud computing resiliency.
H2: Data quality vulnerabilities have negative influence on cloud computing resiliency.

(4)  What are the impact of cybersecurity and data quality capabilities to cloud computing resiliency?

We develop two hypotheses for this question as following:

H3: Cybersecurity capabilities have positive influence on cloud computing resiliency.
H4: Data quality capabilities have positive influence on cloud computing resiliency.

## 3.2 Research Framework

We propose a research framework for cybersecurity and data quality for cloud computing resiliency as follows (Fig. 1):

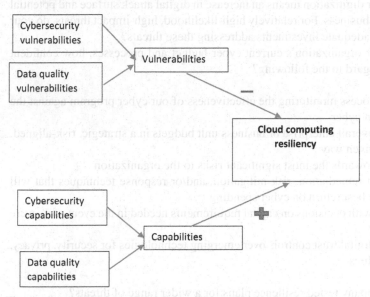

Fig. 1. The proposed research framework for cybersecurity and data quality for cloud computing resiliency

## 3.3 Proposed Items

To address the research questions, we propose a list of items to be used for assessment. Those items are based on the existing literature, and prior studies. Some of the other measurement items will come from established or tested instruments developed from the related research and practical fields.

### Cybersecurity and DQ in Cloud Computing (Capability)

1. Cloud computing provider's ability to protect the *integrity* of my firm's data is high.
2. Cloud computing provider's ability to protect the *confidentiality* of my firm's data is high.
3. Sharing of cloud computing provider's server with other firms' is of great concern.

4. The cloud computing techniques provide sufficient security transfer channel during the process of mass data interchange.

**Budget and Investment for Cybersecurity (Awareness)**

1. More and faster digitization means an increase in digital attack surface and potential for harm to the business. For relatively high-likelihood, high-impact threats, do your company have adequate investments addressing these threats?
2. Regarding your organization's current cyber budget and processes, how confident are you with regard to the following?

    a. Includes process monitoring the effectiveness of our cyber program against the spending on cyber
    b. Linked to overall enterprise or business unit budgets in a strategic, risk-aligned, and data-driven way
    c. Allocated towards the most significant risks to the organization
    d. Focused on remediation, risk mitigation, and/or response techniques that will provide the best return on cyber spending
    e. Integrated with decisions on capital requirements needed in the event of a severe cyber event
    f. Adequate digital trust controls over emerging technologies for security, privacy, and data ethics

3. Have your company tested resilience plans for a wider range of threats?

**Security Policy (Policies)**

1. Are there appropriate Security Policy, Guidelines or Procedures established?
2. Is the existing Security Policy/Guidelines/Procedures adequately enforced?
3. Are users informed of their obligation with regard to the relevant laws, security policy and procedures before being granted access rights?
4. Is the use of strong/complex password policy enforced?
5. Is the use of two-factor authentication enforced for access control?

**Human Aspect of Information Security (Vulnerability)**
Do employees of the company do any of the following?

Internet use

- Installing unauthorized software
- Accessing dubious websites

- Inappropriate use of internet

Social networking site (SNS) use

- Amount of work time spent on SNS is too much
- No award of the consequences of SNS
- Posting about work on SNS

Incident reporting

- Reporting suspicious individuals
- Reporting bad behavior by colleagues
- Reporting all security incidents

Mobile computing

- Physically securing personal electronic devices
- Sending sensitive information via mobile networks
- Checking work email via free network

Information handling

- Disposing of sensitive documents
- Leaving sensitive material unsecured

## 4 Significance

There is limited research on cybersecurity data quality's impact on cloud computing resiliency in the cloud environment, thus our research objectives begin to address this gap.

From the existing literature, it is not clear if a firm that is not aware of its cybersecurity- threats can effectively manage authentication, authorization, data confidentiality, data integrity and non-repudiation concerns of a cloud user wanting to employ a cloud provider. Hence, the first objective of the study is to assess whether a higher level of awareness of the cybersecurity issues in firms lead to better risk management policies for cloud computing. Future research can include collecting data to further test and valid the proposed research framework. The other research objectives can be achieved by using the empirical data to answer the research questions, and test the research hypotheses.

The research framework proposed in this paper is going to make theory contributions by providing a theoretical model to fill the current research gap in cybersecurity and data quality's influence on cloud computing resiliency in cloud computing environment. As for the managerial implications, the future research that build on the research framework of this paper will provide findings that will help businesses to understand the cybersecurity and data quality issues better for cloud computing management, which many of the businesses and managers still have difficulty comprehend.

# References

Almutiry, O., Wills, G., Alwabel, A., Crowder, R., Walters, R.: Toward a framework for data quality in cloud-based health information system. Paper presented at the International Conference on Information Society (i-Society 2013) (2013)

Ani, L.: aCybercrime and national security: the role of the penal and procedural law. Law and Security in Nigeria, 200–202 (2011)

Chen, D., Zhao, H.: Data Security and Privacy Protection Issues in Cloud Computing. Paper presented at the 2012 International Conference on Computer Science and Electronics Engineering (2012)

Goud, N.: Cyber Attack on Tower Semiconductor. Cybersecurity Insiders. http://www.cybersecu rity-insiders.com/cyber-attack-on-tower-semiconductor/ (2020)

Kamara, S., Lauter, K.: Cryptographic cloud storage. Paper presented at the Proceedings of Financial Cryptography: Workshop on Real-Life Cryptographic Protocols and Standardization (2010)

Khan, N., Al-Yasiri, A.: Identifying cloud security threats to strengthen cloud computing adoption framework. Paper presented at the the 2nd International Workshop on Internet of Thing: Networking Applications and Technologies (IoTNAT'2016) (2016)

Ramachandra, G., Iftikhar, M., Khan, F.A.: A comprehensive survey on security in cloud computing. Procedia Comput. Sci. **110**, 465–472 (2017). https://doi.org/10.1016/j.procs.2017. 06.124

Razzaq, A., Hur, A., Ahmad, H.F., Masood, M.: Cyber security: Threats, reasons, challenges, methodologies and state of the art solutions for industrial ap- plications. Paper presented at the 2013 IEEE Eleventh International Symposium on Auton- omous Decentralized Systems (ISADS) (2013)

Schneier, B.: The US has suffered a massive cyberbreach. It's hard to overstate how bad it is. The Guardian. Available at: https://www.theguardian.com/commentisfree/2020/dec/23/cyber-attack-us-security-protocols (2020)

Srinivasamurthy, S., Liu, D.Q., Vasilakos, A.V., Xiong, N.: Security and privacy in cloud computing: a survey. Parallel Cloud Comput. **2**(4), 126–149 (2013)

Strong, D.M., Lee, Y.W., Wang, R.Y.: Data quality in context. Commun. ACM **40**, 103–110 (1997)

Xu, H.: Factor Analysis of Critical Success Factors for Data Quality. Paper presented at the AMCIS (2013)

Xu, H., Horn Nord, J., Brown, N., Daryl Nord, G.: Data quality issues in implementing an ERP. Ind. Manag. Data Syst. **102**(1), 47–58 (2002). https://doi.org/10.1108/02635570210414668

Xu, H., Mahenthiran, S.: Users' perception of cybersecurity, trust and cloud computing providers' performance. Inform. Comput. Secur. **29**(5), 816–835 (2021). https://doi.org/10.1108/ICS-09-2020-0153

Zissis, D., Lekkas, D.: Addressing cloud computing security issues. Future Gener. Comput. Syst. **28**(3), 583–592 (2012). https://doi.org/10.1016/j.future.2010.12.006

# Browser Extension for Detection of Fake News and Disinformation

Lumbardha Hasimi and Aneta Poniszewska-Marańda(✉)

Institute of Information Technology, Lodz University of Technology, Lodz, Poland
lumbardha.hasimi@dokt.p.lodz.pl, aneta.poniszewska-maranda@p.lodz.pl

**Abstract.** Fake news is information usually used to mislead, manipulate or disinform while reaching a certain audience and going viral in rather a short period. Currently it started to be the bigger and bigger problem in Internet, mass media and in everyday life. The pervasive and wide-spreading effect of fake news content is becoming a serious concern of our era. Considering the emergent need for research in this area, our work aims to observe, analyse and propose a solution to the fake news topic. This work presents an Internet browser extension, aiming to notify the user regarding the credibility of the information and carrying out fake news detection. The system as an extension is designed using JavaScript environment, cloud function configured on Google Cloud platform while using Neural Network Model based on TensorFlow library for predictions process on the credibility of the content. The paper also discusses and presents approaches and models on fake news detection, and subsequently security issues on the system's functionality.

**Keywords:** fake news detection · cloud computing · browser plugin · prediction of text · neural networks

## 1 Introduction

Fake news is information usually used to mislead, manipulate or disinform while reaching a certain audience and going viral in rather a short period. It is almost certain that the spread of fake news has had a grave impact on social cohesion, democratic processes and most of all has raised serious concerns among different entities.

In the midst of the great challenge of identifying, analysing, and understanding this phenomenon and many underlying processes, there have been proposed and developed many different approaches and techniques. Although these approaches differ in the choice of algorithmic techniques and adaptation, still there is a common share of techniques in the methodology and deployment [1].

With the recent developments in technology and the internet advances, false information is easier to create, share and spread, making it more complicated to correctly distinguish it from true information. Such news are published on

M. Papadaki et al. (Eds.): EMCIS 2022, LNBIP 464, pp. 209–220, 2023.
https://doi.org/10.1007/978-3-031-30694-5_16

specialized websites, social media platforms or even as podcasts. The multi-modal nature of fake news is making the detection process even more challenging, though it has shown more evidence of the happening of news events and presenting new opportunities to detect features in fake news [2].

Every day, approximately 1.93 billion of people are exposed to information on the leading social media platform. The most recent example of this is the COVID-19 pandemic – almost 80% of consumers in the United States reported having seen fake news on the coronavirus outbreak, highlighting the extent of this issue and the reach fake news can achieve [3].

False information is ubiquitous, and millions of people can be misled in a matter of seconds. Under the commonness of the problem and the number of people it affects, our work aimed to find an appropriate solution that will help people assess information easily. As a result, we proposed a detection system as a browser extension that gets the content of a currently viewed article, sends it to the cloud function, to determine the veracity of the content, and returns the answer in the form of the browser alert.

This solution was intended to be quick, simple, and reliable. The fake news detector in the form of the browser extension provides simplicity, as it requires only a click in the extension to launch the fake news detector. The process of assessing the article content and receiving an outcome takes approximately ten seconds, a feature that needs further enhancement. The reliability of the accuracy of the model used to evaluate the content of the articles reached up to 99%. The fake news detector serves its purpose, providing the user with a short and comprehensible response.

This paper presents the proposal of browser extension for fake news and disinformation detection, analysing the aspects regarding the implementation, design, deployment in the cloud, libraries, algorithms, classification and data processing. The paper is structured as follows: Sect. 2 presents the related work in the field, the existing solutions and the innovations available. Section 3 describes the methodology, basic architecture and model's design. Section 4 deals with the cloud deployment while Sect. 5 showcases the plugin, evaluation and results.

## 2   Related Works

Research was carried out to learn about the current state of the art, existing methods, and research productivity. In general, the literature and existing solutions revolve around the use of Machine Learning and Neural Networks use in fake news detection.

Machine learning as a method concerns ways of finding patterns in data and using it to make estimations [4]. With the help of advanced algorithms in the learning process, it is possible to create models used to classify data provided by the user as fake or not. In the study [5] about the use of machine learning approaches in fake news detection, several algorithms and techniques were analysed and compared. According to the report, among algorithms such as XGboost, Random Forests, Naive Bayes, K-Nearest Neighbours (KNN), Decision

Tree, and SVM, the highest accuracy was obtained with XGboost algorithm, which was higher than 75%.

Neural networks, on the other hand, inspired by the human brain structure might be described as a web of interconnected entities, each of them responsible for a simple computation [6]. Furthermore, there are many possible implementation methods for machine learning and neural networks, for which Python language provides developers with one of the most popular environments. Some of the most commonly known libraries in Python include Scikit-learn, Pandas and TensorFlow.

Existing solutions as a browser extension for fake news detection include *Check-it* [1], *FakerFact* [7], *BRENDA* [8], *TrustedNews* [9], *The Factual* [10]. A significant advantage of the last two solutions is that apart from simple news verification they also display objectivity and credibility expressed as a percentage. Nevertheless, in the case of *The Factual* [10], it was noted that the browser extension often crashes while opening, perceiving also not proper functionality. Furthermore, browser extensions such as *The Factual, TrustedTimes* and *Faker-Fact*, although claimed to support automated fact-checking and being listed in the Google extension store, there is no available documentation of the models used. Moreover, it was not found to have any system which can narrow down the claim within the article using fact-checkworthiness detection and use that claim to detect fake news.

However, *BRENDA* [8] solution as a browser extension for fake news detection provides many feature evidence at the both-word level and the sentence level. It follows a client-server architecture and has a frontend and backend module, where the frontend is a browser extension and the backend is a python Flask server. This way as a solution, it stands out compared to the above-mentioned solutions.

*Check-it* [1], on the other hand, effectively combines a set of diverse signals as a form of the pipeline, to accurately classify fake news articles and inform the user, while ensuring user's privacy and easy experience. The system contains four main components, including matching, similarity checking and comparing, analysing user behaviour in social networks, and classifying linguistic features using different feature engineering processes [1]. Furthermore, most of the works above employ server-side APIs with constant communication, utilize HTTP cookies, request permissions, and require account registration. These actions are taken to have better results and higher accuracy in identifying fake news but may also have a negative impact making users reluctant in using it on browsing routine. Considering everything above, it was decided on browser extension, using the Neural Networks approach. The verification module is implemented with the use of Python language and libraries Pandas and TensorFlow.

## 3   Proposed Fake News and Disinformation Detection Solution

Aiming to tackle the issue of fake news and disinformation detection, this paper presents the proposal of the system on article veracity, that consists of three main components. The browser extension extracts the article content from the HTML file, sends it as a request to the cloud computing system, after receiving a result, it displays to the user the information on article authenticity.

Secondly, the classifying model based on Neural Networks – a trained model that makes predictions on article authenticity based on the parameters put to the model by cloud function. The parameters were obtained in the process of pre-defined data set analysis, whereas Google cloud as a serverless environment reacts to the request sent by the browser extension. In a request, it receives an article content that is passed by the function to the loaded model, and finally, the obtained result is returned from the cloud to the user through the extension.

The proposed solution aims to significantly improve the quality of articles and information that are served to the reader, and this way hinders the fake news spread and dissemination of information. Building the insights of the system, different tools were engaged to reach the objective of veracity and credibility of the content. For the data analysis stage, the pre-processing phase, using Google Cloud Platform through all the available tools and libraries, engaged libraries such as Pandas and Scikit-learn. For the machine learning model, optimized we choose TensorFlow to create and train neural networks, hence carried out by *Tensorboard* library for visualization of learning outcomes. Finally, JavaScript for the plug-in and google cloud employs serverless code calling and parallelization.

To make it possible for the end-user to have a quick warning regarding the content credibility of the article, a browser plug-in was seen as an apt solution. Thus, it is proposed the plug-in acting as a client, whereas the entire classification procedure takes place on the server, namely cloud service supporting parallelization. The classification itself contains artificial intelligence tools carrying out the process based on the obtained data. There are three main files, consisting of *manifest.json, background.js* and *content.js* Manifest.

*Manifest.json* contains important information such as permissions that extension needs, description, or background files which consist of actions it performs. In this case, permissions to access the active tab and scripting were granted. Background.js contains a function that listens for click on the extension's icon and runs the *content.js* file. *Content.js* is the longest file that consists of an HTML parser, function sending a POST request to the server, and function retrieving content from the opened tab. The GET function is responsible for retrieving the webpage in the form of the HTML code utilizing HTTP GET requests. While HTML parser retrieved from the currently opened webpage needs to be parsed, allowing only the relevant article text to be sent to the model. Meanwhile, the POST function sends the data to the server as an HTTP POST request and returns a response from the server, which indicates whether the data sent was assessed as fake or not.

To do so an attempt is made to find a <*article*> tag, and its textual contents then are read. The text is parsed so that all the tags like <*div*>, <*p*> or <*a*> as well as their attributes are removed. The text afterward is formatted, so that all whitespaces are deleted, and the text looks like an article genuinely written and is ready to be sent to the model.

Figure 1 presents the components of the overall system. The plugin acting as a client, allows the whole classification to take place in the server. The client component contains three elements: the browser extension, HTML sender an the receiver allowing the feedback from the server. The architecture focuses largely on the server side, including HTML receiver, data extractor, Fake News detector and decision sender. The data extractor, responsible for the feature extraction process, described above, produces the list of fake articles propagator. The classifier then, with the data extracted from the dataset using the feature process, permits for the final step on the detection of the news through the processes on the detector. Finally, the model was saved with .tf extension for later usage.

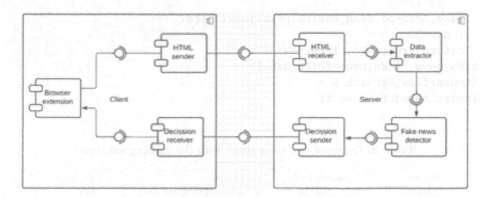

**Fig. 1.** System's architecture for fake detection.

## 4    Dataset and the Pre-processing of the Data

The whole system was tested on the primary dataset using a collection of news articles obtained from the open Machine Learning Repository Kaggle [11]. The data obtained from the Kaggle platform was split into two files: *True.csv* and *Fake.csv*. The dataset that is used is a public fake and real news dataset. The dataset is split into two parts: real news or fake news, almost perfectly balanced with around 20000 real and fake articles respectively. It includes features such as title, content, type of the article and publication date (Table 1).

**Table 1.** Data extraction features

| No. | Attribute | Type | Description |
|-----|-----------|------|-------------|
| 1 | Title | Text | Indicates the title of the Article |
| 2 | Content | Text | Indicates the content of the Article |
| 3 | Publication date | Date | Indicates the date the article was published |
| 4 | Label | Text | Indicates the Article labeled as Fake or True |

The title indicates the title of the article. The content is the body of the article. The subject shows which type of news the article belongs to. Lastly, the date shows the publication date of the article. Considering that the network must analyse all the input text, the article should be split into smaller elements. Such an approach generally increases the speed of data processing and first and foremost, increases the accuracy of predictions [12]. The data is split into words mainly because such a technique is highly effective, considering that it is needed to trace and save the context for every word in the text (Table 2):

```
def fetch_data():
  true_news=pd.read_csv(os.path.join('True.
    csv'))
  fake_news=pd.read_csv(os.path.join('Fake.
    csv'))
  return true_news, fake_news
true_news, fake_news=fetch_data()
%tensorflow_version 2.x
import tensorflow as tf
```

**Table 2.** Overview and comparison with the existing solutions

| Solution | Approach used | Type output | Accuracy |
|---|---|---|---|
| FakerFact | *Deep Learning* | Assessment/Credibility score | 66% |
| BRENDA | *Deep neural network* | Claim and User Feedback False/True | to 86% |
| TrustedNews [13] | *Machine Learning* | Objectivity score | 73% |
| Check-it | *Deep Neural Network* | Classification Fake/True | to 90% |

In the proposed solution, we use a special blacklist of words from the natural language toolkit (*nltk*) library, which makes it possible to exclude the prepositions and words without a high semantic load [13]. Having defined *fetch_data* function with the use of Pandas library it was possible to store data with dataframe objects. For the data pre-processing it was necessary to label positive samples and negative samples, then merge the title with the article content and drop irrelevant data:

```
def reorganize_data(true_news,fake_news):
  fake_news['label']=0
  true_news['label']=1
  dataset=pd.concat([ true_news,fake_news])
  dataset['text'] = dataset['title'] + " " +
    dataset['text']
  dataset = dataset.drop(['title', 'subject',
    'date'], axis=1)
  import sklearn
  from sklearn.model_selection import
```

```
      train_test_split
   x_train,x_test,y_train,y_test =
      train_test_split(dataset['text'],
      dataset['label'],test_size=0.2,
      ran-dom_state = 1)
   return x_train,x_test,y_train,y_test
```

Positive and negative samples were merged into one dataframe object, and then split into training and validation sets. Functions provided by scikit-learn library allowed to shuffle the data and split with a given size of the validation set, namely 20%. To input, this data to the TensorFlow neural network, it is needed to get data converted to python *list_objects*. The main reason to use Neural Network as a text classifier is the huge flexibility and possibility to link with external architectures [15]:

```
import tensorflow_hub as hub
embedding = "https://tfhub.dev/google/
   nnlm-en-dim50/2"
hub_layer = hub.KerasLayer(embedding,
   input_shape=[],
dtype=tf.string, trainable=True)
model = tf.keras.Sequential()
model.add(hub_layer)
model.add(tf.keras.layers.Dense(16,
   activation='relu'))
model.add(tf.keras.layers.Dense(1))
```

The model uses *nnlm-en-dim50*, a pre-trained neural network managing the embedding and tokenizing data. Except for the embedding layer, it contains one more hidden layer and an output layer with a *softmax* activation function (Fig. 2).

```
Model: "sequential"
```

| Layer (type) | Output Shape | Param # |
|---|---|---|
| keras_layer (KerasLayer) | (None, 50) | 48190600 |
| dense (Dense) | (None, 16) | 816 |
| dense_1 (Dense) | (None, 1) | 17 |

```
Total params: 48,191,433
Trainable params: 48,191,433
Non-trainable params: 0
```

**Fig. 2.** The model of pre-trained neural network.

The model was trained with four epochs considering that a longer weight update was not necessary. For the visualization of the learning process, Tensorboard was employed, and after one epoch, the model achieved almost its maximum accuracy (Fig. 3).

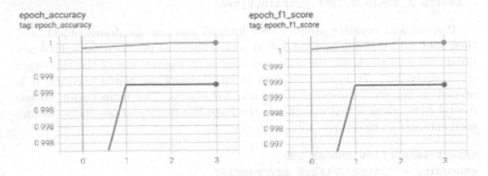

**Fig. 3.** Accuracy and F1-score of trained neural network model.

After three epochs the model reached significant results for the training and validation set (Table 3).

This signifies that clear separation can be observed between positive and negative samples and the architecture is adapted to the classification requirements.

**Table 3.** Accuracy and F1-score results of trained neural network model

|          | Training Set | Validation Set |
|----------|--------------|----------------|
| Accuracy | 100%         | 99.9%          |
| F1-score | 99.9%        | 99.9%          |

## 5  Cloud Deployment

The fake news detection system was deployed through the Google Cloud Function platform, mainly because of the advantages over other systems, namely functions that get triggered when an event is fired, hence terminated after execution of the function. Files such as *variables.index* and *variables.data-00000-of-00001*, were uploaded to Google Storage Bucket before the function was created and configured [15]. The first file stores the list of variable names. The second one stores the actual values of all the variables saved, and the HTTP allows unauthenticated invocations. The memory allocated by the function is set by four gigabytes due to the size of the second file on which the cloud function operates. For the source code of the cloud, there are two files, specifically *main.py* storing

Fig. 4. The proposed extension in Google Chrome [15].

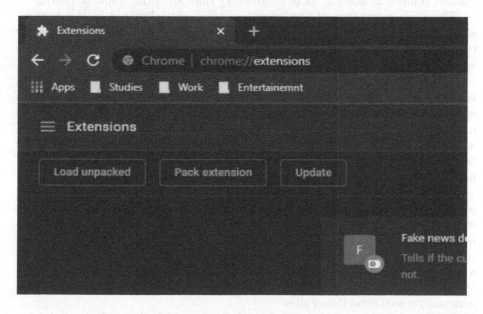

Fig. 5. Proposed extension panel in Google Chrome [15].

the function code and executing when the trigger event happens, and *require-ments.txt*, where libraries required for function execution are declared (Figs. 4 and 5).

To retrieve the article content from the browser, the *is_fake* entry point of the cloud function is called, as a request to HTTP request for the model prediction. The function loads variables from cloud storage and stores them in a form of Blob objects [15]. Following, the variables are passed to the model, which is ready on prediction of the article veracity.

If data sent by the browser extension is not empty, the model proceeds with the prediction and returns a *softmax* result. Otherwise, the function returns information that nothing was sent for prediction. The *softmax* [16] results enable the final stage of cloud function action. Regarding the security, Google Cloud Platform provides a variety of security aspects for the Cloud Functions [17], in this case, the access control based on authentication. To allow all users the possibility of invoking fake news detection function, allowing this way also the full functionality, the access was set public. Furthermore, there is also a full control on permissions request possibilities, meaning agents and entities owe the control on cloud function resources access.

## 6   Discussion and Conclusions

Considering that the dataset contained articles revolving mainly US politics, in the project tests, there was a tendency to have articles with political content classified as fake. Nevertheless, this can be as a result of the fact that such content inflicts, in general, more controversy than any other topic or content.

The main advantage of the offered solution as elaborated in the sections above is the simplicity and the ease of use. It requires no more effort to verify the truthfulness of an article than enabling the plugin, with no need to access any other page, copy URL, or create an account.

There are, however, two main issues that require attention. First of which is the time required to get the result. Although it is very quick and easy to activate the extension itself, having a final output is not processed within the most optimal time. Considering that the existing solutions already offer output within 5 s, it is still not the best user-friendly option. The reason for that is the fact that in the cloud function the model gets downloaded each time the function is called.

The other issue to highlight is the fact that times the HTML cannot be processed even though the site is an actual article. Various sites are built differently when it comes to the HTML structure. In HTML5 an *<article>* tag has been introduced, which in theory should be used to contain articles or longer text contents. Although it is used on most sites, it is still possible to encounter ones with a different structure. Hence, with the implementation of parsing the contents of *<article>* tag, the article cannot be extracted and formatted. It certainly is not common on well-established sites.

Regarding perspectives and future work, some aspects might need further improvement, that is user interface, article recognition, and execution time. The first aspect concerns the way how the browser extension is presented to the user. The interface might be developed further to be more appealing. Additionally, the other aspect regards the fact that the software does not work for websites that do not include an *<article>* tag in their HTML structure. Therefore, the extraction of content shall be added despite the structure. Lastly, the aspect of processing time, to improve that, the model loading needs improvement and re-organization. Enhancing the features mentioned above would lead to better performance and significantly improved version even in comparison to the state of art solutions.

**Acknowledgment.** The publication was created as part of the participation in the project of Polish National Agency for Academic Exchange under the "STER" Programme – Internationalisation of Doctoral Schools" as part of the project "Curriculum for advanced doctoral education & training – CADET Academy of Lodz University of Technology".

# References

1. Paschalides, D., et al.: Check-it: a plugin for detecting and reducing the spread of fake news and misinformation on the web. In: 2019 IEEE/WIC/ACM International Conference on Web Intelligence (WI), pp. 298–302 (2019)
2. Alonso-Bartolome, S., Segura-Bedmar, I.; Multimodal fake news detection. arXiv:2112.04831 [cs] (2021). Accessed 16 Feb 2022
3. Watson, A.: Fake news in the U.S. - statistics & facts, Statista (2022). https://www.statista.com/topics/3251/fake-news/. Accessed 02 Mar 2022
4. Choudhary, M., Jha, S., Prashant, Saxena, D., Singh, A.K.: A review of fake news detection methods using machine learning. In: 2021 2nd International Conference for Emerging Technology (INCET), pp. 1–5. (2021). https://doi.org/10.1109/INCET51464.2021.9456299
5. Khanam, Z., Alwasel, B.N., Sirafi, H., Rashid, M.; Fake news detection using machine learning approaches. In: IOP Conference Series: Materials Science and Engineering, vol. 1099, no. 1, p. 012040 (2021). https://doi.org/10.1088/1757-899X/1099/1/012040
6. Kula, S., Choraś, M., Kozik, R., Ksieniewicz, P., Woźniak, M.: Sentiment analysis for fake news detection by means of neural networks. In: Krzhizhanovskaya, V.V., et al. (eds.) ICCS 2020. LNCS, vol. 12140, pp. 653–666. Springer, Cham (2020). https://doi.org/10.1007/978-3-030-50423-6_49
7. FakerFact. https://www.rand.org/research/projects/truth-decay/fighting-disinformation/search/items/fakerfact.html. Accessed 16 Feb 2022
8. Botnevik, B., Sakariassen, E., Setty, V.: BRENDA: browser extension for fake news detection. In: Proceedings of the 43rd International ACM SIGIR Conference on Research and Development in Information Etrieval, New York, NY, USA, pp. 2117–2120. Association for Computing Machinery (2020). https://doi.org/10.1145/3397271.3401396
9. Trusted News. https://trusted-news.com/. Accessed 16 Feb 2022
10. The Factual - Unbiased News, Trending Topics - The Factual. https://www.thefactual.com/?lp=new. Accessed 16 Feb 2022
11. Fake News—Kaggle. https://www.kaggle.com/c/fake-news/data. Accessed 11 Feb 2022

12. Abdulrahman, A., Baykara, M.: Fake news detection using machine learning and deep learning algorithms. In: 2020 International Conference on Advanced Science and Engineering (ICOASE), pp. 18–23 (2020). https://doi.org/10.1109/ICOASE51841.2020.9436605

13. Kevin, V., et al.: Information nutrition labels: a plugin for online news evaluation. In: Proceedings of the First Workshop on Fact Extraction and VERification (FEVER), Brussels, Belgium, pp. 28–33. Association for Computational Linguistics (2018)

14. Hasimi, L., Poniszewska-Maranda, A.: Ensemble learning-based fake news and disinformation detection system. In: 2021 IEEE International Conference on Services Computing (SCC), pp. 145–153 (2021). https://doi.org/10.1109/SCC53864.2021.00027

15. Abiodun, O.I., Jantan, A., Omolara, A.E., Dada, K.V., Mohamed, N.A., Arshad, H.: State-of-the-art in artificial neural network applications: a survey. Heliyon 4(11), e00938 (2018). https://doi.org/10.1016/j.heliyon.2018.e00938

16. Czyzewski, A., Lech, E., Milosz, A., Kowalski, K.: Cloud computing system - Raport. Lodz University of Technology, Student paper (2022)

17. Multi-Class Neural Networks: Softmax—Machine Learning Crash Course, Google Developers. https://developers.google.com/machine-learning/crash-course/multi-class-neural-networks/softmax. Accessed 16 Feb 2022

18. Securing Cloud Functions—Cloud Functions Documentation, Google Cloud. https://cloud.google.com/functions/docs/securing. Accessed 16 Feb 2022

# Digital Governance

# A Comprehensive Framework for Measuring Governments' Digital Initiatives Including Open Data

Mohsan Ali[✉], Ioannis Zlatinis, Charalampos Alexopoulos, Yannis Charalabidis, and Loukis Euripidis

University of the Aegean, Samos, Greece
{Mohsan,alexop,yannisx,eloukis}@aegean.gr,
icsdm621007@icsd.aegean.gr

**Abstract.** Digital innovation and digital initiatives are generally recognized and considered to be the driving forces behind firm survival and success in the market. This is not the case in the public sector, where digital initiatives have suffered not only from a lack of research trying to explain them but also from a major lack of recognition of their importance. The government's eagerness to introduce more digital initiatives for better e-government plays an important role in advancing the country. Digital initiatives in several governments are being overlooked, even though these days, private companies are competing to provide their customers with the best products and services. In this study, important attributes and sub-attributes of digital initiatives by governments are identified so that we can get a clearer picture of the government's digital initiatives (GDIs). The attributes and sub-attributes are extracted and combined with the new, proposed attributes to make our all-encompassing framework (with the passage of time, maybe more attributes and sub-attributes can be added to the list). The identified attributes are either directly or indirectly related to projects under the government's digital initiatives, which include open data for more transparent and accountable e-governance. The developed framework has been applied to Greece and Pakistan; the authors of the underlying study belong to Pakistan and Greece, and one more reason behind this is to measure the difference between the developing and developed countries' GDIs and to evaluate the governments' digital initiatives (GDIs) with respect to each attribute and sub-attribute. This research will help people understand the problems that governments face with GDIs and maybe provide recommendations for further development initiatives.

**Keywords:** Governments' digital initiatives (GDIS) · Digital Governments · Open data initiatives · Digital innovation · E-government pillars · Greece · Pakistan

## 1 Introduction

Digital government initiatives is the use of information and communications technology (ICT) in the public sector to provide citizens, businesses, and government employees

© The Author(s), under exclusive license to Springer Nature Switzerland AG 2023
M. Papadaki et al. (Eds.): EMCIS 2022, LNBIP 464, pp. 223–241, 2023.
https://doi.org/10.1007/978-3-031-30694-5_17

high-quality oriented services [1–3]. Increased ICT access is vital for bridging the digital gap, promoting effective governance, and advancing sustainable development. Digital government initiatives have improved the delivery of public services, but their overall impact in developing nations such as Pakistan has been hampered by an abundance of rules and a slower adoption rate. Everyone has the right to expect quick services from the government and easy access to information that is correct and often complete [4]. Governments continue to be collectors, consumers, preservers, and creators of primary data and aren't realizing their full potential. Governments must modernize in response to rapid changes in society and the economy, and information technology may aid in this endeavor. In recent years, however, government e-services have not been sufficiently concentrated on citizens.

The 2022 United Nations E-Government Survey [5] reveals that several nations have implemented e-government initiatives and information and communication technologies (ICT) applications for the public in order to enhance public sector efficiencies and streamline governance systems in support of sustainable development. The government's digital initiatives can be motivated by the Sustainable Development Goals (SDGs). Leaders in e-government see innovative technical solutions to boost economic and social sectors that are falling behind. In the current global recessionary climate, the overall conclusion of the 2012 Survey is that governments must rethink e-government and e-governance [5], placing greater emphasis on institutional linkages and among the tiers of government structures to create synergy for inclusive, sustainable development. To reach this goal, it is important to broaden the reach of digital initiatives so that the government can play a transformative role by putting in place procedures and institutions that support sustainable development [5, 6].

The COVID-19 pandemic has proven the relevance of digitalization for the timely and efficient provision of government services. Digital services are basically the fruitful results of digital government initiatives for better service provisions. Countries with a proper road map for digital government initiatives performed well during the pandemic era, and they are confident that these digital initiatives will perform well in the coming days as well. Governments have developed the digital portals, where users may get a plethora of relevant information organized by topic, life cycle, or other chosen usage, are backed by technical advances that promote data exchange and successful optimization of cross-agency governance systems. Services personalization in the digital portals is becoming popular, and more and more countries are changing their content and presentation to suit different initiatives. Digital government initiatives may position the public sector as a demand generator for ICT infrastructure and applications. When government digital initiatives make up a large part of a country's gross domestic product (GDP) and the regulatory environment encourages the growth of ICT manufacturing, software, and related services, the effect will be stronger [4].

However, several developing countries in Asia, such as Pakistan, are not well versed in the provision of government's digital initiatives. There can be several reasons for this, such as technological, the economy, population, and the literacy rate. Governments lack a grasp of how to initiate digital initiatives and what actions must be followed for digital initiatives to be effective. What are the main attributes and sub-attributes of a government's digital initiatives, and how can a government implement them? For instance, how can the government launch a digital initiative to increase transparency in government? The government might be able to give its people digital initiatives that use "open data" to improve transparency. To answer this research question, we developed a framework based on the extraction of determining factors for government digital initiatives, and then compared Pakistani and Greek governments to validate our framework. In this study, we will provide a theoretical and practical framework for successful governments' digital initiatives along with the determining factors, which we call the attributes and sub-attributes of successful governments' digital initiatives (GDIs).

Existing studies discuss government assessment frameworks such as the UN e-government survey [5], Open Data Watch (ODIN), the digital Economy and society index (DESI), and web-portal-based assessments [7], government structure and citizen engagement-based assessments [8], and government accountability [9]. In this study, we will combine all of these indices and define attributes and sub-attributes that evaluate government digital initiatives in conjunction with open data-based initiatives. To the best of our knowledge, our proposed framework includes open data as a digital initiative alongside other government digital initiatives, such as the use of AI and blockchain in government.

The main contributions of this research are:

- To extract and define the attributes and sub-attributes for the GDI's framework development
- To categorize the digital initiatives provisioned by the government as "provided," "not provided," and "partially provided" based on our GDIs framework.
- To determine Pakistan's and Greece's current positions in government digital initiatives with the help of our developed framework
- To provide academics and practitioners a way to evaluate GDIs in any other country in the world.

The words "factor," "pillars," and "attributes" are used as synonyms in this study. The remainder of the article is structured as follows: The second section explains digital initiatives, e-governments, and advancements in ICT, considering research publications and grey documents. In the third section, existing factors to measure the success of digital government initiatives have been discussed. We added more factors (attributes and sub-attributes) to the existing factors to make it a more extensive study. A comparison has been drawn between Greece's and Pakistan's based on our developed framework. The fourth section describes how these two nations approach digital government initiatives and adoption patterns and their success based on the scale of service provision, such as "provided", "not provided", or "partially provided". The final section concludes the research with the overall findings, limitations, and future work.

## 2 Literature Review

E-government is regarded as the initial stage of digital government. The Organisation for Economic Co-operation and Development (OECD) defines digital government as "the use of digital technologies as an integrated part of governments' modernization strategies to create public value" and that it "relies on a digital government ecosystem comprised of government actors, non-governmental organizations, businesses, citizens' associations, and individuals that supports the production of and access to data, services, and content through interoperable technologies". "E-government" is defined as "the use of information and communication technologies (ICTs), and the Internet in particular, by governments to improve government". The digital governments initiatives play a vital role in the development of E-government. The pandemic has given the digital government a new lease on life and helped to define its role, both in the way it delivers digital services and in the new, creative ways it handles crises [10, 11]. The e-government ranking of the countries depends upon the difference factors such as country's gross domestic product(GDP) and its e-government ranking, money is not the only important factor in the development of e- government [12]. According to the 2020 UN e-government survey, Denmark, Estonia, and Korea top the rankings [1]. During the last few years, COVID-19 affected governments, particularly municipalities, to pursue services using innovative technologies, but a vast number of people were unable to access digital services, as published in the UN e-government survey [5].

Most of the time, developing nations are compelled to devote resources to undertaking fundamental changes, such as e-Government, based on models that may not work in settings that are significantly unlike those in the developed world [13, 14]. In underdeveloped countries, a new field of study concerning digital-government-related issues and the usability of e-government websites is expanding, whereas public perspectives (participation) are receiving less attention [15]. Incorporating citizens into the policy-making process is one of the most significant benefits of using e-government services [16, 17]. Even if a new digital initiative or service is developed by the government, the laws related to these digital initiatives and services may slow down the adoption rate in the country [18–20].

Davison et al. (2005) argue that people develop a preference for citizen-centric and responsive e-government websites. In a number of developing countries, e-government websites are not available in native and local languages, indicating that e-government is meant only for a minority of educated individuals. The European Commission published the four key indicators as an e-government benchmark initiative (user centricity, transparency, key-enablers, and cross-border mobility) with the title "eGovernment Benchmark 2020: eGovernment that Works for the People" to measure e-government performance in Europe [7].

Recently, other studies have been published which considers the development and evaluation of e-governments. We used the term "digital government initiatives" for the "e-government development". To measure the governments' digital initiatives(GDIs), different attributes (pillars) and sub-attributes will be extracted from the literature Irawan et al., for example, provide two (2) evaluation models for government websites, with the two models divided into two components: technical and democratic deliberation. The first website evaluation model is the "Qiyuan Fan Website Evaluation Model", which contributes to the e-services section of the pillars (in our study) by providing the financial transactions and the e-procurement sub-pillars of the comparison. The second model is the "Lee-Geiller & Lee Website Evaluation Model", which also contributed to the transparency and e-services sub-pillars. The interoperability of services, coordination at the national level, and error management are some of the sub-pillars of the e-government comparison [21]. The government website evaluation is a key part of the e-government comparison(digital government initiatives) as a whole [17, 22].

As Wu et al. said, "An efficient performance measurement system is essential for controlling, monitoring, and improving service quality in governmental organizations". Their contribution to the comparison is the "Government Structures" pillar, which focuses on the learning and growth of employees, which is an essential measurement of e-government initiatives. More specifically, the "government structures" have staff training and knowledge management, staff satisfaction, and internal communication, which are some of the sub-pillars. Wu et al. also proposed the "Citizen Engagement" pillar with the responsiveness to inquiries and complaints and the political efficacy sub-pillars of the comparison [8].

One of the most important parts of the e-government comparison is the accountability section. Ibrahim et al. have proposed a framework for the evaluation of accountability based on web-based accountability practices; their developed framework uses financial, performance, and political accountability as the main pillars for the evaluation models. We have selected the "accountability" pillar from Ibrahim et al. along with its sub-pillars for our study [9]. The "Use of Disruptive technologies" pillar is constructed based on the cutting-edge technologies for the e-government such as Fintech, block chain, and AI. The use of disruptive technologies is important to develop smart cities, and consequently, in developed countries, the governments are focused on the use of these technologies for digital government initiatives [23]. Finally, the state audit office of Hungary has introduced good governance pillars such as lawmaking, accountability, transparency, economic and financial sustainability, a model organization, and reasonable and effective financial management [24]. These good governance pillars are indirectly related to the digital government initiatives, and we have used them to extract some new sub-pillars for our study, such as accountability and transparency-related sub-pillars.

Different articles discussed the pillars, factors, or attributes to measure the success of digital government initiatives, but in this study, an all-factors-encompassing framework could be useful. For instance, previous studies mentioned the evaluation of government websites, citizen engagements, the role of staff training, accountability, technicalities, and government preferences to evaluate digital government initiatives. However, we evolved with more factors to measure the success of GDIs such as "government structures", "e-services by government", "use of disruptive technologies", "transparency", "accountability", and "citizen engagement" to make this study more comprehensive. Each main pillar has been enriched with sub-pillars as described in the literature, and we have added some more sub-pillars to elaborate more on the digital government initiatives such as "Use of disruptive technologies", where we have introduced some sub-pillars such as use of block chain, AI, and FinTech. In the transparency pillar, we have introduced a sub-pillar for the use of open, big, and linked data for transparent e-government as a digital government initiative. This is the uniqueness of our study among others in that we have considered open data as a digital government initiative for the evaluation of e-governments.

## 3 Proposed Methodology

The entire procedure is depicted in Fig. 1. This study began by extracting the most essential digital government attributes, features, properties, pillars, and sub-dependent attributes. These characteristics are essential for measuring the government's Transparency and accountability in a systematic manner. In Sect. 2 of this research, the attributes or pillars that are directly or indirectly important to digital government are taken from the published literature and policy papers by researchers and governments. Government structures, E-services, the use of disruptive technology, transparency, accountability, and citizen engagement are the six characteristics of digital government. After extracting the six primary attributes of digital government, the next step is to carefully identify sub-attributes associated with each of the six core attributes.

Regarding government initiatives, this list of characteristics is not comprehensive; there may be many more. Several countries may have distinct perspectives, policies, and standards for digital services, for instance. One important point of view is that the government may have missed out on some digital services because of the high cost of technology devices and budget constraints. Each attribute and sub-attribute are exhaustively defined and discussed in the next section. This research can be used in the future to analyze the available services and orientations that are superior in one nation and inferior in another. The Greece and Pakistan have been chosen for the validation and application of our developed framework. We have used the snowball method to search the attributes and sub-attributes for both countries. There was no conflict of interest while deciding about the attributes' and sub-attributes' availability in each country because we just visited the official government websites to find out the relevant information. The reason behind the selection of these countries is that the authors belong to these two countries, and the validation of the framework will be quite easy to perform.

This study will help us trace the two nations' digital government advancements in each field. This is the age of the fourth industrial revolution, and it may be important to find the areas that need more attention to improve governance and, indirectly, make governments more accountable and open. We are aware that several countries are facing financial troubles because of the current global epidemic, making these attributes even more crucial. There may be further obstacles like these outbreaks in the future. There is a unique answer to every situation; it is the utilization of technical solutions to improve the future by putting residents' needs first. In the evaluation of governments' digital initiatives, the government structure such as population, area, ministries, staff training, and internal communication mechanisms are to be measured for each country. This information will help estimate the cost of the digital initiatives and their effectiveness.

E-services are also significant in digital governance. And the government can readily provide a variety of e-services to end-users in a timely and effective manner. Innovation with disruptive technology is also an essential factor to consider while evaluating digital governance. IOT, block chain, FinTech, robotics, cloud-computing, and AI are disruptive technologies to provide the finest services ever to citizens [23].

Transparency in government is a crucial metric for determining the efficacy of the government and its policies. The sub-attributes of transparency are significant and aid in exploring opaque government areas for future improvement. Public sector digital accountability evaluates the effectiveness of previous policies and decisions. This helps

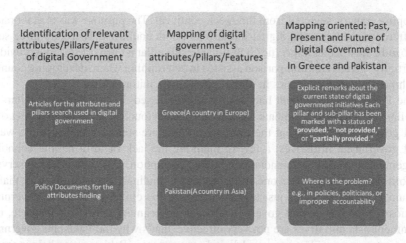

**Fig. 1.** Proposed methodology for extracting government attributes and sub-attributes and exploring their roles in Governments (e.g., in Greece and Pakistan)

improve governance and reduce poverty by making sure that government programs meet their stated goals and the needs of the people they are meant to help. Participation of citizens is essential for effective governance. It allows private individuals and groups to inform, influence, monitor, and assess public choices, procedures, and actions. The primary objective of public involvement is to foster significant public input during the decision-making process. Therefore, public involvement facilitates communication between the public and entities that make decisions. The other important factor is the corruption index in developing countries. Several politicians are creating problems in the way of digital government. Greece and Pakistan are also listed in the corruption index published by World data and Transparency international [25]. Focus must be placed on the digitization of government financial flows in order to build transparent and accountable governments. The Digital Economy and Society Index (DESI) is a second index that uses predefined indicators to measure how well digital services work in European countries. The European Commission has been using the DESI index since 2014 to measure the digital progress in the member states of Europe. The DESI key areas are very helpful for comparing the progress of countries, but we added a few other indicators as well, such as the use of disruptive technology and its sub-attributes in the public sector [26].

## 4 Findings and Experiences

Table 1 explains the pillars of E-government and applied these pillars to Greek and Pakistani government. We used the ministries website, and their initiatives websites to illustrate the current situation in each electronic pillar in e-government. This study helped us to evaluate the digital government initiatives. The countries evaluated in this study were Greece and Pakistan. For the comparison of digital government initiatives, Greece is a developed country, selected from the European continent, and Pakistan is a developing country, selected from the Asian continent. The purpose was to elaborate on the differences between the digital initiatives in both countries. There are three scales used to indicate whether a specific GDIs pillar or sub-pillar is "provided", "not provided", or "partially" given. This comparison assisted in determining where developing countries lag behind developed countries in digital government initiatives.

The Pakistani government had been evaluated based on the identified pillars and sub-pillars of digital governments initiatives. Pakistan touched upon every aspect of the digital government but during the evaluation, there are some obstacles in the way of digital government initiatives: a few of them are the lack of information technology management system, low financial conditions, corruption, less user-oriented services, and political instability. The one more important finding of this study was that the Pakistani government gives just a few datasets to the citizens about the government activities for the transparent government (open government data). For instance, open government data initiatives are not very effective, that's why the Pakistani government is not very transparent. The improved digital government in Pakistan may help in smooth information flow from government to citizens, citizens to government, and also within the government institutions. Consequently, the digital government in Pakistan will help in advancements of administrative activities, improve the economy, and at the same time improve transparency.

The Greek government over the last years is on the right track to achieving those pillars. First of all, in the "E-services by government" section of the pillars, Greece provides good practices, and several services are provided in the online platform of the Greek government but there are not all digitally enabled to the citizens and that is another step that needs to be taken in order to provide quality of service. Moreover, the "Use of Disruptive technologies" part of the evaluation is another good example for the Greek government. There are lots of initiatives that took place in Greece and the government is in the right direction. That also applies to the "Transparency" pillar, based on the specific sub-pillars and indicators the Greek government is on the right track. On the other side, over the last few years, it is noticed a lack of citizen engagement, with no initiatives and actions to encourage citizens to participate in the policy process or to provide feedback on common problems that they are facing. Citizen engagement is a crucial indicator of an efficient digital government, and the Greek government must provide initiatives in order to be productive.

There are several indicators to assess the performance of the digital government, such as digital economy and society index (DESI) designed by European commission (EC), United nations long-standing questionnaire for the e-government assessment. In 2017, the Tallinn Ministerial Declaration developed a monitoring tool "the digital single market vision and broader EU2020 goals". This tool monitors and is used by the EC to provide information about the use of ICT in the public sector [3]. These indicators used some specific areas to assess the progress, but we focused on the depth version of each indicator, such as pillars and sub-pillars. In this article, we extracted the pillars and sub-pillars from the literature and policy documents and added a few of our own developed indicators, such as the use of disruptive technology in digital government initiatives. We compared two countries, one from European and the other from the Asian Continent. We also devised a scale to measure the progress of each pillar and

**Table 1.** Pillars to measure the governments' digital initiatives along with Pakistan and Greece comparisons

| Pillars | Sub-pillars | Description | Greece | Pakistan |
|---|---|---|---|---|
| **1. Government Structures** [8] | 1.1 Ministries | The number of ministries in the country | 19 | 32 |
| | 1.2 Number of services | The number of services provided from the online portal | 1399 | not mentioned |
| | 1.3 Population | The population of the country | 11Million approx | 230M approx |
| | 1.4 Area | The countries' area is in square kilometers | 131,957 km$^2$ | 881,913 km$^2$ |
| | 1.5 Staff training and knowledge management | Training and development practices adopted by the public sector will help in defining policies better. Training also brings measurable changes in knowledge skills, attitude, and social behavior of the employees | [Provided] The Greek government has several institutions for better staff training. However, there are some financial and management constraints | [Partially Provided] The Pakistani governments also train through several programs but due to lack of resources, inadequate financing and technological gadgets, it is difficult to achieve the desired output. e.g., the staff training institute is an example of this initiative |
| | 1.6 Internal communication | Internal communication promotes the valuable role that staff play in communicating your messages and ambitions and ensures that no one misses important information or updates | [Provided] Proper communication within public administration contributes to the efficient operation of the public body and the quality of service to citizens. For achieving this, training seasons and workshops should be done in regular basis. A good example of the procedure is the OECD learning and consultation workshop. | [Partially Provided] The internal communication mechanisms are available in Pakistan. Fewer ministries implemented this for instance, FBR arranged workshops for the improvement of internal communications |
| | 1.7 Government initiatives | The number of government initiatives that took place in a country | [Provided] In the last 3 years the e-government initiatives from the Greek government have been exceptional. The problems raised by the pandemic boost the effort of the Greek public sector. One of the most noticeable initiatives is the online platform of the Greek government | [Provided] The government initiatives are very common and can be found on the Pakistani govern ment official website. These initiatives have several problems (Tax, and financial limits) due to international monetary fund sanctions (IMF) |
| **2. E-services by government** [23, 27] | 2.1 Online request for services | The availability of online portals for the service demand should be active. For instance, few departments book appointime its via a phone call, and some of them use email as well | [Provided] Several services are provided in the online platform of the Greek government. The available services are based on "life events" (birth, insurance, setting up a business, etc.) while the user can also browse all the services per ministry, institution, organization, or independent authority | [Partially Provided] less online environment is observed in the Pakistani government structure. Sometimes, they did not have HR resources to respond to many requests |
| | 2.2 Financial transactions | Financial transactions are very important nowadays. Every person via mobile phone and interlinked bank account as well | [Provided] In the last 2 years, digital transactions have rapidly increased. Citizens use digital transactions in their daily life. However, not every service provides online transactions | [Provided] The financial transactions are common via different banking applications and local mobile services as well. Due to money transfer limits and taxis people use less of these mediums for their business |

(continued)

**Table 1.** (*continued*)

| Pillars | Sub-pillars | Description | Greece | Pakistan |
|---|---|---|---|---|
| | 2.3 E-procurement | E-procurement is very helpful to decide which company was better in previous projects and vice versa. The E-procurement also reduces the biasness | [Provided] The process of requisitioning, ordering and purchasing goods and services is provided by the main portal eprocurement.gov.gr. | [Partially Provided] Not all the departments are following the e-procurement. Pakistani governments usually use the print media for procurement tenders |
| | 2.5 Interoperability of Services | The interoperability of service is important, for example, if any ministry wants to access the national identity database, then it could be feasible for them to make an interoperable transaction within ministries or government departments. Although, governments are working on international/cross-border interoperability of services | [Provided] In the Greek public sector there are many cases of systems that work separately, and without any connection. The important thing is that the online platform of the Greek government has interoperability capabilities as well as some of the central services of the public sector (e.g. e-EFKA) | [Partially Provided] To some extent services are interoperable. For instance, the National Database and Registration Authority (NADRA) has several partners, but other departments are not following true service interoperability |
| | 2.6 Coordination at the national level | Coordination is also important. For instance, in the disaster management concept each department should coordinate for better problem tackling. The Covid-19 was an example, when National health institutions coordinated with other departments | [Provided] The coordination at the national level can be seen by the Greek digital portal gov.gr. The online platform provides coordination to some of the public bodies, but there is not the desired level of support | [Partially Provided] Several departments coordinate to tackle the bigger problems such as in flood times Pakistani several institutions coordinate but in passive mode. They should adopt a proactive mode |
| | 2.7 Content organization and guidelines | The content organization and guidelines for the platforms and portals is very important, otherwise accessibility to information will be reduced | [Partially Provided] Based on the Greek e-Government Interoperability Framework there are directions and standards to be followed by the public agencies at central or local levels. Unfortunately, not all public bodies follow the guidelines | [Not Provided] The paper free project is not truly initiated in Pakistan and that is why content management and organization is not locatable in several institutions |
| | 2.8 Error management | Error and log management should be installed in E-services to make the delivery smooth | [Provided] Most of the ministries are under the "gov.gr" domain. That means backup protocols and error management guideline are implemented on the websites | [Partially Provided] Error management is important to recover the online services in no time. Back up mechanism is implemented in NADRA and some judicial applications |
| | 2.9 Privacy and safety | The platforms and e-portals should use some security measures such as secure certificates and secure URLs | [Provided] Personal data protection guidelines based on GDPR are implemented on most of the ministries' websites. After the authentication process, the transmitted data is deleted from the files of the service provider | [Provided] Privacy and safety measures are observed in a few ministries. Privacy policies are implemented in NADRA. Pakistan is also developing personal data protection rules. |
| | 2.10 Sustainability of e-services | The sustainability of the e-services should be considered while conducting some kind of feedback loop. The social media team of the government must be active to make the E-service improved with the passage of time | [Provided] The digital transformation of the public administration in Greece is the key to the sustainability of e-services within the public sector. The implementation of the digital transformation strategy for the Greek public sector leads to the improvement of existing digital public services as well as the creation of new digital public services for the benefit of citizens | [Partially Provided] The sustainability of e-services is implemented by just a few ministries such as NADRA implements this, but railway, Pakistan international airlines (PIA), Police, and federal investigation authority avoid sustainability of E-services. For instance, few ministries have fewer HR resources to deal with larger service demand |

(*continued*)

**Table 1.** (*continued*)

| Pillars | Sub-pillars | Description | Greece | Pakistan |
|---|---|---|---|---|
| | 2.11 E-Agriculture | E-agriculture is important. For instance, if an area wants an abrupt response against locust attack the E-agriculture portal will help them | [Provided] The Hellenic Republic Ministry of Rural Development and Food website is an important portal for farmers and other stakeholders. This portal provides multiple web services for farmers and vegetable trafficking announcements to e-services for spraying machines | [Provided] The E-agriculture projects are highly encouraged in Pakistan, but it should be standardized for all provinces. One example of e-agriculture is the agriculture department initiative for crops, water and farmer management |
| | 2.12 E-Health | E-health portals played important roles curing the Covid-19 or any other virus which affected the world | [Provided] Based on the ministry of Health website there are numerous initiatives, for instance, the design and implementation of the Greek National eHealth Interoperability Framework (NeHIF) according to European directives, regulations, and international standards | [Provided] The government of Pakistan has built an e-health service13 in the form of personal identification cards that securely save health histories and patient data, enabling clinicians and insurers to make informed decisions based on consistent patient histories. Oladoc, doctHERa, Sehat Kahani, Commission on Science and Technology for Sustainable Development (COMSATS), eVaccsPrime Minister National Health Program, and Aga Khan Development Network e-Health Resource Centre (AKDN eHRC) are digital health initiatives |
| | 2.13 E-Energy | The billing service of electricity, gasses, and other sources should be provided by the E-energy platforms. The electricity shortfalls, and other communications must be made through the use of ICT | [Partially Provided] Although there are platforms for online electricity payment, the Greek public sector does not have a one-stop platform for these procedures. The lack of proactive services in the field of energy is something that must be sorted out | [Partially Provided] Different energy and petroleum ministries use the portals to manage the complaints and other queries. Pakistan lacks the availability of energy sharing mechanisms among the provinces using technology as other countries have |
| | 2.14 eJustice | The jurisdiction in the form of the internet for the criminal cases management. Several institutions can use this technology. For instance, Airport security must be able to use the e justice portal to control the movement of criminals | [Partially Provided] The Hellenic Republic Ministry of Justice has several ICT projects submitted on its website as well as some statistics from the Criminal proceedings from 2016. However, any other data from previous cases are not online | [Partially Provided] The ministry of law has a website for sharing the notifications, but they don't share the previous cases information online. One project for lawyer help online is eJustice (this is a private project) |
| | 2.15 ICT education | ICT education should be provided by the help of some computer labs at an early age | [Provided] The ICT implementation in the education system of Greece is increased over the last few years. The Greek Research and Technology Network is a state-owned enterprise that operates under the authority of the Greek Ministry of Education's General Secretariat for Research and Technology. Its purpose is to offer high-quality infrastructure and services to Greece's academic, scientific, and educational communities, as well as to spread information and communication technology to the general public | [Partially Provided] Pakistan doesn't have such self-initiative for the improvement of ICT education, but other countries' NGOs are working in Pakistan to increase ICT use and education. e.g., UNESCO |

(*continued*)

**Table 1.** (*continued*)

| Pillars | Sub-pillars | Description | Greece | Pakistan |
|---|---|---|---|---|
| 3. Use of Disruptive technologies [5] | 3.1 Internet of things(IoT) | IOT is emerging and several countries are using this for smart cities and countries projects | [Provided] There are some good examples and practices of IoT technologies in Greek government. The Greek state is currently developing the iot.gov.gr portal while also presenting a network guide of 135 sensor stations for the monitoring of environmental indicators | [Provided] There are a lot of examples regarding the use of ICT in Pakistan, but very limited benefits. For instance, a smart city project uses IoT to monitor traffic. However, bad road conditions and other factors hinder the IoT project |
| | 3.2 FinTech | Like private companies, government should improve the use of FinTech for better management of finance using the technology initiatives | [Provided] In the last years, the Greek government made some significant initiatives in the field of finance. Two financial actions are the Digital Step and Digital Leap aimed to provide businesses in all sectors with investments in ICT | [Provided] In order to raise knowledge about financial resource management in the country, the government has launched Financial Literacy Programs for young people |
| | 3.3 Artificial Intelligence (AI)-Robotics | AI and Robotics is emerging, governments should facilitate the youth for Biotechnology and other better projects. Several countries have robots at restaurants and hospitals, e.g., usage of robots to care for COVID-19 patients | [Provided] The Greek government has made the digital transformation as a top priority. Therefore, Greece actively implements European artificial intelligence policy and rules. The Hellenic Ministry of Digital Governance is currently in the final stages of developing its national AI strategy. Another AI policy initiative in Greece is the AI Center of Excellence coordinated by the National Center for Scientific Research "Demokritos" | [Partially Provided] President of Pakistan announced the initiative program for artificial intelligence, cloud computing, and blockchain for the future. Presidential for Artificial Intelligence & Computing (PIAIC). Several other education sectors are working on cutting-edge technologies, but in this regard, government initiative and financing are limited |
| | 3.4 Cloud-Computing | Cloud based services are very important in populated countries | [Provided] Cloud computing is becoming the norm in modern society. One of the bright examples of the operations that took place in Greece is the GRNET research infrastructures including Grid, Cloud, and HPC mentioned in 3.6. GRNET has created and managed its own public IaaS cloud solution, Okeanos, which provides cloud resources to Greek universities | [Provided] PIAIC is also dealing with the cloud computing sector to prepare the generation for the fourth industrial revolution, programs like PIAIC should be increased |
| | 3.5 Blockchain | Blockchain is famous for decentralized version of ledger technology and provides basis for the smart cities and digital oriented countries [23] | [Partially Provided] Distributed ledger technology (DLT) is under the use for the tourism safety, and supply chain management during the covid-19. Greek government also used the DLT to improve health care sector services | [Partially Provided] Pakistan is using the DLT for the remittances purposes in the banking sector to attract the workers in Malaysia. This initiative is forced by the Financial Action Task Force (FATF) to control the terror financing and money laundering |
| 4. Transparency [17, 22] | 4.1 Open Accessibility | The data, products, and services provided by the government should be accessible and transparent on its own | [Provided] Not every data, product or service is provided by the Greek government, but every public service is accessible via the Greek online portal with information about the services | [Partially Provided] Most of the time, the services and products provided by the government are not properly conveyed to the end-users. Unlike in other countries, they do not share the services through a single portal or website. Individuals can access the services and products online, but some may waste a lot of time trying to find them |

(*continued*)

**Table 1.** (*continued*)

| Pillars | Sub-pillars | Description | Greece | Pakistan |
|---|---|---|---|---|
| | 4.2 System availability | The system should be available without any disruption | [Partially Provided] The demand of users is greater than the system capacity. As a result, the Greek digital platform has some availability problems, especially during the pandemic crisis | [Partially Provided] The number of users is greater than the system's capacity. The number of queries is much higher than the available resources. Most of the time, we find the systems are unavailable for certain services, for instance, examination portals in Pakistani school boards and job portals |
| | 4.3 Quality of information, open data, big, and linked data Portals and platforms for E-government | The quality of information should be evaluated before dissemination to avoid any future problems. The portals and platforms also play an important role in transparency of e-government Big data and linked data plays an important role in understanding the semantics of data from different departments of governments in less time. Open data platforms are the need of this technological and fast computing era. For instance, open data is important for government's transparency and accountability. The open government data projects should be initiated for better citizen-oriented governance | [Provided] As mentioned in 3.12 the new open data portal provides quality real-time data, in contrast to the old version that there were no evaluation proceedings for the datasets quality of every dataset is low compared to the new open data portal. Open, big and linked data initiatives can be seen in the Greek open data portal. There are 10 different topics (e.g. Business and Economy, Education and others) with 47 datasets available. The previous version of the open data portal has more than 10000 datasets from 340 public services but the quality of every dataset is low compared to the new open data portal | [Partially Provided] In Pakistan, open data portals and initiatives are still relatively unknown and unproductive. 349 datasets are available. Most of them are following the Berners-Lee 5-star model up to 2 stars. To improve the quality of data, government ministries should stress the benefits of open data quality for better transparency and accountability. One major example regarding the use of big data in Pakistan is for statistical purposes. The Pakistan statistics bureau is using big data technologies. Although, it required more efforts for the use of big data technologies. Open data portal of Pakistan provides the data by 22 organizations and 349 datasets of Pakistan. Just a few ministries are providing the open data. The open data project needs to be extended for better transparency and accountability in the Pakistani government. The open data is just based on data that is given in any format, and no project has been found that looks at data in RDF format |

(*continued*)

**Table 1.** (*continued*)

| Pillars | Sub-pillars | Description | Greece | Pakistan |
|---|---|---|---|---|
| **5. Accountability** [9, 24] | 5.1 Financial | Financial accountability is important to avoid the corruption problems in developing countries. e.g., to control the black money | [Provided] There are some priorities of the Greek fiscal policy. First, the further strengthening of fiscal reliability to reestablish trust in the economy's medium–term prospects and reestablish access to international capital markets. Second, equitable distribution of macroeconomic adjustment costs and assistance to vulnerable households. Gradual adaptation of the fiscal policy mix in order to boost productivity and maintain a sufficient growth rate. Further information in the Greek Ministry of Finance website. | [Provided] Public Expenditure and Financial Accountability Framework (PEFA) has defined seven pillars of PFM(public financial management) which includes budget reliability, transparency of public finances, management of assets and liabilities, policy based fiscal strategy and budgeting, predictability and control in budget execution, accounting and reporting and, external scrutiny and audit. Federal Board of Revenue(FBR) and Finance Division Pakistan. |
| | 5.2 Political | Political accountability should be implemented via ICT for a transparent government. e.g., Panama papers, Pandora papers could be the issues in later stages in case of unfair political accountability | [Partially Provided] Based of the Sustainable Governance Indicators (SGI) Greek executive accountability is in the lower-middle ranks internationally. Candidate lists and agendas are often controlled by political party leadership circles. Citizens are misinformed about government policies because of the prevalence of political and infotainment-focused news | [Partially Provided] In developing countries, political instability and IMF sanctions are two of the main things that weaken them. Political instability has been seen in Pakistan since the beginning. As a result, political accountability is somewhat difficult to achieve. Several organizations are working for political accountability in Pakistan such as National Accountability Bureau (NAB) |
| | 5.3 Performance | The government performance accountability should be evaluated with pre-decided indicators such as GDP, and economy improvements | [Provided] The key indicators of the government performance accountability show a noticeable deficit in 2020, as GDP in the pandemic period of Greece presents negative growth | [Provided] The government's performance accountability is a key factor. The GDP, stock exchange indices, and inflation rate are key performance indicators; better lives for citizens; and, most importantly, citizen satisfaction cannot be overlooked |
| | 5.4 Health | The most important sector of the government. Its accountability and transparency are important by examining the facilities in health sectors | [Provided] National Health System services are provided by a combination of public facilities and a huge number of private providers hired by EOPYYThe pandemic crisis has proved that the creation of an effective network of primary care services to fulfill population requirements is by far the most pressing demand in the health system | [Provided] Ministry of national health services regulations and coordination(NHSRC) is working to make health procedures more efficient. There are other health institutions with the help of China and Iran contributing towards the health initiatives |

(continued)

**Table 1.** (*continued*)

| Pillars | Sub-pillars | Description | Greece | Pakistan |
|---|---|---|---|---|
| | 5.5 Law | The law must be followed by the citizens and politicians for better accountability of government initiatives | [Not Provided] Citizens do not receive adequate knowledge of government policies because the media is either heavily politicized or leans toward entertainment, and individual members of parliament seldom address substantial policy concerns with voters in the electoral districts that they represent | [Partially Provided] The Ministry of law and justice in Pakistan deals with legal accountability. Moreover, Pakistan's supreme court and high courts are also held accountable for the law and justice. There are few courts and few opportunities for justice, and hundreds of thousands of cases go unresolved |
| | 5.6 Technological | The manual processes of the government need to be replaced with technological solutions for better and fast accountability | [Partially Provided] The technological accountability in Greece has significantly increased over the past few years. However, there are many cases of unstable and untrusted systems in the public sector | [Not Provided] Technological accountability is less adopted in Pakistan. A lot of work needs to be based on automatic processes, so till this time, no organization is working for technological accountability. Although technology-based applications are evolving in Pakistan. |
| **6.Citizen Engagement** [17, 28, 29] | 6.1 Political Efficacy | Political efficacy is desirable for the stability of democracy. That is because modern democratic societies tend to provide citizens with the power to influence the action of their government | [Provided] E-participation empowers citizens by ensuring better interactions, increasing access to information and services, and boosting public participation in policy and decision-making using information and communication technology. The Greek government has made some significant steps towards better citizens' engagement with the online Greek platform "gov.gr". However, some basic e-participation features have been missing (e.g., online forum in "gov.gr") | [Partially Provided] Political effectiveness depends upon citizen involvement. There are several programs for political efficacy. One of them is the Citizen Portal for Overall Government. Although every other ministry should have a citizen engagement program for better political efficacy |
| | 6.2 Responsiveness to inquiry/complaints | Ensuring that public service complaints are handled effectively is a key feature of good governance and a good service to deliver | [Provided] When dealing with complaints, each of these public organizations follows its own internal procedures. As a result, there is sometimes uncertainty among the many stakeholders about the efforts being made by other entities and how to maximize efficiency while managing an excessive number of case reports | [Partially Provided] The Prime Minister also listens to the public questions on a live telecast. However, there is a problem with service delivery in Pakistan, where there is a mismatch between population and service availability. Citizen Portal for complaints and tracking system is used, but several cases where never responded because of classism |

(*continued*)

**Table 1.** (*continued*)

| Pillars | Sub-pillars | Description | Greece | Pakistan |
|---|---|---|---|---|
| | 6.3 Direct communication with elected government officials | Bringing issues of importance to the attention of those elected and communicating with government officials is an avenue for every citizen who wants to be heard | [Partially Provided] The number of direct discussions with the citizens for each elected candidate is subjective. However, most elected officials increase discussions with citizens during the election period and limit them during their period of service | [Partially Provided] It depends upon the individual electable. Sometimes it is very difficult to approach the individual national assembly members after the election to ask them about the problem. There is no mechanism for communication |
| | 6.4 Encouragement/promotion of participation | The primary goal of public participation is to encourage citizens to have meaningful influence into decision-making processes | [Partially Provided] The majority of citizens are unaware of government policies. Those who are, however, express their policy views in a variety of ways. Citizens, for example, can engage in an open electronic consultation process on proposed government initiatives, which each ministry must notify and oversee before writing a law | [Not Provided] There is no initiative to raise the involvement of individuals in participatory government. There are a few people who have access to social media, and they control the other person's point of view. Most of the population just have their problems with the inflation rate and daily basis expenses. Controlled social media and printed media is trending in developing countries |
| | 6.5 Sharing the products and outcomes created through collaboration | This will enhance the creativity of the citizen for better thinking to develop the cross-border collaborative products, and services | [Provided] The Hellenic Republic Ministry of Foreign Affairs website presents announcements, statements and speeches in collaboration with other countries. For instance, it presented the third Strategic Dialogue between Greece and the United States | [Provided] One example is the collaboration with China's government, such as CPEC for Gwadar port, the Orange Line Project, and metro bus projects are a few examples of these projects. The Ministry of planning, development and special initiatives deals with these matters |
| | 6.6 E-participation of citizens | Online feedback via portals, mobile applications, google forms, or contact us page, messages, and emails improve the e-participation of the citizens for the improvement of services based on the demands. This will create a citizen-oriented government architecture | [Partially Provided] On the side of the municipalities, some websites provided complaint and suggestion forms, allowing individuals to participate actively. However, more steps need to be done to promote contact between the government and citizens. Citizens' engagement would be increased if forums, communities, and blogs were supported | [Partially Provided] Federal and provincial gateways exist. Locals have protested late wages, traffic congestion, forced conversions, and delayed health, education, judicial, and infrastructure services. The Prime Minister's Performance Delivery Unit created the Pakistan Citizens' Portal app in 2018 to increase citizen participation. The government complaints portal. Portal is available at government agencies |

sub-pillar, such as "provided (if service is available)", "not provided (if service is not available)", and "partially provided" initiatives in digital government as shown in Fig. 2. We transformed "provided (if service is available)", "not provided (if service is not available)", and "partially provided" into 1, 0, and 0.5 real values, respectively. We calculated the mean for each main pillar based on these values, and the results are shown in Fig. 2.

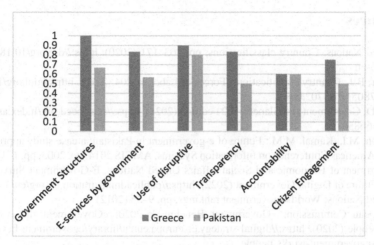

**Fig. 2.** Greece and Pakistan comparison based on their digital initiatives

## 5 Conclusion

In this research article, we devised a framework to compare the digital initiatives of the governments. Pakistan and Greece's initiatives in the field of electronic government are highlighted in the light of our developed framework. In light of our devised plan, a government (e.g., Pakistani and Greek) may elaborate on their weakest and strongest points towards e-government for better transparency and accountability. Other governments may also use this evaluation framework to evaluate and later optimize the futuristic view of e-government. Digital transformation for governments requires governments to consider the systemic use of disruptive technologies such as we tried to elaborate on all the key pillars of digital government. There are several other indicators used to measure the success of digital governments, such as Open Data Watch (ODIN), which provides indexes for each country based on their openness and coverage, but our proposed framework considers the 6 main pillars and several sub-pillars to evaluate the E-government. According to the ODIN score, Pakistan has a score of 43 based on openness and coverage, but Greece has a score of 60. Consequently, our study also proved, based on the GDIs framework, that Greece is much more advanced than Pakistan. In this study, we focused on which sectors of the government need to be improved for better e-government. This framework will show how each pillar and sub-pillar needs more monitoring and development to help governments move up the rankings of different indices like EDGI, DESI, and ODIN and, as a result, give their citizens the best services using digital technologies. Our developed framework is not limited to the assessment of Greece and Pakistan's digital initiatives; the framework is applicable to other countries' digital initiatives assessments as well. More attributes and sub-attributes related to government digital initiatives may emerge in the future, and this framework will be able to adjust for the new attributes and sub-attributes.

# References

1. United Nations: Country classifications, pp. 163–171 (2020). https://doi.org/10.18356/036 ade46-en
2. Silver, B.J.: Country classifications. Forces of Labor, p. 204 (2010). https://doi.org/10.1017/cbo9780511615702.009
3. OECD, Government at a Glance - 2021 edition (2022). https://stats.oecd.org/Index.aspx?QueryId=66856
4. Arfeen, M.I., Kamal, M.M.: Future of e-government in Pakistan: a case study approach. In: 20th Americas Conference on Information Systems, AMCIS 2014, no. 2003, pp. 1–13 (2014)
5. Department of Economic and Social Affairs United Nations, E-Government Survey 2022, The Future of Digital Government (2022). https://publicadministration.un.org/en/
6. United Nations: World e-government rankings, pp. 9–35 (2012)
7. European Commission: eGovernment Benchmark 2020: eGovernment that works for the people (2020). https://digital-strategy.ec.europa.eu/en/library/egovernment-benchmark-2020-egovernment-works-people
8. Wu, J.C.T., Tsai, H.T., Shih, M.H., Fu, H.H.: Government performance evaluation using a balanced scorecard with a fuzzy linguistic scale. Serv. Ind. J. **30**(3), 449–462 (2010). https://doi.org/10.1080/02642060802248017
9. Ibrahim, S.: web-based accountability disclosure: understanding performance and practices in the Malaysian states government, no. 04, pp. 678–699 (2022). https://doi.org/10.17605/OSF.IO/AG7SY
10. Shuja, J., Alanazi, E., Alasmary, W., Alashaikh, A.: COVID-19 open source data sets: a comprehensive survey. Appl. Intell. 1296–1325 (2020). https://doi.org/10.1101/2020.05.19.20107532
11. Manley, L., Ariss, A., Gurin, J.: Sustainable development goals and open data. Digital Development (2015). https://blogs.worldbank.org/digital-development/sustainable-development-goals-and-open-data
12. Charalabidis, Y., Flak, L.S., Pereira, G.V.: Scientific foundations of digital governance and transformation (2022). https://link.springer.com/content/pdf/10.1007/978-3-030-92945-9.pdf
13. Ravšelj, D., Umek, L., Todorovski, L., Aristovnik, A.: A review of digital era governance research in the first two decades: a bibliometric study. Futur. Internet **14**(5) (2022). https://doi.org/10.3390/fi14050126
14. Bertot, J.C., Jaeger, P.T., Mcclure, C.R.: Citizen-centered E-government services: benefits, costs, and research needs, pp. 137–142 (2008)
15. Lourdestorres, V.A.: Public administration - 2011 - torres - performance measurement in Spanish local governments a cross-case comparison study.pdf (2011)
16. Rehman, M., Kamal, M.M., Esichaikul, V.: Adoption of e-government services in Pakistan: a comparative study between online and offline users. Inf. Syst. Manag. **33**(3), 248–267 (2016). https://doi.org/10.1080/10580530.2016.1188570

17. Irawan, B., Hidayat, M.N.: evaluating local government website using a synthetic website evaluation model. Int. J. Inf. Sci. Manag. **20**(1), 449–470 (2022)
18. Lnenicka, M., Nikiforova, A.: Transparency-by-design: what is the role of open data portals? Telemat. Inform. **61**(March), 101605 (2021). https://doi.org/10.1016/j.tele.2021.101605
19. Suri, P.K., Sushil: Effectiveness of strategy implementation and e-governance performance. Eval. Program Plann. **92**(December 2021), 102063 (2022). https://doi.org/10.1016/j.evalprogplan.2022.102063
20. Gonzalo, J., Ruiz, R., Antonio, J., Zarate, A.: Electronic government and digital literacy: temporal comparison in the use of electronic government and digital literacy : temporal comparison in the use of technology in Mexico, no. July (2021)
21. Charalabidis, Y., Alexopoulos, C., Loukis, E.: A taxonomy of open government data research areas and topics. J. Organ. Comput. Electron. Commer. **26**(1–2), 41–63 (2016). https://doi.org/10.1080/10919392.2015.1124720
22. Bello, O., Akinwande, V., Jolayemi, O., Ibrahim, A.: Open data portals in Africa: an analysis of open government data initiatives. Afr. J. Libr. Arch. Inf. Sci. **26**(2), 97–106 (2016)
23. Rizun, N., Ciesielska, M., Pereira, G.V., Alexopoulos, H.C.: Mapping determinants of unintended negative consequences of disruptive technologies use in smart cities. In: Twenty-Ninth European Conference on Information Systems (ECIS 2021) (2021)
24. A series of individual studies entiled: Pillars of Good Governance – Focus on State Audit Office of Hungary as a Supreme Audit Institution. State Audit Office of Hungary (2016). https://www.asz.hu/en/pillars-of-good-governance
25. T. International the global coalition against Corruption: corruption perceptions index (2021). https://www.transparency.org/en/cpi/2021/
26. European Commission: The Digital Economy and Society Index (DESI). Policies (2021). https://digital-strategy.ec.europa.eu/en/policies/desi
27. Ciesielska, M., Rizun, N., Chabik, J.: Assessment of E-government inclusion policies toward seniors: a framework and case study. Telecomm. Policy **46**(7), 102316 (2022). https://doi.org/10.1016/j.telpol.2022.102316
28. Jaeger, P.T., Bertot, J.C.: Designing, implementing, and evaluating user-centered and citizen-centered e-government. Int. J. Electron. Gov. Res. **6**(2), 1–17 (2010). https://doi.org/10.4018/jegr.2010040101
29. Saleh, A.A., Alyaseen, I.F.T.: E-governance system key successful implementation factors. Int. J. Perceptive and Cognitive Computin **8**(1), 40–46 (2022). https://journals.iium.edu.my/kict/index.php/IJPCC/article/view/267%0A, https://journals.iium.edu.my/kict/index.php/IJPCC/article/download/267/163

# Digital Solutions to What?
## - WPR as a Model for Public Servants Seeking a Better Grip on Their Local Digitalization Policy

Marcus Heidlund[✉] and Katarina L. Gidlund

Mid Sweden University, Sundsvall, Sweden
Marcus.heidlund@miun.se

**Abstract.** Public-sector digitalization has gained traction over the years, and with it has come a flood of official documents (policies and grey literature) highlighting what the (digital) future is supposed to look like and proposing a range of digital solutions to inspire action. Such policies and strategic documents propose what will be important in future societies. In this paper, we employ the policy-analysis framework, 'what's the problem represented to be' (WPR), first developed by Bacchi. We conducted a workshop with a Swedish municipality, inviting key actors to work with the idea of digitalization to re-read their digitalization policy in light of the WPR framework. The purpose of this paper is to investigate what surfaced when the policymakers and public servants used WPR to dissect their own digitalization policy. The results show that the key actors' reflections centred around the value of the policy itself, and the WPR framework seemed to enhance their ability to reflect upon the usability of the policy and the work needed to implement and evaluate it. Furthermore, they pinpointed that the digitalization policy appeared rather naïve in terms of contextual factors (lack of recontextualisation on the municipal level) and hindrances (lack of resources to tackle existing hindrances).

**Keywords:** Digitalization · e-Government · policy analysis · WPR

## 1 Introduction

Public-sector digitalization has gained traction in recent years, bringing with it a flood of official documents (policies and grey literature) highlighting what the (digital) future is supposed to look like and proposing a range of digital solutions to inspire action. Such policies and strategic documents illustrate what may be important in a future society, and as Keeney (1996) states, strategic objectives – if enacted properly – provide direction for all decisions within an organisation. Thus, while policy analysis is a rich field in political science, with a variety of available frameworks (for further reading, see Dunn, 2015), there are fewer studies in information systems research analysing the idea of digitalization with the help of established policy-analysis frameworks bridging disciplinary gaps and making interdisciplinary contributions. This study aims to contribute to this area and earlier research such as Savoldelli et al., 2012, Bolgherini, 2007 and Goldkuhl, 2016 who all raise questions on the linkage in between policy and practice also pointing

© The Author(s), under exclusive license to Springer Nature Switzerland AG 2023
M. Papadaki et al. (Eds.): EMCIS 2022, LNBIP 464, pp. 242–250, 2023.
https://doi.org/10.1007/978-3-031-30694-5_18

to the uniqueness of public sector contexts. (Salvoldelli et al., 2012) showed in their study of two decades of e-Government research that one dimension of the 'adoption paradox' is transparent and trustworthy policy decision making processes (Savoldelli, 2012). Bolgherini stress the political, social and cultural nature and claims that these are downplayed by a focus on massive technological intervention (Bolgherini, 2007). Whereas Goldkuhl in his literature review highlights what he calls the policy principle and argues that in public sector the IT artefact is policy-ingrained (Goldkuhl, 2016).

One tool that is increasingly having an impact in political science is Bacchi's (2012) 'what's the problem represented to be' (WPR) framework, and WPR has been used in studies of digitalization policies (see, for example, Sundberg, 2019; Syrstad Høydala and Haldar, 2021; Nyhlén and Gidlund, 2022; Duval Jensen et al., 2022). Sundberg (2019) identified that the overall problem identified with WPR is that digitalization is described as providing certain benefits but the only way to reap these benefits is to adapt to digitalization itself. Sundberg (2019: 8) concludes, "However, since no alternatives to the proposed development are presented, digitalization is ascribed autonomous features, as a solution in search for problems". In a study on digitalization of the Norwegian education system, Syrstad Høydala and Haldar (2021) found that the overall problem to be solved that surfaced was that the education system should use digitalization to provide digitally competent future citizens and workers. Nyhlén and Gidlund (2022) investigated three levels of digitalization policies (EU, national, and sub-national) and conclude that policymaking is trapped in a form of technological determinism, with the answer to every societal problem being assumed to be digitalization in one form or another, without further recontextualisation. Duval Jensen et al. (2022) studied healthcare documentation in Denmark and identified that digital healthcare documentation appears to be the solution to most problems and that a standardised documentation contributing to a lack of individualised healthcare. As these earlier studies indicate, the WPR framework provides insights that might otherwise be hidden during the policymaking process – consciously or unconsciously overlooked – but which may come to the surface during practical work in the different local contexts.

To contribute to the stream of research above, trying to further enlighten the relation in between digitalization policies and their enactment on an overall level and on goal achievement and transparency of what is supposed to be solved in specific, this paper utilizes Bacchi's WPR model but in a slightly different setup. The aim of the study focused on how policymakers and public servants themselves could re-read their digitalization policy with the support of the WPR framework. The study involved a semi-structured workshop in a Swedish medium-sized municipality, with key actors chosen by the head of digitalization. The purpose of this paper was to investigate what surfaced when the policymakers and public servants used WPR to dissect their digitalization policy.

The structure of this paper is as follows. Section 2 presents Bacchi's (2012) What's the problem represented to be (WPR) framework. Section 3 details the method and analysis applied in this paper, while Sect. 4 presents the results and analysis. Finally, Sect. 5 gives the concluding remarks and makes suggestions for future research.

## 2  Theoretical Framework: What's the Problem Represented to Be?

Carol Bacchi (2009) explains that policy has a key role in government and is part of how governing takes place. Bacchi (2009) states that the perception of policy is a good thing that indicates that something needs to be fixed, and if something needs to be fixed there has to be a problem. However, this problem is not always described or even named. In response, Bacchi (2009; 2012) created the 'what's the problem represented to be' approach and describes it as follows: "The 'WPR' approach is a resource, or tool, intended to facilitate critical interrogation of public policies. It starts from the premise that what one proposes to do about something" (Bacchi, 2012: 21). WPR thus identifies not only the problem the policy seeks to solve but also how this problem is represented. A WPR analysis is guided by the following six questions (Bacchi, 2012: 21):

1. What is the problem represented to be in a specific policy?
2. What presumptions or assumptions underpin this representation of the problem?
3. How has this representation of the problem come about?
4. What is left unproblematic in this problem representation? Where are the silences? Can the problem be thought about differently?
5. What are the effects are produced by this representation of the problem?
6. How and where has the representation of the problem been produced, disseminated, and defended? How could it be questioned, disrupted, and replaced?

The first question seeks to clarify the problem addressed by the policy, while the second question targets the underlying premises of this representation of the problem. Question 3 examines the contingent practices and processes that created this under-standing of the problem, thereby highlighting a space for challenge and change (Bacchi, 2012). The fourth question critically examines the gaps and limitations in this repre-sentation of the problem, opening up the possibility of alternatives. Similarly, question 4 opens up a space in which to imagine different futures. Question 5 reflects on the effects or consequences of this problem representation and how it limits what can be talked about as relevant (Bacchi, 2012). Finally, question 6 seeks to increase awareness of the problem representation, also encouraging to think about how the policy could be replaced.

When WPR is used to question, re-read, and analyse one's own policy or proposal, the framework's thinking and ideas then become part of the material to be analysed. Bacchi (2012) argues that policies are not the best way of solving a problem, but they can determine what will be done and what will not. Hence, the purpose of the WPR analysis is to critically examine the conceptual logic and assumptions of public policies. In this way, Bacchi (2012) explains, that the view of the public becomes governed by problematisations rather than policies.

In this study, we used the WPR framework to contribute to the vast field of research into good governance, including the importance of transparency and accountability (Hood and Heald, 2006). More specifically, like Bozeman and Bretschneider high-lighted already in 1986 and Rochelau and Wu (2002) and Wang et al. (2018), i.e. the differences between initiatives in private and public information systems – (or, as better known today, the digitalization of the public versus the private sector). Bozeman and

Bretschneider (1986) described the differences as greater risk-aversion, divided authority, multiple stakeholders with competing goals, short-term budgets related to political management, highly regulated procurement processes, and many links between programmes and organisations driven by legal requirements and other public limitations. Furthermore, a literature review by Agostino et al. (2022) found that discussions of accountability issues in public-sector digitalization have primarily been published in public administration journals (see, for example, Mergel, 2017; Wang et al., 2018). Whereas accounting journals focusing on private-sector digitalization, overlooking the public sector. The review by Agostino et al. stresses the importance of accountability in translations in order to deal with what they describe as more multicentric and blurred processes, and translations and the role of translators (Agostino et al., 2022). This is also why we chose to focus on translators as key actors working to clarify digitalization policies.

## 3  Empirical Material and Methodology

The purpose of this paper was to investigate what surfaced when policymakers and public servants used WPR to dissect their digitalization policy. The research and empirical material were gathered during a semi-structured workshop in a medium-sized Swedish municipality. This municipality, like many other Swedish municipalities, invests substantial resources in digitalization. The participants in the workshop, selected by the municipality's head of digitalization, were five employees working with digitalization questions in either the municipality or one of the municipality companies. The selection process was based around finding suitable employees that work with digitalization both on a strategic but also concrete level in the municipality. The participants role in the municipalities are displayed in table 1 below. The workshop, as explained previously, was developed around Bacchi's WPR framework (described in more detail later in this section), and the theme of the workshop was the digitalization policy.

**Table 1.** Participants role in the municipality

| Participant number | Role in the municipality |
| --- | --- |
| Participant 1 | Head of Digitalization |
| Participant 2 | Chairman of the municipal companies' IT council |
| Participant 3 | Digitalization strategist |
| Participant 4 | Business developer |
| Participant 5 | Unit manager for digital development |
| Participant 6 | Information Security Coordinator |

The objective of the workshop was for the key actors to question and problematise the current digitalization policy. The municipality was about to embark on creating a new policy and they wanted to identify the strengths and weaknesses of the current policy to be taken over into the new iteration. The policy document was selected with the help of the head of digitalization, as the municipality had a wide range of official documents touching on digitalization and this document was deemed the most appropriate for the task at hand. The digitalization policy selected is centred around five goals or perspective namely: digital safety, digital competence, digital leadership, digital sustainability and digital innovation.

The workshop began with the authors giving a brief presentation on policy analysis and then Bacchi's (2012) WPR approach and presenting the six questions (see Sect. 2). The questions were translated into Swedish and presented in everyday language:

1. What problem does the policy identify?
2. What underpinnings are the problem description based upon?
3. Is there anything taken for granted or presented as unproblematic in the policy?
4. Are there points of view other than those described in the policy?
5. What happens if one follow or does not follow the policy?
6. How was the policy created? How could it be replaced?

The question formulations were modified from the original to adapt them to the study context, keeping in mind that the participants had no prior experience with policy analysis. Thus, this modified set of questions was better aligned with the topic of the workshop and to the targeted participants. Prior to the workshop, the participants were sent a PowerPoint document (containing these questions) in which they could take notes throughout the workshop, and they were given a copy of the digitalization policy chosen for discussion. After the WPR presentation, the participants divided themselves in three pairs and went into different rooms to begin their analyses. After 45 min, the particiants were asked to come back to discuss their findings for another 45 min. During the workshop, the participants were asked to type their reflections into the PowerPoint documents, under each of the six questions; and these notes were gathered at the end. As there were only three pairs, most of the empirical material could be presented.

The empirical material gathered for this paper consisted of the notes made by the participants during their analysis sessions, combined with field notes taken by the authors during the workshop and discussions. The material was subject to a directed qualitative content analysis (Hsieh and Shannon, 2005). As the material and workshop were structured by the six WPR questions. This gave that, the workshop was semi-structured, and the participants worked with one question at a time, and this order gently led the participants through an increasingly critical examination. The concept of using WPR for public servants is something that Bacchi (2012) touches on whom state that applying these six questions in relation to one's own policy or proposal allows the policymaker to incorporate their own thinking into the material. As such, in this study, the participants conducted the initial analyses, and their analysis were then re-analysed. This is a slightly different approach to the one described earlier. The advantage of this approach is that the empirical material then included not only the policy itself, but also the key actors' analyses of the same, which provided a richer and more interactive picture of the policy in its context.

# 4   Results and Analysis

As mentioned above, the analysis involved two sets of empirical material: the key actors' notes from their analyses, as guided by the six questions, and the notes taken throughout the workshop to support reflections that the participants made that might not be included in their own notes. The analyses strictly followed the order of the six WPR questions, and the directed qualitative content analysis (Hsieh and Shannon, 2005) ensured that the answers to these questions (expressed as text notes in the Power Points and as verbal expressions in the reflection part of the workshop that were collected as field notes) provided the results.

## 4.1   What is the Problem Represented to Be?

None of the groups identified any clear statements uncovering problems (WPR1) to which digitalization would be the solution in the given policy. However, they did identify a set of implied challenges, such as the "digital leap", explaining that the municipality must commit to digitalization efforts to reap the benefits of enhanced services and increased efficiency, with a competitive race embedded in that. Moreover, the policy included a range of goals and a vision that the municipality wanted to achieve, such as being open, equal, and attractive municipality; being the best school municipality; being a municipality in which it is safe to grow old; decreasing unemployment; increasing occupation; and being environmentally sustainable. However, these goals did not have a clear connection to the challenge on which the policy had been built on. This is in line with the findings of Sundberg (2019), who argues that we must adapt to the digital society in order to reap its benefits, but this is so taken for granted that we miss explaining the mechanisms behind a competitive stance.

The second (modified) WPR question (WPR2) asked about the underpinnings of the problem description. One group referred to the five goals of the policy, and the other two took up analysis of the surroundings, such as identifying the (mega)trends and taking stance in official national and pan-national documents and strategies. It is common in Swedish municipalities to use the goals of the national digitalization strategies. However, as pointed out by some of the participants, these strategies do not translate well to the local challenges and the capacity of local operations, thus contributing to a rather naïve and unreflective enactment of national policies.

The third question (WPR3) asked the participants to identify what had been taken for granted or represented as unproblematic in the policy. Here, the participants reflected on the lack of problem insights that appeared to assume there would be no roadblocks or hindrances to achieving the goals set out in the policy (partly related to the abovementioned unreflected or even naïve stance). Also taken-for-granted was the availability of the resources for achieving the goal exist and will be provided and architecture, and infrastructure is available. This in line with the Nyhlén and Gidlund (2022) finding regarding taken-for-granted(ness) that is reproduced, though in this case it is being assumed that digitalization will occur. In a similar manner, Heidlund and Sundberg (2021) identified that few other alternatives to digitalization were being presented to stakeholders.

The fourth question (WPR4) sought to gauge whether there were points of view other than those presented in the policy. Here, the groups explained that the views taken in the policy were based on national strategies and suggested that the policy lacked a clear connection to the municipality and its core operations. One group asked, "What does the citizen want and need?". This touches on something interesting: while it can be difficult to talk about the citizen in any overarching national policies, as identified by Schou and Hjelholt (2019), this can be beneficial in smaller settings, such as the municipal context.

Question five (WPR5) strayed somewhat from Bacchi's original model but was nonetheless important, as it sought to capture what would happen if one followed (or did not follow) the policy. Here, the participants reflected on that they might not follow the strategy today. They said that there seemed to be no clear assignment of whom should evaluate whether the policy goals had been achieved and what the consequences would be if they had not.

The final question (WPR6) asked the participants how the policy had been created and how it could be replaced. While they were uncertain about the former, they all had ideas about how it could be replaced and made more meaningful. One idea involved the "digital transformation plan", which is narrower and showed how the municipality could benefit from digital transformation grounded in a "citizen first" principle.

As Bacchi (2012) states, a WPR analysis of one's own policy can provide additional insights that can then become part of the material itself. As such, this workshop not only invited a critical examination and identification of the shortcomings of the current policy, more so it also contributed by providing inspiration and learning. During the discussion of these findings, many ideas and lessons emerged with regards to the creation of the new policy. Two key points arose. First, the policy should be more closely tied to the municipalities' own operations and must clarify how digitalization can add value to these. Second, the participants noted the lack of problems identified in the current description, as well as the lack of resources and evaluation protocols; hence, future policy should be more grounded in problems than in vision and goals.

## 5   Concluding Remarks

The purpose of this paper was to investigate what surfaced when policymakers and public servants used WPR to dissect their digitalization policy in order to contribute to earlier research stressing the importance of understanding the political aspects of digitalization in public sector (Salvodelli et al., 2012, Bolgherini, 2007 and Goldkuhl, 2016). As such this study's objective is to add to the existing more theoretical and conceptual contribution by putting forward empirically-based nuances on the enactment of policies in practice. The paper highlights how policymakers and public servants themselves could re-read their digitalization policies with the support of the WPR framework. As shown in earlier research, the process of translations is vital to ensure accountability (Agostino et al., 2022), and tentative results show that emerging critical digital-accountability issues are multicentric accountability, the blurring of accountability roles and boundaries, increasing relevance of translation processes and translators' roles (Agostino et al., 2022).

In this study, the key actors' reflections centred around different aspects of the value of the policy itself. The WPR framework seemed to support their ability to reflect upon the usability of the policy and the work required to implement and evaluate it. The participants indicated that the policy appeared rather naïve in terms of reference to contextual factors (a lack of recontextualisation at the municipal level) and hindrances (a lack resources for tackling existing hindrances). Hence, they stated opinions on that future policies should be narrower in the sense of, targeting specific problems faced by the municipality and connecting the policies to their operational capabilities. Overall it is problematic when digitalization policies present digitalization as a solution to all challenges per se (similar to the findings of Nyhlén and Gidlund, 2022), rather than responding directly to local problems. This is in line with the results by Agostino et al. (2022), which stress the need for critical issue of local translation processes.

Our results indicate that WPR appears to be a promising tool for public servants to analyse their own policy proved to be a success in this context. The participants found the questions and areas of discussion to be useful, and they gained deeper insights into policy construction by using the framework to dissect their own current policy. With policymakers and strategists analysing and questioning their own policies in this way, the next generation of digitalization strategies could move away from the discourse of describing digitalization as the only alternative and such become a goal in itself to provide more contextualised and transparent narratives. As for future research we encourage scholars to pursue this type of empirical work, including the policy makers in the process of analysing their own policies, to substantiate arguments of the importance of political aspects of public sector digitalization. While we choose Bacchi's WPR approach there are of course a multitude of policy tools that can be applied and it would also be interesting to do a similar study in other contexts such as higher-level government (national). This approach could, as Bacchi (2012) states, contribute to ensuring the public is governed by problematisation, rather than policies.

# References

Agostino, D., Saliterer, I., Steccolini, I.: Digitalization, accounting and accountability: a literature review and reflections on future research in public services. Financ. Account. Manage. **38**(2), 152–176 (2022)

Bacchi, C.: Analysing Policy. 1st edn. Pearson Higher Education, Australia (2009)

Bacchi, C.: Introducing the 'what's the problem represented to be?' approach. Engaging with Carol Bacchi: Strategic interventions and exchanges, pp. 21–24 (2012)

Bolgherini, S.: The technology trap and the role of political and cultural variables: a critical analysis of the e-government policies. Rev. Policy Res. **24**(3), 259–275 (2007)

Bozeman, B., Bretschneider, S.: Public management information systems: theory and prescription. Public Adm. Rev. 475–487 (1986)

Dunn, W.N.: Public Policy Analysis. Routledge (2015)

Duval Jensen, J., Ledderer, L., Beedholm, K.: How digital health documentation transforms professional practices in primary healthcare in Denmark: a WPR document analysis. Nurs. Inq. e12499 (2022)

Goldkuhl, G.: E-government design research: towards the policy-ingrained IT artifact. Gov. Inf. Q. **33**(3), 444–452 (2016)

Heidlund, M., Sundberg, L.: How is digitalization legitimised in government welfare policies? An objectives-oriented approach. In: EGOV-CeDEM-ePart-*, pp. 199–207 (2021)

Hood, C., Heald, D.: Transparency: The Key to Better Governance?, vol. 135. Oxford University Press for The British Academy (2006)

Hsieh, H.F., Shannon, S.E.: Three approaches to qualitative content analysis. Qual. Health Res. **15**(9), 1277–1288 (2005)

Keeney, R.L.: Value-focused thinking: Identifying decision opportunities and creating alternatives. Eur. J. Oper. Res. **92**(3), 537–549 (1996)

Mergel, I.: Building holistic evidence for social media impact. Public Adm. Rev. **77**(4), 489–495 (2017)

Nyhlén, S., Gidlund, K.L.: In conversation with digitalization: myths, fiction or professional imagining?. Inf. Polity (Preprint) **27**, 1–11 (2022)

Rocheleau, B., Wu, L.: Public versus private information systems: Do they differ in important ways? A review and empirical test. Am. Rev. Public Adm. **32**(4), 379–397 (2002)

Savoldelli, A., Codagnone, C., Misuraca, G.: Explaining the egovernment paradox: an analysis of two decades of evidence from scientific literature and practice on barriers to egovernment. In: Proceedings of the 6th International Conference on Theory and Practice of Electronic Governance, pp. 287–296 (2012)

Schou, J., Hjelholt, M.: Digitalizing the welfare state: citizenship discourses in Danish digitalization strategies from 2002 to 2015. Crit. Policy Stud. **13**(1), 3–22 (2019)

Sundberg, L.: If digitalization is the solution, what is the problem? In: European Conference on Digital Government, pp. 136-IX. Academic Conferences International Limited (2019)

Syrstad Høydal, Ø.S., Haldar, M.: A tale of the digital future: analyzing the digitalization of the Norwegian education system. Crit. Policy Stud. **16**, 1–18 (2021)

Wang, C., Medaglia, R., Zheng, L.: Towards a typology of adaptive governance in the digital government context: the role of decision-making and accountability. Gov. Inf. Q. **35**(2), 306–322 (2018)

# Concept for an Open Data Ecosystem to Build a Powerful Data Environment

Larisa Hrustek[1]([✉]) [iD], Renata Mekovec[1] [iD], and Charalampos Alexopolus[2] [iD]

[1] Faculty of Organization and Informatics, University of Zagreb, Pavlinska 2, Varazdin, Croatia
lhrustek@foi.unizg.hr

[2] Department of Information and Communications Systems Engineering, University of the Aegean, 83200 Karlovassi, Samos, Greece

**Abstract.** Today, data is seen as the starting point for change and innovation in business and management processes. Since the goal of an open data ecosystem is for stakeholders to actively create and use data, their needs must be clearly identified and defined. An analysis of the ecosystem's key elements, characteristics, and modes of operation is required to create an environment that efficiently harnesses the full potential in the creation, use, and reuse of data. This paper proposes an approach for developing a conceptual model of the open data ecosystem in a given environment based on the identified key elements and their characteristics. The following key elements of the open data ecosystem are: Stakeholders, Data, Infrastructure, and Policy/Governance. The identified elements require detailed engagement and elaboration within the open data ecosystem. By clearly identifying these elements, a strong and resilient open data ecosystem can be analyzed and built in any industry such as agriculture, transport, education, law, finance and other potential field or specific field in the public sector, contributing to long-term growth and development.

**Keywords:** Ecosystem Approach · Open Data Ecosystem · Elements · Open Data

## 1 Introduction

Open data provided an opportunity to establish a participatory society, develop innovative data-based solutions, discover new business opportunities and make better decisions in the public and private sectors [1]. Data is considered a driver of change and innovation in governance and business processes, but the realization of its potential requires building and functioning in a targeted environment called an ecosystem. The ecosystem in a broader sense represents a system in which entities work in collaboration and interdependently, as well as strive to contribute to the ecosystem [2]. In a narrow sense, an ecosystem is an environment with its own specificities and includes stakeholders who contribute to the ecosystem by sharing resources, knowledge and skills, striving to achieve social and business goals. In many domains, the value potential of data is recognized and consequently increasingly dominate by interdependent services and data

M. Papadaki et al. (Eds.): EMCIS 2022, LNBIP 464, pp. 251–263, 2023.
https://doi.org/10.1007/978-3-031-30694-5_19

exchange between different stakeholders, which leads to the emergence of data ecosystems [3]. Different research has been conducted in the open data management field, but knowledge about data ecosystem management and its operation is limited [1, 4, 5], specifically in the various professional field. Nevertheless, the creation, use and reuse of data [1] are necessary for the successful construction and development of a data-driven culture, as one of the strategically important directions in the growing data economy according to European strategy for data [6]. Accordingly, there is a need to research data ecosystems, that is, to identify the key elements, characteristics and ways how to successfully achieve principles that such a system should satisfy. In this research, an ecosystem approach was applied and it was analyzed how the ecosystem elements contribute to the realization of this approach principles. In this paper, the research questions (RQ) are defined as a following:

RQ1: What are the elements of the (open) data ecosystem?
RQ2: What characteristics should the (open) data ecosystem satisfy for the purpose of successful operations?
RQ3: Which identified elements of the (open) data ecosystem have an impact on the realization of the principles of the ecosystem approach?

The contribution of this research stems from the need to create circular data ecosystems and strengthen specific data environments. A circular data ecosystem presents an environment where stakeholders together contribute to building a common field where they operate in a way that shares their own data and uses data from others, they know data needs and upgrade data flows. To create an environment that successfully leverages its full potential in the creation, use, and reuse of data, an analysis of the ecosystem's key elements, characteristics, and modes of operation are essential.

The following part provides an explanation of the ecosystem approach used in this research, a description of the problems observed, and the research design. The third part of the paper provides a literature review with an analysis of the results of existing research. The fourth part provides a presentation of the results, while the fifth part provides a discussion and suggestions for further research.

## 2   Research Approach and Design

### 2.1   The Ecosystem Approach

Originally, the ecosystem approach implies a strategy for the integrated management of land, water and living resources that promotes conservation and sustainable use in an equitable manner [7–9]. This approach seeks to achieve a balance in conservation, sustainable use as well as a fair and equitable sharing of the benefits arising from the use of resources [7–10]. First, this approach came to settle down in natural science such as biological and environmental science, while later it found useful application in different sciences such as information science. The ecosystem approach has become a guide through various theoretical, conceptual and practical aspects of complex systems relevant to socio-ecological management [11]. Each complex system with its own specific, elements, characteristics and objectives are called an ecosystem. An ecosystem is defined as a system of stakeholders, practices, values and technologies in a particular environment, where the mentioned elements are connected and interdependent [12]. The literature mentions 12 principles (P) on which the ecosystem approach is based. They are shortly mentioned as follow: The management objectives of resources are a matter of societal choice (P1), Decentralized management to the lowest appropriate level (P2), Ecosystem managers should consider their effects (P3), Recognizing potential gains from management (P4), Conservation of ecosystem structure and functioning (P5), Ecosystems must be managed within the limits of their functioning (P6), The ecosystem approach should be undertaken at the appropriate spatial and temporal scales (P7), Long-term objectives for ecosystem management (P8), Management must recognize that change is inevitable (P9), Balance between, and integration of, conservation and use of biological diversity (P10), Processes based on information, including scientific and indigenous and local knowledge, innovations and practices (P11), Involvement of all relevant stakeholders (P12) [7, 9, 13]. The operation of the ecosystem based on these principles is significant for the development of a sustainable and strengthened environment, as, for example, a data ecosystem should be. Essential elements and characteristics of some ecosystem contribute to the realization of these principles. The application of the ecosystem approach is necessary for the analysis of the elements and characteristics of a specific ecosystem, as well as the achievability of the principle.

## 2.2 Research Problem and Motivation

According to the ecosystem approach, dependence on other parties is necessary for the strategic management and operation of the ecosystem. Governance institutions, which have largely been initiators and promotors of open data initiatives, have launched open data portals, and are increasingly dependent on other organizations that produce high-quality data. Data management in such organizations is becoming increasingly critical. In addition to existing research on data management within organizations, data management in the ecosystem is more difficult to understand and very limited is known about it [4]. Despite the great interest in the data ecosystem, clear definitions are lacking [1]. The information and data flow between different stakeholders are obscure and their roles are not clearly defined [1, 5]. The publication and exchange of data in the private sector are encouraged [6], but the significant needs for different types of valuable government and private data remain superficial and incomprehensible [1]. The impact of open data in specific sectors of public governance has not yet been deeply understood and assessed [1], as well as initiatives at the level of data ecosystems are missing. Also, it is discussed a lot of issues and concludes that there is a paradox that open data (publication need) potential is decreased due to the provision of already enhanced (well-developed) e-commerce and e-government services [14]. Therefore, the observed problems can be classified in a narrow and broad sense.

Several problems were identified in a narrow sense of the data ecosystem:

- unclear definition of a data ecosystem;
- the lack of a sound identification of the key elements of the data ecosystem and limited knowledge of its characteristics;

- the relationship confusion and unclear data flow between the stakeholders of the ecosystem;
- lack of focus on developing and building the data ecosystem in practice;
- lack of data management knowledge and skills of those involved in the data ecosystem.

In a broader sense, the mentioned problems are reflected in the limited growth and development of the open data field which causes the following problems:

- lack of use and reuse of open data
- questionable sustainability of open data initiatives.

## 2.3   Research Design

The study was divided into two parts. The first part of the research focused on determining the answers to RQ1 and RQ2. The second part of the research provided an overview of the connection of elements to the principles of the ecosystem approach and provided an answer to RQ3.

For the first part of the research, based on the ecosystem approach, the guideline for reporting systematic reviews Preferred Reporting Items for Systematic Reviews and Meta-Analyses (PRISMA) was applied. PRISMA guidelines enable the implementation of research through the identification of relevant literature in three phases: Identification, Screening and Included. Papers from the relevant databases in the investigated research field Web of Science (WoS) and Scopus were included in the investigation process. The

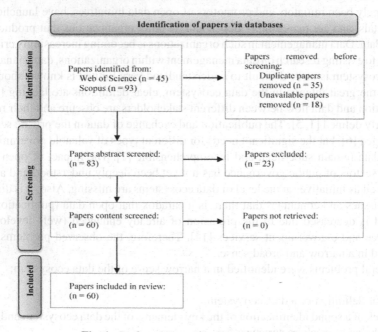

**Fig. 1.** Preferred Reporting Items in research

complex query in the research was: "data ecosystem" OR "open data ecosystem" AND "elements" OR "components".

In the Identification phase (Fig. 1), a list of papers from WoS and Scopus was gathered based on a complex query, and duplicate as well as unavailable papers were removed. In the Screening phase, an analysis of abstracts was made, and papers whose thematically did not correspond to the researched field were excluded from the research. Secondly, papers whose content corresponds to the selected research topic were subjected to a detailed analysis. Focus in literature analysis is placed on papers that deal with the topic of "data ecosystem" or "open data ecosystem", and which in the content include an analysis of the elements or components of the data ecosystem, as well as the characteristics, prerequisites or principles of the ecosystem. The Included phase represents the total number of papers included in the results of this research.

## 3   Literature Review

With the development of appropriate technologies and digital society, data has become a tradable and valuable resource [15]. The establishment of industrial open data ecosystems is still in its infancy with many unknowns and unmapped in how they should be built and operate [16]. Challenges and ambiguity in all their key elements, the development of data ecosystems represent an obstacle [16] and require a research effort in defining the approach. An open data ecosystem is an environment of stakeholders who can mutually exchange, produce and consume data and its provide conditions for the creation, management and maintenance of data exchange initiatives [15]. The conditions that the data ecosystem should satisfy include the multidimensional nature of different perspectives such as economic, technical, operational, legal, social, political and institutional perspectives [15–18].

The previous literature individually deals with the research of different constructs of data ecosystems. Stakeholders in the data ecosystem are mentioned as one of the most important constructs [1, 19–22]. They can simultaneously be producers and/or users of data with different roles in the data ecosystem. The activities they carry out take place under different conditions [15]. The next more significant construct is the technologies, standards and tools used in the data ecosystem [22–24]. They enable networking in the ecosystem, actions on data and visualization of the data environment [25, 26]. Data, as a purposeful resource in the ecosystem, is the construct around which the most significant activities of the data ecosystem are made [27]. Policy/Governance is responsible for defining policies, guidelines and dissemination of data use and reuse activities [4, 28, 29]. Also, their role is significant in providing the infrastructure and support law perspective in data initiatives, such as control of data publishing obligations and rights as well as care of General Data Protection Regulation (GDPR). The concluding reflection points to a set of heterogeneous, dynamic and evolving constructs in the landscape of the data ecosystem [15].

### 3.1   Open Data Ecosystem Elements

The open data ecosystem is an environment that gathers stakeholders at multiple levels [18, 21, 30, 31] and infrastructurally supports all processes related to data [25, 32], which are key resources for creating diverse values in society [30]. The purpose of a data ecosystem is to actively create and use data by stakeholders, therefore their needs should be clearly identified and defined. Stakeholders must know each other's needs in order to be able to contribute to the common goal in the ecosystem such as sharing resources (data), providing support (upgrading of skills and knowledge, infrastructure support), and creating innovations (development of new products, the innovation culture). The ecosystem can be shown through the basic elements and their characteristics with certain contextual conditions that must be met for its process and technological functioning. Key elements of the open data ecosystem are:

1. Stakeholders
2. Infrastructure
3. Data
4. Policy/Governance.

In the literature analysis, emergent entities of the data ecosystem were identified. Entities are identified as elements of the data ecosystem without which it cannot operate. A total of 60 articles were analyzed and the mentioned entities that were partially or completely subject to the treatment of the topic were extracted. One of the most frequently mentioned entities is infrastructure or various infrastructure capacities. As many as 50 articles mention different infrastructure capacities as a significant element in the data ecosystem. Furthermore, 39 articles mentioned stakeholders as an element that is indispensable for the operating of the data ecosystem due to the roles that they, as actors, occupy in the system. The data are highlighted in 38 articles and they are the subject resource of the data ecosystem. Policy/Governance is important for regulatory purposes and is marked in 24 articles.

### 3.1.1 Stakeholders
Stakeholders in the ecosystem, in addition to their primary goal, such as achieving some business result, have an obligation to support and contribute to the development of the data ecosystem [18, 32]. In the open data ecosystem, the contribution of stakeholders is noticeable through active participation, dynamic interaction, support of open data initiatives and the development of a data-based innovation culture in the sector [33]. Open data ecosystems are developed through the adaptation of stakeholders, their feedback loops and dynamic interaction, and the strengthening of other interdependent factors [18]. This process is supported by the proposal of the triple helix model [31]. The inclusion of all stakeholders in the process of knowledge sharing and innovation creation makes this model useful for ecosystem analysis. In one of the studies, the quintuple helix model was presented, which unites all the stakeholders of the observed ecosystem [31]. The same approach is necessary to identify stakeholders that share the same or similar problems, collaborate with similar partners, and have similar needs [34, 35]. To create a sustainable open data environment in a sector, the roles and data needs must be understood [30, 36]. Typical roles of stakeholders in the open data ecosystem are data providers, infrastructure providers, application developers, support service providers, end-users [21].

### 3.1.2 Infrastructure

To enable the proper functioning of all processes and an accessible environment to stakeholders, a complex infrastructure is essential in the background of the open data ecosystem [3, 24, 37]. The technical and technological infrastructure includes programs, tools, services, and data preparation activities [18]. The background activities, such as the development and implementation of programs and tools in the ecosystem, aim to make data available to users on portals or platforms [24]. The infrastructure supports the preparation of the data and its publication in an appropriate form, while the users on the portal or platform should find the data very easily and download it for use, which requires a user-friendly interface, appropriate visualization, and support service [18, 38].

Several activities characterize the importance of this element in the ecosystem. First, (a) the infrastructure for preparing and publishing open data on the Internet is important, (b) cleaning, analyzing, enriching, combining, linking, and visualizing data, (c) searching, finding, accessing, browsing, and evaluating metadata, (d) discussing data and providing feedback to the data provider and other stakeholders [16, 18]. These activities cover a wide range of topics that are critical to the development of the open data ecosystem. Some of them are data audit, dataset selection, address and map data, privacy, licensing, high quality publishing, data access, data discovery, supporting public agencies, engaging data users, promoting reuse, and evaluation [25, 39]. The open data ecosystem requires significant infrastructure capacity and a commitment to improving it in order to build an organized, functioning, and sustainable system.

### 3.1.3 Data

Data is a product of a data ecosystem. Fundamental processes in the data ecosystem are based on and with data, from the generation and collection of data to its availability in the ecosystem [40]. Data in the ecosystem should be easily discoverable, available and known to stakeholders, accessible for download and application, and usable for an operational purpose that provides benefits [41, 42]. The characteristics of datasets can be evaluated based on parameters such as legal, practical, technical, and social aspects [43]. Legal parameters include a machine-readable rights statement, a clear rights statement, licensing and privacy issues. Practical parameters include web accessibility, guaranteed timeliness of data, and quality issues. Technical parameters include appropriate formats and open standards. Social parameters relate to metadata, providing feedback, and promoting data. Open data can be viewed in a variety of ways, and there are several definitions. Rather than provide another formal definition, we prefer to look at the characteristics of what makes data truly open. Finally, the characteristics of data could be included and related to the conceptual model in the following ways: (a) data must be complete, (b) data must be primary, (c) data must be timely, (d) data must be accessible, (e) data must be machine-processable and posted online in a permanent archive, (f) access must be nondiscriminatory, (g) data formats must be non-proprietary, (h) data licensing must be unrestricted and have no usage costs, and (i) data should be as accurate as possible [44].

### 3.1.4 Policy/Governance

Policy/Governance is an element that determines the progress and development direction of data ecosystem initiatives [4, 45]. First, establishing sustainable and strong data ecosystems requires defining policies at the national level and developing them at lower, sectoral levels [2, 17, 46]. Government efforts should focus on promoting better information sharing in the public and private sectors, supporting open innovation for co-creation of products and services through budgeting and infrastructure, and actively engaging in interoperation within the data ecosystem [47, 48].

### 3.2 Open Data Ecosystem Characteristics

The success of the data ecosystem depends largely on the collaboration and communication among stakeholders and their joint interaction. A collaborative environment is important for creating a good climate in the environment and for sharing data knowledge and skills [12, 18]. Infrastructural capacities can be enhanced by sharing technological resources. Certainly, it is necessary to research good practices and analyze what kind of infrastructure is desirable in an ecosystem where multiple stakeholders with different needs operate. Infrastructure capacities should be interoperable and stakeholders should be familiar with their functional features. In the ecosystem, it should be outlined what kind of data is desirable and in what form it is acceptable for further use and reuse, therefore certain rules regarding data should be defined. The data ecosystem should be characterized by a culture of experimentation with data. For an open data ecosystem in a certain industry or sector, it is important to build a development strategy and development models at the management and governance levels. Attempts to develop potential products or services can improve business processes within organizations or contribute to the development of socially useful solutions [42, 47, 49, 50].

## 4 Open Data Ecosystem Conceptualization

The conceptual model of the open data ecosystem (Fig. 2) is presented in terms of the four elements of the data ecosystem and their characteristics to be considered in the detailed elaboration and analysis of specific ecosystems.

Stakeholders in the data ecosystem are important for several reasons. In addition to their primary role in the industry, they also have specific roles in the data ecosystem. They may be data creators or data users, or they may take on other roles. In addition, it is important to identify their potential for creating data, i.e., what useful data they create for other stakeholders in their organization. Identifying data needs, i.e., determining potentially useful data for developing one's business, requires a great deal of attention.

Infrastructure is the most complex element that requires detailed elaboration because it supports all activities in the ecosystem, including activities related to data and activities between stakeholders. The technologies and techniques selected should enable the easiest possible processing and preparation of data so that they are accessible to stakeholders for use and further processing [24]. The data ecosystem should be a secure environment for stakeholders that supports user-friendly features such as a collaborative platform, APIs, licenses, standards, tools and more [24, 51, 52].

Data is a resource that stakeholders in the ecosystem care about. It should be available in an acceptable format and meet certain characteristics and quality standards to be used for an appropriate business or societal purpose.

Policy/Governance is an element that includes the industry as a data ecosystem in the national open data policy. In addition, detailed elaboration at a lower level is desirable. Initiatives should be a priority of government institutions through supporting activities [4, 53, 54].

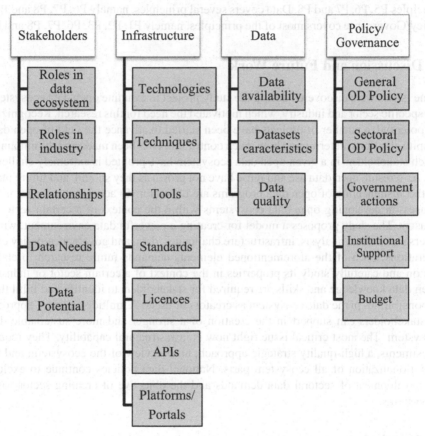

**Fig. 2.** Conceptual Model of Open Data Ecosystem Elements and Characteristics

Proposed elements of open data ecosystem can be explained from perspective of specific sector or a specific industry. For example, stakeholder analysis will explain how to identify roles in the sector and data environment and the potential they have in the creating and using data. In addition, infrastructure element will assess the accessibility and interoperability of technological solutions and principles. Element called data will include identification of what data are available, its quality, and its potential for use in the sector. Element Governance/Policy identify the potential of such an ecosystem and whether government or leading initiatives aim to build and financially support an ecosystem. Future work will include the improvement of proposed model to assess the state of a specific data ecosystem and development of framework. The improvement and framework will be based on the proposed conceptual model in order to assess the identified elements of the data ecosystem. Future research will aim to assess the status of the data ecosystem of individual industries and benchmark at the level of different countries. The data ecosystem should contribute to the fulfillment of the principles of the ecosystem approach through the mentioned elements. Stakeholders are closely related to principles P3, P4, P10, P11 and P12. Infrastructure contributes to the fulfillment of principles P5, P6, P7 and P8. Data covers several principles, namely P6, P7, P8 and P11. Policy/Governance covers most of the principles, namely P1, P2, P3, P6, P7, P8 and P9.

## 5    Discussion and Future Work

Some issues were discovered during the study project to examine open data ecosystems in a specific sector and industry, which motivated the need for this research. Recognizing the potential, a number of projects have been started to advance the field of open data. Despite all efforts, determining how much contribution has been made and the amount to which stakeholders in a given open data ecosystem have profited is extremely challenging. As a result, open data use and reuse have not grown as they should, and future plans for the sustainability of open data programs are uncertain. To address this, our concept focuses on developing open data ecosystems within the context of a certain sector or industry. The study proposes a model for creating a powerful data environment, which covers stakeholder analysis, infrastructure challenges, data, and governance/policy considerations. Each of the aforementioned elements demands future research to elaborate on and carefully study its properties in the context of a certain sector or industry. Open data knowledge and skills are required for stakeholders to identify and fulfil their responsibilities in the data ecosystem as creators or users. A multidisciplinary approach by stakeholders can support in the creation of a stronger and more sustainable data ecosystem. The most critical issue right now is infrastructural capability. They require investments, a high-quality strategic approach, an overview of the ecosystem, and the operationalization of all ecosystem parts. National data policies continue to exclude the development of sectoral data demands and the objective of creating sector-based ecosystems.

## 6    Conclusion

Development-oriented open data efforts should be aligned, created, and operationalized as an ecosystem at the sector or industry level so that actual progress can be evaluated. Existing research frequently mentions the open data ecosystem, but rarely in the context of a sector or industry. These results are not surprising considering the research's complexity, which necessitates industry expertise and abilities connected to the characteristics of the open data ecosystem.

The open data ecosystem is an environment in which stakeholders with different responsibilities but similar interests operate, share, and use specific data, that is enabled by interoperable infrastructure solutions. To create a specialized open data ecosystem, it is required to concentrate on promoting and strengthening a data-driven economy, as well as sharing open data through the deployment and availability of appropriate support technologies. The purpose of data ecosystem development proposed on this paper is to shift stakeholders' roles from passively informed actors to active co-producers of ecosystem services. Future research will concentrate on developing a framework for assessing open data ecosystems based on the proposed conceptual model, which includes important characteristic elements.

**Acknowledgment.** This research is part of the Twinning Open Data Operational - TODO project that has received funding from the European Union's Horizon 2020 research and innovation program under grant agreement No. 857592.

# References

1. Shah, S.I.H., Peristeras, V., Magnisalis, I.: Government (big) data ecosystem: definition, classification of actors, and their roles. Int. J. Comput. Inf. Eng. **14**(4), 102–114 (2020)
2. Yunita, A., Santoso, H.B., Hasibuan, Z.A.: Everything is data: towards one big data ecosystem using multiple sources of data on higher education in Indonesia. J Big Data **9**(1), 91 (2022)
3. Rebstadt, J., Kortum, H., Hagen, S., Thomas, O.: Towards a transparency-oriented and integrating service registry for the smart living ecosystem. Informatik, Germany (2021)
4. van Donge, W., Bharosa, N., Janssen, M.F.W.H.A.: Data-driven government: cross-case comparison of data stewardship in data ecosystems. Gov. Inf. Q. **39**(2), 101642 (2022)
5. Booth, P., Navarrete, T., Ogundipe, A.. Museum open data ecosystems: a comparative study. J. Doc. **78**(4), 761–779 (2021)
6. European Commission, Communication from the Commission to the European Parliament, the Council, the European Economic and Social Committee and the Committee of the Regions. A European strategy for data. https://eur-lex.europa.eu/legal-content/EN/TXT/?qid=159307 3685620&uri=CELEX%3A52020DC0066. Accessed 10 Sept 2022
7. Secretariat of the Convention on Biological Diversity, The Ecosystem Approach. https://www.cbd.int/ecosystem/#:~:text=The%20ecosystem%20approach%20is%20a,three%20o bjectives%20of%20the%20Convention. Accessed 09 Sept 2022
8. Ecosystems Knowledge Network. https://ecosystemsknowledge.net/ecosystem_approach. Accessed 09 Sept 2022
9. Shepherd, G.: The ecosystem approach: five steps to implementation. IUCN (2004). https://www.iucn.org/resources/publication/ecosystem-approach-five-steps-implementation. Accessed 09 Sept 2022
10. Waltner-Toews, D., Kay, J., Lister, N.-M.: The Ecosystem Approach: Complexity, Uncertainty, and Managing for Sustainability, Illustrated Columbia University Press, New York (2008)
11. Lister, N.-M., Waltner-Toews, D., Kay, J.: The Ecosystem Approach: Complexity, Uncertainty, and Managing for Sustainability (2008)
12. van Loenen, B., et al.: Towards value-creating and sustainable open data ecosystems: a comparative case study and a research agenda. JeDEM – eJ. eDemocr. Open Govern. **13**(2), Art. no. 2 (2021)
13. Shepherd, G.: The Ecosystem Approach: Learning from Experience. IUCN (2008)
14. Kassen, M.: Open data and e-government – related or competing ecosystems: a paradox of open government and promise of civic engagement in Estonia. Inf. Technol. Dev. **25**(3), 552–578 (2019)

15. Oliveira, M.I.S., Lóscio, B.F.: What is a data ecosystem? In: Proceedings of the 19th Annual International Conference on Digital Government Research: Governance in the Data Age, New York, NY, USA, pp. 1–9 (2018)
16. Diran, D., Hoppe, T., Ubacht, J., Slob, A., Blok, K.: A data ecosystem for data-driven thermal energy transition: reflection on current practice and suggestions for re-design. Energies 13(2), Art. no. 2 (2020)
17. Gupta, A., Panagiotopoulos, P., Bowen, F.: An orchestration approach to smart city data ecosystems. Technol. Forecast. Soc. Chang. 153, 119929 (2020)
18. Zuiderwijk, A., Janssen, M., Davis, C.: Innovation with open data: essential elements of open data ecosystems. Inf. Polity 19, 17–33 (2014)
19. Hayashi, T., Ishimura, G., Ohsawa, Y.: Structural characteristics of stakeholder relationships and value chain network in data exchange ecosystem. IEEE Access 9, 52266–52276 (2021)
20. Moreno, J., Serrano, M., Fernández, E., Fernández-Medina, E.: Improving incident response in big data ecosystems by using blockchain technologies. Appl. Sci. 10, 724 (2020)
21. Kitsios, F., Papachristos, N., Kamariotou, M.: Business models for open data ecosystem: challenges and motivations for entrepreneurship and innovation. In: IEEE 19th Conference on Business Informatics (CBI), vol. 01, pp. 398–407 (2017)
22. Immonen, A., Palviainen, M., Ovaska, E.: Requirements of an open data based business ecosystem. IEEE Access 2, 88–103 (2014)
23. Osorio-Sanabria, M.A., Amaya-Fernández, F., González-Zabala, M.: Exploring the components of open data ecosystems: a systematic mapping study. In: Proceedings of the 10th Euro-American Conference on Telematics and Information Systems, New York, NY, USA, pp. 1–6 (2020)
24. Pinto, V., Parreiras, F.: Towards a taxonomy for big data technological ecosystem. In: Proceedings of the 22nd International Conference on Enterprise Information Systems, Prague, Czech Republic, pp. 294–305 (2020)
25. Lee, D.: Building an open data ecosystem - an irish experience. In: ACM International Conference Proceeding Series, pp. 351–360 (2014)
26. Alexopoulos, C., Loukis, E., Charalabidis, Y.: A platform for closing the open data feedback loop based on web2.0 functionality. JeDEM – eJ. eDemocr. Open Govern. 6(1), Art. no. 1 (2014)
27. Janev, V.: Ecosystem of big data. In: Janev, V., Graux, D., Jabeen, H., Sallinger, E. (eds.) Knowledge Graphs and Big Data Processing. LNCS, vol. 12072, pp. 3–19. Springer, Cham (2020). https://doi.org/10.1007/978-3-030-53199-7_1
28. Koers, H., Bangert, D., Hermans, E., van Horik, R., de Jong, M., Mokrane, M.: Recommendations for services in a FAIR data ecosystem. Patterns 1(5), 100058 (2020)
29. Sanaei, M., Taslimi, M., AbdolhoseinZadeh, M., Khani, M.H.: A study and analysis of the open government data ecosystem models. Iran. J. Inf. Process. Manage. 34, 609–636 (2019)
30. Lindman, J., Kinnari, T., Rossi, M.: Business roles in the emerging open-data ecosystem. IEEE Softw. 33(5), 54–59 (2016)
31. Kitsios, F., Kamariotou, M., Grigoroudis, E.: Digital entrepreneurship services evolution: analysis of quadruple and quintuple helix innovation models for open data ecosystems. Sustainability 13(21), Art. no. 21 (2021)
32. Runeson, P., Olsson, T., Linåker, J.: Open data ecosystems - an empirical investigation into an emerging industry collaboration concept. J. Syst. Softw. 182, 111088 (2021)
33. Rudmark, D., Andersson, M.: Feedback loops in open data ecosystems. IEEE Softw. 39(1), 43–47 (2022)
34. Hrustek, L., et al.: Towards digital innovation: stakeholder interactions in agricultural data ecosystem in Croatia. Interdiscip. Descr. Complex Syst. 20(2), 190–209 (2022)

35. Hrustek, L., Tomičić Furjan, M., Šalamon, D., Varga, F., Džidić, A., von Loenen, B.: Overview of the open data agricultural ecosystem in Croatia. In: Book of abstracts of the National Open Data Conference (2021). https://www.bib.irb.hr/1148032. Accessed 14 Oct 2022

36. Kassen, M.: Adopting and managing open data: stakeholder perspectives, challenges and policy recommendations. Aslib J. Inf. Manage. 70(5), 518–537 (2018)

37. Pires, F.M., León Quiñonez, L., de Souza Mendes, L.: A cloud-based system architecture for advanced metering in smart cities. In: 2019 IEEE 10th Annual Information Technology, Electronics and Mobile Communication Conference (IEMCON), pp. 1087–1091 (2019)

38. Demchenko, Y., de Laat, C., Membrey, P.: Defining architecture components of the big data ecosystem. In: 2014 International Conference on Collaboration, Technologies and Systems (CTS), pp. 104–112 (2014)

39. Luo, W.: Enterprise data economy: a hadoop-driven model and strategy. In: 2013 IEEE International Conference on Big Data, pp. 65–70 (2013)

40. Zeleti, F.A., Ojo, A.: Capability development in open data-driven organizations. In: Ojo, A., Millard, J. (eds.) Government 3.0 – Next Generation Government Technology Infrastructure and Services. PAIT, vol. 32, pp. 135–171. Springer, Cham (2017). https://doi.org/10.1007/978-3-319-63743-3_6

41. Welle Donker, F., van Loenen, B.: How to assess the success of the open data ecosystem? Int. J. Digit. Earth 10(3), 284–306 (2017)

42. Benitez-Paez, F., Comber, A., Trilles, S., Huerta, J.: Creating a conceptual framework to improve the re-usability of open geographic data in cities. Trans. GIS 22(3), 806–822 (2018)

43. ODI Open Data Certificate. https://certificates.theodi.org/en/. Accessed 30 Aug 2022

44. Charalabidis, Y., Zuiderwijk, A., Alexopoulos, C., Janssen, M., Lampoltshammer, T., Ferro, E.: The open data landscape: concepts, methods, tools and experiences. In: Public Administration and Information Technology, pp. 1–9 (2018)

45. Munshi, U.M.: Data science landscape: tracking the ecosystem. In: Munshi, U., Verma, N. (eds.) Data science landscape. Studies in Big Data, vol. 38, pp. 1–31. Springer, Singapore (2018). https://doi.org/10.1007/978-981-10-7515-5_1

46. Vayena, E., Gasser, U.: Strictly biomedical? Sketching the ethics of the big data ecosystem in biomedicine. 29, 17–39, (2016)

47. Ojo, A., Curry, E., Sanaz-Ahmadi, F.:A tale of open data innovations in five smart cities, vol. 2015, p. 2335 (2015)

48. Freitas, J.A.C., Balaniuk, R., Silva, A.P.B., Silveira, V.S.: The open data ecosystem of federal government: compositions and challenges. Ciencia Inform. 47, 110–132 (2018)

49. Shin, D.-H., Choi, M.J.: Ecological views of big data: perspectives and issues. Telemat. Inform. 32(2), 311–320 (2015)

50. Lockwood, M.: An accessible interface layer for self-sovereign identity. Front. Blockchain 3 (2022)

51. Balaji, V., et al.: Requirements for a global data infrastructure in support of CMIP6. Geosci. Model Dev. 11(9), 3659–3680 (2018)

52. Moreno, J., Gómez, J., Serrano, M.A., Fernandez, E.B., Fernández-Medina, E.: Application of security reference architecture to big data ecosystems in an industrial scenario. Softw. Pract. Exp. 50(8), 1520–1538 (2020)

53. Cue, R., et al.: Data Governance in the Dairy Industry. Anim. (Basel) 11(10), 2981 (2021)

54. Bugbee, K., et al.: Building a data ecosystem: a new data stewardship paradigm for the multi-mission algorithm and analysis platform (MAAP). In: IGARSS IEEE International Geoscience and Remote Sensing Symposium, Yokohama, Japan, pp. 4261–4264 (2019)

# Extending the "Smart City" Concept to Small-to-Medium Sized Estonian Municipalities: Initiatives and Challenges Faced

Karin Amukugo Fröhlich[1]([⊠]) [ID], Ralf-Martin Soe[2] [ID], Pardon Blessings Maoneke[3] [ID], Karishma Jain[1] [ID], Antti Pinomaa[1,4] [ID], and Marko Nieminen[1] [ID]

[1] Aalto University, Espoo, Finland
karin.frohlich@aalto.fi
[2] University of Technology, Tallinn, Estonia
[3] University of Mpumalanga, Nelspruit, Mpumalanga, South Africa
[4] Lappeenranta-Lahti University of Technology, Lappeenranta, Finland

**Abstract.** This study investigated smart city initiatives and challenges faced by small to medium sized municipalities. The literature on smart cities is dominated by findings from large cities yet, both large and small municipalities are expected to contribute towards the fulfillment of the United Nation's 17 sustainable development goals. A mixed method was used in which 35 municipalities were engaged. Study findings suggest that rural municipalities are yet to be fully aware of how they can harness the technology and become smart in the way they operate and serve inhabitants. Nonetheless, cities and towns seem to have an idea of what smart city initiatives to consider. Together, Estonian cities, towns and rural municipalities face several challenges in their efforts to assume "smartness" something that needs to be addressed in a pragmatic way. Municipalities face unique challenges; hence, solutions that work for one municipality may not work for another.

**Keywords:** Digital Governance · Smart City · Smart Village · SDGs · ICTs

## 1 Introduction

The United Nations (UN) estimate that more than 50% of the world population is based in cities that occupy 3% of the earth's surface [1–4]. This population in cities and towns (urbanization) is expected to increase to approximately 66% by 2050 due to birth, migration from rural areas or as new urban settlements are established. There are concerns that the continued population growth will make urban areas difficult to manage while at the same time depleting limited resources that are meant to service the present and future generations. The UN and the European Union (EU) have endorsed the smart city concept as a panacea for managing urban settlements and promoting a sustainable use of resources [1, 2]. The use of disruptive, cutting-edge Information and Communication Technologies (ICTs), Internet of Things (IoTs) and big data analytics are among other technological solutions that are expected to derive the smart city concept [1] and small to medium sized municipalities are also expected to benefit from such

© The Author(s), under exclusive license to Springer Nature Switzerland AG 2023
M. Papadaki et al. (Eds.): EMCIS 2022, LNBIP 464, pp. 264–276, 2023.
https://doi.org/10.1007/978-3-031-30694-5_20

initiatives [5]. However, little is known about challenges faced by small to medium sized municipalities in their endeavor to adopt and use these technologies. Solutions for smart city initiatives are often developed in big cities that have socio, economic and demographic characteristics that are not consistent with those of small municipalities [5]. Thus, factors that enable the development of a smart city such as, high population density, a big service sector and high productivity are not common in small municipalities [6, 7], yet such locations are also expected to succeed in technology adoption and use [5]. Besides, the smart city concept involves digital transformation that will see the adoption of radical and disruptive technologies that are complex with no pre-defined processes [2, 8]. Hence, this study explores challenges faced by small to medium sized municipalities as they adopt the smart city concept. The study uses findings from pilot research that was conducted on Estonian municipalities.

## 2 Literature Review

### 2.1 Smart City Concept

The use of the phrase "smart city" can be traced back to the 1990s when cities started to use ICTs for different purposes such as in electronic government (e-Government) [9, 10]. Today, the smart city concept has become a global phenomenon that is expected to play a pivotal role towards the fulfilment of the UN's 17 sustainable developmental goals (SDGs) by 2030 [1, 10, 11]. Besides, a city that is resilient, safe, inclusive, and sustainable is one of these SDGs [1, 10]. Wang, Luo, Zhang and Furuya [12] goes on to suggest that the desire to attain sustainability through smartness is not specific to cities alone but to all communities including rural areas. Accordingly, the need to attain the 17 SDGs has seen more research effort put in understanding and conceptualizing the smart city concept to all communities – large, small and marginalized communities – as seen with the recent use of phrases such as "smart villages" [1, 4, 12, 13].

Despite all this popularity, there is no single smart city theoretical explanation or definition [4, 10]. For example, the European Commission defines a smart city as a place where the efficacy of traditional networks and services is enhanced by using digital technologies and/or ICTs in a manner that benefit inhabitants and businesses [2, 10]. The European Commission adds that the smart city concept goes beyond the use of ICTs but can also mean "smarter urban transport networks, upgraded water supply and waste disposal facilities and efficient ways to light and heat building", "a more interactive and responsive city administration, safer public spaces and meeting the needs of an ageing population" [10]. In addition, Ismagilova, Hughes, Dwivedi and Raman [11] proposes that smart cities are cities that "use an Information Systems centric approach to the intelligent use of ICT within an interactive infrastructure to provide advanced and innovative services to its citizens, impacting quality of life and sustainable management of natural resources". All these definitions point towards improving the quality of life and simplifying the way communities are managed [4, 10, 12].

### 2.2 Smart City Initiatives and Challenges

The literature identifies different areas where the smart city concept is applicable [14]. For example, in large cities where the use of smart cities is dominant, it is used (smart city initiatives) to improve the lives of inhabitants and manage resources through smart mobility or smart transportation, smart living (public safety and smart buildings), smart environment, smart citizens, smart economy, smart government, and smart architecture and technologies [11, 12]. Successful implementation of these smart city initiatives requires the formulation of a smart city strategy, setting up governance structures, implementing supporting technologies, and identifying and engaging all relevant stakeholders [2, 14]. Furthermore, a community that is adopting a smart city concept must exhibit the following characteristics: high population density, proximity to a big city/be a big city, low remoteness, low unemployment, high productivity, high employment in a service sector and the availability of local university [6, 7]. Successfully harnessing the smart city concept will result in social, economic, and environmental sustainability [10].

However, characteristics for enabling smart city initiatives are not consistent with those of small to medium sized municipalities. For example, Estonia has a population of 1.3 million, 79 sparsely populated municipalities of which 75 municipalities have a population of less than 50,000 [5]. The European Commission classify communities with 10 000 to 100 000 residents as small municipalities. Hence, such municipalities do not meet some of the key requirements for successfully harnessing the smart city concept. Despite being small in nature, marginalized and lacking resources, the EU policy recognize that such communities must attain "smartness" in the way they serve inhabitants and manage resources [16]. There is ongoing debate on whether these small municipalities should assume a scalable approach and equally apply smart solutions developed for bigger cities while others argue that small municipalities have distinct features that warrant a unique set of solution for smartness [12, 16–18]. Regardless of the approach assumed, these small to medium sized municipalities will need to adopt cutting edge technologies that will be used as a means for realizing better quality of life for residents and sustainability [12]. Given the prevailing contextual characteristics of small municipalities, these municipalities are expected to face a number of challenges when harnessing technologies for "smartness" given the vast amount of technologies that need to be considered such as big data, artificial intelligence, IoTs, cloud computing, data analytics and ICTs [1, 2, 10, 11]. The often-reported challenges relate to funding, heterogeneous devices and disintegrated systems, lack of capacity, big data management, information security and, social awareness and acceptance [14, 19–22]. These challenges are discussed below.

**Funding:** Small to medium municipalities tend to have little financial resources for implementing a smart city concept [5] as they are mostly dominated by less productivity and are often agriculture-based [17]. Those aspiring to adopt a smart city concept need to pay for designing, hardware and software, operational and maintenance costs. According to Silva et al. [21], designing costs are related to the initial capital outlay for deploying a smart city. This is followed by the costs for acquiring the necessary hardware and software that includes IoTs and ICTs infrastructure [20]. For example, London invested in 5969 public Wi-Fi hotspots across the city and spent approximately USD 2.32 million on indoor public Wi-Fi for Galleries and Museums [1]. Furthermore, there are operational costs for the day-to-day running of a smart city and maintenance costs that also requires funding. The EU and other international or local organizations offer funding for smart city initiatives as these are seen as funds for infrastructural improvements. For instance,

Helsinki receives funding from the EU for performing pilot testing and experimental tests on smart city research projects [2]. However, Hämäläinen [2] goes on to note that the uncertainty of smart city funding necessitates self-funding as done by Helsinki.

**Heterogeneous Devices and Disintegrated Systems:** The architecture of a smart city is composed of heterogeneous devices and appliances from different vendors that perform a wide range of services [21, 22]. These multi-devices often have incompatible platforms something that make the setting up of a smart city complex and difficult, requiring numerous rounds of pilot testing and experimentations. Furthermore, city data is often gathered within different departments or domains using unique ICTs that are specific to these domains [2]. As such, becoming a smart city is down to the ability of integrating different technologies into one central processing location [21, 22]. Bibri and Krogstie [1] notes that Barcelona's Sentilo and City OS platforms provides a good example of how a city can setup a horizontal information platform that aggregates and standardize open data from heterogenous devices. Barcelona and London, some of the success stories in this area, integrate data from different departments and present this in real-time using visualizations on a dashboard in operation centers [1, 21]. Cities often engage technological companies and research institutes when searching for solutions on integrating heterogenous devices and disintegrated systems for smart cities [2, 21]. However, approaches assumed in London and Barcelona cannot easily be transferred to small municipalities given their lack of resources and supportive institutions [17].

**Lack of Capacity:** Municipalities lack the capacity to deliver technology driven solutions [20] especially small to medium sized municipalities. This include a lack of skills and expertise to effectively harness the technology in a way that is consistent with expectations of smart city initiatives – sustainability and improving the quality of life. The literature suggests that municipalities need to establish an ICT department that is responsible for all technologies that facilitate smart city concept as done by the Finish capital Helsinki [2]. Furthermore, cities engage the private sector when developing new technological solutions for smart city initiatives as revealed by Helsinki City that works with a private organization, Forum Virium Helsinki (FVH) Ltd., for experimenting new technological solutions [2]. According to Silva et al. [21], London's ability to establish an infrastructure to support smart city initiatives is down to the city's engagement with research institutes, among other collaborations. In addition, new academic programs have been developed to equip people with skills for managing the city using big data analytics [1]. Most of these solutions have been found applicable to big cities but they are hardly compatible with rural municipalities, for example, where there is a lack of digital literacy, depopulation, and less connectivity [12, 17].

**Big data Management:** Smart city initiatives are faced with a challenge to manage a huge amount of data that is continuously generated by countless devices [19, 21]. This heterogeneous data from different devices needs to be collected, stored and processed in seamless operations of the city something that presents a huge challenge [21]. Managing big data is expected to be a huge challenge for small to medium sized municipalities given their lack of resources especially supporting infrastructure and digital skills [12, 17]. However, if successfully managed, this data can be used to support the development of various data driven innovation solutions for smart cities as shown in London, Singapore, Helsinki and Barcelona [1, 10, 21, 23]. Bibri and Krogstie [1] notes that London and Barcelona are among the first European cities to use data driven applications to improve the lives of citizens.

**Social Awareness and Acceptability:** There are suggestions that inhabitants may not be privy to smart city technological initiatives for them to accept their use [20]. Even when aware, citizens may still decide against using smart city solutions for different reasons. For example, Bielska et al. [17] notes that rural municipalities lack openness and involvement. Bawany and Shamsi [19] goes on to suggest a need to change social habits and the mindset of inhabitants so that they embrace the use of the technology. For better social awareness, the city of Helsinki made its services accessible using a "one-stop shop" application known as "Service Map" [2]. In addition, there are initiatives for crowd sourcing that aim to engage citizens in addressing city problems. In some cases, citizens are given mechanisms to provide feedback and rate service provision by the city with the intent of using this information to shape service delivery according to citizen' needs [21] something that may promote acceptance. This is complemented by educating citizens to improve their technical skills for easy interaction with various smart city technologies [1].

**Information Security:** Smart city applications rely on gathering personal and sensitive data from citizens using different sets of infrastructure [14, 21, 22]. Thus, smart cities gather data on citizens from heterogeneous devices that operate on unique platforms something that present a security challenge. In addition, there can be no guarantees that all individuals will be security cautious when accessing smart city applications in an environment where attackers have become more skillful. Hence, there is need for a security framework that protect the smart city infrastructure, the privacy and confidentiality of user data [21]. This data needs to be protected during collection, storage, processing, and dissemination [14, 19]. Interestingly, the available solution for aiding information security such as compliance with the General Data Protection Regulation (GDPR) of Europe is seen as a huge challenge for municipalities [2].

## 3 Research Methodology

This study is based on a combination of mixed methods. The assortment included a questionnaire, individual interviews and several workshops with multiple Estonian municipalities. This data was gathered from a pilot project of the idea competition of the Smart City Center of Excellence of the FinEst Twins [24]. The mayor, deputy mayor, development specialists and other experts from municipalities took part in this study. Initially, a questionnaire was sent out to 35 Estonian municipalities. This questionnaire helped to map the most pressing issues of cities, towns, and municipalities. Those who participated in questionnaires (16 Municipalities) were further engaged through interviews to gain an understanding of the nature and magnitude of the challenges faced. This study report findings from interviews. These interviews clarified all the doubts the participant had regarding the questions and collected more elaborated answers on the challenges faced. Thematic analysis was used to analyze the data.

# 4 Results

Sixteen cities, towns, and rural municipalities were engaged in data collection using interviews. Of these municipalities, 7 are classified as towns followed by 5 rural municipalities and 4 cities. The biggest city, in terms of population, in this study – Tallinn city – has a population of 0.5 million while the remaining three cities have population ranging between 50000 and 94000 inhabitants. The researched Estonian towns shows to be small with an average population of 10500 inhabitants. Lastly, rural municipalities that were engaged in this study have an average population of 16000 inhabitants. Furthermore, an average of four participants were interviewed per municipality.

## 4.1 Smart City Initiatives

**Current Smart City Initiatives:** Study findings suggest that there are ongoing smart city initiatives in some of the researched Estonian municipalities. For example, the government city of Võru is involved in a *smart environment* initiative by adopting an automated data management system for managing the energy of real estates. Similarly, a move to have an open and transparent government by conducting online municipality meetings that are open to residents, for example, suggest the use of a *smart government* by the town of Valga. Furthermore, there are suggestions of various *smart living* initiatives being practiced in Jelgava. For example, sensors installed on rivers are used to gather data on water level. This data is then used in flood notifications to those who reside along the riverbanks. Sensors have also been placed on roads to gather data on ice locations so that snow clearance is done on a need basis. In addition, *smart living* is also visible in Jelgava city's use of sensors to monitor climate and the level of carbon dioxide in schools and other public buildings. In this case, data driven decisions of taking lesson breaks and opening windows are assumed if the level of carbon dioxide is too high. Furthermore, Jelgava uses a *smart transportation* system as evidenced by traffic lights whose functionality is determined by the flow of traffic rather that predetermined timeslots.

**Plans on Smart City Initiatives:** Participants were asked to indicate digital solutions that should be implemented in their city by the year 2025 or 2030. The gathered data shows an inclination towards digitalization with the intent of using data driven solutions where IoTs are among the technologies to be used. For example, a participant from Tallin city stated that they expect to be able to visualize data that is gathered using sensors by the year 2025. Expectations are that service delivery will be data driven where real-time feedback from inhabitants will be used to improve city services. In particular, selected Estonian cities, towns and rural municipalities expect to use big data for smart transportation, smart environment, smart living, and smart governance. With reference to smart transportation, the idea is to have *"a self-managing public transport system which enables multimodal mobility that can be completely customized for a person's needs in real time"* or *"smart and remote-controlled traffic management fixtures that enable flexible adjustments of traffic management and support the use of V2X technology. [Furthermore,] a solution for analyzing and modelling traffic [is also required]."* Smart environment initiatives include the use of data to monitor carbon dioxide in the city and *"monitoring energy consumption in real time"* and managing each building's energy consumption.

In addition, a participant from Rakvere town suggested that the town plans to implement smart living initiatives as it was stated that they hope for a *"wider application of the digital solutions in the city's social, cultural, and educational system and in tourism"*. Similarly, suggestions for smart government initiatives were identified from the gathered data. For instance, a participant from Tartu city suggested more use of electronic services (eServices) as it was stated that *"all public services should be available as electronic and mobile services"* and using *"a document management system that allows [one] to monitor the processing of applications"*. Similarly, a participant from Rakvere states that the use of *"smart integrated governance of the city government"* or using *"a system of indicators on governance to monitor different domains"* as stated by a participant from Saaremaa rural further suggest intentions for smart government initiatives. Furthermore, a participant from Valga town suggests the use of crowdsourcing to receive reports of faults by stating that *"a system for residents to notify the local government of [current] problems [should be implemented by 2025]"*.

### 4.2  Challenges Faced by Municipalities

Participants were asked to describe problems their city/rural municipality must address in the next 5 to 10 years. A list of 10 challenges were arrived at following discussions and a challenge-based workshop. The participants, 35 local governments, collectively made some changes to the list of top 10 problems that Estonian cities, towns and rural municipalities are facing when solving these in a smart way. While the Smart City Center of Excellence of the FinEst Twins solely focused on challenges that relate to transport, energy, built environment, governance, and data [24]; this study focuses on findings that are specific to governance and data challenges. Participants indicated that the most common challenges relate to the lack of capacity, data specification and accessibility, big data management, lack of system integration, setting up the infrastructure, information security and big data usage. The nature and extent of these challenges is presented next.

**Lack of Capacity:** Participants cited several issues that relate to the lack of capacity to manage big data. These concerns were raised by participants from six towns, two cities and three rural municipalities. For instance, there is evidence of lacking skills to use big data as participants describe their lack of knowledge on what kind of data and for what purpose must data be gathered. In addition, certain information is being collected, and its volume is constantly increasing something that further complicate the situation as explained by a participant from Tartu city: *"The precise need and extent of data collection is undetermined, i.e., who needs what kind of data. At the same time, we are collecting increasing amounts of information"*. The technical capacity and competency for data processing, analysis, and using big data for governance decisions is also lacking. A participant from Tartu city explains that *"today, the capacity to process and analyze information and use it for governance decisions is small and we have no necessary technical solutions"*. Another participant from the same city also stated that: *"The city of Tartu has never had a Head of Data Management. This is why the situation is somewhat* *chaotic.... We have discussed hiring a manager for information issues."* The lack of human resources with the necessary skills for smart city initiatives were also emphasized by another participant: *"We do not have enough employees who would assemble various data into a single database to improve our administrative capability and the quality of public services".*

**Data is Not Specific and Accessible:** Empirical evidence suggests that cities, towns and rural municipalities mainly rely on data that is captured by Statistics Estonia. However, this data is often not specific to the needs of municipalities. One of the participants explains that: *"Statistics Estonia does not issue a separate set of data for Pärnu alone".* Instead, the data is generalized at country level. Another participant from Lääne clarifies this challenge: *"...A lot of the essential information is collected for the entire county, not for individual rural municipalities... For example, when preparing a public health profile, the data input is available on the county level, but the population profile varies so wildly within the Harju County that it does not characterize the situation in our rural municipality".* Thus, statistics at national level are seen to be too biased towards big cities at the expense of rural municipalities as further comments highlighted that: *"[in] the example of the health profile: county-level data is not accurate enough – the city of Tallinn and the Golden Circle influence it too much"* such that interpretations of the data will be biased towards the cities.

Another challenge is that data is not availed in real-time or when needed. *"The state releases information with a delay that is too long for the data to be used as a basis for plans. We need information about this summer to plan for the next, not information about the summer two years ago."* Similar views were shared by a participant from Saaremaa rural municipality: *"As it stands, we have cooperated with Statistics Estonia, but do not have access to the required information at the necessary time to monitor all the aforementioned areas."* In addition, the private sector does not have direct access to data. Often, the private sector must engage the municipality to have access to data according to a participant from Tartu: *"Most of the collected information is still not accessible as open data. When we have received specific requests for information from the private sector, we have tried to accommodate".*

**Big Data Management:** Participants from different cities and towns cited challenges that suggest difficulties with the management of big data. There appears to be a lack of data standardization and challenges on publishing or organizing data. With reference to data standardization, a participant from the city of Tartu suggested that there is no clarity on what type and format of data needs. Furthermore, there is a lack of knowledge on the value of data to target groups such that the information is availed to the intended users; hence, the need for data standardization. The challenges on publishing or organizing data were raised by participants from Võru town and Tallinn city. For instance, questions were raised on what information should the municipality publish and how this can be done in a more dynamic manner. Focus is on visualizing all the important data in a user-friendly format that could give room for personalized data use by different user groups – residents and city officials. Thus, *"[to] provide information [data] to citizens and organizations that they could use at their discretion and for the creation of new services".* This challenge is exacerbated by the increasing volume of data thereby making it difficult for residents to find the actual data they want. A participant from Tartu city explained that *"due to increasing information volumes, the residents find it more complicated to organize their daily lives and find information about services that are important to them"*. Lastly, there is also a need to consider data requirements for the old population that is not always friendly to the technology: *"the 85+ population is increasing, and we have to consider their needs."*

**Lack of System Integration:** Study findings suggest that cities, towns and rural municipalities face challenges related to a lack of system integration. For example, there are suggestions that some of the data is not available digitally especially data on old files. Even when this data is digitized, there are concerns that the databases are not physically connected something that promote information silos. For example, a participant from Rae rural municipality noted that databases in use are incompatible *"our problem is that different databases are not compatible"*. This is supported by a finding that local governments have individual, need-based systems as stated by a participant from Pärnu city *"every municipality builds a separate system"*. This calls for systems that can collect information from various databases or physically connected databases. This view was shared by a participant from Tallinn city *"[there is a need to] establish data bridges with neighboring municipalities and the urban region of Helsinki to integrate data-based governance, cooperation, and services"*.

In addition to having physically connected databases, the stored data must be compatible with different databases and the data should have *"interlinking"* keywords that make the extraction of related information from different databases easy. A participant from Rae rural municipality stated that *"an important issue is submitting data in a format that would allow it to be used more widely and to be added to different databases. Our goal is to have keywords in documents that help us find and link different information"*. Failure of which, a participant from Valga town explained the cost implications of a disintegrated systems to municipalities that are tasked with the responsibility of providing social and educational services to residents: *"[if] we do not know where people who consume our services live. A person could be an officially registered resident in Estonia and receive benefits [in Valga] while living in Latvia."* Thus, information about a municipality's population is key when allocating resources for service delivery where *"social services and education are the biggest expenses for the municipality"*.

**Setting up the Infrastructure:** Study findings suggest that municipalities are faced with a challenge of setting up the infrastructure and services for enabling data driven solutions in governance. As reported by some participants, there is need for more electronic and mobile services to be set up: *"public services should be remotely accessible as much as possible; however, this is currently not the case. We need consistent development of electronic and mobile services"*. This infrastructure should enable linkage between different service domains, departments and facilitate communication across municipalities. IoTs sensors are among the devices to be setup as one of the participants indicated that: *"It would be great if we could install various sensors in the rural municipality that would provide data in real time. We would like to monitor the energy consumption of buildings and devices and control the temperature in buildings from a distance. For us, improving communication with Latvia is extremely important because many residents* *of the Valga rural municipality and people working here are Latvian citizens".* Furthermore, there are suggestions for other IoT technologies that need to be set up in different locations within the municipality. For instance, there is a need for security cameras in the public space, sensors on rivers to gather data on water level and sensors on roads for gathering data on ice and using smart traffic lights.

**Information Security:** The modern information society needs to contribute more towards Information Technology (IT) security and data protection. Currently, navigating data protection, the GDPR, and cybersecurity regulations is difficult for local governments. These views emanated from the gathered data as it was stated that *"in the modern information society, our IT security (ISKE) and data protection need increasing investments".* Similarly, a participant from Rae rural municipality suggested that complying with the data protection act is another problem that need to be resolved as it was stated that *"another important issue is complying with the requirements for data protection."* Similar views on ensuring information security were shared by a participant from Tallinn city: *"data management must be based on a clear principle that a resident's information belongs to them and we must guarantee lawful and secure data management".*

**Big Data Use:** Study findings suggest a slow adoption and use of big data in decision making. Thus, even though the municipalities are gathering data that is important for decision making, these municipalities do not go on to use the data for decision making. This is emphasized in feedback that was given by a participant from Elva town who stated that *"we collect a lot of data but do not use it much when making decisions. We mostly 'follow our guts'".* It is not clear if this is related to a lack of capacity or just a traditional practice.

## 5   Discussion and Conclusion

This study investigated challenges associated with digitalization when small to medium sized municipalities are adopting a smart city concept. The literature on smart cities is mainly focused on the success stories of leading adopters with little focus paid on small cities and towns. Interestingly, the smart city concept is regarded as one of the solutions for meeting the 17 SDGs, goals that should be met by both cities, towns and rural municipalities. Study findings showed that Estonian cities and town municipalities are engaged in various smart city initiatives such as smart transport, smart living, smart government, and smart environment. These cities and towns are also keen on using various technologies, emerging technologies included, to attain a smart city status. Nonetheless, little data emanated from the engaged rural municipalities on what smart initiatives they are currently and plan to implement. Rural municipalities that showed interest in the smart city or smart village appear to be emulating what is being done in large cities and towns. Otherwise, a participant from Rae rural municipality appear to summaries the general status of smart village adoption by stating that *"so far, we have concentrated on smaller things that definitely function. We would need a huge leap...".* These findings suggest a void in the implementation of smart city or smart village initiatives in rural municipalities given that the smart city concept has been dominated by large cities [12,

17]. It can be said that rural municipalities are struggling to conceptualize "smartness" within their context hence the need for more research on how such areas can best benefit from the use of technology.

Furthermore, study findings shows that municipalities face several challenges that include the lack of capacity, data specification and accessibility, big data management, lack of system integration, setting up the infrastructure, information security and big data usage. For example, there is a lack of knowledge on the kind and purpose of data that must be gathered, a lack of technical capacity and competency to collect, process, analyze and use data in decision making. The literature suggest that such challenges can be resolved by pilot testing smart city solutions, training users and establishing a department that is dedicated to ICT issues and/or engaging local research institutions and technological companies [1, 2, 21]. These suggestions appear more suited for large towns and cities that have a better service sector according to study findings by Duygan et al. [6] and Yigitcanlar et al. [7]. In this study, the town of Rakvere is taking advantage of a better service sector by working with the University of Tartu in developing a solution that uses satellite data in developing and planning cities. Viimsi rural municipality shows the use of a private sector organization, KPMG, in assessing the implementation of its information systems. However, a small service sector in rural areas suggest that rural municipalities are exposed to a limited base for consultations when compared to their counterparts. Furthermore, challenges faced by rural municipalities are compounded by the lack of relevant reference case studies on how such municipalities can successfully harness the smart village concept. This points to the need of a bottom-up approach when adopting the smart city or smart village concept. The use of a bottom-up approach in implementing smartness across cities, towns and rural areas is supported in the literature [5, 12, 17, 18]. It is important to point out that this study findings suggest a top-down approach is being used to implement the smart city concept. For example, Statistics Estonia appears to be the central organization that gathers and disseminate data to different municipalities. This data appears to be limited and does not meet all the information needs of municipalities as mentioned by participants from Lääne and Rae rural municipality. Again, a participant from Valga town further emphasizes the need for a localized solution for implementing a smart city solution by reporting challenges faced due to towns and cities' system disintegration. The current system does not seem to permit the collating of different forms of data across cities, towns and rural areas as required by municipalities. On the other hand, it is accurate that small municipalities lack resources (skills and human resources) something that may warrant a top-down approach to implementing smartness. However, if assumed, such an approach should be based on a facilitative role otherwise this may result in biased initiatives that do not conform to interests of the intended target/rural areas [16].

**Study limitations and Future Research:** The current study combine views of cities, towns and rural municipalities, and went on to consolidate the most common challenges. While this approach identified the most important challenges, it may have unintentionally disregarded concerns from rural municipalities as these have unique characteristics compared to those of cities and towns. Hence, more research effort is needed on challenges faced by rural municipalities and how they can assume "smartness" within their contexts.

# References

1. Bibri, S.E., Krogstie, J.: Smart eco–city strategies and solutions: the cases of Royal Seaport, Stockholm, and Western Harbor, Malmö, Sweden. Urban Sci. **4**(1), 1–42 (2020)
2. Hämäläinen, M.: A framework for a smart city design: digital transformation in the Helsinki smart city. In: Ratten, V. (ed.) Entrepreneurship and the Community. CMS, pp. 63–86. Springer, Cham (2020). https://doi.org/10.1007/978-3-030-23604-5_5
3. Shao, Q.G., Jiang, C.C., Lo, H.W., Liou, J.J.: An assessment model of smart city sustainable development: integrating approach With Z-DEMATEL and Z-TOPSIS-AL (2022)
4. Wirsbinna, A., Grega, L.: Assessment of economic benefits of smart city initiatives. Cuadernos Econ. **44**(126), 45–56 (2021)
5. Agriesti, S.A.M., Soe, R.M., Saif, M.A.: Framework for connecting the mobility challenges in low density areas to smart mobility solutions: the case study of Estonian municipalities. Eur. Transp. Res. Rev. **14**, 32 (2022)
6. Duygan, M., Fischer, M., Pärli, R., Ingold, K.: Where do smart cities grow? The spatial and socio-economic configurations of smart city development. Sustain. Cities Soc. **77**, 103578 (2022)
7. Yigitcanlar, T., Kenan, D., Luke, B., Kevin C. D.: What are the key factors affecting smart city transformation readiness? Evidence from Australian cities. Cities **120** (2022)
8. Kutty, A.A., Kucukvar, M., Abdella, G.M., Bulak, M.E., Onat, N.C.: Sustainability performance of European smart cities: a novel DEA approach with double frontiers. Sustain. Cities Soc. (2022)
9. Bibri, S.E.: A novel model for data-driven smart sustainable cities of the future: the institutional transformations required for balancing and advancing the three goals of sustainability. Energy Inform. **4**(1), 1–37 (2021)
10. Shamsuzzoha, A., Nieminen, J., Piya, S., Rutledge, K.: Smart city for sustainable environment: a comparison of participatory strategies from Helsinki, Singapore and London. Cities **114**, 103194 (2021)
11. Ismagilova, E., Hughes, L., Dwivedi, Y.K., Raman, K.R.: Smart cities: advances in research— An information systems perspective. Int. J. Inf. Manage. **1**(47), 88–100 (2019)
12. Wang, Q., Luo, S., Zhang, J., Furuva, K.: Increased attention to smart development in rural areas: a scientometric analysis of smart village research. Land **11**(8), 1362 (2022)
13. Bibri, S.E., John, K.: A novel model for data-driven smart sustainable cities of the future: A strategic roadmap to transformational change in the era of big data. Future Cities Environ. **7**(1) (2021)
14. Eckhoff, D., Wagner, I.: Privacy in the smart city—applications, technologies, challenges, and solutions. IEEE Commun. Surv. Tutor. **20**(1), 489–516 (2017)
15. Hämäläinen, M., Tyrväinen, P.: Improving smart city design: a conceptual model for governing complex smart city ecosystems. In: Bled eConference. University of Maribor Press (2018)
16. Szalai, Á., Varró, F., Szabolcs, F.: Towards a multiscalar perspective on the prospects of 'the actually existing smart village'–a view from Hungary. Hung. Geogr. Bull. **70**(2), 97–112 (2021)
17. Bielska, A., Stańczuk-Gałwiaczek, M., Sobolewska-Mikulska, K., Mroczkowski, R.: Implementation of the smart village concept based on selected spatial patterns–a case study of Mazowieckie Voivodeship in Poland. Land Use Policy **104**, 105366 (2021)
18. Cowie, P., Leanne, T., Koen, S.: Smart rural futures: will rural areas be left behind in the 4th industrial revolution? J. Rural. Stud. **79**, 169–176 (2020)
19. Bawany, N. Z., Shamsi, J.A.: Smart city architecture: vision and challenges. Int. J. Adv. Comput. Sci. Appl. **6**(11) (2015)

20. Khan, H.H., et al.: Challenges for sustainable smart city development: a conceptual framework. Sustain. Dev. **28**(5), 1507–1518 (2020)
21. Silva, B.N., Khan, M., Han, K.: Towards sustainable smart cities: a review of trends, architectures, components, and open challenges in smart cities. Sustain. Cities Soc. **38**, 697–713 (2018)
22. Zhang, K., Jianbing, N., Kan, Y., Xiaohui, L., Ju, R., Xuemin, S.S.: Security and privacy in smart city applications: challenges and solutions. IEEE Commun. Mag. **55**(1), 122–129 (2017)
23. Chen, Y., Silva, E.A.: Smart transport: A comparative analysis using the most used indicators in the literature juxtaposed with interventions in English metropolitan areas. Transp. Res. Interdisc. Perspect. **10**, 100371 (2021)
24. Soe, R., Sarv, L., Gasco-Hernandez, M.: Systematic mapping of long-term urban challenges. Sustainability **14**(2), 817 (2022)

# Tools for Calculating the ICT Footprint of Organisations: Adaptation of a European Study

Guillaume Bourgeois[1,2]([✉]), Kassandra Bigot[1,2], Vincent Courboulay[1,2], and Benjamin Duthil[1,2]

[1] IT, Image, Interaction Laboratory (L3I), University of la Rochelle, 17000 La Rochelle, France
guillaume.bourgeois@univ-lr.fr
[2] EIGSI, 17041 La Rochelle, France

**Abstract.** In the face of global warming caused by greenhouse gas (GHG) emissions, the IT industry is responsible for 3–4% of global carbon dioxide emissions. Organisations aiming for a green IT strategy need to do more than just improve their environmental impact. Carbon footprinting is a valuable decision support tool that allows organisations to measure and communicate the environmental impact of their activities. To do this, they need tools that can calculate, track and report their greenhouse gas emissions and guide actions to reduce and offset these emissions. The aim of this article is to present a specific tool for calculating an organisation's digital carbon footprint, called WeNR Light. It is directly based on the WeNR2021 study.

**Keywords:** Sustainable Information Systems · carbon footprint calculator · individual actions · Green Computing · eco-responsible · ecological impact · sustainable development · energy consumption

## 1 Introduction

Firstly, one of the biggest challenges facing the world today is climate change, because of its impact on people and the environment. In the 1970s, governments around the world became aware of the need to address sustainable development. In 1983, the World Commission on Environment and Development (WCED) took on the mission of uniting nations in the pursuit of sustainable development and popularised the term "sustainable development" with the publication of the Brundtland Report [6]. Since the Rio Declaration, Agenda 21 and the United Nations Framework Convention on Climate Change, society has become more sensitive to the environment, including greenhouse gases (GHGs). Later, in 2015, the 2030 Agenda for Sustainable Development (United Nations, 2015) proposed a set of 17 Sustainable Development Goals (SDGs) to be achieved by 2030, including SDG13 (Climate Action - Acting Urgently to Combat Climate

M. Papadaki et al. (Eds.): EMCIS 2022, LNBIP 464, pp. 277–290, 2023.
https://doi.org/10.1007/978-3-031-30694-5_21

Change and its Impacts). More recently, in 2019, the European Commission (EC) proposed the European Green Deal, which includes 50 specific actions to combat climate change, with the aim of making Europe the first climate neutral continent by 2050 [3].

In addition, the main factor contributing to climate change is global warming, which is measured by the concentration of greenhouse gas (GHG) emissions released into the atmosphere. For organisations seeking to achieve their climate neutrality goals, the first step is to determine their current environmental performance in terms of their carbon footprint (CF) [19]. Then, based on an analysis of the current situation, organisations can propose action plans to reduce or even offset their GHG emissions [26]. The most important regulatory framework for accounting for greenhouse gas emissions is the Greenhouse Gas Protocol (2004), which defines the carbon footprint as the total quantity of greenhouse gas emissions, generally expressed in carbon dioxide equivalents (CO2e), resulting directly or indirectly from an organisation's activities [27].

This is why the concept of sustainable computing has gained new media visibility in recent months due to the increasing place of digital technology in our lives and activities [18], and the growing awareness of the role of human activity in global warming. On the intangible side, the IT sector (all sectors combined) now accounts for nearly 4% of global GHG emissions. Thus, climate change and sustainable development must be taken into account in all areas [29].

In summary, organisations, as the main users of digital technologies [25], must play an important role in the convergence of climate and social emergencies and technological transformation. To reach the awareness phase of the digital environment and social impact, we need to move to the next phase: the measurement phase. Indeed, as long as measurement is not at the heart of management, organisations will blindly experiment and fail to act on the critical levers. To remedy this, and as a direct legacy of the WeGreenIT work published by WWF in 2018. ISIT (Institute for Sustainable IT) in France, Switzerland and Belgium has conducted the only free study to measure the quantitative and qualitative footprint of information systems, WeNR 2021.

The WeNR method allows everyone to use a simplified process to assess the carbon footprint of their IT assets. The work of collecting and compiling the data sources and assessing the reliability of the values is carried out by a team of WeNR engineers with the aim of disseminating a common reference framework within the community. In this respect, WeNR is a big step forward [28].

However, data collection for those who wish to participate in the work can be complicated. For example, the amount of refrigerant gas injected into the chiller each year, the total power consumption of the server room, is often difficult to obtain. This difficulty should not be an obstacle to participation in the WeNR process.

To remedy this problem, the WeNR Light was born. The WeNR Light version is a much lighter version replacing the long questionnaire with a much smaller one of about ten questions to be filled in directly on the site and allowing a quick result of the equivalent number of kilograms of $CO_2$ that an employee produces per year for an organisation. However, there is an uncertainty rate close to 10–15%.

We will begin this article by detailing the general WeNR 2021 study that was used and its main results, drawing conclusions and strategies for recommendations. We will then show the declination of the WeNR tool into a lighter version, WeNR Light, and the conclusions that can be drawn from it, discussing them at the end of the article.

## 2   Study WeNR

In this chapter we present the results of the WeNR2021 study, which we will use to create our WeNR Light tool from the same database.

### 2.1   Presentation of the Sample

**Fig. 1.** Firms distribution by size.

The sample studied represents 62 French (82%), 9 Belgian (11%) and Swiss 4 (7%) companies, for a total of 1,309,604 jobs, distributed among different sizes of companies: VSEs (10%), SMEs (18%), MSEs (36%), and LMEs (36%). The companies in the sample are established in different sectors of activity (other specialized, scientific and technical activities: 26%; public administration: 23%; financial and insurance activities: 12%; production and distribution of electricity, gas, steam and water: 7%; transport and warehousing: 7%; administrative and support service activities: 6%; trade, repair of motor vehicles and motorcycles: 6%; other: 4%; legal, accounting, management, architectural, engineering, controlling, and technical analysis activities: 4%; real estate activities: 3%; and manufacture of food products, beverages, and tobacco products: 1%), either in the public domain (27%) or in the private domain (74%) (Fig. 1).

## 3  Interpretation

We will detail the different interpretations that can be made of the results of the study.

**Facts**

The main number to remember is the digital carbon footprint of a European employee which represents 265 kg CO2e per year (220 working days). It would take the planting of 66 trees per year to offset it.

**Focus on Equipment**

First of all, for the distribution of GHGs by equipment area, in their life cycle, the manufacturing of office equipment is responsible for 77% of the GHG emissions of the organizations participating in the study. Overall, nearly 90% of GHG emissions come from manufacturing (office equipment and data centers combined) [24].

Otherwise, 7% of the office equipment of the organizations studied is purchased reconditioned. This is a significant area of improvement for organizations wishing to reduce their environmental footprint. After use within the organization, an average of 34% of office equipment is reconditioned or given a second life. This is also a lever for improvement to reduce the environmental impact of an IT asset.

In terms of lifespan, monitors and network equipment stand out with a lifespan of 7 years. In the study, smartphones and tablets have a lifespan of 4 years (global average of 18 to 36 months depending on the country), which shows that organizations are committed to extending the lifespan of their equipment. The longevity of equipment is a very encouraging sign that needs to become more widespread. To continue their efforts, organizations that are already well on their way to a digitally responsible approach can implement key best practices such as equipment repurposing and reuse. This exercise can significantly reduce the environmental footprint of the Information System.

**Focus on Maturity**

Regarding maturity, the participating entities demonstrate maturity on the subject of responsible digital with an average score of 59%. Digital responsibility is expanding and moving out of the environment of expert organizations, which are already far ahead on the subject. Less mature entities are taking up the issue and are seeking to evaluate the footprint of their information systems from the very beginning of their DR approach. The democratization of the deployment of Green Procurement approaches and the ability of organizations of all sizes to work to implement good practices to reduce the social and environmental footprint of their Information System play a determining role and is very encouraging [7].

**Recommendations**

The recommendations that we can make are:

– Develop a Sustainable IT Strategy

Develop a responsible digital strategy based on measuring the environmental footprint of the Information System with WeNR. The deployment of the NR approach allows for a 10 to 20% reduction in the environmental impact of the information system [5].

– Extend the life of equipment

From the moment of purchase, select equipment that is easy to repair and to upgrade within the organization or outside: second life, reconditioning... It is essential to extend the life of equipment to reduce its environmental footprint.

– Eco-design of digital services

While the environmental footprint of equipment is essentially a result of manufacturing, usage will have a direct impact on its lifespan. Thus, designing digital services that are more virtuous from an environmental, social and economic point of view will have a direct impact on the quantity of equipment consumed by the organization [2].

## 4    WeNR Light Application

In this section, we will describe how we have adapted the WeNR 2021 study into the WeNR Light online tool:

### 4.1    Goal

The Carbon Footprint Estimator combines a data model with a user interface. It allows people to measure their carbon emissions and track methods to reduce them. It is called an online estimator because it exists online. Self-management is about measuring your carbon footprint or taking action to minimise it. Many estimators are available for public and private use to measure carbon footprints or emissions. This is to provide users with the ability to manage their own habits [1].

To use the information effectively, it must be relevant to the employee's daily activities. When collecting information, professionals need to consider the context - in particular, carbon footprint calculators can be useful tools for decision-makers to understand climate change and raise environmental awareness. Acquiring information in this way is particularly effective when the information comes from a professional context [4].

While there is some literature on personal carbon footprint calculators in everyday contexts as we can see in Table 1, there is no research on specific tools for professional contexts. This paper attempts to initiate and track greenhouse gas emission reductions from the WeNR2021 study by providing an open source carbon footprint calculator relevant to a professional domain to quickly measure carbon impacts within an organization and drive action. The National Climate Action Simulator (NGC) was chosen as a solid foundation for business applications. The data model is based on the WeNR2021 study and the scope is limited to ICT [8].

**Table 1.** Comparison of carbon footprint Calculator.

| Name | Accessibility | Scalability | Broad scope of application |
|---|---|---|---|
| Carbon Footprint (CF, 2020) | ●●● | ● | ●●● |
| Carbon Fund (CFund, 2021) | ●● | ●● | ●● |
| Cool climate (CoolCalifornia, 2021; Simpson, 2009) | ●● | ● | ●● |
| GHG Protocol Calculator (GHG Protocol, 2021) | ● | ● | ●●● |
| myclimate (Foundation myclimate, 2021) | ●●● | ●● | ●● |
| Simple Carbon Calculator (National Energy Foundation, 2017) | ●●● | ●● | ●● |
| Simplified GHG Emissions Calculator (SGEC, 2020) | ●●● | ●● | ●●● |
| Terrapass Calculator (Terrapass, 2021) | ●●● | ● | ●● |
| CA-CP (CA-CP, 2020) | ●● | ●● | ●● |
| SIMAP (SIMAP, 2020) | ● | ●● | ●● |

The objective of the calculator is to enable users to identify the main sources of ICT-related emissions. Special focus on user engagement and adoption. One of the issues addressed in this article is to make users feel guilty and increase positive influence by providing social comparisons and contextual suggestions.

### 4.2   Source Release

The French non-profit ABC has created a calculator called Nos Gestes Climat en 2020. It was inspired by a calculator created in 2010 by ADEME, a French public agency. In addition to the inspiration, Nos Gestes Climat also received input from Avenir Climatique, a French NGO dedicated to improving carbon accounting and the transition to a low-carbon society. The final product was published as an open-source application and supported by ADEME and the beta.gouv.fr incubator. NGC was created with Avenir Climatique and ADEME as advisors. They worked on a project to facilitate access to carbon footprint information, as well as on the need to transform awareness into active engagement. Many people, organisations and institutions are discussing climate change and a low carbon transition. This includes schools, businesses, governments and even families. Until recently, the public had access to climate information, including emission factors, via an online database maintained by ADEME, but it was difficult for non-specialists to use [1].

"Nos Gestes Climat" is a fork of the futur.eco project (copied from the code, then modified to serve another purpose), which is itself a fork of mon-entreprise.fr (my-company.fr), mon-entreprise.fr (my-company.fr) is produced by the French public service incubator beta.gouv.fr and URSSAF. This forked chain shows an interest in open source code for public services, where working on professional fields (social contributions) allows working on very different subjects. This success prompted the my-company team to outsource the core of the project to the TypeScript publi.codes library, which allows users to write public interest models in a new French programming language, and expose these algorithms as well as simple tables to collect. The necessary input goes directly to the web page. Models developed using publi.codes can be fully documented in the web application itself. In UCS, users can easily access the account details [1].

NGC is divided into two code bases, the user interface and the data model represented by a yaml file. The model repository is where contributions are mainly made by the Datagir and ABC teams, but also includes anyone who finds bugs, has problems or wants to incorporate better local models. The use of publi.codes ensures that template updates automatically update end-user forms without any additional effort from the developer [9].

Since the first deployment of the project in June 2020, over 600 issues have been resolved or prioritised in the GitHub repository, which also includes the UCS roadmap and detailed release notes [22]. The user interface itself is coded in Javascript/Typescript and is rapidly evolving to help users understand their footprints in an interactive, mobile and graphical way, and easily share their results on social networks to engage in conversations to move away from changes in personal social debate at the level and want action at the political level. The user experience of the form follows the design principle that there should only be one question on the screen at a time (even in desktop mode), rather than the traditional collection of 10+ entries on a single page. Questions are grouped by consumer category, which also allows users to pause between sets of questions [17].

### 4.3   Advantages of the Organisational Design Tool

We chose the source version because it is faster than implementing and deploying from scratch. The final open source version will be made available on the GitHub platform so that future development can take place within other organisations [14].

The second advantage of WeNR light is the way users are engaged. User engagement is essential to ensure that the carbon footprint estimates via the calculator have an impact on user habits. Features such as simulation speed, interface and user-specific recommendations are carefully considered in the variants.

### 4.4   Designing the Declination

In order to build the declination, we chose to take the data from the WeNR study and we clustered them in order to have several possible scenarios, i.e. several homogeneous groups. Once this cluster was created, we performed a multivariate linear regression to give a weight to each question. The "our climate actions" tool enabled us to implement this method.

The image below represents the clustering into several homogeneous groups of organisations:

## 4.5   Implementation

The image below shows how the WeNR2021 study and its offshoot, the WeNR Light, have been implemented (Figs. 2 and 3):

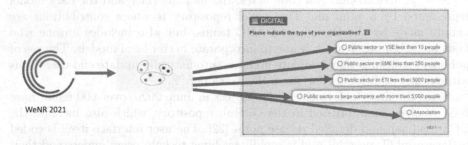

**Fig. 2.** Grouping of WeNR data by organisation type.

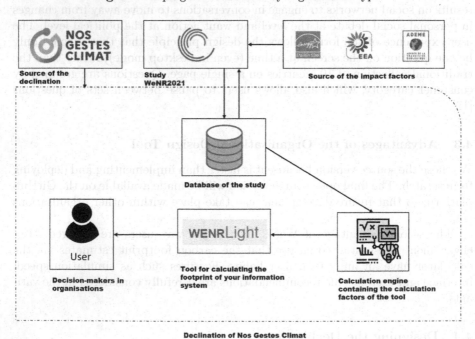

**Fig. 3.** Overall operation of the tool.

Unlike other calculators, WeNRLight is based on a mathematical model of the WeNR 2021 study that is a mutlivariate linear regression on multiple groups of data in order to simplify the WeNR model, allowing us to obtain the most accurate results possible by limiting the number of questions asked.

## 5   Achievement

We will describe the implementation of the WeNR Light, we will find the list of the 9 questions we have to answer and the information, if the decision maker does not have any values, a default one is proposed. These default answers are an average that represents the type of organisation (Fig. 4).

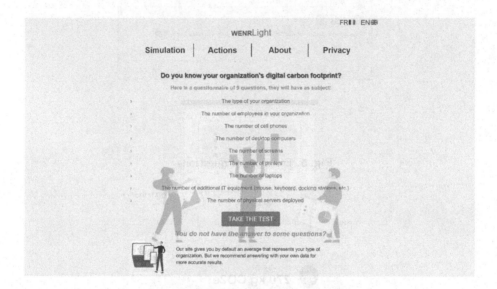

**Fig. 4.** Home page.

In order, the most relevant questions extracted from the WeNR study are

- The number of staff in your organisation,
- The number of laptops,
- The number of desktop computers,
- The number of screens,
- The number of computer accessories (mouse, keyboard, docking stations, etc.),
- The number of mobile phones,
- The number of printers (printer/copier, copier, etc.),
- The number of physical servers deployed in your organisation.

The first thing to do is to click on the START button to begin the questionnaire.

For the questions, we have to choose the corresponding answer among the six proposals. Ergonomics being very important to guide the decision-makers, the small information logo represented by an i in white on a blue background. We will then have a frame appearing in the foreground with an explanatory text on the definitions (Figs. 5 and 6).

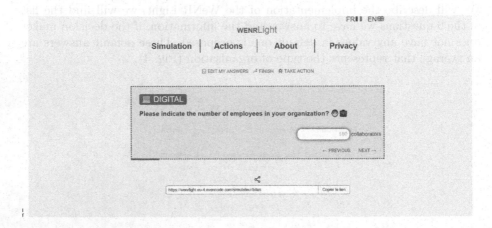

**Fig. 5.** Example of questions.

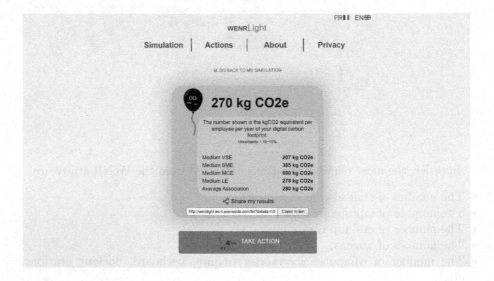

**Fig. 6.** End page.

The second most important page of the site, it allows us to have our equivalent consumption of kilograms of $CO_2$ per employee per year. A calculation will be made according to the answers to the questionnaire and the result will be written in the address bar of your browser. This result will be read out and indicated in large print to the right of the image of a balloon filled with greenhouse gases.

Here are the average values for each type of organisation:

- VSE: 207 kg CO2e,
- SME: 385 kg CO2e,
- MCE: 600 kg CO2e,
- LE: 278 kg CO2e,
- Association: 280 kg CO2e.

In an even smaller box you will see a small explanatory text followed by the uncertainty rate.

Further down you will see the average value for your type of organisation, allowing you to see where you stand (remember, however, that there is some uncertainty) [12].

As for good practices to improve your carbon score, a page has been created that allows you to know more in detail the action related and how to act well in the organisation to reduce its digital carbon footprint [30].

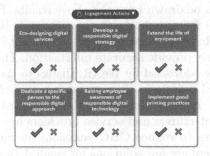

**Fig. 7.** Recommendations page.

Figure 7 gives an overview of the practical actions that companies can take after completing the calculator. For example, extending the lifetime of their equipment, these are all extracted from the best practices guide of the French government's interministerial digital department [30].

Regarding the collection of data, the site does not collect data, it will just monitor some small trivial information such as the pages consulted, the time spent and the address of the page at the end of the simulation containing the total digital carbon footprint [13].

# 6   Discussion

Firstly, it is important to have a sufficiently accurate and up-to-date inventory for the entire range covered by WeNR. The free GLPI software is so popular in the community that we recommend it. We recommend including its average power consumption or carbon footprint in the checklist when purchasing equipment. This will greatly facilitate your future WeNR evaluation [11].

For server rooms, the use of a communication PDU (Power Distribution Unit) is the best method. In the absence of these readings, theoretical values are recommended as part of the method [20].

Electricity in mainland France is particularly decarbonised. We recommend using the value from the EcoInvent database (0.119 kgCO2e), although it corresponds to peak production. This value is very low compared to the European or world average due to the high share of nuclear in the French energy mix. It is important to keep in mind that this creates other problems that do not translate into $CO_2$ (no decommissioning, waste storage, etc.) [21].

We also recommend publishing the data obtained in order to raise awareness of climate issues among users, who are often well-intentioned. Finally, the WeNR methods are designed to work independently, but WeNR can help you implement or interpret the results, as well as provide relevant advice.

## 7   Conclusion

Several conclusions can be drawn from the above results. Firstly, we found that there are few solutions for estimating an organisation's ICT carbon footprint, and the solutions that have been implemented are mainly focused on households and transport, and rarely provide accurate information. The transparency issues mentioned are partially addressed by WeNR, thanks to the thorough sharing of sources and factors used [10]. Secondly, our research related to green IT revealed that ICT needs to pay more attention to its energy consumption. While most large data centre manufacturers and service providers are aware of the need to act and adopt measures to reduce GHG emissions at various stages [15], the research shows that there are few solutions for small data centres [23].

Thus, the results of this study confirm that there is still much to be done in this area and that application services are needed to help companies reduce their costs and emissions. In carbon reduction efforts, communication is as important as the policies and actions themselves: to involve as many of the company's stakeholders as possible, public organising activities need to be carried out. These activities will increase the number of interested people [22].

It is also important to broaden and improve the data sources used. In the future, WeNR aims to become the reference tool to support the deployment of sustainable IT approaches and reduce the environmental impact of European organisations' information systems over time. ISIT announces the release of a new version in 2022 in addition to the WeNR Light, the WeNR Plus, which will provide benchmarking and actionable sustainable IT insights for participating ISIT member organisations and the WeNR 2022 and 2023 campaign.

Finally, WeNR Plus will use the WeNR models and calculations to provide a more comprehensive and detailed report in terms of quantity, quality and comparison with peer organisations, but most importantly by analysing the impact of policy decisions. In addition, the analytical tools provided will help define an action plan for developing a responsible digital strategy. Finally, as different cloud providers liberalise their carbon impact, the next version of WeNR will include the implementation of APIs to quantify the greenhouse gas emissions of cloud systems.

# References

1. Auger, C., Hilloulin, B., Boisserie, B., Thomas, M., Guignard, Q., Rozière, E.: Open-source carbon footprint estimator: development and university declination. Sustainability **13**(8), 4315 (2021). https://doi.org/10.3390/su13084315
2. Foogooa, R., Dookhitram, K.: A self green ICT maturity assessment tool for SMEs. In: 2014 IST-Africa Conference Proceedings, Le Meridien Ile Maurice, Pointe Aux Piments, Mauritius, pp. 1–9 (2014). https://doi.org/10.1109/ISTAFRICA.2014. 6880671
3. Belkhir, L., Elmeligi, A.: Assessing ICT global emissions footprint: trends to 2040 and recommendations. J. Clean. Prod. **177**, 448–463 (2018). https://doi.org/10. 1016/j.jclepro.2017.12.239
4. Thackray, H., Kor, A., Pattinson, C., Earle, L.: Audit of an organisation's ICT systems for flexible working. Undefined (2018). Accessed 13 June 2022
5. Montbroussous, B., Berthoud, F., Feltin, G., Moreau, G., Schaeffer, J.: Calculer le bilan Carbone de votre parc informatique avec EcoDiag, un service EcoInfo, p. 10 (2019)
6. Valls-Val, K., Bovea, M.D.: Carbon footprint assessment tool for universities: CO2UNV. Sustain. Prod. Consum. **29**, 791–804 (2022). https://doi.org/10.1016/ j.spc.2021.11.020
7. Pandey, D., Agrawal, M., Pandey, J.S.: Carbon footprint: current methods of estimation. Environ. Monit. Assess. **178**(1–4), 135–160 (2011). https://doi.org/10. 1007/s10661-010-1678-y
8. Cheung, C.W., Berger, M., Finkbeiner, M.: Comparative life cycle assessment of re-use and replacement for video projectors. Int. J. Life Cycle Assess. **23**(1), 82–94 (2017). https://doi.org/10.1007/s11367-017-1301-3
9. Teehan, P., Kandlikar, M.: Comparing embodied greenhouse gas emissions of modern computing and electronics products. Environ. Sci. Technol. **47**(9), 3997–4003 (2013). https://doi.org/10.1021/es303012r
10. Berkhout, P.H.G., Muskens, J.C., Velthuijsen, J.W.: Defining the rebound effect. Energy Policy **28**(6), 425–432 (2000). https://doi.org/10.1016/S0301-4215(00)00022-7
11. Bourgeois, G., Courboulay, V., Duthil, B.: Deployment of a campaign to measure the ICT carbon footprint experimentation in French-speaking Europe (2021)
12. Court, V., Sorrell, S.: Digitalisation of goods: a systematic review of the determinants and magnitude of the impacts on energy consumption. Environ. Res. Lett. **15**(4), 043001 (2020). https://doi.org/10.1088/1748-9326/ab6788
13. Zhang, C., Khan, I., Dagar, V., Saeed, A., Zafar, M.W.: Environmental impact of information and communication technology: unveiling the role of education in developing countries. Technol. Forecast. Soc. Chang. **178**, 121570 (2022). https:// doi.org/10.1016/j.techfore.2022.121570
14. Berthoud, F., Parry, M.: Évaluation des impacts environnementaux de l'informatique. Terminal. Technologie de l'information, culture and société, no. 106–107 (2010). https://doi.org/10.4000/terminal.1794

15. Mzabi, A.E., Khihel, F.: Green IT une solution informatique pour l'environnement. Rev. Afr. Manage. **3**(2) (2018). https://doi.org/10.48424/IMIST.PRSM/ram-v3i2.11178

16. Malmodin, J., Moberg, Å., Lundén, D., Finnveden, G., Lövehagen, N.: Greenhouse gas emissions and operational electricity use in the ICT and entertainment media sectors. J. Ind. Ecol. **14**(5), 770–790 (2010). https://doi.org/10.1111/j.1530-9290.2010.00278.x

17. Zhou, X., Zhou, D., Wang, Q., Su, B.: How information and communication technology drives carbon emissions: a sector-level analysis for China. Energy Econ. **81**, 380–392 (2019). https://doi.org/10.1016/j.eneco.2019.04.014

18. Jones, N.: How to stop data centres from gobbling up the world's electricity. Nature **561**(7722), 163–166 (2018). https://doi.org/10.1038/d41586-018-06610-y

19. Kern, E., Dick, M., Naumann, S., Hiller, T.: Impacts of software and its engineering on the carbon footprint of ICT. Environ. Impact Assess. Rev. **52**, 53–61 (2015). https://doi.org/10.1016/j.eiar.2014.07.003

20. Lean ICT: Towards Digital Sobriety: our new report. The Shift Project, 05 March 2019. https://theshiftproject.org/en/article/lean-ict-our-new-report/. Accessed 13 June 2022

21. Arushanyan, Y., Ekener-Petersen, E., Finnveden, G.: Lessons learned - review of LCAs for ICT products and services. Comput. Ind. **65**(2), 211–234 (2014). https://doi.org/10.1016/j.compind.2013.10.003

22. Malmodin, J., Lundén, D., Moberg, Å., Andersson, G., Nilsson, M.: Life cycle assessment of ICT. J. Ind. Ecol. **18**(6), 829–845 (2014). https://doi.org/10.1111/jiec.12145

23. Andrae, A.S.G., Edler, T.: On global electricity usage of communication technology: trends to 2030. Challenges **6**(1), 117–157 (2015). https://doi.org/10.3390/challe6010117

24. Grimm, D., Weiss, D., Erek, K., Zarnekow, R.: Product carbon footprint and life cycle assessment of ICT - literature review and state of the art. In: 2014 47th Hawaii International Conference on System Sciences, Waikoloa, HI, pp. 875–884 (2014). https://doi.org/10.1109/HICSS.2014.116

25. Guillaume, B., Benjamin, D., Vincent, C.: Review of the impact of IT on the environment and solution with a detailed assessment of the associated gray literature. Sustainability **14**(4), 2457 (2022). https://doi.org/10.3390/su14042457

26. Set Science-Based Emission Reduction Targets—UN Global Compact. https://www.unglobalcompact.org/take-action/action/science-based-target. Accessed 13 June 2022

27. Hu, J.-L., Chen, Y.-C., Yang, Y.-P.: The development and issues of energy-ICT: a review of literature with economic and managerial viewpoints. Energies **15**(2), 594 (2022). https://doi.org/10.3390/en15020594

28. WeNR—Information System footprint measurement tool. WeNR. https://wenr.isit-europe.org/. Accessed 13 June 2022

29. Gonzalez, A., Chase, A., Horowitz, N.: What we know and don't know about embodied energy and greenhouse gases for electronics, appliances, and light bulbs, p. 12 (2012)

30. Guide de bonnes pratiques numérique responsable pour les organisations. https://ecoresponsable.numerique.gouv.fr/publications/bonnes-pratiques/. Accessed 07 Sept 2022

# Designing an API for the Provision of Public Service Information Based on CPSV-AP

Christos Pappis[1], Dimitris Zeginis[2,3]($\boxtimes$), Efthimios Tambouris[3], and Konstantinos Tarabanis[2,3]

[1] Hellenic Open University, Patras, Greece
std142986@ac.eap.gr
[2] Centre for Research and Technology - Hellas, 6th km. Charilaou-Thermi RD, Thessaloniki, Greece
{zeginis,kat}@iti.gr
[3] University of Macedonia, Egnatia 156, 54636 Thessaloniki, Greece
{tambouris,zeginis,kat}@uom.edu.gr

**Abstract.** The provision of public services (PS) is at the heart of public authority operations as it directly affects citizens' lives and the prosperity of society. Part of PS provision is publishing PS descriptions in online catalogues to inform citizens and promote transparency. The European Union has developed the Core Public Service Vocabulary Application Profile (CPSV-AP) as a PS data model to facilitate PS catalogue creation and promote semantic interoperability. The use of CPSV-AP is increasing within the European Union leading to a number of relevant PS descriptions that are also potentially beneficial for the implementation of the Single Digital Gateway (SDG) regulation. However, the use of heterogeneous, some times not very popular technologies (e.g. RDF) to provide access to such data hinders their further uniform exploitation. Towards this problem definition, Application Programming Interfaces (APIs) can provide a uniform, easily-consumable way of serving such data hiding any heterogeneity of the underlying technologies. The aim of this paper is to design and implement an API to serve data based on CPSV-AP facilitating the implementation of the SDG and the further exploitation of data at added value services and tools e.g. chatbots. The paper describes the requirements, use cases, design choices and implementation details of the API. We anticipate that the proposed API will help towards the further adoption of CPSV-AP by facilitating and homogenizing the consumption of relevant data.

**Keywords:** eGovernment · API · public service · CPSV-AP

## 1 Introduction and Motivation

The provision of public services (PS) is the main task of most public authorities and a major part of their operations in total. PS provision is usually the main

M. Papadaki et al. (Eds.): EMCIS 2022, LNBIP 464, pp. 291–304, 2023.
https://doi.org/10.1007/978-3-031-30694-5_22

task for the implementation of a governmental policy targeting the fulfillment of citizens' needs. As PS provision is a major part of public sector operations, the opening of information on PSs, i.e. publishing structured PS descriptions, highly contributes to the realization of the Open Government vision, promoting transparency and trust between public administrations and citizens.

Every eGovernment system that provides PSs is based on an underlying PS model [15]. PS models and standards are considered as the main enablers for promoting interoperability and quality in the PS provision domain. Embracing the need for a standard PS model, the European Commission has developed the Core Public Service Vocabulary (CPSV) [9] that enables the description of public services.

The adoption of CPSV is expected to facilitate semantic interoperability of public service descriptions throughout Europe. This will be beneficial for the implementation of the Single Digital Gateway (SDG) regulation [14], since the European coordinator of the SDG has to collect the descriptions of public services from European public administrations in one unique portal. However, the use of heterogeneous technologies (e.g. SQL, RDF) to store the data produced based on CPSV-AP as well as the unfamiliarity with some of the technologies related to the CPSV-AP (e.g. RDF) may hinder the collection of such data by the SDG as well as their exploitation by added value services and tools.

Towards this problem definition, APIs can provide uniform access to data described based on CPSV-AP hiding any heterogeneity of the underlying technologies [20]. Additionally, the use of most CPSV-AP data requires skills and tooling (e.g. RDF, SPARQL) that are less widespread than some other web technologies (e.g. JSON, Javascript). For example, there are many Javascript visualization libraries that consume JSON data (e.g. D3.js, charts.js), while there are just a few that consume RDF and their functionality is limited.

In order to unleash the full potential of CPSV-AP described data there is a need to standardize the interaction (i.e. input, output and functionality) in a way that facilitates: i) the development of reusable services and tools and ii) the automatic collection of public service descriptions from different catalogues. This paper describes the requirements, use cases, design choices and implementation details of an API that aims to exploit and unify CPSV-AP data stored using heterogeneous technologies while also making data available in a structure and format that is familiar to a larger group of developers. The API offers a uniform way to access the data, thus enabling the development of generic services and tools that can be reused across datasets and even cross-boarder.

The rest of the paper is organized as follows, Sect. 2 presents relevant background and related work, Sect. 3 describes the methodology followed for the development of the API, Sect. 4 presents the use cases and requirements that need to be covered by the API, Sect. 5 describes the design decisions for the API, Sect. 6 presents the implementation details, Sect. 7 describes some exaples scenarions using the API and finally Sect. 8 concludes the paper.

## 2    Background Work

This section presents some background work needed to understand the context of the paper and the developed API.

### 2.1    The Core Public Service Vocabulary (CPSV)

The Core Public Service Vocabulary (CPSV) [9] (Fig. 1) is a data model for describing public services and the associated life and business events (e.g. "having a baby", "starting a new business"). The two main classes of the model are the "Public Service" and the "Public Organization" that is responsible for offering the service. Except from these two classes there are many other describing other aspects of the public service including the required input to execute the service (i.e. "Evidence"), the "Output" of the service, the "Cost" of the service, the "'Channel" through which the service is offered, the "Contact Point" of the service etc.

Based on CPSV a linked data application profile CPSV-AP was also developed. An Application Profile is a "specification that re-uses terms from one or more base standards, adding more specificity by identifying mandatory, recommended and optional elements to be used for a particular application, as well as recommendations for controlled vocabularies to be used" [13]. For example, CPSV-AP reuses classes and the properties of existing vocabularies including FOAF [4], DCT [11] and DCAT [1] as well as other Core Vocabularies e.g. CPOV. It also proposes the use of controlled vocabularies to populate the properties e.g., for the names of countries [17] and languages [18]. The more recent version of CPSV-AP is version 3.0 and was published in 2022.

The classes and properties of CPSV-AP are categorized as mandatory or optional. The minimum requirements of a public service description, in order to comply with CPSV-AP, is to provide at least information on the mandatory properties of the mandatory classes. Optional classes can still have mandatory properties for which information should be provided when an optional class is used.

The CPSV-AP belongs to a set of Core Vocabularies that have been developed by the European Comission in order to conceptualise the public service provision domain. These vocabularies include, apart from the CPSV-AP, the Core Criterion and Core Evidence Vocabulary (CCCEV) [6], Core Person Vocabulary (CPV) [7], the Core Public Organisation Vocabulary (CPOV) [8] and the Core Business Vocabulary (CBV) [5].

Currently there are many known uses of CPSV-AP throughout Europe [12]:

- Belgium used the CPSV-AP to harmonise public services data from different regional sources and centralise them into a common system. Additionally, the region of Flanders adopted a slightly updated version of the CPSV-AP as their regional model for describing public services.
- The Estonian Ministry of Economic Affairs extended CPSV-AP to address local needs, as well as to cover the public service life-cycle. They also developed a service to harvest data based on their extension.

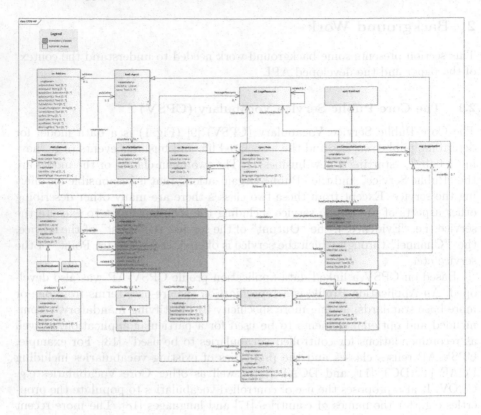

**Fig. 1.** The CPSV-AP v3.0 model [9].

- Finland used the CPSV-AP to create a national data model for describing public services. In addition, they ran a pilot with Estonia for which they used the CPSV-AP and tools to create a cross-border catalogue of public services.
- The Region of Epirus in Greece used the CPSV-AP to model a subset of their public services catalogue and used the suggested procedure to turn them into linked data.
- Ireland used CPSV-AP to create a national data model tailored to their needs for describing public services.
- The Agency of Digitisation in Italy created a national data model that extends CPSV-AP to include country-specific characteristics. They are using the model in a national catalogue of public services. Additionally, they use CPSV-AP at local level and specifically at the Autonomous Province of Trento to describe public services through a distributed Content Management System.
- Portugal created also a national data model that adapts the CPSV-AP to include country-specific characteristics. They are implementing their model in a national catalogue of public services.

- Slovakia is working on a national data model based on the CPSV-AP to support the mapping of the Slovakian central meta information system of public administration to the CPSV-AP.
- Spain is working on a pilot with Portugal for which they are using the CPSV-AP to create a cross-border catalogue of public services.
- The Netherlands reused certain CPSV-AP tools to develop a simple way of creating public service descriptions matching the Dutch national data model.

## 2.2  Semantic Web and Linked Data

The term semantic web refers to an extension of the current "web of documents" in order to build a "web of linked data". In order to achieve the vision of linked data, a set of principles have been proposed to "publish data on the web in such a way that it is machine-readable, its meaning is explicitly defined, it is linked to other external datasets, and can in turn be linked to from external datasets" [3]. Semantic Web and linked data are empowered by technologies such as RDF [10] and SPARQL [21] in order to publish data on the web.

More specifically, RDF is a W3C standard model for data interchange on the Web. It uses URIs to name things and their relationships that are expressed as "triples" $<X, Y, Z>$ (e.g., $<$ PublicService1, hasCompetentAuthority, Organization1$>$). RDF has many serialization formats such as RDF/XML [23], Turtle [22] and JSON-LD [24]. The JSON-LD is the most recently developed serialization that aims to be lightweight and easy for humans to read. It is based on the successful JSON format and is ideal for APIs. SPARQL is a language to express queries across diverse RDF data sources.

Many researchers suggest that the linked data paradigm is ideal for Public Sector Information publishing, enabling the potentiality of breaking the bureaucratic silos [16]. Towards this direction, CPSV-AP incorporates linked data as an underpinning technology. Thus, public service descriptions, that are modeled using CPSV-AP, can be published as linked data and become part of the linked data cloud.

However, linked data technologies, apart from its benefits, puts some significant challenges [25] [16] hindering the adoption of relevant vocabularies such as CPSV-AP. For example, using linked data requires skills and tooling (e.g. RDF, SPARQL) that are less widespread than some other web technologies (e.g. JSON, Javascript). Thus, there is a need to standardize the interaction (i.e. input, output and functionality) with linked data in a way that facilitates the development of reusable services and tools.

## 2.3  APIs for Public Service Descriptions

The Application programming interfaces (APIs) are essential enablers of the transformation towards digital governments due to their modularity, reusability and scalability [20]. The APIs are the connection nodes that standardise the interactions at digital architectures and thus foster interoperability through data transfer.

In the context of public service cataloging, EU has already identified cases [19] where either there is limited use of relevant APIs or cases where a CPSV-AP based API will be beneficial. For example a case where there is a limited use of APIs, is the Autonomous Province of Trento. They have established a common datastore[1] of concepts (e.g. Public Services) which is used by several local authorities. The data exchanged among local authorities is shared by means of a REST API and published in JSON format. In order to map the shared data structure to other data models, such as the CPSV-AP, a mapping tool has been developed so that data is transformed from JSON to JSON-LD format.

Another case that is quite mature but there is still no relevant API is the Flemish Information Agency. They have developed a model, called OSLO-Dienstencataloog[2], which extends concepts of CPSV-AP in order to uniform their public services. On top of this model they also provide a JSON-LD context that can be shared between APIs.

## 3   Methodology

This section presents the methodology that has been followed in order to develop an API for the provision of public service information based on CPSV-AP:

- Define the use cases and the corresponding requirements that need to be covered by the API (Sect. 4).
- Define the design criteria and decisions for the API (Sect. 5).
- Implement the API (Sect. 6).
- Apply the API at some relevant scenarios (Sect. 7).

The following sections describe in detail these steps.

## 4   Use Cases and Requirements

The main requirement of the API is to support Create, Read, Update, Delete (CRUD) operations on all the CPSV-AP concepts. Additionally, there is also a need for basic search functionality of public services based on some criteria that have been identified by the bibliography [15]. The main use cases that need to be supported by the API are depicted in a UML use case diagram (Fig. 2) and are the following:

- Retrieve instances of CPSV-AP concepts (e.g. public service). The API should support retrieving either all the instances or a specific instance (e.g. based on its id). The description of the instances should include all the relevant properties defined by CPSV-AP.

---

[1] http://ontopa.opencontent.it/openpa/classes.
[2] https://data.vlaanderen.be/doc/applicatieprofiel/dienstencataloog/.

– Search for Public Services. This includes: i) searching based on the competent authority of the service, ii) searching based on the input/output of the service, iii) searching based on the event (i.e. life event or business event) related to the public service and iv) search for dependent public services (based on the properties "requires" and "related" of CPSV-AP).
– Create a new instance of a CPSV-AP concept.
– Update an existing instance of a CPSV-AP concept.
– Delete an existing instance of a CPSV-AP concept.

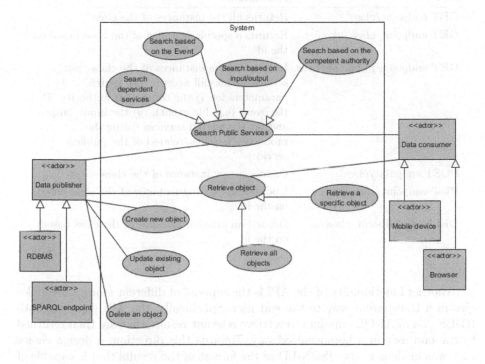

**Fig. 2.** UML Use Case diagram of the API

Mainly there are two actors/systems that are expected to use the API. The "data publisher" that is responsible for publishing and maintaining the public service descriptions using the API. The "data publisher" can use any storage technology e.g. RDBS or SPARQL endpoints. The "data consumer" can be any application (e.g. mobile device or browser) that retrieves or searches for public service descriptions in order to present them to the end user.

## 5    API Design

Based on the use cases and requirements described at Sect. 4 the main functionality of the API is to support CRUD operations on all the classes of CPSV-AP, but also support searching of public services based on specific criteria. The architectural approach that has been selected for the design of the API is the REST, since it offers many advantages including the simplicity, scalability, speed, and ability to handle all data type through simple HTTP requests. The functionality of the API, in terms of supported REST calls, is depicted at Table 1.

**Table 1.** Supported REST API calls

| API call | Description |
|----------|-------------|
| GET endpoint/class | Returns all the instances of the class |
| GET endpoint/class/id | Returns a specific instance of the class based on the id |
| GET endpoint/public_services | Returns all the instances of the class "public service". This call accepts also search parameters for: i) the competent authority, ii) the event (e.g. life event), iii) the input/output and iv) the related services (using the properties requires/related of the public service) |
| POST endpoint/class | Creates a new instance of the class |
| PUT endpoint/class/id | Updates an existing instance of the class based on the id |
| DELETE endpoint/class/id | Deleted an existing instance of the class based on the id |

Another functionality of the API is the support of different storage technologies in a transparent way to the end user. Specifically, the API supports both RDBS and SPARQL enpoints to retrieve relevant results that are then returned to the end user in a homogenized way. Towards this direction, a design choice that was made is to use JSON-LD as the format of the results that is capable of "blending" relational data with RDF data using also the CPSV-AP model.

## 6   API Implementation

The API is implemented in PHP language making use of the Symfony 6.0 web application framework that supports the Model-View-Controller architecture. In such architectures the Model component undertakes the management of the data (i.e. storage, processing and retrieval), the View component manages the presentation of information to the end user and the Controller component is the interface between Model and View components. The code of the developed API follows this approach.

Additionally, the developed API uses the api-platform 2.6 framework that facilitates the building of web APIs, the Doctrine platform that supports the abstracted use of data bases and the Composer platform for dependency management. The code of the API is available at Github[3].

The code is organized in a way where a separate source file is created for each of the CPSV-AP classes and contains information about the way to handle each class. For example, Listing 1 presents the code that defines the URI of the relevant CPSV-AP class (i.e. Public Service) as well as the supported REST API calls, while Listing 2 presents the code that defines the supported search parameters for the Public Service class.

**Listing 1.** Supported API calls for the class PublicService

```
#[ApiResource(
    iri: 'http://purl.org/vocab/cpsv#PublicService',
    collectionOperations: ['get','post'],
    itemOperations: ['get','put','delete'])]
```

**Listing 2.** Search parameters for the class PublicService

```
#[ApiFilter(SearchFilter::class, properties: [
    'identifier' => 'exact',
    'name' => 'exact',
    'description' => 'partial',
    'hasCompetentAuthority.id' => 'exact',
    'hasCompetentAuthority.preferredLabel' => 'exact',
    'isGroupedBy.id' => 'exact',
    'isGroupedBy.name' => 'exact',
    'produces.id' => 'exact',
    'produces.name' => 'exact',
    'requires.id' => 'exact',
    'requires.name' => 'exact',
    'related.id' => 'exact',
    'related.name' => 'exact',
    'hasInput.id' => 'exact',
    'hasInput.name' => 'exact'])]
```

The returned results of the API are in JSON-LD format that extends JSON by adding linked data annotations. In this way the returned results are also compatible with CPSV-AP.

The API supports the connection both to relational databases and RDF SPARQL endpoints. With the appropriate configuration the API can retrieve (using the GET method) data both from relational databases and SPARQL endpoints and return them in a unified way. Specifically, the GET method executes the appropriate SELECT statements on the relational database and the SPARQL endpoint, collects the results from both sources, merges them and returns the result. Currently, this functionality is supported only for the Public Service and the Public Organization classes. Additionally, the current implementation supports only the GET method for the SPARQL endpoints, while all

---

[3] https://github.com/chpappis/cpsv-php-api.

the methods (GET, POST, PUT, DELETE) are supported for the relational databases.

Figure 3 presents a part of the API implementation description for the "Public Service" and the "Rule" classes.

**Fig. 3.** Part of the description of the API

# 7   Using the API: Example Scenarios

An instance of the API has been deployed in order to execute some example scenarios. The instance of the API is available at the URL: http://pappis.gr/api. The API is connected at two data sources:

- A relation database with CPSV-AP data related to the public service "Getting a Greek Passport" [2].
- The SPARQL endpoint http://data.dai.uom.gr:8890/sparql that contains CPSV-AP based linked data descriptions of Public Services. The SPARQL endpoint uses the Virtuoso open-source RDF database.

The following scenarios are executed using the Postman application for calling the API and getting the results.

Scenario 1: search public services based on the name of the public organization (i.e. using the property hasCompetentAuthority of CPSV-AP) that offers them. For example, the following API call retrieves all the public services that are offered by the Public Organization with name "Passport Issuer".

```
http://pappis.gr/api/public_services?
    hasCompetentAuthority.preferredLabel=Passport Issuer
```

An excerpt of the returned result is presented at Fig. 4. The excerpt contains two public services with id 1 and 3 that are offered by the Public Organization with name "Passport Issuer" (some of the API results are in Greek since the relational database and the SPARQL endpoint exploited contain data in Greek).

```
{
    "@context": "/api/contexts/PublicService",
    "@id": "/api/public_services",
    "@type": "hydra:Collection",
    "hydra:member": [
        {
            "@id": "/api/public_services/1",
            "@type": "http://purl.org/vocab/cpsv#PublicService",
            "id": 1,
            "identifier": "Ps0001",
            "name": "Έκδοση Διαβατηρίου",
            "description": "Διαβατήριο είναι το έγγραφο που εκδίδεται ...",
            "processingTime": "P0Y0M3DT0H0M0S",
            "keyword": "Έκδοση, Διαβατήριο, Ελληνική Αστυνομία, ...",
            ....
        },
        {
            "@id": "/api/public_services/3",
            "@type": "http://purl.org/vocab/cpsv#PublicService",
            "id": 3,
            "identifier": "Ps0003",
            "name": "Έκδοση Πιστοποιητικού",
            "description": "Υπηρεσία Έκδοσης Πιστοποιητικών",
            "processingTime": "P0Y0M1DT0H0M0S",
            "keyword": "Έκδοση,Πιστοποιητικό",
            ...
        },
        ...
    ]
}
```

**Fig. 4.** Excerpt of the result of Scenario 1

Scenario 2: search of public services based on the life event they belong. For example, the following API call retrieves all the public services that belong to the life event "Traveling abroad" with id = 1. The searching is also possible using the life event name (i.e. isGroupedBy.name) but in this case the greek name of the event is required as input.

```
http://pappis.gr/api/public_services?isGroupedBy.id=1
```

Scenario 3: search of public services based on their input/output. For example, the following API call retrieves the public service that produces as output the "Passport with 3 years duration" with id = 1. The searching is also possible using the output name.

```
http://pappis.gr/api/public_services?produces.id=2
```

Scenario 4: search of dependent public services (i.e. using the properties "requires" and "related" of CPSV-AP). For example, the following API call retrieves the public service that requires the execution of the public service "Certificate issuance" with id = 4. The searching is also possible using the public service name.

```
http://pappis.gr/api/public_services?requires.id=4
```

An excerpt of the returned result is presented at Fig. 5. The excerpt contains one public service with id = 1 that requires for its execution the public service with id = 4.

```
{
    "@context": "/api/contexts/PublicService",
    "@id": "/api/public_services",
    "@type": "hydra:Collection",
    "hydra:member": [
        {
            "@id": "/api/public_services/1",
            "@type": "http://purl.org/vocab/cpsv#PublicService",
            "id": 1,
            "identifier": "Ps0001",
            "name": "Εκδοση Διαβατηρίου",
            "description": "Διαβατήριο είναι το έγγραφο που εκδίδεται από ...",
            "processingTime": "P0Y0M3DT0H0M0S",
            "keyword": "Εκδοση, Διαβατήριο, Ελληνική Αστυνομία, Ταξίδι, ...",
            "requires": [
                {
                    "@id": "/api/public_services/4",
                    "@type": "http://purl.org/vocab/cpsv#PublicService",
                    "id": 4,
                    "identifier": "Ps0004",
                    "name": "Εκδοση Βεβαίωσης",
                    "description": "Υπηρεσία Εκδοσης Βεβαιώσεων",
                    "processingTime": "P0Y0M2DT0H0M0S",
                    "keyword": "Εκδοση,Βεβαίωση",
                    ...
                }
            ],
            ...
        }
    ...
}}}
```

Fig. 5. Excerpt of the result of Scenario 4

# 8   Conclusion

The Core Public Service Vocabulary Application Profile (CPSV-AP) has been proposed as a European PS data model to facilitate PS catalogue creation and promote semantic interoperability in the PS information provision. Its broader use will also be beneficial for the implementation of the Single Digital Gateway regulation. However, the use of heterogeneous, some times not very popular technologies (e.g. RDF) to provide access to CPSV-AP data hinders their further uniform exploitation.

Towards this direction, this paper proposes the development of an API to provide a uniform, easily-consumable way of serving CPSV-AP data hiding any heterogeneity of the underlying technologies and also facilitating the automatic collection of PS descriptions from multiple catalogues. The API currently provides CRUD operations on all the concepts of CPSV-AP and also basic search functionality of public services. The API is designed in a flexible way enabling the easy addition of more functionalities e.g. searching based on other more advanced criteria. Additionally, the API supports both relational databases and partially SPARQL endpoints enabling also their integration (only for public services and public organizations). In a future version of the API the authors aim at extending the functionality for the SPARQL endpoints by supporting also the PUT, POST and DELETE methods as well as by supporting all the CPSV-AP concepts at the GET.

We anticipate that the proposed API will help towards the further adoption of CPSV-AP by facilitating the development of relevant added value services and applications as well as the automatic collection of PS descriptions from various catalogues throughout Europe. This will provide multiple benefits to citizens including the provision of integrated information about PS throughout Europe, the increase of the quantity/quality of (mobile) applications that use CPSV-AP compliant data and the personalization of PS based on the citizen's profiles. All the above can lead to the reduction of bureaucracy and time spend to identify and execute public services that in turn can foster transparency and trust between the public administration and the citizens.

**Acknowledgment.** Part of this work was funded by the European Commission, within the H2020 Programme, in the context of the project inGov under Grant Agreement Number 962563 (https://ingov-project.eu/).

# References

1. Albertoni, R., Browning, D., Cox, S., Beltran, A.G., Perego, A., Winstanley, P.: Data catalog vocabulary (dcat) - version 2. W3C Recommendation (2020). https://www.w3.org/TR/vocab-dcat-2/
2. Antoniadis, P., Tambouris, E.: Passbot: a chatbot for providing information on getting a greek passport. In: ICEGOV 2021, pp. 292–297. Association for Computing Machinery, New York, NY, USA (2021). https://doi.org/10.1145/3494193.3494233
3. Bizer, C., Heath, T., Berners-Lee, T.: Linked data-the story so far. semantic services, interoperability and web applications: emerging concepts, pp. 205–227 (2009)
4. Brickley, D., Miller, L.: Foaf vocabulary specification (2004). https://xmlns.com/foaf/0.1/
5. Core Vocabularies Working Group: Core business vocabulary. EU Semic Recommendation (2022). https://semiceu.github.io/Core-Business-Vocabulary/releases/2.00
6. Core Vocabularies Working Group: Core criterion and core evidence vocabulary. EU Semic Recommendation (2022). https://semiceu.github.io/CCCEV/releases/2.00/
7. Core Vocabularies Working Group: Core person vocabulary. EU Semic Recommendation (2022). https://semiceu.github.io/Core-Person-Vocabulary/releases/2.00

8. Core Vocabularies Working Group: Core public organisation vocabulary. EU Semic Recommendation (2022). https://semiceu.github.io/CPOV/releases/2.00
9. Core Vocabularies Working Group: Core public service vocabulary application profile (cpsv-ap). EU Semic Recommendation (2022). https://semiceu.github.io/CPSV-AP/releases/3.0.0
10. Cyganiak, R., Wood, D., Lanthaler, M.: Rdf 1.1 concepts and abstract syntax. W3C Recommendation (2014). https://www.w3.org/TR/rdf11-concepts/
11. DCMI Usage Board: Dcmi metadata terms (2020). https://www.dublincore.org/specifications/dublin-core/dcmi-terms/
12. DG-DIGIT/ISA-Programme: About core public service vocabulary application profile. https://joinup.ec.europa.eu/collection/semantic-interoperability-community-semic/solution/core-public-service-vocabulary-application-profile/about. Accessed 18 Oct 2022
13. DG-DIGIT/ISA-Programme: D04.01 core public service vocabulary application profile 2.0. European Commission, Technical Report (2016)
14. European Union: Establishing a single digital gateway to provide access to information, to procedures and to assistance and problem-solving services. Regulation (EU) 2018/1724 of the European Parliament and of the Council (2018)
15. Gerontas, A., Peristeras, V., Tambouris, E., Kaliva, E., Magnisalis, I., Tarabanis, K.: Public service models: a systematic literature review and synthesis. IEEE Trans. Emerg. Top. Comput. **9**(2), 637–648 (2021). https://doi.org/10.1109/TETC.2019.2939485
16. Mouzakitis, S., et al.: Challenges and opportunities in renovating public sector information by enabling linked data and analytics. Inf. Syst. Front. **19**(2), 321–336 (2016). https://doi.org/10.1007/s10796-016-9687-1
17. Publications Office of the European Union: The country authority table (2022). https://publications.europa.eu/resource/dataset/country
18. Publications Office of the European Union: The language authority table (2022). https://publications.europa.eu/resource/dataset/language
19. PwC EU Services: APIs for cpsv-ap based catalogue of services (2019)
20. Vaccari, L., et al.: Application programming interfaces in governments: Why, what and how. European Commission - JRC science for policy report (2020)
21. W3C: Sparql 1.1 overview. W3C Recommendation (2012). https://www.w3.org/TR/sparql11-overview/
22. W3C: RDF 1.1 turtle: Terse RDF triple language. W3C Recommendation (2014). https://www.w3.org/TR/turtle/
23. W3C: RDF 1.1 xml syntax. W3C Recommendation (2014). https://www.w3.org/TR/rdf-syntax-grammar/
24. W3C: Json-ld 1.1 a json-based serialization for linked data. W3C Recommendation (2020). https://www.w3.org/TR/json-ld11/
25. Zeginis, D., Kalampokis, E., Roberts, B., Moynihan, R., Tambouris, E., Tarabanis, K.: Facilitating the exploitation of linked open statistical data: JSON-QB API requirements and design criteria. SemStats (2017)

# An AI-Enhanced Solution for Large-Scale Deliberation Mapping and Explainable Reasoning

Nikos Karacapilidis[1]([⊠]) [iD], Dimitris Tsakalidis[2] [iD], and George Domalis[2] [iD]

[1] University of Patras, 26504 Rio Patras, Greece
karacap@upatras.gr
[2] Novelcore, 26500 Patras, Greece

**Abstract.** This work aims to respond to the profound lack of dialogue between citizenship and policy making institutions by proposing a novel solution that enables the transition to inclusive, transparent, accountable and trustworthy deliberation practices. The proposed solution builds on cutting-edge AI tools and technologies to develop a sustainable digital platform, and bridges theories from the fields of argumentation and digital democracy. It may transform scattered islands of emerging knowledge and practices, as well as fragmented discussion threads, into an integrated and coherent dialogue, and provides mechanisms for expanding this dialogue and converting it into tangible actions. Much attention is paid to issues related to knowledge extraction, knowledge graph-based representation of large-scale deliberation, argument mining, aggregation and visualization, as well as to explanation and awareness services about the evolution and outcome of a deliberation.

**Keywords:** Digital Governance · Digital Democracy · Citizen-centric e-Governance · Artificial Intelligence in Government · conceptual framework

## 1 Introduction

Citizens worldwide are increasingly worried about the socio-political crises and conflicts emerging around the world and seek new ways to exercise their social responsibility. In such settings, they ask to be engaged in democratic and inclusive discussions about how these crises and conflicts can be prevented, mitigated and even resolved in new ways, such as through mobilization of grassroot collective actions. However, current methods to engage citizens in fruitful deliberations, gain trust, and harness their commitment to act are problematic; without the practical means to engage them into policy making, a meaningful and effective joining-up of bottom-up citizen-led initiatives and top-down policy making has not yet been realized.

At the same time, political science research highlights the need of balance as well as communication, interaction and exchange of knowledge between 'democracy' (democratic institutions, consultations with citizens) and 'technocracy' (specialized knowledge of experts and policy makers), as they are complementary, each of them needing inputs

© The Author(s), under exclusive license to Springer Nature Switzerland AG 2023
M. Papadaki et al. (Eds.): EMCIS 2022, LNBIP 464, pp. 305–316, 2023.
https://doi.org/10.1007/978-3-031-30694-5_23

from the other, while both making significant but different contributions to the design of effective and socially acceptable public policies (Androutsopoulou et al., 2018). In particular, participants in democratic processes need extensive knowledge and expertise on the social problems they are dealing with. On the other hand, experts dealing with important social problems tend to ignore important aspects of public policies, such as their impact on employment, social inequalities, and quality of life. To reduce these negative tendencies, experts need inputs from democratic political processes concerning the values of citizens and other stakeholder groups, as well as their diverse perspectives and approaches.

The approach proposed in this paper aspires to address the above issues by unleashing the power of democratic and participatory processes towards the aggregation of ideas and the co-creation of efficient and effective solutions to multi-dimensional societal problems. It aims to exploit and meaningfully integrate internal and external data, by considering all the operational stakeholders as key co-creators of value information-knowledge-action chains, thus sustaining and inspiring better-informed collaboration towards innovative actions. The proposed approach creates a novel deliberation solution to transform scattered islands of emerging knowledge and practices, and fragmented discussion threads into an integrated and coherent dialogue and provides mechanisms for expanding this dialogue and converting it into tangible actions.

The key contributions of the proposed solution are: (i) novel knowledge extraction algorithms to yield factual and affective knowledge; (ii) a knowledge graph-based representation of large-scale deliberation enriched with state-of-the-art natural language understanding, argument mining, aggregation and visualization mechanisms to turn unstructured user-generated content into knowledge and actions, and (iii) explanation and awareness services about the evolution and outcome of deliberation to enable better informed collaboration, augment sense making and increase transparency of the overall process.

The remainder of this paper is structured as follows: Sect. 2 reports on background issues concerning large-scale online deliberation methods and tools, argumentation and social knowledge mining, group decision making in large-scale deliberations, and knowledge graphs for deliberation mapping. Section 3 describes the proposed solution for the facilitation and enhancement of large-scale deliberations, along with its potential and expected impact. Finally, Sect. 4 outlines concluding remarks and future work directions.

## 2 Background Issues

### 2.1 Large-Scale Online Deliberation Methods and Tools

Current deliberation platforms are rudimental in the way they structure data, scarcely support evidence-based reasoning, lack features to enhance personal understanding and situational awareness, and hardly support effective deliberation and decision-making. If we look at *social media solutions*, a wide research literature demonstrates how online dialogue on these platforms is prone to toxic behaviors such as biased and un-supported information, rumors, misinformation, hate speech and echo chambers effects. These technologies are therefore inapt to promote public discussion and fail to enable the realization of constructive attitude, informative and rational dialogue, civility and equality.

On the other hand, *participatory democracy solutions* such as *Consul, Democracy OS, Loomio* and *Decidim* have demonstrated large adoption in supporting a variety of democratic processes, such as solicitation of ideas on public issues, community voting, and participatory budgeting. While this second category of solutions is able to promote active change in specific policy making contexts, and provides a much more constructive and inclusive environment to promote citizens engagement in collective decision making, it shares some of the weaknesses of social media; it provides simple discussion features and hardly supports evidence-based thinking since discussion data is neither presented nor collected in a way that makes it easy for people or machines to make sense of the knowledge embedded in the dialogue. Moreover, when the discussion scales, it is hard for participants to grasp the status and progress of the deliberation.

To address these shortcomings, *issue-centric solutions* such as *Kialo, Deliberatorium, Cohere, DebateGraph* and *The Evidence Hub* enable people to interact by building deliberation maps that are made up of interlinked questions, answers and arguments. Such tools help communities be much more systematic and complete in their deliberations about complex topics, enhance evidence-based dialogue, build common ground, support the development of shared understanding of complex problems and improve the quality of online argumentation (De Liddo *et al.*, 2012). However, the uptake and impact of these solutions is hindered by a lack of usable and intuitive interfaces for online dialogue.

## 2.2 Argumentation and Social Knowledge Mining

Argumentation mining lies between natural language processing, argumentation theory and information retrieval, aiming to automatically detect the arguments expressed in a deliberation process, their individual or local structure and the interactions between them. The main goal of argumentation mining is to automatically extract arguments from generic textual corpora, in order to provide structured data for computational models of argument and reasoning engines. Recent advances in Computational Argumentation and Natural Language Processing (NLP) enable the development of novel methods that may capture arguments and inform stakeholders about the evolution of a deliberation through contextualization, representation and aggregation of argumentation in diverse contexts (Cabrio and Villata, 2018).

Argumentation mining systems developed so far adopt a pipeline architecture through which they process unstructured textual documents and produce as output a structured document, where the detected arguments and their relations are annotated so as to form an argument graph (Lippi and Torroni, 2016). Such a pipeline consists of three basic subtasks, namely argumentative sentence detection, argument component boundary detection, and argument structure prediction. There are many similarities between these subtasks that are typically addressed by prominent Machine Learning (ML) and NLP techniques. Approaches to argumentation mining adopt either a *discourse-level perspective*, aiming to analyze local argumentation structures, or an *information-seeking perspective*, aiming to detect arguments that are relevant to a predefined topic. Consequently, such approaches call for a subsequent argumentation aggregation step, which can aggregate similar arguments for the same topic.

As far as argument aggregation is concerned, a variety of models have been already proposed based on *argument-wise* and *framework-wise* methods (Bodanza *et al.*, 2017).

In the former, individually supported arguments are aggregated by a voting mechanism, while in the latter the aggregation comes from merging the individually supported criteria or different argumentation frameworks through a collectively decided method, depending on the specific argumentation context under consideration. The framework-wise approach is considered more efficient in the context of deliberative democracy, while the argument-wise approach could be the most efficient one in the context of a debate among experts. In a similar research line, contextualized word embeddings that classify and cluster topic-dependent arguments have been recently proposed in the literature. Two of the most popular approaches are *Embeddings from Language Models* (Peters *et al.*, 2018) and *Bidirectional Encoder Representations from Transformers (BERT)* (Devlin *et al.*, 2018). Contrary to traditional word embeddings, these approaches calculate the embeddings for a sentence dynamically, by considering the context of a target word. This generates word representations that better match the specific sense of the word in a sentence.

## 2.3 Group Decision Making in Large-Scale Deliberations

New technological paradigms such as social networks, e-participation, e-democracy and e-marketplaces enable the participation of big numbers of stakeholders in the decision-making process. Consequently, these paradigms make it possible to obtain more and more subjective and objective data. At the same time, the group decision making process is characterized by the following: (i) the scale of groups participating in the process has become much larger than before, varying from dozens to thousands; (ii) people involved in the process come from different organizations and in most cases have different backgrounds, interests and constraints; (iii) individuals can express opinions at different times or places, while the final solution is no longer attributable to a single decision maker, but rather to a large-scale group making decisions jointly (Karacapilidis, 2014).

State-of-the-art approaches attempt to address the following major challenges (Tang and Liao, 2019): *(i) reduction of the decision makers' dimension:* clustering analysis is the most widely used method so far, aiming to reduce the complexity and cost of the associated problems, as well as to identify common opinion patterns (e.g. clusters with similar opinions and a spokesman who represents each cluster); *(ii) weighting and aggregating decision information:* the development of a reasonable method that considers the diverse characteristics of individuals and subgroups to determine weights is very crucial, making simple aggregation strategies such as arithmetic average or weighted average not appropriate; *(iii) management of participants' behavior:* existing studies often adopt a social network analysis perspective to investigate the consensus reaching process and the associated detection and elimination of conflicts among decision makers (Liu *et al.*, 2019); *(iv) cost management:* diverse consensus models with minimum cost have been already proposed to address this challenge, which is associated to the feedback mechanism of the whole process; *(v) knowledge distribution and information increase:* this concerns the diverse social relationships that may exist among decision makers as well as the consideration of additional information such as trust and reputation of them.

Most prominent tools and technologies build on concepts and techniques from Artificial Intelligence and Operational Research to enable a sophisticated data analysis, while also discovering patterns of data and inferring data content relationships and rules from them (Karacapilidis *et al.*, 2014). Such tools and technologies certainly facilitate diverse aspects of decision making. Although there exist certain limitations in their suitability, they may aid users to make better and faster decisions. However, there is still room for further developing the conceptual, methodological and application-oriented aspects of the problem. One critical point that is still missing is a holistic perspective on the issue of large-scale group decision making. This originates out of the growing need to develop applications by following *a more human-centric (not problem-centric) view*, in order to appropriately address the requirements of contemporary knowledge-intensive settings. Such requirements stem from the fact that decision making has also to be considered as *a social process* that principally involves human interaction. The structuring and management of this interaction requires the appropriate technological support.

### 2.4   Knowledge Graphs for Deliberation Mapping

Knowledge Graphs (KGs) facilitate the storage and representation of knowledge in a direct and expandable way (Wang *et al.*, 2014). Recent KG-based approaches can represent knowledge extracted from either structured (e.g., tabular and matrix data) or unstructured data (e.g., media and textual data). The advantage of KGs against the classical knowledge bases is that they generally perform better in data-intensive environments, since they allow for: (i) easier data schema expansion and alternation, (ii) better knowledge extraction and representation, (iii) masking of the underlying data complexity, (iv) integration of knowledge from external sources (e.g., Wikipedia), and (v) exploitation of graph algorithms. In addition, KGs can effectively represent both information related to the relations between entities and information that concerns each individual entity (Lin *et al.*, 2017).

With respect to the representation of the knowledge existing in such graphs, *KG embeddings* have been recently adopted. KG embeddings provide low-dimensional dense vectors, which incorporate important information related to the entities (i.e. nodes) and relations (i.e. edges) of a KG. Most important, KG embeddings assist traditional ML models in performing a list of tasks more accurately. Such tasks may concern entity classification, inference of relations, network analysis and prediction of links between the entities of a KG.

KGs have already been applied to several practical domains including question-answering, language models, entity matching, chatbots, dialog systems, recommendation engines, fraud detection, and prediction of future research collaborations (Wang *et al.*, 2017). Furthermore, they excel in real world applications, where complex data from multiple sources can only be processed together, aiming to gain important insights. As far as the implementation and the utilization of a KG are concerned, several well-tried and mature programming libraries and tools exist in the literature. For instance, the *Neo4j* database provides the user with already implemented graph and ML algorithms, thus enabling the construction of robust and production-ready KGs. Graph databases can be seamlessly used along with widely used ML frameworks such as *TensorFlow, PyTorch* and *scikit-learn* to build meaningful ML models and pipelines.

# 3 The Proposed Solution

## 3.1 Research Methodology

The development of the proposed solution follows the *Design Science* paradigm, which seeks to create innovations that define the ideas, practices, technical capabilities, and products through which the analysis, design, implementation, and use of Information Systems can be effectively and efficiently accomplished (Hevner *et al.*, 2004). This paradigm has been extensively adopted in the development of Information Systems in order to address what are considered to be *wicked problems*, i.e., problems characterized by unstable requirements and constraints based on ill-defined contexts, complex inter-actions among issues of the problem, inherent flexibility to change design processes and artifacts, and a critical dependence upon human cognitive and social abilities to produce effective solutions. At the same time, our approach is in line with the *Action Research* paradigm, which aims to contribute both to the practical concerns of people in a problematic situation and to the goals of social science by joint collaboration within a mutually acceptable ethical framework (Rapoport, 1970). As such, it concerns the improvement of practices and strategies in the particular cognitively complex environment under consideration, as well as the acquisition of additional knowledge to improve the way stakeholders address issues and solve problems (Checkland and Holwell, 1998).

## 3.2 Conceptual Architecture

The proposed platform offers a holistic and modular solution that securely hosts and effectively supports large-scale deliberation processes. All the individual modules are designed to be built on top of a cloud service system, configured to be aligned with the needs of all types of stakeholders. This modular approach constitutes the backbone of our solution, which is capable of thoroughly addressing the complexity of deliberative processes, while also enhancing trust, transparency and legitimacy of policy making.

Guided by advancements in (Explainable) AI, ML, NLP, Graph Theory and Argu-mentation, our human-centric approach will produce an efficient and scalable platform that can support deliberation processes of different models and at all levels, from local to global. The proposed technical solution, whose *three-layer architecture* is illustrated in Fig. 1, ensures the seamless integration (at both a conceptual and a technical level) and interoperability of diverse components and services. It enables a *synergy of human and machine reasoning* towards facilitating and augmenting the participation and delib-eration of diverse types of stakeholders in structured discursive interactions. In addi-tion, it exploits *rich semantics at machine level* to enable the meaningful incorpora-tion and orchestration of interoperable services, aiming to reduce the inherent data-intensiveness of the context under consideration. In particular, the proposed solution seamlessly integrates:

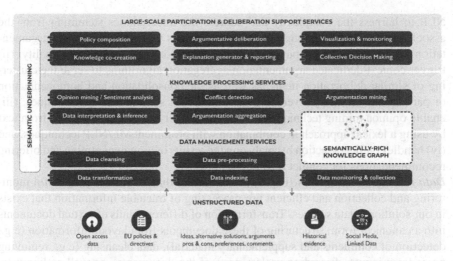

**Fig. 1.** The architecture of the proposed large-scale deliberation platform.

- *Participation and deliberation support services* (Fig. 1, top layer) that (i) support incremental formalization of argumentative deliberation, i.e., a stepwise and controlled evolution from a mere collection of individual ideas and resources to the production of highly contextualized and interrelated knowledge artifacts; (ii) augment sense-making through advanced visualization and monitoring dashboards that offer an informative and user-friendly overview of a deliberation process in terms of participants' engagement and contributed knowledge, while also providing insights about the structure, evolution and dynamics of a deliberation process; (iii) provide advanced knowledge exchange and co-creation functionalities by offering a deliberation environment that supports interpretation of diverse knowledge items and their interrelationships; (iv) are geared towards facilitating collective decision making and consensus building through innovative virtual workspaces that enable participants rank alternative solutions; (v) aid policy makers discover and elucidate key points relevant to the deliberation and accordingly compose evidence-informed policies and practices, and (vi) aid citizens and other stakeholders to get explanations and reporting about the inferential process of the underlying AI algorithms and decision making mechanisms, thus promoting trust in the deliberation outputs.
- *Knowledge processing services* (Fig. 1, middle layer) that enable a sophisticated analysis on the associated textual content of deliberative processes. Building on prominent AI techniques, this set of services facilitates the comprehension of the structure and meaning of argumentative deliberation by breaking down the input received into a machine-readable format. By understanding a set of linguistic and structural cues, our solution enables a precise interpretation of the corresponding texts and their transformation into actionable, measurable and easily accessible knowledge, thus augmenting the quality of human-computer interaction. Items identified populate and are meaningfully linked with the platform's knowledge graph to support a sophisticated representation of deliberation entities and their dynamics. This set of services builds on and extends state-of-the-art ML frameworks and neural architectures for

NLP to harness the complexity and address the uncertainties stemming from the associated data and knowledge. It includes services that (i) deploy novel argumentation mining mechanisms paying particular attention to argumentation quality; (ii) trigger context-dedicated argumentation aggregation algorithms to serve the clustering of similar deliberation items, as well as the consolidation and homogenization of participants' input by leveraging prominent text summarisation techniques; (iii) enable opinion mining techniques to identify, extract and process opinions from text by using a lexical approach in combination with prominent ML/NLP techniques, and (iv) handle conflict detection by analyzing the results of argument mining and offering recommendations for conflict resolution.

- *Data management services* (Fig. 1, bottom layer) that enable the purposeful monitoring and collection and efficient pre-processing of tractable information that exists in our solution's data sources. Transformation of different kinds of textual documents into a canonical form, structuring of these documents from layout information (e.g., detection of comments and supplementary material), data cleansing (e.g., removing noise from inputs, discarding useless parts of the documents), as well as linguistic annotations that facilitate data indexing are some of the functionalities foreseen in this category of services.

### 3.3 Advancements of the Proposed Solution

With respect to *large-scale online deliberation*, the proposed solution provides two main technological advancements to the state of the art:

- *Improved Sense Making.* Large scale deliberations are hard to monitor and make sense of. The proposed platform will develop powerful deliberation analytics and visual interfaces to make sense and assess the state, progress and quality of a deliberation process, as well as alerts that guide users to the parts of the discussion where they can offer most.
- *Improved Evidence-Based Thinking.* Large scale discussions often produce shallow content and low-quality debate. The proposed solution takes a knowledge-based approach to improve deliberation quality. The foreseen deliberation platform will provide a series of features and services to recommend scientific literature to participants during the debate. In this way, the proposed solution fosters evidence-based thinking and more informed discussion, which improve the overall quality of the deliberation.

As far as *argumentation mining and aggregation* are concerned, the proposed solution will shape novel ways of supporting and facilitating online deliberations: (i) the foreseen framework will employ automatically extracted arguments in order to improve decision and policy making and support strategic actions; (ii) it will adopt a joint discourse-level and information-seeking perspective paying much attention to argumentation clustering and argumentation aggregation procedures for the context under consideration, and (iii) it will employ novel argumentation mining pipelines paying particular attention to argument quality, while also facilitating the creation of explanations that disclose how the information on which the machine relies to make its own decisions is retrieved and interpreted (Karacapilidis *et al.*, 2017).

The proposed solution will also advance *large-scale decision-making support technologies*, by adopting a knowledge-based decision-making view, enabled by the meaningful accommodation of the results of the argumentation mining and aggregation processes. According to this view, decisions will be considered as pieces of descriptive or procedural knowledge referring to an action commitment. In such a way, the decision-making process is able to produce new knowledge, such as evidence justifying or challenging an alternative or practices to be followed or avoided after the evaluation of a decision, thus providing a refined understanding of the problem under consideration. On the other hand, in a decision-making context the knowledge base of facts and routines alters, since it has to reflect the ever-changing external environment and internal structures of citizen assemblies. Knowledge management activities such as knowledge elicitation, representation and distribution influence the creation of the decision models to be adopted, thus enhancing the decision-making process.

The abovementioned synergy of decision-making and knowledge management will be further strengthened in the proposed platform by the incorporation of features enabling decision makers to perform argumentation on the issues raised. Many collaborative decision-making problems have to be solved through dialoguing and argumentation among a group of people. In such contexts, conflicts of interest are unavoidable and support for achieving consensus and compromise is required. Independently of the model used for decision making, argumentation is valuable in shaping a common understanding of the problem. It can provide the means to decide which parts of the information brought up by the decision makers will finally be the input to the model used. Moreover, argumentation may stimulate the participation of citizens and decision makers and encourage constructive criticism. To address the above category of requirements, a user-friendly argumentative deliberation-based decision-making support environment will be developed (Christodoulou *et al.*, 2016).

The proposed solution integrates novel mechanisms to aggregate citizens' and subgroups' opinions into collective positions. Different aggregation functions will be tested to assess the robustness of the results obtained. These mechanisms will comply with the foreseen transition model for the scaling of an ongoing deliberation. In addition, the proposed solution can further elaborate the consensus reaching process through the development of new models in which consensus is measured through aggregated collective opinions at both the intra-subgroup and the inter-subgroup levels. Aiming to augment the explainability and interpretability of the models and data involved in the overall large-scale group decision making process (Samek *et al.*, 2019), the proposed solution will also develop and integrate in the foreseen deliberation framework a dedicated explanation mechanism that will benefit the user in terms of *justification* (exposing the reasoning behind a decision may help the user decide how much credence to give in it), *user involvement* (allowing the user to add her knowledge and inference skills to the overall decision process), and *system acceptance* (in that the system's functionality is fully visible and its suggestions are adequately justified).

Finally, the proposed solution will employ a novel *knowledge graph* to model stakeholders' knowledge and interactions jointly. We plan to advance current knowledge aggregation methods that are based on neural architectures such as attention mechanisms and Graph Neural Networks (GNNs). The foreseen advancement will build on large-scale pre-training via transformers and variants like BERT models. Large-scale pre-training models will aid the acquisition (and injection in the KG) of factual knowledge. Considering its overall objectives, the proposed solution will rethink the way of knowledge aggregation in an efficient and interpretable manner. Specifically, it enables the construction of dynamic knowledge graphs, together with novel mechanisms to capture the dynamics of a deliberation, thus addressing limitations of traditional knowledge representation and reasoning by meaningfully monitoring and analyzing the temporal dimension. It will advance current approaches to thoroughly address the scalability issue, which is certainly crucial in large-scale knowledge graphs. The ubiquitous trade-off between computational efficiency and model expressiveness will be addressed through transformer-based models to encode graph entities, relations and path sequences, as well as GNNs to aid the learning of connectivity structure under an encoder-decoder framework. For the analysis of deliberation data, the foreseen KG builds on a graph-based text representation, namely *graph-of-docs* (Giarelis *et al.*, 2020). The proposed knowledge graph structure and related advancements also serves explainability purposes, aiming to aid stakeholders build a complete and informed mental model of the inferential process of the underlying machine learning algorithms and the knowledge-based decision-making support system and promote trust for its outputs. The proposed explanations generator engine adopts a human-in-the-loop approach towards the development of interactive interfaces to support model interpretability and inference explainability.

## 4   Conclusions

This paper has described a novel solution that adopts a pluralist and bottom-up approach to increase the quality of deliberation and its ability to influence public policy. The proposed solution aims to facilitate and augment the scaling of this approach through an AI-enhanced digital democracy platform that builds on prominent ML/NLP technologies to enable lay and expert stakeholders exchange and reform their opinions, co-create actionable solutions, and collectively reach decisions in a highly transparent and trustful way. We argue that this solution will contribute to the improvement of the quality of democracy nowadays, which demands the active involvement and effective participation of citizens in policy making from the design to the implementation phase. It can be viewed as a digital transformation tool that is an essential enabler of a socially cohesive society, where all individuals and groups have a sense of belonging, participation, inclusion, recognition and legitimacy.

The proposed solution has been shaped through long and fruitful collaboration among diverse types of stakeholders (representing academia, citizens and civil society, government and public authorities, and ICT-focused SMEs), through which a series of rich application scenarios have been sketched and analyzed. The main limitation of our study is that though the proposed approach has undergone a first level assessment and validation by experienced practitioners, which has been highly positive, its application has to be carefully planned by taking into account the capacity and available resources of diverse organizations. Future work directions include the full implementation and integration of the proposed solution's modules and services, as well as the collection of feedback through its assessment in diverse deliberation settings, ranging from a local to an international level. Its application will be evaluated through a set of dedicated Key Performance Indicators, focusing on the usefulness and ease of use of the proposed approach.

# References

Androutsopoulou, A., Karacapilidis, N., Loukis, E., Charalabidis, Y.: Combining technocrats' expertise with public opinion through an innovative e-participation platform.IEEE Trans. Emerg. Top. Comput. **2018** (2018). https://doi.org/10.1109/TETC.2018.2824022

Bodanza, G., Tohmé, F., Auday, M.: Collective argumentation: a survey of aggregation issues around argumentation frameworks. Argument Comput. **8**(1), 1–34 (2017)

Cabrio, E., Villata, S.: Five years of argument mining: a data-driven analysis. In: Proceedings of the 27th International Joint Conference on Artificial Intelligence (IJCAI-18), pp. 5427–5433. AAAI Press (2018)

Checkland, P., Holwell, S.: Action research: Its nature and validity. Syst. Pract. Action Res. **11**(1), 9–21 (1998)

Christodoulou, S., Karacapilidis, N., Tzagarakis, M.: Exploiting alternative knowledge visualizations and reasoning mechanisms to enhance collaborative decision making. In: Tweedale, J.W., Neves-Silva, R., Jain, L.C., Phillips-Wren, G., Watada, J., Howlett, R.J. (eds.) Intelligent Decision Technology Support in Practice. SIST, vol. 42, pp. 89–106. Springer, Cham (2016). https://doi.org/10.1007/978-3-319-21209-8_6

Devlin, J., Chang, M.-W., Lee, K., Toutanova, K.: BERT: pre-training of deep bidirectional transformers for language understanding. arXiv preprint arXiv:1810.04805 (2018)

Giarelis, N., Kanakaris, N., Karacapilidis, N.: On a novel representation of multiple textual documents in a single graph. In: Czarnowski, I., Howlett, R.J., Jain, L.C. (eds.) IDT 2020. SIST, vol. 193, pp. 105–115. Springer, Singapore (2020). https://doi.org/10.1007/978-981-15-5925-9_9

Hevner, A.R., March, S.T., Park, J., Ram, S.: Design science in information systems research. MIS Q. **28**(1), 75–105 (2004)

Karacapilidis, N., Malefaki, S., Charissiadis, A.: A novel framework for augmenting the quality of explanations in recommender systems. Intell. Decis. Technol. J. **11**(2), 187–197 (2017)

Karacapilidis, N., Christodoulou, S., Tzagarakis, M., Tsiliki, G., Pappis, C.: Strengthening collaborative data analysis and decision making in web communities. In: Proceedings of the 23rd International World Wide Web Conference (WWW2014), Companion Volume - Workshop on Web Intelligence and Communities, Seoul, Korea, 7–11 April 2014, pp. 1005–1010 (2014)

Karacapilidis N. (ed.): Mastering Data-Intensive Collaboration and Decision Making: Cutting-Edge Research and Practical Applications in the Dicode Project. Studies in Big Data Series, vol. 5, Springer, Heidelberg (2014). https://doi.org/10.1007/978-3-319-02612-1

De Liddo, A., Sándor, Á., Shum, S.B.: Contested collective intelligence: rationale, technologies, and a human-machine annotation study. Comput. Support. Coop. Work **21**(4–5), 417–448 (2012)

Lin, H., Liu, Y., Wang, W., Yue, Y., Lin, Z.: Learning entity and relation embeddings for knowledge resolution. Proc. Comput. Sci. **108**, 345–354 (2017)

Lippi, M., Torroni, P.: Argumentation mining: state of the art and emerging trends. ACM Trans. Internet Technol. **16**, 2 (2016). Article 10. https://doi.org/10.1145/2850417

Liu, B.S., Zhou, Q., Ding, R.X., Palomares, I., Herrera, F.: Large-scale group decision making model based on social network analysis: trust relationship-based conflict detection and elimination. Eur. J. Oper. Res. **275**(2), 737–754 (2019)

Peters, M., et al.: Deep contextualized word representations. In: Proceedings of the 2018 Conference of the North American Chapter of the Association for Computational Linguistics, vol. 1, pp. 2227–2237 (2018)

Rapoport, R.N.: Three dilemmas in action research. Hum. Relat. **23**(6), 499–513 (1970)

Samek, W., Montavon, G., Vedaldi, A., Hansen, L.K., Müller, K.-R. (eds.): Explainable AI: Interpreting, Explaining and Visualizing Deep Learning. Springer, Heidelberg (2019). https://doi.org/10.1007/978-3-030-28954-6

Tang, M., Liao, H.: From conventional group decision making to large-scale group decision making: what are the challenges and how to meet them in big data era? A state-of-the-art survey. Omega 102141 (2019). https://doi.org/10.1016/j.omega.2019.102141

Wang, Q., Mao, Z., Wang, B., Guo, L.: Knowledge graph embedding: a survey of approaches and applications. IEEE Trans. Knowl. Data Eng. **29**(12), 2724–2743 (2017)

Wang, Z., Zhang, J., Feng, J., Chen, Z.: Knowledge graph embedding by translating on hyperplanes. In: Proceedings of AAAI 2014, pp. 1112–1119 (2014)

# Digital Services and Social Media

# Fairness Issues in Algorithmic Digital Marketing: Marketers' Perceptions

Veronika Pavlidou[1]([⊠]) [iD], Jahna Otterbacher[1,2] [iD], and Styliani Kleanthous[1,2] [iD]

[1] Open University of Cyprus, Latsia, Cyprus
design@caliber.com.cy
[2] CYENS Centre of Excellence, Nicosia, Cyprus

**Abstract.** Data-driven algorithms are becoming more prevalent in our everyday lives, automating various decisions that can impact access to opportunities and resources. For this reason, much research concerns the ethical and social consequences of algorithms. Algorithmic *digital marketing* represents a key means by which people encounter and/or are affected by algorithms in daily life. While previous work has considered the perceptions of consumers in this realm, little work to date has considered those of marketing professionals. The current work presents an exploratory study of marketers, aiming to uncover their general views on the use of algorithmic marketing, as well as their own practices and if/how they address the ethical concerns surrounding algorithmic marketing. The findings underscore the need for ethical regulations and guidelines to be defined for fair practices in the context of algorithmic digital marketing.

**Keywords:** digital marketing · marketers' perceptions · fairness in algorithmic processes · micro-targeting · ethics in digital marketing

## 1 Introduction

The field of artificial intelligence (AI) has significantly expanded in recent years; from 2010 to 2021, the total number of AI publications doubled (AI Index Report, 2022) and new applications of AI are transforming the way we work and live. Likewise, they are evolving in almost every business sector. The great benefit of AI for business is the possibility of processing a huge volume of data and interpreting it in real-time, to save time and increase revenue. Businesses in general, are increasingly looking for ways to put AI technologies to work to improve their productivity, profitability and business results. Lately, industry and researchers have adopted the term "cognitive technologies" when referring to narrow AI, implying that the technologies take the automation to a new level of "human-like" thought.

There has been a progression of AI from Cognitive Assistance such as automation of repeatable tasks or process automation (e.g., email automation, calendar and scheduling assistant, auto-responsive, etc.) to Cognitive Insights, highly data-dependent applications of AI for predicting behaviour (e.g., customer buying patterns), and finally, to Cognitive Engagement. This newest form of application involves companies' engagement with

M. Papadaki et al. (Eds.): EMCIS 2022, LNBIP 464, pp. 319–338, 2023.
https://doi.org/10.1007/978-3-031-30694-5_24

people inside and outside of business (e.g., digital customer support agents like Amelia, which is recognized as the Most Human AI™) (Mussomeli, Neier, Takayama, Sniderman, & Holdowsky, 2019). One of the branches of narrow AI consists of algorithms based on machine learning (ML) that have had great influence on marketing practices. In particular, algorithmic marketing is a process that is automated to such a degree that it can be steered by setting a business objective in a marketing software system (Katsov, 2001). Moreover, those new applications have biases, intentional or otherwise, and these shape and constrain individuals' lives (Winter, 2014).

Today, most consumers conduct online research before making a purchase, and this fundamental change in buying behaviour forces marketers to adapt their business marketing strategies for the digital age. Marketers (i.e., advertisers) always try to reach their target audience based on demographics (e.g., gender, age, race, ethnicity), preferences, and by using cognitive biases (e.g., authority bias, scarcity bias, etc.), which influence potential consumer behaviour and decisions in order to increase sales. Marketers are aware that people tend to run from commercial ads, knowing very well that the number of users who adopt ad-blocking is increasing rapidly all over the world (Tudoran, 2019). Thus, they try to find more creative ways to get customers' attention e.g., by using cognitive biases. As (Nadler, 2017) states, marketers identify consumers' cognitive and affective biases and target their vulnerabilities. Here, there is the risk to cross the line from ethical to unethical practices.

Thus, it is important to understand what impact the interactions among the stakeholders or users (customers/consumers and marketers) have in all stages of algorithmic digital marketing and their perceptions of algorithmic processes since user perception and belief are created, amplified, or reinforced by algorithms and vice versa. Additionally, from a user's perspective, the algorithm is a "black box," and it is not possible for the user to know how the computation is completed. At the same time, algorithms themselves do not have the affordances that would allow users to understand them or how best to utilize them to achieve their goals (Shin & Park, 2019).

The marketing system should always be of service to people (Murphy, Laczniak, & Harris, 2016), and this statement makes it obvious that marketers should always behave ethically. Companies and marketers must adjust to the constantly challenging digital economy, and those that desire to enter global competition should pay attention to customer benefits and business fairness in order to achieve sustainability (Anggadwita & Martini, 2020). Fairness in organizational practices can foster various sources competitive advantage and hence improve organizational performance (Yeolman & Mueller Santow, 2016).

However, as a first step, we need to better understand marketers' perceptions of fair, ethical decisions in digital marketing practices and how they relate to the usual moral approaches used by marketers today. To this end, our research addresses these issues by using qualitative methods to understand the impact of digital marketing on people's lives. Understanding this impact is important because moral awareness gives the ability to people to identify the ethical aspects of their decisions, and helps to shape marketers' attitudes toward their practices. This article aims to contribute to the debate about the impact of digital marketing on people's lives and to encourage digital marketers to actively apply and pursue algorithmic fairness that will undoubtedly help to build a trust relationship and therefore effective communication between the stakeholders (e.g. consumer-brand) in digital marketing processes.

## 2 Related Work

### 2.1 Interactions Between Stakeholders in Algorithmic DM

To ground our study of marketers' perceptions of ethical issues in algorithmic digital marketing, we first consider the stakeholders involved in the marketing process, i.e., consumers, brands, marketers, and the relationship between them. Moreover, we consider the perceptions they have of digital marketing practices (i.e., data-driven algorithms, microtargeting, cognitive marketing) that have impacted traditional marketing, businesses, the tourism sector, and consumers, as well as their moral approaches to these practices.

Consumer-brand relationship quality is a comprehensive concept that reflects the intensity, depth, continuity, and effect of the relationship between a consumer and brand (Lee & Jin, 2019). Today, consumers often have concerns regarding privacy or ethical issues, and they often feel violated when they sense they have no control over the algorithmic processes that involve their personal data during their browsing (Pavlidou, et al., 2021). As (Korolova, 2011) states, one of the big concerns users have when they share personal information on social networking sites is that the service does not "sell" their personal information to advertisers. Moreover, previous research (Pavlidou, et al., 2021) that examined users' perceptions regarding their online activities and fairness in algorithmic targeted marketing, revealed that users have ethical concerns regarding the microtargeting or detailed targeting such as FB Lookalike or Special Ad Audiences tools. In other words, users often believe that they are being tracked and monitored without their consent and also that marketers manipulate them by creating a sense of scarcity intentionally. Additionally, the study of (Herder, E., & Dirks, S., 2022) that examined similar variables such as privacy concerns, trust and vulnerability showed that many people report concerns of privacy and consider commercially targeted advertising to be unethical, as they exploit people's vulnerabilities. Furthermore, the authors cited the danger of filter bubbles that bear several risks, such as the manipulation of people's decision-making, when they have access only to algorithmically selected advertisements.

According to (Dwork et al., 2012), fairness issues were detected in machine learning (ML) known as group fairness or individual fairness; for example, Lookalike and Special Ad Audiences tools can create similarly biased target audiences from the same source audiences (AIES, 2022). It is known that microtargeting audience selection practice is one of the more opaque processes where criteria selection relies on inputs resulting from machine learning (ML) workings of proprietary software. Thus, it holds potential risks of stereotyping and discrimination. As (Birner, N., Hod, S., Kettemann, M. C., Pirang, &., Stock, F., 2021) state, there is a fairness issue in ad delivery where a lack of privacy and user self-determination can be detected, and in general, personalization becomes problematic when the incentives of the consumer and the firm are not aligned (Calo, 2013).

Moreover, consumers have concerns regarding different persuasive techniques that marketers use to promote their products/services. These persuasive techniques used by marketers to convince consumers to buy are based on cognitive biases such as Scarcity (Pavlidou, et al., 2021), Authority or Consensus biases. The presence of unyielding cognitive biases makes individual decision-makers susceptible to manipulation by those able to influence the context in which decisions are made (Hanson & Kysar, 1999). As (Calo, 2013) states, consumers differ in their susceptibility to various forms of persuasion. Some consumers, for instance, respond to consensus bias that causes people to see their own behavioral choices and judgments as relatively common and appropriate to existing circumstance, while viewing alternative responses as uncommon, deviant (Ross, 1976). Others bristle at following the herd but instead find themselves reacting to scarcity or another frame. But in any case, these cognitive biases affect us all with uncanny consistency and unflappable persistence (Hanson & Kysar, 1999).

In a social exchange relationship, both parties must feel they are being respected by the other in order for future social contracting to take place (Wasieleski & Gal-Or, 2008). Creating fairer approaches to marketing and being perceived as fair is necessary for developing good customer relationships and increased loyalty (Nguyen, 2016). That means representing products in a clear way in advertising and to reject manipulations and sales tactics that harm customer trust (Sheth & Naresh, 2010). However, since marketing practices continue to use different sales tactics or cognitive biases many questions come up about how fair these campaigns and practices are. For instance, by creating a temporary product scarcity – either unintentionally or deliberately – a product provider can increase overall demand and stimulate customer enthusiasm over a specific period, leading to improved overall market performance (Shi, Li, & Chumnumpan, 2020).

However, based on research on user perceptions (Pavlidou, et al., 2021) people believe that it is not fair to use scarcity to convince them to purchase the product and people who feel that marketers intentionally created scarcity seem to do not trust the online ads with offers messages. Moreover, people who believe that marketers create misleading scarcity seem to do not trust the online ads with offers or scarcity effect. Many times, to attract the consumer's attention, marketers may resort to use different types of scarcity that could be not fair to their customers, especially, when it concerns online advertising since technology has no inherent morality and the way in which it is utilized is what really matters (Bergman, 1997). Moreover, the consumer of the future will be increasingly mediated, and the firm (marketers) of the future increasingly empowered to capitalize on that mediation in ways that are both fair and suspect (Calo, 2013). Unfortunately, there is to date little literature on these issues, especially on digital marketers' perceptions of fairness in digital marketing processes. Therefore, the purpose of this research is to fill this gap, enriching the literature regarding marketers' views on the ethics of digital marketing practices.

## 2.2   Ethical Concerns in AI and DM

As (Giovanola and Tiribelli, 2022a) argue, fairness as an ethical value is articulated in a distributive and socio-relational dimension. It comprises three main components. *Fair equality of opportunity* holds that all individuals should own the same amount of material resources (Ali, 2022) or information. *Equal right to justification* calls for non-discrimination against individuals and social groups. Finally, *fair equality of relationship* means that there is a mutual respect of each person's interests and desires (Giovanola & Tiribelli, 2022a).

In the Fair Machine Learning literature, the goal is to develop models that are built to make unbiased decisions (e.g., classifications) or predictions (Wing, 2018). According to (Binns, 2018), fairness as used in the fair machine learning community is best understood as a placeholder term for a variety of normative egalitarian considerations that all come back to the concept of equal treatment of all in society. In any case, the development of methods to define, measure and ensure fairness in predictive models is an active area of research, and represents a promising direction for more ethical automated decision making (Tubella, 2022).

In the marketing literature, fairness may be considered as an important element to Ethical Marketing. Specifically, Ethical Marketing is the application of ethics into the marketing process where marketing practices emphasize transparent, trustworthy, and responsible personal and/or organizational marketing policies and actions that exhibit integrity as well as fairness to consumers and other stakeholders (Murphy, Laczniak, & Harris, 2016). According to the research of (Shin & Park, 2019), in which the authors investigated how trust is related to fairness, transparency, and accountability in algorithm-based services people unanimously agreed that fairness is critical factor in algorithms. In this study (Shin & Park, 2019) utilized a triangulated mixed method design and showed that while personalized results have great benefits to certain users, other users may find the results unfair, depending on what characteristics are perceived in personalized experiences.

Hence, fairness is an ethical principle that speaks to how we treat one another in our social and economic interactions (Yeolman & Mueller Santow, 2016). As one of the core values of ethical marketing, it is important in building relationships and enhancing consumer confidence in the integrity of marketing (Murphy, Laczniak, & Harris, 2016). Specifically, fairness in Digital Social Marketing should be associated with morality, impartiality and uprightness (Nguyen, Steve Chen, Sharon Wu, & Melewar, 2015). Moreover, fairness, privacy, autonomous choices may be important rights or entitlements of individual consumers/citizens, but they are also the quintessential building blocks of a free digital society (Helberger, 2016).

It is well accepted that marketing practices should aim to be fair and ethical; therefore, digital marketers should focus not only on their benefits, but also how they benefit consumers or other stakeholders and, hence, the society as a whole. Unfortunately, there are few such guidelines for digital marketers. Towards this effort, we need to understand how today's digital marketers perceive fairness issues in their digital marketing practices, and what are their ethical principles or moral approaches.

## 2.3 Digital Marketing Practices and Concerns of Marketers

For some marketers, it is natural to operate ethically according to their personal principles. But for others, the principals of ethics such as honesty, integrity, loyalty, and fairness are not so important as compared to profit. According to (Murphy, Laczniak, & Harris, 2016), unethical marketers abuse and exploit their customers by regularly extracting valuable personal information from their clients and not protecting the data files adequately or by selling it. Most of the information gathered by online marketers ends up in the hands of data aggregators, who create enhanced consumer profiles available for additional re-sale (Murphy, Laczniak, & Harris, 2016).

In general terms, the marketers use digital marketing practices, such as search engine marketing (SEM), content marketing, influencer marketing, content automation, e-commerce marketing, campaign marketing, and social media marketing, social media optimization, e-mail direct marketing, etc. that have in common the using of micro-targeting or detailed targeted advertising and persuasive techniques (Bala et al., 2018). Targeted advertising is the practice of monitoring people's online behavior and using the collected information to show people individually targeted advertisements (Herder & Dirks, 2022). These two techniques based on findings of previous research have fairness issues such as privacy, distrust, misleading, and manipulation (Pavlidou, et al., 2021). Specifically, the microtargeting technique or detailed targeted advertising can serve as an effective way to deliver relevant information to citizens. But it can also be used as a persuasive tool and impact people's intentions (Zarouali & Dobber, 2020). Influencer Marketing or Digital influencer marketing practices have the fake follower problem since machine learning algorithms cannot detect all the fake accounts, and where influencers try to boost their own follower counts via unethical means like buying fake followers (Anand, et al., 2022). In general, Influencer Marketing considered as dispersed, nonlinear, and sometimes ephemeral advertising formats (Asquith Fraser, 2020) where digital influencers do not adequately disclose whether their review or endorsement has been paid for or if they have a financial relationship with an advertiser. This creates a lack of transparency and ability for the consumer to recognise content that is in fact paid for. As (Einstein, 2016) characteristically says in his book, "a world where there is no real content: everything we experience is some form of sales pitch".

Moreover, according to (Hanson & Kysar, 1999) there is a serious problem of 'market manipulation' defined as the utilization of cognitive biases to influence peoples' perceptions. Cognitive biases such as scarcity, Consensus or Authority biases - social influence (e.g., 'social proof'—informing users of others' behaviour—and shopping with others) (Browne & Swarbrick, 2017). These cognitive biases might harm consumers' trust, decrease loyalty and sense of fairness. Based on the research of (Mathur, et al., 2019), 'Scarcity' refers to the category of dark patterns that signal the limited availability or high demand of a product, thus increasing its perceived value and desirability. Digital market manipulation recognizes that vulnerability is contextual and a matter of degree and specifically aims to render all consumers as vulnerable as possible at the time of purchase (Calo, 2013). While the Digital Marketing is crucial to every business there are concerns regarding consumer's privacy, unethical marketing practices, and manipulation.

## 2.4   Current Work

All these issues could be avoided if marketers use ethical marketing practices.

If they would recognize that they have a natural duty to treat others fairly (G. Graaf, 2006). Ethical marketing practices (product-, pricing-, place-, and promotion-related ethics) affect brand loyalty through the mediators of the consumer-brand relationship and perceived product quality (Lee & Jin, 2019) and most marketing decisions have ethical ramifications whether business executives realize it or not (Laczniak & Murphy, 1991). Considering previous business ethics research, consumers continue to demand more high-quality products, and they display a preference for brands that are socially reputable even at higher prices when evaluating similar products. However, it is not that simple since there is a very fine line between informing, nudging and outright manipulation (Zarouali & Dobber, 2020).

To this end, it is important to study not only consumer perceptions that help a better understanding of what motivates consumers to engage with brands/companies (Pavlidou et al., 2021), but also those of marketers to provide insight into digital marketing stakeholders' perceptions regarding fairness in Algorithmic Digital Marketing, and moreover, to enrich the literature regarding digital marketers' perceptions on DM practices and ethical issues such as fairness.

Specifically, the current research questions are:

RQ1. *How do marketers perceive the impact of algorithmic digital marketing?*
RQ2. *How are marketers making fair/ethical decisions in marketing practices? Specifically, what factors do they consider or assess regarding their particular marketing practices?*

**Table 1.** Demographic characteristics of participants.

| Gender | Percent |
|---|---|
| Male | 75 |
| Female | 25 |
| Occupation | Percent |
| Digital Marketing Employee | 56 |
| Agency Owners | 25 |
| Freelance Marketers | 19 |
| Age | Percent |
| 30 plus | 75 |
| 24–29 | 25 |

## 3  Design/Methodology/Approach

To explore the above research questions, a grounded theory approach was considered as appropriate, and it was used in the data analysis, facilitated by the NVivo software. According to (Charmaz, 2006) grounded theory methods consist of systematic, yet flexible guidelines for collecting and analyzing qualitative data, with the goal of constructing theories 'grounded' in the data themselves. Our research approach involved the conduct of in-depth interviews. Thus, we recruited marketing specialists by means of our personal and business network from different business sectors in Cyprus with an average of

10 years' work experience. Sixteen semi-structured interviews were carried out (Fig. 1). The interviews, which had an average duration of 1 h, were recorded and transcribed.

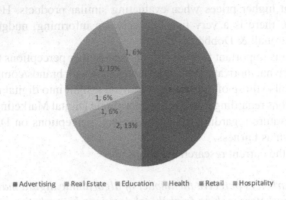

■ Advertising  ■ Real Estate  ■ Education  ▒ Health  ■ Retail  ■ Hospitality

**Fig. 1.** Industry in which participants work.

Cyprus, a growing international hub due to its strategic location, was chosen as a case study. Many individuals and companies are attracted to Cyprus given its competitive income tax rate and personal tax incentives. Cyprus represents a constantly growing tech hub within the European Union with 64 percent of core innovation service enterprises (Eurostat, 2022) It is center of many international IT companies that use Cyprus as a hub for software development and digital marketing, such as Microsoft, Oracle, SAP, IBM and lately, global companies such as NCR, Kardex, and Wargaming, to name a few. Since we live in digital era where rapidly accelerating e-commerce transactions is our everyday reality, it is important to explore the perceptions of marketing practitioners regarding fairness in algorithmic digital marketing.

Initially, an interview guide was prepared, as a key part of the semi-structure interview protocol. Its goal was to ensure each interview covered, with the participants, topics related to fairness in algorithmic digital marketing. After piloting, the final set of questions was developed (Table 2). The aim of these unstructured questions was to allow the interviewees to guide the conversation and discuss topics important to them, at the same time scoping the discussion in line with the purpose of the current research. During the online interview registration, the participants (Table 1) were asked at first to provide their informed consent to participate in the study, answer demographic questions, and then to select their preferable date and time for the interview. A follow-up phone call was made to ensure that a link for the online interview was received by the participants and that all details, such as the permission to voice record, using of camera, interview duration, etc. were clarified.

**Table 2.** Open-ended questions used in the semi-structured interviews.

| |
|---|
| How would you describe today's consumer? |
| How would you describe today's marketer? |
| Do consumers want relationships with brands? |
| Do you follow steps in marketing campaigns? |
| Who decides in your company what is wrong or right in a marketing campaign or advertisement? |
| What marketing strategies do you use usually? (e.g., social networks, paid media advertising, email marketing, influencers) Do you use persuasive techniques? |
| Do you believe that scarcity effect (e.g., limited product/time/quantity) is always fair to be used? |
| Do you use data collection techniques? |
| Do you use microtrageting? |
| Do you like how the Facebook algorithm works? |
| How do you create target audience? |
| How would you order hierarchically (*the best interest of your client, the best interest of a consumer or *the best interest of the company you work for) |

## 4   Data Analysis

Each participant transcript was read, and open coding was begun upon a second reading. The first stage of the grounded theory approach involved coding as many categories as possible from the data. Next, a codebook was created, which included a comprehensive list of all codes, properties of each code, and emic phrases. A node is a collection of references about a specific theme, place, person, or other area of interest (Bazeley, 2007). The next step, axial coding (integration and dimensionalization), was used to make connections by comprehensively examining codes to identify relationships among the open codes (Lindlof and Taylor, 2002). In other words, thematic analysis was applied by using an inductive approach to describe the different perceptions regarding fairethical decisions in marketing practices. The final step involved selective coding, which begun with the identification of a core variable that encompasses the data. After the core variable was identified (i.e., marketer perceptions on fair-ethical decisions in marketing practices), selective coding was used by rereading the transcripts and selectively coding any data that pertained to the core variable.

The main themes identified in the data were visualized by using the feature of NVivo Explore Diagram. This feature was used to get sense of the important items connected to main findings such as: the perceptions of DM impact (Sect. 4.1), Digital Marketing usual online strategies (Sect. 4.2) and moral approaches to DM practices (Sect. 4.3).

## 4.1 The Perceptions of DM Impact

In the data concerning the first theme, participants explained the role of IT in Digital Marketing, and specifically, the transition from traditional marketing to digital. They also commented on its impact on people's lives, particularly in the tourism sector, since it is a vital economic sector of Cyprus, business and related issues e.g., brands awareness, consumers behaviour and communication. Moreover, the new forms of marketing such as a sensory marketing and collaborative marketing were mentioned. Digital Marketing usual online strategies were named and briefly explained. Nonetheless, topics such as Cognitive Marketing through its tools (microtargeting, data-driven algorithms, persuasive technique (scarcity), influencer marketing) along the fairness issues were emphasized and explained in detail (Fig. 2).

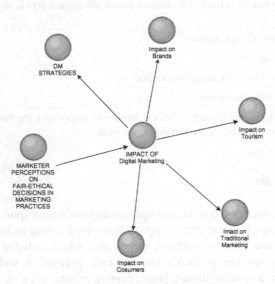

**Fig. 2.** Impact of Digital Marketing.

**Impact on Traditional Marketing.** The transition from traditional to digital/electronic marketing is not easy. As marketing becomes more online and ever more automated IT comes to play a major role in digital marketing. So, today's marketers must be 'multifunctional' (i.e., multi-skilled) and need to have technical background since all digital marketing activities are based on collected data. All interviewees noted that traditional marketing is 'a dead end' and marketers have to upgrade their technological knowledge to be able to move on to digital. Several participants pointed out that digital marketers analyze numbers and not people's moods as traditional marketers used to do. But, perhaps the biggest impact comes from the fact that "the traditional marketer had to come forward while a digital one may not even be recognizable because they are behind a computer and sells a service and so it is much safer. A digital marketer can play the smart one once the results are not so visible and customers may have some technological barrier to understand. So, they can easily cheat", as described by participant (P1). Traditional or digital marketers' mindset is focused on the consumer, and the aim is to sell and grow the business by keeping customers satisfied and happy. But an app/web developer involved in digital marketing practices is focused around the impersonal end-user and have different mindset, "more mathematical or algorithmic logic, they do not talk about ethics or anything related to morality in their practices," as participant (P8) said.

**Impact on Tourism.** It is generally accepted that IT helped Tourism to improve services, enhance efficiency, and reduce costs. Data-driven insights probably help managers to have greater confidence in their decisions or to have more accurate predictions. However, according to participants the customer loyalty, which was a big and very important part for the tourism sector in Cyprus has disappeared since repeaters (travellers who prefer to return to familiar travel destinations) are not exist anymore. As participant (P16) said: "We have destroyed an important part of ourselves…of our country. There is no tourism without a tour operator, today traveller must do everything by himself, and slick persons often take advantage of it and there is nobody to complain… Nowhere. Is this the "evolution" of hotel marketing that we need?". Moreover, there is an'illusion of control' created by changing the business model from B2B to B2C of hotel services. Hotel owners think that by this way they can increase revenue and they will control the market game themselves. P16 noted: "The big tour operators control us; the big booking providers control us. This is where the end of traditional tourism begins. Instead of tour operators, booking providers have the control."

**Impact on Brands.** Today, IT is considered as the backbone of many businesses since consumers have moved dramatically toward online channels, and companies/brands have responded in turn. Thus, they had to invest a lot of money by employing digital marketers or by paying for digital marketing services. Unfortunately, no one was prepared for this challenge. According to participants. There is a lack of technological knowledge such as data management, and experience in entrepreneurs as well as in digital marketers. Many of new digital marketers are not qualified therefore they offer superficial and unprofessional services to brands. As a participant (P2) characteristically said: "In the land of the blind, the one-eyed man is king. This is the impression I have." Therefore, both external and internal communication of company have many issues (i.e. conflicts, unclear directions, misunderstandings etc.). Moreover, those digital marketers who work with commission don't have a moral sense, "money talks… That's all I got to say", said P2. Therefore, the issue of brand trust is one that comes up today. It seems that nowadays the product name overcomes the brand name.

**Impact on the Consumer.** All participants believe that although IT empowers consumers, and algorithms are able to make recommendations, at the same time, it reshapes and makes it harder for consumers to determine their course. There is an abundance of choices, thus today's consumer has become confused, demanding, and with little brand loyalty. "To fill this gap marketers constantly need to provide information about products and adapt more to the consumer's needs and marketers' work became harder", stated participant (P11). Lately, the consumer's behaviour changed, he or she is more vulnerable, worried, much more sensitive and reacts immediately to any changes in the environment (e.g., TV news, government announcements, etc.). Also, by shopping online, consumers became suspicious, careful, and sceptical, demanding to see value for money. "Especially millennium consumers are very difficult to predict", said P10. Regarding the new tendency on plain and minimal websites and e-shops design, two participants explained "today's consumer gets tired of graphics and ads" (P12). "Consumers are not smart, so to convert them into buyers we need to simplify everything" (P2). Moreover, some participants believe that the motto'the customer is always right' does not apply online. The company's policy and new regulations have priority today. Participants acknowledged that lately there is an issue with a customer loyalty. However, DM Agency owners are not concerned much about it comparing to digital marketing employees or marketing consultants. They believe that the successful data-driven campaign results are more important.

## 4.2 Digital Marketing Usual Online Strategies

As mentioned, marketers use different digital marketing practices that have in common the use of microtargeting and persuasive techniques or cognitive marketing. Their perceptions on these issues are explored below.

**Perceptions of Data Driven Algorithms in Market Research.** Marketers greatly benefit from using AI in market research since it lets them exploit their data for effective business decisions. All participants agreed that it is great to have immediate insights and build a marketing strategy quicker by using data from Meta, Google analysis, Google data studio etc.). Although the benefits of algorithms and the way how they work are known and familiar to marketers, many participants expressed their concerns about the tracking, censorship, opaqueness (results numbers), its complexity and vagueness, and its constantly changes. As participant (P12) said: "FB algorithm sometimes scares me a bit both as a consumer (more) and as a marketer. On the one hand, it is useful for easy targeting. On the other hand, it is scary since it monitors every movement of mine." It is known that data driven algorithms help marketers to improve the relevance of their audience by demographics and previous behaviour to make tailor-made offers and deliver them, often in real time. However, marketers mentioned that for better performance of their ads they need to engage and interact constantly as profile accounts as well. Participant (P7) said: "I feel that it is unfair that we need to do non-stop feeding of algorithms by payments and engagements, just to make our posts to be seen". The social networks algorithms prioritize which content a user sees in their feed based on relevancy and there is phenomenon that most of users do not have the ads' function turned off so they get too many ads and can not control them. Participant (P16) suggests: "It would be more ethical and fairer to pay a small fee in using social media platforms than to continue to live in madness of advertisements while all these platforms gain billions". It is known that algorithm outputs are opaque probably because of technical and social reasons. Many of participants mentioned that unfortunately, social media platforms (for example FB) give only few variables and instructions nothing more, "it is just a black box", they said. Participant (P13): "It does not bother me as long as I know that it is none of marketers knows how it works" (Fig. 3).

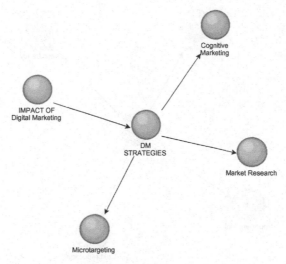

**Fig. 3.** Perceptions of DM online strategies.

**Perceptions of Microtargeting.** Concern about lack of transparency on algorithmic microtargeting (e.g., paid ads analysis), arose as well, as several participants mentioned that they worried about the lack of understanding of how an algorithm works and hence, they worry about results validity. And, since there is nothing can be done in this direction from their side, they have no choice, but to trust the systems. All of participants believe that microtargeting is essential marketing strategy used for target ads at lower cost by several audience exclusions. They say that they would prefer to go broadly and do not exclude audience by preferences in their marketing campaigns, but the cost would be too high. Participant (P16) said: "Marketing Science says'yes' to microtargeting, but ethics say'no'". All of participants mentioned that microtargeting has to be combined with any of persuasive techniques (e.g. scarcity) for better results, but as they stated it cannot be used in all products since it can become dangerous. "It is excellent tool to promote coffee but not to be used in political arena, there it will be unethical", Participant (P8), "that is why FB does not allow political advertising and use a content filtering for it (Fig. 4)."

**Perceptions of Cognitive Marketing.** For marketers the term 'Cognitive Marketing' generally means making people think positively about the brands they promote; thus, they try to create meaningful, and at the same time effective, ads by using cognitive biases such as Authority bias or Scarcity bias. Authority bias used in influencer marketing, where someone who people trust, influences people's purchasing decisions. Most of the participants believe that Influencer marketing it is just a temporary trend, and that influencer industry now is over-saturated and more competitive. Some of the participants believe that this kind of marketing is not suitable for expensive products and as a tool is not effective in all countries because of different culture in each location. However, all of the participants believe that there is always a risk to lose the customer's trust because influencers tend to promote similar products of different brands at the same time, therefore choosing the right influencer for the campaign is a difficult and time-consuming task. Moreover, some of the participants believe that an influencer could be a danger

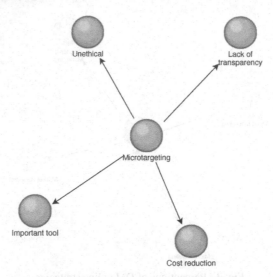

**Fig. 4.** Perceptions of microtargeting.

in case they do not follow the provided instructions and guidelines. All participants mentioned that people today look for authenticity and quality of product/service to be convinced.

Regarding scarcity bias several participants believe that it is unfair to use scarcity effect in online marketing practices constantly because it makes people psychologically stressed. On the other hand, "people need a push to move forward, so scarcity effect should be used rationally and with conscience while a non-aggressive ad can be used all year round", stated participant (P9). Participant (P1) said: "The scarcity is fair when it falls into the context of marketing…it is an essential part of usual marketing practices". However, Participant (P8) pointed out that sometimes scarcity used, in a total unethical way, e.g., 'Happy Hour' event in different video games where marketers attract customers by offering them something extra at certain times of the day or at the middle of night by keeping young people and kids awake.

On the one hand, participants acknowledged that scarcity an absolute 'must' especially if there is a need for immediate sale (for example during the sales period or obsolete inventory), moreover ad with scarcity performs much better than a plain one. On the other hand, Participant (P8) explained: "Scarcity helps with sales but not with communication with customers and it does not help to build a brand trust. People don't embrace this technique because if they use the offer ones, they will always expect discounts afterwards". All interviewees mentioned that scarcity is good to be used in online marketing practices only in case of low-cost products, products in stock for immediate selling since it may have negatively impacts on the brand image.

### 4.3 Moral Approaches to DM Practices

It is known that ethical actions are those that bring out the best in ourselves and others, and by this way benefit all members of the society. And since societies are those that determine what is right and acceptable, ethical/moral actions could be considered as actions that follow societal values. So, in this part of data, participants explained their moral approaches and values regarding information technology in digital marketing practices in the society. Specifically, the ethical principles such as honesty, caring, professionalism, policy following etc. Moreover, the participants explain how they realize what is the best interest of the customer, client and employer. Although to trust technology people need to feel confident that their activities online are safe, secure, and the technology is not opaque or complex in its implementation, participants mentioned that they trust technology, and specifically algorithmic content filtering system for its accuracy (Fig. 5).

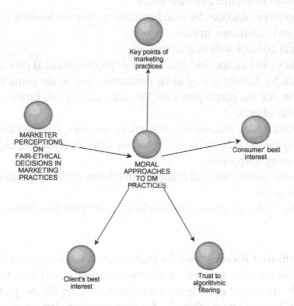

**Fig. 5.** Moral approaches to DM practices.

**Consumers' Best Interests.** In digital marketing, marketers play the role of mediators between their clients and consumers, and in ethical marketing, marketers try to achieve the well-being of everyone involved. However, it was observed that the participants as digital marketers do not act in the best interests of the consumer. They mostly focus on insight analysis that help them to create a data-driven marketing strategy quickly. In other words, they do not see consumers as persons, so they don't matter to them. However, they pay close attention to general consumers' satisfaction level by analyzing consumers' feedback and reviews through machine learning tools such as sentiment analysis.

**Client's Best Interests.** All the participants said that acting in the best interests of the client is important. Effective communication by having win-win relationship is also important. According to the participants a satisfied client will always bring another client.

**Key Points of Marketing Practices.** Most of the participants mentioned that during the online marketing practices they usually do not refer to the specific policy or ethical code but they follow their own personal ethical principles and they consider general factors such as (Fig. 6):

- Being environmentally conscious (by ensuring sustainable development and meet environmental standards).
- Being ethical in a personal sense but the ethical/moral responsibility should weigh on brand or marketing department managers.
- Using persuasive techniques such as scarcity or authority in moderation.
- Being honest with consumers or customers.
- Using non-deceptive practices by avoiding false and/or misleading information or by putting terms and conditions in case.
- Having personal contact with consumers.
- Following policy only in specific cases such as pharmaceutical products or tobacco.
- Professionalism by finding a way to be productive and at the same time effective in communication. As one participant said "We take care of two kinds of customers: the customer and his clients".
- Managing reputation by monitoring consumers' reviews. All the participants believe that reviews are a good tool for quality measure so the specific machine learning tools employed by many companies.
- Idea validation by getting second opinions and doing online experiments to see what does / does not work.
- Trustworthiness by keeping promises to clients and "treat the clients with kid gloves" (P1).

**Trust in Algorithmic Filtering.** All the participants mentioned that they have partial trust in algorithmic filtering systems of different platforms and they believe that algorithms work well. As one of participant mentioned: "We let the the platform to control and decide about the ad content ethically, for example when it detects alcohol, it makes an error to put the age over 21 to run". However, marketers feel that they have no control on algorithms "we are actually in the hands of the algorithm". On the one hand marketers see algorithms as "insurance valve" and "the only tool that can handle all these countless online transactions". On the other hand, the censorship and lack of transparency worry them.

## 5  Discussion

Machine learning, algorithms, and AI have impacted many applications relying on all sorts of data, and these applications are used as communication tools in Digital Marketing to increase personalization, create content, improve automation, analyze data sets, utilize chatbots etc. Thus, the question arises as to what happens to the relationship of trust between marketer and consumer. The traditional marketing approach was more personal; typically, marketers were reaching out not globally, but to local consumers and it was easy to have person-to-person relationships. The transition to digital marketing had undoubtedly increased the distance between the marketer and consumer, since the relationship between them nowadays is based on online impersonal communication. Taken together, the results of the current study indicate that there is a kind of negative impact of Digital Marketing on Tourism, on Brand Trust, and on Consumer Loyalty.

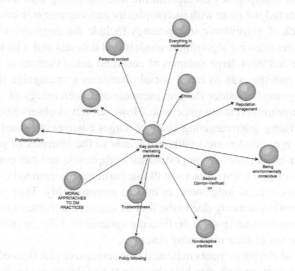

**Fig. 6.** Key Points of marketing practices

One of the reasons of this negative impact is related to the fact that consumers believe that not all marketers behave fair and ethically (Pavlidou, et al., 2021).

In summary, the results show that digital marketers seem to be largely indifferent to the issues of algorithmic fairness such as privacy, tracking, manipulations (persuasive techniques by using cognitive biases). They care mostly about the real time results, highest possible number of views and sales even if the results seem to them complex and vague. Therefore, any digital marketing technique e.g., Microtargeting or Cognitive biases – Influencers, Scarcity, Authority etc. marketers going to apply is considered from the profit side, but not from the ethical one. In other words, digital marketers seem to not act in the best interests of the consumer since they do not refer to any specific policy or ethical code, and just follow their own personal ethical principles along the trust in algorithmic filtering.

Like any empirical study, there are a number of limitations associated with our research. First, the sample is small, as we recruited individuals with significant experience in the industry. Secondly, we used a case study approach, with participants all being based in Cyprus. It is obvious in their responses that the region has particular characteristics, e.g., emphasis on the tourism sector. Nonetheless, it provided an opportunity to examine the impact of digital marketing on more traditional practices, and how the marketers themselves view this.

# 6  Conclusion

This study examined marketers' perceptions about fairness issues in algorithmic digital marketing. The analysis revealed a number of concerns regarding Digital Marketing impacts on the traditional way of marketing and moral approaches to Digital Marketing practices. Lack of transparency on algorithmic microtargeting also raised concerns, as all interviewees noted the issue with its complexity and vagueness. Factors contributing to the overall lack of algorithmic transparency include the cognitive impossibility for humans to interpret massive algorithmic models and datasets and a lack of appropriate tools to visualise and track large volumes of code and data (Tsamados, 2021).

It is known that the way to make social change is to recognize that algorithmic digital marketing must consider the requirements and preferences of all stakeholders (marketers and consumers) at the same time. However, it was observed that participants (marketers), by using microtargeting strategy, target customers (based on profession, region, income, age, gender, etc.) without regards to the interests of prospective customers. Good business ethics are important for companies, and can even lead to them achieving higher sales. Despite this, most of our participants seem to be unfamiliar with terms like fairness, ethical judgement, or ethical responsibility. They primarily aim to avoid trickiness and dishonesty due to the fear of negative self-image or bad reputation in their own community, and just let the filtering systems of different platforms to decide what is ethical or not in their marketing practices.

Although all of the participants indicated that persuasive practices of cognitive marketing such as scarcity or authority bias should not be false or misleading, they did not report taking any responsibility as marketers for the ads content, target audience or persuasive practices used in Digital Marketing. The aim of this work is to emphasize the importance of marketers' ethical decisions, their impacts on our everyday life, and to prevent uncertainty about responsibility. Having in mind that the future of personalization is where advertisers will develop ever more powerful and reality-bending ways to make sure their products are seen (Pariser, 2011) governments have a powerful regulatory role to play. There is a need for a broader discussion on what *fair marketing practices* are in the context of microtargeting, and more generally, the use of algorithms in cognitive marketing (persuasive practices). Furthermore, frameworks are needed to guide design choices, to regulate the reaches of algorithmic systems, and to ensure proper data stewardship (Dignum, 2018). As demonstrated by the findings of our study, it seems unlikely that digital marketers will actively pursue means to enforce ethical and fair algorithmic digital marketing themselves, thus, frameworks/guidelines and regulation will need to be made easy to understand and apply and/or enforce.

# References

Aler Tubella, A., Barsotti, F., Kocer, R.G., et al.: Ethical implications of fairnessinterventions: what might be hidden behind engineering choices? Ethics Inf. Technol. **24**, 12 (2022). https://doi.org/10.1007/s10676-022-09636-z

Alì, N.: Fair equality of opportunity and the place for individualmerit in a liberal democratic society. Braz. Polit. Sci. Rev. **16**(1), e0003 (2022). https://doi.org/10.1590/1981-382120220010002

Anand, A., Dutta, S., Mukherjee, P.: InfluencerMarketing with Fake Followers (March 16, 2022). IIM Bangalore Research Paper No. 580. SSRN: https://ssrn.com/abstract=3306088 or https://doi.org/10.2139/ssrn.3306088

Nadler, A., McGuigan, L.: An impulse to exploit: the behavioral turnin data-driven marketing. Crit. Stud. Media Commun. **35**(2), 151–165 (2018).https://doi.org/10.1080/15295036.2017.1387279

Anggadwita, G., Martini, E. (Eds.): Digital economy for customer benefitand business fairness. In: Proceedings of the International Conference on Sustainable Collaboration in Business, Information and Innovation (SCBTII 2019), Bandung, Indonesia, 9–10 October 2019 (1st ed.). Routledge (2020)

Asquith, K., Fraser, E.M.: A critical analysis of attempts to regulate native advertising and influencer marketing. Int. J. Commun. **14**, 5729–5749 (2020)

Bazeley, P.: Qualitative Data Analysis with NVivo. Sage Publications, London (2007)

Bala, Madhu and Verma, Deepak, A Critical Review of Digital Marketing (October1, 2018). M. Bala, D. Verma (2018). A Critical Review of Digital Marketing. International Journal of Management, IT & Engineering, 8(10), 321–339., Available at SSRN: https://ssrn.com/abstract=3545505

Bergman, A.: Ethique et gestion. In: Encyclop edie de Gestion, Economica Paris (1997)

Birner, N., Hod, S., Kettemann, M.C., Pirang, A., Stock, F. (Eds.): Increasing fairness in targeted advertising [The Ethics of Digitalisation Clinics Report, 1]. Alexander von Humboldt Institute for Internet and Society (2021). https://graphite.page/fair-targeted-ads

Calo, R.: Digital market manipulation. SSRN Electron. J. **82**(4) (2013)

Charmaz, K.: Constructing Grounded Theory: A Practical Guide Through Qualitative Analysis, 2 edn. Sage, Thousand Oaks, CA, Taylor Francis, 2016 (2006). ISBN 1317235657, 9781317235651

Dwork, C., Hardt, M., Pitassi, T., Reingold, O., Zemel, R.: Fairnessthrough awareness. In: Proceedings of the 3rd Innovations in Theoretical Computer Science Conference on - ITCS 2012 (2012). https://doi.org/10.1145/2090236.2090255

Einstein, M.: Black ops advertising: Native ads, content marketing, and thecovert world of the digital sell, p. 8. OR Books, New York, NY (2016)

Giovanola, B., Tiribelli, S.: Weapons of moral construction? On the value of fairness in algorithmic decision-making. Ethics Inf. Technol. **24**(1), 1–13 (2022a). https://doi.org/10.1007/s10676-022-09622-5

Glaser, B., Strauss, A.: The Discovery of Grounded Theory: Strategiesfor Qualitative Research. Aldine, Chicago (1967)

Graaf, G.: The autonomy of the contracting partners: an argument for heuristic contractarian business ethics. J. Bus. Ethics, **68**, 347–361(2006)

Infocredit Group. (2022). Cyprus Business Landscape Overview 2016–2021: A Statistical Analysis Report

Hanson, J.D., Kysar, D.A.: Taking Behavioralism Seriously: The Problemof Market Manipulation. Harvard Law School (1999)

Herder, E., Dirks, S.: User attitudes towards commercial versus political microtargeting. In: Adjunct Proceedings of the 30th ACM Conference on User Modeling, Adaptation and Personalization, pp. 266–273, July 2022

Helberger, N.: Profiling and targeting consumers in the Internet of Things–Anew challenge for consumer law (2016). SSRN 2728717

Lee, J.Y., Jin, C.H.: The role of ethical marketing issues in consumer-brand relationship. Sustainability MDPI **11**(23), 1–21 (2019)

Korolova, A.: Privacy violations using microtargeted ads: a case study. J. Priv. Confid. **3**(1) (2011). https://doi.org/10.29012/jpc.v3i1.594

Laczniak, G.R., Murphy, P.E.: Fostering ethical marketing decisions. J. Bus. Ethics **10**, 259–271 (1991). https://doi.org/10.1007/BF00382965

Lee, J.-Y., Jin, C.-H.: The role of ethical marketing issues in consumer-brand relationship. Sustainability **11**, 6536 (2019). https://doi.org/10.3390/su11123653

Mathur, A., Acar, G., Friedman, M.J., Lucherini, E., Mayer, J., Chetty, M., Narayanan, A.: Dark patterns at scale: findings from a crawl of 11 K shopping websites. Proc. ACM on Hum. Comput. Interact. **3**(CSCW), 1–32 (2019)

Murphy, P.E., Laczniak, G.R., Harris, F.: Ethics in Marketing: International casesand perspectives

Nguyen, B., Steve Chen, C.-H., Sharon Wu, M.-S., Melewar, T.C.: Ethical marketing. Ethical Soc. Mark. Asia 55–79 (2015)

Proceedings of the 2022 AAAI/ACM Conference on AI, Ethics, and Society. Association for Computing Machinery, New York, NY, USA

Perrault, R., et al.: The AI Index 2019 Annual Report", AI Index Steering Committee, Human-Centered AI Institute, Stanford University, Stanford, CA, December 2019. (c) 2019 by Stanford University, "The AI Index 2019 Annual Report" is made available under a Creative Commons Attribution- NoDerivatives 4.0 License (International). https://creativecommons.org/licenses/by-nd/4.0/legalcode

Riefa, C., Clausen, L.: Towards Fairness in Digital Influencers' Marketing Practices. J. Eur. Consum. Mark. Law (EuCML)(April 12, 2019). 8 (2019). SSRN: https://ssrn.com/abstract=336 4251

Ross, L., Greene, D., House, P.: The false consensus effect: an egocentricbias in social perception and attribution processes. J. Exp. Soc. Psychol. **13**(3), 279–301 (1977). https://doi.org/10.1016/0022-1031(77)90049-X

Sheth, Jagdish; Malhotra, Naresh (2010). Wiley International Encyclopedia ofMarketing —— Ethical Marketing and Marketing Strategy

Shi, X., Li, F., Chumnumpan, P.: The use of product scarcity in marketing. Eur. J. Mark. **54**(2), 380–418 (2020)

Strauss, A., Corbin, J.: Basics of Qualitative Research: Grounded Theory Procedures and Techniques. Sage Publications, Beverly Hills, CA (1990)

The AI Index 2022 Annual Report by Stanford University is licensed underAttribution-NoDerivatives 4.0 International. To view a copy of this license. http://creativecommons.org/licenses/by-nd/4.0/

Tsamados, A., Aggarwal, N., Cowls, J., et al.: The ethics of algorithms: key problems and solutions. AI Soc. **37**, 215–230 (2022). https://doi.org/10.1007/s00146-02101154-8

Tudoran, A.A.: Why do internet consumers block ads? New evidence from consumer opinion mining and sentiment analysis. Internet Res. **29**(1), 144–166 (2019). https://doi.org/10.1108/IntR-06-2017-0221

Will Browne and Mike Swarbrick Jones. (2017). What works in e-commerce - ameta-analysis of 6700 online experiments. Qubit Digital Ltd.

Wasieleski, D.M., Gal-Or, M.: An enquiry into the ethical efficacy of theuse of radio frequency identification technology. Ethics Inf. Technol. **10**(1), 27–40 (2008). https://doi.org/10.1007/s10676-008-9152-z

Winter, J.S.: Surveillance in ubiquitous network societies: normative conflicts related to the consumer in-store supermarket experience in the context of the Internet of Things. Ethics Inf. Technol. **16**, 27–41 (2014). https://doi.org/10.1007/s10676-0139332-3

Zarouali, B., Dobber, T., De Pauw, G., de Vreese, C.: Using a Personality Profiling Algorithm to Investigate Political Micro-targeting: Assessing the Persuasion Effects of Personality-Tailored Ads on Social Media. Communication Research (2020)

# The Importance of Platforms to Achieve Digital Maturity

Tristan Thordsen[✉] ⓘ and Markus Bick ⓘ

ESCP Business School Berlin, Heubnerweg 8-10, 14059 Berlin, Germany
{tthordsen,mbick}@escp.eu

**Abstract.** An important "building block" for an organization's digital maturity
seems to be the implementation of platforms. In general, platforms as software or
hardware infrastructures through which users and organizations can create applications, services, and communities, can trigger so-called network effects. These
effects act as a catalyst for the organization's digital maturity, by positively impacting key factors such as IT infrastructure, collaboration or innovation within the
firm. The importance of implementing platforms for organizations pursuing the
goal of raising their digital maturity is thus widely accepted. However, there is a
lack of knowledge concerning the type of platforms that is beneficial for an organization's desired digital transformation. Based on the qualitative content analysis of
24 Digital maturity models (DMMs) that are frameworks prescribing key elements
and a model path for an organization's digital maturity, we establish a first comprehensive overview of ten different platform types relevant for digital maturity.
The most prominent types are: IT-Infrastructure, collaboration, and value chain
platforms. Using an established framework, we were able to classify the differing
platform concepts into "platforms with selectively open interfaces" and "platforms with N-sided market infrastructures". Based on these insights, we derive
a first working definition of the term platform in the context of digital maturity.
We thus contribute to the advancement of both research fields and their overlap.
Furthermore, we offer managers guidance in the interpretation and application of
DMMs – in due consideration of different platform types.

**Keywords:** Digital Maturity · Platforms · Digital Transformation · Maturity
Models

## 1 Introduction

Most scholars and practitioners believe that being 'digital' is a key priority for businesses
to stay competitive in the era of digital transformation [1]. As an effort to conceptualize
this abstract idea, the construct of *digital maturity* has emerged. It designates "what a
company has already achieved with regard to transformation efforts" [2]. To satisfy the
omnipresent need of assessing and ultimately raising the digital maturity of an organization, so-called digital maturity models (DMMs) have been designed. These frameworks
define key elements for the digital maturity of organizations and prescribe a concrete
path along different maturity stages [3].

M. Papadaki et al. (Eds.): EMCIS 2022, LNBIP 464, pp. 339–351, 2023.
https://doi.org/10.1007/978-3-031-30694-5_25

For today's organizations, the implementation of platforms is widely considered as important building block in achieving digital maturity [4]. In general, platforms as software or hardware infrastructures through which users and organizations can create applications, services, and communities [5], can trigger so-called network effects [e.g., 6]. These effects act as a stimulant for the organization's digital maturity, by positively impacting key elements such as *IT infrastructure*, *collaboration* or *innovation* within the firm [7]. Even though the outstanding relevance of implementing platforms for organizations pursuing the goal of raising their digital maturity is largely recognized within the academic community, there is a lack of knowledge concerning the type of platforms that is beneficial for an organization's desired digital transformation.

This is a significant problem as the platform literature proposes a great variety of distinct platform types for differing areas of application within and across organizations and ecosystems: e.g., *internal platforms* in contrast to *industry platforms*. *Internal platforms* are infrastructures that are confined to the boundaries and the use of one single firm, whereas *social media platforms* dispose of open interfaces with a potentially unlimited pool of external complementors including firms and individuals [8].

Given this knowledge gap, there is a great uncertainty, especially among managers, seeking to keep pace with the ongoing digital transformation, regarding the definition of the term platform and the relevance of its respective types for this endeavor. DMMs, as popular practical tools, prescribing a variety of key elements for an organization's digital maturity, including platforms, can provide valuable insights on the nature and context of differing platform types relevant for this transformation process. We thus derive the following research questions for this study:

1. *In the context of present DMMs, what are the different platform types addressed as relevant for an organization's digital maturity?*
2. *In the context of digital maturity, what definition of the platform concept can be derived?*

In this study, we seek to paint a comprehensive picture of the differing platform types, that are deemed significant for an organization's digital maturity – based on the contents of present DMMs. In doing so, we further underline the fact that the concept of platforms and digital maturity are indeed closely interrelated. To provide a foundation for future research at this point of intersection of the two research fields, we derive a first working definition of platforms in the context of digital maturity. While investigating on this matter, we thus contribute to the advancement of the research field of digital maturity and platforms accordingly.

From a practical perspective, we inform managers on the meaning of platforms in the context of their organization's digital maturity and provide them with an overview of the relevant platform types. By implementing a suitable platform concept, practitioners can add an important building block to increase their organization's level of digital maturity.

To reach our aims, we will first outline current definitions of digital maturity, DMMs, the term platform and its overarching classifications in a general context, to then show how these concepts are interrelated. Then, by executing a systematic literature review, we shall identify all relevant DMMs for a business context of the past 10 years covering the term of platform according to a general definition. After having amassed the literature pool, we will follow Mayring's [9] qualitative content analysis approach for inductive category creation to identify the differing platform concepts and their frequency of distribution within the models. Subsequently, we will discuss the findings. Then, outline theoretical and practical contributions of this paper. Following, we present a working definition of platforms in the context of digital maturity and present and research agenda, consisting of potential research questions while proposing suitable research approaches for further investigation. Finally, we provide respective limitations.

## 2　Research Background

### 2.1　Digital Maturity Models

Today, we look back on eleven years of digital maturity research [10]. In this context, we refer to the following established definitions in the IS community:

*Digital Maturity* describes "the status of a company's digital transformation" – designating "what a company has already achieved with regard to transformation efforts" [2]. Here, efforts encompass both implemented changes from an operational perspective and acquired capabilities with regards to the management of the organization's transformation endeavor. In line with current beliefs of the academic community, [11] see a positive relationship between digital maturity and business performance as highly likely. It is suggested that the assessment of an organization's maturity is one of the key factors in the process of achieving a higher level of firm performance [12].

*Digital Maturity Model* denotes normative reference models which are utilized in organizations to assess the status quo of its digital maturity and provide concrete measures to increase its level [13]. Through a catalogue of benchmark indicators, various stages of maturity are framed in terms of evolutionary levels [14]. Current models display in average four to six evolutionary stages on the path to digital maturity [13]. DMMs are mainly designed for the following fields of application. The largest stake of the models serves a general business context [15, 16]. Manufacturing [17, 18] constitutes the target sector of another substantial portion of present DMMs. Remaining models address a broach spectrum of business fields such as IT [19], telecommunications [20] and Education [21, 22].

In general, every digital organization starts with a carefully formulated companywide *digital strategy* [23–25]. Intermediary stages towards digital maturity are characterized by more radical changes in the firm's IT infrastructure, processes, culture, and hierarchy. In this context, *platformization*, the implementation of platforms, is considered a key factor for an organization's digital maturity.

Typically, the DMMs' final stages are marked by a new orientation of the organization focusing on the customer. This novel approach is enabled by the data-driven enterprise. To maintain the organization's digital maturity, a continuous anticipation and adaptation to the increasingly dynamic business environment is necessary. Given the simple nature of maturity models and their particular practical relevance and value, a plethora of DMMs has emerged over the last decade [26].

## 2.2 Platforms

The notion of *platform* was established by the tech industry in the early 2000s to label digital intermediaries connecting information, goods, and persons [27]. A platform can be described as software or hardware infrastructure through which users, organizations, and institutions can create applications, services, and communities [5]. Such a network provides common standards, interfaces, and instruments to leverage core technologies, which enable data and knowledge sharing, collaboration, and foster innovation within or across firms [7]. Platforms are catalysts for these key factors for achieving digital maturity [28]. The phenomenon of platforms can be observed at different levels and in several organizational settings: within one firm, across supply-chains, or across entire ecosystems.

Despite the relevance of platforms for the management discipline, the research agenda has been limited and divided. Gawer [8] proposes a theoretically sound integrative framework for platforms that combines the two dominant, yet distinct theoretical perspectives. Instead of conceptualizing platforms either as markets – following the economic perspective – or as infrastructures – according to the engineering perspective, Gawer [8] unifies these differing approaches, portraying platforms through the organizational lens. Gawer's [8] unified framework suits the context of digital maturity as organizational phenomenon perfectly. It underlines the essential characteristics of different platform types while taking into consideration the particular setting of an organization. The following figure outlines Gawer's [8] classification of platforms (Fig. 1).

|  | Internal platform | Supply-chain platform | Industry platform |
|---|---|---|---|
| Level of analysis | • Firm | • Supply-chain | • Industry ecosystems |
| Platform's constitutive agents | • One firm<br>• Its constituent sub-units | • Assembler<br>• Suppliers | • Platform leader<br>• Complementors |
| Technological architecture |  | • Modular design<br>• Core and periphery |  |
| Interfaces | • Closed interfaces<br>Interfaces specifications are shared within the firm, but not disclosed externally | • Interfaces selectively open<br>Interface specifications are shared exclusively across the supply-chain | • Open interfaces<br>Interface specifications are shared with complementors |
| Accessible innovative capabilities | • Firm capabilities | • Supply-chain's capabilities | • Potentially unlimited pool of external capabilities |
| Coordination mechanisms | • Authority through managerial hierarchy | • Contractual relations between supply-chain member organizations | • Ecosystem governance<br>○ In the special case of multi-sided markets: exclusively through pricing |
| Examples | • Black and Decker (machine tools)<br>• Sony Walkman (consumer electronics) | • Renault-Nissan (automotive manufacturing)<br>• Boeing (aerospace manufacturing) | • Facebook (social networking)<br>• Google (Internet search and advertising)<br>• Apple iPhone and Apps (Mobile) |

**Fig. 1.** Classification of platforms according to Gawer [8]

Following the above unified framework, three overarching types of platforms can be identified. Internal platforms refer to a set of subsystems and *closed interfaces* that are accessible only within one firm. Here, innovative capabilities, collaboration and knowledge sharing are confined to the firm's boundaries. Examples for such platforms include the *Sony Walkman* and *Black and Decker machine tools*. Platforms with *selectively open interfaces* are an advancement of internal platforms. Their added value is shared across the entire supply chain of a certain organization. Therefore, they are also called supply chain platforms [8]. These types of platforms are particularly popular in the automotive industry. Industrial platforms display *N-sided market infrastructures* that connect buyers and suppliers [29]. They are much more complex as they represent entire ecosystems consisting of multiple organizations. Accordingly, innovative capabilities, collaboration and knowledge sharing are potentially unlimited [30]. Nowadays, entire business models are based on the service of providing digital platforms. In line with previous years, in 2020, seven of the top ten companies are taken by providers of such platforms [31].

### 2.3 Platforms and Digital Maturity

*Platformization* describes the process of a stepwise transformation, where "the IT silo structure is transformed to a platform-oriented infrastructure" [32]. Experts claim that platforms are necessary to enhance the quick connect capability of internal and external interactions and thus to harness market opportunities in today's dynamic business environment. The added value of a platform is determined by the amount and intensity of interactions on that a platform [33]. Through platformization, business networks can be transformed into digital ecosystems. Digital ecosystems are environments that allow time- and location-independent, "flexible and demand-driven collaboration" [30]. Due to these network effects, platforms are considered a key factor for an organization's digital transformation journey [e.g., 6]. In this context driving factors for digital maturity, such as *IT infrastructure*, *collaboration* or *innovation*, are boosted by the implementation of platforms [7].

## 3 Research Design

Our research goal is to identify the differing platform concepts that are addressed in current DMMs and thus deemed relevant for an organization's digital transformation journey. Furthermore, we seek to identify the respective business contexts in which the differing platform concepts are covered. Ultimately, we aim at gaining a better understanding of the platform phenomenon and its types in the context of DMMs and to derive avenues for future scientific investigations.

An explorative research design seems to be the appropriate approach for this given subject matter. The method of choice is a systematic literature review. In this endeavor, we draw on the insights of previous works [1] and further optimize the search strategy by adding new search terms and extending the relevant timeframe to encompass the past eleven years of DMM literature (2011–2021). This timeframe is especially relevant, as to our best knowledge, the first DMM was developed in 2011 [34]. Subsequently, we shall identify the models encompassing any notion of platforms to then outline the respective business foci of the DMMs. As a next step, we will conduct an inductive qualitative content analysis according to Mayring [9] to analyze the respective platform concepts and how they relate to the differing business contexts. In the following, we are going to lay out the details of our modus operandi.

### 3.1 Literature Search Methodology

Based on the guidelines of Vom Brocke et al. [35], we focused on a eleven-year period (2011 to 2021), searching in ten leading IS journals (Senior Basket of 8 plus BISE and

MISQE), four major IS conferences (AMCIS, ECIS, ICIS, PACIS) and two complementary databases (Business Source Premier and Google Scholar). In the selection of
databases and outlets, we draw on the experience of [36], who have previously engaged
in the analysis of existing DMMs. The keywords of this systematic literature review
were designed according to the PICO criteria (Population, Intervention, Comparison
and Outcomes). Kitchenham and Charters [37] deem these parameters as particularly
suitable for a literature review in the discipline of IS. Furthermore, we identified synonyms and alternative spellings for the search phrases by consulting both experts and
literature. Search strings for Business Source Premier are e.g.: (("maturity model" OR
"stages of growth model" OR "stage model" OR "change model" OR "transformation
model" OR "grid")) AND (("Transformation" OR "Digit*")). In addition to the existing catchphrases identified by Thordsen et al. [1], we added "index", "matrix", "eval-
uat*", "framework", "quotient", "industry 4.0", "readiness", and "assess*". A detailed
overview of the literature search process can be found in Fig. 2.

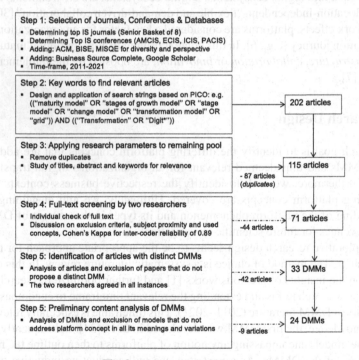

**Fig. 2.** Literature search process

The search was first performed focusing on titles, abstracts, and keywords. It resulted
in a total of 202 papers, of which 87 were duplicates. After a full-text screening of the
remaining 115 papers, we further excluded 44 works that did not address a business
context. This step was at first completed by each author individually. Then, we discussed
our screening and agreed on the papers to be included. Cohen's Kappa indicating the
inter-coder reliability was at 0.89 [38]. Finally, our pool of literature comprised 71

articles. These included two publications from leading IS journals, 27 conference articles, and 42 papers of other journals or publications of consultancies. We included consultancy reports and practitioners' works as a majority of DMMs come from these backgrounds and we seek to paint a comprehensive picture of this research field. Within the 71 papers, we could identify 33 DMMs. A preliminary screening of all 33 publications reveals that the concept of platform is addressed in a total of 24 studies –forming our final literature pool. This step was performed based on the presence of all varieties and meanings of the platform concept as it has been described earlier in the text. As a last step, we identified the industry focus of the DMMs for further investigation on the relationship between the addressed platform type and potential area of application. Here the inter-coder reliability was at 0.85. The following table provides an overview of the analyzed models with their respective business focus (Table 1).

As indicated, 24 models address the platform idea as it is described by [5]. Taking a closer look at the respective publication dates of the DMMs, it can be noted that the platform concept remains relevant along the past eleven years. In line with the findings of previous studies, the large majority of DMMs identified in our literature pool addresses a general business context (13) and thus has no industry focus. Seven DMMs concentrate solely on the manufacturing sector, whereas logistics, education, auditing, and IT are the target sectors of one DMM of the literature pool respectively.

## 3.2 Qualitative Content Analysis

For our DMM analysis we applied Mayring's [9] qualitative content analysis approach for inductive category creation. It is an established qualitative research approach success-fully utilized in various IS studies for the analysis of systematic literature reviews [e.g., 56]. Mayring's [9] qualitative content analysis procedure allows an unbiased, transparent and methodologically controlled content analysis of qualitative data and thus ensures qualitative rigor. The first step of this content analysis consists in a frequency analysis. Within this procedure, we identify different passages within the studies' terminology and compare their frequency of occurrence with each other. In this context, the use of comprehensive category systems, including all relevant elements of the relevant concept is of special importance. Following this approach, we highlighted all text passages refer-ring to the concept of "platform" in its simplest form – as it is defined by Casilli and Posada [5]: software or hardware infrastructure through which users, organizations, and institutions can create applications, services, and communities. Once we have identified all instances referring to this concept, we investigated on the respective contexts and the number of DMMs that the platform construct was mentioned in. In the analysis, we thus unitized and coded every identified idea of platforms to abstract codes with similar context and meaning. In total, we formed ten categories. For this procedure, we used the QCAmap-Software (qcamap.org). After approximately 50% of the content analysis, we revised and refined the category system [9, 57]. Two researchers performed the initial coding process independently. The intercoder-reliability was at 90%. Remaining cases were clarified during a discussion with a third researcher [57]. In a next step, our goal was to increase the level of abstraction of the previously identified classifications to further reduce the number of sets. Consulting the foregoing platform classifications according to [8], through deductive reasoning, we were able to further cluster the platforms into:

*platform with selectively open interfaces (also Supply Chain Platform)* and *platform with N-sided market infrastructure (also Industry Platform)*. Figure 3 depicts the results of our first analysis.

**Table 1.** Literature pool

| ID | Study | Business focus | Year |
|----|-------|----------------|------|
| 1 | [39] | General | 2011 |
| 2 | [34] | General | 2011 |
| 3 | [40] | General | 2012 |
| 4 | [16] | General | 2015 |
| 5 | [41] | Logistics | 2015 |
| 6 | [42] | Manufacturing | 2015 |
| 7 | [43] | Manufacturing | 2016 |
| 8 | [44] | General | 2016 |
| 9 | [45] | General | 2016 |
| 10 | [6] | General | 2016 |
| 11 | [17] | Manufacturing | 2018 |
| 12 | [46] | Education | 2019 |
| 13 | [47] | General | 2019 |
| 14 | [48] | General | 2019 |
| 15 | [49] | General | 2019 |
| 16 | [28] | Manufacturing | 2020 |
| 17 | [10] | General | 2020 |
| 18 | [26] | General | 2020 |
| 19 | [50] | General | 2020 |
| 20 | [51] | Auditing | 2020 |
| 21 | [52] | Manufacturing | 2020 |
| 22 | [53] | IT | 2020 |
| 23 | [54] | Manufacturing | 2020 |
| 24 | [55] | Manufacturing | 2021 |

## 4  Findings

The differing concepts of platforms that we were able to detect in present DMMs are displayed in Fig. 3. Moreover, Fig. 3 shows the broader classification of these platforms according to by Gawer [8]. Nine of the 24 DMM studies name platforms in the context of *IT or digital infrastructure* (see number in parentheses). The same frequency of naming applies to *collaboration*, followed by *value/supply chain* with seven, and *knowledge* with six instances. *Social media* appears in five models, whereas *e-commerce* is named four times, *innovation* and *manufacturing* three and finally *CRM* and *business model* two times. The frequency analysis also shows that the overarching classification *platform with selectively open interfaces (also Supply Chain Platform) is* addressed more than three times as frequently and by a significantly larger number of DMMs than the classification *platform with N-sided market infrastructure (also Industry Platform).*

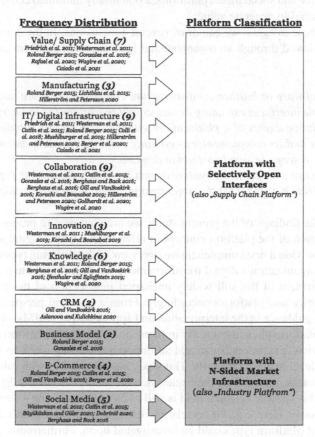

**Fig. 3.** Overview of the platform types in present DMMs

## 5    Conclusion and Working Definition

In the last decade of DMM research, the dominating perspective on platforms is a that of a modular software and hardware infrastructure across the firm and its suppliers. Internal platforms, however, are not addressed by current DMMs and thus do not seem beneficial for an organization's digital journey. The positive network effects of platforms with selectively open interfaces on collaboration, knowledge sharing, and innovation are appreciated by the present studies. These platforms are confined to the firm and its suppliers. In this setting, especially the benefit of platforms designed for the value chain management and manufacturing are widely considered. These platforms limit the access to internal data from the outside and thus protect sensible information. However, the sharing and innovation potential is thus also confined. Present DMMs also acknowledge the advantages of platforms with N-sided market infrastructures that form entire data ecosystems. Particularly in the context of business models, based on the exchange of data, e-commerce and social media platforms, a potentially unlimited ecosystem of users and firms is crucial to generate value.

Based on these insights, we can draw several conclusions. In the context of digital maturity and viewed through an organizational lens, the term platform can be defined as:

1. *Modular software or hardware infrastructure composed of a core and a periphery with multiple interfaces extending across supply chains and/ or industry ecosystems.*
2. *The constitutive agents of a platform are either a firm and its suppliers or a platform leader and its complementors - creating value by enabling and making use of economies of scope in supply or/and in demand.*
3. *As the platform's access to innovating agents increases, the diverse capabilities and network effects beneficial for the organization's digital transformation increase.*

Based on the findings of the present study, we are confident to propose a valid first working definition of the platform concept in relation to a firm's digital maturity. Furthermore, we provide a first comprehensive overview of the differing types of platforms relevant to an organization's digital transformation. We thus contribute to the advancement and refinement of the still widely uncharted intersection of the research fields of digital maturity and platforms accordingly. From a practical perspective, we offer managers with guidance in the interpretation and application of DMMs – in due consideration of the platform phenomenon. By implementing a suitable platform type from the provided overview, practitioners can add an important building block to increase their organization's level of digital maturity. This study has provided us with valuable insights regarding platforms, underlining their relevance for an organization's digital maturity. Nevertheless, this specific research field is still at its very beginning – further scientific inquiry is necessary. Here, e.g., the correlation between the organization's industry focus and the relevant platform type could be investigated upon. Furthermore, it could be of added value to analyze at which level of digital maturity, the present DMMs suggest implementing platforms.

We acknowledge the limitations of this paper. Of course, the keywords and catchphrases of the literature search, as well as the databases and outlets, can be further extended to complement the existing pool of DMMs. Nevertheless, we are confident to further stimulate the academic discussion and to encourage future investigations on the subject matter.

## References

1. Thordsen, T., Murawski, M., Bick, M.: How to measure digitalization? A critical evaluation of digital maturity models. In: Hattingh, M., Matthee, M., Smuts, H., Pappas, I., Dwivedi, Y.K.,

Mäntymäki, M. (eds.) I3E 2020. LNCS, vol. 12066, pp. 358–369. Springer, Cham (2020). https://doi.org/10.1007/978-3-030-44999-5_30

2. Chanias, S., Hess, T.: How digital are we? Maturity models for the assessment of a company's status in the digital transformation. Manag. Rep./Institut für Wirtschaftsinformatik und Neue Medien **2**, 1–14 (2016)

3. Pavel, E.V., Kudryashova, T.V., Bykova, P.A.: Research on the digital maturity of mechanical engineering companies in Russia. In: International Scientific and Practical Conference "Russia 2020-a New Reality: Economy and Society (ISPCR 2020), pp. 408–411. Atlantis Press (2021)

4. Pauli, T., Fielt, E., Matzner, M.: Digital industrial platforms. Bus. Inf. Syst. Eng. **63**(2), 181–190 (2021). https://doi.org/10.1007/s12599-020-00681-w

5. Casilli, A., Posada, J.: The platformization of labor and society. Society and the internet: how networks of information and communication are changing our lives, pp. 293–306 (2019)

6. Gill, M., VanBoskirk, S.: The digital maturity model 4.0. Benchmarks: digital transformation playbook (2016)

7. Teece, D.J.: Dynamic capabilities and (digital) platform lifecycles. In: Entrepreneurship, Innovation, and Platforms. Emerald Publishing Limited (2017)

8. Gawer, A.: Bridging differing perspectives on technological platforms: toward an integrative framework. Res. Policy **43**, 1239–1249 (2014)

9. Mayring, P.: Qualitative content analysis: theoretical foundation, basic procedures and software solution (2014)

10. Aslanova, I.V., Kulichkina, A.I.: Digital maturity: definition and model. In: 2nd International Scientific and Practical Conference "Modern Management Trends and the Digital Economy: from Regional Development to Global Economic Growth" (MTDE 2020), pp. 443–449. Atlantis Press (2021)

11. Eremina, Y., Lace, N., Bistrova, J.: Digital maturity and corporate performance: the case of the Baltic states. J. Open Innov. Technol. Market Complex. **5**, 54 (2019)

12. Bititci, U.S., Garengo, P., Ates, A., Nudurupati, S.S.: Value of maturity models in performance measurement. Int. J. Prod. Res. **53**, 3062–3085 (2015)

13. Williams, C., Schallmo, D., Lang, K., Boardman, L.: Digital maturity models for small and medium-sized enterprises: a systematic literature review. In: ISPIM Conference Proceedings, pp. 1–15. The International Society for Professional Innovation Management (ISPIM) (2019)

14. Becker, J., Niehaves, B., Poeppelbuss, J., Simons, A.: Maturity models in IS research (2010)

15. Westerman, G., Bonnet, D., McAfee, A.: Leading Digital: Turning Technology into Business Transformation. Harvard Business Press (2014)

16. Catlin, T., Scanlan, J., Willmott, P.: Raising your digital quotient. https://www.mckinsey.com/business-functions/strategy-and-corporate-finance/our-insights/raising-your-digital-quotient

17. Colli, M., Madsen, O., Berger, U., Møller, C., Wæhrens, B.V., Bockholt, M.: Contextualizing the outcome of a maturity assessment for Industry 4.0. Ifac-papersonline **51**, 1347–1352 (2018)

18. Gajsek, B., Marolt, J., Rupnik, B., Lerher, T., Sternad, M.: Using maturity model and discrete-event simulation for Industry 4.0 implementation. Int. J. Simul. Model. **18**, 488–499 (2019)

19. Isaev, E.A., Korovkina, N.L., Tabakova, M.S.: Evaluation of the readiness of a company's IT department for digital business transformation. Бизнес-информатика (2018)

20. Ochoa-Urrego, R.-L., Peña, J.-I.: Digital maturity models: a systematic literature review. In: ISPIM Conference Proceedings, pp. 1–15. The International Society for Professional Innovation Management (ISPIM) (2020)

21. Jugo, G., Balaban, I., Pezelj, M., Begicevic Redjep, N.: Development of a model to assess the digitally mature schools in Croatia. In: Tatnall, A., Webb, M. (eds.) WCCE 2017. IAICT, vol. 515, pp. 169–178. Springer, Cham (2017). https://doi.org/10.1007/978-3-319-74310-3_19

22. Ðurek, V., Kadoic, N., Reðep, N.B.: Assessing the digital maturity level of higher education institutions. In: 41st International Convention, pp. 671–676 (2018)

23. Matt, C., Hess, T., Benlian, A.: Digital transformation strategies. Bus. Inf. Syst. Eng. **57**, 339–343 (2015)
24. Ochoa, O.L.: Modelos de madurez digital:¿ En qué consisten y qué podemos aprender de ellos?/Digital maturity models: what are they and what can we learn from them? Boletín de estudios económicos **71**, 573 (2016)
25. Chanias, S., Hess, T.: Understanding digital transformation strategy formation: insights from Europe's automotive industry. In: PACIS, p. 296 (2016)
26. Büyüközkan, G., Güler, M.: Analysis of companies' digital maturity by hesitant fuzzy linguistic MCDM methods. J. Intell. Fuzzy Syst. **38**, 1119–1132 (2020)
27. Evans, D.S., Hagiu, A., Schmalensee, R.: Invisible Engines: How Software Platforms Drive Innovation and Transform Industries. The MIT Press (2008)
28. Rafael, L.D., Jaione, G.E., Cristina, L., Ibon, S.L.: An Industry 4.0 maturity model for machine tool companies. Technol. Forecasting Soc. Change **159**, 120203 (2020)
29. Iansiti, M., Levien, R.: The Keystone Advantage: What the New Dynamics of Business Ecosystems Mean for Strategy, Innovation, and Sustainability. Harvard Business Press (2004)
30. Aulkemeier, F., Iacob, M.-E., van Hillegersberg, J.: Platform-based collaboration in digital ecosystems. Electron. Mark. **29**(4), 597–608 (2019). https://doi.org/10.1007/s12525-019-003 41-2
31. Forbes: The 100 largest companies in the world by market capitalization in 2020 (in billion U.S. dollars). https://www-statista-com.revproxy.escpeurope.eu/statistics/263264/top-companies-in-the-world-by-market-capitalization/
32. Bygstad, B., Hanseth, O.: Transforming digital infrastructures through platformization (2018)
33. Alt, R., Zimmermann, H.D.: Electronic markets on platform competition. Electron Markets **29**, 143–149 (2019). https://doi.org/10.1007/s12525-019-00353-y
34. Westerman, G., Calméjane, C., Bonnet, D., Ferraris, P., McAfee, A.: Digital transformation: a roadmap for billion-dollar organizations. MIT Center for Digital Business and Capgemini Consulting, vol. 1, pp. 1–68 (2011)
35. vom Brocke, J., Simons, A., Niehaves, B., Riemer, K., Plattfaut, R., Cleven, A.: Reconstructing the giant: on the importance of rigour in documenting the literature search process. In: ECIS, vol. 9, pp. 2206–2217 (2009)
36. Thordsen, T., Bick, M.: Towards a holistic digital maturity model. In: ICIS 2020 (2020)
37. Kitchenham, B., Charters, S.: Guidelines for performing systematic literature reviews in software engineering (2007)
38. Kane, M., Crooks, T., Cohen, A.: Validating measures of performance. Educ. Meas. Issues Pract. **18**, 5–17 (1999)
39. Friedrich, R., Le Merle, M., Grone, F., Koster, A.: Measuring Industry Digitization: Leaders and Laggards in the Digital Economy. Booz & Co., London (2011)
40. Westerman, G., Tannou, M., Bonnet, D., Ferraris, P., McAfee, A.: The digital advantage: how digital leaders outperform their peers in every industry. MITSloan Management and Capgemini Consulting, MA, vol. 2, pp. 2–23 (2012)
41. Berger, R.: The digital transformation of industry. The study commissioned by the Federation of German Industries (BDI), Munich (2015)
42. Lichtblau, K., et al.: Industrie 4.0 Readiness. IMPULS-Stiftung for mechanical engineering, plant engineering, and information technology (2015)
43. Gonzalez, A.A., et al.: Digitale Transformation-Wie Informations-und Kommunikationstechnologie etablierte Branchen grundlegend verändern. Abschlussbericht des vom Bundesministerium für Wirtschaft und Technologie geförderten Verbundvorhabens. IKT-Wandel (2016)
44. Berghaus, S., Back, A.: Stages in digital business transformation: results of an empirical maturity study. In: MCIS 2016, Cyprus, p. 22 (2016)

45. Berghaus, S., Back, A., Kaltenrieder, B.: Digital maturity & transformation report 2016. Crosswalk AG (2016)
46. Ifenthaler, D., Egloffstein, M.: Development and implementation of a maturity model of digital transformation. TechTrends **64**, 302–309 (2020)
47. Muehlburger, M., Rueckel, D., Koch, S.: A framework of factors enabling digital transformation (2019)
48. Nguyen, D.K., Broekhuizen, T., Dong, J.Q., Verhoef, P.C.: Digital readiness: construct development and empirical validation (2019)
49. Korachi, Z., Bounabat, B.: Towards a maturity model for digital strategy assessment. In: Ezziyyani, M. (ed.) AI2SD 2019. AISC, vol. 1105, pp. 456–470. Springer, Cham (2020). https://doi.org/10.1007/978-3-030-36674-2_47
50. Berger, S., Bitzer, M., Häckel, B., Voit, C. (eds.): Approaching Digital Transformation-Development of a Multi-dimensional Maturity Model (2020)
51. Dobrinić, D.: Digital maturity of auditing companies in the Republic of Croatia (2020)
52. Hillerström, M., Petersson, I.: Measuring Digital Maturity in the CNC Manufacturing Industry: A Maturity Evaluation Model (2020)
53. Gollhardt, T., Halsbenning, S., Hermann, A., Karsakova, A., Becker, J.: Development of a digital transformation maturity model for IT companies. In: 2020 IEEE 22nd Conference on Business Informatics (CBI), pp. 94–103. IEEE (2020)
54. Wagire, A.A., Joshi, R., Rathore, A.P.S., Jain, R.: Development of maturity model for assessing the implementation of Industry 4.0: learning from theory and practice. Prod. Plan. Control 1–20 (2020)
55. Caiado, R.G.G., Scavarda, L.F., Gavião, L.O., Ivson, P., de Mattos Nascimento, D.L., Garza-Reyes, J.A.: A fuzzy rule-based industry 4.0 maturity model for operations and supply chain management. Int. J. Prod. Econ. **231**, 107883 (2021)
56. Osterrieder, P., Budde, L., Friedli, T.: The smart factory as a key construct of industry 4.0: a systematic literature review. Int. J. Prod. Econ. **221**, 107476 (2020)
57. Krippendorff, K.: Content Analysis: An Introduction to its Methodology. Sage Publications (2018)

# Can Process Mining Detect Video Game Addiction Through Player's Character Class Behavior?

Maxime Guénégo and Rébecca Deneckère[✉]

Centre de Recherche en Informatique, Université Paris 1 Panthéon-Sorbonne, Paris, France
rebecca.deneckere@univ-paris1.fr

**Abstract.** For more than fifteen years, we have witnessed the effervescence of video games, first considered by a handful of scholars this hobby knew how to penetrate most homes. Video games are no longer necessarily considered a simple hobby, players have been able to take advantage of this enthusiasm to make a business, the birth of e-sport, community channels, and various monetized videos are proof of this. However, the news shows us that for some people video games can become a real addiction and bring with it its share of the problem.

Since video games are by nature computer programs, we may wonder if it is not possible to retrieve information about their users in real-time to detect any addictive behavior and thus protect the player. In this paper, we will use process mining techniques to analyze player logs to try to find a behavioral pattern of addiction.

**Keywords:** Addiction · Video Games · Process Mining

## 1 Introduction

A video game is usually defined as an electronic game with a user interface allowing a playful human interaction by generating visual feedback [1]. The video game player has devices to act on the game and perceive the consequences of his actions in a virtual environment. According to this definition of a video game, its history can begin around 1950 with OXO (1952), Tennis for Two (1958), or Spacewar! (1962), which is the most commonly accepted date. Pong (1972), is the first game whose gameplay is catchy and addictive enough to make it a known success with a public audience.

There are many benefits that video game players get from engaging in their chosen activity. These can be educational, social, and/or therapeutic [2]. However, excessive use can pose a health threat to some people (basically young people but addicts can be found in all categories of people). The most important risk concerns networked games and in particular multiplayer role-playing games. Detecting these addictions can be a real challenge to help addicts as soon as possible. Besides traditional techniques used by health professionals (time consumed playing for instance), we investigate the possibility to use user logs to identify addiction by looking at the player's character behavior. In network games, players are often playing a character belonging to a specific class, like

M. Papadaki et al. (Eds.): EMCIS 2022, LNBIP 464, pp. 352–364, 2023.
https://doi.org/10.1007/978-3-031-30694-5_26

Knight, Mage, Hunter, Druid, Warrior, etc. The player activities are completely different as each class has different objectives in the group (kill enemies, collect treasures, treat injuries, and so on). This means that each class will have a specific behavior and that we have to study each class separately from the other.

We believe that we should be able to detect an addict's behavior by looking at the activity logs of a user belonging to a specific class. Thus, our main research question is the following: "**RQ: Can we identify a video game addiction from activity logs?**". This research question is decomposed into two other questions:

- RQ1: "Does it exists tools to detect addiction to video games?" and
- RQ2: "Can process mining help to identify addiction in class logs?"

Section 2 will give some background works on video-games addiction and process mining. Section 3 shows an experiment to try to demonstrate that it is possible to make a correlation between addicted players and their logs. We conclude in Sect. 4.

## 2   Background

This section will give some insights into the two domains studied in this work: video game addiction and process mining.

### 2.1   Video Game Addiction

Historically, the term addiction only involved addictions to toxic substances (drugs) but today we can mention addictions such as sexual addiction, compulsive shopping, gambling, and so on. Among these new addictions, we find the addiction to video games, officially an addiction since June 18, 2018, when the World Health Organization (WHO) listed this addiction as a disease just like addictions to drugs, tobacco, or alcohol. The WHO gives us the following definition of a video games addiction: *"Gaming disorder is characterized by a pattern of persistent or recurrent gaming behavior ('digital gaming' or 'video-gaming'), which may be online (i.e., over the internet) or offline"*[1] It also gives details about the characteristics of this addiction, like an impaired control over gaming, an increasing priority given to gaming to the extent that gaming takes precedence over other life interests and daily activities, and the continuation or escalation of gaming despite the occurrence of negative consequences.

Currently, the majority of the literature works in the field of video game addiction is Asian and deals mainly with adolescents. Yet the popularity of video games continues to grow (especially online video games). According to [3], the increase in this popularity has consequences on the population as it also develops an addiction. The particularity of this pathology is that it is difficult to diagnose because in normal times the gaming activity is not problematic. In addition, the use of video games may in some cases have beneficial effects, which is seen as inconceivable for "traditional" addictions such as

---

[1] WHO Definition of Gaming disorder: https://icd.who.int/browse11/l-m/en#/http%253a% 252f%252fid.who.int%252ficd%252fentity%252f1448597234.

cocaine or alcohol (even if there are studies that show that marijuana can help people reduce nausea and vomiting during chemotherapy, improve appetite in people with HIV/AIDS, and reduce chronic pain and muscle spasms [4, 5]).

The Diagnostic and Statistical Manual of Mental Disorders [6] worked on provisional criteria from studies on gambling addictions. The DSM-5 Checklist (DSM5) is an 11-item questionnaire that measures the degree (mild, moderate, severe) to which an individual meets diagnostic criteria for a substance use disorder.

There are currently several measurement tools to try and assess the intensity of addiction. However, these tools do not distinguish between game styles or the environment [7].

- The Internet Addiction Test (IAT) [8], in 1998, assess Internet addiction in general.
- The Problem Video game Playing (PVP) [9], in 2002, is based on several criteria of addiction defined by the DSM [6].
- The Game Addiction Scale (GAS) [10]: Created in 2009 with two versions of 7 and 21 items, also based on [6]. The objective here is to distinguish the playing time and the intensity of the addiction. However, the 21-item version has not been validated.
- The Questionnaire for Measuring Intensity of Addictive Behaviors (QMICA) [11], in 2010, aims at identifying the co-addictions (compensation of one addiction by another).
- The Internet Gaming Disorder (IDG) [12], in 2014, consists of 20 items distributed within the Griffiths model.

Currently, one of the most used models is the Griffiths one. Dr. Mark Griffiths [13, 14] has been working on substance-free addictions since 1987. He then began his research on gambling before turning to video games, putting in parallel dependent games and excessive games. This notion is important because excessive play may result in inordinate enthusiasm but it can remain an added value for the patient's life. On the other hand, a dependent game is not exciting and is harmful to the patient's life. He then concludes that two people can play a lot but without having the same result. To be able to distinguish between these two behaviors, Griffiths' model is based on 6 components found in behavioral addictions. The components of the model are Salience (the game becomes the main activity or you think about it even when you don't play), Tolerance (the gear system the more you play the more you have to play to be satisfied), Withdrawal symptoms (a psychological or physiological problem when you can't play), Changing mood (It is by playing that you manage to control your negative mood), Conflict (problem in personal relationships to keep a playtime, or deletion of other activity to keep the play time), Relapse (failure to stop the game or increase the playing time when a new extension arrives or when a new game is released).

Today we consider that the most addictive video games are the «Massively Multiplayer Online Role-Playing Games» (MMORPG) [15]. The reason for a massive addiction to this type of game is that we often get rewards (reward circuit) which cause a craving and a desire for «more and more» in the game. One can easily make the parallel with the winnings obtained in games to scratch or chance. In some countries such as Belgium, video games that offer random rewards (called loot boxes) for money have seen their «Pan European Game Information» (PEGI) ranking go from +13 years to +

18 years as if the game were a game of money. In addition, another factor to be taken into account in MMORPG-type video games is membership in a group/community/guild. Joining a group of players dependent on each other to progress in the virtual adventure gives a sense of social obligation to always play more. Finally, a well-known business model in the world of video games is Free to play (F2P), by which we mean a free game that has an internal store to the game to buy bonuses (experience boosts, cosmetics, extra characters, etc.). This internal store is the only source of revenue for the game publisher, but it remains very prolific. For example, the Fortnite game generated more than $300 million in 200 days after its launch on IOS [16].

## 2.2 Process Mining

In recent years, the sum of digital data continues to increase with more than 33 zettabytes in 2018 [17] and The 2020 Data Attack Surface Report estimates data stored in the cloud will reach 100 zettabytes by 2025 [12]. This increase is not going to stop, on the contrary. Faced with this, the computer sciences have matured in the exploitation of this data, thus creating many promising research frameworks. In this context, no company can ignore business intelligence (BI). BI is defined as a set of tasks such as the collection, storage, processing, processing, and operation of data to assist in decision-making. The objective is to better understand the activity and the sector in which we are located (for example a company). In other words, collecting data is nothing new for the IT sector, but it is so important that this collection is almost systematic and crucial. Controlling your data is then a real competitive advantage.

Process mining [18] is a mixture of Business Process Management (BPM) and Business Process Intelligence (BPI). BPM [19] is a process-oriented field, which provides a global view of business processes. Among other things, it allows us to see how these processes (and the business that composes them) interact with each other. BPI is a recent discipline that emerged thanks to the emergence of data but also and especially to their varieties [20]. According to [15] the BPI can be described as a set of forecasting, control, analysis, audit, and optimization tools serving business users but also IT services to manage the quality of the organization's processes.

In process mining, the process search is based on the analysis of event logs. The objective is then to analyze them to extract new business processes that would not have been prescribed but used since defined via the collected data. Once identified, they can be monitored and/or optimized. There are several types of objectives in process mining: process discovery, conformance checking, enhancement, and recommendation. Following [15], the main advantages of these techniques are objectivity (we discover the real processes because they are based on event logs), compliance (between the formalized process and the actual process), speed, the predictions and simulations, and the transparency (just like objectivity, processes being extracted directly from logs it is possible to go into detail of a process knowing who did what, when and how). There are a lot of tools that offer process mining techniques [21, 22].

## 3 Experiment

To answer our research questions, we made an experiment (a method of data collection designed to test hypotheses under controlled conditions, to eliminate threats to internal validity [23]) with players of a video game classified as highly addicting, an MMORPG[2] called World Of Warcraft (WoW) which is one of the most played MMORPG in the world (millions of players, distributed in 244 countries and territories).

The process used in this experiment was as follows (Fig. 1).

- Our first step was to define a survey that will allow us to define if the person was addicted or not. Then we selected a specific group in the game and submitted the survey to our players.
- The second step was to collect the logs of these players to use in process mining tools.
- We were then able to make a cross-analysis to see if predictions about addiction via the recovered logs were possible.

**Fig. 1.** Method used

### 3.1 Survey

We used the method preconized in [24] for the survey part. The process is composed of several steps (Fig. 2) that we detail in this section.

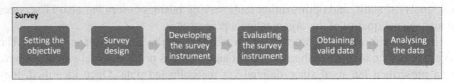

**Fig. 2.** Survey Process

**Setting the Objective.** Our goal is to be able to identify addicts and non-addict people in a specific group population of WoW players.

**Survey Design.** We ask for information at one fixed point in time, which means that our survey is *cross-sectional*, it gives a snapshot of the problem. The survey is a self-administered questionnaire where each participant is reached personally to increase the answering rate.

---

[2] MMORPG: Massively Multiplayer Online Role-Playing Game.

**Developing the Survey Instrument.** We identified the questions following several categories. The first one allows defining the respondent characteristics. The others are defined to identify the possible player addiction. The Griffiths model [13] is the most widespread model to wonder about the addiction of someone and we based our questionnaire on its 6 categories, as shown in Fig. 3.

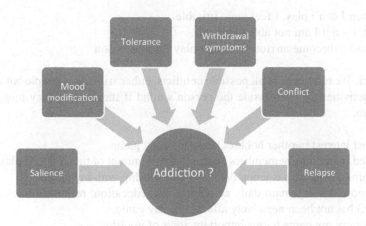

**Fig. 3.** Griffiths model elements (based on [13])

- *Respondent characteristics.* It is always useful to identify some characteristics of the respondent to be able to link answers to some profile.

  a. Gender (Male / Female)
  b. Age
  c. Pseudonym in the video game

- *Salience.* We wonder here about the importance of the playing activity compared to other usual ones in the person's mind.

  a. long sessions of games often interfere with sleep
  b. I usually think about my next game session when I do not play
  c. I think the game has become the most time-consuming activity in my life

- *Mood modification.* We report if the mood of a person changes when playing.

  a. I never play games to feel better
  b. I play to help me deal with any negative feelings I may have
  c. I play to forget everything that bothers me.

- *Tolerance.* This is more about the time spent in the game and if this one increase.

a. I significantly increased play time during the last year
b. I have to spend more and more time playing
c. I often think that a full day is not enough to do everything I need to do in the game

- *Withdrawal symptoms.* These are the mood symptoms that declare themselves when the person is not able to play as usual.

a. When I don't play, I feel more irritable
b. I feel sad if I am not able to play
c. I tend to become anxious if I can't play for any reason

- *Conflict.* We refer here to all possible conflicts, either with other people but also with other activities, or even inside the person's mind if they realize they might have a problem.

a. I lost interest in other hobbies because of my game
b. I lied to my family members because of the amount of time I spend playing
c. I think my game has jeopardized my relationship with my partner
d. I know that my main daily activity (work, education, responsibilities at home, etc.) has not been negatively affected by my game.
e. I believe my game harms important areas of my life.

- *Relapse.* When a person has realized he's an addict, succeeds to escape this addiction but comes back to it again.

a. I would like to reduce my playing time, but it is difficult
b. I don't think I could stop playing
c. I often try to play less, but I see I can't

The interviewees were able to answer on a Likert scale with the following values: Strongly disagree, Disagree, no opinion, Agree, Strongly agree.

**Evaluating the Survey Instrument.** We checked if the questions were understandable by asking some other people what they were understanding exactly. We made several testing to check the effectiveness of the follow-up procedures, and the reliability and validity of the questionnaire.

**Obtaining Valid Data.** On World of Warcraft, players are organized in guilds. Guilds participate in various events available in the game. These events require 40 players. So the logs that we were able to retrieve were flagged under the name of a specific guild (player group). We choose a specific guild called Inglorious, composed of 49 persons with a middle age of 31. Each of them was personally reached out and asked to fill out the form, and 43 accepted to do it.

**Analyzing the Data.** Once the questionnaire has been filled, we gave a ponderation to each of the answers as follows: Strongly agree *5, Agree *4, No opinion *3, Disagree

*2, and Strongly disagree *1. As a result, people having a score superior to 60 will be considered addicts, which was the case for 15 respondents. These results were communicated to the players who (quite honestly) validated the founding. This analysis answers the first research question.

## 3.2   Process Mining

We followed the first steps of the PM2 methodology [25] as shown below (Fig. 4).

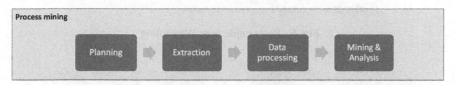

**Fig. 4.** Process mining methodology

**Planning.** The objective is to discover process models in the WoW game, to be able to compare the behaviors of addicts or non-addict people.

**Extraction.** There is a website[3] that provides a tool to record the logs of a WoW group to analyze the actions done by each player. Via this site, it is possible to analyze all the uploader data for a group, and offers different player-by-player metrics (who did the most damage, the most care, which skill was executed, and so on). The real objective of this site is to offer metrics on who does better and thus offer competition between players. We downloaded one complete month of data for the chosen guild whose members answered our survey.

**Data Processing.** The obtained CSV files only give two columns: the timestamp and a text chain (Fig. 5). This text chain follows several patterns (Fig. 6) that can be defined as follows: P1 (Name - Skill - Target – Amount), P2 (Name - Class - Skill – Target), P3 (Name - Win - Bonus - Source – Skill), P4 (Name - Swings-At – Target). We decomposed each of these text chains to create several useful columns.

---

| 1 | Time | Event |
|---|------|-------|
| 2 | 00:00:03 | Stalfos Charge étourdissante Géant de lave 2 Immune |
| 3 | 00:00:03 | Stalfos casts Charge on Géant de lave 2 |
| 4 | 00:00:04 | Stalfos gains 15 Rage from Stalfos's Charge |
| 5 | 00:00:04 | Stalfos's Posture de combat fades from Stalfos |
| 6 | 00:00:04 | Stalfos gains Posture berserker from Stalfos |
| 7 | 00:00:04 | Stalfos casts Posture berserker |
| 8 | 00:00:04 | Stalfos swings at Géant de lave 2 |
| 9 | 00:00:04 | Stalfos Melee Géant de lave 2 318 Glancing |
| 10 | 00:00:06 | Stalfos casts Tourbillon |
| 11 | 00:00:06 | Stalfos Tourbillon Géant de lave 2 *682* |
| 12 | 00:00:06 | Stalfos Tourbillon Géant de lave 1 *534* |
| 13 | 00:00:07 | Stalfos gains Rafale from Stalfos |
| 14 | 00:00:07 | Stalfos's Rafale is refreshed by Stalfos |

**Fig. 5.** Log sample from Warcraftlog

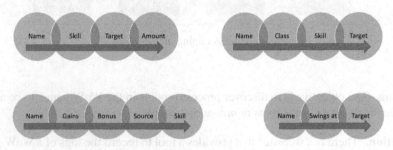

**Fig. 6.** Logs patterns

Another problem had to be solved in the logs export. We can export the logs of a fight (a fight is a very current event and does not represent much at the scale of the game session). To have a sufficiently consistent measurement we had to export a large number of CSV files to have the equivalent of a game time in hours. For example, extracting a player's logs for a game time of 2h23 on 16 October 2019 requires 98 CSV files. We then had to merge all the CSV files into a single file.

**Mining and Analysis.** We used Disco to discover the generic process model of the game players. However this was not significant as, in WoW, each player is a member of a specific class and we know that the player's role behavior is strongly dependent on this class. We then decided to analyze each class behavior separately and discovered the corresponding process models. This gives some insights into the real sequences between the various possible activities of each class member, which can help players to adapt their behavior when sharing a session with other people.

### 3.3 Cross Analysis

In this part, we tried to identify some relationship between the results of the survey and the process models obtained in the logs.

Let's take the example of the Warrior class. We had 6 players playing this role in our dataset. We used process mining techniques to identify the behavior of the class and the behavior of each of these 6 persons for all the quests. We then compared the models and the questionnaire results. Figure 7 shows the behavior obtained for the most addicted player in the group.

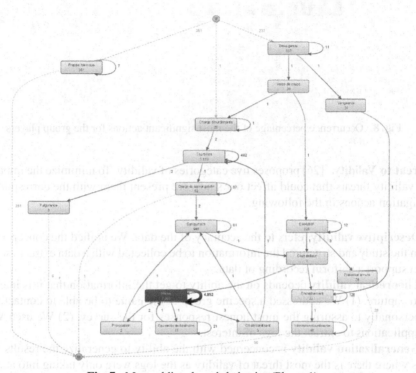

**Fig. 7.** Most addict player's behavior (Player 1)

However, when looking at the generic warrior discovered process model or all the behaviors of the class players, no conclusion can be made. The behaviors are quite different from one player to the other and we found that addict players don't play in the same way as the other addicts. The same conclusion can be drawn with all the different classes of the game.

Disco software provides several metrics such as the occurrence of each event. We extracted these metrics to analyze them (Fig. 8). Let's consider the first Melee event. This event is the basic attack of the Warrior class and means that as long as the player is in range he will attack automatically. We note however that the distribution between automatic attacks and non-automatic events is different depending on the players. Indeed, player 1 and player 2 scores mean that they spent this percentage of time letting their character attack alone. Unfortunately, looking at these metrics doesn't allow us to identify a link with addiction: players 1, 3, or 5 are addict players in this group but some of their metrics are similar to the non-addict players 2, 4, or 6.

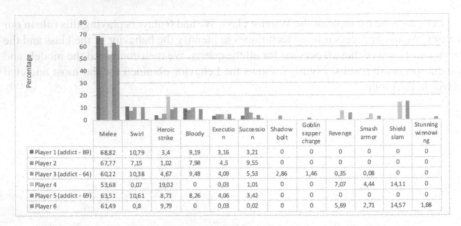

**Fig. 8.** Occurrence percentage of the most significant actions for the group players

| | Melee | Swirl | Heroic strike | Bloody | Execution | Succession | Shadow bolt | Goblin sapper charge | Revenge | Smash armor | Shield slam | Stunning winnowing |
|---|---|---|---|---|---|---|---|---|---|---|---|---|
| ■ Player 1 (addict - 89) | 68,82 | 10,79 | 3,4 | 9,19 | 3,16 | 3,21 | 0 | 0 | 0 | 0 | 0 | 0 |
| ■ Player 2 | 67,77 | 7,15 | 1,02 | 7,98 | 4,5 | 9,55 | 0 | 0 | 0 | 0 | 0 | 0 |
| ■ Player 3 (addict - 64) | 60,22 | 10,38 | 4,67 | 9,48 | 4,09 | 5,53 | 2,86 | 1,46 | 0,35 | 0,08 | 0 | 0 |
| ▩ Player 4 | 53,68 | 0,07 | 19,02 | 0 | 0,03 | 1,01 | 0 | 0 | 7,07 | 4,44 | 14,11 | 0 |
| ■ Player 5 (addict - 69) | 63,51 | 10,61 | 8,71 | 8,26 | 4,06 | 3,42 | 0 | 0 | 0 | 0 | 0 | 0 |
| ■ Player 6 | 61,49 | 0,8 | 9,79 | 0 | 0,03 | 0,02 | 0 | 0 | 5,69 | 2,71 | 14,57 | 1,68 |

**Threat to Validity.** [26] proposes five categories of validity. To minimize the impact of the validity threats that could affect our study, we present them with the corresponding mitigation actions in the following.

- **Descriptive validity** refers to the accuracy of the data. We unified the concepts used in the study and structured the information to be collected with a data extraction form to support a uniform recording of data.
- **Theoretical validity** depends on the ability to get the information that it is intended to capture. (1) We addressed a specific group in the game to be able to contact them personally to assuring the most honest responses for the survey. (2) We used WoW applications to extract the logs of interest.
- **Generalization validity** is concerned with the ability to generalize the results. This is where there is the most threat of validity as the logs were only taking into account the actions made for specific quests, not all the data from each player when he was playing the game (the tool used to record the logs had to be started by the player when a quest is launched to record its statistics, which he doesn't do if there is no quest on the run). Moreover, concentrating the dataset on only one group restricts the number of players in each class, which doesn't give enough data to use correctly process mining techniques.
- **Evaluative validity** is achieved when the conclusions are reasonable given the data. (1) Two researchers studied the results independently to identify potential analysis differences. (2) At least two researchers validated every conclusion.
- **Transparency validity** refers to the repeatability of the research protocol. The research process protocol is detailed enough to ensure it can be exhaustively repeated.

## 4   Conclusion

In this work-in-progress paper, we focused on the study of addiction in video games using process mining techniques as a detection tool. We studied and specified what is an addiction and more precisely what is a video game addiction. As explained, addictions without an active chemical substance are different from others based on a 100% behavioral addiction. We then analyzed event logs on a video game known as particularly addictive (WoW). With a questionnaire, on a set of forty people playing this game, we were able to identify people as addicts or not. After exporting and analyzing their logs we cross-checked these results with the questionnaire responses to identify whether one or more metrics would be viable to say whether a person is addicted or not.

Unfortunately, according to this analysis, no available metrics were relevant. The process model obtained for each player is quite different from the others and no specific deviation can be found. Some addict players have very high metrics, as not addict payers. Some addict players have low rates, as not addict players. It was then impossible to define a typical behavior pattern that would highlight the addiction of one player compared to another.

Based on this initial analysis, we will extend our data set by collecting more data on this game by contacting other players of other groups. Process mining uses statistical techniques, so it is necessary to work on a lot of data to have objective results. It may also be appropriate to move towards another analysis type. In this type of game we know that players try to reproduce a cycle of action that is as optimal as possible depending on their goal (e.g.: doing the most damage, giving the most care, and so on). A second hypothesis would be to identify the optimal process and measure the gap between this prescribed process and the player's behavior. We could see if the addict players manage to better respect the optimal path or not and thus can highlight one or more metrics to identify these players.

# References

1. IGI Global Dictionary: What is video game. https://www.igi-global.com/dictionary/video-game/31541
2. Buss, S., Becker, D., Daniels, M., Nöldeke, G., Blumtrit, C., Striapunina, K.: Economy Compass, Statista (2019)
3. Casati, F., Dayal, U., Sayal, M., Shan, M.: Business process intelligence. Comput. Ind. **53**, 321–343 (2004)
4. Griffiths, M.D., Meredith, A.: Videogame addiction and its treatment. J. Contemp. Psychother. **39**, 247–253 (2009)
5. van der Aalst, W.M.P., Weske, M. (eds.): Business Process Management: BPM 2003, vol. 2678. Springer, Heidelberg (2003). https://doi.org/10.1007/3-540-44895-0
6. Van der Aalst, W.M.P.: Process Mining: Data Science in Action. Springer, Heidelberg (2016). https://doi.org/10.1007/978-3-662-49851-4
7. Petry, N., Rehbein, F., Ko, C.-H.: Internet gaming disorder and the DSM-5. Curr. Psychiatry Rep. **17**(9) (2014)
8. Nelson, R.: Fortnite Revenue on iOS Hits $300 Million in 200 Days, sensor tower blog (2018)
9. Billieux, J., Deleuze, J., Griffiths, M.D., Kuss, D.J.: Internet Gaming Addiction: The Case of Massively Multiplayer Online Role-Playing Games. Textbook of Addiction Treatment: International Perspectives. Springer, Milano (2015)
10. Tiwari, A., Turner, C.J., Majeed, B.: A review of business process mining: state-of-the-art and future trends. Bus. Process Manag. J. **14**(1) (2008)

11. Mahendrawathi, E.R., Astuti, H.M., Nastiti, A.: Analysis of customer fulfilment with process mining: a case study in a telecommunication company. Procedia Comput. Sci. J. **72**, 588–596 (2015)

12. Jensen, B., Chen, J., Furnish, T., Wallace, M.: Medical marijuana and chronic pain: a review of basic science and clinical evidence. Curr. Pain Headache Rep. **19**(10), 1–9 (2015). https://doi.org/10.1007/s11916-015-0524-x

13. Décamps, G., Battaglia, N., Idier, L.: Élaboration du Questionnaire de mesure de l'intensité des conduites addictives (QMICA): évaluation des addictions et co-addictions avec et sans substances. Psychol. Fr. **55**(4), 279–294 (2010)

14. Whiting, P.F., et al.: Cannabinoids for medical use: a systematic review and meta-analysis. J. Am. Med. Assoc. (JAMA) **313**(24), 2456–2473 (2015)

15. Tejeiro Salguero, R.A., Bersabé Morán, R.M.: Measuring problem video game playing in adolescents, research report (2002)

16. Saunders, J.B., et al.: Gaming disorder: Its delineation as an important condition for diagnosis, management, and prevention. J. Behav. Addict. **6**(3), 271–279 (2017)

17. Plessis, C., Altintas, E., Guerrien, A.: Addiction aux jeux vidéo en ligne: étude comparative des outils de mesure en langue française. Ann Med Psychol (Paris) (2018)

18. Griffiths, M.: Addiction sans drogue, quand le cerveau a le goût du jeu. Adolescence J. **79**(1), 51–55 (2012)

19. Kandell, J.J.: Internet addiction on campus: the vulnerability of college students. CyberPsychology Behav. **1**(1) (2009)

20. Griffiths, M.D., Kuss, D.J., Demetrovics, Z.: Social networking addiction: an overview of preliminary findings. In: Behavioral Addictions: Criteria, Evidence and Treatment (2014)

21. American Psychiatric Association: Diagnostic and statistical manual of mental disorders: DSM-5. Washington, D.C: American Psychiatric Association. (Under Alcohol Use Disorders) (2013)

22. Lemmens, J.S., Valkenburg, P.M., Peter, J.: Development and validation of a game addiction scale for adolescents. Media Psychol. **12**, 1 (2009)

23. Sheppard, V.: Research Methods for the Social Sciences: An Introduction (2002)

24. Kitchenham, B.A., Pfleeger, S.L.: Personal opinion surveys. In: Shull, F., Singer, J., Sjøberg, D.I.K. (eds.) Guide to Advanced Empirical Software Engineering, pp. 63–92. Springer, London (2008). https://doi.org/10.1007/978-1-84800-044-5_3

25. van Eck, M.L., Lu, X., Leemans, S.J.J., van der Aalst, W.M.P.: PM$^2$: a process mining project methodology. In: Zdravkovic, J., Kirikova, M., Johannesson, P. (eds.) CAiSE 2015. LNCS, vol. 9097, pp. 297–313. Springer, Cham (2015). https://doi.org/10.1007/978-3-319-19069-3_19

26. Thomson, S.B.: Qualitative research: validity. JOAAG **6**(1) (2011)

# Evaluating Greek Government Digital Distance Learning Policies in Higher Education for the Covid-19 Period

Niki Kyriakou, Nikolaos Kompos, Euripidis Loukis$^{(\boxtimes)}$, and Theodoros Leoutsakos

University of the Aegean, 83200 Karlovasi, Samos, Greece
eloukis@aegean.gr

**Abstract.** The Greek government rapidly after the outbreak of the Covid-19 pandemic in order to mitigate its spread adopted for all universities a central policy of shutdown, and at the same time of continuing their educational activities through asynchronous and synchronous online teaching, based on e-learning platforms and online educational material. This study aims to evaluate these government policies, based on public policy evaluation theory, with respect to both their direct outputs, meant as educational resources (technological and human) provided to the students, as well as educational outcomes. For this purpose, evaluation data have been collected through a survey of 269 undergraduate students of the Department of Information and Communication Systems Engineering of the University of Aegean. The results show that the participants in this survey were neutral to satisfied with the digital educational resources (technological and human) provided to them during the Covid-19 period. With respect to the educational outcomes the participants perceive a slightly level of understanding the online lectures in comparison with the traditional face-to-face ones, but a similar level of concentration. Finally, the extent of online participation of them in the exams of the theory and the laboratories has been large to very large.

**Keywords:** distance learning · digital learning · e-learning · synchronous e-learning · asynchronous e-learning · policy evaluation · higher education · university

## 1 Introduction

Distance education is the learning process in which instructors and learners are not physically present at the educational institution (Kaplan and Haenlein 2016). The flexibility of distance education curricula allows more learners to participate in the educational process, as it reduces the barrier of limited time imposed by personal responsibilities and commitments, and geographical distance, and provides access to the educational process to people from many different geographical areas and socio-economic backgrounds (Oblinger 2000; Masson 2014). Distance education programs can be both innovative (Masson 2014), and just as effective as the traditional face-to-face learning programs (Nguyen 2015), especially if the former is conducted using digital means (e-learning

© The Author(s), under exclusive license to Springer Nature Switzerland AG 2023
M. Papadaki et al. (Eds.): EMCIS 2022, LNBIP 464, pp. 365–380, 2023.
https://doi.org/10.1007/978-3-031-30694-5_27

technologies), and the instructor has specialized knowledge and experience in online teaching methods, which are quite different from the traditional ones (Masson 2014). Furthermore, most textbooks are available in digital forms. Also, all e-learners have equal access to the digital education, regardless of demographic factors, such as socio-economic status, place of residence, gender, origin, age, or tuition (Casey and Lorenzen 2010). Digital distance education, in comparison with the traditional education, enables learners to learn in the most appropriate for each of them way and pace, to follow the courses according to their needs and background, focusing on and spending more time in subjects in which they have weaker knowledge (Kirtman 2009). Through asynchronous and synchronous digital distance learning, students can have continuous access to the educational material at the time and pace they desire, and also can have collaboration (e.g., for group assignments and projects) flexibly (Masson 2014). Digital distance education can be highly beneficial for learners and instructors during a pandemic, as there they do not have to move to the educational institution and have direct contact, which contributes to the mitigation of the transmission of the disease (Masson 2014).

However, in addition to the above advantages of digital distance education, asynchronous and synchronous, there are also some disadvantages. According to Al-Saleh (2013), the lack of direct interaction, cooperation, and communication between the learner and the instructor, makes it more difficult for learners to ask questions and additional information from the instructor, which has negative impact on the quality of the education of the former. In addition, weaknesses and problems of the technological infrastructure in educational institutions, especially with respect to the high speed and availability access of the students and instructors to the educational content, as well as to synchronous e-learning sessions, may lead to further problems in the education quality (Jawida et al. 2019).

The Greek government rapidly after the outbreak of the Covid-19 pandemic in order to mitigate its spread adopted a central policy for all universities of shutdown, and at the same time of continuing their educational activities through asynchronous and synchronous online teaching, based on e-learning platforms and online educational material (Bao 2020; Crawford et al. 2020). In particular, in early March 2020 a Legislative Act titled 'Urgent Measures for Handling the Negative Consequences of the Appearance of Covid-19' was issued by the Greek government (Government Gazette A, 55, March 11th, 2020), which in Article 12 included the shutdown of all Greek universities, and at the same time the continuation of their educational activities using digital distance learning methods. A few days later (16/3/2020) the Ministry of Education issued a relevant Administrative Circular titled 'Application of Distance Learning in Higher Education Institutions', which included guidelines for the practical implementation of digital distance learning, using not only asynchronous e-learning methods (upload educational content to an electronic platform), but also synchronous ones. For the universities that already had e-learning platforms with sufficient capacity this should start immediately, while the remaining ones were given quite strict deadlines for upgrading their e-learning platforms within a short time period, and also the option of using e-learning platforms that had been offered by Google and Microsoft.

It is quite important to evaluate the application of these digital distance learning policies of the Greek government for the universities during the Covid-19 period, which have been extensively and intensively debated (and there has been strong confrontation among politicians about them), and investigate the degree of their success or failure, and also identify their strengths and weaknesses. In general, it is important to gain as much knowledge as possible from the application of these digital distance learning policies of the Greek government in the universities during the Covid-19 period, which can be quite useful for making the required improvements of them. This is going to be highly beneficial, as digital distance learning is expected to be used extensively in the future in higher education, as both instructors and students have become more familiar with this education method: a) for postgraduate programs and continuous education; b) for the provision of cross-departmental courses (attended by students of several different departments, which are located in different cities); c) for the provision of higher education to special groups of students who cannot participate in the 'traditional' face-to-face learning processes.

This study aims to make a contribution in this direction: it evaluates these government policies of adopting digital distance learning in higher education during the Covid-19 period, based on public policy evaluation theory (Adelle and Weiland 2012; Wollmann 2016; Vedung 2017; Bundi and Trein 2022) (outlined in Sect. 2.1), with respect to both their direct outputs, meant as educational resources (technological and human) provided to the students, as well as educational outcomes. So, the main research objectives of this study are:

i)   to assess the quality of the main educational resources provided to the students during the Covid-19 period: the quality of the e-learning platform and the quality of 'e-instruction' (i.e. teaching the distant students by instructors though the e-learning platform);
ii)  to assess the educational outcomes of the digital distance learning during the Covid-19 period: level of understanding the online lectures as well as of concentration on them, in comparison with the traditional face-to-face teaching, and finally extent of online participation in the online exams.

For this purpose, evaluation data have been collected through a survey of undergraduate students of the Department of Information and Communication Systems Engineering of the University of Aegean, Greece. The statistical analysis of their responses allows drawing interesting and useful conclusions concerning their degree of satisfaction with the technological and human educational resources provided to the students during the Covid-19 period, as well as the level of understanding the online lectures as well as of concentration on them, and finally the extent of their participation in a critical element of the university courses: the examinations, which had been conducted online as well.

In the following Sect. 2 a brief review of representative relevant literature is provided, while in Sect. 3 the method and data of this study are described, followed by the results in Sect. 4 and finally the conclusions in Sect. 5.

## 2 Literature Review

### 2.1 Public Policy Evaluation

Since the policy interventions of modern state has become quite costly (consuming large amounts of taxpayers' money), complex to implement and also have high impacts on the economy and the society, their comprehensive and rational evaluation is an imperative, so there has been extensive research and practical effort in this direction (Adelle and Weiland 2012; Wollmann 2016; Vedung 2017; Bundi and Trein 2022). Public policy evaluation can be defined as 'careful retrospective assessment of the merit, worth, and value of administration, output, and outcome of government interventions, which is intended to play a role in future, practical action situations' (Vedung 2017). It is usually conducted after the end of the implementation of a public policy (ex-post evaluation) in order to assess its impact as well as the degree of attainment its objectives, and also identify weaknesses and possible improvements, and in general gain relevant knowledge, which can be useful for future relevant decision-making and for the design and implementation of similar policies in the future. Furthermore, it can be conducted also before the implementation of a public policy (ex-ante evaluation) in order to assess on one hand its costs and on the other hand its impacts, and examine whether the latter are worthy of the former, and also to assess and compare alternative courses of action.

A public policy requires some 'inputs' (including usually financial and human resources), which are used by the government agency(ies) responsible for this policy in order to produce some direct 'outputs'; these outputs affect the target group (e.g., some citizens or firms) of the public policy and produce some first-level 'outcomes' (meant as impacts and changes in their situation/behavior), and possibly some second-level elements, etc.; this 'structure' of a public policy is shown in Fig. 1; therefore the general methodology of the evaluation of a public policy has to follow this structure, and focus on these three main elements of it (inputs, outputs and outcomes) as well as the relationships among them (Vedung 2017).

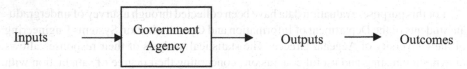

**Fig. 1.** The structure of public policy and its evaluation methodology

### 2.2 Evaluation of Digital Distance Learning in Higher Education

Some research has been conducted for the evaluation of the digital distance learning policies adopted by governments during the Covid-19 period in higher education. It is useful to review the most representative of these studies. Fabriz et al. (2021), based on a survey of 3056 students and 396 instructors from a large German university, investigated whether the dominance of synchronous or asynchronous online teaching and learning in higher education during the Covid-19 period has affected the whole experience of students as well as their performance/results. In addition, it examined how well these two online teaching and learning methods satisfy students' basic psychological needs for autonomy, competence, and relatedness suggested by self-determination theory. The results suggest that students who were taught mainly through synchronous methods reported more self-centered activities, such as feedback, than students taught mainly through asynchronous method. In contrast, teachers perceived fewer differences between these two teaching methods (synchronous and asynchronous online learning), especially concerning students' feedback activities.

A qualitative case study, assisted by an online survey, has been conducted by Irfan et al. (2020) aiming to identify the barriers that arise during online learning in the mathematics domain in higher education. The data was collected through an online survey consisting of 27 structured questions concerning basic skills challenges, teaching and learning challenges, and university challenges. Twenty-six Professors from universities in Sumatra, Java, Kalimantan, and Sulawesi who teach mathematics participated in the research. The results of this study reveal that all teachers used a Learning Management System (LMS)-based website as a means of online teaching: The learning management system-based platform is the most widely used (google class and Edmodo), while video conferencing is the second choice (Zoom and Skype). It has been concluded that there have been significant obstacles, such as the limitations of writing mathematical symbols and the limited basic capabilities of the system and multimedia software to support online learning.

Baxter and Hainey (2022) explore views of students of a UK higher education institution concerning distance online delivery. Students were asked about their views on distance learning and the psychological impact it had on students and students' studies. The research provided students with an opportunity to reflect on whether the practice of providing distance education continues to provide students with a beneficial learning experience. The research adopted a case study methodology using questionnaires; in total, 894 students completed the questionnaire. The survey findings showed that some participants felt that distance learning was beneficial for immediate feedback, motivational support, and encouragement. The negative findings identified consequences for feeling isolated and unmotivated and a preference for face-to-face delivery.

However, further research is required concerning the digital distance learning in the higher education during the Covid-19 period, in different national contexts and thematic disciplines, in order to obtain extensive knowledge about various aspects of it, which can provide a strong basis for reaching higher levels of maturity of it.

## 3 Method and Data

The method of this study was based on the theory of public policy evaluation outlined in Sect. 2.1, and especially on the general methodology of public policy evaluation shown in Fig. 1. We have focused our evaluation:

a) on the outputs that have been provided to the students as part of this government digital distance learning policy during the Covid-19 period: the technological resources (the e-learning platform) provided to them as well as the human resources (the 'e-instruction', meant as distant teaching by instructors though the e-learning platform) provided to them;

b) and on the educational outcomes of the digital distance learning during the Covid-19 period: the level of understanding the online lectures as well as the level of concentration on them, in comparison with the traditional face-to-face teaching and also the extent of online participation in the online exams (as the exams constitute the final and highly important stage of a university course, and the extent of students' participation in these online exams is significantly affected by - and is a good measure of - the level of learning they have achieved through the online teaching of theory and labs during the semester - if students feel that they have not gain sufficient knowledge through the online teaching they will probably have lower propensity to participate in the online exams;

Furthermore, we have examined the relationships between the above policy out-puts and outcomes (in order to investigate which of the former affect the latter). Our evaluation method is shown below in Fig. 2.

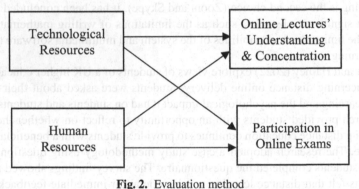

**Fig. 2.** Evaluation method

Data were collected through a questionnaire (provided in the Appendix) from students of the Department of Information and Communication Systems Engineering of the University of the Aegean after the first wave of coronavirus. It included questions concerning the degree of satisfaction of students with:

– the technological resources provided them for digital distance learning: "The quality of the e-learning platforms", "The response of the technical support to problems/malfunctions of the distance learning platforms" and "The information provided to students about the use of the various distance learning platforms";
– the human resources: the online theory teaching, and the online conduct of the laboratories (it should be mentioned that as mentioned above the context of our study was a Department of Information and Communication Systems Engineering, so the laboratories of the courses concerned programming and/or use of sophisticated software, therefore it was possible to be conducted online); and also "The consistency of the

teachers and the response to their obligations", "The encouragement of students by teachers for active participation in the courses" and "The academic secretariat of the Department";

– the educational outcomes: the level of understanding the online lectures as well as the ease of concentration on them, in comparison with the traditional face-to-face teaching; and also the extent of online participation in the online exams of the theory and in the laboratories.

Also, the questionnaire included in the beginning questions about some respondent's demographic characteristics: gender, year of birth, year of study and previous use of video conferencing platforms before the start of this digital distance learning in the Covid-19 period.

We received valid questionnaires from a sample of 269 students. The sample consisted of 73.98% men, and the mean age was 23.21 (Std = 2.60) years, while the average of their year of study was 3.94 (Std = 2.52) years of study. Most of them (81.78%) had used a videoconferencing platform before.

Using these data we tested the above variables for normality using the Kolmogorov-Smirnov and Shapiro-Wilk normality tests; according to the results the hypothesis of normality of distribution is rejected for all variables. As a result, the non-parametric Mann-Whitney U test and the non-parametric Spcarman correlation coefficients were used to investigate possible correlations among them.

## 4  Results

### 4.1  Descriptive Statistics

In Table 1 we can see the descriptive statistics (average and standard deviation) of the variables that measure students' degree of satisfaction with the technological and human resources provided to them as part of the digital distance learning policies during the Covid-19 period.

We can see that the respondents are neutral to satisfied, however being closer to the latter, with the online theory teaching (Average = 3.65, Std = 1.03) and the online conduct of the laboratories (Average = 3.65, Std = 1.09) during the Covid-19 period. Therefore, students seem to be rather satisfied with the online conduct of the theory teaching and the laboratories, but only to some degree, and definitely not completely satisfied. The reasons for this can be identified by examining the average values with the next six variables shown in the same Table 1. We can see that the respondents are on average satisfied, or between neutral and satisfied, but closer to the latter, with "the response of the technical support to problems/malfunctions of the e-learning platforms" (Average = 4.10, Std = 0.93), "the information provided to the students on the use of the e-learning platforms" (Average = 3.75, Std = 0.96) and "the quality of the e-learning platforms (technical problems)" (Average = 3.68, Std = 0.99), which all concern the technological resources provided to the students as part of the digital distance learning policies during the Covid-19 period. On the contrary there is a lower degree of satisfaction, between neutral and satisfied but closer to the former, with "the encouragement of students by the teachers for active participation in the courses" (Average = 3.31, Std =

**Table 1.** Descriptive statistics for the variables measuring students' satisfaction with the technological and human resources provided to them during the Covid-19 period.

| Variable (Satisfaction with) | Mean | Std |
|---|---|---|
| Online theory teaching | 3.65 | 1.03 |
| Online laboratories conduct | 3.65 | 1.09 |
| The consistency of the teachers and the response to their obligations | 3.26 | 1.25 |
| The quality of the e-learning platforms (technical problems) | 3.68 | 0.99 |
| The encouragement of students by the teachers for active participation in the courses | 3.31 | 0.97 |
| The response of the technical support to problems/malfunctions of the e-learning platforms | 4.10 | 0.93 |
| The information provided to the students about the use of the e-learning platforms | 3.75 | 0.96 |
| The academic secretariat of the department | 3.13 | 1.25 |

0.97), "the consistency of the teachers and the response to their obligations" (Average = 3.26, Std = 1.25) and "the academic secretariat of the department" (Average = 3.13, Std = 1.25), 54.65%), which all concern the human resources provided to the students as part of the digital distance learning policies during the Covid-19 period. These indicate the lack of experience and knowledge: a) on one hand of the teaching staff about effective methods and practices of online conduct of theory teaching and laboratories, as well as ways of motivating and encouraging the remote students to be more engaged, participate and not remain passive; and b) on the other hand of the staff of the academic secretariat about the administration of this online conduct of the courses and the effective communication with the remote students. On the contrary, the technical staff managed to provide a high-quality operation of the e-learning platforms, as well as of technical support of them concerning the management and recovery of their problems/malfunctions.

In Table 2 we can see the descriptive statistics of educational outcomes' variables, which concern the level of understanding the online lectures as well as of the ease of concentration on them, in comparison with the traditional face-to-face teaching. We can see that on average there is disagreement to neutrality (however closer to the latter) to the statement "I can understand better the online lectures than the traditional face-to-face ones" (Average = 2.82, Std = 0.83); also, there is on average neutrality to the statement "It is easier for me to concentrate on the online lectures than in the traditional face-to-face ones" (Average = 2.99, Std = 1.55). These findings indicate that the participants perceive a slightly lower level of understanding the online lectures in comparison with the traditional face-to-face ones, and a similar level of concentration.

**Table 2.** Descriptive statistics for the understanding and concentration variables.

| Variable | Mean | Std |
|---|---|---|
| I can understand better the online lectures than the traditional face-to-face ones | 2.82 | 1.28 |
| It is easier for me to concentrate on the online lectures than in the traditional face-to-face ones | 2.99 | 1.55 |

In Table 3 we can see the descriptive statistics of. We can see a high to very high level of "Participation in online exams of theory" (Average = 4.57, Std = 0.83), as well as "Participation in online exams of laboratories" (Average = 4.22, Std = 1.21). This extensive participation reflects on one hand the need and pressure that the students feel to pass as many courses as possible during these difficult times, and not lag behind in their studies, but on the other hand it also reflects their perception that they had gained sufficient knowledge through the online teaching, so they had a good chance of passing the exams.

**Table 3.** Descriptive statistics for the participation in online exams' variables.

| Variable | Mean | Std |
|---|---|---|
| Participation in online exams of theory | 4.57 | 0.83 |
| Participation in online exams of laboratories | 4.22 | 1.21 |

## 4.2 Effects of Demographic Characteristics

Next, we examined whether students' satisfaction levels with online teaching and online laboratories conduct, as well as with overall online courses organization (measured through the average of remaining six variables that measure students' satisfaction with these important aspects of the organization and technology of the online courses), is affected by gender. For this purpose, the non-parametric Mann-Whitney U-test was performed for a significance level of $\alpha = 0.05$, and the results are shown in Table 4. It can be concluded that the gender of the respondent does not affect "Satisfaction with online theory teaching" ($Z = -1.183$, $p = 0.237 > 0.05$), "Satisfaction with online laboratories' conduct" ($Z = -1.106$, $p = 0.269 > 0.05$) and "Satisfaction with online courses' organization and technology" ($Z = -1.513$, $p = 0.130 > 0.05$).

Accordingly, in order to investigate whether the above three satisfaction variables are affected by prior use of videoconferencing platforms (before the start of the digital distance learning in the Covid-19 period) we performed similar non-parametric Mann-Whitney U-tests, and their results are shown in Table 5. It can be concluded that the prior use of videoconferencing platforms does not affect "Satisfaction with theory teaching" ($Z = -1.443$, $p = 0.149 > 0.05$), "Satisfaction with laboratories" ($Z = -1.608$, $p = 0.108 > 0.05$) and "Satisfaction with courses' organization and technology" ($Z = -1.324$, $p = 0.185 > 0.05$) (Table 4), for $\alpha = 0.05$.

**Table 4.** Results of non-parametric Mann-Whitney U-tests – effect of gender on satisfaction with online theory teaching, satisfaction with online laboratories' conduct, satisfaction with online courses' organization and technology.

|  | Mann-Whitney U-test | |
| --- | --- | --- |
|  | Z | p |
| Satisfaction with only theory teaching | -1,183 | ,237 |
| Satisfaction with online laboratories' conduct | -1,106 | ,269 |
| Satisfaction with online courses' organization and technology | -1,513 | ,130 |

**Table 5.** Results of non-parametric Mann-Whitney U-tests – Effect of prior use of video conferencing platforms -on satisfaction with online theory teaching, satisfaction with online laboratories' conduct, satisfaction with online courses' organization and technology.

|  | Mann-Whitney U-test | |
| --- | --- | --- |
|  | Z | p |
| Satisfaction with online theory teaching | -1,443 | ,149 |
| Satisfaction with online laboratories conduct | -1,608 | ,108 |
| Satisfaction with online courses' organization and technology | -1,324 | ,185 |

Finally, we investigated whether these three satisfaction variables are affected by respondent's age and year of study; for this purpose, we calculated Spearman's Rho non-parametric correlation coefficients, using the significance level of $\alpha = 0.05$, and the results are shown in Table 6. We can see that age has a statistically significant weak positive correlation with "Satisfaction with online theory teaching" (Rho = 0.200, p = 0.001 < 0.05) and "Satisfaction with online laboratories' conduct" (Rho = 0.149, p = 0.017 < 0.05); at the same time, there is no statistically significant correlation with the "Satisfaction with courses' organization and technology" (Rho = 0.087, p = 0.154 > 0.05). With respect to the year of study, it shows a statistically significant weak positive correlation with "Satisfaction with online theory teaching" (Rho = 0.178, p = 0.004 < 0.05) and "Satisfaction with laboratories" (Rho = 0.139, p = 0.027 < 0.05), while there is no statistically significant correlation with "Satisfaction with online courses' organization and technology" (Rho = 0.076, p = 0.213 > 0.05). Therefore, it is concluded that as age and year of study increase, satisfaction with online theory teaching and online laboratories' conduct tends to increase weakly, probably because of the increasing maturity and overall ability of students to cope with the inherent disadvantages of distance education (lack of co-location of learners and instructors); on the contrary, the satisfaction with the organization and technology of the online courses is not affected by age and year of study.

**Table 6.** Results of non-parametric Spearman's Rho calculations of age and year of study with satisfaction with online theory teaching, satisfaction with online laboratories' conduct, satisfaction with online courses' organization and technology.

|  |  |  | Age | Year of Study |
|---|---|---|---|---|
| Spearman's rho | Satisfaction with online theory teaching | Rho | ,200 | ,178 |
|  |  | p | ,001 | ,004 |
|  |  | N | 266 | 266 |
|  | Satisfaction with online laboratories' conduct | Rho | ,149 | ,139 |
|  |  | p | ,017 | ,027 |
|  |  | N | 254 | 254 |
|  | Satisfaction with online courses' organization and technology | Rho | ,087 | ,076 |
|  |  | p | ,154 | ,213 |
|  |  | N | 269 | 269 |

The same analysis was made for the educational outcomes' variables that measure the perceived level of understanding of the online lectures as well as of ease of concentration on them, in comparison with the traditional face-to-face teaching. We found than male students have a higher level of both in comparison with female students; also, both have weak statistically significant correlation with age and year of study, but no statistically significant correlation with the prior use of videoconferencing platforms.

### 4.3 Relationships Between the Policy Outputs and Outcomes

Finally, we investigated the effects of the examined policy outputs (technological and human resources) on the policy outcomes: a) on the average of the perceived level of understanding the online lectures and the perceived level of ease of concentration on them, in comparison with the traditional face-to-face teaching, and b) the average of the degrees of participation in the online theory exams and in the online laboratories' exams. For this purpose, we calculated Spearman's Rho non-parametric correlation coefficients of these two average variables with the abovementioned eight policy output variables; the results are shown below in Tables 7 and 8 (the statistically significant values are shown in bold).

We can see that all the examined policy outputs (technological and human resources), with the only exception of the academic secretariat of the Department) have positive effects on the average of the perceived level of understanding the online lectures and the perceived level of ease of concentration on them, in comparison with the traditional face-to-face teaching (level of educational outcomes). The quality of (degree of students' satisfaction with) the online theory teaching and the online laboratories conduct have the strongest effects, followed by the level of encouragement of students by teachers for active participation in the courses; this indicates that the quality of the human resources provided to the students during the Covid-19 period (= the 'e-instruction', meant as

**Table 7.** Spearman's Rho non-parametric correlation coefficients of the average level of understanding the online courses and concentrating on them with the policy output variables

| Policy outcome variable | Spearman's rho | Sig |
|---|---|---|
| The consistency of the teachers and the response to their obligations | ,280 | ,000 |
| The quality of the e-learning platforms | ,263 | ,001 |
| The encouragement of students by teachers for active participation in the courses | ,377 | ,000 |
| The response of the technical support to problems/malfunctions of the e-learning platforms | ,295 | ,000 |
| The information provided to the students on the use of the e-learning platforms | ,258 | ,000 |
| The academic secretariat of the Department | ,106 | ,175 |
| Satisfaction with the online theory teaching | ,450 | ,000 |
| Satisfaction with the online laboratories conduct | ,417 | ,000 |

distant teaching by instructors though the e-learning platform) affects most the level of the educational outcomes of these digital education policies in higher education.

**Table 8.** Spearman's Rho non-parametric correlation coefficients of the average degree of participation in the theory and laboratories exams variable with the policy output variables

| Policy outcome variable | Spearman's rho | Sig |
|---|---|---|
| The consistency of the teachers and the response to their obligations | ,015 | ,807 |
| The quality of the e-learning platforms | ,135 | ,027 |
| The encouragement of students by teachers for active participation in the courses | ,078 | ,203 |
| The response of the technical support to problems/malfunctions of the e-learning platforms | ,202 | ,001 |
| The information provided to the students on the use of the e-learning platforms | ,107 | ,081 |
| The academic secretariat of the Department | ,068 | ,266 |
| Satisfaction with the online theory teaching | ,117 | ,050 |
| Satisfaction with the online laboratories conduct | ,155 | ,011 |

We can see that the response of the technical support to problems/malfunctions of the e-learning platforms, the quality of (degree of students' satisfaction with) the online laboratories conduct and the quality of the e-learning platforms have the strongest positives effects on the extent of students' participation in the online exams, followed by the quality (degree of students' satisfaction with) the online theory teaching as well as the information that had been provided to them about the use of the e-learning platforms.

# 5 Conclusion

The Greek government rapidly after the outbreak of the Covid-19 pandemic, in order to mitigate its spread, adopted a central policy for all universities of shutdown, and at the same time of continuing their educational activities using digital distance learning technologies and methods. In the previous sections has been presented an evaluation of these government policies; it has been based on public policy evaluation theory, which distinguishes between direct policy outputs and policy outcomes. In this direction have been evaluated both the direct outputs of these policies (i.e. the educational resources, both technological and human ones, provided to the students), and its educational outcomes (level of understanding the online lectures as well as of concentration on them, in comparison with the traditional face-to-face teaching, and also extent of online participation in the online exams). Evaluation data have been collected through a survey of 269 undergraduate students of the Department of Information and Communication Systems Engineering of the University of Aegean.

The results of the analysis of the data we collected indicated a moderate to good level of success of these policies with respect to the educational resources provided to the students during the Covid-19 period, which however was not a complete success; this difficult (but absolutely necessary) undertaking worked to some extent, but had also some important weaknesses, which concerned "the encouragement of students by the teachers for active participation in the courses", "the consistency of the teachers and the response to their obligations" and "the academic secretariat of the department". These weaknesses reflect the lack of experience and knowledge of the teaching staff about effective methods and practices of online conduct of theory teaching and laboratories, as well as ways of motivating and encouraging the remote students to be more engaged, participate and not remain passive. Also, they reflect the lack of experience and knowledge of the administrative staff of the academic secretariat about the administration of this online conduct of the courses and the effective communication with the remote students.

Therefore, since as mentioned in the Introduction digital distance learning is expected to be used extensively in the future in higher education (e.g. for postgraduate programs and continuous education, for providing shared courses attended by students of several geographically remote departments), it is necessary to provide sufficient training to the teaching staff about the effective conduct of digital distance learning, and especially the motivation, encouragement and engagement of the remote students. Also, it is necessary to provide sufficient training to the administrative staff about the effective administration of this digital distance learning, in which the students are not physically present in the same geographical location.

With respect to the educational outcomes the results indicate a satisfactory level of success of these policies: the participants perceive a slightly lower level of understanding the online lectures in comparison with the traditional face-to-face ones, and a similar level of concentration. Furthermore, there has been extensive participation in the most critical element of these online courses, their online examinations of the theory and the laboratories of the courses, which probably reflects a positive students' perception about their learning from this digital conduct of the courses: their feeling that they had gained sufficient knowledge through the online teaching for having a good chance of being successful in the exams.

Our study has two main limitations. The first limitation is that it is dealing only with the learners' perspective, but not with the instructors' perspective (who are more experienced, so they can provide a more substantial and in-depth evaluation of these digital distance learning policies in higher education during the Covid-19 period, and identification of weaknesses that have to be addressed). So, it is necessary to conduct similar evaluation research in the future from the instructors' perspective, using both quantitative methods (questionnaire-based) as well as qualitative ones (e.g. based on interviews and focus-groups). The second limitation is that our study has been based on a survey of undergraduate students of a Department of Information and Communication Systems Engineering, whose instructors and students are quite familiar with the use of digital technologies, and also its laboratories can be conducted online. So, it is necessary to conduct similar evaluation research in other types of departments, in which instructors and students have lower familiarity with the use of digital technologies, and also it is not possible to conduct the laboratories online (e.g. this might need more sophisticated simulation approaches).

## Appendix: Questionnaire

1) Gender (Male, Female, Other)
2) Year of birth
3) Year of study (1, 2, 3, 4, 5, 6, 7, 8, 9, 10, >10)
4) Were you using video conferencing platforms before the start of this digital distance learning in the Covid-19 period? (Yes, No)
5) How satisfied are you with the organization of remote digital:

    a. theory teaching (Very Disappointed, Disappointed, Neutral, Satisfied, Very Satisfied)

    b. conduct of laboratories (Very Disappointed, Disappointed, Neutral, Satisfied, Very Satisfied)

6) How satisfied are you with:

    a. The consistency of the teachers and the response to their obligations (Very Disappointed, Disappointed, Neutral, Satisfied, Very Satisfied)

    b. The quality of the e-learning platforms (technical problems) (Very Disappointed, Disappointed, Neutral, Satisfied, Very Satisfied))

    c. The encouragement of students by the teachers for active participation in the courses (Very Disappointed, Disappointed, Neutral, Satisfied, Very Satisfied)

    d. The response of the technical support to problems/malfunctions of the e-learning platforms (Very Disappointed, Disappointed, Neutral, Satisfied, Very Satisfied)

    e. The information provided to the students about the use of the e-learning platforms (Very Disappointed, Disappointed, Neutral, Satisfied, Very Satisfied)

    f.  The academic secretariat of the Department (Very Disappointed, Disappointed, Neutral, Satisfied, Very Satisfied)

7) To what extent do you agree or disagree with the following statements:

    a.  I can understand better the online lectures than the traditional face-to-face ones (Totally Disagree, Disagree, Neutral, Agree, Totally Agree)

    b.  It is easier for me to concentrate in the online lectures than in the traditional face-to-face ones (Totally Disagree, Disagree, Neutral, Agree, Totally Agree)

8) To what extent did you participate on the remote online examinations of:

    a.  the theory (Not at All, To a Small Extent, To a Moderate Extent, To a Large Extent, To a Very Large Extent)

    b.  the laboratories (Not at All, To a Small Extent, To a Moderate Extent, To a Large Extent, To a Very Large Extent)

# References

Adelle, C., Weiland, S.: Policy assessment: the state of the art. Impact Assess. Project Appraisal **30**(1), 25–33 (2012)

Al-Saleh, B.A.: Critical issues in e-learning distance education model. In: Proceedings of the Third International Conference for e-Learning via Distance Learning, Riyadh, Saudi Arabia (2013)

Bao, W.: COVID-19 and online teaching in higher education: a case study of Peking University. Hum. Behav. Emerg. Technol. **2**, 113–115 (2020)

Baxter, G., Hainey, T.: Remote learning in the context of COVID-19: reviewing the effectiveness of synchronous online delivery. J. Res. Innov. Teach. Learn. vol. ahead-of-print, no. ahead-of-print (2022)

Bundi, P., Trein, P.: Evaluation use and learning in public policy. Policy Sci. **55**, 283–309 (2022)

Casey, A.M., Lorenzen, M.: Untapped potential: seeking library donors among alumni of distance learning programs. J. Libr. Adm. **50**(5–6), 515–529 (2010)

Crawford, J., et al.: COVID-19: 20 countries' higher education intra-period digital pedagogy responses. J. Appl. Learn. Teach. **3**, 1–20 (2020)

Fabriz, S., Mendzheritskaya, J., Stehle, S.: Impact of synchronous and asynchronous settings of online teaching and learning in higher education on students' learning experience during COVID-19. Front. Psychol. **12**, 733554 (2021)

Irfan, M., Kusumaningrum, B., Yulia, Y., Widodo, S.A.: Challenges during the pandemic: use of e-learning in mathematics learning in higher education. Infinity **9**(2), 147–158 (2020)

Jawida, A., Tarshun, O., Alyane, A.: Characteristics and objectives of distance education and e-learning—a comparative study on the experiences of some Arab countries. J. Arab. Lit. **6**, 285–298 (2019)

Kaplan, A.M., Haenlein, M.: Higher education and the digital revolution: about MOOCs, SPOCs, social media, and the Cookie Monster. Bus. Horiz. **59**(4), 441–450 (2016)

Kirtman, L.: Online versus in-class courses: an examination of differences in learning outcomes. Issues Teach. Educ. **18**(2), 103–115 (2009)

Masson, M.: Benefits of TED talks. Can. Fam. Physician **60**(12), 1080 (2014)

Nguyen, T.: The effectiveness of online learning: beyond no significant difference and future horizons. MERLOT J. Online Learn. Teach. **11**(2), 309–319 (2015)

Oblinger, D.G.: The Nature and Purpose of Distance Education. The Technology Source. Michigan: Michigan Virtual University (2000)

Vedung, E.: Public Policy and Program Evaluation. Routledge, New York (2017)

Wollmann, H.: Utilization of evaluation results in policy-making and administration. Croatian Comp. Public Adm. **16**(3), 433–458 (2016)

# The Determinants of Technology Acceptance for Social Media Messaging Applications – Fixed-Effect Questionnaire Design

Pawel Robert Smolinski[1], Monika Mańkowska[2(✉)], Barbara Pawlowska[3], and Jacek Winiarski[3]

[1] University of Gdansk, Gdansk, Poland
p.smolinski.674@studms.ug.edu.pl
[2] WSB University, Gdansk, Poland
mon.mankowska@gmail.com
[3] Faculty of Economics, University of Gdansk, Gdansk, Poland
{barbara.pawlowska,jacek.winiarski}@ug.edu.pl

**Abstract.** Technology acceptance of social media and instant messaging applications is an important area of research. However, a common framework consistent with fixed-effects technology acceptance models is lacking. Researchers disagree on which variables are most useful in which circumstances, and consequently there is no unified item pool. Researchers are forced to use different variables and design technology acceptance questionnaires from scratch, resulting in unacceptable variability and lack of generalizability across studies. Our paper aims to address this problem by reviewing the literature, selecting the best variables and items, and creating a state-of-the-art questionnaire that can be used by authors and future researchers in the field of social media technology acceptance studies.

## 1 Introduction

*Instant messaging* (IM) applications allow users to maintain interpersonal contacts, share opinions or exchange impressions. Through them, users can maintain relationships regardless of distance and reach people with whom a face-to-face meeting would be impossible. The number of IM applications available is constantly increasing. This phenomenon has been recognized by several researchers, and a lot of research is currently being done in this area [1, 2]. The number of entries resulting from a search for the term "instant messenger" on portals such as Google Scholar, ResearchGate.com or Scopus.com clearly shows this. Many questions arise, such as:

- What methods should be used to assess the usability of instant messaging applications?
- How to design, construct, and test a research tool?
- Which variables to select?

© The Author(s), under exclusive license to Springer Nature Switzerland AG 2023
M. Papadaki et al. (Eds.): EMCIS 2022, LNBIP 464, pp. 381–396, 2023.
https://doi.org/10.1007/978-3-031-30694-5_28

These are only a few of the problems that a researcher who studies the phenomenon, which is the contemporary popularity of instant messaging, must find a solution to.

All instant messaging applications, regardless of the details of its functionality, are classified as IT software. The first known tool for testing the acceptance of an IT application was the Technology Acceptance Model (TAM). In the course of scientific development, many other methods emerged, such as the Unified Theory of Acceptance and Use of Technology (UTAT) and its variants UTAUT2 and UTAUT3 [3].

These models have already been used to measure technology acceptance of instant messengers. An example is research [4] using the theory of planned behavior (TPB), TAM, and the flow theory. Results demonstrate that users' perceived usefulness and perceived enjoyment (TAM variables) positively influence their attitude towards IM applications, which in turn impacts their satisfaction. The results of [5] also support that perceived enjoyment and perceived usefulness are positively associated with user satisfaction. Issues such as the influence of technology acceptance model factors, social influence factors and demographic factors on instant messaging adoption in the workplace are presented in [6]. The researchers found social influence to be a more important factor in determining IM adoption than perceived usefulness and perceived ease of use. They also showed that gender and age did not impact the adoption of IMs. The acceptance of IMs in the workplace is studied by [7]. The author added an additional, new variable: "curiosity about other people" to the TAM model. The obtained results prove that perceived usefulness was not significantly important; however, perceived enjoyment, social norms, curiosity about other people, and perceived ease of use were all important prerequisites for IM usage outside the workplace.

## 2   Research Problem and Hypothesis

Technology acceptance models have been applied to many instant messaging applications, however, the variety of variables and constructs used by researchers is quite large (see next section for a comprehensive literature review). It is difficult to compare the results of studies from the same field when they use different models, variables, and measurement tools. We believe that technology acceptance research can benefit from identifying a common set of variables and operationalizing them in a unified questionnaire. Therefore, we present the first research problem:

*Is it possible to identify the most robust technology acceptance variables and select questionnaire items that can be used consistently across different instant messaging applications?*

The first research aim leads us to the second problem, which has never been fully addressed in the literature: technology acceptance models can be viewed as either fixed effect models or random effect models. The fixed effect view assumes that there is only one true technology acceptance and that, once identified, it can explain phenomena from a variety of domains (e.g., variable A explains behavioral intentions to use social media applications as well as blockchain adoption). This view is at the core of technology acceptance research and corresponds to the Davis F.D. conceptualization of TAM as a collection of variables that are general enough to explain the widest range of IT applications [8]. The same fixed-effect view is characteristic for the UTAUT model [3]. However, the recent meta-analysis results [9, 10] suggest that there is too much heterogeneity in the different technology acceptance models to grant the assumption of fixed effects. Variable A can explain well the intention to use social media, but not the intention to adopt blockchain. The need for random effects becomes clear. Random effects do not assume a constant effect that varies only slightly between different IT applications, but many effects. Random effects do not describe one technology acceptance that applies to all and explains both social media and blockchain, but one technology acceptance for a particular social media application and the second for a particular blockchain application (and they are certainly different). The existence of random effects is the reason (albeit often unconscious) researchers add new constructs for technology acceptance tailored to a specific IT application e.g., [11] added two new UTAUT variables to specifically explain technology acceptance of e-learning software).

However, the concept of fixed-effect technology acceptance may not be completely out of the air. For specific IT applications, such as the one we address in this paper, there is variability, but there may be hope for homogeneity. For example, WhatsApp might be perceived as a more useful messaging application (perceived usefulness) than Meta Messenger, but this variability might be generated by the underlying common effect for all messaging applications. Of course, this depends on the size of the variance and whether it is small enough to confirm the assumption of an underlying homogeneous effect.

We can formulate our second research problem as follows:

*Can social media TAM variables be measured as fixed effects?*

In other words, we want to select questionnaire items that are good indicators of TAM variables regardless of the type (brand) of social media messaging application. We can express this problem in a testable hypothesis:

*H: Questionnaire factor loadings are reliably invariant across different social media applications.*

In the first section, we present our approach to item selection and the conceptualization of new technology acceptance variables for social media research. In the second section, we test whether item invariance can be detected across different social media applications, confirming the fixed effect hypothesis at the variable level. Finally, in the appendix, we present a final set of robust items that can reliably assess technology acceptance of instant messaging applications under different circumstances.

## 3  Research Design and Methodology

In order to build the questionnaire for the pilot studies, six stages of design and analysis presented in Fig. 1 were identified.

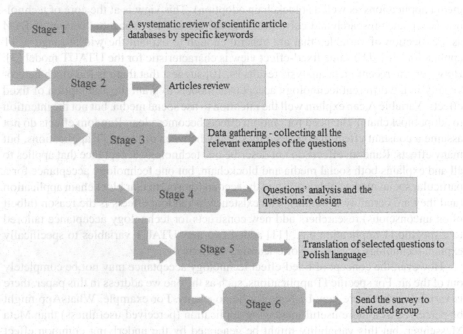

**Fig. 1.** Research stages

Work on examining the acceptance of technology in the case of messaging apps began with a literature review in this area (stage 1). The focus was on articles on technology acceptance research in social media, including messaging apps using the TAM or UTAUT models. A scoping literature search was conducted using the following electronic journal databases: Scopus, Science Direct, Emerald, IEEE, Springer, Taylor & Francis, Wiley, and Google Scholar. This scoping review, or scoping study, synthesized exploratory keywords aimed at mapping key concepts, types of evidence and gaps in research by systematically searching, selecting and synthesizing existing knowledge. We specifically searched for the following keywords and terms, including: "social media + TAM", "social media+UTAUT model:, "Facebook+TAM model", "Facebook+UTAUT model", "Whatsapp+TAM model", "Whatsapp+UTAUT model:, "Instagram +TAM model", "Instagram+UTAUT model", "Snapchat+TAM model", "Snapchat+UTAUT model", "Telegram+TAM model", "Telegram+UTAUT model", "Signal+TAM model", "Signal+UTAUT model", "TikTok+TAM model", "TikTok+UTAUT model", "Discord+TAM model", "Discord+UTAUT model", "Twitter+TAM model", "Twitter+UTAT model".

Table 1 presents a numerical list of the effects of the review of article databases in the studied area.

**Table 1.** Number of results according to combinative keywords search used for the methodology in electronic journal databases

| Keywords | Database | | | | | | |
|---|---|---|---|---|---|---|---|
| | Science Direct | Taylor& Francis | Springer | Wiley | Emerald | Scopus | IEEE |
| Social media+TAM model OR UTAUT model | 3 914 | 3 227 | 4 122 | 2 631 | 3 815 | 47 | 54 |
| Facebook+TAM model OR UTAUT model | 1 290 | 726 | 591 | 435 | 1 189 | 25 | 23 |
| Whatsapp+TAM model OR UTAUT model | 212 | 113 | 131 | 67 | 184 | 2 | 5 |
| Instagram +TAM model OR UTAUT model | 284 | 166 | 106 | 62 | 288 | 1 | 4 |
| Snapchat+TAM model OR UTAUT model | 64 | 35 | 21 | 21 | 74 | 0 | 0 |
| Telegram+TAM model OR UTAUT model | 27 | 32 | 12 | 24 | 18 | 1 | 1 |
| Signal+TAM model OR UTAUT model | 788 | 126 | 0 | 0 | 989 | 0 | 190 |
| TikTok+TAM model OR UTAUT model | 20 | 4 | 0 | 5 | 5 | 1 | 1 |
| Discord+TAM model OR UTAUT model | 38 | 122 | 26 | 4 | 18 | 1 | 1 |
| Twitter+TAM model OR UTAUT model | 747 | 409 | 364 | 228 | 652 | 3 | 10 |

At length, we identified almost 28 7960 articles and conference papers in the first step of the search. To better focus our base, we filtered out articles published before 2010. From this original compilation of literature, we then identified articles that included survey-based research. After the initial analysis, around 600 articles were identified. We were interested in the variables used in the research and the questions that made up the questionnaires. In the database of the collected articles, particular attention was paid to review articles containing an overview of research into the application of technology acceptance models in the area of social media. The literature review was compiled, publications were systematically analyzed, using strategic and critical reading methods [12, 13]. We found the article by Al-Qaysi et al. [14], which served as a road map for the authors. This article is a review article identifying research based on the survey method in the field of social media research. It consists of the list about 60 articles, most of which used the survey method in the study. Some of these articles contained questions that authors asked to respondents. After a detailed analysis of the identified articles and with the use of the snowball method, we gained access to further studies using the questionnaire to measure the acceptance of technology in social media. On this basis, we created a database of 68 articles that we used to build the research tool. These articles were reviewed in detail and 21 of them were used to build the final tool.

*Data gathering - questions*
The systematic review mentioned in previous paragraph consists of 68 selected articles. To build the initial list of questions, we defined the following exclusion/inclusion criteria:

- Is the article related to general available, popular social media application?
- Does the article include the questionnaire with TAM/UTAUT related questions?

First step at the stage 3 of the pilot research design was to eliminate the articles without questions included in the paper and not related to popular social media generally available for users. As an example, articles related to network usage, web 2.0 and web-based communities were removed as there were not related to any software that can be utilized by the global society. The 21 articles related to social media tools were identified.

To build the final question database (stage 4) with TAM/UTAUT questions, all the 21 articles were reviewed. 825 questions related to 16 social media applications are identified. For each of selected items we captured details about the construct from UTAUT/TAM, social media name, article details and the question itself. All the captured details were grouped by the construct name – in total, we identified 87 UTAUT/TAM variable names (see Supplementary materials for a list of all variable names and items identified in the literature).

Created database was a subject of further analysis.

Building the final pilot questionnaire requires reduction number of questions and eliminate or merge the variables of UTAUT/TAM model, to keep the model sustainable and consistent. For this part of stage 4 of the design, the following exclusion/inclusion criteria were specified:

- Is this question duplicated?
- Does the defined software have messaging/communication functionality?
- Is the question related to communication/messaging aspects between users (not marketing or announcements)?
- Does the question fit into Likert scale?

The research is focused on communication aspect of social media, so the first step of selecting relevant questions was eliminating non-communication/messaging related tools. As an example, questions related to YouTube were eliminated, as text-based communication is not the main usage model of the application. Moreover, questions related to internal corporate blogs were removed because this kind of communicator is not widely available for general usage. On the limited number of questions (283), we looked for duplicated or similar questions and rephrase or remove them from the database.

Social media are powerful communication medium with high impact on social lives [15, 16] and it offers variety of use cases [22, 23], including e-commerce [19], marketing [2, 20], advertising [21], education and learning [22, 23], as well as exchanging messages and information between users [24]. To achieve the sustainable number of questions, we decided to select text-based communication between users as a focal point of the research. Questions related to Social Media Ads, online teaching, marketing, job offering were removed from the database. The last step of database clean-up was removing the questions which do not fit into the Likert Scale, for example, questions with Yes/No answers.

To create the final base for survey, the UTAUT/TAM constructs shall have the unified names [1]. Selected questions were grouped by the variables and variable names were reviewed. As a conclusion some of the variable names were unified, for example "Performance Efficiency" related questions were classified as "Perceived Usefulness", "Satisfaction", "Entertainment value", "Perceived playfulness" was translated into "Perceived Enjoyment". During the analysis, two additional constructs as model extension were identified and are used in the questionnaire - "Security & Privacy" and "Technology Attachment". "Security & Privacy" questions were taken from other variables named "Trustworthiness", "Online privacy" and "Intrusiveness tolerance". "Technology attachment" is combination of "Affinity with computer", "Technology-fit", "Computer playfulness".

Final list of the questions contains 86 items and contains 8 technology acceptance constructs:

1. Perceived Ease of Use
2. Perceived Enjoyment
3. Social Influence
4. Behavioral Intention
5. Attitude
6. Perceived Usefulness
7. Security & Privacy
8. Technology attachment

*Building the questionnaire*
Stage 4 of creating the questionnaire contained selection of the most used social media application, taking into consideration communication aspect of the media. Based on the information collected on the popularity of IM applications [25], we decided to qualify nine applications for the study. Selected applications which are includes in final version of questionnaire:

1. Meta Messenger
2. WhatsApp,
3. Instagram DM
4. Tik Tok
5. Twitter
6. Snapchat
7. Signal
8. Telegram
9. Discord

Research was designed to be taken in Poland, so all the questions were translated to Polish. The survey was configured and distributed using the Unipark online tool. Survey was available from 24.03.2022–24.04.2022.

## 4  Participants

97 respondents completed a designed questionnaire. The respondents were university students from northern Poland. Among the respondents, the majority are Meta Messenger users (92 out of 97). The second most popular messaging app is Instagram (74 of 97), followed by WhatsApp (56 of 97), SnapChat (36 of 97) and Discord (37 of 97). Other IM applications have user bases of less than 30 users, in our study.

# 5   Questionnaire Analysis

*Data analysis scheme*
To select items that are the best indicators for UTAUT variables, we created a data analysis scheme, which we divided into three stages: item analysis, exploratory factor analysis, and reliability analysis. Figure 2 presents the proposed scheme. We excluded messaging applications with a small user base (<30) due to the impracticality of conducting a factor analysis with samples smaller than 30 observations. The analysis includes only Meta Messenger, Instagram, WhatsApp, SnapChat and Discord. All results were calculated using laavan package in R programming language framework. Our data is available with supplementary materials.

*Item analysis*
In the first stage, for each of the 8 operationalized UTAUT variables and each of the 5 messaging applications, we conducted a classic item analysis. We selected three descriptive statistics that we considered key indicators of item goodness: standard deviation, range and kurtosis.

Standard deviation provides information about the variance in respondents' answers. Items that have a small standard deviation do not discriminate between respondents, since most give very similar answers. Therefore, if an item's standard deviation was lower than 1 (1 point on the Likert scale) or at least one messaging app, we removed this item from further analysis.

Most of the small standard deviations are for responses to items concerning Meta Messenger. We observed a very strong tendency to rate Meta Messenger positively as a messaging app. For example, item PU7, indicating a positive attitude toward the app, (*Using the indicated communicator makes it easier for me to keep in touch with others*) - has a standard deviation of 0.88 and a mean of 6.71 for Meta Messenger. It suggests a strong tendency of respondents to agree with this characteristic of this particular messaging app. For the other messaging applications, this tendency was weaker, and we considered it insignificant (e.g., item statistics for item PU7 and Discord are $SD = 2.05$ and $M = 5.03$). 13 items were removed from the analysis due to small standard deviation.

Range provides information about the spectrum of responses. Since we used a 7-degree Likert scale, we should expect a range equal to 6 for representative items. A range lower than 6 could indicate that an item does not capture the complete possible spectrum. For example, item SI7 (*My friends also use the indicated messenger*) - had a range equal to 3 for Messenger and Instagram, indicating that participants selected only responses between 5 and 7 (a strong tendency to agree). We found 9 items in our dataset that had a range below 6 and removed them from the analysis accordingly.

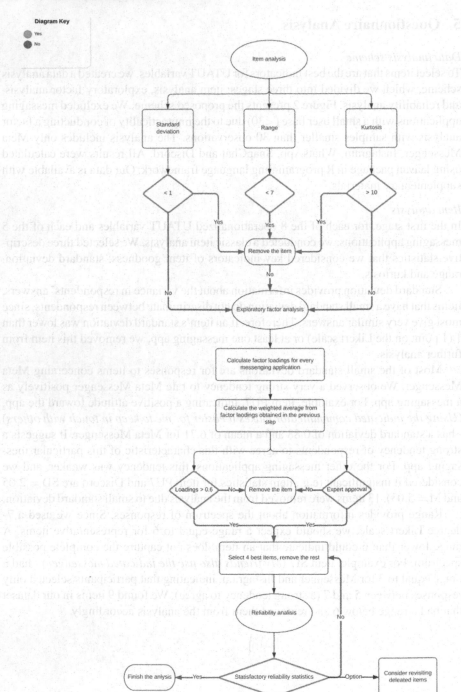

**Fig. 2.** Data analysis flowchart

Kurtosis is a distribution measure. Distributions with large (positive) kurtosis are characterized by a concentration of responses around a single value. This is an undesirable characteristic for a well-discriminating item. For example, item SI3 (*Most of my friends use the indicated communicator*) - had a kurtosis of 30.57, suggesting an extreme concentration of responses around the mean. We removed 13 items with extreme kurtosis (above 10) from the analysis. Again, the majority of items with extreme kurtosis occurred for social influence variable and for Meta Messenger and Instagram messaging applications.

*Exploratory factor analysis*
In the second stage, we conducted an exploratory factor analysis for each of the 8 variables and each of the 5 messaging applications. Exploratory factor analysis measures which items are the best indicators for the UTUAT variables. The basic statistics of factor analysis are factor loadings. We can interpret a factor loading as a correlation between an item and a latent variable (UTAUT variable). The larger the factor loading, the better the item functions as a statistical indicator. Due to the small sample size for some messaging applications (SnapChat and Discord) and the ordinal nature of our variables (Likert scale), we decided to use a non-parametric variant of exploratory factor analysis with a polychromatic correlation matrix and WLS as an estimator.

For each of 5 messaging applications, we conducted 9 exploratory factor analyses (one for each variable), giving us 45 factor analysis results. In order to quantify the association strength between an item and the UTAUT variable, we decided to aggregate factor loadings for 5 messaging applications using a factor loading weighted average. We used the proportion of users in the sample as weights (e.g.: Messenger has a weight of 92/97–0.948, WhatsApp 56/97–0.577). For each variable, we selected 4 items with the highest factor loading and lowest weighted factor variance (homogeneity).

*Reliability analysis*
In the third stage, we measured reliability of selected items. For each 8 variables, we calculated a reliability coefficient [26] using weighted factor lodgings from the previous stage. The reliability coefficient is formulated as:

$$\omega = \frac{\left(\sum \lambda_i\right)^2}{\left(\sum \lambda_i\right)^2 + \sum \left(1 - \lambda_i^2\right)}$$

where is an average-weighted loading.

Table 2 presents the reliability and descriptive statistics for the final selection of items.

*Invariance results*
Table 2 shows that the variances of the weighted factor loadings are almost nonexistent. This is true only for the final selection of items. Other combinations of items, rejected by our analysis, would have resulted in significantly larger variance. This proves that invariance can be achieved at the item level but does not mean that it is not possible to create questionnaires that do not measure technology acceptance homogeneously across different IT applications. For this reason, we encourage future researchers to use our questionnaire, as it is consistent with the fixed effect hypothesis.

**Table 2.** Factor and reliability analysis results

| Latent construct | Item | Weighted Factor Loading (variances) | McDonald's Omega |
|---|---|---|---|
| Perceived usefulness (PU) | PU1 | 0.715 (0.0039) | 0.866 |
| | PU6 | 0.849 (0.0038) | |
| | PU11 | 0.756 (0.0041) | |
| | PU12 | 0.821 (0.0104) | |
| Social Influence (SI) | SI4 | 0.656 (0.0122) | 0.787 |
| | SI7 | 0.777 (0.0097) | |
| | SI8 | 0.688 (0.0301) | |
| | SI9 | 0.650 (0.0118) | |
| Perceived enjoyment (PE) | PE1 | 0.831 (0.0039) | 0.886 |
| | PE4 | 0.873 (0.0026) | |
| | PE11 | 0.749 (0.0020) | |
| | PE13 | 0.793 (0.0080) | |
| Technology attachment (TA) | TA2 | 0.792 (0.0650) | 0.871 |
| | TA3 | 0.820 (0.00) | |
| | TA5 | 0.817 (0.0033) | |
| | TA10 | 0.741 (0.0081) | |
| Perceived ease of use (EU) | EU7 | 0.831 (0.0061) | 0.888 |
| | EU9 | 0.762 (0.0093) | |
| | EE11 | 0.801 (0.0108) | |
| | EU13 | 0.866 (0.0033) | |
| Security aspects (SA) | SA2 | 0.717 (0.0054) | 0.859 |
| | SA3 | 0.813 (0.0091) | |
| | SA5 | 0.806 (0.0027) | |
| | SA7 | 0.770 (0.0043) | |
| Attitude (A) | A1 | 0.844 (0.0053) | 0.871 |
| | A2 | 0.798 (0.0023) | |
| | A3 | 0.849 (0.0072) | |
| | A4 | 0.673 (0.0010) | |
| Behavioral intention (BI) | BI1 | 0.743 (0.0038) | 0.878 |
| | BI2 | 0.821 (0.0103) | |
| | BI4 | 0.855 (0.0077) | |
| | BI5 | 0.784 (0.00) | |

## 6   Conclusions and Further Research

Applying an extensive literature review, we were able to select items that have reliably measured technology acceptance variables in previous works on social media messaging applications [1, 27, 28]. In addition, we combined these items into 7 variables that best reflect technology acceptance of social media [29]. We further added the eighth variable, reflecting the user's psychological dependence on a messaging application. These items were analyzed extensively to:

- select the best items that can be used by future researchers of social media technology acceptance,
- prove that technology acceptance is invariant at the construct level and within a specific IT application (social media messaging).

The final product of our research is a questionnaire that can be used by other researchers to investigate technology acceptance of social media applications. Our questionnaire unifies the common constructs of technology acceptance in a comprehensive and reliable framework. It guarantees that the measured variables are invariant and satisfy the fixed effect hypothesis.

Future researchers can use our questionnaire to deepen the understanding of social media usage and to test the structural relationships between the proposed variables. The problem of fixed/random effects in technology acceptance models is far from solved. We have demonstrated that the construction of an invariant questionnaire is possible. Further research should investigate whether the TAM variables are invariant at the structural equation (SEM) level. To prove the invariance, a technology acceptance model (TAM) of major social media applications must be created (Facebook-Messenger, Instagram, WhatsApp, Twitter, etc.). Besides model invariance, researchers can also use technology acceptance for comparative SWOT analysis and product design analysis of individual IM applications. Such analyses can be useful in determining flaws and desirable functionalities of those applications. They may also provide insight into which TAM variables contribute the most to the market position of a certain social media.

## Appendix: Final Questionnaire

PE1: The <application> is useful in my social life.
PE6: Using the <application> improves the quality of my relationships with others.
PE11: Using the <application>improves my social skills.
PE12: I use the <application> because I want to keep up to date with information.

SI4: My friends find the <application> useful for sharing knowledge and information
SI7: My friends also use the <application>
SI8: I communicate with my friends mainly using the <application>
SI9: People close to me recommend the <application> as the best one

PE1: The use of the <application> is interesting
PE4: Using the <application> makes me happy
PE11: Using the <application> supports the development of my interests
PE13: Using the <application> triggers my curiosity

TA2: Using the <application> is one of the most important things I do every day
TA3: I feel that I am not up to date when I have not used/used <application> for a while
TA5: I would feel uncomfortable if I did not use the <application> regularly
TA10: When I use the <application> I am not aware of the passage of time

EU7: I believe that the <application> is easy to configure
EU9: It is easy to navigate the interface of the <application>
EU11: It is easy to customize the <application> to suit your needs
EU13: I find that using the <application> to communicate with others is easy

SA2: I believe it is safe to use <application> and send confidential data with it
SA3: I believe that my social profile data in the <application> is safe
SA5: I believe that the <application> sufficiently protects my privacy
SA10: I believe that the <application> provides an adequate level of security for my messages

A1: I like to use the <application>
A2: I think communication with friends using the <application> is good
A3: I like to communicate with friends through the <application>
A4: I think it is good to have an account on the <application> to connect and interact with people

BI1: I will be recommending/recommending the <application> to others
BI2: I intend to use the <application> in the future
BI4: I intend to use the <application> as my main tool for online communication
BI5: I intend to use the <application> to communicate with my friends

## Supplementary Materials

Link to the repository with discussed data: https://osf.io/ec2aj/?view_only=9437a625c
dea4ebdab2129f148c31c41.

## References

1. Idemudia, E.C., Raisinghani, M.S., Samuel-Ojo, O.: The contributing factors of continuance usage of social media: an empirical analysis. Inf. Syst. Front. 20(6), 1267–1280 (2016). https://doi.org/10.1007/s10796-016-9721-3
2. Assimakopoulos, C., Antoniadis, I., Kayas, O.G., Dvizac, D.: Effective social media marketing strategy: Facebook as an opportunity for universities. Int. J. Retail Distrib. Manag. 45(5), 532–549 (2017). https://doi.org/10.1108/IJRDM-11-2016-0211
3. Venkatesh, V., Davis, F.: A theoretical extension of the technology acceptance model: four longitudinal field studies. Manag. Sci. 46(2), 186–204 (2001). https://doi.org/10.1287/mnsc.46.2.186.11926
4. Lu, Y., Zhou, T., Wang, B.: Exploring Chinese users' acceptance of instant messaging using the theory of planned behavior, the technology acceptance model, and the flow theory. Comput. Hum. Behav. 25(1), 29–39 (2009). https://doi.org/10.1016/j.chb.2008.06.002

5. Wang, W., Ngai, E.W.T., Wei, H.: Explaining instant messaging continuance intention: the role of personality. Int. J. Hum.-Comput. Interact. **28**(8) (2011). https://doi.org/10.1080/104 47318.2011.622971

6. Glass, R., Li, S.: Social influence and instant messaging adoption. J. Comput. Inf. Syst. **51**(2), 24–30 (2010)

7. Rouibah, K.: Social usage of instant messaging by individuals outside the workplace in Kuwait: a structural equation model. Inf. Technol. People **21**(1), 34–68 (2008). https://doi.org/10.1108/09593840810860324

8. Davis, F.D., Bagozzi, R.P., Warshaw, P.R.: User acceptance of computer technology: a comparison of two theoretical models. Manag. Sci. **35**, 982–1003 (1989). https://doi.org/10.1287/mnsc.35.8.982

9. Hwang, J.S., Lee, H.J.: A meta-analysis of advanced UTAUT variables in the ICT industry: an analysis of published papers in Korean journals. Int. J. Innov. Comput. Inf. Control **14**(2), 757–766 (2018)

10. Tamilmani, K., Rana, N.P., Dwivedi, Y.K.: Consumer acceptance and use of information technology: a meta-analytic evaluation of UTAUT2. Inf. Syst. Front. **23**(4), 987–1005 (2021)

11. Smolinski, P.R., Szóstakowski, M., Winiarski, J.: Technology acceptance of MS teams among university teachers during COVID-19. In: Themistocleous, M., Papadaki, M. (eds.) Information Systems: EMCIS 2021, pp. 346–361. Springer, Cham (2022). https://doi.org/10.1007/978-3-030-95947-0_24

12. Matarese, V.: Using strategic, critical reading of research papers to teach scientific writing: the reading–research–writing continuum. Supporting Res. Writing 73–89 (2013). https://doi.org/10.1016/B978-1-84334-666-1.50005-9

13. Renear, A.H., Palmer, C.L.: Strategic reading, ontologies, and the future of scientific publishing. Science **325**(5942), 828–832 (2009). https://doi.org/10.1126/science.1157784

14. Al-Qaysi, N., Mohamad-Nordin, N., Al-Emran, M.: Employing the technology acceptance model in social media: a systematic review. Educ. Inf. Technol. **25**(6), 4961–5002 (2020). https://doi.org/10.1007/s10639-020-10197-1

15. Al-Rahmi, W.M., Zeki, A.M.: A model of using social media for collaborative learning to enhance learners' performance on learning. J. King Saud Univ. – Comput. Inf. Sci. **29**(4), 526–535 (2017). https://doi.org/10.1016/j.jksuci.2016.09.002

16. Rauniar, R., Rawski, G., Yang, J., Johnson, B.: Technology acceptance model (TAM) and social media usage: an empirical study on Facebook. J. Enterp. Inf. Manag. **27**(1) (2014). https://doi.org/10.1108/JEIM-04-2012-0011

17. Chintalapati, N., Daruri, V.S.K.: Examining the use of YouTube as a learning resource in higher education: scale development and validation of TAM model. Telemat. Inform. **34**(6), 853–860 (2017). https://doi.org/10.1016/j.tele.2016.08.008

18. Fedorko, I., Fedorko, R., Gavurova, B., Bacik, R.: Social media in the context of technology acceptance model. Entrepreneurship Sustain. Issues **9**(1), 519–528 (2021). https://doi.org/10.9770/jesi.2021.9.1(32)

19. Cha, J.: Shopping on social networking web sites. J. Interact. Advert. **10**(1), 77–93 (2009). https://doi.org/10.1080/15252019.2009.10722164

20. Lowe, B., D'Alessandro, S., Winzar, H., Laffey, D., Collier, W.: The use of Web 2.0 technologies in marketing classes: key drivers of student acceptance. J. Consum. Behav. **12**(5), 412-422 (2013). https://doi.org/10.1002/cb.1444

21. Lin, C.A., Kim, T.: Predicting user response to sponsored advertising on social media via the technology acceptance model. Comput. Hum. Behav. **64**, 710–718 (2016). https://doi.org/10.1016/jchb.2016.07.027

22. Barn, S.S.: 'Tweet dreams are made of this, who are we to disagree?' Adventures in a #Brave New World of #tweets, #Twitter, #student engagement and #excitement with #learning. J. Mark. Manag. **32**(9–10), 965–986 (2016). https://doi.org/10.1080/0267257X.2016.1159598

23. Ifinedo, P.: Students' perceived impact of learning and satisfaction with blogs. Int. J. Inf. Learn. Technol. **34**(4), 322–337 (2017). https://doi.org/10.1108/IJILT-12-2016-0059

24. Tan, X., Qin, L., Kim, Y., Hsu, J.: Impact of privacy concern in social networking web sites. Internet Res. **22**(2), 211–223 (2012). https://doi.org/10.1108/10662241211214575

25. Hootsuite Digital 2021 Poland report. https://datareportal.com/reports/digital-2021-poland

26. McDonald, R.P.: Test Theory: A Unified Treatment. Psychology Press (1999). https://doi.org/10.4324/9781410601087

27. Akram, M.S., Albalawi, W.: Youths' social media adoption: theoretical model and empirical evidence. Int. J. Bus. Manag. **11**(2), 22 (2016). https://doi.org/10.5539/ijbm.v11

28. Kwon, S.J., Park, E., Kim, K.J.: What drives successful social networking services? A comparative analysis of user acceptance of Facebook and twitter. Soc. Sci. J. **51**(4), 534–544 (2014). https://doi.org/10.1016/j.soscij.2014.04.005

29. Lee, Y., Kozar, K.A., Larsen, K.R.: The technology acceptance model: past, present, and future. Commun. Assoc. Inf. Syst. **12**(1), 752–780 (2002). https://doi.org/10.1037/0011816

# Conceptual Model of User Experience for Personalization

Elena Kornyshova[✉] and Eric Gressier-Soudan

CEDRIC, Conservatoire National des Arts et Métiers, Paris, France
{elena.kornyshova,eric.gressier_soudan}@cnam.fr

**Abstract.** Designers of Information Technology (IT) devices and Information Systems (IS) are more and more concerned about providing better conditions of use: more efficient interactions, enjoyable interfaces, and personalization. These aspects are studied within the concept of User Experience (UX). UX is necessarily specific and should be dynamically adapted to a given user or group of users. Personalized user experience links with the user's characteristics, user's mood, and user's expectations but also with the targeted object under a UX design process. In addition, in many cases, multiple devices are involved in user interactions. In the context of museum devices, visitors use tablets, interactive screens, geolocation sensors, headphones, etc. These devices are also supporting the IS dedicated to visiting applications. In general, the IS is not shared among users or very few of them. To provide personalized UX, we advocate building a shared IS supported over users' devices and back-office servers. Thus, we can introduce a conceptual model to help UX personalization. Our model proposal is illustrated through Man-Museum Interactions literature.

**Keywords:** User Experience · Conceptual Model · Personalization · Man-Museum Interaction

## 1 Introduction

Information and Communication Technologies (ICT) overwhelm everyday life to provide an enhanced form of living. Digital extensions to our common senses are flourishing essentially based on a more connected world. Connections encompass humans, objects, homes, cars, pets, cities, organizations, industry, banks to offer a smarter life. The key concept of this new era is interaction. As a wide concept, it is fostered through Information systems engineering, Communications, and more and more efficient Technologies to deliver an improved User Experience (UX).

We believe that valuable user experiences need to be designed carefully with all dimensions of user context, thus, UX is implicitly personalized. Due to the changing nature of user context, especially when emotions are entering the loop of the design process, UX is necessarily specific and should be dynamically adapted to a given user or group of users. Dynamicity should consider time, space, and the user's profile. We assume that the profile is not uniform, it is a time-dependent concept. It is evolving through time

M. Papadaki et al. (Eds.): EMCIS 2022, LNBIP 464, pp. 397–406, 2023.
https://doi.org/10.1007/978-3-031-30694-5_29

and is influenced by all sorts of occurring events. Personalized user experience links with user characteristics, mood, and expectations. It also links with the object under consideration by the UX design process.

In addition, in almost all cases, multiple supports (devices and/or associated IS) take part in a user experience. This is the case with devices used in museums. We aim to provide the adaptation of these devices to different visitors, for instance, if a child is close to a screen, this one should show images attractive to the child. If the visitor is an adult, the presentation on the same screen could be more serious. For an elderly person, the font size could become bigger. The personalization could be done at an individual level, but also for a persona (user type). The same issue is present in organizations when users (employees) connect to different tools in their workplace and could have a personalized representation of available data.

To obtain this personalization, data about users and user experience should be stored and shared between different devices and supported by the IS of the organization. Despite numerous works on UX, we have not identified a conceptual model allowing to structure the required data. Thus, the goal of this paper is to present a UX conceptual model reflecting the different UX dimensions and used to personalize UX. In our work, we consider user experience only supported by digital technologies.

To validate our model, we have been interested in heritage applications; the Museum came rapidly to our mind as a convenient use case. Museums offer emotional visits, and most of the applications developed for museums could be, by design, obsoletes at the time they are launched. The missing point is evolution. Visitors change and what they feel too is changing. Artifacts move from one place to another (sometimes to another museum). Temporary events are programmed to underline an artist or a piece of work… This is a very preliminary list of the kind of evolution a museum should face. Any change is a risk for the launched application because it can be unable to reconfigure to take into account changes. Applications for museums should be designed in a different way. The challenge is to provide a model that supports changes over time and that the applications that rely on this model can evolve accordingly to these changes.

The paper is organized as follows. Section 2 introduces related works. In Sect. 3 we present the UX conceptual model. In Sect. 4, we illustrate this model with the Man-Museum interactions literature. We conclude the paper and give our future research in Sect. 5.

## 2 Related Works

In this section, we present works related to user experience in general and applied to Man-Museum interactions.

### 2.1 User Experience and Its Representation

As defined in ISO 9241–210, "User Experience is a person's perceptions and responses that result from the use or anticipated use of a product, system or service." (definition from [1]). All works on UX agree on the complexity and richness of this term [2–5]. [2] defines three facets in UX: "beyond the instrumental", "emotion and affect", and the "experiential" (which means context-awareness and temporality). [3] enumerates different definitions of UX. [4] shows results of a Systematic Literature Review on aspects and dimensions of UX with the main goal of UX evaluation. The authors have identified five dimensions: values, user needs experience, brand experience, technology experience, and context. [5] presents a survey on the UX nature to obtain a shared definition that converges on UX as "dynamic, context-dependent and subjective" [5]. They detail different kinds of experience: product, system, service, and object experiences.

The most detailed generic definition of UX is done in [1]. The authors present a product-oriented model of user experience. It includes the following dimensions: human perception (senses, cognition and affects, and responses), product (product sensors and product responses), experience context, and temporality of experience [1].

Considering the UX representations, the authors of [6] suggest and validate a mathematical model of UX in the case of dynamic adaptive video streaming. The authors of [7] develop a simplified model of User Experience to explicitly link UX with usability and Human-Computer Interactions. A temporal model of the UX lifecycle is highlighted in [8] with an explanation of different UX phases.

More detailed works on personalized UX are [9] and [10]. [9] presents a three-layers contextual gameplay experience model linking the player (with his experience corresponding to player characteristics and internal influences) to the game system (playability), and external influences (called contextual gameplay experience). External influences include spatial, temporal, social, and cultural influences. In [10], the authors detail a UX model, which includes product features having an apparent product character for each user. The user is subjected to different consequences of the apparent product character depending on the situation. The authors apply their model to augmented reality in the case of urban heritage tourism.

## 2.2 User Experience in the Context of Man-Museum Interactions

User experience with application to Man-Museum interactions is presented in [11–18]. The authors of [11] present a study made at the Acropolis Museum and, in parallel, in social media networks. The goal of this research is to explore personalization in the museum experience. [12] analyzes different approaches to understand visitor behavior and defines the following perspectives: socio-cultural, cognitive, psychological orientation, physical, and environmental. [13] doesn't focus on visitors' behaviour but on visits and visitors' motivations to explain why people are coming to museums. The expectations are compared with the visit itself. This work leads to a classification of visitors: Explorers, Facilitators, Experience Seekers, Hobbyists, Rechargers, Respectful Pilgrims, and Affinity Seekers. [14] presents a framework architecture to support three visit phases (pre-visit, on-site, and post-visit). This framework contains three models: visitor model, site model, and visit model. [15] suggests using recommendation systems to take care of visiting styles in addition to user interests to improve the quality of museum visits. [16] details a multi-sensory approach to design the museum experience. Several works detail serious games developed to improve Man-Museum Interaction, such as [17, 18].

We observed multiple works mentioning the necessity to have a shared vision of UX, suggesting definitions and aggregating information about different aspects of UX.

However, from the literature review, we have not identified a conceptual model covering the different dimensions of UX and allowing to personalize UX in a distributed environment. In the next section, we present a UX conceptual model.

## 3   Conceptual Model of User Experience

Figure 1 depicts the UX conceptual model. This model features different UX-related dimensions. Senses, affects, responses, and context are the basic components that could be considered as input/output to feed the user experience. User experience is related to the corresponding objects (which are used in a specific experience) and subjects (that we foresee larger than the concept of a simple user). It is also connected to a device which is represented by an ICT component used in the experience. User experience could be expected (by the user or by designers) and lived during the experience. In the following, we explain all these concepts.

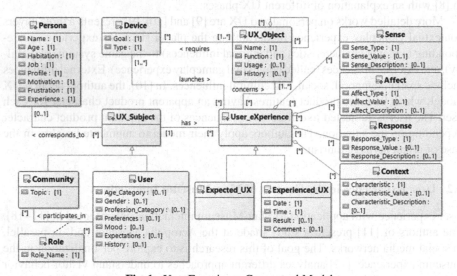

**Fig. 1.** User Experience Conceptual Model.

**User_Experience.** UX expresses an experience of a user toward an object. A user experience is unique for a user and for a time slot, but it could be associated with one or with a set of objects. This is the core concept as we consider that users communicate with objects through user experience. The UX concept is composed of four related concepts: sense, affect, response, and context (the four *composition* links) each of them representing a detailed taxonomy of possible elements. UX is a combination of different possible values of the taxonomy elements. The user experience *concerns* at least one object but could be associated with many objects. Each object is associated with multiple user experiences. Each user experience is associated with only one UX subject (the *has* association).

**Sense.** The typology of senses is taken from [1] which distinguishes the following types: exteroceptive (external to the organism stimuli), proprioceptive ("spatial body orientation"), interoceptive ("stimuli produced within the organism"), and chronoception (sense of time). Different senses are sight, hearing, taste, smell, touch, thermic, pain, and so on. **Affect.** Affects (lived emotions, values, etc.) as cognitive processes "link external stimuli information with brain... in order to reach an interpretation of the stimuli on their semantic and aesthetic character" (from [1]). **Response.** Human responses include physiological, motor, and motivational affects [1]: temperature sensation, respiration change, cardiovascular change, posture, gesture, mimic, voice, etc. **Context.** The context factors are inspired from [19] as better reflecting the context content: external context (like weather, season, time), organizational context, etc. For each of them, the model allows to identify the *Type* and to specify a *Value* or a *Description* of the given characteristic. For instance, for a cardiovascular change, we can register the corresponding value, or for an emotion or a value (as an affect), we can give a description.

**User Experience.** As in [20], we consider two kinds of UX: expected and really experienced (the *inheritance* associations). The instances of **Experienced_UX** store data about real UX together with data about this experience: *Date, Time*, and possibly *Result* (for instance, a "like") and *Comment* if they are left by subjects. In addition, a UX subject has expectations concerning his/her future user experience. Thus, an instance of an **Expected_UX** could be defined.

**UX_Subject.** UX_Subject could be a User of a Community (the *inheritance* associations). Both could *correspond to* a concept of Persona mainly used to characterize users in this field. A UX subject *has* multiple "user experiences".

**User.** "A user is a human who is targeted to utilize a product" [1]. Users have different parameters describing them: *Age_category, Gender, Profession_category, Preferences, Mood, Expectations*, and *History*. Only data authorized by the user could be stored respecting the General Data Protection Regulation (GDPR) rules.

**Community.** A community represents a group of users identified within a friendship network and is mainly characterized by a *Topic* and people *participating in* it [21] expressed with an association class **Role**. Users can have various roles in different communities (*Role_Name*).

**Persona.** The Persona term is related to a type of user often an imaginary one but deduced from data gathered during the exploration of users. [22] defines a Person as a "representation of the most common users, based on a shared set of critical tasks." A Persona describes different characteristics, needs, and behavior of this typical user. This concept is central to the users' representation in UX-related approaches. It includes generally demographic and biographical data: personal, technical, relationship, and opinion information [23]. The authors of [24] present a detailed ontology-based user characterization using the concept of Persona to personalize UX applications depending on context. We take the most important characteristics in our model (*Name, Age, Habitation, Job, Profile, Motivation, Frustration*, and *Experience*), but this list could be extended if needed in a given case. A persona could be associated with a unique user or with a community.

**UX_Object.** Each UX is related to at least one UX Object. We use this generic term to group products, services, systems, or objects of experience from [5]. Users can experience a whole group of objects, for instance, from a museum room without differentiating concrete objects (the *composition* link on the UX_Object concept). Different attributes characterize UX Objects. An object is not only the physical object itself but a set of relevant data about it: history of creation and of evolutions, author(s), way of production, uses, civilization it belongs to, maintenance events, etc. These aspects are revealed using ICT devices, augmented reality for example. It can include variations in colors, texture, sounds, forms, and so on. We define the main attributes: *Name, Function, Usage,* and *History.* This list could be extended obviously.

**Device.** A device supports the user experience itself. The nature of the device could be different from communication devices (smartphones, tablet computers, glasses, VR masks) to sensors used to capture gestures, movements, eye tracking, and so on. At least one device is required to represent a UX Object, but it could be done by multiple devices. [1] enumerates types of sensors that could be associated with a consumer product: physical, logical sensors, sensors capturing external factors (contact, range, vision sensors), internal factors (like heat monitoring), and so on. We characterize devices by two main attributes: *Goal* and *Type.* An object *requires* to have supporting devices, each device could support one or more object(s). A device could contribute to multiple user experiences (the *launches* association). The user experience should have at least one associated device (for our purpose, we do not consider UX without any technology-based support).

This conceptual model of User Experience aims at highlighting different UX-related dimensions that we can store in IS for further data utilization to personalize the experience. In this manner, data about UX are centralized and standardized. It helps also to compare the planned and real experience. In addition to these practical needs, the introduced conceptual model contributes to several challenges (based on [25]): to formalize UX knowledge through different concepts and their relationships; to develop a shared representation of UX concepts; to make UX knowledge reusable in different projects and contexts; to support the creation of UX models applied to various fields, and to check and validate the existing UX models or other representations.

## 4    UX Conceptual Model in Man-Museum Interactions Literature

Museum IS are often database-oriented (collection management systems – [26]). In [26], the author suggests an approach to integrate five museum legacy information systems applied to the case of the National Palace Museum in Taiwan. [27] presents an Internet-of-Things architecture to design smart museums. Except for [14] and [20], we did not identify any other work dealing with a conceptualization of UX in the context of Man-Museum interactions. [14] presents the visitor, site, and visit models. [20] presents a model of the Visitor-Player experience. In this work, the visit game personalization is thought as a resolution of puzzle intrigues between external components (playability and context) and players' expectations.

We apply the UX conceptual model to Man-Museum Interactions as they are considered in the current research literature to identify related works depending on the established concepts.

**User Experience.** The main research sources dealing with UX in museums are [11–16]. User experience in museums is considered as containing three phases: pre-visit, on-site visit, and post-visit [14]. [20] details the notion of museum visit experience to define balanceable visit games. This work also distinguishes between expected and really lived experiences.

**Sense, Affect, Response, Context.** These UX components are less studied in the literature. [16] presents a multi-sensory transformation approach to enhance the museum visit experience. [20] defines four groups of contexts: museal, temporal, cultural, and social.

**UX Subjects.** Museum users are studied through the "Persona" concept. "Personas are detailed descriptions of imaginary people constructed out of well-understood and highly specified data about real people" [28]. Different approaches consider individual visitors [29–32], groups [14, 33], or both [34, 35]. Several works detail user characteristics. The visitor model form [14] includes a visitor profile (demographics and preferences), his/her state, together with the number of visitors in a group. [36] gives an overview of identified user characteristics from literature such as user profile (age, gender, education, skills, and so on) or user preferences (related or not to the museum context). Several approaches also consider the feedback of the user after visiting the museum.

**UX Objects.** Different museum objects, art pieces, etc. A group of objects can be an exhibition room or a logical group of objects like Lavoisier Lab (https://www.arts-et-met iers.net/musee/visitor-information) representing a set of objects which have a real additional value when presented together. Several works mention objects attributes as factors for visit personalization: available multimedia information (graphical, video, audio, etc.) regarding the artworks [34]; multimedia collection containing digital reproductions of sculptures, educational videos, audio guides, textual and hypermedia documents with a description of authors and sculptures [37]. In addition, several museums and other cultural heritage institutions detail various characteristics about museum information considered during user experience: museum map [29, 34], repository of cultural heritage data [30], and exhibitions' locations [31].

**Devices.** The most detailed typology of devices used in museums is given in [38]. The author enumerates more than 50 devices grouped into 13 categories like handling devices, viewing devices, projection devices, etc. [36] summarizes devices used in different projects dealing with museum user experience: PDA (personal digital assistant) devices with RFID (Radio Frequency Identification) tags [34, 35], mobile devices [11, 14, 29], glasses [32], sensors [32], etc.

In addition to the identification of main concepts, the application of the UX conceptual model to Man-Museum Interactions allowed us to identify the following open issues: (i) lack of UX conceptual models formalized to represent Man-Museum interactions; (ii) UX components are under-explored and personalization mechanisms are still limited for Man-Museums Interactions; and (iii) devices used in Museums are not related to their information system.

# 5 Conclusion and Future Works

Currently, different kinds of organizations start to apply UX to promote their activities and to improve the relationships with their users: customers, visitors, and clients. For example, it can be refined as customer experience or brand experience [5]. To help them in establishing an IS for implementing shared data about UX, we presented in this paper a conceptual model allowing to store data about UX and to share information about how UX could be personalized for a given user. The proposed conceptual model offers a shared representation of UX and allows checking the completeness of the UX-related concepts of a real application. We have applied this model to UX in Man-Museums Interactions.

In our future research, we will develop a generic method for engineering unique personalized UX, which could be applied not only in the case of visits in museums, but also in the case of employees' experience within digital workplaces. We foresee the engineering of adaptable, personalized experience as situational, thus depending on different characteristics of the context.

# References

1. Bongard-Blanchy, K., Bouchard, C.: Dimensions of user experience - from the product design perspective. Journal d'Interaction Personne-Système, Association Francophone d'Interaction Homme-Machine (AFIHM) 3(1) (2014)
2. Hassenzahl, M., Tractinsky, N.: User experience – a research agenda. Behav. Inf. Technol. 25(2), 91–97 (2006)
3. Law, E., Roto, V., Vermeeren, A., Kort, J., Hassenzahl, M.: Towards a shared definition of user experience. In: Proceedings of CHI 2008, pp. 2395–2398 (2008)
4. Zarour, M., Alharbi, M.: User experience framework that combines aspects, dimensions, and measurement methods. Cogent Eng. 4(1) (2017)
5. Law, E., Roto, V., Hassenzahl, M., Vermeeren, A., Kort, J.: Understanding, scoping and defining user experience: a survey approach. In: Proceedings of CHI 2009, USA, pp. 719–728 (2009)
6. Liu, Y., Dey, S., Ulupinar, F., Luby, M., Mao, Y.: Deriving and validating user experience model for DASH video streaming. IEEE Trans. Broadcast. 61(4), 651–665 (2015)
7. Jetter, H.-C., Gerken, J.: A simplified model of user experience for practical application. In: 2nd COST294-MAUSE International Open Workshop "User eXperience - Towards a unified view" (NordiCHI 2006), pp. 106–111 (2007)
8. Pohlmeyer, A., Hecht, M., Blessing, L.: User experience lifecycle model continue [Continuous User Experience]. In: Proceedings of BWMMS 2009 (2009)
9. Engl, S., Nacke, L.E.: Contextual influences on mobile player experience – a game user experience model. Entertainment Comput. J. 4(1), 83–91 (2013)
10. Han, D.-I., Tom Dieck, M.C., Jung, T.: User experience model for augmented reality applications in urban heritage tourism. J. Herit. Tour. 13(1), 46–61 (2018)
11. Antoniou, A., et al.: Capturing the visitor profile for a personalized mobile museum experience: an indirect approach (2016)
12. Goulding, C.: The museum environment and the visitor experience. Eur. J. Mark. 34(3/4), 261–278 (2000)
13. Falk, J.H., Dierking, L.D.: The Museum Experience Revisited. Routledge (2016)

14. Kuflik, T., Wecker, A.J., Lanir, J., Stock, O.: An integrative framework for extending the boundaries of the museum visit experience: linking the pre, during and post visit phases. Inf. Technol. Tourism **15**(1), 17–47 (2014). https://doi.org/10.1007/s40558-014-0018-4
15. Lykourentzou, I., et al.: Improving museum visitors' quality of experience through intelligent recommendations: a visiting style-based approach. In: Intelligent Environments (Workshops) (2013)
16. Harada, T., Hideyoshi, Y., Gressier-Soudan, E., Jean, C.: Museum experience design based on multi-sensory transformation approach. In: Proceedings of the DESIGN 2018, Dubrovnik, Croatia (2018)
17. Gressier-Soudan, E., Pellerin, R., Simatic, M.: Using RFID/NFC for pervasive serious games: the PLUG experience. In: Near Field Communications Handbook, pp. 279–304 (2011)
18. Guardiola, E.: CULTE gameplay prototype: deliverable of project CULTE (2015)
19. Kornyshova, E., Deneckère, R., Claudepierre, B.: Towards method component contextualization. Int. J. Inf. Syst. Model. Des. **2**(4) (2011)
20. Astic, I.: Adaptation dynamique des jeux de visite pour les musées: contribution à l'équilibrage de l'expérience du visiteur joueur. Ph.D. thesis, Paris, France (2018)
21. Zhang, Z., Li, Q., Zeng, D., Gao, H.: User community discovery from multi-relational networks. Decis. Support Syst. **54**(2), 870–879 (2013). https://doi.org/10.1016/j.dss.2012.09.012
22. Tomlin, W.C.: What's a persona? In: UX Optimization. Apress, Berkeley (2018). https://doi.org/10.1007/978-1-4842-3867-7_2
23. Aquino, P.T., Filgueiras, L.V.L.: User modeling with personas. In: Proceedings of the 2005 Latin American Conference on Human-Computer Interaction (CLIHC 2005), pp. 277–282. Association for Computing Machinery, New York (2005). https://doi.org/10.1145/1111360.1111388
24. Skillen, K.-L., Chen, L., Nugent, C.D., Donnelly, M.P., Burns, W., Solheim, I.: Ontological user profile modeling for context-aware application personalization. In: Bravo, J., López-de-Ipiña, D., Moya, F. (eds.) UCAmI 2012. LNCS, vol. 7656, pp. 261–268. Springer, Heidelberg (2012). https://doi.org/10.1007/978-3-642-35377-2_36
25. Kornyshova, E., Deneckère, R.: Decision-making ontology for information system engineering. In: Parsons, J., Saeki, M., Shoval, P., Woo, C., Wand, Y. (eds.) ER 2010. LNCS, vol. 6412, pp. 104–117. Springer, Heidelberg (2010). https://doi.org/10.1007/978-3-642-16373-9_8
26. Wu, S.-C.: Systems integration of heterogeneous cultural heritage information systems in museums: a case study of the National Palace Museum. Int. J. Digit. Libr. **17**(4), 287–304 (2015). https://doi.org/10.1007/s00799-015-0154-2
27. Chianese, A., Piccialli, F.: Designing a smart museum: when cultural heritage joins IoT. In: Third International Conference on Technologies and Applications for Smart Cities, UK (2014)
28. Pruitt, J., Adlin, T.: The persona lifecycle: keeping people in mind throughout product design, 1st edn. Interactive Technologies, p. 744 (2006)
29. van Hage, W.R., Stash, N., Wang, Y., Aroyo, L.: Finding your way through the Rijksmuseum with an adaptive mobile museum guide. In: Aroyo, L., et al. (eds.) ESWC 2010. LNCS, vol. 6088, pp. 46–59. Springer, Heidelberg (2010). https://doi.org/10.1007/978-3-642-13486-9_4
30. Noor, S., Martinez, K.: Using social data as context for making recommendations: an ontology based approach. In: Proceedings of the 1st Workshop CIAO (2009)
31. Bohnert, F., Zukerman, I.: Personalised pathway prediction. In: De Bra, P., Kobsa, A., Chin, D. (eds.) UMAP 2010. LNCS, vol. 6075, pp. 363–368. Springer, Heidelberg (2010). https://doi.org/10.1007/978-3-642-13470-8_33
32. Damala, A., Stojanovic, N., Schuchert, T., Moragues, J., Cabrera, A., Gilleade, K.: Adaptive augmented reality for cultural heritage: ARtSENSE project. In: Ioannides, M., Fritsch, D.,

Leissner, J., Davies, R., Remondino, F., Caffo, R. (eds.) EuroMed 2012. LNCS, vol. 7616, pp. 746–755. Springer, Heidelberg (2012). https://doi.org/10.1007/978-3-642-34234-9_79

33. Luyten, K., Van Loon, H., Teunkens, D., Gabriëls, K., Coninx, K., Manshoven E.: ARCHIE: disclosing a museum by a socially-aware mobile guide. In: 7th International Symposium on Virtual Reality, Archaeology and Cultural Heritage (2006)

34. Ghiani, G., Paternò, F., Santoro, C., Spano, L.D.: UbiCicero: a location-aware, multi-device museum guide. Interact. Comput. **21**(4), 288–303 (2009)

35. Amato, F., Chianese, A., Mazzeo, A., Moscato, V., Picariello, A., Piccialli, F.: The talking museum project. Procedia Comput. Sci. **21**, 114–121 (2013)

36. Chou, S.-C., Hsieh, W.-T., Gandon, F.L., Sadeh, N.M.: Semantic web technologies for context-aware museum tour guide applications. In: AINA 2005, vol. 1 (2005)

37. Al Hudar, S.: Approaches to design museum visit game adaptable to the context: a literature review. Master thesis, CNAM, Paris, France (2019)

38. Recouvreur, J.: Typology of Digital Devices for Museums (2019). https://julienr.pro/docs/typology-of-digital-devices-for-museums.pdf. Accessed Oct 2022

# Emerging Computing Technologies and Trends for Business Process Management

# Mining Contextual Process Models Using Sensors Data: A Case of Daily Activities in Smart Home

Ramona Elali[1]([✉]), Elena Kornyshova[2], Rébecca Deneckère[1], and Camille Salinesi[1]

[1] Paris 1 Panthéon Sorbonne, Paris, France
{ramona.elali,rebecca.deneckere,camille.salinesi}@univ-paris1.fr
[2] Conservatoire National des Arts et Métiers, Paris, France
elena.kornyshova@cnam.fr

**Abstract.** Different techniques are used by companies to enhance their processes. Process mining (PM) is one of these techniques that relies on the user activity logs recorded by information systems to discover the process model, to check conformance with the prescribed process, to enhance the process, and to recommend or guess the next user activity. From another hand, many contextual factors such as time, location, weather, and user's profile influence the user activities. However, PM techniques arc mainly activity-oriented and do not take into consideration the contextual environment. Our main goal is to enrich process models obtained using process mining technics with contextual information issued from sensors data and to construct contextual process models for a better process discovery, conformance checking, and recommendations. In this paper, we test the feasibility to integrate events logs with sensor logs to provide meaningful results. We use existing datasets with events and sensors logs about daily activities in Smart Home to construct a process model enriched by contextual information.

**Keywords:** Process Mining · Process Model · Context · Contextual Process Model · Sensors Log · Events Log · Smart Home

## 1 Introduction

While the rapid evolution of Information Systems (IS) is taking its rise in all domains, data turns out to be the most powerful and silent weapon that can change the world since it can be used to get insights, make decisions, increase revenues, etc. Process mining (PM) [1] is one of the techniques that helps in processing the available data to get better knowledge. PM main objectives are to discover the process model, to check its conformance with the current process, to enhance the process, and to finally recommend to the user the next activity by relying on activity logs recorded from IS [1]. However, PM does not take into consideration the contextual background behind the user activity. It discovers the process model based on the user activity logs only. In fact, users are not activity-oriented and there are many external factors such as time, location, profile, etc. that can affect the activity selection. Hence, with the existence of Artificial Intelligence

© The Author(s), under exclusive license to Springer Nature Switzerland AG 2023
M. Papadaki et al. (Eds.): EMCIS 2022, LNBIP 464, pp. 409–425, 2023.
https://doi.org/10.1007/978-3-031-30694-5_30

tools, it would be easier to access the contextual environment behind an activity while relying on different types of sensors. Since the smart home is a rich context environment, it would be interesting to study it in order to explore how contextual information can affect user activity directly or indirectly.

The enrichment of the process model by additional data had already been handled with semantic process models [3]. In this work, the authors discussed the benefits and the capability of a semantic process model. Several works study how process mining could be enhanced using semantic data organized into ontologies [4–6]. Our goal is to improve process models by semantics issued from sensor data to build a contextual process model.

In this paper, we explore Smart Home datasets to check how likely contextual information can affect user activities and to construct a contextual process model. The datasets used in this report were provided by BP-Meets-IoT Challenge [7] and contain data about everyday life and home activities. It is composed of 2 main simulated datasets: a dataset that consists of the activity log of a single living inside a home (DS1) coupled with the corresponding sensor logs, and a second one that consists of the activity log of two livings (DS2) also coupled with the sensor logs. All logs were provided in XES format.

The next section will present the background. Section 3 will focus on data exploration and Sect. 4 will present the research questions and the method used to construct a contextual process model. In Sect. 5, we describe the results and the discovered process model. Related works are detailed in Sect. 6. We conclude in Sect. 7.

## 2 Background

Hereafter, we describe the background of the research fields.

Event logs are considered the most important source of information and the major input for the mining techniques. Usually, there is a difference between the existing prescribed business process model which is provided by the organization, and what the user really does to complete their tasks in the actual process. In fact, events logs are the base of the process mining technique which permits the discovery of the actual business processes, the conformance verification with the existing prescribed processes, and the enhancement of the process model [19].

PM is a Business Process Management evolving technology where the main objective is to discover, check the conformance, and enhance the process models that are based on event logs [1]. PM focuses on the generated activities from the business processes so it can be used as a recommendation technique to direct the user on which next activity to follow according to his current activity [20, 21]. PM focuses on activity-oriented models. PM has shown that the actual processes that are extracted from event logs can be different from the prescribed business process. In contrast to PM models that are created as a sequence of steps that don't support variability [22]. According to [23] to properly understand research processes, it is essential to trace them. The collected traces depend on the process model established, which must be as accurate as possible to comprehensively record the traces. Still, the major drawback in tracing processes is finding an adequate modeling language that covers all the aspects needed when analyzing these traces. [23] presents five types of process models from the information systems engineering domain to use to represent processes: Activity-oriented process models, Product-oriented process models, Decision-oriented process models, Context-oriented process models, and Strategy-oriented process models. In addition, different process models' annotations were described in [1], that are used to represent the process after the execution of a process discovery algorithm: Transition Systems, Petri Nets, Workflow Nets, YAWL, BPMN, EPCs, Causal Nets, and Process Trees.

Typically, PM techniques do not take into consideration the context behind the user activity. In [24, 25], they proposed a contextualization methodology based on the process to be able to construct models on the fly while taking into consideration the situation behind them. Although, in [3] they have discussed and pointed out the benefits and the capability of a semantic process model. Hence, it would be interesting to step to build a contextual process model.

## 3 Dataset Exploration

In this work, we have relied on two main datasets DS1 and DS2 that describe the daily habits of two individuals living in a Smart Home (DS1 for the individual 1 and DS2 for individual 2). Both datasets were recorded for 21 consecutive days between 16 March 2020 and 6 April 2020 from 0:00 am to 11:59 pm. Each dataset contains the person activities logs and the sensor logs. Initially, the activity logs in DS1 contain 4068 event records while the sensors log contains 34571 event records. And the activity logs in DS2 contain 28238 event records and the sensors log contains 39304 event records.

The following subsections present both activity and sensor logs (correspondingly Subsects. 3.1 and 3.2) and data classification in order to identify groups of different elements of both logs (Subsect. 3.3).

### 3.1 Activity Logs

The activity logs are composed of a set of traces. Each trace contains a set of events grouped under a case name. Each record in the activity logs is characterized by different attributes as illustrated with an extraction in Fig. 1. These attributes are described with: *Case Name:* categorizes a set of activities under a specific goal; *Trace Id:* groups a set of events; *Event Id:* indicates the unique Id that distinguishes every record inside the dataset; *Activity Name:* describes the activity that is taking place; *Resource Id:* describes the person who's doing the activity; *Timestamp:* indicates the date and time when the

| ActivityName | ResourceId | Timestamp | Transition | EventId | CaseName | TraceId |
|---|---|---|---|---|---|---|
| go_wardrobe | 1 | 2020-03-16 00:00:00+00:00 | complete | 4071 | sleeping_BP | 435 |
| get_clothes | 1 | 2020-03-16 00:00:00+00:00 | start | 4072 | sleeping_BP | 435 |
| get_clothes | 1 | 2020-03-16 00:01:00+00:00 | complete | 4076 | sleeping_BP | 435 |
| change_clothes | 1 | 2020-03-16 00:01:00+00:00 | start | 4077 | sleeping_BP | 435 |
| change_clothes | 1 | 2020-03-16 00:05:00+00:00 | complete | 4080 | sleeping_BP | 435 |
| go_bathtub | 1 | 2020-03-16 00:05:00+00:00 | complete | 4081 | sleeping_BP | 435 |

**Fig. 1.** Event Logs Samples.

activity has occurred; *Transition:* indicates if the event record is a start or a complete activity.

### 3.2 Sensors Log

The sensor logs correspond to a set of events. Each record in the sensor logs is characterized by some fields as depicted in Fig. 2.

| SensorName | ResourceId | Timestamp | EventId | Value |
|---|---|---|---|---|
| position | 1 | 2020-03-16 00:01:00+00:00 | 18 | 1202 852 |
| position | 1 | 2020-03-16 00:06:00+00:00 | 24 | 1184.75270036629 848.33325125897€ |
| water_use | 0 | 2020-03-16 00:07:00+00:00 | 25 | 10 |
| position | 1 | 2020-03-16 00:07:00+00:00 | 27 | 1246 638 |
| water_use | 0 | 2020-03-16 00:31:00+00:00 | 51 | 1 |

**Fig. 2.** Sensor Log Samples.

The sensor logs are characterized by the following attributes: *Event Id:* which indicates the unique Id that distinguishes every record inside the dataset; *Activity Name:* describes the sensor type that was triggered; *Resource Id:* describes the resource that is triggering the sensor. It's either the person living inside the home or it's automatic by the system; *Timestamp:* indicates the date and time when the sensor event has occurred; *Value:* indicates the value range of the sensor. Noting that each sensor has a different set of values according to the sensor type.

### 3.3 Data Classification

We classified the main elements of the datasets (activities and sensors) into different groups. As a result, we were able to identify 53 activities and 14 sensors that were provided in [7] and that are listed below in Table 1 and Table 2.

In addition, we have found that each activity can be grouped into a set of different activity types. The provided dataset contains already defined categories of activities. However, these categories are defined by the authors with regards to goals. We aim at grouping activities regarding their nature; thus, it will allow us to avoid having the same activity classified into multiple categories as it is done in the initial dataset description [7]. We have defined 13 activity types such as activities specific to the bathroom or the kitchen etc. as shown in Fig. 3. Also, we have classified the sensors into different categories. We have defined 7 categories as shown in Fig. 4. Note that each sensor acts differently from another sensor and each sensor has its own range of value.

**Table 1.** Home Activities List in the Dataset.

| Activity Name | | | |
|---|---|---|---|
| brush_teeth | go_kitchen_shef | dress_up_outdoor | go_windows |
| change_clothes | go_kitchen_sink | drink_water | go_workplace |
| Clean | go_outside | eat_cold_meal | have_bath |
| close_windows | go_shoe_shelf | finish_walk | interact_with_man |
| do_exercise | go_wardrobe | get_bread | lower_blinds |
| dress_down_outdoor | go_wc | get_clothes | open_windows |
| get_food | pack_goods | go_fridge | go_exercise_place |
| get_food_from_fridge | put_plate_to_sink | go_entrace | go_dining_table |
| get_glass | raise_blinds | go_computer_chair | go_computer |
| get_water | rest_in_chair | go_chair | Work |
| go_bathroom_sink | sleep_in_bed | wc_flush | wash_hands |
| go_bathtub | switch_computer_off | wc_do | walk_outside |
| go_bed | switch_computer_on | wash_dishes | use_the_computer |
| go_book_shelf | | | |

**Table 2.** Sensors List in the Dataset.

| Sensor Name | | | |
|---|---|---|---|
| air_Condition | position | food | fridge_door_contact |
| blinds | power_use | home_Aired | home_presence |
| cooked_food | pressure_bed | windows | water_use |
| unwashed_dishes | temp | | |

Subsequently, we have identified the different locations or positions where a sensor can be linked to, or an activity can take place. The different positions are listed in Fig. 5.

While categorizing the activities, the sensors, and the positions, we relied on the provided dataset. However, we tried to define the groups in a generic way to allow extension when it's used in different contexts, countries, or cultures.

**Fig. 3.** Activity Types Grouping.

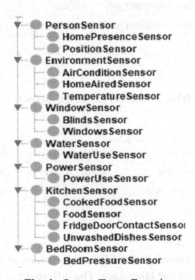

**Fig. 4.** Sensor Types Grouping.

# 4    Research Questions and Proposed Approach

The research methodology used in this proposal has been detailed in [2]. We explain below the research questions specific to the work presented in this paper and the main proposed approach.

## 4.1    Research Questions

The natural behavior of a living person tends to be variable and doesn't stick to a fixed schedule or a routine to perform its daily living activities. We believe that the person will act or adjust his daily activities and tasks according to the contextual environment that can affect him directly or indirectly. For instance, the person during the weekend performs different activities than on the weekdays. In addition, on a rainy day, the person will exercise indoors while when it's a sunny day, he can go outside for a walk. We believe that it is possible to identify the links between context data and user activities.

Therefore, in this paper, our research questions are the following: **Question 1:** *Does the contextual environment affect user activities?* **Question 2:** *Can links between sensors' data and activity logs be automatically identified?*

**Fig. 5.** Positions.

## 4.2    Proposed Approach

As mentioned above, we relied on DS1 and DS2 datasets. For the first question, we used both of them while for the second question, we used only DS1. The method used in this paper is illustrated in Fig. 6.

It includes data transformation as a first step. Then, the method contains the next three steps (Activity and Sensor Mapping, application of the Apriori Algorithm [8], and the Process Model Discovery using Disco) which could be applied separately or in parallel. As a final step, we create a contextual process model using the outcome of the previous steps.

**Data Transformation.** Data transformation consists of data cleansing and data manipulation according to our needs. The activity logs and the sensors log contained some

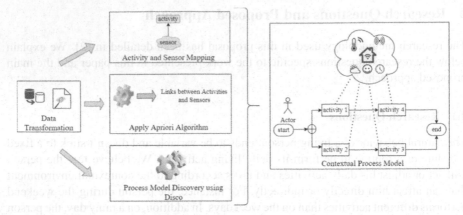

**Fig. 6.** Approach Overview.

noise, in addition to some duplicated records. A data cleansing was established on those logs to remove noise and duplicated data. Hence, we obtained 3154 records in the activity logs and 4332 records in the sensors log in DS1, and in DS2 we obtained 23166 event records. For data manipulation, we have transformed both logs from XES format to CSV format. Then, we have added two additional attributes for both logs: the *Position* attribute and the *Day* attribute. For the *Position* attribute, we have annotated each activity record and each sensor event record with a position value that indicates the actual location where the activity has taken place or the location where the sensor should be positioned relying on Fig. 5. For instance, the *Position* attribute for an activity related to *wash_dishes* or *unwashed_dishes* should be the *kitchen_sink*. As for the *Day* attribute, we have annotated the records from Day1 to Day21 to group all the events that are linked to a specific day. For example, all the activities records and the sensor events records that have occurred between *2020-03-16 00:00:00 + 00:00* and *2020-03-16 23:59:00 + 00:00* are annotated by *Day1*.

**Activity Mapping with Sensors.** To answer the first question, we have done a mapping between the sensors and the activities according to the *Position* and the *Timestamp* attributes. Hence, we obtained 112 correspondences between the sensors and the activities that will be described in Subsect. 4.1 in Fig. 8.

**Apriori Algorithm Application.** To answer the second question, we have applied Apriori [8] which is an Association rule mining technique. Hence, association rule mining main's goal is to find the hidden relationships between different items. It is commonly used for marketing purposes such as in Market Basket analysis to identify the items that are frequently bought together. Association rule mining allows us to find frequent patterns, causal structures, and associations [8, 9]. We used association rule mining because it allows us to find the rules that show us how an appearance of a specific item will allow the occurrence of other items. In our case, an item represents either an activity or a sensor.

Therefore, we will use the association rule mining technique since it's a rule-based technique to find the causal structures between the activities and the sensors as we believe that there is a hidden relationship between the contextual environment and the user behavior. Association rules are composed of an antecedent and a consequent and is represented by if–then statements. The Apriori algorithm is one of the top algorithms in the rule mining technique [10] and it allows us to find the relationships between the sensors and the activities. In order to apply it to our dataset, we first had to combine both logs in a single log file sorted according to the Timestamp attribute. Then, we had to transform the activities and sensors records into a transactional records list which is supported by Apriori as an input parameter. We obtained 410 transaction records containing the activities and sensors as transaction items. In addition to the transactional list, Apriori needs additional parameters such as the minimum support and confidence. Since we want to obtain strong rules with good confidence, we set the confidence value to 80% in all the experiments. We did 3 experiments as shown in Table 3, and the value of minimum support was set through the process of trial and error.

**Table 3.** Apriori Experiments.

| Experiment | Minimum support | Confidence | # of obtained rules |
|---|---|---|---|
| First Experiment | 0.15 | 80% | 56 rules |
| Second Experiment | 0.05 | 80% | 8565 rules |
| Third Experiment | 0.03 | 80% | 34623 rules |

After analyzing, manually, the generated rules, we found that the result of the second experiment is more realistic due to the number of generated rules in addition to the minimum support which is not very low. In Subsect. 5.2, we will present a set of the generated rules from the second experiment.

**Process Model Discovery Using Disco.** This part allows us to complete the answer on the first question. We have used the process mining tool Disco [11] to obtain the process models in order to know the difference between the habits of the weekend and weekdays and how likely a resource profile would affect the process model. We mapped the Day attribute in the activity logs to the Case attribute in Disco and we applied the process discovery on the activity logs. The results are presented in Subsect. 5.3.

**Contextual Process Model Creation.** Based on the results of the previous steps, we generated a contextual process model (described in Subsect. 5.4).

## 5   Results Analysis and Contextual Process Model

In this section, we analyze the results of our experiment.

## 5.1   Activity Mapping with Sensors Result

We believe that context elements can affect user activities directly or indirectly. Since the sensors' data represent the contextual environment, we did a mapping using human reasoning between the sensors and activities. Figure 8 represents the established mapping, to show that each activity can have one (or more) triggered sensor(s), or, in the contrary, that a sensor can affect one or more activities. As an instance, when *unwashed_dishes* sensor value is greater than 0 then the *wash_dishes* activity might take place. In addition, the *water_use* and the *position* sensors values will be modified accordingly. The *position* sensor value will be set to the *kitchen_sink* while the *water_use* sensor will indicate the water usage. Plus, the *put_plate_to_sink* activity will trigger the *unwashed_dishes* sensor value which will cause the occurrence of other activities. For the mapping, we used the colored sensors from Fig. 7 to illustrate the different sensor categories.

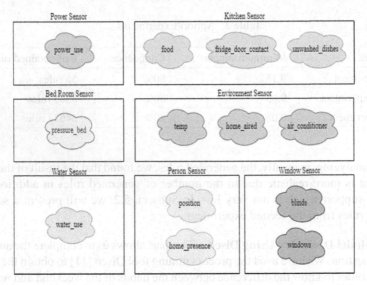

**Fig. 7.** Sensors Grouping.

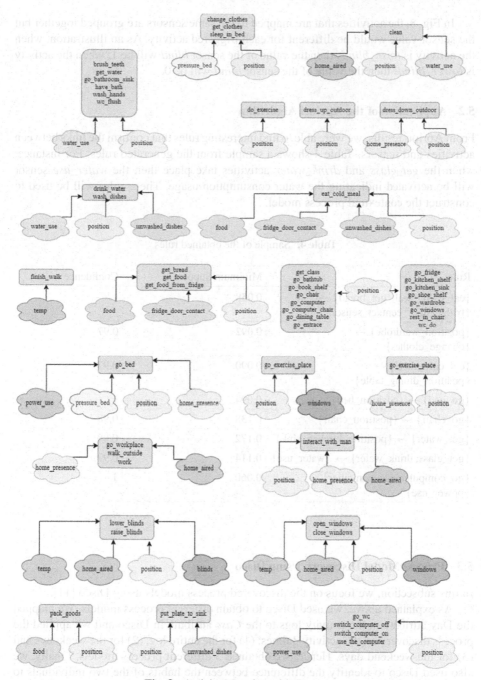

**Fig. 8.** Activity Mapping with Sensors.

In Fig. 8, the activities that are mapped to the same sensors are grouped together but the sensor value would be different for each triggered activity. As an illustration, when the activity is *raise_blinds* then the value of the sensor *blind* will be 1; when the activity is *lower_blinds* then the value of the sensor *blind* will be 0.

## 5.2   Application of the Apriori Algorithm

From Apriori results, we were able to find interesting rules that confirm the links between activities and sensors. Table 4 shows a sample from the generated rules. For instance, when the *get_glass* and *drink_water* activities take place then the *water_use* sensor will be activated indicating the water consumption usage. These rules will be used to construct the contextual process model.

**Table 4.** Sample of the obtained rules.

| Rule | Minimum Support | Confidence |
|------|-----------------|------------|
| {eat_cold_meal_ get_bread} → {fridge_door_contact_sensor} | 0.058 | 1 |
| {position_wardrobe} → {change_clothes} | 0.092 | 0.97 |
| {eat_cold_meal} → {position_dining_table} | 0.090 | 0.97 |
| {go_bed} → {position_bed} | 0.094 | 1 |
| {go_chair} → {position_chair} | 0.138 | 0.98 |
| {get_water} → {position_kitchen_sink} | 0.172 | 0.88 |
| {get_glass, drink_water} → {water_use} | 0.114 | 0.97 |
| {go_computer, use_computer} → {power_use} | 0.060 | 1 |

## 5.3   Process Model Discovery Using Disco

In this subsection, we focus on the discovered process models using Disco [11].

As explained above, we used Disco to obtain different process models. We mapped the Day attribute in the activity logs to the *Case* attribute in Disco and we applied the process discovery on the activity 3 times: (1) for the entire log, (2) for the weekdays and (3) for the weekend days. Hence, we obtained 3 different process models. Finally, we also used Disco to identify the difference between the habits of the two individuals to obtain 2 different process models. Figure 9 represents two extractions for the discovered process models using the weekdays and weekend activity logs.

Based on the obtained models, we can directly deduce that the process model discovered from the weekdays event records is different from the process model discovered

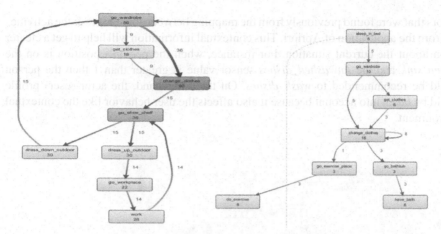

**Fig. 9.** Extractions of the Obtained Process Models for Weekdays (left) and Weekend (right).

from the weekend event records. Human living tends to execute different types of activities between the weekdays and the weekends due to their work schedule, the time, the country, the weather, etc.

As expected, the activity *go_to_work* is missing from the weekend process model and we can find replacement activities as *go_exercise_place* or *do_exercise*, which is not the case in the rest of the week, the individuals having not enough time to exercise after spending hours at the workplace. Moreover, the highest number of events that occurred during the entire week is during the weekends on Sundays and Saturdays because the person tends to spend more days at home. This difference confirms that the date affects the user activities.

We have discovered two process models using the event logs of each person separately. It showed us how likely the activity process model would be different between two persons living inside the same home. We noticed clearly that resource 2 (the second individual) does not go to any workplace. In addition, it seems that resource 2 is responsible for the wash_dishes activity, which does not appear at all in the process model of resource 1 (the first person). This difference between the two process models relates to how the person's profile such as age, gender, character, and hobbies can also affect the user's behavior while enacting his daily activities. Thus, these relations between profile, sensor data and activities would provide more precise information to construct contextual process models.

## 5.4 Contextual Process Model

We showed that the contextual elements such as time, location, profile, etc. affect directly or indirectly the user activities. It would be interesting to annotate the process model with contextual information. This will enrich the process model by providing more accurate information about the activity current situation which will help in better decision making and insights. Figure 10 shows a process sample that was extracted from the weekend process model from DS1. The process model was annotated by the information of the sensors that were found previously from the mapping between the sensors & the activities and from the application of Apriori. This contextual information will help to get a clearer vision about the current situation. For instance, when the person's position is on the *kitchen sink* and the *unwashed_dishes* sensor value is greater than 1 then the person should be recommended to *wash_dishes*. On the other hand, the actor (user) profile should be taken into account because it also affects the user behavior like the contextual environment.

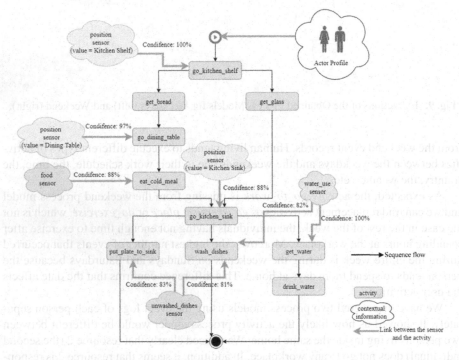

**Fig. 10.** Contextual Process Model.

# 6   Related Work

Multiple research works have already identified the importance of putting context into process models.

In [4], they described scenarios that illustrate how the process mining could be enhanced by using semantically annotated event logs. The authors of [5] described a semantic process mining approach allowing to enrich event logs using semantic data organized in an ontology. In [6], the industrial benefits and challenges of semantic process mining are analyzed. In [3], the authors presented the benefits of semantic annotation for process modeling.

In [12], the authors have considered 4 main types of contexts: the context that is directly linked to the process instance, the context that is related to the process overall, the social context that is linked to how people interact with others, and the external context that is affected by external factors such as weather, economic climate, etc. They conclude that contextual information should be used in the construction of process models.

The authors of [12, 14] have presented the necessary core building elements to enable semantic process mining which focuses on ontologies to find the link between the generated events and the real concepts they presented in the ontologies.

In [15], the authors presented an approach to filtering and abstracting event logs using ontologies and cluster analysis in the healthcare domain. Hence, their approach consists of incorporating data mining with process mining techniques to create contextual process mining.

[16] presents a framework for a knowledge-based abstraction of event logs, and the output of the framework which is the abstracted traces will be given as input to the semantic process mining technique.

The researchers in [17] introduced a framework considered an ontology-based system that supports the development of semantic process mining techniques.

The authors in [18] applied process mining methods to event logs of the activities of daily living of the elderly inside a Smart Home but they didn't take into consideration the contextual data.

All the mentioned previous works acknowledge that the contextual information related to each specific event will enhance the process mining techniques. Thus, we studied the scenario of daily living inside a Smart Home in order to build a process model using the sensors data, because we believe that the daily activities of a human living can be affected by different factors and can be simulated to build a contextual process model.

# 7  Conclusion

In this work, we propose an approach to enrich the process models mined from the activity logs with contextual data. We applied this approach to a case study by exploring a dataset related to Smart Home activities.

In the explored case study, we were able to find automatically and manually the links between the sensors and the activities. We highlighted the differences between the mined process models whenever contextual information (*weekdays* Versus *weekends*), or the user profile (*resource 1* Versus *resource 2*) is changing. The multiplicity of the different mined process models, each related to a specific context, suggests the importance of constructing contextual process models. A lot of other contextual information can provide better knowledge since a person's activities would be different in different circumstances such as time, location, country, culture, weather, etc. Using contextual process models would allow us to offer better recommendations to the users by contextually recommending the best-suited activity at a specific time and place.

However, the case study and the provided dataset were quite simple as they didn't offer any detailed information about the user profile or other context characteristics. The dataset is of small size and lacks data; the only provided characteristics were timestamp and resource Id.

In future work, we aim to be able to discover the contextual process model automatically and we plan to work on a larger dataset to extract more links between the sensors and activities to be able to guide the user on the fly.

# References

1. Van Der Aalst, W.M.: Process mining: data science in action. Springer, Heidelberg (2016). https://doi.org/10.1007/978-3-662-49851-4
2. Elali, R.: An intention mining approach using ontology for contextual recommendations. In: Proceedings of the Doctoral Consortium Papers Presented at the 33rd International Conference on Advanced Information Systems Engineering (CAiSE 2021), Melbourne, Australia, June 28–July 2 2021. CEUR Workshop Proceedings, vol. 2906, pp. 69–78 (2021). https://ceur-ws.org/
3. Lautenbacher, F., Bauer, B., Seitz, C.: Semantic business process modeling - benefits and capability. AAAI Spring Symposium - Technical Report, pp. 71–76. (2008)
4. Alves de Medeiros, A.K., Van Der Aalst, W.M.: Process mining towards semantics. In: Dillon, T.S., Chang, E., Meersman, R., Sycara, K. (eds.) Advances in Web Semantics I. LNCS, vol. 4891, pp. 35–80. Springer, Heidelberg (2008). https://doi.org/10.1007/978-3-540-89784-2_3
5. Okoye, K., Tawil, A.R.H., Naeem, U., Lamine, E.: Discovery and enhancement of learning model analysis through semantic process mining. Int. J. Comput. Inf. Syst. Ind. Manag. Appl. 8(2016), 93–114 (2016)
6. Ingvaldsen, J.E., Gulla, J.A.: Industrial application of semantic process mining. Enterp. Inf. Syst. 6(2), 139–163 (2012)
7. Koschmider, A., Leotta, F., Serral, E., Torres, V.: BP-Meets-IoT 2021 Challenge Dataset (2021)
8. Zaki, M.J., Meira, W.: Data Mining and Analysis: Fundamental Concepts and Algorithms. Cambridge University Press, Cambridge (2014)
9. Aher, S.B., Lobo, L.M.R.J.: A comparative study of association rule algorithms for course recommender system in e-learning. Int. J. Comput. Appl. 39(1), 48–52 (2012). https://doi.org/10.5120/4788-7021
10. Agrawal, R., Srikant, R.: Fast algorithms for mining association rules. In: Proc. 20th Int. Conf. Very Large Data Bases, VLDB, vol. 1215, pp. 487–499 (1994)
11. Disco: https://fluxicon.com/disco/
12. Van Der Aalst, W.M., Dustdar, S.: Process mining put into context. IEEE Internet Comput. 16(1), 82–86 (2012). https://doi.org/10.1109/MIC.2012.12
13. de Medeiros, A.K.A., Van Der Aalst, W.M., Pedrinaci, C.: Semantic process mining tools: core building blocks (2008)
14. de Medeiros, A.K.A., et al.: An outlook on semantic business process mining and monitoring. In: Meersman, R., Tari, Z., Herrero, P. (eds.) OTM 2007. LNCS, vol. 4806, pp. 1244–1255. Springer, Heidelberg (2007). https://doi.org/10.1007/978-3-540-76890-6_52
15. Rabbi, F., Fatemi, B., Maccaull, W.: Analysis of patient pathways with contextual process mining (2022)
16. Leonardi, G., Striani, M., Quaglini, S., Cavallini, A., Montani, S.: Towards semantic process mining through knowledge-based trace abstraction. In: Ceravolo, P., van Keulen, M., Stoffel, K. (eds.) SIMPDA 2017. LNBIP, vol. 340, pp. 45–64. Springer, Cham (2019). https://doi.org/10.1007/978-3-030-11638-5_3
17. Okoye, K., Islam, S., Naeem, U., Sharif, M.S., Azam, M.A., Karami, A.: The application of a semantic-based process mining framework on a learning process domain. In: Arai, K., Kapoor, S., Bhatia, R. (eds.) IntelliSys 2018. AISC, vol. 868, pp. 1381–1403. Springer, Cham (2019). https://doi.org/10.1007/978-3-030-01054-6_96

18. Theodoropoulou, G., Bousdekis, A., Miaoulis, G., Voulodimos, A.: Process mining for activities of daily living in smart homecare. In: 24th Pan-Hellenic Conference on Informatics (PCI 2020). Association for Computing Machinery, New York, NY, USA, pp. 197–201 (2020). https://doi.org/10.1145/3437120.3437306

19. Diaz, O.E., Perez, M.G., Lascano, J.E.: Literature review about intention mining in information systems. J. Comput. Inf. Syst. **61**(4), 295–304 (2021). https://doi.org/10.1080/08874417.2019.1633569

20. Khodabandelou, G., Hug, C., Deneckère, R., Salinesi, C.: Process mining versus intention mining. In: Nurcan, S., et al. (eds.) BPMDS/EMMSAD -2013. LNBIP, vol. 147, pp. 466–480. Springer, Heidelberg (2013). https://doi.org/10.1007/978-3-642-38484-4_33

21. Compagno, D., Epure, E.V., Deneckere-Lebas, R., Salinesi, C.: Exploring digital conversation corpora with process mining. Corpus Pragmatics **2**(2), 193–215 (2018). https://doi.org/10.1007/s41701-018-0030-6

22. Khodabandelou, G.: Contextual recommendations using intention mining on process traces: doctoral consortium paper. In: IEEE 7th International Conference on Research Challenges in Information Science (RCIS), pp. 1–6. IEEE (2013)

23. Hug, C., Salinesi, C., Deneckère, R., Lamassé, S.: Process modeling for humanities: tracing and analyzing scientific processes. In: Revive the Past: Proceedings of the 39th Annual Conference on Computer Applications and Quantitative Methods in Archaeology, pp. 245–255 (2011)

24. Kornyshova, E., Deneckere, R., Claudepierre, B.: Towards method component contextualization. Int. J. Inf. Syst. Model. Design **2**(4), 49–81 (2011)

25. Kornyshova, E., Deneckere, R., Claudepierre, B.: Contextualization of method components. In: 2010 Fourth International Conference on Research Challenges in Information Science (RCIS), pp. 235–246. IEEE (2010)

# Business Process Automation in SMEs

Sílvia Moreira[1](✉) [iD], Henrique S. Mamede[2,3] [iD], and Arnaldo Santos[3] [iD]

[1] UTAD, Vila Real, Portugal
al75401@alunos.utad.pt
[2] INESC TEC, Lisbon, Portugal
[3] Universidade Aberta, Lisbon, Portugal

**Abstract.** Business Process Automation has been gaining increasing importance in the management of companies and organizations since it reduces the time needed to carry out routine tasks, freeing employees for other more creative and exciting things. The use of process automation seems to be a growing trend in the business's operational restructuring, combined with digital transformation. It can be applied in the most varied business areas. Organizations from any sector of activity can also adopt it. Given these benefits, the granted success in transforming business processes would be expected. However, 30 to 50% of automation initiatives with Robotic Process Automation technology fail. In this work, a set of guidelines will be proposed that will constitute, after validation, a framework capable of guiding organizations, with a focus on SMEs, in the procedure of automating their processes, thus obtaining the maximum return of this transformation.

**Keywords:** Business Process Automation · Framework · Robotic Process Automation · Business Process Management · Digital Transformation

## 1 Introduction

The COVID-19 pandemic context accelerated Digital Transformation (DT) because companies and organizations have reinvented how they act in the market to continue providing services and products to customers and users [1]. There are about 22.6 million Small and Medium-sized Enterprises (SMEs) in European Union. Of those, 34% have already adopted digital technology, 24% recognize that they need a digital tool, 10% consider adopting advanced digital technology, and 8% have the same perception as the previous group. However, due to a lack of digital literacy or funding, they do not know which one will be the most appropriate [2].

Business process digitalization can involve automation. For Zaoui and Souissib [3], DT is a current theme with enormous importance for companies in all sectors of activity because it changes the relationship with customers, suppliers, and human resources and alters the value creation process. Gartner [4] mentioned that in 2021 (given the pandemic context), the use of process automation technologies would be an essential topic (90% of large companies worldwide would adopt this technology by 2022), with increasing importance for companies and the economy in general. These process automation technologies will automate critical business processes, freeing employees from the manual effort for other tasks [4].

© The Author(s), under exclusive license to Springer Nature Switzerland AG 2023
M. Papadaki et al. (Eds.): EMCIS 2022, LNBIP 464, pp. 426–437, 2023.
https://doi.org/10.1007/978-3-031-30694-5_31

Business Process Management (BPM) becomes essential for the constant improvement of the efficiency and effectiveness of an organization [5]. Chakraborti et al. [6] define BPM as a multidisciplinary area that includes business process management, modelling, automation, execution, control, measurement, and optimization. BPM sees the company as a whole, involving activity flows (process workflow), systems and people involved, internal and external to the organization [6]. Wewerka and Reichert [7] recognize the importance of BPM in the business process digital lifecycle, involving all its participants and information systems. Due to the evolution of changing contexts, the authors point out that business processes must be adaptable, efficient, and with low costs; in short, companies increasingly need a greater degree of business process automation to remain competitive.

In the definition of the concept of BPM, suggested by Wewerka and Reichert [7], process mining appears as a sub-discipline of BPM where the discovery of business processes, compliance verification, and data analysis are carried out. This sub-discipline also helps discover processes for automation. It can contribute to BPM projects by surveying the as-is scenario, in short, the current state of the organization's business processes [7]. BPM can be adopted in numerous areas present in the organizational structure [8]. Despite its broad spectrum of use in the business area and the processes eligible for adoption, it is always essential to check its adaptability, suitability, and the return that comes from this change since this adoption can translate into a paradigm shift in the way the business works.

New technologies allied to BPM allow new ways of working, affecting humans, which can cause some fears regarding jobs and the lack of knowledge of the technology itself [5]. The technologies capable of being disruptive, changing and improving business processes, are Blockchain, Robotic Process Automation (RPA), Artificial Intelligence (AI) [5, 9], IoT, process mining, reality virtualization, and 4D printing [9].

Chakraborti et al. [6], Stravinskienė and Serafinas [5], and Wewerka and Reichert [7] mention RPA technology as being essential in BPM and Business Process Automation (BPA). They complement each other [7]. While BPM provides companies with actual knowledge about their business processes and workflows, RPA is concerned with developing bots whose function is to imitate human interaction with information systems [7]. According to Ahuja and Tailor [10], the term RPA came into force in the vocabulary of organizations in 2000, yet, its development began in 1990. Bhatnagar [11], Chakraborti et al. [6], Siderska [12], Syed et al. [13], Siderska [14], Wewerka and Reichert [7], and Puica [15] consider RPA an advanced/emerging technology that through software automates routine and rule-based tasks, previously performed by humans. Despite all the benefits and positive effects that process automation with RPA brings to organizations, applying this type of technology is not always successful. According to Stravindkiené and Serafinas [5], 30 to 50% of RPA initiatives fail.

Given the general benefits of adopting business process automation tools, there are various reasons automation can fail. Stands out in the assessment of the process of digitalization, its documentation, its maturity, in short, the assessment of digital readiness; there is also the doubt in choosing the ideal process to automate; lack of knowhow about the automation technology in the organization; and lack of knowhow about the implementation methodology. These reasons and the references are presented in Table 1.

**Table 1.** Reasons for automation to fail.

| Reason | References |
| --- | --- |
| The assessment of digital readiness | Romão et al. [16]; Hofmann et al. [17]; Siderska [12]; Syed et al. [13]; Siderska [14]; Yatskiv et al. [18]; Flechsing et al. [19]; Ng et al. [20]; Puica [15]; Sobczak [21] |
| Doubt in choosing the ideal process to automate | Ansari et al. [22]; Asquith and Horsman [23]; Auth et al. [24]; Mishra et al. [25]; Romão et al. [16]; Sobczak [21]; Chakraborti et al. [6]; Sobczak [26]; Hindel et al. [27]; Hofmann et al. [17]; Leite et al. [28]; Patri [29]; Rizk et al. [30]; Siderska [12]; Syed et al. [13]; Yatskiv et al. [18]; Choi et al. [31]; Flechsing et al. [19]; Sobczak [32]; Puica [15] |
| Lack of knowhow about the automation technology in the organization | Ansari et al. [22]; Dey and Das [33]; Mishra et al. [25]; Reddy et al. [34]; Sobczak [26]; Dechamma and Shobha [35]; Hindel et al. [27]; Hofmann et al. [17]; Patri [29]; Rizk et al. [30]; Sethi et al. [36]; Yatskiv et al. [18]; Choi et al. [31]; Flechsing et al. [19]; Zhang et al. [37]; Sobczak [21] |
| Lack of knowhow about the implementation methodology | Ansari et al. [22]; Sobczak [26]; Hindel et al. [27]; Hofmann et al. [17]; Sobczak [32]; Zhang et al. [37]; Sobczak [21] |

There are still gaps in the implementation of the automation process, like a lack of guidelines/roadmaps for RPA adoption; definition of formal techniques for choosing target processes for automation; definition of metrics for measuring the benefits achieved with RPA; and definition of critical factors for success in automating and their implications, among others gaps in the literature. The most cited gaps and the references are listed in Table 2.

**Table 2.** Gaps in the implementation of the automation process.

| Gap | References |
| --- | --- |
| Lack of guidelines/roadmaps for RPA adoption | Mishra et al. [25]; Sobczak [26]; Ahmad and Van Looy [9]; Enríquez et al. [38]; Hofmann et al. [17]; Siderska [12]; Syed et al. [13], Flechsing et al. [19]; Wewerka and Reichert [7]; Puica [15]; Sobczak [21] |
| Definition of formal techniques for choosing processes for automation | Ansari et al. [22]; Mishra et al. [25]; Ahmad and Van Looy [9]; Patri [29]; Syed et al. [13]; Choi et al. [31]; Puica [15] |
| Definition of metrics for measuring the benefits achieved with RPA | Mishra et al. [25]; Hofmann et al. [17]; Syed et al. [13]; Stravinskienė and Serafinas [5] |
| Definition of critical factors for success in automating and implications | Sobczak [26]; Ahmad and Van Looy [9]; Hofmann et al. [17]; Syed et al. [13]; Siderska [14] |

Combining the factors that contribute to the failure in the procedure of process automation and some existing weaknesses in research in this area, one can assume that, a roadmap (framework) that guides this entire process, from the as-is study, the to-be study, adoption, and follow-up (guidelines/roadmaps for process automation), is necessary.

So, this work aims to propose creating a framework (guidelines) that can answer the main research question: *what methodological support can be given to SMEs for successfully adopting process automation tools?* Based on the previous question, five research questions were formulated:

1. How to typify the processes to be automated?
2. What are business process automation's main barriers, risks, and limitations?
3. What are the leading technologies/tools to support business process automation?
4. What guidelines exist to support process automation in an organization?
5. How to evaluate the success of business process automation?

## 2 Framework/Guidelines Design

### 2.1 Design Science Research

The Design Science Research (DSR) methodology, using the model developed by Peffers et al. [39], was chosen to accomplish the goal of this work, whose main objective is creating and evaluating new technological artefacts that help organizations deal with technological problems [40]. So, Table 3 illustrates the goal of this work, using the DSR model of Peffers et al. [39]. It should be noted that the mentioned artefact will be the final product of this research, the support framework for SMEs to carry out the automation procedure of their processes minimizing the risks involved to maximize the positive effects of this change.

### 2.2 Formal Methodology for Business Process Automation (FM4BPA)

Business Process Automation (BPA) and Business Process Management (BPM) are related. At least four pillars form BPM and BPA ecosystem (Fig. 1): Human Resources, Business Process, Technology and Business Strategy. In these four sectors, there are some concerns that we must address in the FM4BPA:

1. Business Strategy – Is automation proper for the business? What does the organization have to do to apply automation? Will automation change the business model?
2. Human Resources – Human is the essential capital in the business core. Human Resources must be involved in the procedure of BPM and BPA from the beginning;
3. Business Processes – Materialization of business strategies. List of existent processes? Which one is a candidate for automation? Is there already BPM?
4. Technology – How automate? Which tool does the enterprise have to use to maximize outcomes?

**Table 3.** DSR Model of Peffers et al. [39] applied to this work.

| Activity of the DSR Model | Description |
|---|---|
| 1. Problem identification and motivation | Understand why automation fails; How to help SMEs in the BPA process so they achieve success in the procedure |
| 2. Define the objectives for a solution | A framework/guidelines will be proposed and validated for automation in SMEs |
| 3. Design and development | *"Formal Methodology for Business Process Automation"* (FM4BPA) will be the artefact developed to overcome the difficulties of SMEs in their BPA procedure. FM4BPA will describe all steps and provide tools for successful automation |
| 4. Demonstration | FM4BPA will be applied in SMEs that are in automation process or intend to start one. At least five companies will be invited to use FM4BPA in their automation procedure |
| 5. Evaluation | Questionnaires (with close answer questions) and case studies will be conducted with SME employees to collect data to analyze the suitability of the developed artefact. Case studies will help validate if all the steps are in the FM4BPA, and questionnaires will measure the success of the artefact. The possible answers will be defined on a scale of 1 to 5. The success will be calculated with the medium of the responses. If the result is less than 2, FM4BPA is not adequate. Suppose the result is between 2,01 and 4; FM4BPA is partially adequate and helpful. If the result is more than 4 FM4BPA is adequate and helpful |
| 6. Communication | This work will be communicated in the form of papers submitted to relevant conferences in the area and scientific journals, culminating with the thesis document |

These pillars have relations and are interdependent. They must be in the same direction. BPM and BPA have cycle characteristics. The study of processes and their management are a continuous effort in the organization to optimize its resources [5]. The automation focus appears on the execution stage in the BPM cycle (see Fig. 2).

Five phases form the BPM cycle [6, 7]: Design, Modelling, Execution, Monitoring and Optimization. These phases are cycle and sequential (Fig. 2).

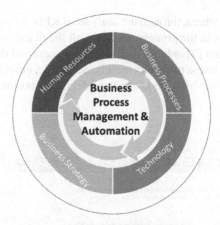

**Fig. 1.** BPM and BPA ecosystem.

1. Design – Identify the current business process and design of the proposed change and break the process down into multiple tasks;
2. Modelling – What if analysis to model change is proposed on multiple variables?; Model process using suitable BPM software?; As-is model to be in the future (to-be scenario);
3. Execution – Use of software to track and manage; Execute the process, or put a system in place; Integration and automation;
4. Monitoring – Keep track of completion level of changes and effectiveness; Monitor and analyze the system; KPIs;
5. Optimization – Identify problems and opportunities and apply changes for more significant cost savings and efficiency; Make changes to the business process to improve it.

Van der Aalst [41] proposed that BPM be implemented in three phases: design or redesign of the process, its implementation, and its execution and adjustment. Other authors follow roughly the same lifecycle, with more or less detail than others.

In the design phase, the understanding and analysis of the functioning of the process are carried out, which tasks are done, the roles of its stakeholders, that is, the process mining, and the description of the current state (as-is model) of the organization [7, 9, 42]. In the second phase of the BPM cycle, modelling is responsible for the graphic description, through BPMN (Business Process Management Notation), of the to-be scenario if improvements have been detected to be carried out in the previous phase [42]. In this description, weaknesses are analyzed, tasks are restructured, weaknesses are eliminated, and opportunities for improvement are mentioned [42].

The execution phase follows, the modelling implementation occurs, and the characterized tasks are automated [42]. The monitoring phase is responsible for verifying compliance with the established objectives. It detects errors, anomalies, or deviations from the objectives, through the verification of the Key Performance Indicators (KPI) that have been defined to quantify the impact of automation [42]. The cycle ends with the optimization phase, where, through the analysis of KPIs, new measures are drawn up for future improvements in the process, always with the objective of optimization [42].

Wewerka and Reichert [7] indicate that, as a consequence of the constantly changing context of our world, business processes must be adaptable, efficient, and with low costs. In short, companies increasingly need a greater degree of business process automation to remain competitive.

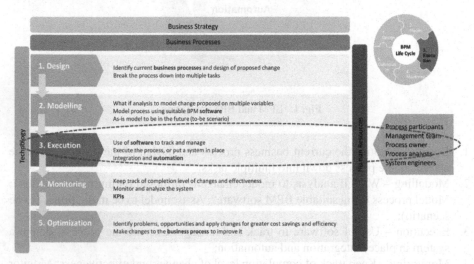

**Fig. 2.** BPM cycle with the four pillars from their ecosystem [5–7, 41]

Figure 2 complements the BPM cycle with four pillars of the ecosystems and calls our attention to the nomination of five actors (pillar of Human Resources) important to the automation procedure: process participants, management team, process owner, process analysis, and system engineers.

*Integration and automation* are mentioned in the execution (third step of the BPM cycle). This is related to integration with other systems or processes related to the study process. Automate the process with a software tool. At this point, a *Formal Methodology for Business Process Automation* (FM4BPM) appears (Fig. 3).

The automation life cycle includes six steps [43]: 1. *Identify processes for automation*; 2. *Design and optimization of the process*; 3. *Verification of the digital readiness*; 4. *Selection of the automation technology/tool*; 5. *Automation implementation*, and 6. *Solution governance*. Step 2 is a cycle task since optimization and design of process must be a concern, with the high performance of automation as a goal.

In each of the steps, some issues must be considered (Fig. 4). So, in step 1 we have to verify the scenario as-is and project a to-be, and choose processes that are more suitable for automation. In this step, we want to propose a *checklist with the characteristics of candidates for automation*. With this tool, we can help the correct identification of processes for automation. As an output of step 1, we have the delivery of the ranking list of the processes to automate. In this phase, we think it is important and necessary to include stakeholders to help the process mining in the organization.

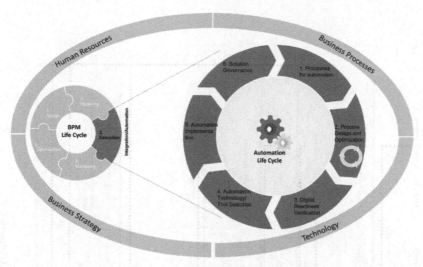

**Fig. 3.** Formal Methodology for Business Process Automation (FM4BPA)

One of the most critical and time-consuming tasks in step 2 is to analyze the process and optimize (remember the cycle characteristic and its goal). The detail of the process and its optimization can be found in the output of step 2.

It is considered essential to define also the KPIs (Key Performance Indicators) because we must know how automation affects the process (compared to as-is scenarios and KPIs before automation).

Digital readiness is the concern in step 3. We want to develop a *checklist with the digital characteristics for automation.* Is the process digitalized? If not, is it possible? What are the changes? If a process is ready for automation, the organization can choose the automation technology/tool (step 4 of FM4BPA).

FM4BPA defines a *matrix that relates processes automation requirements versus technology characteristics.* Step 4 delivers a *list of process requirements* and *the tool for the implementation of the automation.* In step 4, system engineers must be included in the IT governance and architecture-related tasks.

The focus in *automation implementation* (step 5 of FM4BPA) is the materialization of the choices made in the previous steps, the process, its design, its study, its requirements and a computer tool to support automation. At least six tasks are essential in this step: focus on the responsibilities of each human resource; installation and configuration of the technology/tool; training of the participants in processes; automation process design in the tool; implementation of solution; and test.

Lastly, step 6 *solution governance* aims to *monitor and analyze the effectiveness of the adopted solution* (with evaluation dashboards with KPIs (before and after automation) and the list of processes requirements); to concern the *security/RGPD and privacy* of the solution; *dealing with errors and exceptions,* and *scalability and balance resources.*

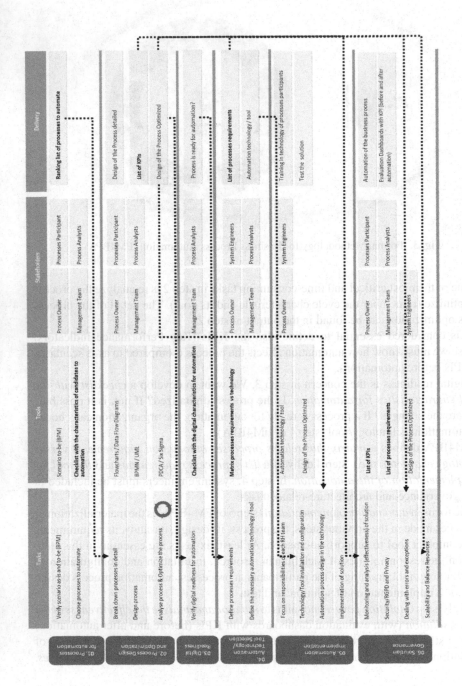

**Fig. 4.** FM4BPA detail with tasks, tools, stakeholders and deliveries

# 3 Final Considerations

Process automation using supporting technologies has been an excellent driving factor in the digital transformation of organizations [30]. It is a significant technological evolution in which emerging software platforms are already in a very acceptable state of maturation and are already scalable, reliable, and resilient [11].

The benefits and the challenges of adopting process automation technologies were well stated in this paper. We concluded that there are several theoretical benefits that automation can add to organizations. We believe that the challenges are being addressed. Namely, the question of discovering the critical processes for automation and the concern with the social and human aspects that technologies can influence becomes evident. This concern is significant because a process that transforms the paradigm of organizational functioning cannot forget its human capital. Human resources are a significant part of the core ecosystem.

This document draws attention to those challenges. With the research and investigation into the issue of guidelines/roadmap for the adoption of automation, covering the entire procedure, especially in SMEs, it is intended that part of the challenges demonstrated here will be overcome.

# References

1. Kergroach, S., Bianchini, M.M.: OECD: The digital transformation of SMEs. OECD Publishing (2021).https://doi.org/10.1787/bdb9256a-en
2. Statista: SMEs in Europe - Statistics and Facts. Published by D. Clark. https://www.statista.com/statistics/1254442/digitalization-adoption-among-eu-smes/
3. Zaoui, F., Souissib, N.: Roadmap for digital transformation: a literature review. Procedia Comput. Sci. **175**, 621–628 (2020). https://doi.org/10.1016/j.procs.2020.07.090
4. Gartner: Gartner says worldwide robotic process automation software revenue to reach nearly $2 billion in 2021. https://www.information-age.com/rpa-revenue-reach-nearly-2-billion-2021-gartner-123491711/
5. Stravinskienė, I., Serafinas, D.: Process management and robotic process automation: the insights from systematic literature review. Manag. Organ. Syst. Res. **86**(1), 87–106 (2021). https://doi.org/10.1515/mosr-2021-0006
6. Chakraborti, T., et al.: From robotic process automation to intelligent process automation. In: Asatiani, A., et al. (eds.) BPM 2020. LNBIP, vol. 393, pp. 215–228. Springer, Cham (2020). https://doi.org/10.1007/978-3-030-58779-6_15
7. Wewerka, J., Reichert, M.: Robotic process automation - a systematic mapping study and classification framework. Enterp. Inf. Syst. **17**(2), 1986862 (2021). https://doi.org/10.1080/17517575.2021.1986862
8. IBM: RPA is essential for small and midsized enterprises to gain a competitive advantage. https://www.techtarget.com/
9. Ahmad, T., Van Looy, A.: Business process management and digital innovations: a systematic literature review. Sustainability **12**, 6827 (2020). https://doi.org/10.3390/su12176827
10. Ahuja, S., Tailor, RK.: Performance evaluation of robotic process automation on waiting lines of toll plazas. NVEO Nat. Volatiles Essent. Oils 10437–10442 (2021). https://www.nveo.org/index.php/journal/article/view/2945

11. Bhatnagar, N.: Role of robotic process automation in pharmaceutical industries. In: Hassanien, A.E., Azar, A.T., Gaber, T., Bhatnagar, R., F. Tolba, M. (eds.) AMLTA 2019. AISC, vol. 921, pp. 497–504. Springer, Cham (2020). https://doi.org/10.1007/978-3-030-14118-9_50

12. Siderska, J.: Robotic process automation - a driver of digital transformation? Eng. Manag. Prod. Serv. **12**(2), 21–31 (2020). https://doi.org/10.2478/emj-2020-0009

13. Syed, R., et al.: Robotic process automation: contemporary themes and challenges. Comput. Ind. **115**, 103162 (2020). https://doi.org/10.1016/j.compind.2019.103162

14. Siderska, J.: The adoption of robotic process automation technology to ensure business processes during the COVID-19 pandemic. Sustainability **13**(14), 8020 (2021). https://doi.org/10.3390/su13148020

15. Puica, E.: How is it a benefit using robotic process automation in supply chain management? J. Supply Chain Cust. Relatsh. Manag. **2022**, 1–11 (2022). https://doi.org/10.5171/2022.221327

16. Romão, M., Costa, J., Costa, C.: Robotic process automation: a case study in the banking industry. In: 14th Iberian Conference on Information Systems and Technologies (CISTI) (2019). https://doi.org/10.23919/CISTI.2019.8760733

17. Hofmann, P., Samp, C., Urbach, N.: Robotic process automation. Electron. Mark. **30**(1), 99–106 (2019). https://doi.org/10.1007/s12525-019-00365-8

18. Yatskiv, N., Yatskiv, S., Vasylyk, A.: Method of robotic process automation in software testing using artificial intelligence. In: International Conference on Advanced Computer Information Technologies (ACIT) (2020). https://doi.org/10.1109/ACIT49673.2020.9208806

19. Flechsing, C., Anslinger, F., Lash, R.: Robotic process automation in purchasing and supply management: a multiple case study on potentials, barriers, and implementation. J. Purch. Supply Manag. **28**(1), 100718 (2021). https://doi.org/10.1016/j.pursup.2021.100718

20. Ng, K., Chen, C., Lee, C., Jiao, J., Yang, Z.: A systematic literature review on intelligent automation: aligning concepts from theory, practice, and future perspectives. Adv. Eng. Inform. **47**, 101246 (2021). https://doi.org/10.1016/j.aei.2021.101246

21. Sobczak, A.: Robotic process automation as a digital transformation tool for increasing organizational resilience in Polish enterprises. Sustainability **14**(3), 1333 (2022). https://doi.org/10.3390/su14031333

22. Ansari, W., Diya, P., Patil, S.: A review on robotic process automation - the future of business organizations. In: 2nd International Conference on Advances in Science and Technology (ICAST 2019) (2019). https://doi.org/10.2139/ssrn.3372171

23. Asquith, A., Horsman, G.: Let the robots do it! – taking a look at robotic process automation and its potential application in digital forensics. Forensic Sci. Int.: Rep. **1**, 100007 (2019). https://doi.org/10.1016/j.fsir.2019.100007

24. Auth, G., Czarnecki, C., Bensberg, F.: Impact of robotic process automation on enterprise architectures. In: INFORMATIK 2019: 50 Jahre Gesellschaft für Informatik – Informatik für Gesellschaft (2019). https://doi.org/10.18420/inf2019_ws05

25. Mishra, S., Devi, K.S., Narayanan, M.B.: People and process dimensions of automation in business process management industry. Int. J. Eng. Adv. Technol. **8**(6), 2465–2472 (2019). https://doi.org/10.35940/ijeat.F8555.088619

26. Sobczak, A.: Developing a robotic process automation management model. Informatyka Ekonomiczna **2**(52), 85–100 (2019). https://doi.org/10.15611/ie.2019.2.06

27. Hindel, J., Cabrera, L., Stierle, M.: Robotic process automation: hype or hope? In: 15th International Conference on Wirtschaftsinformatik (2020). https://doi.org/10.30844/wi_2020_r6-hindel

28. Leite, G., Albuquerque, A., Pinheiro, P.: Process automation and blockchain in intelligence and investigation units: an approach. Appl. Sci. **10**(11), 3677 (2020). https://doi.org/10.3390/app10113677

29. Patri, P.: Robotic process automation: challenges and solutions for the banking sector. Int. J. Manag. **11**(12), 322–333 (2020). https://doi.org/10.34218/IJM.11.12.2020.031

30. Rizk, Y., et al.: A conversational digital assistant for intelligent process automation. In: Asatiani, A., et al. (eds.) BPM 2020. LNBIP, vol. 393, pp. 85–100. Springer, Cham (2020). https://doi.org/10.1007/978-3-030-58779-6_6

31. Choi, D., R'bigui, H., Cho, C.: Candidate digital tasks selection methodology for automation with robotic process automation. Sustainability 13(16), 8980 (2021). https://doi.org/10.3390/su13168980

32. Sobczak, A.: Robotic process automation implementation, deployment approaches and success factors – an empirical study. Entrepreneurship Sustain. Issues 8(4), 122 (2021). https://doi.org/10.9770/jesi.2021.8.4(7)

33. Dey, S., Das, A.: Robotic process automation: assessment of the technology for transformation of business processes. Int. J. Bus. Process Integr. Manag. 9(3), 220–230 (2019). https://doi.org/10.1504/IJBPIM.2019.100927

34. Reddy, K.N., Harichandana, U., Alekhya, T., Rajesh, S.M.: A study of robotic process automation among artificial intelligence. Int. J. Sci. Res. Publ. 9(2), 392–397 (2019). https://doi.org/10.29322/IJSRP.9.02.2019.p8651

35. Dechamma, P.J., Shobha, N.S.: A review on robotic process automation. Int. J. Res. Eng. Sci. Manag. 3, 103162 (2020). https://www.ijresm.com/Vol.3_2020/Vol3_Iss5_May20/IJRESM_V3_I5_60.pdf

36. Sethi, V., Jeyaraj, A., Duffy, K., Farmer, B.: Embedding robotic process automation into process management: case study of using tasks. AIS Trans. Enterp. Syst. (2020). https://doi.org/10.30844/aistes.v5i1.19

37. Zhang, C., Issa, H., Rozario, A., Søgaard, J.: Robotic process automation (RPA) implementation case studies in accounting: a beginning to end perspective. Account. Horiz. (2021). https://ssrn.com/abstract=4008330

38. Enríquez, J.G., Jiménez-Ramírez, A., Domínguez-Mayo, F.J., García-García, J.A.: Robotic process automation: a scientific and industrial systematic mapping study. IEEE Access 8, 39113–39129 (2020). https://doi.org/10.1109/ACCESS.2020.2974934

39. Peffers, K., Tuunanem, T., Rothenberger, M.A., Chatterjee, S.: A design science research methodology for information systems research. J. Manag. Inf. Syst. 24, 45–77 (2007). https://doi.org/10.2753/MIS0742-1222240302

40. Hevner, A.R., March, S.T., Park, J., Ram, S.: Design science in information system research. MIS Q. 28, 75–105 (2004). https://doi.org/10.2307/25148625

41. van der Aalst, W.M.P.: Business process management: a comprehensive survey. Int. Sch. Res. Notices (2013). https://doi.org/10.1155/2013/507984

42. De Ramon Fernandez, A., Ruiz Fernandez, D., Sabuco Garcia, Y.: Business process management for optimizing clinical processes: a systematic literature review. Health Inform. J. 26(2), 1305–1320 (2020). https://doi.org/10.1177/1460458219877092

43. Mamede, H.S.: Automatização de Processos com RPA. FCA, Lisboa (2021)

# An Integrated Approach Using Robotic Process Automation and Artificial Intelligence as Disruptive Technology for Digital Transformation

Anderson Araújo[1]([⊠]) [iD], Henrique S. Mamede[2] [iD], Vítor Filipe[1,3], and Vitor Santos[2]

[1] University of Trás-os-Montes e Alto Douro, Quinta de Prados, 5000-801 Vila Real, Portugal
`anderson.s.araujo@inesctec.pt`
[2] INESC TEC, Universidade Aberta, R. da Escola Politécnica nº 147, 1269-001 Lisbon, Portugal
[3] INESC TEC – INESC Tecnologia e Ciência, 4200-465 Porto, Portugal

**Abstract.** Digital transformation is a phenomenon arising from social, behavioral and habitual changes due to global economic and technological development. Its main characteristic is adopting disruptive digital technologies by organizations to transform their capabilities, structures, processes and business model components. One of the disruptive digital technologies used in organizations' digital transformation process is Robotic Process Automation. However, the use of Robotic Process Automation is limited by several constraints that affect its reliability and increase the cost. Artificial Intelligence techniques can improve some of these constraints. The use of Robotic Process Automation combined with Artificial Intelligence capabilities is called Hyperautomation. However, there is a lack of solutions that successfully integrate both technologies in the context of digital transformation. This work proposes an integrated approach using Robotic Process Automation and Artificial Intelligence as disruptive Hyperautomation technology for digital transformation.

**Keywords:** Robotic Process Automation · Artificial Intelligence · Digital Transformation · Hyperautomation

## 1 Introduction

Digital Transformation (DX) is a phenomenon arising from social, behavioral and habitual changes as a result of global economic and technological development [1, 2]. Nadkarni and Prügl [3] argue that digital transformation consists of the *"adoption of disruptive digital technologies on the one side and actor-guided organizational transformation of capabilities, structures, processes and business model components on the other side"* [3, p. 236].

In this sense, Ribeiro et al. [4], Bu et al. [5], and Daptardar [6] claim that Robotic Processes Automation (RPA) is an excellent tool for innovation and disruptive technology that enable digital transformation in organizations. There are many advantages to

M. Papadaki et al. (Eds.): EMCIS 2022, LNBIP 464, pp. 438–450, 2023.
https://doi.org/10.1007/978-3-031-30694-5_32

implementing RPA in an organization embedded in a DX process. Daptardar [6] argues that *"RPA gets organizations past the halfway point. It permits the organization to mechanize undertakings without waiting to finish each change. When done appropriately, RPA can help departments of the organization gain interval ground while hanging tight for their chance on the need list"* [6, p. 889].

However, RPA has some constraints. Complex workflows, processes with many exceptions, immature systems and unstable environments which require workflow changes are not recommended for RPA applications [7–9]. Nevertheless, to deal with RPA constraints, various authors propose an integrated approach between RPA and others research topics such as Business Process Management (BPM) [10, 11] and Artificial Intelligence (AI) [8, 12, 13].

However, some authors, such as Daptardar [6] and Hartmann [14], claim that RPA/AI solutions are still in the early stages of their evolution. Only a few large companies use such a solution in their DX processes. One of the reasons the authors give is that automated robots are often limited in changes and new developments, which leads to reduced flexibility and is more error-prone. In addition, there is a lack of solutions that successfully integrate both technologies in the context of digital transformation [6, 8, 12–14].

Therefore, this work proposes an integrated approach using RPA and AI as disruptive Hyperautomation technology for digital transformation. As a result, we hope to pave the way for enabling the adaptability of RPA to handle changes, making it less error-prone and, consequently, reducing maintenance efforts and costs.

In order to do that, in the following sections, we will describe the problem statement (Sect. 2) and discuss the theoretical background on DX, RPA, AI and Hyperautomation (Sect. 3). After that, in Sect. 4, we will describe the proposed solution. Finally, in Sect. 5, we will point out some final considerations.

## 2 Problem Statement

According to Siderska [7], RPA is at the forefront of organizations' digital transformation process. Bu et al. [5] claim that RPA is an excellent tool for innovation and disruptive technology that enable digital organizational transformation. In addition, Turcu and Turcu [15] also state that RPA is essential in adequately shifting human resources functions for organizations' digital transformation processes.

Indeed, currently, several business processes, when executed, have activities that can be performed by pieces of software called bots automatically without the need for direct human interference. RPA replaces human actions performed in Graphical User Interfaces (GUI) when performing a specific task in a business process by a software (bot) that imitates them [10].

There are many advantages to implementing RPA in an organization. Ribeiro et al. [4], Siderska [7], and Madakam et al. [16], for example, claim that RPA increases the ability of employees to deal with more work processes, reduces errors in data analysis in order to provide a more assertive decision-making process, reduces the performance of repetition activities by employees, which encourages a focus on more creative work and problem solving, promotes the standardization of processes, which increases the quality of services and products, reduces costs and brings more reliability to the production chain.

However, automated robots are often limited to deal with changes and new developments, which leads to them having reduced flexibility and makes them more error-prone. Nevertheless, workflow changes and software updates are becoming increasingly common due to the dynamic environment in which these solutions work. Moreover, the cost of implementing and maintaining RPA is high [4, 7, 8, 10]. As a result, despite of be an excellent tool for innovation and disruptive technology that enables digital transformation in organizations, RPA has not been a widely used approach in organizations' DX processes [6, 14, 17]. Hartmann [14] claims that *"many things are already feasible today with RPA and AI, although AI, in particular, is still in the early stages of its evolution. Nevertheless, only a few companies are relying on digital transformation. The deployment of one or both technologies can be described as early adopters. This applies specifically to larger companies with sufficient financial resources and human capital and the possibility of coping with setbacks easier"* [14, p. 54]. Figure 1 illustrates this scenario.

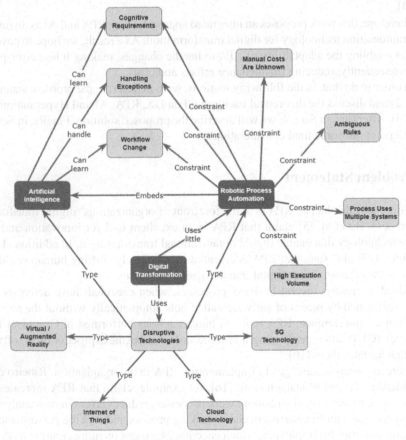

**Fig. 1.** Conceptual map of the problem statement.

Several authors propose solutions using AI and RPA together to solve that problem. Yatskiv et al. [12] claim that *"the best way to overcome the mentioned challenge is to apply RPA together with other innovative solutions such as Machine Learning"* [12, p. 502]. Nakano [13] claims that *"in this process of interconnectedness, Artificial Intelligence (AI) and Robot Process Automation (RPA) are disruptive and promise emerging technologies"* [13, p. 51]. In addition, Hartmann [14] claims that *"whereas the areas of application for RPA are rather limited to repetitive, non-complex tasks, AI can cover a broader spectrum"* [14, p. 52]. Bornet et al. [18] go further and claim that RPA/IA solutions can provide a new renaissance for our society, making it more human and reinventing what we call "work". That is what the authors call "Intelligent Process Automation" or "Hyperautomation". However, these approaches were developed to solve a specific problem in a specific context. There is a lack of solutions that successfully integrate both technologies (AI and RPA) in the context of digital transformation.

Thereby, considering what has been discussed so far, we have then summarized our Research Problem (RP):

- RPA has not been widely used in organizations' DX processes due to its constraints in dealing with changes and new developments.

  Thus, our Research Question (RQ) can then be summarized:
- How can we leverage RPA as a widely used approach in organizations' DX processes?

  With the basis of our research questions and research problem, we summarized our research goal (RG):
- Defining an integrated approach using RPA and AI improves RPA's adaptability to handle changes and makes it less error-prone.

  As part of the research goals, we can then summarize three hypotheses (H):
- The hypothesis (H1) is that AI techniques enables RPA solutions to learn from changes.
- The hypothesis (H2) is that AI solutions can be integrated into RPA solutions.
- The hypothesis (H3) is that an integrated approach using RPA and AI can leverage RPA as a digital disruption technology widely used in organizations' DX processes.

## 3 Background

### 3.1 Introduction

In this section, we will discuss the main topics related to this research: Digital Transformation (DX), Robotic Process Automation (RPA), Artificial Intelligence (AI), and Hyperautomation.

The primary purpose of this section is to present the concepts and the main characteristics of the topics mentioned above, as well as to share the results of other studies closely related to what is being carried out. This section aims to establish the importance of the study and works as a benchmark for comparing the results with other findings [19].

## 3.2 Digital Transformation

Nowadays, the term Digital Transformation (DX) is in evidence. According to Hess et al. [1], DX is the transformation *"concerned with the changes digital technologies can bring about in a company's business model, ... products or organizational structures"* [1, p. 124]. In this context, we are concerned about how new digital technologies and business models associated with them impact the existing business models and the value propositions of goods and services. According to Silva Neto and Chiarini [2], *"digital platforms have emerged as a new technical and organizational element capable of changing the dynamics of consolidated socioeconomic models"* [2, p. 1]. In this sense, we are dealing with the change in the dynamics of the customer-supplier relationship phenomena. Moreover, society is experiencing new ways of consuming, buying, communicating, and offering products and services. That is, organizations are experiencing a multidimensional transformation [1].

Hess et al. [1] claim that these changes are not limited to "products, business processes, sales channels or supply chains, but all business models are being reformulated and often overturned". It is about the disruption of business models caused by the impact and effect of new digital technologies, as presented in Udovita [19].

According to Bradley et al. [20], digital disruption *"is the effect of digital technologies and business models on a company's current value proposition and its resulting market position"* [20, p. 1]. The authors claim that digital disruption can change an organization's business models faster than any force in history and reshape the market.

In this way, Nadkarni and Prügl [3] conceptualize digital transformation in the context of adopting disruptive digital technologies by organizations but also consider an organizational transformation of capabilities, structures, processes, and business approach. One of these disruptive technologies used in the organizations' DX processes is RPA.

RPA is becoming an essential element of organizations' business operations. According to Bu et al. [5], RPA is an excellent tool for innovation and disruptive technology that enable digital transformation in organizations. The authors claim that *"RPA has a wide range of applications in various industries such as healthcare and pharmaceuticals, financial services, outsourcing, retail, telecommunications, energy and utilities, real estate, and fast-moving consumer goods"* [5, p. 30]. In addition, Turcu and Turcu [15] also state that RPA is essential in adequately shifting human resources functions for organizations' digital transformation processes.

## 3.3 Robotic Process Automation

Currently, RPA is an emerging technology when it comes to business process automation. It is about replacing human actions performed in Graphical User Interfaces (GUI) when executing a specific task in a business process or workflow by a software (robot) that imitates them. In this context, due to the rapid advance of digital transformation in organizations, the digitisation of business processes is increasing evidence. This fact has positioned technologies such as RPA at the forefront of organizations' DX process [7].

There are many advantages to implementing RPA in an organization. According to Siderska [7] and Madakam et al. [16], RPA:

- increases the ability of employees to deal with more work processes;
- reduces errors in data analysis in order to provide a more assertive decision-making process;
- reduces the performance of repetition activities by employees, which encourages a focus on more creative work and problem-solving;
- promotes the standardization of processes, which increases the quality of services and products;
- reduces costs and brings more reliability to the production chain;
- follows regulatory compliance rules and provides an audit trail history.

However, RPA has some constraints. Complex workflows and processes with many exceptions are not recommended for RPA applications. Siderska [7] presents some criteria to identify suitable processes for RPA, such as processes with low cognitive requirements; processes with no need to access multiple systems, given that RPA is applied to existing applications; processes and tasks that are performed relatively frequently. Kaarnijoki [8] also proposed a set of criteria to identify suitable processes for RPA. Table 1 shows these criteria.

**Table 1.** Criteria for RPA.

| Criteria | Description |
| --- | --- |
| High volume | The process has a high volume of transactions or is performed frequently |
| The process uses multiple systems | The process involves accessing multiple systems that otherwise could not be easily integrated |
| Stable environment | Mature systems and environments remain the same every time process is executed |
| The process can be broken down into unambiguous rules | The process can be broken down into exact step-by-step rules which have no room for misinterpretation |
| Minimum need for exception handling | A highly standardized process with little need for handling exceptions |
| Manual costs are known | The cost structure of the current process is known, and ROI can be calculated for the RPA solution |
| Low cognitive requirements | The process does not require judgement or complex interpretation skills |

Furthermore, van der Aalst et al. [9] present a graphical representation to understand the relevance of RPA and to identify the suitable processes considering the proposed by Siderska [7] and Kaarnijoki [8]. Figure 2 shows the graphical representation proposed by van der Aalst et al. [9].

444     A. Araújo et al.

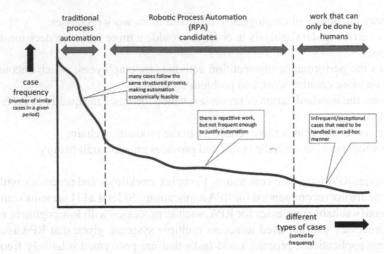

**Fig. 2.** Positioning RPA. (Source: van der Aalst et al. [9]).

Considering the content presented in Table 1 and Fig. 2, it is evident that, from an economic perspective, the RPA candidate's processes are those where there is repetitive work but not frequent enough to justify the traditional process automation. On the other hand, when we have infrequent and exceptionalities cases, it must be done by humans. In addition, König et al. [10] claim that *"RPA is limited in that many techniques required to successfully implement it lay outside its scope. This includes gathering the necessary information for automation enactment, dealing with exceptions during the execution of automated processes, and managing process automation on an organizational level"* [10, p. 132].

To deal with RPA constraints, various authors propose an integrated approach between RPA and other research topics. Liermann et al. [21] propose a algorithm-driven decision-making solution in the automated process. Moreover, Martins et al. [22], propose to use Machine Learning (ML) to create cognitive RPA.

Yatskiv et al. [12] and Kaarnijoki [8] propose an integrated approach between RPA and Artificial Intelligence (AI) in order to use AI capabilities to complement RPA. Furthermore, Nakano [13] claims that *"in this process of interconnectedness, Artificial Intelligence (AI) and Robot Process Automation (RPA) are disruptive and promise emerging technologies"* [13, p. 51]. The author presents a study on AI and RPA's effects on the accounting and auditing fields in Japanese society. Patel et al. [23] integrated Machine Learning (ML) and RPA techniques to customize an automated email response bot. Hartmann [14] presents the chances and risks of RPA and AI for process optimization within the supply chain. Bellman and Göransson [24] present a framework to bridge RPA and AI. Finally, Parchande et al. [25] integrated RPA with a ML technique to develop a contractual employee management system. These approaches were developed to solve a specific problem and in a specific context.

## 3.4 Artificial Intelligence

The term Artificial Intelligence (AI) was formally proposed at a conference at Dartmouth University in 1956 [According to Hu and Jiang [26] *"The Dartmouth Conference of 1956 proposed that 'every aspect of learning or any other feature of intelligence can be so precisely described that a machine can be made to simulate it.' The Conference was the moment that artificial intelligence gained its name, its mission, its first success and its major players, and is widely recognised as the birth of artificial intelligence"* [27, p. 241]. According to Nunes et al. [27], AI can be defined as the *"ability of a system to correctly interpret external data, learn from that data, and use that learning to achieve specific goals and tasks through flexible adaptation"* [28, p. 2].

AI is one of the most promising areas of computer science, and AI techniques are expected to solve complex problems faster and more accurately than humans, including problems for which currently there are no solutions. It is about the ability of machines and software to make decisions based on the analysis of a set of received information. Moreover, AI solutions must be able to learn by accumulating knowledge and successfully applying it.

For the last seven decades, AI has experienced a long development process that has brought us to the current scenario where AI algorithms and ML techniques have been successfully used in several business areas, such as commerce, industry, and digital services. In this scenario, ML has been a critical component of organizations' digital transformation. The basic idea behind ML is to teach machines how to learn from data more efficiently through AI techniques embedded in reasoning techniques such as statistics, probabilities, use cases, etc. [4, 28]. These AI algorithms enable organizations to develop solutions to extract, classify information, associate, optimize, group, predict, identify patterns, etc.

According to Grekousis [29], there are four main categories of learning methods:

- supervised (labelled data);
- unsupervised (unlabeled data);
- semi-supervised (labelled data only for a small portion of the training dataset); and
- reinforcement (via a system of sensors interacting with a dynamic environment to obtain and analyze data on the fly).

In order to select the appropriate learning category, we need to consider the given problem. Grekousis [29] also claims two categories of learning: shallow and deep learning. The main difference between the two categories lies in the depth of the analysis.

Finally, it is essential to highlight the increase in the number of applications that use AI, ML and RPA technologies, including a combination of these and other related technologies. According to Ribeiro et al. [4], *"given the scope of the applicability of AI, RPA has gradually been adding, to its automation features, implementations of algorithms or AI techniques applied in certain contexts (e.g., Enterprise Resource Planning, Accounting, Human Resources) to classify, recognise, categorise, etc. In recent years, some academic studies have been published as challenges and potential, as well as case studies of the applicability of RPA and AI"* [4, p. 53].

## 3.5 Hyperautomation

According to Bornet et al. [18], *"Intelligent Automation (IA) or 'Intelligent Process Automation', is a new notion, officially coined in 2017 by Institute of Electrical and Electronics Engineers. More recently, IA has been given different names, including Hyperautomation (by Gartner), Integrated Automation Platform (by Horses For Sources), and Cognitive Automation (by several sources)"* [18, p. 25].

It is about to apply the power of AI capabilities in robotics in order to meet most of the needs both can't satisfy individually. Hyperautomation combines AI capabilities with robotics capabilities and complements each other by connecting these capabilities. Figure 3 shows the positioning of Hyperautomation with other recent technology concepts proposed by Bornet et al. [18].

Madakan et al. [30] claim that *"Hyperautomation involves cutting-edge technologies such as artificial intelligence, machine learning and others to automate business operations, processes, services and thus complement human talent"* [31, p. 2]. It is an advanced version of automation followed by RPA. But the point of Hyperautomation is not to replace humans because when you handle repetitive tasks in an automated way you free up your workforce and allow them to focus on more strategic and cognitive tasks.

Martins et al. [22] propose an RPA application, which in real-time, dynamically detects objects in software applications interface. To do that, the authors trained a Convolution Neural Network (CNN) by using several interfaces and menus and used them to classify software interfaces in real-time. Furthermore, a developed software takes automated actions like moving the mouse pointer, editing text, clicking, etc.

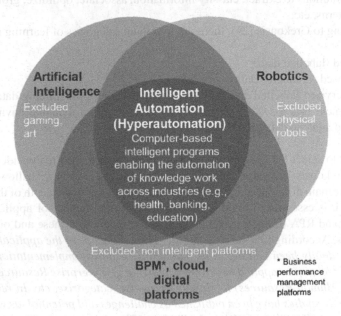

**Fig. 3.** Positioning RPA. (Source: van der Bornet et al. [18]).

The authors claim the techniques *"based on deep learning is capable of detecting objects in real-time, classify them, with outstanding accuracy, and take dynamic actions"* [23, p. 1].

Finally, Jha et al. [31] point that *"Process Automation has the potential to bring great benefits for businesses and organizations especially in the financial services industry where businesses are information-intensive and experience rich data flows. This was achieved mainly via Robotic Process Automation (RPA), but the increased complexity of the Machine Learning (ML) algorithms increased the possibility of integrating classic RPA with Artificial Intelligence (AI)"* [32, p. 1].

## 4   Proposed Solution

The proposed solution is to develop an integrated approach using RPA and AI that improves RPA adaptability to handle changes and makes it less error-prone, in order to leverage RPA as a digital disruption technology widely used in organizations' DX processes.

To do this, we need to analyze some AI techniques to define how these techniques, such as those proposed by Martins et al. [22], Jha et al. [31], O. I. Abiodun et al. [32], Somvanshi et al. [33], Charbuty and Abdulazeez [34] or Kuo and Huang [35], can be applied in RPA solutions in order to mitigate some of their constraints, especially to improve RPA adaptability. This is **Artifact 1**. Secondly, we need to define a framework for applying Artifact 1 in the organization's DX processes. This is **Artifact 2. Artifacts 1 and 2** make up the RPA/AI (Hyperautomation) solution. Figure 4 illustrates this scenario.

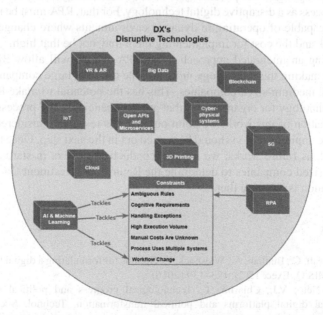

**Fig. 4.** Integration between AI techniques and RPA in the context of DX.

In the next step, we will demonstrate the use of the artefact to solve one or more instances of the problem by performing simulations and case studies. For this, we will analyze the impact of RPA solutions (RPA solutions without embedded AI techniques) in organizations' DX processes and the impact of **Artifacts 1 and 2** (RPA/AI solutions) in organizations' DX processes.

Next, we will compare the objectives of the solution to actual observed results from using the RPA/IA solution in the demonstration. The objective is to demonstrate how well the developed artefact supports the solution to the problem. For this, we will define success indicators and metrics, run the solution, measure them, and analyze the results.

The next step is to compare the results of the impact of RPA solutions (RPA solutions without embedded AI techniques) in organizations' DX processes with the analysis of the impact of the RPA/AI solution in organizations' DX processes. **Artifacts 1 and 2** (RPA/AI solution) are expected to perform better than RPA (RPA solution without embedded AI techniques).

## 5  Conclusions

Considering the current scenario, in which society is experiencing the emergence of new ways of consuming, buying, communicating, and offering products and services, organizations undergo a multidimensional transformation. DX is one of the critical pieces of this puzzle. However, before adopting disruptive digital technologies such as RPA, AI or ML, organizations need to consider the benefits and risks these technologies can bring.

This paper highlights the possibility to RPA be used in organizations' digital transformation process as a disruptive digital technology. For that, RPA must be more reliable and must be capable of operating in dynamic environments where changes are part of the daily work and the cost for implementing them may not be that high.

So, defining an integrated approach using RPA and AI will allow RPA to take a step forward, making this technology more reliable to use in large companies, startups, and small and medium-sized companies. This has the potential to make RPA a highly disruptive technology for organizations that are experiencing a DX process.

As a limitation of this work, we point out that this research considered a literature review and the implementation should be carried out in the next step. Once this approach is operational, as future works, we suggest conducting surveys in startups and small and medium-sized companies to determine the Return On Investment (ROI) that using Hyperautomation is bringing them.

## References

1. Hess, T., Matt, C., Benlian, A., Wiesböck, F.: Options for formulating a digital transformation strategy. MIS Q. Exec. **15**(2), 123–139 (2016)
2. Da Silva Neto, V.J., Chiarini, T.: Technological progress and political systems: non-institutional digital platforms and political transformation. Technol. Soc. **64**, 101460 (2021)

3. Nadkarni, S., Prügl, R.: Digital transformation: a review, synthesis and opportunities for future research. Manag. Rev. Q. **71**(2), 233–341 (2020). https://doi.org/10.1007/s11301-020-00185-7

4. Ribeiro, J., Lima, R., Eckhardt, T., Paiva, S.: Robotic process automation and artificial intelligence in industry 4.0–a literature review. Procedia Comput. Sci. **181**, 51–58 (2021)

5. Bu, S., Jeong, U.A., Koh, J.: Robotic process automation: A new enabler for digital transformation and operational excellence. Bus. Commun. Res. Pract. **5**, 29–35 (2022)

6. Daptardar, S.: A review-the golden triangle of RPA, AI and digital transformation. Int. Res. J. Mod. Eng. Technol. Sci. **3**, 887–891 (2021)

7. Siderska, J.: Robotic process automation – a driver of digital transformation? Int. J. Inf. Manag. **12**, 21–31 (2020)

8. Kaarnijoki, P.: Intelligent automation-assessing artificial intelligence capabilities potential to complement robotic process automation. M.S. thesis. Faculty of Engineering and Natural Sciences, Tampere University of Technology. Tampere, Finland (2019)

9. van der Aalst, W.M.P., Bichler, M., Heinzl, A.: Robotic process automation. Bus. Inf. Syst. Eng. **60**(4), 269–272 (2018). https://doi.org/10.1007/s12599-018-0542-4

10. König, M., Bein, L., Nikaj, A., Weske, M.: Integrating robotic process automation into business process management. In: Asatiani, A., et al. (eds.) BPM 2020. LNBIP, vol. 393, pp. 132–146. Springer, Cham (2020). https://doi.org/10.1007/978-3-030-58779-6_9

11. Mendling, J., Decker, G., Hull, R., Reijers, H.A., Weber, I.: How do machine learning, robotic process automation, and blockchains affect the human factor in business process management? Commun. Assoc. Inf. Syst. **43**, 19 (2018)

12. Yatskiv, N., Yatskiv, S., Vasylyk, A.: Method of robotic process automation in software testing using artificial intelligence. In: 10th International Conference on Advanced Computer Information Technologies (ACIT), pp. 501–504. IEEE, Deggendorf, Germany (2020)

13. Nakano, M.: Artificial intelligence and robotic process automation for accounting and auditing in society 5.0. J. Soc. Sci. **89**, 51–61 (2022)

14. Hartmann, F.: Evolving digitisation: chances and risks of robotic process automation and artificial intelligence for process optimization within the supply chain. B.A. thesis, Berlin School of Economics and Law, Berlin, Germany (2018)

15. Turcu, C.E., Turcu, C.O.: Digital transformation of human resource processes in small and medium sized enterprises using robotic process automation. Int. J. Adv. Comput. Sci. Appl. **12**(12), 70–75 (2021)

16. Madakam, S., Holmukhe, R.M., Jaiswal, D.K.: The future digital work force: robotic process automation (RPA). JISTEM-J. Inf. Syst. Technol. Manag. **16** (2019)

17. Houy, C., Hamberg, M., Fettke, P.: Robotic process automation in public administrations. Digitalisierung von Staat und Verwaltung (2019)

18. Bornet, P., Barkin, I., Wirtz, J.: Intelligent automation: welcome to the world of hyperautomation: learn how to harness artificial intelligence to boost business and make our world more human (2021)

19. Udovita, P.: Conceptual review on dimensions of digital transformation in modern era. Int. J. Sci. Res. Publ. **10**, 520–529 (2020)

20. Bradley, J., Loucks, J., Macaulay, J., Noronha, A., Wade, M.: Digital vortex: how digital disruption is redefining industries. Global Center for Digital Business Transformation: an IMD and Cisco Initiative (2015)

21. Liermann, V., Li, S., Waizner, J.: Hyperautomation (automated decision-making as part of RPA). In: Liermann, V., Stegmann, C. (eds.) The Digital Journey of Banking and Insurance, Volume II, pp. 277–293. Springer, Cham (2021). https://doi.org/10.1007/978-3-030-78829-2_16

22. Martins, P., Sá, F., Morgado, F., Cunha, C.: Using machine learning for cognitive robotic process automation (RPA). In: 15th Iberian Conference on Information Systems and Technologies (CISTI), pp. 1–6. IEEE (2020)
23. Patel, M., Shukla, A., Porwal, R., Kotecha, R.: Customised automated email response bot using machine learning and robotic process automation. In: 2nd International Conference on Advances in Science and Technology (ICAST). SSRN, Maharashtra, India (2019)
24. Bellman, M., Göransson, G.: Intelligent process automation: building the bridge between robotic process automation and artificial intelligence. M.S. thesis. School of Industrial Engineering and Management, Kth Royal Institute of Technology. Stockholm, Sweden (2019)
25. Parchande, S., Shahane, A., Dhore, M.: Contractual employee management system using machine learning and robotic process automation. In: 5th International Conference On Computing, Communication, Control And Automation (ICCUBEA), pp. 1–5. IEEE, Pune, India (2019)
26. Hu, S., Jiang, T.: Artificial intelligence technology challenges patent laws. In: International Conference on Intelligent Transportation, Big Data and Smart City (ICITBS), pp. 241–244. IEEE, Changsha, China (2019)
27. Nunes, T., Leite, J., Pedrosa, I.: Automação Inteligente de Processos: Um Olhar sobre o Futuro da Auditoria Intelligent Process Automation: An Overview over the Future of Auditing. In: 5th Iberian Conference on Information Systems and Technologies (CISTI). IEEE, Sevilla, Spain (2021)
28. Ray, S.: A quick review of machine learning algorithms. In: International Conference on Machine Learning, Big Data, Cloud and Parallel Computing (COMITCon), pp. 35–39. IEEE, Faridabad, India (2019)
29. Grekousis, G.: Artificial neural networks and deep learning in urban geography: a systematic review and meta-analysis. Comput. Environ. Urban Syst. **74**, 244–256 (2019)
30. Madakam, S., Holmukhe, R.M., Revulagadda, R.K.: The next generation intelligent automation: hyperautomation. J. Inf. Syst. Technol. Manag. **19** (2022)
31. Jha, N., Prashar, D., Nagpal, A.: Combining artificial intelligence with robotic process automation—an intelligent automation approach. In: Ahmed, K.R., Hassanien, A.E. (eds.) Deep Learning and Big Data for Intelligent Transportation. SCI, vol. 945, pp. 245–264. Springer, Cham (2021). https://doi.org/10.1007/978-3-030-65661-4_12
32. Abiodun, O.I., Jantan, A., Omolara, A.E., Dada, K.V., Mohamed, N.A., Arshad, H.: State-of-the-art in artificial neural network applications: a survey. Heliyon **4**(11), e00938 (2018)
33. Somvanshi, M., Chavan, P., Tambade, S., Shinde, S.V.: A review of machine learning techniques using decision tree and support vector machine. In: International Conference on Computing Communication Control and Automation (ICCUBEA). IEEE, Pune, India (2016)
34. Charity, B., Abdulazeez, A.: Classification based on decision tree algorithm for machine learning. J. Appl. Sci. Technol. Trends **2**, 20–28 (2021)
35. Kuo, P., Huang, C.: A high precision artificial neural networks model for short-term energy load forecasting. Energies **11**(1), 213 (2018)

# Modern Warehouse and Delivery Object Monitoring – Safety, Precision, and Reliability in the Context of the Use the UWB Technology

Krzysztof Hanzel[✉] [iD]

Department of Telecommunications and Teleinformatics, Faculty of Automatic Control, Electronics and Computer Science, Silesian University of Technology, Akademicka 16, 44-100 Gliwice, Poland
Krzysztof.Hanzel@polsl.pl

**Abstract.** This document assesses six technologies – including visual or radio methods – currently used to monitor shipments and traffic in warehouses and logistics centers. The work presents eight important aspects from the point of view of systems monitoring the location of people and loads in areas where logistic operations are carried out. Among them, you can distinguish factors such as the accuracy of determining the distance aimed at improving the positioning of objects in warehouse spaces, the frequency of acquiring items that affect the safety and management of human resources, and the effectiveness of the system in unfavorable conditions in the form of objects on the line of sight. The document's conclusion indicates the preferential technology and the proposed application possibilities within industry 4.0. At the same time, the analysis showed that the existence of a universal method is currently impossible. Still, the dissemination of radio technologies such as UWB is a new opening in the aspect of warehouse management.

**Keywords:** industry 4.0 · warehouse management · package tracking · logistics

## 1 Introduction

Transportation of goods, despite the crisis caused by COVID-19, continues to grow. Forecasts indicate a further increase in merchandise trade volume, and thus also in global and intercontinental shipping, as well as in last-mile transport. A big challenge in such transportation is constant supervision and detailed monitoring of individual batches of goods in the context of the increasingly frequent transport of general cargo. Monitoring more comprehensive and automated logistics centers is also a significant challenge, especially with rising energy prices. Also, ensuring the safety and optimization of employees' work, especially in the Western European market, where an employee's supply significantly shrinks, becomes a significant problem.

These and many other challenges make you check how you can monitor warehouse spaces and freight so that it is as precise and reliable as possible, not energy-consuming, and simultaneously qualitatively competitive with the currently used methods. These

M. Papadaki et al. (Eds.): EMCIS 2022, LNBIP 464, pp. 451–462, 2023.
https://doi.org/10.1007/978-3-031-30694-5_33

considerations lead to a comparison of several leading monitoring methods, both in the global frame of reference and confined spaces such as warehouses, production halls, the interior of container ships, etc. The analysis of technologies and methods that can be used to determine the location of people and objects in warehouse centers should begin with determining the factors based on which the optimal strategy will be selected.

At this stage, it should be noted that the set of 8 factors responsible for the technology division has a large impact on the target choice and should be made dependent on a specific application. Their detailed description in the context of the paper's topic is presented. The study considered six technologies (including vision and radio) that were analyzed in the context of their use in warehouses and distribution centers. A detailed description of the methods and their potential with the adopted coefficients is presented. Based on the analysis, it was decided to conduct a study of the UWB technology, which showed the most significant potential applicability in the context of low-energy monitoring of warehouse spaces.

## 1.1 Methods of Assessing the Quality of Positioning Systems

The proposed system for monitoring storage space can be characterized in the following areas:

- System application place (indoor/outdoor/mixed)
- Energy demand (AC/battery/computing power)
- Accuracy and precision
- Operating range (local, global, scalability)
- Price
- Required infrastructure
- Communication and other sensing possibilities
- Working conditions & ease of use

The selection of parameters is based on an analysis of the literature and a review of the most frequently raised weaknesses of the systems [1–3]. It was also decided to place a detailed description of the parameters near the tables they refer to in the next chapter.

## 1.2 Systems Included in the Analysis and Their Discussion

Based on the study of the systems currently used on the market and of potentially good solutions enabling the achievement of the set goals, six methods were selected, which were then analyzed based on the criteria presented above. The systems finally considered were:

- Vision recognition [4–7]
- Barcodes [8–10]
- GPS (Global Positioning System) [11–14]
- UWB (Ultra-Wideband) [3, 15–18]
- LoRaWAN (LoRa Wide Area Network) [19–21]
- RFID (Radio-frequency identification) [22–24]

To prepare the characteristics in the best possible way, the features of the systems are listed in eight tables, each table for one of the features. To simplify the description of each technology, an identifier is assigned in the first table, represented in the following.

The first area to be dealt with is the system application place. This parameter defines the capabilities of the system application. A modern system should enable tracking in the door-to-door system with the simultaneous possibility of interoperability within the warehouse space, sea freight process, road transport, etc. Assigning subsequent identifiers in this process, with the simultaneous change of monitoring systems, leads to errors resulting from the human factor. At the same time, monitoring both shipments and a fleet of specialized vehicles (forklifts, self-propelled warehouse vehicles, indoor cargo vehicles) is an essential aspect of security. For example, the ability to detect potential collisions among staff, monitor unforeseen events among employees and respond faster in case of a need for cooperation. A detailed comparison of systems in the context of this parameter is presented in Table 1.

**Table 1.** System application place.

| System | Description |
|---|---|
| Vision systems (1) | The system characteristics largely assume a static infrastructure with known camera locations and a constant power supply. The ability to monitor objects within the range of the camera limits the operation to a relatively small area. Possible identification problems |
| Barcode analysis (2) | However, position information provided during scanning is limited to a specific item, e.g., a conveyor belt or a mobile scanner. Reliable identification is possible as long as the label is not broken |
| GPS (3) | Limited to open-air spaces or equipped with additional systems imitating the GPS signal. The necessity of communication and identification on the end tag side |
| UWB (4) | The exact position is only available if infrastructure exists in a given location. The ability to track the distance (point to point) using a mobile terminal anywhere |
| LoRaWAN (5) | The position determination accuracy depends on the network density in a given area, the operating environment, and the distance from the system nodes. There are currently around 88,000 system nodes scattered around the world, with a range of about 10–15 km. The determination of the position is possible when the marker has access to a minimum of 3 of them |
| RFID (6) | It is required to keep relatively short distances during scanning. It requires dense infrastructure (for passive tags) or a constant power supply to all system nodes |

Another proposed parameter is energy demand – a factor gaining importance in the continuous increase in electricity prices. It makes it possible to determine whether the planned system can take the form of mobile devices in the form of wearable IoT or it will be an energy-consuming system, forcing a constant power supply from the mains. At the same time, attention should be paid to the frequency of charging devices and the risk of semi-portable infrastructure, i.e., one that formally works on battery power, but in the operation process, the battery life does not coincide with the employee's working time unit, so it is necessary to replace the cells or their recharging during work A comparison of the systems in this respect is shown in Table 2.

**Table 2.** Energy demand.

| System | Description |
|--------|-------------|
| (1) | The imaging device itself consumes a relatively large amount of energy (especially if it has an appropriate infrared or visible light illumination). In the case of the desire to transmit identification, automated decision-making, and storage of data provided by the system on the part of the shipping company/warehouse, it is also necessary to provide power for the infrastructure, as well as for computing power and data warehouses |
| (2) | The average energy consumption level includes, for example, a scanner on a belt feeder and a system for video processing of the data obtained in this way. A mobile scanner's consumption is lower and limited to the PDA device's battery power. The need to consider the energy required, e.g., printing labels |
| (3) | Depending on the application, relatively high GPS radio power consumption is required in continuous monitoring. The system consumes less energy if an item is delivered within a specific time interval (every 30 min, for example) |
| (4) (5) | Very low energy consumption on the side of the dimensioning infrastructure, the ability to work on battery power. The low complexity of the positioning algorithm also results in low energy consumption. Mobile dimensioning devices have the power consumption of a standard PDA |
| (6) | Possibility to use passive tags not equipped with batteries (supply via induction). Battery-powered active tags might also create the system. Mobile dimensioning devices have an energy consumption slightly higher than standard PDAs |

Accuracy and precision are other parameters taken into account. These are two independent position measurement attributes that depend on the system used. The first one – accuracy – tells us about the overall quality of determining the position. The system with this feature can indicate the distance range from the tag to search for the located object. The second factor – precision, illustrates the certainty of finding a given object with the search area, e.g., on a heatmap. If both factors provide high quality, we can accurately indicate the desired object regarding its place and distance. Their comparison in the context of systems is presented in Table 3.

Operating range – this parameter is similar to the application place, but it expresses the possible location area in a broader context. We can distinguish direct systems requiring contact at a distance of a few cm, systems with a wider range of operations within one locating point, or global systems that allow us to determine the location regardless

**Table 3.** Accuracy and precision.

| System | Description |
|---|---|
| (1) | Depends on the resolution of the camera and the monitored area. The precision depends on the object's size and the classification algorithm's quality. Typically, the accuracy in industrial systems is around one to several meters |
| (2) | Very high accuracy in determining the object's position in the case of scanners on conveyor belts (the precision depends on the object's size to the label). In the case of handheld scanners, the position is limited to, e.g., the area of operation of a given scanner or manual operator integration |
| (3) | Dependent on location and environmental conditions. Usually in the range from a few to several meters |
| (4) | Precision corresponds to the standard deviation of 20 cm. However, the accuracy of the position depends on the quality of the infrastructure calibration and the working environment |
| (5) | The position determination accuracy depends on the network density in a given area, the operating environment, and the distance from the system nodes. It can range from 20–200 m |
| (6) | It is a system heavily dependent on technology (active/passive) and the density of reference points. In one of the more favorable cases, the accuracy is at the level of 3 m [25], but the real cases allow for estimation in the range of 6–7 m [22] |

of the infrastructure we manage. The operating range of the analyzed systems is shown in Table 4.

**Table 4.** Operating range.

| System | Description |
|---|---|
| (1) | In local operation, a large number of cameras also necessitates the use of a local data processing center. Extending the system largely requires local infrastructure, cabling, and computing resources |
| (2) | It can work globally (e.g., a manual scanner) but does not provide the location. In another variant, it only works stationary. It is possible to precisely locate, for example, a package on a conveyor belt, but it is not allowed to do so at any time during storage or transport |
| (3) | Global coverage without the need to invest in additional infrastructure (not counting the required receiver). The range is limited inside buildings where GPS simulation systems can be used |

(*continued*)

**Table 4.** (*continued*)

| System | Description |
|--------|-------------|
| (4) | The local range. However, relatively low energy consumption allows for quick adaptation to larger areas. Requires more than one reference point to obtain a position |
| (5) | The metropolitan range. The availability of areas with a higher density of LoRaWAN system transmitters limits the system. Requires more than one reference point to obtain a position |
| (6) | Local scope. The RTLS variant requires a high concentration of reference points and an active marker. Suppose it is only necessary to notify the presence. In that case, passive tags can be used to confirm their presence within the antenna range (e.g., presence information within several dozen meters), but it requires low interference and good access for waves |

Another factor taken into account is price. It considers two components: the cost of implementation (including initial investments in infrastructure and hardware) and operating costs (related to consumables, licenses, software, and maintenance). The estimated list of prices, based on the analysis of available solutions, is presented in Table 5.

**Table 5.** Price (based on commercial systems)

| System | Description |
|--------|-------------|
| (1) | A single node (cam + wiring) is about $100, but the entire system to the warehouse, including network stuff and computing servers, is around $50–$100 thousand. In this case, operating costs should also include electricity costs. We can omit server costs when we outsource video processing, but there will be extra cloud computing power to pay for |
| (2) | Depends on the type and class of the device. One personal scanner costs around $100, while one scanner on a belt feeder can cost from several to several thousand dollars, depending on the quality and scanning (e.g., scanning individual codes in a designated area on one plane or a multi-scanner that allows you to handle multiple labels at once) on five or even six sides of the package) |
| (3) | A single device costs a dozen or so dollars. In addition, there is a position information transmission system (e.g., in the form of an industrial sim card). The cost of the server monitoring the position of shipments and the costs of the network infrastructure should be added to this, which depends on the number of devices and the scale of activity, but is cheaper than systems processing video images |
| (4) | A single device currently costs $20–$30. However, reference infrastructure is also required. Position information exchange may be based on communication between nodes (mesh) or another protocol (WLAN, industrial sim). Additionally, infrastructure for position monitoring and analysis should be included |

(*continued*)

**Table 5.** (*continued*)

| System | Description |
|---|---|
| (5) | Depending on the infrastructure model. If we use public infrastructure, the cost is for each module (currently around $10) and subscription, allowing the use of public infrastructure (from $3,000 to $50,000, depending on the level of support). Additionally, infrastructure for position monitoring and analysis should be included |
| (6) | The cost of the tag ranges from $0.10–$2 to $15–$20 depending on whether it is passive, sticker, sew-on component, or active with a small battery. The reader costs $200–$500 depending on the presence of the screen and other functions, e.g., the form of communication with the database or additional location based on different technologies |

A factor that cannot be overlooked when implementing significant complex investments is the required infrastructure. Depending on the form of infrastructure (centralized or distributed), it may be necessary to purchase local modules (endpoints of the location system) that can be connected to cloud services. It may also be required to buy a central infrastructure in the form of computing and aggregating systems, decision-making systems, etc. Some systems require a similar or very similar infrastructure, which is included in Table 6.

**Table 6.** Required infrastructure.

| System | Description |
|---|---|
| (1) | Technology prefers a centralized infrastructure, the main component of which is a server that performs video analysis. It often forms an extended star topology, where smaller units analyze data from a given sub-area and send information about events to the chief supervisor |
| (2) (6) | The infrastructure is limited to single points with a declared position (e.g., on a conveyor belt) or operating in a given area (e.g., a handheld scanner in a given section of a warehouse) that connect to a centralized database |
| (3) (4) (5) | Single devices, for example, use the Internet to connect to the server or cloud services, transmitting information about the position and status |

Communication and other sensing possibilities verify whether there is computing capacity on the part of the located object and the possibility of feedback (in the form of a message, warning) or sensing the parameters of the positioned object (e.g., temperature measurement of transported goods, verification if the object is in motion, etc.). The description of the implementation (and its availability) of communication is presented in Table 7.

The last factor taken into account is working conditions & ease of use. It is information on what environmental conditions a given system can operate and the level of complexity of the service. For example, some solutions require direct contact, being in

**Table 7.** Communication and other sensing possibilities.

| System | Description |
|---|---|
| (1) | The lack of communication possibilities, however, allows, for example, a visual assessment of the condition of the shipment and archiving its appearance |
| (2) | Unable to communicate |
| (3) | Unable to communicate. Possible communication is realized with different technology |
| (4) | Full two-way communication enables transmitting position information and additional data from possible sensors. It also allows the reception of messages to, for example, change the parameters of heating, tightness, etc. |
| (5) | In some variants, two-way communication is possible, but usually, communication is in the marker-reference point direction. The transmission speed is limited to 27 kbit/s |
| (6) | They offer both one-way and two-way communication, depending on the technology. Theoretically, performing two-way communication using passive RFID tags (such as saving the last transaction on a payment card) is possible. Still, active RFID tags are often used for this purpose due to the increased range of operation |

sight, under the open sky, etc. At the same time, the level of complexity of service is essential in determining the cost of implementing a new employee or functional expansion of the proposed system. A summary of the capabilities of the proposed systems in relation to this parameter is presented in Table 8.

**Table 8.** Working conditions & ease of use.

| System | Description |
|---|---|
| (1) | The system is practically maintenance-free, but its preparation is a tailor-made solution. It requires visibility conditions (LOS) and good lighting conditions (lighting of the halls, maneuvering areas, etc.). The service is limited to a system of notifications and alerts about the detection of individual objects, motion detection, etc. The risk is the possibility of the real-time video preview, which may reveal sensitive data in the event of a leak |
| (2) | Due to its limited capabilities, the system is straightforward, and its implementation and subsequent implementation of employees do not generate additional costs. It requires direct contact with the parcel or the appearance of the parcel in the scanning area |
| (3) (4) (5) | The system requires assigning an identifier to a particular tag and attaching it to the shipment (container, pallet). The tag itself must be pre-configured and, e.g., assigned to a cloud account of a given company |

*(continued)*

**Table 8.**  (*continued*)

| System | Description |
|---|---|
| (3) | Additionally, it requires sky visibility for proper operation |
| (4) (5) | The system does not require visibility conditions (it works in NLOS). However, the very close proximity of metal objects to the antenna may disturb the transmission quality |
| (6) | Depending on the tag type, it requires its association with a given shipment (passive tag) or initial configuration (active tag) |

## 2  Analysis of the Prepared Statement

From the analysis performed, it can be seen that there is a huge discrepancy between the positioning systems used in trade and logistics, and in particular, among the systems that can be used to track a shipment from the manufacturer of the goods through logistics hubs and centers up to "last mile" delivery. It is impossible to select an undisputed leader regarding the entire process. Still, it can be pointed out that its elements – are due to the constant development of wireless technologies and their advantage over solutions, e.g., video. The steps involved in covering long distances in the open air are indeed unrivaled regarding satellite navigation technologies (whether GPS or other commonly available technologies). However, in the context of logistics centers, warehouses, and their immediate vicinity, the matter is more complex due to the lack of GPS signal.

For retail customers, cheap solutions will most often be the leading one because monitoring parcels worth several or several dozen USD is challenging to implement with systems exceeding their value (even assuming that these systems will be returnable). However, if we consider small and medium-sized enterprises, the situation is no longer so obvious.

Both reducing errors in logistics, ensuring constant monitoring of transport conditions and quality, as well as precise step-by-step tracking of deliveries allow us to believe that the systems enabling two-way communication – such as UWB – will be the future of forwarding in this area. In addition, the UWB technology is the only one of the discussed technologies that allow for two-way communication while meeting the RTLS requirements, which also allows for implementation in the warehouse as an internal system of communication and warning about danger.

Based on these dependencies, it is proposed that the system under which logistics is carried out should be based on a layered model. In this case, on par with the currently used solution, it also uses the approach based on UWB technology. It can be used as a support or transition period, as presented in Fig. 1.

The advantage of the proposed solution is interoperability (e.g., within a ship whose position is determined using GPS, there is also the possibility of visualizing the position of containers and determining, e.g., the humidity prevailing in them). It can also increase safety (the forklift operator no longer has to be warned about a dangerous event) by the security guarding the video surveillance. Still, it can be done by a system that will automatically inform about a potential collision or even turn off the drive. Finally, there

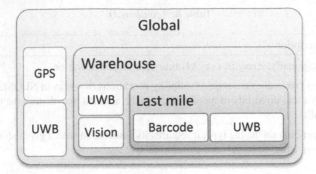

**Fig. 1.** Proposition of coexistence of a UWB-based system at various stages of storage and shipping.

is also a guarantee of delivery, where the courier, even if by mistake scans the wrong label, will be informed when leaving the parcel that the wrong one is the package was removed from the delivery truck.

The current expansion of UWB technology remains a question. It indicates two main trends – the development of mobile applications, the implementation of the latest flagship smartphones of brands such as Apple [26] or Google [27], and the trend focused on automotive technology. Of course, the interoperability of these two approaches is also possible. So, for example, Apple is considering cooperation with BMW [28], where cars are to be opened using virtual keys stored on the brand's phones and communicating with vehicles using UWB technology. None of these approaches fit in with market solutions aimed directly at industry and transport, but more and more companies are offering such solutions commercially on a smaller scale [29]. Moreover, the intensified development of this technology in the above-mentioned areas allowed for its gradual miniaturization and cost reduction.

## 3 Summary and Conclusions

The article presents several requirements and how the latest technologies used in warehouses and distribution centers deal with them. As has also been shown, no one-size-fits-all method can meet the requirements of large-scale shipment monitoring and security, and positioning within warehouses. Nevertheless, it has been shown that many warehouse requirements can be met with radio technologies, which allow for constant location monitoring, with simultaneous, two-way communication and low energy consumption. The technology that attracted particular attention is UWB, which both meets the requirements presented above and is currently strongly developed in the context of industry and consumer solutions. In addition, there are more and more commercial solutions on the market that introduce this technology for use within the discussed topic.

**Acknowledgment.** This research was partially funded by European Social Funds, project no POWR.03.02.00-00-I007/17-00 "CyPhiS – the program of modern PhD studies in the field of Cyber-Physical Systems"; and by Polish Ministry of Science and Higher Education by Young

Researchers funds Faculty of Automatic Control, Electronics, and Computer Science, Silesian University of Technology.

# References

1. Lewczuk, K., Załęski, A.: Selected aspects of indoor positioning based on AIDC elements in warehouse facilities. J. KONES 25(4), 555–562 (2018). https://doi.org/10.5604/01.3001. 0012.8005
2. Löffler, C., Riechel, S., Fischer, J., Mutschler, C.: Evaluation criteria for inside-out indoor positioning systems based on machine learning. In: Proceedings of the 2018 International Conference on Indoor Positioning and Indoor Navigation (IPIN), pp. 1–8 (2018)
3. Hanzel, K., Paszek, K., Grzechca, D.: Possibilities of using data from the UWB system for the validation of ADAS subsystems. In Networking Issues in Innovative Applications Based on Cyber-Physical Systems Paradigm; Wydawnictwo Politechniki Śląskiej, pp. 108–124 (2020). ISBN: 978-83-7880-736-0
4. Ng, Z.Y.: Indoor-positioning for warehouse mobile robots using computer vision. UTAR (2021)
5. Yan, T.: Positioning of logistics and warehousing automated guided vehicle based on improved LSTM network. Int. J. Syst. Assur. Eng. Manag. (2021).https://doi.org/10.1007/s13198-021-01243-3
6. Aravindaraj, K., Rajan Chinna, P.: A systematic literature review of integration of Industry 4.0 and warehouse management to achieve sustainable development goals (SDGs). Clean. Logist. Supply Chain 5, 100072 (2022). https://doi.org/10.1016/j.clscn.2022.100072
7. El-sayed, M.E., Youssef, A.W., Shehata, O.M., Shihata, L.A., Azab, E.: Computer vision for package tracking on omnidirectional wheeled conveyor: case study. Eng. Appl. Artif. Intell. 116, 105438 (2022). https://doi.org/10.1016/j.engappai.2022.105438
8. Jonnalagadda, V.: A package tracking application based on software agents (2012)
9. Jia, C., Huang, J., Gao, Q., Luo, S.: Application of barcode technology in warehouse management of printing and packaging enterprises. In: Zhao, P., Ouyang, Y., Xu, M., Yang, L., Ren, Y. (eds.) Applied Sciences in Graphic Communication and Packaging. LNEE, vol. 477, pp. 533–541. Springer, Singapore (2018). https://doi.org/10.1007/978-981-10-7629-9_66
10. Istiqomah, N.A., Sansabilla, P.F., Himawan, D., Rifni, M.: The implementation of barcode on warehouse management system for warehouse efficiency. J. Phys. Conf. Ser. 1573, 012038 (2020). https://doi.org/10.1088/1742-6596/1573/1/012038
11. Crato, N.: How GPS works. In: Figuring It Out, pp. 49–52. Copernicus, Berlin, Heidelberg (2010). https://doi.org/10.1007/978-3-642-04833-3_12
12. GPS.Gov: GPS Accuracy. https://www.gps.gov/systems/gps/performance/accuracy/. Accessed 28 Jan 2020
13. Chambers, A., Scherer, S., Yoder, L., Jain, S., Nuske, S., Singh, S.: Robust multi-sensor fusion for micro aerial vehicle navigation in GPS-degraded/denied environments. In: Proceedings of the 2014 American Control Conference, pp. 1892–1899 (2014)
14. Grzechca, D., Tokarz, K., Paszek, K., Poloczek, D.: Using MEMS sensors to enhance positioning when the GPS signal disappears. In: Nguyen, N.T., Papadopoulos, G.A., Jędrzejowicz, P., Trawiński, B., Vossen, G. (eds.) ICCCI 2017. LNCS (LNAI), vol. 10449, pp. 260–271. Springer, Cham (2017). https://doi.org/10.1007/978-3-319-67077-5_25
15. Hanzel, K., Paszek, K., Grzechca, D.: The influence of the data packet size on positioning parameters of UWB system for the purpose of tagging smart city infrastructure (2020). https://doi.org/10.24425/BPASTS.2020.134173

16. Hanzel, K., Grzechca, D.: Increasing the security of smart cities of the future thanks to UWB technology. In: Themistocleous, M., Papadaki, M. (eds.) Information Systems (EMCIS 2021). LNBIP, vol. 437, pp. 585–596. Springer, Cham (2022). https://doi.org/10.1007/978-3-030-95947-0_41

17. IEEE Standard for Low-Rate Wireless Networks--Amendment 1: Enhanced Ultra Wideband (UWB) Physical Layers (PHYs) and Associated Ranging Techniques. https://standards.ieee.org/standard/802_15_4z-2020.html. Accessed 13 Apr 2021

18. Analysis of the Scalability of UWB Indoor Localization Solutions for High User Densities. https://www.researchgate.net/publication/325626565_Analysis_of_the_Scalability_of_UWB_Indoor_Localization_Solutions_for_High_User_Densities. Accessed 6 Feb 2020

19. The Things Network. https://www.thethingsnetwork.org/map. Accessed 2 Nov 2022

20. Haxhibeqiri, J., De Poorter, E., Moerman, I., Hoebeke, J.: A survey of LoRaWAN for IoT: from technology to application. Sensors **18**, 3995 (2018). https://doi.org/10.3390/s18113995

21. Wong, M.A., Lau, T., Alsayaydeh, A.J., Shahrom, H.H., Pembuatan, U.T.M.M.: Portable warehouse environmental monitoring using LoRaWAN. Int. Rev. Red Cross **95**, 383–413 (2019)

22. How Accurate Can RFID Tracking Be? https://www.rfidjournal.com/question/how-accurate-can-rfid-tracking-be. Accessed 24 Oct 2022

23. Ni, L.M., Liu, Y., Lau, Y.C., Patil, A.P.: LANDMARC: indoor location sensing using active RFID. Wirel. Netw. **10**, 701–710 (2004). https://doi.org/10.1023/B:WINE.0000044029.06344.dd

24. Saab, S.S., Nakad, Z.S.: A standalone RFID indoor positioning system using passive tags. IEEE Trans. Ind. Electron. **58**, 1961–1970 (2011). https://doi.org/10.1109/TIE.2010.2055774

25. Seco, F., Plagemann, C., Jiménez, A.R., Burgard, W.: Improving RFID-based indoor positioning accuracy using gaussian processes. In: Proceedings of the 2010 International Conference on Indoor Positioning and Indoor Navigation, pp. 1–8 (2010)

26. Zafar, R.: IPhone 11 Has UWB With U1 Chip - Preparing Big Features For Ecosystem. Wccftech (2019)

27. Google Has Added an Ultra-Wideband (UWB) API in Android. Xda-Dev (2021)

28. What's the Deal with Ultra Wideband Technology and What Will It Do for Your Car? https://www.bmw.com/en/innovation/bmw-digital-key-plus-ultra-wideband.html. Accessed 2 Nov 2022

29. Ultra-Wideband for Indoor Positioning – RTLS by Infsoft. https://www.infsoft.com/basics/positioning-technologies/ultra-wideband/. Accessed 2 Nov 2022

# Enterprise Systems

# The Effects of Artificial Intelligence in the Process of Recruiting Candidates

Lasha Abuladze[1]([✉]) and Lumbardha Hasimi[2]

[1] University of Economics in Bratislava, Petržalka, Slovakia
Lasha.abuladze@euba.sk
[2] Comenius University in Bratislava, Bratislava, Slovakia
Lumbardha.hasimi@uniba.sk

**Abstract.** AI-based solutions have found a great application in filling the gap and enhancing the massive recruiting processes. With the recent developments, the role of gamification in the overall managerial processes, especially in recruiting has proven to be crucial. AI as a powerful tool towards the challenges the hiring process faces, appears as contradictory in a number of issues. In this paper, we have observed and analyzed the advantages and disadvantages of AI in recruiting, followed by a proposed model for resume screening based on keywords and phrases against job description. Furthermore, a case has been presented and assessed regarding results and implications of AI-based tools, namely machine learning models in a simple scenario of hiring process.

## 1 Introduction

Human resource management (HRM) is constantly looking for innovative technologies to increase efficiency and effectivity in recruitment and selection. Recent technological advancements have enabled it to move from a traditional style of operation to a modern, digitized one in some key areas of their recruitment process [1]. The use of AI technologies is currently described and understood by different researchers and authors as an emerging trend in recruitment and selection activities, which has brought new opportunities and challenges for the enterprise instances. Using artificial intelligence represents one of the biggest evolutions in HRM. With the help of AI, specifically the Machine Learning as a subset of AI, HRM has gained opportunities to automate some basic and important areas of recruitment and selection that were previously done manually by human [2–4].

### 1.1 Theoretical Background

HRM is constantly looking for innovative technologies to increase efficiency and effectivity in recruitment and selection. Recent technological advancements have enabled HRM to move from a traditional style of operation to a modern, digitized one in some key areas of their recruitment process [3]. The process of finding and recruiting candidates electronically has brought new aspects into the recruiting scene. Most popular methods of e-recruitment are:

M. Papadaki et al. (Eds.): EMCIS 2022, LNBIP 464, pp. 465–473, 2023.
https://doi.org/10.1007/978-3-031-30694-5_34

- Social media sourcing. Social platforms like Facebook, LinkedIn, Glassdoor, Instagram are widely used by companies to attract, search and recruit candidates online. It is also a great way and an effective tool to spread a positive name among candidates and build the brand status of the employer.
- Employer's website. It provides information about new job opportunities in the company, allows applicants to apply for it and collects their data.
- Job boards. Websites used by HRM to advertise and promote job offers while job seekers can use it to find new vacancies and job opportunities.
- Online interviews. Communication technologies (Skype, Teams, Zoom, G-meet) allow HRM to interview candidates online without using any physical space.
- AI powered software. Software, that use humankind intelligence to find, screen candidates and simplify the entire process of recruitment for the HRM.

In today's recruitment use of artificial intelligence became one of the biggest evolution and trend for HRM. With the help of AI, HRM has gained opportunities to automate some basic and key areas of recruitment and selection that were previously done manually by humans. According to [5] approximately 70% of companies will adopt some form of AI by 2030, and countries that can establish themselves as AI leaders could reap up to 20 to 25% more economic benefits than current levels.

### 1.1.1 Artificial Intelligence - Its Use in the Recruitment Process

AI as a sub field of computer science is focused on training computers to perform initially designed human tasks. While seeking to imitate and therefore improve the human intelligence, it compares natural human intelligence to artificial intelligence. As a model of science, it aims and targets alleviating the human physical and mental labour using computational intelligence behavioral models with the goal of developing reasoning, learning, decision making, and performing complex issues that can be executed by human brain [6, p.]. We have witnessed in the recent years the expansion in the unprecedented impact of AI in recruitment processes. Enormous software in today's recruitment markets find AI- based solutions, has a great hand to helping employers mitigate the overall recruitment process. The AI systems up to date are available across wide range functions. Specifically speaking, AI has found a great application in: Candidate Sourcing, Engagement Candidate Tracking, CV Screening, Pre-Employment Assessments, Video Interviewing etc. Recruitment AI encompasses a wide array of technologies functioning at different points in the recruitment process [7].

AI-based solutions help employers scan a considerable number of apps for the best possible candidates. In fact, this is one of the most widely used forms of AI recruitment solutions today [7]. The use of AI in the recruiting, has completely redefined the relationship between employer and the applicant, not only in the process of selection, but in the overall employer experience. AI tools like Chatbot provide applicants with new features and enhanced employer practices. The entire assessment process, such as interview scheduling, customized candidate profiles, customized offers, reference checking are all tasks easily handled by AI-imbued applications. Despite the wide use of AI, only 10% of companies currently use it in a high context, whereas 36% of organizations expect to make full use of AI in the future [8]. Some of the commonly known AI applications adopted by big companies include:

- ATS and CRM Systems Applicant Tracking Systems (ATS) – platforms that offer the recruiters the chance to track every stage of hiring process, from the initial contact to the end of hiring
- Candidate Relationship Management (CRM)
- CV/Resume Screeners CV screening
- Conversational Agents Recruitment conversational agents, or chatbots
- Pre-Employment Assessments
- AI powered interviewing

In the case of resume screeners, AI powered recruitment innovation addresses the issue of high-volume data processing. Trained models screen and detect the characteristics in the content given, through keywords and main expressions that are crucial for the criteria. Whereas for the chatbots, the models are designed to mimic human conversational skills, using NLP technique to analyze and reply effectively. AI-powered interviewing analyze facial expressions, evaluating on basis of the tone, language, non-verbal measures, while applicating standardized process [9].

AI technologies used in recruitment are seen as highly capable and objective tools. However, the practical use and benefit of AI to support recruitment are opposing seen in different perspectives [7]. Among many concerns regarding the benefits of AI in the recruitment processes researchers have demonstrated records of inaccurate results as well as issues with the disability problem. In the case of the assessment of the disability fairness, the issue has gathered quite little attention. The impact of AI aside, the structural issues affecting people with disabilities in gaining and maintaining employment is a complex and ongoing concern. Therefore layering a complex system of automated assessment of candidates risk complicating the situation and expanding the risk of harm [10]. Through implementation of AI in the business strategies the recruiters can identify candidates and effortlessly get data on persona. Most of the repetitive tasks disappear because of the efficiency and use of a living soul the recruiters free to deal with more strategic issues.

Although the AI is designed to overcome the bias during the selection process, the technique itself can still suffer the bias inherited from their programming and data sources. How much this affects the AI systems of recruitment, it is a prone to many questions and complexities levels within the primary functions of systems. In the case of primary sources of bias such as name, age, gender etc., such data unbiasedly pass through the support of AI [8, 11, 12].

Despite the contradictive attitudes towards benefits of AI, many studies argue that AI recruiting does not inherently conflict with human rights [13]. Although on the issue of conflicting with ethical principles, highly depends on the conditions under AI tools are used and trained. However, it is of a crucial importance to derive possible implications and responsibilities in terms of human rights standards in context of AI recruiting [8]. This alongside with bias issues of AI methods, represents a new gap that requires further research and standardization.

Majority of companies which adopt AI-enabled recruiting are unfamiliar with the implementation process. It is especially important to identify certain categories of talent pool, before applying the AI-based tools. In massive applications scenarios, AI tools are applied as cost-effective solutions [14].

Alongside cost-efficiency, AI in the recruitment helps on evaluation process, processing of data, ranking and qualification screening, reduction of administrative and routine tasks, communication, the speed of the overall process carrying etc. Most of all, at the bias context AI allows the equal opportunity to all candidates, leaving no room for human tendency. As an output, AI tools impact business competitiveness [6], while help to gain better insight into the talent and recruitment process.

## 2 Proposed Methodology

In the theoretical part of research, we defined e-recruitment, artificial intelligence with their characters based on articles, research papers from different authors and researchers. With help of deduction and analysis methods, we screened, filtered existing knowledge, and explained mutual relations between e-recruitment and artificial intelligence. By using methods of logic and induction, we compared and synthesized existing knowledge and have found limited or no information, knowledge on the implications of AI-based solutions in recruiting processes [15].

While analyzing the current state-of-art in the field, we have summarized research gaps and current issues, trying to answer the research questions of this study:

1. What are the implications of AI-based solutions in recruiting processes?
2. At what extent, the machine learning models can affect the performance and result of the screening process?

Hence, the main goal of the research is to answer the research questions stated, while investigating and exploring the effects of AI-based solutions implemented in recruiting processes. Namely, the importance of time-consuming and number-efficient processing of resume datasets.

In the practical part of the research, to elaborate further what has been discussed above, we have implemented a simple model for *resume* screening using machine learning, based on text classification and keyword similarities found within resumes or candidate profiles. Another element that would enhance the screening process, is the readability of the resume. Hence, a proposed readability score as a metrics of screening and classification would add an asset. The dataset that is used in this paper is *"Resume dataset"* available at Kaggle.com [16] containing over 2000 resumes from various fields and categories (Table 1).

**Table 1.** Dataset used to train the model.

| Label | Description |
| --- | --- |
| ID | Unique identifier and file name for the respective pdf. |
| Resume_html | Contains the resume data in html format as present while web scrapping. |
| Resume_str | Contains the resume text only in string format. |
| Category | Category of the job the resume was used to apply. |

The dataset is separated into training and test sets by a 75/25 split, respectively. After importing relevant modules, and loading the dataset, a pre-processing of the data was conducted, using stemmer function from nltk library, namely, Porter Stemmer [17] Algorithm as a text normalizing algorithm.

**Table 2.** The pre-processing of data

| Function | Description |
| --- | --- |
| Stop words | Tokenizing the input words into individual tokens and stored it in an array. StopWords [8] |
| Tokenization | Converting the corpus to a vector of token counts. Count Vectorizer (sklearn) |
| Lemmatization | Transform the corpus of text into a list of words and assign words to lemmas. |

Considering the importance of the dataset and the pre-processing phase, throughout the process, the data had undergone the cleaning stage, tokenization, stemming and lemmatization, as seen in the Table 2. To ignore the case of letters, all words were converted to lower case, whereas to avoid words without high semantic load, we used a special blacklist of words from the natural language toolkit (*nltk*) library. To avoid overload of the dictionary with equal words in various forms, and increase the overall accuracy, the stemmer-function from the '*nltk*' library was found to be highly effective.

## 3  Implementation and Results

The proposed solution uses various techniques with the aim of achieving automated screening of candidate's resume that mainly focuses on the content of the resume, namely focusing on the keywords used throughout the content. For this reason, the feature extraction stage covers keyword extraction followed by the next stage which includes the similarity computation. In the extraction stage we perform the extraction of keywords, phrases, and related parameters to match candidates with the job description of the company.

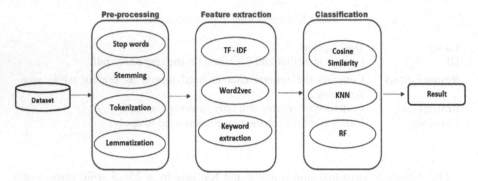

**Fig. 1.** The architecture and activity overflow of the implemented model

As seen in the Fig. 1, the architecture of the proposed model – the basic concept consists of three phases, namely the pre-processing, data extraction and final classification.

**Tf-Idf.** In the second phase, Tf-idf, was used to calculate the terms. As a technique it ranks each word in accordance with the number of times it appears within a document [18–20]. This term is then normalized by dividing by the number of words in the entire group. Considering that words with shorter segment tend to hold more weight. Furthermore, the inverse document frequency ranks the words based on the individuality against the specific segment. Meaning that words are separated towards their use in a section of the text, from the global used words (Fig. 2).

```
wordfrequencydist = nltk.FreqDist(total_Words)
mostCommon = wordfrequencydist.most_common(50)
print(mostCommon)
from sklearn.model_selection import train_test_split
from sklearn.feature_extraction.text import TfidfVectorizer
from scipy.sparse import hstack
required_Text = resumeData['structured_resume'].values
required_Target = resumeData['Category'].values
word_vectorizer = TfidfVectorizer(sublinear_tf=True,stop_words='english',max_featu
res=1500)
word_vectorizer.fit(required_Text)
WordFeatures = word_vectorizer.transform(required_Text)
```

**Fig. 2.** Feature extraction and vectorizing

**Word2Vec.** Considering that the relationship between the words is important for our model, the Word2vec model was an effective solution. Word2vec is the most common technique as a word embedding model that processes text. Its effectiveness lies on the ability to group together vectors of similar words [21]. A word2vec model can be generated with two different approaches as a Skip-gram and Common Bag of Words. The skip-gram model takes a certain word and tries to predict its surrounding word or context words, whereas in the case of Common Bag of Words, the context of the words is taken and predicted to the target word (Fig. 3).

```
wordscores = Calculate_Word_Scores(phraseList)
keywordresume = Generate_resume_Keyword_Scores(phraseList, wordscores)
sortedKeywords = sorted(keywordresume.items(), key=operator.itemgetter(1), reverse=True)
totalKeywords = len(sortedKeywords)
rake = Rake("db\\Stop_Word_List.txt")
keywords = rake.run(text)
print (keywords)
```

**Fig. 3.** Keyword generating and printing after extraction phase

**K-Nearest Neighbors Algorithm (k-NN).** Nearest neighbors is a supervised learning model which is used for classification and regression analysis [20]. KNN does not have parameters we can change or optimize in order to achieve better performance. This model is used to identify the resumes that are nearest matching the given job description. First, to have similar parameter the open-source library "gensim" was taken into account, generating so the summary of the provided text in the provided word limit.

Final output on running similarity on words and phrases, with the percentage score reached against parameters defined (Fig. 4).

```
# import cosine similarity
from sklearn.metrics.pairwise import cosine_similarity

#similarity score
matchpercentage = cosine_similarity(count_matrix)[0][1]
matchpercentage = round(matchpercentage*100,2)
print('Resume matches {} % score to job description! ',cosine_similarity(count_matrix))
```

Resume matches 83.26 % score to job description!

**Fig. 4.** Cosine similarity output

Considering that the core functionality of the model is to find out the the level at which a resume fits within a job description, the final technique engaged in this case for the similarity measure is the cosine similarity. Cosine similarity is applied to calculate the similarity between vectors and measure a cosine of angle between the corresponding vectors. In our case the count matrix checked containing 0 and 1 values, was implemented to check on the similarity and have the final output in the percentage as the final result on the candidates resume profile.

## 4 Conclusion

In the theoretical part of the research, we used the methods of deduction and analysis to screen, filter existing knowledge and then defined and explained e-recruitment, artificial intelligence with their mutual relationships, characteristics, elements, principles, and potential benefits. To find a research gap, we used logical and inductive methods, through which we compared, synthesized existing information, knowledge, and found the research gap on the implications of AI-based solutions in recruiting activities.

Based on the AI powered recruitment process, we found that using of AI powered software in recruitment process can reduce recruiters' time to screen and shortlist applicants by seeking special keywords in their resumes and comparing, matching specific job requirements with applicants' skills, knowledge, and experience. It can improve quality of hiring process by quickly and efficiently finding the most suitable candidates, while reducing the human bias, preventing similarity attraction effect confirmation bias, halo effect, demographic discrimination, and other factors. However, despite the minimizations on the human bias, the machine learning models can be only as good as human, in terms of the model training and parameter setting. This means that human bias can be inherited indirectly to the AI-based model itself, what leaves huge room for future scientific and research endeavors.

AI-based models can positively impact the overall performance and outcome of the recruitment process, but then again there are limitations and challenges to consider before using it in the recruitment process. Programming limitations, ability to inherit human bias, certain dependencies, lack of human judgement, lack of know-how are all elements that need to be considered and backed up. By answering research questions, we succeeded to fulfil main goal of our research, in investigating and exploring the effects of AI-based solutions implemented in recruiting processes. Namely, the importance of time-consuming and volume-efficient processing of resume datasets.

In addition to all the provided information and knowledge, our research has several barriers and limitations in the application of its results. In our case we focused on the profile of the candidate, taking into account only the skill attribute and the keywords extracted from the resume. Future work could consider more thoroughly the other attributes of the candidates' profile, such as the sentiment score of the overall content using both sentiment analysis and recommendation systems concept.

**Acknowledgement.** This research was supported by Slovak Academy of Sciences VEGA project No. 1/0623/22 Virtualization in people management – employee life cycle in businesses in the era of digital transformation.

# References

1. Kirovska, Z., Josimovski, S., Kiselicki, M.: Modern trends of recruitment - introducing the concept of gamification. J. Sustain. Dev. **10**, 55–65 (2020)
2. Mantello, P., Ho, M.-T., Nguyen, M.-H., Vuong, Q.-H.: Bosses without a heart: socio-demographic and cross-cultural determinants of attitude toward Emotional AI in the workplace, AI & Soc, November 2021. https://doi.org/10.1007/s00146-021-01290-1
3. Pissarides, C., Bughin, J.: Embracing the New Age of Automation | by Christopher Pissarides & Jacques Bughin, Project Syndicate, 16 January 2018. https://www.project-syndic ate.org/commentary/automation-jobs-policy-imperatives-by-christopher-pissarides-and-jac ques-bughin-2018-01. Accessed 27 Oct 2022
4. Joy, M., Assistant, J.: An investigation into gamification as a tool for enhancing recruitment process. Ideal Res. **3** (2017)

5. Bughin, J., Hazan, E., Lund, S., Dahlström, P., Wiesinger, A., Subramaniam, A.: Skill shift: automation and the future of the workforce | VOCEDplus, the international tertiary education and research database. https://www.voced.edu.au/content/ngv%3A79805. Accessed 27 Oct 2022

6. Al-Alawi, A.I., Naureen, M., AlAlawi, E.I., Naser Al-Hadad, A.A.: The role of artificial intelligence in recruitment process decision-making. In: 2021 International Conference on Decision Aid Sciences and Application (DASA), Sakheer, Bahrain, pp. 197–203, December 2021. https://doi.org/10.1109/DASA53625.2021.9682320

7. FraiJ, J., László, V.: A literature review: artificial intelligence impact on the recruitment process. Int. J. Eng. Manag. Sci. **6**(1), Art. no. 1 (2021). https://doi.org/10.21791/IJEMS. 2021.1.10

8. Hunkenschroer, A.L., Kriebitz, A.: Is AI recruiting (un)ethical? A human rights perspective on the use of AI for hiring, AI Ethics, July 2022. https://doi.org/10.1007/s43681-022-001 66-4

9. Armstrong, M., Landers, R., Collmus, A.: Gamifying Recruitment, Selection, Training, and Performance Management: Game-Thinking in Human Resource Management, 2016, pp. 140–165 (2016). https://doi.org/10.4018/978-1-4666-8651-9.ch007

10. Nugent, S.E., et al.: Recruitment AI has a disability problem: questions employers should be asking to ensure fairness in recruitment, SocArXiv, preprint, July 2020. https://doi.org/10. 31235/osf.io/emwn5

11. Obaid, I., Farooq, M.S., Abid, A.: Gamification for recruitment and job training: model, taxonomy, and challenges. IEEE Access **8**, 65164–65178 (2020). https://doi.org/10.1109/ ACCESS.2020.2984178

12. Hunkenschroer, A.: How to improve fairness perceptions of AI in hiring: the crucial role of positioning and sensitization. AIEJ **2**(2) (2021). https://doi.org/10.47289/AIEJ20210716-3

13. Fritts, M., Cabrera, F.: AI recruitment algorithms and the dehumanization problem. Ethics Inf. Technol. **23**(4), 791–801 (2021). https://doi.org/10.1007/s10676-021-09615-w

14. Black, J.S., van Esch, P.: AI-enabled recruiting: what is it and how should a manager use it? Bus. Horiz. **63**(2), 215–226 (2020). https://doi.org/10.1016/j.bushor.2019.12.001

15. Korn, O., Brenner, F., Börsig, J., Lalli, F., Mattmüller, M., Müller, A.: Defining Recrutainment: a Model and a Survey on the Gamification of Recruiting and Human Resources (2017)

16. snehann bhawal, Resume Dataset. https://www.kaggle.com/datasets/snehaanbhawal/resume-dataset. Accessed 07 Oct 2022

17. Chandra, R.: Co-evolutionary Multi-task Learning for Modular Pattern Classification, presented at the Neural Information Processing (ICONIP 2017), PT VI, 2017, vol. 10639, pp. 692–701 (2017). https://doi.org/10.1007/978-3-319-70136-3_73

18. Keita, Z.: Text data representation with one-hot encoding, Tf-Idf, Count Vectors, Co-occurrence Vectors and…, Medium, 06 April 2021. https://towardsdatascience.com/text-data-representation-with-one-hot-encoding-tf-idf-count-vectors-co-occurrence-vectors-and-f1bccbd98bef. Accessed 12 July 2022

19. Paullada, A., Raji, I.D., Bender, E.M., Denton, E., Hanna, A.: Data and its (dis)contents: a survey of dataset development and use in machine learning research. Patterns **2**(11), 100336 (2021). https://doi.org/10.1016/j.patter.2021.100336

20. Wendland, A., Zenere, M., Niemann, J.: Introduction to text classification: impact of stemming and comparing TF-IDF and count vectorization as feature extraction technique. In: Yilmaz, M., Clarke, P., Messnarz, R., Reiner, M. (eds.) EuroSPI 2021. CCIS, vol. 1442, pp. 289–300. Springer, Cham (2021). https://doi.org/10.1007/978-3-030-85521-5_19

21. McCormick, C.: Word2Vec Tutorial - The Skip-Gram Model. http://mccormickml.com/2016/ 04/19/word2vec-tutorial-the-skip-gram-model/. Accessed 12 July 2022

# Information System Security
# and Information Privacy Protection

# Security Risk Assessment of Blockchain-Based Patient Health Record Systems

Nedaa B. Al Barghuthi[1]([⊠]) [iD], Huwida E. Said[2]([⊠]) [iD], Sulafa M. Badi[3] [iD], and Shini Girija[2] [iD]

[1] Higher Colleges of Technology, Sharjah, UAE
Nedaa.albarghuthi@hct.ac.ae
[2] Zayed University, Dubai, UAE
{huwida.said,shini.girija}@zu.ac.ae
[3] The British University in Dubai, Dubai, UAE
sulafa.badi@buid.ac.ae

**Abstract.** Blockchain technology is receiving greater attention for enhancing the security of patient records systems; however, it is not a panacea, as many security risks have been found in these healthcare applications. This study conducts a state-of-the-art analysis of emerging risks in blockchain-based patient health record systems, their severity level, impact, and the corresponding countermeasures against them. In addition, we conclude our observations and indicate how blockchain security vulnerabilities may develop in the future. This study aims to promote more research on blockchain security challenges by offering researchers insights into future security and privacy developments in blockchain-based patient health record systems.

**Keywords:** Blockchain · Electronic health records (EHR) · Patient health records (PHR) · risks · impact · countermeasures · privacy · security

## 1 Introduction

The COVID-19 pandemic has worn out medical personnel, overburdened institutions, adversely affected and marginalized sizable population segments, and reduced demand for and access to non-COVID-19-related medical care [1]. Interoperability, lengthy procedures, delays in diagnosis and treatment, information-sharing delays, high operating expenses, long insurance processing times, and control, privacy, and security issues are just a few difficulties facing current healthcare systems. With the advent of blockchain technology, a distributed and decentralized ecosystem will be possible, ultimately securing and safeguarding critical medical data [2]. For example, an innovative decentralized record management system called MedRec was proposed by Azaria et al. [3], providing patients with a secure means to access an immutable medical log to store treatment details using blockchain technology.

The development of blockchain technology has created new research opportunities in some fields, including medical data preservation, data integrity, patient ownership of

M. Papadaki et al. (Eds.): EMCIS 2022, LNBIP 464, pp. 477–496, 2023.
https://doi.org/10.1007/978-3-031-30694-5_35

their data, simple medical data exchange, and efficient medical insurance claims [4, 5]. However, several studies [6–8] have concentrated on the security features of blockchain-based healthcare due to the growing demand for patient data and its associated security and privacy issues. These studies have paid little attention to the impact, severity level, and relevant countermeasures in the healthcare arena. Such a gap makes it challenging to properly tackle security threats in blockchain-based patient health record systems (BPHRS). Our research aims to identify potential security risks in BPHRS, analyze their severity level and impact, and identify the corresponding countermeasures available to lessen these dangers and secure BPHRS. The three main research questions that underpin this study are as follows:

*RQ1: What are the emerging security risks in blockchain-based patient health record systems (BPHRS)?*
*RQ2: What are the severity levels and impacts of these risks?*
*RQ3: What are the recommended countermeasures to mitigate these risks?*

This paper is organized as follows: The background of blockchain technology is described in the next section. The methodology is presented in section three. The study's findings are described in section four. We summarize the results and study limitations and suggest areas for future investigation.

## 2   Background: Blockchain Technology

A blockchain collects chronologically ordered, publicly accessible records called blocks [9]. The information is encrypted using cryptography to protect user privacy and prevent data manipulation. Since the information is managed and stored in a decentralized ledger, no single central authority makes all the decisions. Instead, a consensus of all the network's participating nodes, which are dispersed around the globe, is used to make most choices [10]. Security, transparency, decentralization, immutability, and distribution are some of the distinctive characteristics of blockchain technology. Blockchain does not rely on centralized, trustworthy entities to process data transactions. Therefore, no intermediary third party is required to audit and confirm the data exchanges [11]. According to their characteristics and network behavior, blockchains can be classified into public, private, and hybrid [12] (Table 1).

**Table 1.** Features of different kinds of blockchains.

|  | Public | Private | Hybrid |
|---|---|---|---|
| Type of database | Decentralized | Partially decentralized | Partially decentralized |
| Definition | Anyone can join and complete transactions on this permissionless distributed ledger [10] | A permissioned blockchain network functions in a private setting, such as a closed network, or is managed by a single identity [11] | It allows businesses to build private, permission-based, and public permission-less systems [10] |
| Advantages | Trustable, secure, and transparent [12] | Faster transactions and scalable [12] | Safe and cost-effective [12] |
| Disadvantages | Scalability issues and high energy consumption [2] | Trust-building issues, lower security, and centralization [2] | Lack of transparency and less incentive [3] |
| Examples | Ethereum [10] | Hyperledger [10] | Ripple [11] |

# 3  Methodology

Using the search terms (TS = "Healthcare" or "Risks" or "Assessments" or "Counter-measures" AND TS = "Blockchain"), we performed a literature search using the Web of Science (WoS) and Scopus databases, establishing a time constraint from 2017 and beyond, and obtained 18 results. The IEEE and Science Direct search engines produced 20 and 13 papers, respectively, which were used to retrieve the supplemental material for the study. Thus, for the systematic literature review, 51 articles published between 2017 and 2022 were found and reviewed for inclusion and exclusion. The inclusion criteria included the study's publishing period (2017–2022) and applicability to blockchain-based healthcare systems. A PRISMA diagram is shown in Fig. 1 to illustrate the steps the researchers performed to identify relevant published materials and choose whether to include or exclude them. These steps include identification, screening, eligibility, and final inclusion.

# 4  Findings

This section discusses the findings of the systematic literature review organized according to the three research questions, RQ1, RQ2, and RQ3.

## 4.1  RQ1: What Are the Emerging Security Risks in BPHRS?

The healthcare industry faces challenges and inefficiencies, including fraud, erroneous healthcare data, a lack of stakeholder participation, and privacy and security concerns. Blockchain is seen as a logical technological solution for solving these issues and short-falls [13–15]. However, significant problems must be resolved before a safe BPHRS is effectively deployed. We present the outcome of our systematic literature review in Fig. 2, which represents a taxonomy of the risks associated with BPHRS based on its features and network behavior. The most significant risks related to BPHRS are technical, threat/security, privacy, organizational, and regulation. The terms risk register, risk profile, and risk treatment are used to provide detailed explanations of each of the risks. A risk register is utilized to detect possible risks associated with a project or an enterprise. An organization's risks are analyzed in a risk profile to determine their severity and likelihood. Risk treatment is selecting and implementing actions to reduce the risk [39].

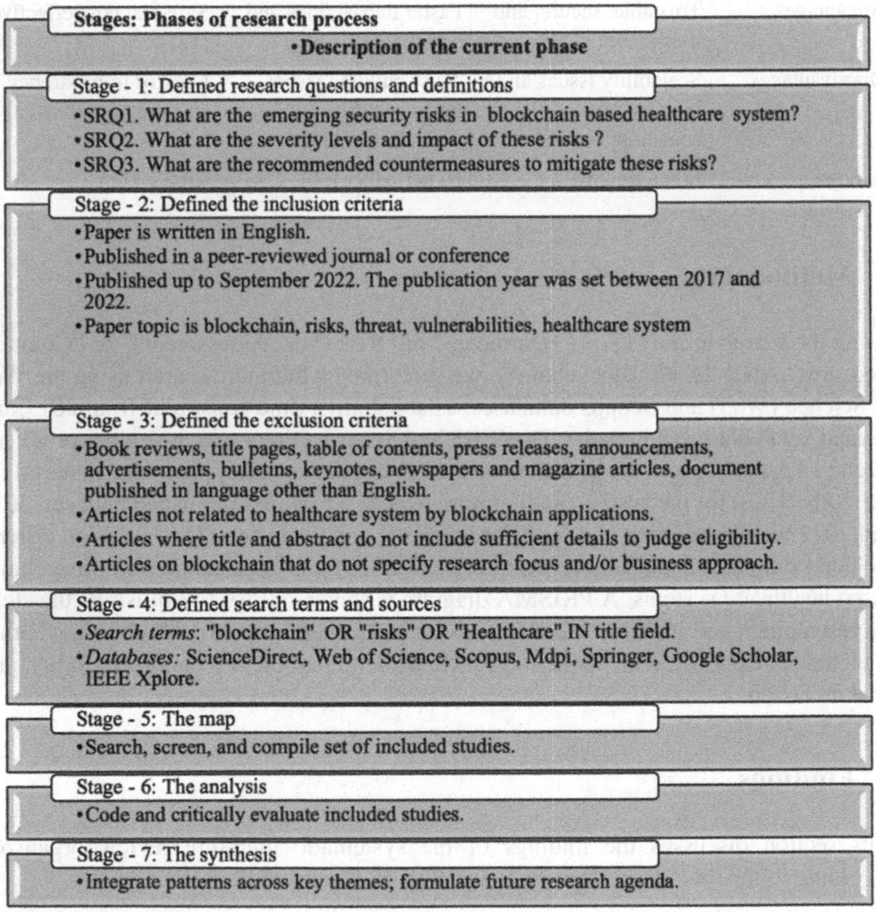
Stages: Phases of research process
•Description of the current phase
Stage - 1: Defined research questions and definitions
•SRQ1. What are the emerging security risks in blockchain based healthcare system?
•SRQ2. What are the severity levels and impact of these risks ?
•SRQ3. What are the recommended countermeasures to mitigate these risks?
Stage - 2: Defined the inclusion criteria
•Paper is written in English.
•Published in a peer-reviewed journal or conference
•Published up to September 2022. The publication year was set between 2017 and 2022.
•Paper topic is blockchain, risks, threat, vulnerabilities, healthcare system
Stage - 3: Defined the exclusion criteria
•Book reviews, title pages, table of contents, press releases, announcements, advertisements, bulletins, keynotes, newspapers and magazine articles, document published in language other than English.
•Articles not related to healthcare system by blockchain applications.
•Articles where title and abstract do not include sufficient details to judge eligibility.
•Articles on blockchain that do not specify research focus and/or business approach.
Stage - 4: Defined search terms and sources
•*Search terms*: "blockchain" OR "risks" OR "Healthcare" IN title field.
•*Databases:* ScienceDirect, Web of Science, Scopus, Mdpi, Springer, Google Scholar, IEEE Xplore.
Stage - 5: The map
•Search, screen, and compile set of included studies.
Stage - 6: The analysis
•Code and critically evaluate included studies.
Stage - 7: The synthesis
•Integrate patterns across key themes; formulate future research agenda.

**Fig. 1.** Research phases

A. **Technical risks**

Before implementing a blockchain, several technical risks to its fundamental functions must be assessed and mitigated. The technical analysis concentrates on the characteristics of the created blockchain-based system, including its applications, the Blockchain it uses, and the consensus algorithm it employs [24]. The most prominent technical risks are scalability, smart contract bugs, poor consensus mechanism, and high energy consumption. As the number of nodes increases, validating every node and every transaction becomes a **scalability** [25] challenge. Data duplication makes it difficult to scale blockchain networks in the healthcare industry [25]. The poor consensus mechanism is mainly due to the lack of proper selection of consensus protocols [28]. Smart contract bugs occur due to poor contract code that generates an invalid result [31]. The Proof of Work (PoW) consensus mechanism used by the blockchain network requires considerable energy. Blockchains consume high energy levels because, no matter how many miners are on the network, blocks can only be added to the chain at set times. Most Ethereum-based healthcare blockchain uses this consensus algorithm, leading to **high energy consumption** [27]. In addition, several other technical risks are associated with BPHRS, as listed in Table 2.

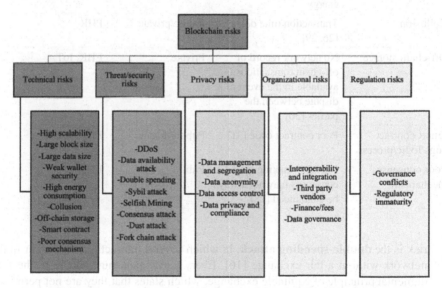

**Fig. 2.** Taxonomy diagram of blockchain risks of the healthcare system

B. **Threat/security Risks**

Even though Blockchain is considered safe and there is no participation by third parties, attacks such as double spending [16], consensus attack [17], Sybil attack [18], DDoS [17], and others have become a serious issue, especially in healthcare. When hacking, many cybercriminals aim directly at customers' financial information stored in their wallets. Hackers often try to boost their earnings by generating network congestion and unnecessary mining blocks. The most prominent security

**Table 2.** List of technical risks in BPHRS

| Risk register | Risk profile | Blockchain types | Healthcare domain |
|---|---|---|---|
| High scalability | Difficulty in scale to a large number of transactions [25] | Public | Covid trace tracking and PHR |
| Large block size | Maximum number of transactions that can be added to a block at once [20] | Public/private | EHR |
| Large data size | Difficult to handle data with high temporal resolution [6] | Public | EHR |
| Weak wallet security | Poor key management [27] | Public/private | EHR |
| High energy consumption | The PoW consensus mechanism requires a considerable amount of energy | Public | E-healthcare App |
| Collusion | Transaction time delay [26, 29] | Public/private | PHR |
| Off-chain storage | No network record of off-chain transactions is available in the event of a dispute between the parties [30] | Private | EHR, IoT |
| Smart contract bugs/logic/process | Poor contract code [31] | Public/private | EHR, IoT |
| Poor consensus mechanism | Decision-making by consensus may not always be guaranteed [28] | Public | EHR |

risk is the **double-spending attack**, in which several transactions can occur in the network without a fair exchange [16]. Every participant must adhere to the fundamental principle of equitable exchange, which states that they are not permitted to discover more messages about other participants' inputs than they would while implementing the consensus protocol [6]. A **consensus attack** occurs in Blockchain when a group of miners or a single miner controls more than 50% of the network's mining hash or computer [17]. Attackers use a 51% attack to reverse transactions on a blockchain and hinder the process of storing new blocks. During a **Sybil attack**, the attacker disrupts information flow, blocks the trustworthy nodes, and refuses to receive or send information after false identities are recognized by the blockchain system [17]. In a **DDoS attack**, an attacker can fill up blocks with spam transactions if they submit many blockchain transactions to the network, causing valid transactions to sit in "*mempools.*" If legitimate transactions are not included in blocks, they are not added to the ledger, and the Blockchain will not be able to function [1]. Table 3 illustrates the list of security risks that disrupt the proper functioning of BPHRS.

**Table 3.** List of threat/security risks in BPHRS

| Risk register | Risk profile | Blockchain types | Healthcare domain |
|---|---|---|---|
| DDoS attacks | Transaction flooding [17] | Public | EHR, PHR |
| | Dust transactions [20] | Private | EHR |
| Data availability attack | Receive an erroneous block by concealing the malicious part of the block from other nodes [21] | Public | E-healthcare App |
| Double spending | Double spending without fair exchange [16] | Public | EHR, PHR |
| Sybil attack | Run several fake nodes [18] | Public | EHR |
| Selfish mining | Allowing nodes with more than 51% computational power to reverse transactions in a blockchain [20] | Public | PHR |
| Consensus attack (51% attack) | The majority of the mining power is controlled by entities [17] | Private | EHR, PHR |
| Dust attack | Dust transactions [22] | Public | EHR |
| Fork chain attack | A fork on the Blockchain and more than one chain exist [23] | Public | EHR |

## C. Privacy Risks

When patient records in BPHRS are shared with other organizations without the data owners' consent for research or medication advertising, serious data privacy problems arise. Maintaining the integrity and confidentiality of outsourced data leads to a significant burden on stakeholders and blockchain nodes in computation and communication [37]. Table 4 shows the list of privacy risks in BPHRS.

**Table 4.** List of privacy risks in BPHRS

| Risk register | Risk profile | Blockchain types | Healthcare domain |
|---|---|---|---|
| Data management and segregation | Poor data management results in an information overload [37] | Public/private | EHR |
| Data anonymity | Leakage of sensitive patient information [37] | Public/private | EHR, PHR |
| Data access control | Unauthorized access to medical data [38] | Public/private | EHR, PHR |
| Data privacy and compliance | Compliance issues with privacy laws such as HIPPA and GDPR [38] | Public/private | EHR |

D. **Organizational risks**

Information about patients can be shared securely with healthcare organizations via Blockchain. Blockchain technology has helped organizations by making it easier to manage the clinical trials required for drug trials. Since copies of the shared ledger are stored across users' devices, Blockchain allows organizations to keep and back up medical insurance [7, 31]. However, investigating Blockchain's internal and external organizational challenges should be considered. Table 5 presents the list of organizational risks associated with BPHRS.

**Table 5.** List of organizational risks in BPHRS

| Risk register | Risk profile | Blockchain types | Healthcare domain |
|---|---|---|---|
| Interoperability and integration | Occurs due to a lack of trust between parties, and a lack of open standards [7, 31] | Public/private | PHR |
| Third-party vendors | Risks of sensitive information leakage due to the involvement of third-party vendors [32] | Public/private | PHR |
| Finance/Fees | A large number of transactions and fraud activities contribute to the high cost of these services [13] | Public/private | EHR |
| Data governance | Lack of guidelines and standards to control the accuracy, security, and use of sensitive data [31] | Public/private | EHR, E-healthcare App |

E. **Regulation risks**

Regulation risks such as governance conflicts and regulatory immaturity were other significant risks identified within BPHRS (see Table 6). The majority of BPHRS are created to be Health Insurance Portability and Accountability Act (HIPAA) [33] and General Data Protection Regulation (GDPR) compliant [34]. The implementation of these regulations has been hampered by decentralization and a lack of involvement from reliable third parties. However, because these regulations will link various social, economic, and healthcare systems, patients and service providers may find it difficult to follow the applications' results in the absence of a legal or compliance code, which results in governance conflicts [35]. Regulatory immaturity involves difficulties in defining the rules that will consider the cooperation of diverse stakeholders to develop an entire ecosystem that also considers the current regulatory system [36].

**Table 6.** List of organizational risks in BPHRS

| Risk register | Risk profile | Blockchain types | Healthcare domain |
|---|---|---|---|
| Governance conflicts | Difficult to follow the applications' results in the absence of a legal or compliance code [35] | Public/private | EHR, E-healthcare App |
| Regulatory immaturity | Problems in defining the rules that will consider the cooperation of diverse stakeholders to develop an entire ecosystem [36] | Public/private | PHR |

## 4.2  RQ2: What are the Severity Levels and Impacts of These Risks?

An in-depth analysis of the risks' immediate impact and severity level in blockchain-based healthcare systems is conducted. The projected harm or unfavorable outcome from exposure to the risk is known as risk severity (also known as risk impact). Using an ordinal scale is one of the most popular techniques to describe risk severity. Low, moderate, high, and severe are the most typical qualitative values on an ordinal scale [39]. Table 5 lists the impacts of all security risks in BPHRS. Table 7 illustrates the effects of emerging security risks in BPHRS.

A. **Impact of technical risks**

Scalability, consensus, smart contract bugs/logic, and transaction time delay/real-time are the most severe technical risks [25, 29]. BPHRS has a scalability issue that forces users to pay considerable fees and wait hours for transaction approval, delaying the provision of services [25]. The systems cannot manage millions of healthcare records in real time due to transaction time delays [29], and there is a probability that a medical history error will occur. The lack of records may cause treatment to be delayed [29]. Blockchain demands a tremendous amount of computer power, which is energy-intensive; it is estimated that Bitcoin mining alone uses 0.5% of the world's electrical supply [28].

B. **Impact of threat/security risks**

The most severe vulnerability risks that expose healthcare data to hackers and cyberattacks are consensus attacks and double spending [16, 17]. Transaction data integrity is compromised during a 51% attack assault, and the network's resources are depleted. The availability of services and the integrity of the data, which are crucial for healthcare applications, are adversely affected [17]. The possibility of double spending undermines the ledger's credibility. Numerous dangers can result in double spending, including Sybil-based double spending and 51% attacks, among others [16]. DDoS attacks have a high severity level, which can immediately interrupt network operations and prohibit access to essential data [20]. The patients and the medical staff may be unable to converse or exchange information because of this attack. Massive data requests block the server. As a result, the attack impacts demand and response generation [20].

C. **Impact of privacy risks**

Important security and privacy issues are brought up by introducing a single interoperable platform to make all healthcare data available in one place. Recent cyberattacks like WannaCry and the breach of medical data at Anthem are evidence of this [41]. When medical data is uploaded to the cloud to be shared in a healthcare blockchain, it can raise essential privacy issues that previous studies have largely ignored. For example, in cloud blockchain networks, hackers can become curious about medical resources and steal sensitive patient data without the patients' permission [20].

D. **Impact of organizational risks**

Interoperability, integration, and data governance are the most severe organizational risks. Premier Healthcare Alliance estimates that a lack of interoperability costs 150,000 lives and US$18.6 billion annually [40]. Most EHR products now available on the market impose restrictions on the open exchange of patient data across different product platforms. Although blockchain technology is intended to be more secure than traditional methods of data exchange, a lack of industry standards may make it difficult for devices to communicate with one another [31]. Industry standards are essential to the success of the healthcare blockchain market as it evolves [7].

E. **Impact of regulation risks**

Governance conflicts have a significant impact on how well BPHRS operates. Major security regulations must be followed, which apply to EHR contents [35]. For instance, anyone can access the data on a blockchain, and no one is responsible for ensuring its availability or security. Users are the data controllers under GDPR; however, Blockchain's immutability cannot erase or modify their data. Who should be held responsible for breaking the rules and regulations is a crucial concern for regulators in governance [31].

**Table 7.** Summary table for impacts of all the emerging security risks

| Risk register | Risk Impact | Blockchain types |
|---|---|---|
| High scalability | Increase in processing needs across the entire BPHRS infrastructure [25] | Public |
| Large block size | Unprocessed patient data, including genomic, critical organs, and others, resulting in unnecessary operating costs [20] | Public/private |
| Large data size | Issues with handling multi-dimensional medical data and high computational costs [6] | Public |
| Weak wallet security | If the key is stolen, it puts both patients' sensitive data and finances in jeopardy [27] | Public/private |
| High energy consumption | Critical performance degradation of patient healthcare systems [28] | Public |
| Collusion | Unable to handle millions of healthcare records in real-time [26, 29] | Public/private |
| Off-chain storage | Introduces a single point of failure, which continuously limits the availability of medical records [30] | Private |
| Smart contract bugs/process/logic | User revocation is expensive and results in a significant blockchain computation overhead. [31] | Public/private |
| Poor consensus mechanism | Impact on how consensus decisions are made [28] | Public |
| DDoS attacks | Massive medical transaction backlogs and higher mining fees [19] | Public/private |
| Data availability attack | Prompt diagnosis and treatment would be delayed [21] | Public |
| Double spending attack | Blocks specific IP addresses and transactions between various hospitals on the blockchain network [18] | Public |
| Sybil Attack | Targets sensitive data such as personal information, insurance details, and patient medical records [18] | Public |
| Selfish mining | If the patient's treatment record transactions are reversed, it could pose a significant threat to the patient [20] | Public |

(*continued*)

**Table 7.** (*continued*)

| Risk register | Risk Impact | Blockchain types |
|---|---|---|
| Consensus attack | Threatens the integrity of medical data on the Blockchain [17] | Private |
| Dust attack | Unavailability of patients' records during the treatment [22] | Public |
| Fork chain attack | A potential threat to the accuracy and integrity of medical data [23] | Public |
| Data management and segregation | Problems in managing and storing enormous numbers of EHRs locally and communicating secure data [37] | Public/private |
| Data anonymity | Conceals the actual identity of the nodes accessing the data [37] | Public/private |
| Data access control | Lack of authorization and distribution of medical records among healthcare providers [38] | Public/private |
| Data privacy and compliance | Raises concerns about compliance with international privacy and security laws, including the GDPR and HIPAA [38] | Public/private |
| Interoperability and integration | Problems in sharing medical data across many blockchain-based BPHRS [7, 31] | Public/private |
| Third-party vendors | Risks of sensitive information leakage are considered when a patient shares part of their medical records with an authorized third party [32] | Public/private |
| Finance/Fees | Insurance frauds and medical trials without planning contribute to high transaction fees [13] | Public/private |
| Data governance | Lack of framework for all healthcare stakeholders is not available [31] | Public/private |
| Governance conflicts | Both patients and service providers may find it difficult to follow the applications' results in the absence of a legal or compliance code [35] | Public/private |
| Regulatory immaturity | Unsure of responsibility for breaking privacy rules and regulations [36] | Public/private |

## 4.3   RQ3: What are the Recommended Countermeasures to Mitigate These Risks?

Here, we outline the current security and privacy-preserving methods and detection techniques that BPHRS can apply. Table 8 presents the list of countermeasures to mitigate BPHRS risks.

A. **Countermeasures for technical risks**

   The technical study examines the features of the developed blockchain-based system, including its applications, the Blockchain it uses, and the consensus algorithm it employs. A variety of techniques are investigated to address scalability problems (such as permissioned blockchains, the lighting protocol, delegated proof of stake, and directed acyclic graphs) [25, 44]. The Practice Byzantine Fault Tolerance (PBFT) algorithm instead of the PoW consensus algorithm can be used to solve scalability issues since it is better suited for BPHRS [25]. Segregated Witness restricts block sizes to 1MB, which minimizes DDoS attacks because forged blocks with larger sizes would be checked out and thrown away [45].

   Ethereum attempted to tackle the security limitations of proof of work, the lower danger of centralization, and high energy consumption using the Proof of Stake (PoS) mechanism [42]. To reduce data size, metadata is stored in a blockchain, and its sensitive and significant data is stored in a separate storage system such as the cloud [14].

   The techniques mainly used for tackling weak wallet security are using a multi-level authentication method when accessing wallets or generating wallet keys. In addition, we might use multi-signature wallets and cold wallets and not share the private keys of wallets with anyone [45]. Estimable PoW estimates how much work has been done and if the corresponding agreement reached a consensus [46]. IoT sensors can measure a patient's health conditions in real-time, which can be used in public blockchains such as the Ethereum environment [47].

B. **Countermeasures for threat/security risks**

   Increased authentication that permits pairing with blockchain blocks is required to reduce double spending attacks, which calls for more confirmations. It is also possible to apply non-interactive non-knowledge proof (NIZK), which aids in spotting anomalies in blockchain systems and allows for the addition of detection criteria to the network, making it impervious to fraudulent and early detection [4]. When nodes surpass a specified threshold, the power monitoring tool should impose restrictions to ensure that no single miner or mining pool has more than 50% of the network hash rate [25]. This helps to track node computing power continually to protect against consensus attacks. It is common practice to detect DoS/DDoS using anomaly detection techniques and reactive defense strategies. While unsupervised learning is frequently used for anomaly and novelty detection, machine learning (ML) techniques are now being utilized to predict harmful and legitimate traffic. Fee- and age-based designs are reactive defense strategies [42]. The mempool accepts an incoming transaction in the fee-based architecture if it pays the minimum relay and mining fees [19]. By only taking transactions that will be added to the Blockchain via mining, the main goal of this approach is to thwart an attacker's plan of attack. The authors calculated the inputs or parent transactions for each incoming transaction in an age-based process and set the "average age" variable to zero [19]. By randomly requesting/sampling portions of the block from the malicious node, Coded Merkle Tree (CMT) was developed to help light nodes identify data availability attacks [21]. Anti-Dust is offered to defend against various dust attacks effectively [22]. Through PBFT consensus, communication with peers can be done directly, reducing the chance of forgery and eliminating financial costs [28]. To protect from fork chain attacks, users should ensure the nodes they connect to are reliable to prevent multiple forks [23]. Selfish mining can be reduced by a backward-compatible protection method in which the fork resolution strategy ignores blocks not released in time [43]. The smart contracts should be designed with formal verification, which checks that a computer program executes as per the standard specification anticipated by the stakeholders [48].

C. **Countermeasures for privacy risks**

BPHRS must develop privacy policies to guarantee that only the patient and healthcare professionals can access patient medical records with the patient's express authorization. Healthblock is used to prevent security risks observed in widely used systems for intelligent healthcare and to strengthen the resiliency of healthcare data management systems [51]. Town Crier maintains anonymity using encrypted variables while allowing smart contracts to leverage data from sources beyond the Blockchain [19]. Ancile uses advanced cryptographic algorithms and smart contracts in an Ethereum-based blockchain for increased access control and data encryption [50]. No direct personal data should be stored on the Blockchain to ensure privacy. Some methods for dealing with this involve adding a cryptographic hash to the chain [38].

D. **Countermeasures for organizational risks**

The effectiveness of blockchain systems depends on organizational security controls for blockchains. As a result, we intend to examine countermeasures for corporate risks associated with BPHRS. Interoperability and integration can be controlled by building future capabilities, training, funding, and setting a suitable regulatory framework for blockchain adoption in the healthcare sector [6]. Smart contracts could implement agreements to secure agreements from healthcare professionals and patients before granting third-party vendors access to their content [31]. The system would be more effective if disintermediation led to lower transaction costs and near-real-time processing [13]. Organizations should agree on a framework for defining the data, size, and format that will be saved to solve data governance issues. This framework should be familiar to all healthcare stakeholders [6].

E. **Countermeasures for regulation risks**

All stakeholders in the healthcare industry should agree on a framework for specifying the data, size, and format that organizations will save to overcome **regulatory immaturity** issues [6]. To ensure that blockchains comply with national and international laws, the legislative frameworks must be evaluated and the required changes enacted. This may reduce **governance conflicts** by gaining certification from the International Standardization Authority, which will facilitate the rapid and secure development of BPHRS [49].

**Table 8.** List of countermeasures for all emerging security risks

| Risk register | Risk severity level | Risk treatment |
|---|---|---|
| Large scalability | Severe | Use the PBFT algorithm instead of the PoW consensus algorithm [25] |
| Large block size | High | Minimize the block size to 1 MB by Segregated Witness [45] |
| Large data size | Medium | Separate storage areas into the cloud [14] |
| Weak wallet security | Medium | Implement multi-level authentication, wallet keys, multi-signature wallets, and cold wallets [45] |
| High energy consumption | High | Introduce a hybrid consensus algorithm based on the PBFT algorithm, and the POS algorithm [42] |
| Collusion/Transaction time delay | Severe | Estimable PoW [46] |
| | High | Use IoT sensors [47] |
| Off-chain storage | Low | Applying masking blocks [30] |
| Poor consensus mechanism | Severe | Use PBFT consensus [28] |
| Smart contract bugs/logic/process | Severe | Verifying the logic of the intelligent contract programs within the Blockchain [48] |
| DDoS attacks | High | Anomaly detection methods [19] |
| | High | Fee-based design and age-based design [19] |
| Data availability attack | Medium | Coded Merkle Tree (CMT) [21] |
| Double spending | Severe | Non-interactive non-knowledge proof (NIZK) [4] |
| Sybil attack | High | Pure PoW consensus protocol [18] |
| Selfish mining | Low | Backward-compatible protection approaches [43] |
| Consensus attack | Severe | Power monitoring tool [25] |
| Dust attack | High | Anti-Dust [22] |
| Fork chain | High | Use reliable nodes [23] |
| Data management and segregation | Low | Healthblock to strengthen the resiliency of healthcare data management systems [51, 52] |

*(continued)*

**Table 8.** (*continued*)

| Risk register | Risk severity level | Risk treatment |
|---|---|---|
| Data anonymity | High | Town Crier maintains anonymity [19] |
| Data access control | Severe | Use Ancile for increased access control [50] |
| Data privacy and compliance | Severe | Add a cryptographic hash to the chain [38] |
| Interoperability and integration | Severe | Set the suitable regulatory framework for blockchain adoption in the healthcare sector [6] |
| Third-party vendors | Medium | Smart contracts [31] |
| Finance/Fees | high | Disintermediation techniques [13] |
| Data governance | medium | A standard framework for defining the data, size, and format [6] |
| Governance conflicts | medium | Legislative frameworks need to be evaluated [6] |
| Regulatory immaturity | high | Gaining certification from International Standardization Authority [49] |

## 5   Discussion

A systematic review of published blockchain-based healthcare systems literature identified the critical area where Blockchain may be used to address data management and access control problems in the EHR. Blockchain technology can reduce costs while increasing the process quality and efficiency in many different areas of healthcare. According to the research, private blockchains are less vulnerable to security risks than public blockchains. The network's scalability, technological risks, rising transaction fees, and security and privacy threats are the ongoing problems that must be resolved for a safe and effective BPHRS. Numerous studies and real-world applications offer countermeasures against these hazards. The best solutions identified are the PBFT algorithm and the PoS consensus protocol, which reduce the overhead of scalability, transaction delay issues, and transaction cost to a large extent [25, 28, 46]. However, there are still difficulties and unresolved research problems with developing reliable and efficient security solutions that can guarantee the proper operation of BPHRS. There is still a regulatory issue with defining the rules and conditions of usage for all parties interested in the BPHRS. One of the primary potential techniques for adopting a blockchain into various healthcare areas is to create a compliance code with consistent standards, standardizations, and international legislation. BPHRS would benefit from adopting AI-based methods like machine learning and deep learning with Blockchain to improve clinical trial verdicts, medical research, and treatment processes.

# 6    Conclusion

This research examined emerging risks related to BPHRS and identified numerous technological security and vulnerability issues. To conduct an in-depth analysis, the authors reviewed 51 publications using PRISMA's inclusion and exclusion criteria in response to the RQs. We provided an overview of BPHRS security and identified vulnerabilities, threats, and viable countermeasures for security specialists and researchers. This study mainly concerns the severity level and impact of these hazards on the patient record system. To forecast the potential harm caused by these threats and confirm whether the current technology is sufficient to survive persistent hacking, it is essential to evaluate the severity level and impact of security and privacy concerns in BPHRS. The study found that, compared to public blockchains, private blockchains are less susceptible to security risks. The ongoing issues that must be fixed for a secure and reliable BPHRS include the network's scalability, technological hazards, increased transaction fees, and security and privacy threats. Future work on BPHRS will center on more secure architecture, creating a robust consensus mechanism, a standard regulatory framework, and a more thorough smart contract detection.

# References

1. Behnke, R.: How Blockchain DDoS Attacks Work (2022). Halborn.com. https://halborn.com/how-blockchain-ddos-attacks-work/
2. Marbouh, D., et al.: Blockchain for COVID-19: review, opportunities, and a trusted tracking system. Arab. J. Sci. Eng. **45**(12), 9895–9911 (2020). https://doi.org/10.1007/s13369-020-04950-4
3. Azaria, A., Ekblaw, A., Vieira, T., & Lippman, A.: MedRec: using blockchain for medical data access and permission management. In: 2016 2nd International Conference on Open and Big Data (OBD), pp. 25–30. IEEE, August 2016
4. Alsunbul, A., Elmedany, W., Al-Ammal, H.: Blockchain application in healthcare industry: attacks and countermeasures. In: 2021 International Conference on Data Analytics for Business and Industry (ICDABI), pp. 621–629. IEEE, October 2021
5. Kumar, T., Ramani, V., Ahmad, I., Braeken, A., Harjula, E., Ylianttila, M.: Blockchain utilization in healthcare: key requirements and challenges. In: 2018 IEEE 20th International conference on e-health networking, applications and services (Healthcom), pp. 1–7. IEEE, September 2018
6. Attaran, M.: Blockchain technology in healthcare: challenges and opportunities. Int. J. Healthc. Manag. **15**(1), 70–83 (2020). https://doi.org/10.1080/20479700.2020.1843887
7. Onik, M.M.H., Aich, S., Yang, J., Kim, C.S., Kim, H.C.: Blockchain in Healthcare: Challenges and Solutions. Big Data Analytics for Intelligent Healthcare Management, pp. 197–226. Academic Press, Cambridge (2019)
8. Ismail, L., Materwala, H.: Article; a review of blockchain architecture and consensus; protocols: use cases, challenges, and solutions. Symmetry **11**(10), 1198 (2019). https://doi.org/10.3390/sym11101198
9. Nakamoto, S.: Bitcoin: A Peer-to-Peer Electronic Cash System (2008). Accessed 20 Aug 2018. https://bitcoin.org/bitcoin.pdf
10. Rifi, N., Rachkidi, E., Agoulmine, N., Taher, N.C.: Towards using blockchain technology for eHealth data access management. In: 2017 4th International Conference on Advances in Biomedical Engineering (ICABME 2017), pp. 1–4. IEEE (2017)

11. Vacca, A., Di Sorbo, A., Visaggio, C., Canfora, G.: A systematic literature review of blockchain and smart contract development: techniques, tools, and open challenges. J. Syst. Softw. **174**, 110891 (2021)
12. Morkunas, V.J., Paschen, J., Boon, E.: How blockchain technologies impact your business model. Bus. Horiz. **62**(3), 295–306 (2019)
13. Noon, A.K., Aziz, O., Zahra, I., Anwar, M.: Implementation of Blockchain in Healthcare: A Systematic Review. In 2021 International Conference on Innovative Computing (ICIC), pp. 1–10. IEEE, November 2021
14. Chen, Y., Ding, S., Xu, Z., Zheng, H., Yang, S.: Blockchain-based medical records secure storage and medical service framework. J. Med. Syst. **43**(1), 1–9 (2018). https://doi.org/10.1007/s10916-018-1121-4
15. Abunadi, I., Kumar, R.: Blockchain and business process management in health care, especially for COVID-19 cases. Secur. Commun. Netw. **2021**, 1–16 (2021). https://doi.org/10.1155/2021/2245808
16. Khan, S.N., Loukil, F., Ghedira-Guegan, C., Benkhelifa, E., Bani-Hani, A.: Blockchain smart contracts: applications, challenges, and future trends. Peer-to-Peer Netw. Appl. **14**(5), 2901–2925 (2021). https://doi.org/10.1007/s12083-021-01127-0
17. Wang, H., Wang, Y., Cao, Z., Li, Z., Xiong, G.: An overview of blockchain security analysis. In: Cyber Security: 15th International Annual Conference, CNCERT 2018, pp. 14–16 August (2018), Revised Selected Papers 15, pp. 55–72. Springer Singapore (2019)
18. Iqbal, M., Matulevičius, R.: Exploring Sybil and double-spending risks in blockchain systems. IEEE Access **9**, 76153–76177 (2021)
19. Hasanova, H., Baek, U.J., Shin, M.G., Cho, K., Kim, M.S.: A survey on blockchain cybersecurity vulnerabilities and possible countermeasures. Int. J. Netw. Manag. **29**(2), e2060 (2019)
20. Jabarulla, M., Lee, H.: A Blockchain and artificial intelligence-based, patient-centric healthcare system for combating the COVID-19 pandemic: opportunities and applications. Healthcare **9**(8), 1019 (2021). https://doi.org/10.3390/healthcare9081019
21. Mitra, D., Tauz, L., Dolecek, L.: Overcoming Data Availability Attacks in Blockchain Systems: LDPC Code Design for Coded Merkle Tree (2021). arXiv preprint arXiv:2108.13332
22. Wang, Y., Yang, J., Li, T., Zhu, F., Zhou, X.: Anti-dust: a method for identifying and preventing Blockchain's dust attacks. In: 2018 International Conference on Information Systems and Computer Aided Education (ICISCAE), pp. 274–280. IEEE, July 2018
23. Ploder, C., Spiess, T., Bernsteiner, R., Dilger, T., Weichelt, R.: A Risk Analysis on Blockchain Technology Usage for Electronic Health Records. Cloud Computing And Data Science, pp. 1–16 (2021). https://doi.org/10.37256/ccds.222021777
24. Wright, S.: Technical and legal challenges for healthcare blockchains and smart contracts. In: 2019 ITU Kaleidoscope: ICT for Health: Networks, Standards, and Innovation (ITU K) (2019)
25. Panda, S.K., Jena, A.K., Swain, S.K., Satapathy, S.C. (Eds.): Blockchain Technology: Applications and Challenges. Springer International Publishing, Cham (2021). https://doi.org/10.1007/978-3-030-69395-4
26. Griggs, K.N., Ossipova, O., Kohlios, C.P., Baccarini, A.N., Howson, E.A., Hayajneh, T.: Healthcare blockchain system using smart contracts for secure automated remote patient monitoring. J. Med. Syst. **42**(7), 1–7 (2018). https://doi.org/10.1007/s10916-018-0982-x
27. Beinke, J., Fitte, C., Teuteberg, F.: Towards a stakeholder-oriented blockchain-based architecture for electronic health records: design science research study. J. Med. Internet Res. **21**(10), e13585 (2019). https://doi.org/10.2196/13585
28. Wu, Y., Song, P., Wang, F.: Hybrid consensus algorithm optimization: a mathematical method based on POS and PBFT and its application in Blockchain. Math. Probl. Eng. 2020 (2020)

29. Siyal, A.A., Junejo, A.Z., Zawish, M., Ahmed, K., Khalil, A., Soursou, G.: Applications of blockchain technology in medicine and healthcare: challenges and future perspectives. Cryptography **3**(1), 3 (2019)

30. Castillo, J.: Blockchain: a decentralized solution for secure applications (doctoral dissertation, university of texas at San Antonio) (2022)

31. Xiong, H., Chen, M., Wu, C., Zhao, Y., Yi, W.: Research on progress of blockchain consensus algorithm: a review on recent progress of blockchain consensus algorithms. Futur. Internet **14**(2), 47 (2022). https://doi.org/10.3390/fi14020047

32. Esmaeilzadeh, P.: Benefits and concerns associated with blockchain-based health information exchange (HIE): a qualitative study from physicians' perspectives. BMC Med. Inform. Decis. Mak. **22**(1), 1–18 (2022)

33. Gostin, L.O., Levit, L.A., Nass, S.J. (Eds.): Beyond the HIPAA privacy rule: enhancing privacy, improving health through research (2009)

34. Wachter, S.: Normative challenges of identification in the Internet of Things: privacy, profiling, discrimination, and the GDPR. Comput. Law Secur. Rev. **34**(3), 436–449 (2018)

35. Nguyen, D., Pathirana, P., Ding, M., Seneviratne, A.: Blockchain for secure EHRs sharing of mobile cloud based E-Health systems. IEEE Access **7**, 66792–66806 (2019). https://doi.org/10.1109/access.2019.2917555

36. Min, M., et al.: Learning-based privacy-aware offloading for healthcare IoT with energy harvesting. IEEE Internet Things J. **6**(3), 4307–4316 (2019). https://doi.org/10.1109/jiot.2018.2875926

37. Bernal Bernabe, J., Canovas, J., Hernandez-Ramos, J., Torres Moreno, R., Skarmeta, A.: Privacy-preserving solutions for blockchain: review and challenges. IEEE Access **7**, 164908–164940 (2019). https://doi.org/10.1109/access.2019.2950872

38. Sookhak, M., Jabbarpour, M.R., Safa, N.S., Yu, F.R.: Blockchain and smart contract for access control in healthcare: a survey, issues and challenges, and open issues. J. Netw. Comput. Appl. **178**, 102950 (2021)

39. Lagrama, E.R.C.: Preventing Disaster: Quantifying Risks at the UP Diliman University Library (2009)

40. DeVore, S., Champion, R.W.: Driving population health through accountable care organizations. Health Aff. **30**(1), 41–50 (2011)

41. Ghafur, S., Grass, E., Jennings, N.R., Darzi, A.: The challenges of cybersecurity in health care: the UK national health service as a case study. Lancet Digit. Health **1**(1), e10–e12 (2019)

42. Rodrigues, B., Stiller, B.: Cooperative signaling of DDoS attacks in a blockchain-based network. In: Proceedings of the ACM SIGCOMM 2019 Conference Posters and Demos, pp. 39–41 (2019)

43. Zhang, R., Preneel, B.: Publish or perish: a backward-compatible defense against selfish mining in bitcoin. In: Handschuh, H. (ed.) CT-RSA 2017. LNCS, vol. 10159, pp. 277–292. Springer, Cham (2017). https://doi.org/10.1007/978-3-319-52153-4_16

44. Singh, S., Sanwar Hosen, A.S.M., Yoon, B.: Blockchain security attacks, challenges, and solutions for the future distributed IoT network. IEEE Access **9**, 13938–13959 (2021)

45. Wen, Y., Lu, F., Liu, Y., Huang, X.: Attacks and countermeasures on blockchains: a survey from layering perspective. Comput. Netw. **191**, 107978 (2021)

46. Hsueh, C., Chin, C.: EPoW: solving blockchain problems economically. In: 2017 IEEE SmartWorld, Ubiquitous Intelligence Computing, Advanced Trusted Computed, Scalable Computing Communications, Cloud Big Data Computing, Internet of People and Smart City Innovation, SmartWorld/SCALCOM/UIC/ATC/CBDCom/IOP/SCI, 2017, pp. 1–8 (2017)

47. Pham, H.L., Tran, T.H., Nakashima, Y.: A secure remote healthcare system for a hospital using blockchain smart contract. In: Proceedings of the IEEE Globecom Workshops, pp. 1–6 (2018)

48. Hewa, T.M., Hu, Y., Liyanage, M., Kanhare, S.S., Ylianttila, M.: Survey on blockchain-based intelligent contracts: technical aspects and future research. IEEE Access **9**, 87643–87662 (2021)

49. Pinter, K., Schmelz, D., Lamber, R., Strobl, S., Grechenig, T.: Towards a multi-party, blockchain-based identity verification solution to implement clear name laws for online media platforms. In: Business Process Management: Blockchain and Central and Eastern Europe Forum. BPM 2019. LNBIP, vol. 361, pp. 151–165. Springer, Cham (2019). https://doi.org/ 10.1007/978-3-030-30429-4_11

50. Dagher, G.G., Mohler, J., Milojkovic, M., Marella, P.B.: Ancile: privacy-preserving framework for access control and interoperability of electronic health records using blockchain technology. Sustain. Cities Soc. **39**, 283–297 (2018)

51. Zaabar, B., Cheikhrouhou, O., Jamil, F., Ammi, M., Abid, M.: HealthBlock: a secure blockchain-based healthcare data management system. Comput. Netw. **200**, 108500 (2021)

52. Papadaki, M., Karamitsos, I., Themistocleous, M.: Covid-19 digital test certificates and Blockchain. J. Enterp. Inf. Manag. **34**, 993–1003 (2021). https://www.researchgate.net/pub lication/353272635_ViewpointCovid-19_digital_test_certificates_and_blockchain

# Domain and Semantic Modeling in the Context of Interactive Systems Development: User and Device Cases

Saulo Silva[✉] [iD] and Orlando Belo [iD]

ALGORITMI R&D Centre/LASI, University of Minho, Braga, Portugal
saulo.silva@ifg.edu.br, obelo@di.uminho.pt

**Abstract.** The development of interactive systems requires tools and knowledge from a number of domains. The combination of Software Engineering with Human Factors aspects assists designers and engineers in designing error free, safe systems for humans' operators. This paper explores the use of multidimensional knowledge representation as means to narrow the gap between design and implementation of interactive systems. Two models are discussed for delivering heavyweight ontologies, by the means of Ontology–Driven Development. This might enable the development of software artifacts, with potential of improving system specification, information sharing and reusability. The theoretical approach is discussed; a literature revision in the topic is included, along with taxonomy enabling the description of user and device's characteristics. Finally, the ontological development is presented and as well as analyzed.

**Keywords:** Ontology-based Development · Requirements Analysis · User Interface · Interactive Systems

## 1 Introduction

When Information Systems (IS) are interactive, that is, operated by humans, they must include a fundamental component, namely the User Interface (UI), which is defined as the part of the system that humans come into contact physically, perceptually and conceptually [1]. Usually, user interfaces that provide graphical elements for controlling the system – for instance, buttons, screens, text boxes, and sliders, among others – are recognized as Graphical User Interfaces (GUI). User Interfaces seek providing system representation for human operators, exposing the control logic behind the system under control. Humans make use of the user interface for carrying out her or his goals with the computational system. Hence, problems in user interfaces have the potential of compromising such goal achievement. For computer systems with critical functions, such as medical devices, energy or aeronautical systems, referred to as interactive critical systems or High Assurance (HA) systems, problems in user interface design may cause losses of different nature, such as financial, environmental or social, which can be characterized as accidents [2].

M. Papadaki et al. (Eds.): EMCIS 2022, LNBIP 464, pp. 497–514, 2023.
https://doi.org/10.1007/978-3-031-30694-5_36

Developing efficient user interfaces for IS requires knowledge from different domains, such as Software Engineering (SE), Ergonomics, Psychology, or Human Factors, for delivering error-free, trustworthy and reliable interfaces. User interface experts use information from those domains for perceiving the most relevant aspects for an ideal system use, providing the best solutions in terms of computer controls suitable for human expectations, and consequently averting gaps that could induce to interaction errors. In this context, modeling can play an important role in assisting designers during user interface development. Model–Driven Development (MDD) is a software development paradigm that makes use of information present in models for reasoning about problems and solutions regarding aspects being implemented in software artifacts. Hence, user and device modeling, for instance, might provide important insights about the description of a potential system user and the computational system in use, respectively.

Another approach for information organization relies on knowledge representation or on ontologies, for representing explicit specifications of conceptualization [3]. For [4], ontologies might have different roles in Software Engineering field. They may be a conceptual basis for software components specification and development. Ontology–Driven Development (ODD) is a software engineering vision considering the need of semantic constraints for supporting development processes, "through which software becomes a direct projection of a semantic definition" [5]. Those constraints are valuable assets for requirement analysis activities, with the potential of reducing costs and assuring the ontological adequacy of the user interface of an information system.

Despite the specificities of ODD and MDD approaches, ontologies may be seen as models. ODD might be considered as a particular type of MDD and might assist in describing problem domain at development time, with potential of reducing the cost of conceptual analysis or assisting in the constraints/requirements phase by providing relations between knowledge elements extracted from application domains. By providing human factors knowledge in the context of user interface development processes, this paper presents a work that seeks narrowing the gap between design and implementation of user interfaces. Our strategy is discussing the necessity and usefulness of human and device knowledge in user interfaces development, describing a multidimensional knowledge representation model related with (abstracts) human and device models, and presenting a human device ontology based on multidimensional models. The remaining part of this paper is organized as follows: Sect. 2 presents the background related with ontologies and knowledge representation. Section 3, approach ontology development, reveals the human device ontology we developed, and discusses how Interactive Systems development can benefit from ontologies. Section 4 highlights results jointly with an ontological analysis; and finally, Sect. 5, presents some remarks and conclusions, and a few research lines for future work.

## 2 Related Works

According with [6], a (abstract) human model can be considered as a set of human parameters, represented by variables, which can be employed for describing users of a product. They differentiate a user model declaration from a user model instantiation, whereas the first relates with the establishment of user variables, and the second relates with the creation of a user profile. The model is described using a machine and human-readable format. User models are relevant for a number of disciplines. They provide human factors information about users that can be considered during design project. Due to this, a number of standards have been created to provide user-modeling guidelines.

A different user-modeling paradigm is based upon context-aware information. In [7] we find an example of that approach. Authors propose a learning mechanism using historical context-aware information gathered from user for reflecting user models. The parameters considered in such models seek assisting systems in dealing with inter-user variance. According with [8], it refers to the variance of user parameters that must be recognized in system specification, which requires that the design cope with different types of users in the system's users. The intra-user variance refers to monitoring internal user knowledge that is necessary for performing the task. User knowledge is especially relevant in the design of interactive systems. It supports different types of analysis that have the potential of preventing unwanted user interface interactions.

Device models are also relevant in designing interactive systems, since it demands rigorous design of user interface structure and behavior for avoiding hazards. Device information can provide valuable understanding of user interface parameters, which can be considered in the analysis. This is especially relevant for interactive systems with critical functions, such as critical interactive systems, or high assurance systems.

Device modeling allows investigating user interface structural and behavioral parameters that are presented as actions that users are allowed to perform for achieving their goals and keeping performance. The interface structure is useful for reasoning about the user interface adequate working, in terms of structure's individual components (for instance, buttons, displays). It is motivated by the need of reasoning about the quality of this structure [9]. For instance, device based user interface individual components, such as buttons, when pressed, causes the activation or deactivation of internal system processes, and displays, which are employed for monitoring changes in internal variables of systems. On the other hand, user interface behavior is useful for reasoning about the adequate user interface working behavior, resulting in the individual components coordination of such systems, being motivated by the need for reasoning about the quality of this behavior – sets of user interface components, such as panels, which are a collection of components like displays and buttons that have different working processes intending the control of different output signals.

The relevance of user and system modeling in the development of interactive systems is undeniable. It has the potential of highlighting crucial and critical user and device aspects that might contribute for positive outcomes such as user satisfaction, improving efficiency and productivity during the interaction process, along with decreasing costs. For achieving that, a crucial phase in this process is the analysis of the models, i.e., providing assurance that it complies with the system specification. Semantic modeling and analysis can play an important role in this task, as models with semantic constraints have the potential in assisting designers and system engineers during the requirements phase of interaction system development. Knowledge Engineering is a discipline closely related with SE that involves methods for knowledge representation [10]. According with [11], agreeing on a single definition of Knowledge is difficult due to its level abstractness. While their work provides a range of different definitions, in the current work, we adopt the definition given by [12], which consider knowledge as interrelations between entities and relationships, highlighting that "relationships between entities provide meaning, at the same time that entities derive their meaning from their relationships". Knowledge representation is then a research field that makes use of modeling activity for creating knowledge bases. In its turn, knowledge bases are useful for representing conceptualizations that are assumed existing in a certain domain of interest, along with the relations between them. Those conceptualizations and their respective relations represent commitments about that domain and might be used as a reusable knowledge repository. The level of commitments depends on the technology for representing knowledge. From the knowledge viewpoint, [13] defines ontology as the most useful organization of knowledge about a particular domain, evolving from the need of satisfying formal postulates of signification. This is achieved by the use of object-attribute-value (O-A-V) triplets, which in ontology theory represents facts about objects and their attributes [14].

Ontologies might be described according with the level of expressiveness and granularity, as graphically illustrated in Fig. 1 [15][16]. Top-level ontologies refer to general and high-level (abstracts) concepts like space, time, and object, among others and provide higher expressiveness, at the cost of lower granularity. Examples can be found in the works of [17] and [18]. Domain ontologies represent concepts that belong to a generic domain or part of the world, such as politics, biology or computer science [19]. Task ontologies describe knowledge about a task or activity, such as initiating a system procedure, ending a system mode selection, or engaging a specific operation mode. Example of task ontology can be found in the work of [20]. Application ontologies, which provide higher granularity, are given as specializations of both domain and task ontologies for a specific case application [21]. Apart from the potential of reusability at development time for reducing the cost of conceptual analysis, ontologies might also offer another advantage in the context of the software development life cycle, as observed in other types of formal specifications, namely allowing designers and engineers to devote time in conceptual modeling (planning) phase prior to development phase. This has the potential of preventing unnecessary overload on software development life cycle implementation and maintenance phases. For the reader interested in covering a wider list of ontologies applications in the context of the Software Engineering life cycle, the works of [10] and [22] provide a categorization of ontologies as well as enumerate other approaches for using ontologies in the context of SE.

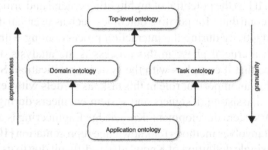

**Fig. 1.** Top-level, domain, task and application ontology in relation with granularity and expressiveness (Source: [15] and [16]).

The literature covers multiple strategies on Ontology development, according with [23] and [24]. It is important highlighting their main differences, as well agreeing on at least one. Ontology development approaches differs on their formalism level. Some approaches base their methodology on elaborating "tasks" or "principles" for ontology development (for instance, the methodology proposed by [25]). Other strategies focus on narrowing the distance from ontology development to other engineering disciplines, incorporating a defined life cycle for it (for instance, the work of [26] in establishing well-defined and iterative phases for ontology development). Following the same principle, the work of [27] comprises the creation of the *Methontology framework* intending standardizing the whole ontology development life cycle, transforming the process in engineering development.

[5] states that more importantly is agreeing on iterative and structured process, which expresses an ontology life cycle, with well-defined processes for each activity for building and maintaining the ontology, such as:

- Domain determination – the first important task in ontology development is considering the domain whose syntactic and semantic knowledge are modeled into a knowledge model.
- Terms enumeration – important expressions, which describe the domain or scope and can contain underlying concepts terms, properties related with those terms, and others, are enumerated [25].
- Definition of class and hierarchy – by examining the objects in previously defined concepts, intends developing the first idea of a class and its subclasses (i.e., a hierarchy). The approaches might be top-down (classes are first defined and then specialized in its subclasses), bottom-up (most relevant subclasses are enumerated and then grouped in different classes) or a combination of both with no loss for the result.
- Class properties definition – also referred to as slots, relates with definition of the internal structure of the class. All subsequent classes inherit the slots of their super classes.
- Domain and range of properties – define the requirements for the information slots hold, i.e., its allowed values (also referred to as facets definition). A facet describes values characteristics such as type, domain and cardinality and therefore has the potential of enforcing constraints for slots-values. For instance, slots-value *type* describes what types of values are allowed in the slot. Accepted types range from *String, Number, Boolean, Enumerated* or *Instance*. Slots-value *domain* refers to the classes describing the slot, which is hence referred to as the range of a slot. Slot-value *cardinality* is a property that refers to the quantity of values assigned to a slot. Some slot values require a single cardinality for correctly describing certain aspects of the reality (e.g., an animal has one body). Multiple-cardinality describes any number of values (e.g., the number of species of the animal kingdom).
- Properties instances – finally individual instance(s) are created by i) choosing a class, ii) creating individual instance of that class, and iii) filling all the defined slots of that class [25].
- Relations Definition – the main goal of identifying and characterizing relations (also referred to as *connections* or *associations*) is uncovering associations expressed between the characteristics and groups. [28] explains this identification as related

with uncovering associations between the characteristics and groups, while characterization follows the identifications and characterization of relations in a way that they could allow the relational knowledge being utilized in a near future for making useful inferences. For instance, the relation *is_part_of*, holds an association between two entities such as *software_button* and *user_interface*, in which it is built in.

- Data properties – its goal is twofold. They are used for connecting classes with specific data values, supporting the parameters that compose the models. Data properties are also used for restricting classes named *Defined Classes* under necessary and sufficient conditions.

We can find other examples of works in the literature related with ontologies in the context of SE life cycle, such as the work of [28] who proposed the IDEF5 ontology development method for promoting knowledge sharing and terminology standardization. [29] proposed the Definitions for User Experience Experimental Terms, a collection of terms interconnected into semantic network and followed by related axioms created in the context of Suggested Upper Merged Ontology (SUMO) ontology framework. [30] described GUMO, a user model ontology formalized in OWL language, capturing user important dimensions for use in user-adaptive systems, such as *Mental State, Personality* and *Contact Information*. Their ontology is revisited later, in the context of the semantic web ontology language [31]. [32], which described ontologies classifications and proposed taxonomy for assisting Software Engineering and Technology (SET) experts for understanding the linking between ontologies with some software development aspects. Other author, [33], presented a conceptualization for user interface development through ontological modeling, which provided descriptions of the user interface itself for a given architecture, for instance, in the description of components, relationships between components, among others. [34] presented a formal user interface representation describing both interaction and components aspects, while also discussed how the formal representation of user interface might benefit its development. In the context of Knowledge Management (KM), [35] presented OntobUMf, an ontology-based Framework for modeling user behavior. She proposed a model for user behavior and implemented a classifier for ranking users based on their knowledge level. [36] proposed a semantic framework in the context of the human system integration (HSI) research field. The main goal is extending the overall modeling capabilities in human-machine systems for improving human representation inside the overall system view. The work of [21] presented a particular implementation of user interface ontology tailored for the software development in avionics domain, where it might play a role as documentation resource. The approach made use of requirements ontology for supporting the creation of software test cases. Finally, [37] provided example of ontology creation for web user interfaces since the specification to the functional fragments.

## 3   The Design Approach

In this paper we propose cataloguing information on User and System domains for formalizing software models. This information includes the characteristics of the models, along with its descriptors (Table 1). The goal of the taxonomy is assisting in answering the following questions:

- What human and device characteristics are important considering for user interface development?
- What are the literature references that support these human factors and device characteristics?

For assisting this task, the taxonomy presents a set of ten (10) descriptors, such as name, description, unit of measurement, value space, taxonomy group, data type, how to measure/detect, reference/source, relations and comment (Table 1).

**Table 1.** Index of descriptors in the taxonomy.

| Descriptor name | Descriptor summary |
| --- | --- |
| *Name* | A characteristic must present unique information that describes it |
| *Description* | Additional information for detailing the characteristic |
| *Unit of measurement* | The unit of measurement of the characteristic |
| *Value space* | The value space of the characteristic (*nominal, ordinal, interval, ratio, absolute*) |
| *Taxonomy group* | Which category the characteristic belongs to |
| *Data type* | Which kind of data the characteristic presents (*string, enumeration, list/vector, real, integer*) |
| *How to measure/detect* | How the characteristic information is measured or detected (type of measurement method) |
| *Reference/source* | Which are the references/standards that supports the characteristic |
| *Relations* | If the characteristics is related with any other |
| *Comment* | Further explanations/observations from the given characteristic |

Table 2 presents a summary for user domain composed by a set of twenty-five (25) descriptors (for instance, related with physical body conditions), along with its seven (7) groups (encapsulating demographic, anthropometric, interaction related states, hearing, visual, motor and knowledge/experience data).

An excerpt (first level) composed by six (6) device descriptors, alongside its three (3) groups is presented (Table 3). The complete work extends up to five other levels providing classification for components, devices and systems.

The next tasks in ontology creation are supported by the taxonomy, whose main role is ensuring required information for the ontology development tasks.

- *Domain determination* – is performed by considering the need to know *Device* and *User* aspects. The literature of human-machine interaction provides a number of measurements for estimating the quality of that interaction. In this context, one of the most widely accepted measurements considered is provided by the Ergonomics of human-system *interaction's* chapter of the International Organization for Standardization

**Table 2.** User characteristics and groups.

| Group name | Characteristic name |
|---|---|
| Demographic data | Age |
| | Gender |
| Anthropometric data | Weight |
| | Stature |
| Interaction related states | Perceived stress |
| | Stress and reaction time |
| | Perceived Fatigue |
| | Perceived Attitude towards computer |
| | Motivation |
| Hearing parameters | Hearing |
| | Hearing @ 500Hz |
| | Hearing @ 1kHz |
| | Hearing @ 2kHz |
| | Hearing @ 4kHz |
| | Background Noise |
| Visual parameters | Visual acuity (and sensitivity) |
| | Color perception |
| | Field of vision |
| Motor parameters | Contact grip |
| | Finger precision |
| | Hand precision |
| | Arm precision |
| | Pinch grip |
| | Clench grip |
| Knowledge and Experience | Semantic User's Knowledge and Experience |

(ISO), which is standard ISO 9241–11. In this particular case the construct of *Usability* considers *User* and *Equipment* component, among others, as central for evaluating man-machine interaction aspects [38].

- *Terms enumeration* – important terms are enumerated for describing the domain or scope based on elements from the main taxonomy structure, i.e., taxonomy group names and taxonomy characteristics names.
- *Class and hierarchy definition* – a top-down approach might provide the first idea of a class and its subclasses (i.e., hierarchies), given by examining the objects in the previously defined concepts from domain or scope enumeration.

**Table 3.** Excerpt of Device characteristics and groups.

| Group name | Characteristic name |
|---|---|
| Component Type | Input |
| | Output |
| Device Type | Generic Device |
| | Specialised Device |
| System Type | Application System |
| | Operational System |

- *Class properties definition* – definition of class properties (also referred to as slots) is given by examining the taxonomy-defined hierarchy.
- *Domain and range of properties* – the descriptors *Unit of* measurement, *Value space* and *Data type* of the taxonomy assist in defining facets.
- *Properties instances – ultimately, individual instance(s) of class hierarchy and characteristics might be created.*
- *Definition of relations* –based on the analysis of user and device models previously presented, the authors propose an extension of the relations framework provided by [39]. A list of 33 tailored relations along with their inversion relation names, are presented (Table 4) in the context of the human and device ontology.

**Table 4.** Types and names of relations.

| Relation type | Relation name | Inverse relation name |
|---|---|---|
| Compositional | | |
| | Component of | Has component |
| | Has component | Component of |
| | Element of | Has element |
| | Has element | Element of |
| | Member of | Type |
| | Has member | Member of |
| | Part of | Has part |
| | Has part | Part of |
| | Gender of | Has gender |
| Spatial | | |
| | Interacts with | Operator |

(continued)

**Table 4.** (*continued*)

| Relation type | Relation name | Inverse relation name |
|---|---|---|
| | Push | Is pushed |
| | Pull | Is pulled |
| | Press | Is Pressed |
| | Turn | Is Turned |
| Role | | |
| | Instrument | Is resource |
| | Operator | Interact with |
| | Resource | Is instrument |
| | Input | Output |
| | Output | Input |
| General | | |
| | Describes | Represents |
| | Represents | Describes |
| | Type | Member of |
| | Has attribute | Attribute of |
| | Has Gender | Gender of |
| Dependency | | |
| | Depends on | Affected by |
| | Presumption for | Affected by |
| Influence | | |
| | Influence on | Interact with |
| | Is opposing | Is supporting |
| | Is supporting | Is opposing |
| | Affect | Depends on |
| Rate | | |
| | Age of | Has age |
| | Measure of | Has measure |
| | Rate of | Has rate |

The set of relation types presented, albeit useful for reasoning about the main characteristics of the proposed ontology, when depleting its capability of capturing the totality of properties and relations from existing models, can be extended and specialized for supporting additional and more representative semantics. Ontologies are on-going work, i.e., they require evolution with updated entities, properties and relations, and so are these relations classification. Additionally, we identified 132 mappings between relations and their source and destinations classes.

The goals of data properties are twofold. They are used for connecting classes to specific data values, supporting the parameters that compose the models, and also restrict classes named Defined Classes under necessary and sufficient conditions. We identified 25 data properties in the context of the current ontology (Table 5).

**Table 5.** Data property in the *User System Ontology*.

| Data Properties |
| --- |
| hasClenchGripValue |
| hasPinchGripValue |
| hasArmPrecisionValue |
| hasHandPrecisionValue |
| hasFingerPrecisionValue |
| hasContactGripValue |
| hasColourPerceptionValue |
| hasFieldOfVisionValue |
| hasVisualAcuityValue |
| hasHearingValue |
| hasBackgroundNoiseValue |
| hasHearing500HzValue |
| hasHearing4kHzValue |
| hasHearing2kHzValue |
| hasHearing1kHzValue |
| hasMotivationValue |
| hasPerceivedAttitudeTowardsComputerValue |
| hasPerceivedFatigueValue |
| hasStressAndReactionTimeValue |
| hasPerceivedStressValue |
| hasStatureInCM |
| hasGenderFM |
| hasWeightInKg |
| hasAgeInYears |
| hasSemanticKnowledgeAndExperienceValue |

Next, a domain model is provided for representing the modeled objects (conceptual classes), as depicted in Fig. 2. Domain modeling organizes knowledge around the basic concepts that is investigated. The model is illustrated with a class diagram, which use classes and interfaces for capturing details about the entities that make up your system and the static relationships between them [40]. The reader will notice that the Device portion is condensed for space constraints.

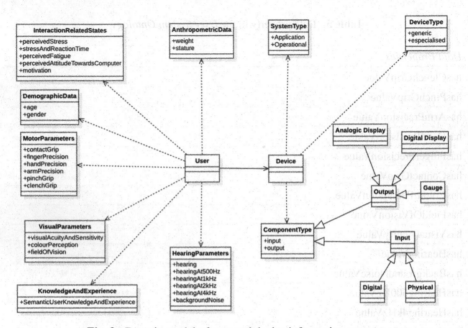

**Fig. 2.** Domain model of user and device information system.

The modeling is also supported by the conceptual model for the User portion of the Ontology, which is depicted in Fig. 3.

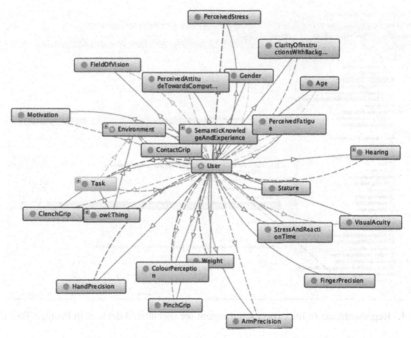

**Fig. 3.** Conceptual model from the User Ontology.

## 4 Results

The User Device Ontology is developed in OWL language, based on RDF and RDF Scheme, and implemented in Protégé Tool [25], presenting classes, properties and instances. The present ontology is heavy weighted (as opposite to those using controlled vocabularies), i.e., its internal structure represents semantic knowledge enriched with value and logical constraints. This particular type of ontology also has the potential of assisting designers in the requirements phase [32, 41]. Figure 4 depicts a fragment of the *User Device Ontology*, highlighting the restrictions that ties all the ontology parts together, i.e., the classes, sub-classes, class properties, range of properties, relations and data properties. This particular instance highlights the *User* requirement for operating *Devices* of the type *Specialized Device*. The restriction demonstrates the requirement of a user with *Semantic Knowledge and Experience* rated as value *3*, which according with [42] represents a domain expert user. The semantics of the restriction might also be expressed by the formal assertion in First Order Logic (FOL) notation:

$$\forall (SpecalisedDevice \in Devices) \exists User$$

$$\ni hasSemanticKnowledgeAndExperienceValue(User) \equiv 3$$

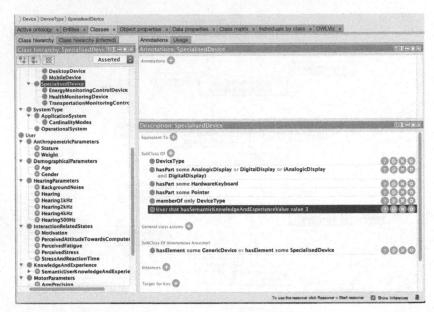

**Fig. 4.** Representation of the expert requirement for specialized devices in Protégé Tool [25]

A snippet of the OWL code for the User Device Ontology is provided in Fig. 5, contemplating the definition of classes and sub-classes.

```
1     <!-- http://www.semanticweb.org/saulo/ontologies/2020/9/User-Ontology/Age -->
2
3     <owl:Class rdf:about="http://www.semanticweb.org/saulo/ontologies/2020/9/User-Ontology/Age">
4         <rdfs:subClassOf rdf:resource="http://www.semanticweb.org/saulo/ontologies/2020/9/User-Ontology#User"/>
5         <rdfs:subClassOf>
6             <owl:Restriction>
7                 <owl:onProperty rdf:resource="http://www.semanticweb.org/saulo/ontologies/2020/9/User-Ontology/ageValue"/>
8                 <owl:someValuesFrom rdf:resource="http://www.w3.org/2001/XMLSchema#integer"/>
9             </owl:Restriction>
10        </rdfs:subClassOf>
11    </owl:Class>
12
```

**Fig. 5.** Fragment of OWL code, for the User Device Ontology.

Ontologies are evaluated according with different viewpoints (e.g., development quality, mathematical correctness, metrics). A *reasoner* might provide knowledge consistency checking, based on checking reflexive, transmission and redundancy properties of ontologies, providing contradictions implicit in the definitions, if any exist. In the case of the proposed ontology, logical consistency is evaluated against the HermiT OWL reasoner [43], and all detected classes are given as satisfiable. Additionally, descriptive metrics for the ontology's asserted classes are presented in Table 6.

**Table 6.** Descriptive metrics for the ontology's asserted class hierarchy

| Name | Metric |
| --- | --- |
| Axiom | 422 |
| Logical axiom count | 253 |
| Declaration axioms count | 168 |
| Class count | 93 |
| Object property count | 46 |
| Data property count | 25 |
| Individual count | 4 |
| Annotation Property count | 1 |
| SubClassOf | 126 |
| EquivalentClasses | 26 |

# 5   Conclusion and Future Work

This presented an on-going version of user device ontology tailored for assisting the design of interactive systems. The ontology aims contributing for the improvement of user interface design, considering semantic knowledge from two domains: human factors and devices. By providing a heavy weighted ontology involving knowledge from the explored domains, we intent assisting designers in the constraints/requirements phase of user interface development, with potential of narrowing the gap between user interface design and implementation. Therefore, we achieved the goals of the work, namely, we discussed the necessity and usefulness of human knowledge during the user interfaces development; we described a multidimensional knowledge representation model related with a (abstract) human model and device model and we presented an ontology based on the multidimensional models.

Future research includes extending the ontology with the complete four-domain models taxonomy, which are part of a research strand involving the design of an interactive systems evaluation platform, and submitting it for a journal. Also, we intent extending the validation method for the ontology, i.e., converting the protégé ontology for Alloy notation, which will allow testing instances and validate formal assertions about the represented knowledge. Counter-examples might be useful on the validation of mathematical/logical assertions. Additionally, moving to a realistic case study will allow the authors inform and validate the use of such ontologies in the development of user interfaces for interactive systems.

**Acknowledgments.** This work was supported by Conselho Nacional de Pesquisa (CNPq), COMPETE: POCI-01–0145-FEDER-007043, and FCT – Fundação para a Ciência e Tecnologia within within the R&D Units Project Scope: UIDB/00319/2020. We also acknowledge the support of Instituto Federal de Educação, Ciência e Tecnologia de Goiás (IFG).

# References

1. Benyon, D., Designing Interactive Systems: A Comprehensive Guide to HCI, UX and Interaction Design (2014)
2. Leveson, N.: Engineering a Safer World: Systems Thinking Applied to Safety. MIT press, Cambridge (2011)
3. Gruber, T.: A translation approach to portable ontologies. Knowl. Acquisition **5**, 199–220 (1993)
4. Hesse, W.: Ontologies in the Software Engineering Process. In: EAI, pp. 3–16 (2005)
5. Tanasescu, V. (2004). An ontology–driven life–event portal. Master. Comput. Sci.
6. Kaklanis, N., et al.: Towards standardisation of user models for simulation and adaptation purposes. Univ. Access Inf. Soc. **15**(1), 21–48 (2014). https://doi.org/10.1007/s10209-014-0371-2
7. Moon, A., Choi, Y.I., Lee, B.S.: Context-aware user model for personalized services. In 2008 Third International Conference on Digital Information Management, pp. 858–863. IEEE (2008)
8. Eason, K.D.: Ergonomic perspectives on advances in human-computer interaction. Ergonomics **34**(6), 721–741 (1991)
9. Dix, A., Finlay, J., Abowd, G.D., Beale, R.: Human-Computer Interaction. Pearson Education, London (2003)
10. Happel, H.J., Seedorf, S.: Applications of ontologies in software engineering. In: Proceedings of Workshop on Sematic Web Enabled Software Engineering (SWESE) on the ISWC, pp. 5–9 (2006)
11. Jakus, G., Milutinovic, V., Omerovic, S., Tomazic, S.: Concepts, Ontologies, and Knowledge Representation. Springer, Cham (2013)
12. Milligan, C., Halladay, S.: The realities and facilities related to knowledge representation. In: IPSI Belgrade, Proceedings of the IPSI-2003 Montengro Conference, Sveti Stefan, Montenegro (2003)
13. Chan, C.W.: The knowledge modeling system and its application. In: Canadian Conference on Electrical and Computer Engineering 2004 (IEEE Cat. No. 04CH37513), vol. 3, pp. 1353–1356. IEEE (2004)
14. Gašević, D., Djuric, D., Devedžic, V.: Model Driven Engineering and Ontology Development. Springer, Cham (2009)
15. Guarino, N.: Formal ontology in information systems. In: Proceedings of the first International Conference (FOIS'98), vol. 46. June 6–8, Trento, Italy. IOS press (1998)
16. Beißwanger, A.E. Developing Ontological Background Knowledge for Biomedicine (Doctoral dissertation) (2013)
17. Lenat, D.B.: CYC: a large-scale investment in knowledge infrastructure. Commun. ACM **38**(11), 33–38 (1995)
18. Miller, G.A., Beckwith, R., Fellbaum, C., Gross, D., Miller, K.J.: Introduction to WORDNET: an on-line lexical database. Int. J. Lexicogr. **3**(4), 235–244 (1990)
19. Clark, P.: Some Ongoing KBS/Ontology Projects and Groups (2000)
20. Musen, M.A., Gennari, J.H., Eriksson, H., Tu, S.W., Puerta, A.R.: PROTEGE-II: computer support for development of intelligent systems from libraries of components. Medinfo **8**(Pt 1), 766–770 (1995)
21. Tan, H., Adlemo, A., Tarasov, V., Johansson, M.E.: Evaluation of an application ontology. In: Proceedings of the Joint Ontology Workshops 2017 Episode 3: The Tyrolean Autumn of Ontology Bozen-Bolzano, Italy, September 21–23, 2017, vol. 2050. CEUR-WS (2017)

22. Alonso, J.B.: Ontology-based software engineering: engineering support for autonomous systems. Integrating Cognition+ Emotion+ Autonomy, 8–35 (2006)
23. Corcho, O., Fernández-López, M., Gómez-Pérez, A.: Methodologies, tools and languages for building ontologies: where is their meeting point? Data Knowl. Eng. **46**(1), 41–64 (2003)
24. Staab, S., Studer, R. (eds.): IHIS, Springer, Heidelberg (2009). https://doi.org/10.1007/978-3-540-92673-3
25. Noy, N.F., McGuinness, D.L.: Ontology development 101: a guide to creating your first ontology. 2001 (2004). http://protege.stanford.edu/publications.
26. Falbo, A.: Integração de Conhecimento em um Ambiente de Engenharia de Software (Doctoral dissertation. Universidade Federal do Rio de Janeiro, Rio de Janeiro), Tese de Doutoramento (1998)
27. Fernández-López, M., Gómez-Pérez, A., Sierra, J.P., Sierra, A.P.: Building a chemical ontology using methontology and the ontology design environment. IEEE Intell. Syst. **14**(1), 37–46 (1999)
28. Benjamin, P.C., et al.: IDEF5 Method Report. Knowledge Based Systems, Inc (1994)
29. Niles, I., Pease, A.: Linking lixicons and ontologies: mapping wordnet to the suggested upper merged ontology. In: Ike, pp. 412–416 (2003)
30. Heckmann, D., Schwartz, T., Brandherm, B., Schmitz, M., von Wilamowitz-Moellendorff, M.: Gumo – the general user model ontology. In: Ardissono, L., Brna, P., Mitrovic, A. (eds.) UM 2005. LNCS (LNAI), vol. 3538, pp. 428–432. Springer, Heidelberg (2005). https://doi.org/10.1007/11527886_58
31. Heckmann, D., Schwarzkopf, E., Mori, J., Dengler, D., Kröner, A.: The user model and context ontology GUMO revisited for future web 2.0 extensions. Contexts Ontol. Representation Reasoning, 37–46 (2007)
32. Ruiz, F., Hilera, J.R.: Using ontologies in software engineering and technology. In: Calero, C., Ruiz, F., Piattini, M. (Eds.) Ontologies for software engineering and software technology, pp. 49–102. Springer, Heidelberg (2006). https://doi.org/10.1007/3-540-34518-3_2
33. Shahzad, S.K.: Ontology-based user interface development: user experience elements pattern. J. UCS **17**(7), 1078–1088 (2011)
34. Paulheim, H., Probst, F.: A formal ontology on user interfaces-yet another user interface description language? Position Paper. In: CEUR Workshop Proceedings, vol. 747, pp. Paper-9. RWTH (2011)
35. Razmerita, L.: An ontology-based framework for modeling user behavior—A case study in knowledge management. IEEE Trans. Syst. Man, Cybern.-Part A: Syst. Humans **41**(4), 772–783 (2011)
36. Orellana, D.W., Madni, A.M.: Human system integration ontology: enhancing model based systems engineering to evaluate human-system performance. Procedia Comput. Sci. **28**, 19–25 (2014)
37. Engelschall, R.S.: Hierarchical User Interface Component Architecture. Doctoral Thesis, Universit Augsburg. Germany (2018)
38. ISO/IEC (International Organization for Standardization) (1998). Standard 9241: Ergonomic Requirements for Office Work with Visual Display Terminals (VDT)s, Part 11. Guidance on Usability. https://www.iso.org/obp/ ui/#iso:std:iso:9241:-11:ed-1:v1:en. Accessed 20 Jan 2019
39. Štorga, M., Marjanović, D., Andreasen, M.M.: Relationships between the concepts in the design ontology. In: 16th International Conference on Engineering Design–ICED 07 (2007)
40. Pilone, D., Pitman, N.: UML 2.0 in a Nutshell. O'Reilly Media, Inc (2005)
41. Decker, B., Ras, E., Rech, J., Klein, B., Hoecht, C.: Self-organized reuse of software engineering knowledge supported by semantic wikis. In: Proceedings of the Workshop on Semantic Web Enabled Software Engineering (SWESE), p. 76. ISWC Galway, Ireland (2005)

42. Karwowski, W.: International Encyclopedia of Ergonomics and Human Factors, -3, vol. Set. CRC Press (2006)
43. Glimm, B., Horrocks, I., Motik, B., Stoilos, G., Wang, Z.: HermiT: an OWL 2 reasoner. J. Autom. Reason. **53**(3), 245–269 (2014)

# A Review of Internet of Things (IoT) Forensics Frameworks and Models

Khalil Al-Hussaeni$^{(\boxtimes)}$, Jarryd Brits, Meghna Praveen, Afra Yaqoob,
and Ioannis Karamitsos

Rochester Institute of Technology, Dubai, UAE

{kxacad,jtb4246,mk7898,aay4689,ixkcad1}@rit.edu

**Abstract.** The abundance of data generated and processed by the ubiquitous number of Internet of Things (IoT) devices around the world is a promising enabler for civil, criminal, and digital investigations. Such data can be used as valuable artefacts to support IoT forensic investigations. This paper provides a survey of IoT forensics models presenting recent advances in the field and identifying opportunities for future research. Through this survey, we identified the challenges in IoT devices and protocols and the solution of the IoT frameworks. More specifically, we examine 18 IoT forensics frameworks in hardware and software architectures in which the challenges were addressed. The future research directions of IoT forensics lie in the implementation and testing of frameworks and models proposed in the latest research. Such research should focus on forensic soundness, data privacy and data carving.

**Keywords:** Blockchain · Digital Forensics · Fog Computing · Internet of Things · IoT Forensics · Privacy · Smart City · Smart Homes · Smart Wearables

## 1 Introduction

The Internet of Things (IoT) consists of a global network of interconnected devices that communicate through various network and communication protocols. Wireless Sensor Networks, Radio Frequency Identification (RFID), and other networking technologies are some of the protocols used by IoT devices for communication and collaboration purposes [1]. With the rapid growth in Internet and wireless communications, more IoT devices are developed, produced and connected globally. IoT devices are used daily by individuals and organizations in various sectors, including Smart environments, Smart homes, Smart vehicles, Smart power grids, and the healthcare sector [2]. The trend of IoT devices in recent years has seen continuous growth, as reported by the International Data Corporation (IDC), which predicted a 17% compound annual growth rate (CAGR) with individuals and organizations spending $698.6 Billion in 2015 up to $1.3 trillion in the year 2019 [3].

As digital device usage has increased, so have crimes that take advantage of these devices. Because of this, forensics investigators need to analyze and examine multiple digital devices for potential digital forensic evidence. Digital forensics is a forensics

© The Author(s), under exclusive license to Springer Nature Switzerland AG 2023
M. Papadaki et al. (Eds.): EMCIS 2022, LNBIP 464, pp. 515–533, 2023.
https://doi.org/10.1007/978-3-031-30694-5_37

method used by investigators to analyze electronic data in digital devices for potential evidence relating to a crime or an incident that is being investigated [4, 6]. As the digital field continues to expand with the invention of different devices and technologies, so does the field of digital forensics. Digital forensics can be divided into the following general categories: Network, Cloud, and IoT forensics.

Network forensics is a field of digital forensics where the main focus is the analysis of network traffic and communications for probable evidence. The majority of the gathered evidence is for analyzing cyber-crimes that used networks for malicious activities such as DDOS attacks, malware, identity theft, and other related crimes. Some of these tools include an Intrusion Detection System (IDS) that aids in the detection of cyber-attacks occurring within the network, and firewalls, which log network traffic and can then be used to analyze malicious network traffic that passes through a network [5].

Cloud forensics is a rapidly growing field in digital forensics as the use of cloud computing increases across organizations and enterprises. Cloud forensics relies on both network and digital forensics, as the data is stored in the cloud. There are a few challenges faced in this discipline as the physical location of the cloud storage might be in a completely different jurisdiction while the entity using it is in another, thus raising the concern of legal jurisdiction on the investigative evidence stored in the cloud [5].

IoT forensics is another subfield of digital forensics that focuses on the forensics of IoT technologies. IoT forensics is a newer field of digital forensics, yet it is becoming increasingly popular as the usage of IoT devices and systems has increased drastically. IoT forensics combines the subfields of digital forensics, and cloud forensics in this discipline due to IoT's method of storing data in both itself (the device) and in the cloud [5]. The components involved in an IoT forensic investigation are illustrated in Fig. 1.

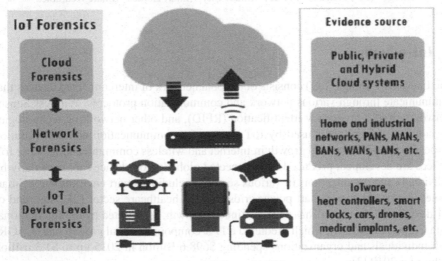

**Fig. 1.** IoT Forensics [1]

This paper surveys the challenges faced in IoT forensics and the solutions proposed by researchers to address these challenges. With the rise in IoT devices around the world in various areas, industries, and locations it is paramount that IoT forensics is expanded upon and consolidated into standard everyday practice by forensic investigators, along with comprehensive frameworks and implemented tools and techniques to carry out Digital forensics activities. IoT Digital forensics is critical to obtaining justice and security in the near future's digital society.

Specifically, the aim of this study is to provide a systematic review of the literature on IoT forensics to answer the following research questions (i) What are the challenges faced when conducting digital investigations that involve forensic analysis of IoT devices? (ii) What is the current state of research that address these challenges? and (iii) What are the major gaps in IoT forensics that can be addressed by future research?

This paper critically surveys eighteen different existing models in the literature, between 2006 and 2022. Among these models, we have identified ten challenges faced in IoT forensics, some of which are common. Moreover, this survey presents two practical recommendations based on the surveyed models. Building on top of that, we have formulated four important future directions that we hope other researchers and engineers may take further.

The remaining of the paper is organized as follows. Section 2 outlines the methodology followed for the systematic literature review, Sect. 3 provides a brief overview of the challenges. Section 4 presents a critical review of the different solutions proposed to one or more of the challenges presented in Sect. 3. In Sect. 5, we provide our recommendations based on some of the solutions surveyed and summarizes key future research directions in IoT forensics. Section 6 concludes this work.

## 2   Methodology

A systematic literature review methodology is used, involving a process for searching existing body of related work, evaluating and analyzing the literature on findings. The systematic literature review follows three steps. The first step is to formulate the research questions presented in the Introduction section. The second step is to identify paper with certain keywords, mainly "IoT", "Internet of Things", "Forensics" and "Digital Forensics". The third step involves identifying the data sources and setting inclusion and exclusion criteria. The inclusion criteria used are (a) papers that propose, implement, or test an IoT forensics model or framework, (b) papers that discuss a particular challenge or issue with IoT forensics, (c) guidelines and technical reports proposed by standardization bodies, and (d) journal and conference publications. The search conducted looked into papers published between 2006 and 2022. Papers that passed the inclusion criteria were prioritized by their citation count and skimmed through to assess their relevance to the research defined questions. Exclusion criteria used are (a) papers that provided general information without any in-depth analysis, and (b) non-English publications.

## 3   Challenges in IoT Forensics

### 3.1   Heterogeneity in IoT Devices and Protocols

One of the primary challenges faced in IoT forensics is the distribution of an overwhelming number of heterogeneous devices across networks. IoT devices make use of different communication protocols, such as ZigBee, WiFi, and LoRa, and run different embedded operating systems with different file systems and formats [8, 9]. The variation in the technical specifications of each of these devices and protocols makes it difficult to have a common and standardized form of forensic acquisition. The heterogeneity in IoT devices and employed protocols result in extreme difficulty in analyzing evidence in a coherent and timely manner.

### 3.2 Closed Source and Proprietary Software

Most of the well-known forensic acquisition and analysis tools and frameworks focus on host-based operating systems like common Linux distributions and Windows. Due to the heterogeneity of the IoT devices, it is not a surprise that when examining an IoT incident, the forensically relevant data are found in different file formats [10]. Adding to the complexity, many of these file formats are proprietary and used by closed-source software, resulting in difficult analysis or reverse engineer. Much of this analysis is time consuming and often impossible due to lack of cooperation from software vendors and the lack of a single forensic analysis tool or standard that addresses this challenge.

### 3.3 Privacy

Collecting information from a shared suspect machine affects the privacy of the innocent users of the system. There is plenty of research works in preserving the privacy of individuals in the collected evidence [9]. Nonetheless, this body of work only focuses on preserving privacy during collection and storage of evidence. However, privacy is also a concern during analysis and while making correlations.

### 3.4 Miniaturization of Devices

With the advantages of miniaturization of IoT devices, there comes a few disadvantages, such as computational and storage limitations. Implementing a strong cryptographic stack and using protocols such as HTTPS is computationally intensive with lots of overhead for an IoT application; this makes IoT devices the low-hanging fruit for attackers. An example of such an attack is the Mirai Botnet infection. The Mirai malware affects IoT devices that use the ARC processor. Due to miniaturization, this processor runs a minimal version of Linux, which, when left exposed on the Internet with default credentials, can be infected by Mirai [11].

### 3.5 Cloud Limitations

Most IoT applications leverage the power of the cloud for storage and processing. Retrieving forensic evidence from the cloud can pose a challenge when the cloud Service Providers (CSPs) hesitate to provide access, share information or cooperate with the investigators [12]. In cases where the CSPs do cooperate with the investigators, there are challenges in forensic acquisition on different cloud architectures such as Software as a Service (SaaS), Platform as a Service (PaaS) and Infrastructure as a service (IaaS).

Forensic imaging is comparatively easier on IaaS than on SaaS and PaaS due to intellectual property rights concerns and the sheer size of the cloud platform in SaaS and PaaS [12].

### 3.6 Evidence of Physical Tampering

IoT devices are easier to be physically tampered with due to miniaturization. They can easily be stolen, replaced, or disconnected. Most IoT devices are supplemented with a remote-control platform, which is often ignored in security audits and often left exposed physically and online.

In 2018, Martin Hron, a security researcher at Avast, discovered over 32,000 MQTT servers exposed on the Internet with no password protection using the Shodan search engine [13]. This left multiple homes and businesses at risk, as these devices could be easily discovered and tampered with both physically and remotely. Remote commands could be issued to disable a device or change its configuration and state. Disconnecting a device makes it extremely difficult to track when it was physically stolen; this also leads to the loss of any volatile data that it holds. In such cases, it is also tough to identify if the cause of disconnection was malicious or legitimate, like a component failure [12].

### 3.7 Volume and Volatility of IoT Evidence

Most IoT applications involve real-time processing of data. This requires that a large volume of data is stored and processed in a very short period, with minimal buffer delays. The real-time nature of processing is volatile and provides little to no data evidence for IoT forensic investigations. Real-Time Operating System (RTOS) is an operating system that is used by many IoT devices where real-time data processing is of the essence [8].

### 3.8 Challenges in Legal Procedures

When implementing the newly-proposed forensic techniques such as Next-Best-Thing (NBT) and Last-on-Scene (LoS) algorithms [14], the scope of the investigation cannot be determined beforehand. They can only be determined during the investigation. However, in most jurisdictions, digital warrants require the scope of the investigation to be determined when applying for it [15]. Therefore, to use these newly-proposed forensic techniques, legal reforms should be made to allow for the acquisition of evidence from newly identified sources throughout the investigation.

Another common challenge faced, not just IoT forensics, but across all digital forensic investigations, is dealing with cross-border jurisdiction challenges [15]. This challenge is predominant in IoT due to the distributed nature of IoT devices, their users, and cloud providers.

### 3.9 Maintaining the Chain of Custody

One of the most difficult challenges and most important aspects of digital forensics, or even just forensics in general, is maintaining chain of custody. It is crucial to formulate cases against the accused and to follow through with an airtight case, allowing for proper justice to be served where applicable.

However, as many forensic analysts have learnt, maintaining chain of custody for given evidence, especially in IoT forensics, is increasingly difficult to do [16]. One of the reasons for this aspect being such a challenge is because chain of custody relies on many different factors, whether optional or mandatory. For instance, chain of custody may rely on evidence of physical tampering, volume, and volatility of IoT evidence and legal procedures. This list is not exclusive, as the challenges incorporated into the chain of custody could include all other challenges mentioned in this section.

### 3.10 Identification of Evidence

This challenge is quite standard amongst all subfields of study in any security-related area. Whether investigators are trying to identify malicious or benign activity or the identify relevant information, current computers are incapable of making that distinction autonomously without the aid of hints and helpers. Usually, these hints and helpers are in the form of hashes of malware or digital signatures; however, helpers will not function for most digital forensic investigations unless the evidence searched for is a malware/virus executable.

Although all data on a system may be considered evidence in a court of law, the value of that evidence is also critical. In other words, that piece of evidence must have some intrinsic value in an investigation, and it is this distinction that is challenging and must be faced in an IoT infrastructure.

## 4  Overview of IoT Forensics Models (2006–2022)

Section 4 presents in brief some of the basic IoT forensics models over the past years (2006–2022) and has adapted to the challenges mentioned in the previous section.

### 4.1  FoBI Framework

Al-Masri et al. [12], have proposed a framework called Fog Based IoT forensic Framework (FoBI), where forensic intelligence is decentralized into a fog node, enabling forensic evidence to be gathered from the interaction between the IoT device and the fog node. Fog computing is an architecture where the storage and computing resources are decentralized to nodes in a "fog layer", which are closer to the data generation point than the cloud infrastructure. This framework addresses the challenges of forensic acquisition due to the volume and volatility of IoT data, heterogeneity of IoT devices and their data, and the miniaturization of IoT devices as the data is stored and acquired from the fog nodes. It also addresses the challenges of data acquisition from cloud providers, as the fog nodes store and process data before sending it over to the cloud, hence providing local access and control of the data and preventing the need to obtain an approval or the cooperation of the cloud service providers.

The FoBI framework is based on the DFRWS investigative model that was proposed in the 1st Digital Forensics Research workshop [12].

The FoBI framework looks ideal at first glance; however, it has several limitations. Firstly, the proposed framework was not implemented or tested for forensic soundness. This will make any evidence acquired by this framework inadmissible in court. Secondly, setting up a fog layer with FoBI nodes can be expensive for large organizations and doesn't provide a significant advantage over manually implementing forensic acquisition after detecting attacks using some of the well-known SIEM or endpoint management software suites available in the market.

## 4.2 FSAIoT

The Forensic State Acquisition from the Internet of Things (FSAIoT) [8] is a framework that focuses on extracting forensically relevant data by monitoring the state changes of an IoT device. The primary component used in this framework is a Forensic State Acquisition Controller (FSAC) [8].

Unlike most other proposed frameworks that were not implemented or tested, the feasibility of acquiring state data using FSAIoT was tested in a home automation environment. The tests were conducted using openHAB as the FSAC.

The motivation for this framework was to address the heterogeneity challenge of IoT devices; however, during their tests, they noticed that the states of some devices could not be retrieved until the device registers an event.

## 4.3 Application-Specific Digital Forensics Investigative Model

Zia et al. [17] proposed a new model of forensic investigation called the Application-Specific Digital Forensics Investigative Model. This model incorporates the four phases of the NIST 800-86 guide [18]. Special focus is placed on the curated list of application-specific IoT artefacts identified during the collection phase. The model has three components:

1. The application-Specific Forensics component applies an appropriate forensic extraction method after identifying the forensics issues unique to a particular IoT application.
2. The Digital Forensics component identifies physical tampering, such as wireless or RF inference or malicious code injection.
3. The Forensics Process component closely follows the NIST process to produce legally admissible evidence using the outputs of the previous two components.

Table 1 summarizes the list of IoT artefacts of forensic value based on the IoT application curated by the authors of [17].

Identifying application-specific artefacts helps narrow down the scope of an investigation, thus assisting investigators in prioritizing their search when investigating a crime in a heterogeneous IoT environment.

## 4.4 BIFF

Blockchain is a system of storing information in a way such that the integrity of the stored information is preserved without the interference of any third-party monitoring system.

**Table 1.** Applications specific IoT artefacts

| IoT Application | Application Specific Artefacts |
|---|---|
| Smart Home | Time log revealing the schedule of the people in the home, mobile synced data, wifi connection and usage data, other device usage details. |
| Smart City | Weather and traffic forecast data, wireless access points, GPS synced data such as navigation data, active routes, saved locations, smart-vehicle synced data, parking details, other connected device details. |
| Smart Wearables for healthcare | Patient vitals such as heart rate, respiratory rate, blood pressure and body temperature, patient activity, doctor visit or appointment details. |

Le et al. [19] have proposed a chain of custody system called Blockchain-based IoT Forensic Framework (BIFF), which leverages the power of blockchain to automatically preserve the chain of custody of forensic evidence. BIFF uses Merkle Trees for identity verification of a DW when an evidence entry is added to the ledger. The performance of a blockchain system is of concern only when the use case involves frequent modifications. However, BIFF's use case is to ensure that there are no modifications to the ledger to maintain the Chain of Custody of digital evidence, which makes this a feasible framework in terms of cost of performance. Overall, this is a promising framework that addresses the challenge of maintaining the Chain of Custody of digital evidence and also some aspects of the privacy challenges in IoT forensics.

### 4.5 DFIF-IoT

Kebande and Ray [20] proposed a new IoT forensics framework that provides a combination of three modules that should be completed together for an overall conclusive digital forensic investigation. DFIF-IoT comprises three different approaches to forensic investigations, namely proactive process, IoT forensics process, and a reactive process.

Kebande and Ray [20] highlight the most prominent difference between their proposed framework (DFIF-IoT) and other related frameworks: its inclusion and ability to implement Digital Forensic Readiness.

Overall, DFIF-IoT is ideal as a standard guide and approach to conducting a digital forensic investigation in an IoT-based environment. The fact that the concurrent processes adhere to an already existing standard from 2015 (ISO/IEC 27043) adds further legitimacy to the proposed framework's effectiveness and "admissibility in a court of law", as suggested in [20].

### 4.6 FAIoT

Zawoad and Hasan [21] propose a new forensics aware model for IoT (FAIoT) devices and environments aimed at tackling the prominent challenge of heterogeneity and distributed nature of IoT infrastructures. In addition to the submission of potential evidence, this "centralized trusted evidence repository" [21] is also used to store secure logging schemes to maintain evidence reliability. The secure logging scheme is, in short, a method of securely logging data in cloud environments as Logging as a Service (LaaS).

FAIoT consists of three main components that, when combined, offer the ability to easily and efficiently retrieve potential evidence from the system. These three components are the Secure Evidence Preservation Module, the Secure Provenance Module, and the API Access.

### 4.7 Privacy Preserving Forensic Model

Almolhis and Haney [9] proposed a privacy preserving forensic methodology by suggesting changes to each step in the ISO/IEC 27037 guide.

In the first two steps of Identification and Collection, the authors promote prioritizing the sources of evidence based on the sensitivity of the data present in the artefact rather than the sensitivity of the investigation conducted.

In the third step of Acquisition, the ISO/IEC 27037 allows for partial acquisition when the suspect device's storage is too large or the system is critical. The final step of Preservation remains the same as the recommended guidelines by the ISO/IEC 27037.

### 4.8 FEAAS

Dorai et al. [22], studied the forensic artefacts produced by an iPhone that controls Nest IoT devices. They built a command line tool called Forensic Evidence Acquisition and Analysis System (FEAAS) to automate their acquisition, analysis, and reporting tasks. Using this tool, the authors acquired a list of Nest artefacts which are tabulated in Table 2.

**Table 2.** Artefacts acquired by the FEAAS tool

| IoT Application | Application Specific Artefacts |
|---|---|
| Info.plist | iPhone build version, IMEI number, phone number, last backup date, UDID, list of applications installed on the device |
| Status.plist | Device backup date and status |
| GooseEvent-Logging | Network type with timestamps, geo-fence data in FenceEvent and FenceReport |

On experimenting with multiple IoS devices, it was found that the paths of the artefacts are universal to all iPhones where the Unique Device Identifier (UDID) of the device will only change. Although not implemented into the FEAAS tool, the authors also managed to reconstruct deleted videos from home.nest.com using the Chrome cache in the iPhone using the "chromagnon" tool. Recovering deleted data from device cache is a very useful technique to overcome the challenges of recovering deleted data from the cloud provider which is next to impossible.

The downside of this research and the FEAAS tool is that it is limited only to iPhone devices that interface with Nest IoT devices. It is difficult to keep up with such changes which reduces the scalability of the FEAAS tool.

## 4.9 Generic IoT Forensics Framework

Li et al. [7] propose a model for digital forensics in the realm of IoT devices. Their model is quite similar to FAIoT in terms of general components and purpose, yet it defines, in more detail, sub-parts of the different stages.

This generic framework integrates the three main stages of digital forensic investigation: Identification, Preservation, Analysis, and Presentation. The main components are categorized as Offence classification, IoT device Identification, Evidence Preservation, IoT Forensic Analysis and presentation. In evidence preservation, Li et al. [7] consider the characteristics in Table 3 as the main areas of potential data to be extracted for the preservation of evidence in a forensic environment and investigation. Again, it is imperative to use the correct tools to extract and preserve such evidence, specifically to acknowledge the integrity and the chain of custody.

**Table 3.** Hardware Characterization for IoT devices

| Hardware Features | Capabilities | Evidence Sources |
| --- | --- | --- |
| Processor | 48 MHz to 2 GHz | Cache |
| Microcontroller | Less than 52 MHz | RAM and Flash |
| Memory | 5 MB to 128 GB | Chip Memory |
| Card slots | MicroSD | SD-based card |
| Display | LCD to HD | - |
| Camera | Still, Video | DF/SD Card, ex memory |
| Interfaces | SPI, I2C | Buil0in RAM/SRAM |
| JTAG | JTAG scanner | - |
| Input interface | RS232 to Keyboards | - |
| Voice input | Voice recognition | EEPROM, ex-storage |
| Positioning | GPS receiver | External storage |
| Wireless | IdDa, BT, WiFi, NFC | - |
| Battery | Li-on Polymer | - |

In IoT Forensic Analysis and Presentation, we analyze the IoT-specific devices and configurations. Li et al. [7] suggest that some general approaches to the analysis and presentation may be to attempt to "reconstruct the IoT crime/event scenes". Table 4 shows software characteristics for IoT devices.

Li et al. [7] tested their approach using a DIY Amazon Echo device built with a Raspberry Pi.

## 4.10 FIF-IoT

Hossain et al. [23] develop of a blockchain-based framework to address the issue of the soundness of extracted logs. Their framework utilizes a decentralized, distributed, public

**Table 4.** Software Characterization for IoT devices

| Software Features | Capabilities | Evidence Sources |
| --- | --- | --- |
| OS | Closed | Android, iOS etc |
| Secure Coding | Programming languages | RAM and FS |
| Logs | Event logs, data logs | FX, ex-storage |
| PIM | Calendar, list | FS, storage |
| Applications | Application data | Cache, RAM, FS, ex-S |
| Data type | Application-based | Cache, RAM, FS, ex-S |
| Call | Logs | File systems |
| Data Processing | Depends | Cache, RAM, FS, ex-S |
| Email | Applications | Ram and FS |
| Protocols | Communication | Cache, RAM, FS, ex-S |
| Web | Web browser | Cache, RAM, FS, ex-S |
| Web | Network services | Cache, RAM, FS, ex-S |

ledger that contains a continuous list of records of transactions between IoT devices, cloud services, and users.

The proposed framework, FIF-IoT, analyses IoT-based systems to collect the components various interactions.

There are multiple types of interactions occurring throughout IoT-based systems such as Things to Users, Thing to Cloud, Things to Things.

1. Things to users (T2U): interactions between users and the IoT devices themselves through various user interfaces
2. Thing to Cloud (T2G): interactions between the IoT devices with the cloud where it stores and publishes it content
3. Thing to Thing (T2T): interactions between IoT devices among themselves

This proposed model goes through various phases, including the transaction creation, insertion of transactions in a blockchain ledger, and the investigation phase. After these steps are completed, the investigators begin the investigation phase by analyzing and examining the evidence [23].

### 4.11 Consortium-IOF

Ryu et al. [24] introduces a framework for IoT forensics using consortium blockchain technology. In this framework, IoT technology is used as the main source for evidence gathering and communications, while the consortium blockchain technology is used in the process of evidence management.

The framework was built with few assumptions, such as members of the consortium blockchain are national and international legislation offices, heterogeneous devices compose the perception layer of the blockchain technology, and trusted key generator-verifier is available.

The framework is divided into four different layers, each perform different roles and responsibilities. The four layers are Edge-IOF, Fog-IOF, Consortium-IOF, and Cloud Storage.

Due to the significance of evidence integrity and chain of custody for admissibility, this framework utilizes a lightweight signcryption process to preserve confidentiality, integrity and non-repudiation.

In addition to these four layers, heterogeneous devices make up the perception layer of the proposed model. As privacy of device owners and investigators is a major concern in this method, a programmable hash function (PHF) from lattices is used to secure the signatures and identities, which are then mapped to a pseudo-identity. The previously mentioned signcryption process is carried out by using PHF ensuring that the IoT system is composed of legitimate devices. This promotes confidentiality, integrity, and non-repudiation of the record stored in the blockchain transactions [24].

## 4.12  FIF-SHE

Goudbeek et al. [25] propose a new forensic framework for Home Automation Systems (HAS) in which systems and infrastructure are built upon IoT devices. Primarily, the authors attempt to address the lack of existing forensic procedure and framework for IoT devices and HAS systems.

The HAS framework consists of seven major stages: preparation off-site, search for HAS on-site, preserve the HAS, understand the specific HAS, check the security level, locate and acquire evidential data, and process analyses of seized data.

Upon analyzing this framework, we notice that the framework is a general forensic procedure that is almost identical to NIST's Incident response model. That said, the framework provides in-depth guide towards a comprehensive investigation of IoT devices deployed in a HAS environment.

## 4.13  GFP for NOR Flash

Li et al. [26] propose a method of retrieving forensically sound evidence from ESP devices, specifically from the storage units on such devices. ESP microcontrollers have risen in popularity in recent years, especially in IoT devices. This is primarily due to the fact that they are small, both in dimensions and volumetrically.

In order to address this need, Li et al. [26] have created a General Forensics Process (GFP) for NOR flash data retrieval, specifically aimed at ESP devices. To do so, they created a 3D printed module for the WROOM module to interface with.

Overall, Li et al. [26] was able to interface with ESP devices and retrieve forensically sound evidence from the flash storage built into the ESP devices, preserving the integrity of evidence along with chain of custody. Though, due to the lack of standardization the authors suggest a new regulation to prevent the hiding of SPI pins, especially in the case of NOR flash storage in microcontrollers. Further research is still needed to detail the specifics of implementing such regulations and the supporting confidentiality preservation methods.

### 4.14 SCADA IIoT

There are currently a variety of problems in Industrial IoT (IIoT) environments, specifically with regards to Supervisory Control and Data acquisition (SCADA) Systems. Some of these problems include the complex integration into IIoT, multiple access points, the lack of post forensics models and tools, lack of live forensics, lack of non-repudiation, and limited logging. The main five procedural steps are *Examination, Identification, Collection, Analysis, and documentation.* Yet, while these steps may be identical because of the nature and components used in IIoT, specifically SCADA systems, the same tools and methods are inadequate for reliable and forensically sound evidence gathering and analysis.

Malik et al. [27] do not create a specific solution and platform to address the aforementioned problems in IIoT; however, they present various possible solutions to the problems on a theoretical and hypothetical basis.

Malik et al. [27] propose solutions to some current problems in digital forensics with regards to IIoT environments and SCADA systems. However, the proposed solutions also come with their own set of challenges. It is unclear whether these identified solutions are realistically capable of being implemented for digital forensic investigation at a reasonable cost of resources. Consequently, the authors mention that the overlining problem in such areas is the complete lack of standardization amongst machines and devices. Their solutions would require further research, investigation, and development to test the viability of each one and the effectiveness of the proposed solution.

### 4.15 MIoT Forensics

There exist various common challenges in digital forensics for IoT infrastructures and devices; some challenges are more prevalent in certain areas of IoT. Liu et al. [16] focus on challenges and solutions for the medical IoT forensics field. Their research indicates that some of the more Medical Field related challenges are difficulties in acquiring adequate amounts of forensic evidence that is said to be relevant and reliable, as well as big data IoT data analysis. Liu et al. [16] propose a Holistic, Medical centric IoT forensics model that is composed of four "spaces": Cyber space, Social space, Physical space, and Psychological space. Their model is purely theoretical in nature, and, thus, it is easier to implement in the vast majority of MIoT infrastructures regardless of their heterogeneity. Below, we elaborate on the four spaces.

With the new holistic model of MIoT forensics, the scope of potential evidence and, subsequently, the scope of investigation is no longer limited to the collection of digital evidence. The limitations of digital forensic investigation of MIoT devices prevent a comprehensive analysis of incidents. This is mainly because MIoT devices, naturally defined to be in the Cybers pace scope, are able to interact with all other defined scopes.

### 4.16 Forensics Enabled Through SecLaaS

IoT heterogeneity is serious challenge for forensic investigators. Luckily, the literature contains emerging technologies and standards to address this challenge.

One of the prominent proposed solutions is by [28], where a cloud-based approach was developed. Although other research work discussed in this survey also talk about using cloud services to aid in the use of IoT forensics, their purpose or implementation has been shown to be limited.

The work in [28] focuses on the logging ability of cloud services, primarily SaaS, PaaS, and IaaS, with various degrees of control and insights. Much like IoT forensics, some of the challenges with cloud services include their logging capabilities, such as volatility of logs, multi-tenancy, accessibility of logs, and decentralization.

Zawoad et al.'s proposed solution in [28], called SecLaaS, works in a similar way to the FAIoT model. Logs are exposed to a variety of possible attacks, such as log modification, privacy violation, and repudiation by user and Cloud Service Provider (CSP). SecLaaS extracts a variety of log types and stores them in a persistent log database for offline analysis. The SecLaaS scheme provides an entry into another database called the "proof database" that keeps its own records to preserve the integrity of the logs much in the same way that checksum for downloaded content works.

Overall, this proposed method of logging and accessing logs allows the preservation of confidentiality amongst cloud users. Moreover, when integrated with some current IoT infrastructures that utilize cloud services, SecLaaS can be considered a promising solution for the future of IoT infrastructure as we move towards more cloud centric applications.

### 4.17 IoT Forensics with Data Reduction

Due to the enormous amount of data generated by IoT devices, Quick and Choo [29] introduces a model of IoT forensics with data reduction processes. The primary goal is to minimize the volume of data to be used for digital forensics investigations whilst maintaining the original metadata. As the volume of IoT devices increases, so does the amount of data used for forensics investigation.

The major concern with the increase of data volume is time and resources it would take to analyze the data in an efficient and forensically sound manner. To address this concern, in Quick and Choo's model [29], data reduction is used to reduce the size of the data used for forensics investigation.

In their model, Quick and Choo [29] used the Data Reduction by Selective Imaging (DRbSI) method to reduce the volume of the data subsets created during the preservation phase and Quick Analysis to analyze data relevant to the scope or objectives of the investigation. In their study, they conducted a test on the M57 datasets, which are collections of forensics datasets intended for research purposes [29]. The data source of the forensics evidence was at a volume of 498 GB. After using DRbSI, the resulting volume decreased to 4.25 GB of extracted data and forensic files.

This proposed model considers some of the challenges faced by investigators during IoT forensic investigation with regards to the volumes of generated evidence and its privacy and confidentiality. Upon evaluating this model, DRbSI has proven to be a well implemented model.

To conclude this section, we summarize the body of work presented in this section. Table 5 compares the models and frameworks surveyed in this section based on the challenges they address, whether they were implemented and tested by the authors, and if they take forensic soundness into account.

**Table 5.** Comparison of IoT Forensics frameworks and models

| Name of the proposed approach | Challenges in IoT Forensics Models | | | | | | | | Implementation & testing | Forensics Soundness |
|---|---|---|---|---|---|---|---|---|---|---|
| | Heterogeneity in IoT Devices | Protocols | Cloud Challenges | Privacy | Chain of Custody | Closed Source | Proprietary software | Volume and Volatility of IoT Evidence | | |
| FoBi [12] | ✓ | ✓ | ✓ | | | | | | Yes | No |
| FSAIoT [8] | ✓ | ✓ | | | | | | | Yes | No |
| Application Digital Model [17] | ✓ | ✓ | | | | | | | Yes | No |
| BIFF [19] | | | | ✓ | ✓ | | | | No | Yes |
| DFIF-IoT [20] | | | | | ✓ | | | | No | Yes |
| FAIoT [21] | ✓ | | ✓ | | ✓ | ✓ | ✓ | | No | No |
| Privacy Preserving Model [9] | | | | ✓ | | | | | No | No |
| FEAAS [22] | ✓ | ✓ | | | | ✓ | ✓ | | Yes | No |
| Generic IoT Forensics Framework [7] | ✓ | ✓ | | | | | | | Yes | No |
| Smart Home IoT Forensics [2] | ✓ | ✓ | | | | | | | Yes | Yes |
| FIF-IoT [23] | | | | ✓ | | | | | Yes | Yes |
| Consortium IoT Framework [24] | | | | ✓ | ✓ | | | | Yes | Yes |
| FIF-SHE [25] | ✓ | ✓ | | | | | | ✓ | Yes | Yes |
| GFP for NOR Flash [26] | | | | | | | | ✓ | Yes | Yes |
| SCADA IIoT [27] | ✓ | ✓ | | | | | | ✓ | No | No |
| MIoT [16] | ✓ | ✓ | | | | | | | No | No |
| Forensics Enabled SecLaaS [28] | | ✓ | | | ✓ | | | ✓ | Yes | No |
| IoT Forensics with Data reduction [29] | | ✓ | | | ✓ | | | ✓ | Yes | Yes |

# 5   Recommendations

We provide potential future work based on the surveyed methods in Sect. 4. Particularly, we suggest two enhancements by extending or combining some of the surveyed models and frameworks, where combined models can complement one another.

## 5.1   FAIoT Hybrid Blackbox Alternative

The FAIoT framework that was proposed and developed by Zawoad and Hasan [21] shows plenty of promise in digital forensics investigations. However, this framework suffers from the problem of the significantly costly HDFS system that is required to handle the huge amount of data. The HDFS system could also be considered difficult to maintain, especially in an enterprise setting depending on the deployment of this framework. If the framework is deployed by a service provider, then cost becomes less of an issue for the enterprise, however privacy concerns rise.

Another potential problem is any form of Man-In-the-Middle attacks, whether during the initial data/evidence transfer phase from IoT devices or the API fetch requests phase instigated by the analyst. Note that for the evidence gathering phase, Zawoad and Hasan in [21] do not explicitly mention HTTPS. HTTP2.0 or TLS. Nevertheless, we assume that these protocols are implied, or some form of encryption mechanism is involved in the evidence transfer phase.

Due to the above identified concerns, we propose an extension to the FAIoT framework. We suggest a hybrid blackbox. The hybrid blackbox has two main components instead of the singular "secure evidence preservation module". One component is essentially identical to the originally proposed module. However, the second component locally stores similar evidence using various compression methods. This data storage module would be updated less frequently with logs to give forensic analyst insight on potential malicious activities.

## 5.2   SCADA IIoT Possible Solution

Although the SCADA IIoT Model [27] offers solutions to the challenges presented in Sect. 4 the main IoT challenges, described in Sect. 3, remain. Our potential solution would be to utilize the rise in popularity of ESP devices and incorporate them into an IIoT environment, eventually replacing the standard SCADA systems. While at this point we do not have detailed information on the feasibility and cost of implementation of such a solution, challenges, such as complex integration and multiple access points are still addressed when combined with the various other solutions and frameworks presented in this survey, such as the FAIoT Model in conjunction with our own proposed hybrid model.

## 5.3   Key Future Research Directions

From the comparison in Table 5, we see that much of the challenges in IoT forensics, listed in Sect. 3, still remain unaddressed. The solutions proposed for some of these challenges are either not implemented or not tested for usability and forensic soundness. Through this survey, we have identified the following key research gaps:

1. Lack of forensic soundness in all the acquisition and analysis models and frameworks proposed in the existing literature. Forensic soundness is crucial for evidence admissibility in court.
2. Lack of implementation and testing of proposed solutions. Much of the solutions in the current literature are purely theoretical or modifications to existing standards. Indeed, some solutions were implemented and tested, however, their scope was limited to only certain device types or scenarios.
3. The issue of data privacy is prevalent in all fields of computing. From our survey, we found very little focus in the literature on addressing privacy concerns associated with IoT forensics.
4. Data carving is an essential component of computer system forensics. However, as mentioned in Sect. 3, data carving is very challenging due to the volatility and volume of data generated by IoT devices. Through this survey we were unable to find any research that addresses this concern.

# 6 Conclusion

This paper surveyed the existing literature to identify challenges in the field of Internet of Things Forensics along with their proposed solutions. Eighteen models and frameworks were critically evaluated to identify open challenges and future research directions. Most of these solutions were purely procedural, while others had various implementations that contributed to increases the reliability and effectiveness for IoT forensics.

Considering the lack of progression throughout existing research work, we believe that the majority of IoT forensics research is limited to hypotheses and future implementation and testing. Most of these proposed solutions lack proper scientific validation and are not tested for forensic soundness, yet, they claim forensic soundness.

We believe that more focus should be directed toward the implementation and validation of proposed IoT forensics techniques. By doing so, we may be able to turn theoretical solutions into usable tools.

# References

1. Stoyanova, M., Nikoloudakis, Y., Panagiotakis, S., Pallis, E., Markakis, E.K.: A survey on the internet of things (IoT) forensics: challenges, approaches, and open issues. IEEE Commun. Surv. Tutor. **22**, 1191–1221 (2020)
2. Servida, F., Casey, E.: IoT forensic challenges and opportunities for digital traces. Digit. Investig. **28**, S22–S29 (2019)
3. Zhou, W., Jia, Y., Peng, A., Zhang, Y., Liu, P.: The effect of IoT new features on security and privacy: New threats, existing solutions, and challenges yet to be solved. IEEE Internet Things J. **6**, 1606–1616 (2018)
4. Paul Joseph, D., Norman, J.: An analysis of digital forensics in cyber security. In: Bapi, R.S., Rao, K.S., Prasad, M.V.N.K. (eds.) First International Conference on Artificial Intelligence and Cognitive Computing. AISC, vol. 815, pp. 701–708. Springer, Singapore (2019). https://doi.org/10.1007/978-981-13-1580-0_67
5. Koroniotis, N., Moustafa, N., Sitnikova, E.: Forensics and deep learning mechanisms for botnets in internet of things: a survey of challenges and solutions. IEEE Access **7**, 61764–61785 (2019)

6. Lartey, J.: Man suspected in wife's murder after her Fitbit data doesn't match his alibi. The Guardian, 25 April 2017. https://www.theguardian.com/technology/2017/apr/25/fitbit-data-murder-suspect-richard-dabate. Accessed 24 Apr 2022

7. Li, S., Choo, K.K.R., Sun, Q., Buchanan, W.J., Cao, J.: IoT forensics: Amazon echo as a use case. IEEE Internet Things J. 6(4), 6487–6497 (2019)

8. Meffert, C., Clark, D., Baggili, I., Breitinger, F.: Forensic state acquisition from internet of things (FSAIoT) a general framework and practical approach for IoT forensics through IoT device state acquisition. In: Proceedings of the 12th International Conference on Availability, Reliability and Security, pp. 1–11 (2017). https://doi.org/10.1145/3098954.3104053

9. Almolhis, N., Haney, M.: IoT forensics pitfalls for privacy and a model for providing safeguards. In: International Conference on Computational Science and Computational Intelligence (CSCI), pp. 172–178 (2019). https://doi.org/10.1109/CSCI49370.2019.00036

10. Lutta, P., Sedky, M., Hassan, M., Jayawickrama, U., Bastaki, B.B.: The complexity of internet of things forensics: a state-of-the-art review. Forensic Sci. Int.: Digit. Invest. 38, 301210 (2021). https://doi.org/10.1016/j.fsidi.2021.3012102666-2817

11. Antonakakis, T., et al.: Understanding the mirai botnet. In 26th USENIX security symposium (USENIX Security 2017), pp. 1093–1110 (2017). https://doi.org/10.5555/3241189.3241275

12. Al-Masri, E., Bai, Y., Li, J.: A fog-based digital forensics investigation framework for IoT systems. In: 2018 IEEE International Conference on Smart Cloud (SmartCloud), pp. 196–201 (2018)

13. Martin, H.: Are smart homes vulnerable to hacking? Avast, 16 August 2018. https://blog.avast.com/mqtt-vulnerabilities-hacking-smart-homes. Accessed 21 Apr 2022

14. Oriwoh, E., Jazani, D., Epiphaniou, G., Sant, P.: Internet of things forensics: challenges and approaches. In: 9th IEEE International Conference on Collaborative Computing: Networking, Applications and Worksharing, pp. 608–615 (2013)

15. Alabdulsalam, S., Schaefer, K., Kechadi, T., Le-Khac, N.A.: Internet of things forensics – challenges and a case study. In: Peterson, G., Shenoi, S. (eds.) DigitalForensics 2018. IFIP Advances in Information and Communication Technology, vol. 532, pp. 35–48. Springer, Cham (2018). https://doi.org/10.1007/978-3-319-99277-8_3

16. Liu, R., Sasaki, T., Uehara, T.: Towards a holistic approach to medical IoT forensics. In: 2020 IEEE 20th International Conference on Software Quality, Reliability and Security Companion (QRS-C), pp. 686–687 (2020)

17. Zia, T., Liu, P., Han, W.: Application-specific digital forensics investigative model in internet of things (IoT). In ARES 2017: Proceedings of the 12th International Conference on Availability, Reliability and Security (2017)

18. Kent, K., Chevalier, S., Grance, T.: Guide to integrating forensic techniques into incident response (2006). https://www.nist.gov/publications/guide-integrating-forensic-techniques-incident-response. Accessed 21 Apr 2022

19. Le, D.P., Meng, H., Su, L., Yeo, S.L., Thing, V.: BIFF: a blockchain-based IoT forensics framework with identity privacy. In: TENCON 2018 IEEE Region 10 Conference, pp. 2372–2377 (2018). https://doi.org/10.1109/TENCON.2018.8650434

20. Kebande, V.R, Ray, I.: A generic digital forensic investigation framework for internet of things (IoT). In: 2016 IEEE 4th International Conference on Future Internet of Things and Cloud (FiCloud), pp. 356–362 (2016). https://doi.org/10.1109/FiCloud.2016.57

21. Zawoad, S., Hasan, R.: FAIoT: towards building a forensics aware eco system for the internet of things. In: 2015 IEEE International Conference on Services Computing, pp. 279–284 (2015)

22. Dorai, G., Houshmand. S., Baggili, I.: I know what you did last summer: your smart home internet of things and your iPhone forensically ratting you out. In Proceedings of the 13th International Conference on Availability, Reliability and Security, pp. 1–10 (2018)

23. Hossain, M., Karim, Y., Hasan, R.:FIF-IoT: a forensic investigation framework for IoT using a public digital ledger. In: 2018 IEEE International Congress on Internet of Things (ICIOT), pp. 33–40 (2018)
24. Ryu, J.H., Sharma, P.K., Jo, J.H., Park, J.H.: A blockchain-based decentralized efficient investigation framework for IoT digital forensics. J. Supercomput. **75**(8), 4372–4387 (2019)
25. Goudbeek, A., Choo, K.K.R., Le-Khac, N.A.: A forensic investigation framework for smart home environment. In: 2018 17th IEEE International Conference on Trust, Security and Privacy in Computing and Communications (2018)
26. Li, Z., Ren, H., Chou, E., Liu, X., McAllister, C.D.: Retrieving forensically sound evidence from the ESP series of IoT devices. IEEE Internet Things J. **9**, 13144–13152 (2022)
27. Malik, V.R., Gobinath, K., Khadsare, S., Lakra, A., Akulwar, S.V.: Security Challenges in industry 4.0 SCADA systems – a digital forensic prospective. In: 2021 International Conference on Artificial Intelligence and Computer Science Technology (ICAICST) (2021)
28. Zawoad, S., Dutta, A.K., Hasan, R.: Towards building forensics enabled cloud through secure logging-as-a-service. IEEE Trans. Dependable Secure Comput. **13**(2), 148–162 (2015)
29. Quick, D., Choo, K.K.R.: IoT device forensics and data reduction. IEEE Access **6**, 47566–47574 (2018). https://doi.org/10.1109/ACCESS.2018.2867466

# Innovative Research Projects

# Local Energy Markets: A Market Transformation Survey Towards Segments of Interest

Evgenia Kapassa(✉) ⓘ and Marios Touloupou ⓘ

Department of Digital Innovation, University of Nicosia, Nicosia, Cyprus
{kapassa.e,touloupos.m}@unic.ac.cy

**Abstract.** Given the rapid growth of distributed energy generation and the anticipated rise in demand-response, one of the most important issues is how to integrate
the energy resources into the existing market. The energy sector has undergone
a considerable transformation during the past years due to changing legislation,
technologies and consumer behavior. Therefore, this work explores the patterns
that form the energy market transformation, measures awareness regarding energy
market and gain insights on energy market tools and services, through a survey
implemented within the innovative project PARITY. Moreover, the knowledge
gained from the participants is used to identify segments of interest. The main
findings from the market transformation survey include among other topics, that
the energy market is envisioned to be entirely based in DERs by the end of 2035,
while the primary barrier for the latter are the lack of EU and national regulations, as well as the inadequate technological design, lack of standardization, and
interoperability. The survey also revealed that smart energy contracts, load forecasting mechanisms, personalized profile models and dynamic pricing schemes
are some of the services that can increase the productivity and profitability within
local energy markets. Finally, this report is based on research carried out in early
2022, drawing from professionals within the H2020 project PARITY and were
from a variety of industries with an emphasis on energy production, analysis, and
consulting.

**Keywords:** Local Energy Market · Market Penetration · Market
Transformation · Survey · Segments of Interest

## 1 Introduction

The unifying objective of reaching net-zero greenhouse gas emissions by the year 2050
was reinforced when the European Parliament proclaimed a state of climate and environmental emergency in November 2019 [1]. In that sense, the most polluting aspect
of human activity is the energy sector, which is responsible for 33% of the greenhouse
gas emissions in the atmosphere [2] and could be considered as the first factor towards
a cleaner energy. The repercussions of Russia's invasion to Ukraine are a second factor
driving the shift to cleaner energy. According to Fisher [3], Russia provides 40% and

27%, respectively, of the natural gas and oil that the European Union (EU) imports. This conflict exposed the EU's reliance on Russian resources, highlighting the importance of diversifying energy supply toward renewables for achieving energy stability [4, 5, 34–37].

However, the latter demands the adaptation of grid infrastructures, in order to handle the capacity of renewable Distributed Energy Resources (DERs) [6, 7]. The idea of an energy community is growing in this context [8, 38], usually offering advanced services, such as self-consumption, energy storage, peer-to-peer trading and management of DERs [9, 39]. An example is the HORIZON 2020 project named PARITY [10], which also implements a model of an energy community, named Local Energy Market (LEM). In order to encourage the incorporation of more DERs into the energy system, LEMs have emerged as a key strategy and are designed to encourage small energy prosumers to trade energy with one another towards supply and demand balance [12, 40, 41].

Thus, the current work's contribution is focused on identifying the patterns that form the energy market transformation, measuring awareness related to LEMs and gaining insights on relevant tools and services, through a qualitative data analysis. The remainder of this paper is organized as follows: Sect. 2 provides the research methodology, while Sect. 3 presents the data analysis and results. Section 4 discusses the knowledge gained, focusing on identifying segments of interest within the LEMs. Finally, Sect. 5 concludes the paper.

## 2 Research Methodology

In this paper, the methodology followed to investigate the energy market transformation in LEMs is based on quantitative research [12], through the use of a survey, as a mean of data collection. Even though surveys are used in a wide range of areas, market research is one of their primary applications. The rationale behind the selection of the researcher lies in the fact that survey research supports the investigation of characteristics and opinions of a group of people in a specific matter (i.e., local energy market transformation). Through survey research the current work will understand the participants' needs, preferences and perspectives on the concept of LEM and will identify factors and barriers to the progress of local renewable energy development.

To that end, the survey was distributed primarily to the PARITY project experts, including distributed system operators, aggregators, retailers etc. Moreover, we included a secondary target group of experts outside PARITY, with an understanding of emerging technologies in the area of energy. The survey was administered to participants within European countries and the total responses were 50. The distribution of the participants is presented in Fig. 1. The questionnaire survey was divided into four sections: (a) Demographics, (b) Energy transformation insights, (c) Market Change and (d) Market Segments. The detailed presentation of the questionnaire as well as the (anonymous) data outcomes are open source and can be found in Zenodo [13].

# 3   Data Analysis and Results

The main results related to the energy market transformation elaborated from the questionnaire, are presented below. Initially, the background knowledge and the familiarity of respondents with the LEM technologies are investigated. Then, energy market transformation insights, as well as market segments are presented.

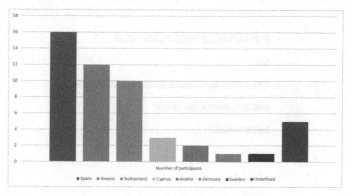

**Fig. 1.** Questionnaire survey distribution

## 3.1   Preliminary Data Analysis

The preliminary data of the survey was related to the demographics and was used to understand the participants' background. The majority of the respondents were male. Regarding their age, range between 30 and 39 are over-represented in the sample, young adults aged 18–24 are under-represented, while people above 60 are not represented. The education background was also investigated, with the majority of the participants holding a master's degree (55,3%), while an important part holds a Doctorate (29,8%). An important parameter of the current survey is that most of the respondents have experience in emerging technologies in the energy sector, such as renewable energy, microgrids, smart meters, blockchain and energy storage. Specifically, the majority of the respondents (51,1%) has a very strong understanding of such technologies, while 6,4% do not really understand their benefits.

## 3.2   Energy Transformation Insights

This section provides a review related to the energy transformation the energy market is currently facing. To begin with, data shows that the majority of the respondents know from where their final energy consumption is coming from, while a small percentage (8,5%) is not. Most energy consumption is coming from residual mix (59,6%), pure renewable energy sources are coming next with 17%, energy coming from power stations burning fossil fuels represent the 14,9% of the responses, while none of the respondents consumes energy coming exclusively from nuclear sources. An interesting aspect regarding the origin of the energy consumed nowadays is depicted also in Fig. 2, where it is evident how each country mostly consumes energy. From the data coming from the conducted survey, we can say that Sweden, Switzerland, Austria, Germany and Cyprus are moving rapidly towards greener energy consumption since all of the energy consumed is coming either from residual mix or from pure renewable energy sources. On the other hand, Spain and Greece are still consuming some amount of energy coming from power stations burning fossil fuels.

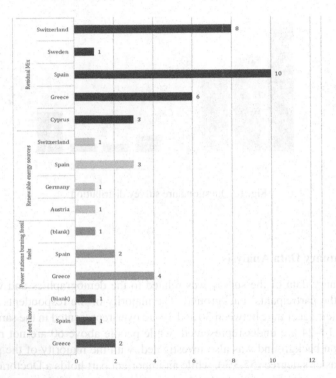

**Fig. 2.** Main source of energy consumption per country

In relation to their energy bill, most respondents state that they are satisfied in terms of costs (51,1%). A significant percentage though does not find its current energy bill reasonable (33,3%). It is worth mentioning though, that the majority of people consuming energy from residual mix are feeling satisfied with their energy bill. The large majority of the sample (85,1%) considers that the risk is higher with energy systems based on more centralized energy 'generation', compared with the 8,5% of the total sample which thinks the risk is lower, and the 4,3% which thinks is the same. Trying to identify why the majority of the sample thinks the risk is higher, we tried to categorize possible risks electricity markets based on more centralized generation are facing. Most of the participants state that the highest risk in more centralized energy markets are the emissions and/or the air pollution. The second risk in line is identified as medium by the majority of the respondents and is related to the energy availability and supply risk of the centralized markets, while also cyberattack, bulk generation of energy and the risk of unjusting pricing schemes are considered as medium risks as well (Fig. 3).

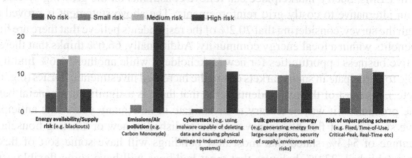

**Fig. 3.** Main risks of centralized energy generation

Regardless their belief related to the risk of the more centralized energy markets, the respondents were asked in how many years they anticipate the energy market to be based on DERs. The response was distributed almost equally, making it difficult to extract a clear prediction. A slightly higher percentage believes that by the end of 2035 the energy market will be based mostly on DERs, although there were some comments from the respondents stating that they anticipate it later (2050 onwards) or never. Additionally, it is believed that especially EVs will be widely applied by 2025, but in a rural area, centralized power plants (e.g., hydro, gas) will still play a significant role in electricity production. Based on the respondents, the anticipation is due to the fact that EU and national regulations and legislations are considered as the obstacles having the highest impact in the fast adoption of LFMs, as shown in Fig. 4. Insufficient technological design, lack of standardization as well as lack of interoperability between equipment and stakeholders are obstacles that are also having a medium impact.

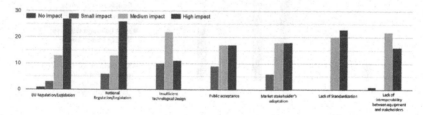

**Fig. 4.** Impact of obstacles in the fast adoption of LFMs

## 3.3 Market Change

Moving on to the next section of the survey, it is becoming evident that a number of changes have taken place within the global energy market. Thus, this section provides a review related to the energy market change, focusing on what would motivate people to participate in such markets.

Since local energy communities are entering the hype, the adoption of peer-to-peer (P2P) electricity trading will turn individual consumers from passive to active participants in LEMs. Such a marketplace can relieve constraints on the growing system and offer an alternative to costly grid reinforcements. The above statement is proved also through the survey, considering that 70,2% of the respondents believe that there are financial benefits within a local energy community. Additionally, 66,6% thinks that there are attractive business opportunities for new stakeholders, while another 66,6% find it reasonable to participate in such markets due to the increased investment in DERs (e.g. PVs, EVs etc.). As most of the respondents agree that there is a significant financial benefit for the prosumers, it was necessary to investigate in how many years it is anticipated to have some sort of flexible energy assets established in new or existing households. Percentage of 54,3% stated that most new buildings will have some sort of flexible assets established, 23,9% believes that most buildings will have some flexible assets, while only 8,7% thinks that all buildings will provide such capabilities (Fig. 5). The most important concepts that are foreseen to change the current energy market are the EVs, the large-scale centralized renewable energy generation as well as the distributed energy generation (Fig. 6). Local energy systems and infrastructure were characterized as of medium importance, while the majority of the respondents agree that large-scale fossil fuel power generation will have no importance at all in the future. Consequently, distributed and renewable enabled energy market that will mostly be based on DERs is becoming a reality.

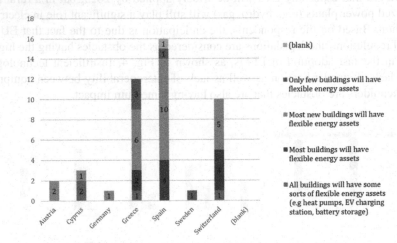

**Fig. 5.** Adoption of flexible energy assets within households per country in 5 to 10 years

## 3.4 Market Segments

At the final stage of this survey, the participants were asked questions related to specific energy market tools and services. Thus, this section provides a review related market segments and possible trends. The participants were asked to indicate which tools and

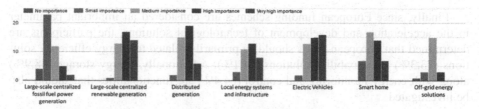

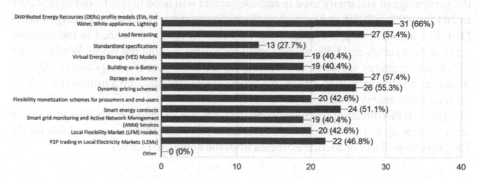

**Fig. 6.** Technological concepts of high importance for the next years

services they think may boost the profitability and efficiency. As it was anticipated, the majority of the respondents believe that DERs personalized profile models (66%), load forecasting mechanisms (57,4%), dynamic pricing schemes (55,3%) and smart energy contracts (51,1%) are the services that have the ability to accelerate the profitability and efficiency of DERs (Fig. 7).

**Fig. 7.** Technological Services that will boost the profitability and efficiency of DERs

Moreover, considering that smart energy contracts could accelerate the profitability and efficiency of DERs, the respondents were asked if they find blockchain enabled smart contracts useful for that matter. The majority of the sample (66,7%) responded yes, leading to the outcome that blockchain technology and smart contracts specifically is a promising technology and could be a trend within LEMs (Fig. 8).

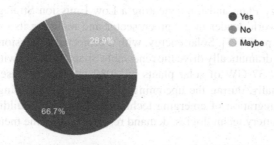

**Fig. 8.** Blockchain and smart contracts for LEM/LFM

Finally, since European funding schemes are considered an important parameter in the acceleration and development of technological solutions, the participants are determined that European funds should be primarily related to energy efficiency solutions (73,3%) and flexibility solutions (71,1%). Additionally, energy storage (68,9%) along with energy monitoring and control (64,4%) are functionalities that should also be investigated.

## 4 Discussion

Despite the advent of the global pandemic and the economic contraction, decarbonization strategies continued to be declared. Despite the absence of a clear opportunity for green infrastructure growth in the economic stimulus initiatives implemented in reaction to COVID-19, sustainable energy demand remained robust as renewables and storage saw decreasing prices and increasing Capital Expenses (CapEx) [14]. Also, it is expected that the percentage of electricity used in end-use sectors will need to rise from roughly 20% in 2015 to 40% in 2050, as low-carbon electricity becomes the dominant energy carrier [16]. For instance, as it is evident from the research results described in the previous section, heat pumps and EVs, would spread widely throughout the world. While the use of renewable energy would be accountable for a sizable part of energy use in industry, buildings and transportation, renewable power would still make up a little under 60% of all renewable energy consumption.

Towards this direction, after analyzing the market in the previous section, the current section is going to provide segments of interest based on information available on the literature as well as knowledge extracted from the conducted survey.

### 4.1 Segments by Country

**Spain – Country Profile and Forecasts.** Solar energy has become a major source of electricity production, and its share in Spain is rising. In the forthcoming years of 2020–2025, the renewables market in Spain is projected to develop at a CAGR of more than 6%. Indicators like promoting government policies and efforts to satisfy increasing power demand with clean energy sources and reduce reliance on fossil fuels, are projected to play a major role in market development [17]. Moreover, the Spanish government took ambitious measures, such as eliminating the sun tax in 2018, establishing a National Energy Technology Plan, and implementing a Low Emission Strategy 2050, to place the country as a world leader in the green sector and reducing its carbon footprint at an unprecedented pace [17]. Solar energy, with a projected expansion of more than 35 GW by 2030, will dramatically drive the renewable Spanish sector, with the government planning to install 37 GW of solar plants by 2030 in order to raise renewable share by 74%. Additionally, during the upcoming years, Spain is focusing on solar energy generation and integration of emerging technologies in smart buildings (e.g., energy storage through battery technologies, demand response, real time metering data etc.).

**Switzerland – Country Profile and Forecasts:** Switzerland's renewable energy market is projected to expand at a CAGR of more than 3% between 2020 and 2025 [18].

Increasing concerns over greenhouse gas pollution, which cause environmental harm and global warming, are forcing the government to search for cleaner electricity generation alternatives. This is expected to fuel the Swiss renewable energy market. However, Switzerland's stagnant power demand is likely to limit the nation's renewable energy market [17]. It is worth mentioning though, Switzerland signed an agreement in 2017 to limit greenhouse gas emissions by half by 2030, expected to open up a number of potential prospects for the Swiss green energy industry in the upcoming years. Additionally, the country's growing installed capacity for solar and hydro power is expected to boost the Switzerland renewable energy market. Moreover, Apox proceeded with the installation of a solar plant of over 6,000 photovoltaic modules on the Glarus Alps site, which covers a surface of 10,000 square meters [19]. The plant can generate 2.7 gigatons of energy per year, which is equivalent to the demand of 600 households. As a consequence of the above, the Switzerland green market is projected to be powered by rising renewable energy installed capacity during the upcoming years period.

**Greece – Country Profile and Forecasts:**  As stated in the 2020 Greek Energy Market Report [20], Because of the COVID-19 epidemic, gross energy usage fall in 2020. Greece has a high degree of solar energy and 2.9 GW of solar thermal systems deployed, with photovoltaic installations accounting for the majority of RES in the country. [20, 21]. Solar PV generated 25,43% of RES electricity and 8.2% of total electricity, due in part to the 512 GWh provided by the Special Photovoltaic Rooftop Program. Additionally, the Hellenic Energy Exchange, which offers high-quality, open, and socially responsible services to environmental market players, was a pivotal move for the region. Working against the EU target image and combining its economy with neighboring nations, not to mention providing low-income families with grants covering up to 60% of the construction costs of solar water heaters, Greece illustrates a strong penetration of renewable energy in the sector as well as an attempt to implement European and national strategies [21]. RES objectives are set to allow a cumulative contribution of 20% to overall final energy consumption, among other items. RES is expected to account for 75% of total energy generated in Greece by the end of 2025, while by 2030, overall, RES capacity will increase by 61–64%, with PV having the largest share in Greece [20].

**Sweden – Country Profile and Forecasts:**  During the projected timeline of 2020–2025, the Swedish renewable energy market is estimated to grow at a CAGR of more than 2%. Government efforts to reduce greenhouse gas emissions in the country are expected to propel the Swedish clean energy industry. However, problems in operating renewable energy plants are likely to restrict Sweden's renewable energy market. Sweden plans to target 100% clean energy power generation by 2040 and 0% greenhouse gas emissions by 2045. These goals are likely to provide a great potential for the Swedish green energy industry. Furthermore, the country's growing renewable energy installed capacity and future developments are expected to boost the Swedish renewable energy market. The European Union has set goals for reducing their carbon footprint, and Sweden is increasing its renewable installation capacity to meet those targets [22]. Regardless, Sweden requires innovative ways to achieve the above-mentioned aggressive policy target of achieving 100% clean energy by 2040. IRENA has therefore suggested four tailor-made options [23] focused on a holistic approach to solve the country's unique difficulties in scaling up Variable Renewable Energy (VRE) in collaboration with the Swedish Energy Agency. The solutions are highly linked with emerging technologies such as batteries, IoT technologies, EV smart charging, as well as blockchain and smart contract technologies. Additionally, the solutions envision innovative market design based on ancillary services, local markets and time of use tariffs.

## 4.2  Segments by Type

**Solar PV:**  Solar PV, or solar photovoltaic energy, is a form of solar energy that uses photovoltaic technology to turn sunlight into electricity. Solar power is probably the most trustable and sustainable green energy source. While hydropower remains the most cost-effective renewable energy source, steadily dropping prices have rendered solar PV the primary market for investment. Particularly, in a variety of countries around the globe, solar PV-generated electricity nowadays cost competitive with fossil fuels [24]. Companies are increasingly shifting towards the use of renewable energy sources for electricity production, especially solar energy and wind power, as a result of strict government legislation and regulation related to carbon emissions. This is expected to boost the solar PV market's growth in the future years [24].

**Solar Thermal - Heating and Cooling:**  Solar thermal has the possibility to be a significant source of heating and cooling in Europe as an incredibly convenient heating source based on a basic principle improved by cutting-edge technologies [25]. As technology has improved, solar thermal is becoming not only a better alternative for more conventional applications like domestic hot water processing, but also a potential approach for modern and more advanced applications like industrial process heat. Moreover, the provision of district heating facilities will boost the viability of centralized solutions such as large-scale sustainable heating and cooling systems, such as solar-thermal. Several thermal heating and cooling technologies, such as solar-thermal, might be much affordable at scale, but delivery requires a district heating grid. The district heating system, on the other hand, is not always financially viable to construct [25].

**Wind Power:**  Wind power capacity continues to expand rapidly, fueled by an established track record and precedent from the previous decade, as well as low levelized cost of energy (LCOE). International Wind Energy Council (GWEC) study [26] states that despite the global offshore wind industry having a great year in 2021 in terms of new capacity, it is still expected to fall behind of the net zero goals set by the International Energy Agency by 2030. The trade association also stated that it was extremely likely to increase installation projections this year as nations sought to modernize their energy infrastructure in reaction to the volatility of fossil fuel prices, which was made worse by the invasion of Ukraine [27].

**Bio-Energy:**  Bio-energy (or biomass) is a significant renewable energy source that will help Europe achieve its climate goals in 2020 and 2030, when renewable energy sources must account for 32% of total energy demand in the EU [28]. Regardless of the fact that international investment in biomass and waste-to-energy fell by 29% from 2014 to 2015, this sector ranks third behind wind and solar energy. Bioenergy has the potential to play a major role in meeting the EU's renewable energy goals by 2030 and beyond [29]. According to the European Commission's sustainability scenario, gross inland bioenergy consumption would range from 170 to 252 Mtoe (i.e., million or mega tonnes of oil equivalent) by 2050 [28]. Bioenergy can also act as a versatile carrier, allowing for higher shares of renewable energy sources such as wind and solar power in power grids [28].

## 4.3  Segments by End-Use

**Building Sector:** Gas heaters, lighting, central air conditioning, refrigeration, and electric and gas water heating are the facilities that use the most electricity in the residential sector [30]. In comparison, the reliability picture for light bulbs, as well as electric and gas water heating use, shows less effectiveness so far. Lighting provides many opportunities for productivity improvements, both because removing incandescent lamps with fluorescent bulbs saves a considerable amount of electricity and because the first cost tends to decline. Fluorescent and incandescent lighting, air-conditioning, office appliances, supply/return fans, and bundled heating are the main electricity appliances in the commercial sector [31].

**Transportation Sector:** Transportation is critical to the global economy because it facilitates the flow of people and global commerce. However, it comes at an expense, since it is a big source of carbon due to the current reliance on fossil fuels. Although it accounts for one-third of global energy consumption, it is also the field with the least amount of renewable energy usage but with the greatest potential [32]. There are some preferred green alternatives for some modes of transportation, but not all [33]. Despite considerable progress in energy efficiency, especially in road transport, global energy demand in the transport sector has steadily increased over the last decade, owing primarily to the increasing number and scale of vehicles on the world's roads. Even if all announced policy initiatives are adopted, the transportation industry is projected to increase GHG emissions by 60% by 2050, owing primarily to increased freight and non-urban transportation. If no steps are taken, road transport is expected to account for at least 70% of GHG emissions by 2050 [33]. Thus. it is clear that green energy strategies for the road transport sector must be incorporated in a broader system of measures that also decrease demand for transportation facilities, change transportation modes, and improve vehicle quality [32].

**Industrial Sector:** Industry accounts for more than one-quarter of global energy-related Carbon Dioxide ($CO_2$) emissions, and manufacturing activities account for a further 8% of global $CO_2$ emissions [29]. Iron and steel, aluminum and plastics account for more than 85% of manufacturing energy and process-related pollutants, and account for more than two-thirds of overall industrial energy consumption. There are actually only a few commercially feasible options for reducing $CO_2$ emissions at scale in these manufacturing industries [33]. According to the "buildings section" of this deliverable, electricity usage in commercial buildings accounts for almost two quads of annual energy use. The ITP Best Practices Program does not discuss the construction aspects of energy use in the sector and focuses primarily on process and plant utility energy use. Finally, there are many opportunities in the lighting, HVAC, and building covering [33].

## 5 Conclusions

Under the scope of this work, a high-level market penetration took place, exploring the patterns that form the energy market transformation, measure awareness regarding energy market and gain insights on energy market tools and services through a survey implemented within the PARITY project [9]. Within this work, we considered market penetration as a measurement on how much PARITY related products and services are being used by potential customers. Additionally, within the market penetration phase, we evaluated potential customers' opinions on the concept of LEMs, to identify factors and barriers to the progress of local renewable energy development based on their knowledge and what would motivate people to participate in such markets. The current market needs were also identified and analyzed. Finally, after evaluating the research outcomes, segments of interest were presented.

In a summary, the market transformation survey, revealed among others that the energy market is anticipated to be fully based in DERs by the end of 2035, while it is forecasted that by the end of 2025 eV will be widely applied. To that end, the main obstacles for the faster energy transformation are the lack of EU and national regulations and legislations, as well as the insufficient technological design, lack of standardization and lack of interoperability between equipment and stakeholders. Furthermore, distributed, and renewable enabled energy market that will mostly be based on DERs seems that is becoming a reality. Finally, a main outcome of the market transformation survey is linked with future trends. Specifically, the survey revealed that DERs personalized profile models, load forecasting mechanisms, dynamic pricing schemes and smart energy contracts are the services that can accelerate the profitability and efficiency of DERs.

The knowledge gained from the current survey as summarized before, it will help the PARITY project to succeed and be established within the market, but at the same time it provides research outcomes relevant to energy policy makers, energy operators, organizations as well as researchers in this area. The following benefits were provided during the discussion of this work: (a) gain holistic view of the market, (b) potential customers retention, (c) identify gaps and give actionable insights, (d) gain a competitive advantage.

**Acknowledgements.** This work has been partially supported by the PARITY project, funded by the European Commission under Grant Agreement Number 864319 through the Horizon 2020.

## References

1. European Union declares a climate emergency. https://climateemergencydeclaration.org/eur opean-union-declares-a-climate-emergency/. Accessed 07 Sept 2022

2. Hua, W., Zhou, Y., Qadrdan, M., Wu, J., Jenkins, N.: Blockchain enabled decentralized local electricity markets with flexibility from heating sources. In: IEEE Transactions on Smart Grid (2022)

3. EU reveals its plans to stop using Russian gas. https://www.bbc.com/news/science-enviro nment-61497315, https://www.cnbc.com/2022/03/02/russia-ukraine-war-lessons-for-global-energy-markets.html. Accessed 07 Sept 2022

4. Kapassa, E., Themistocleous, M., Quintanilla, J. R., Touloupos, M., Papadaki, M.: Blockchain in smart energy grids: a market analysis. In: Themistocleous, M., Papadaki, M., Kamal, M. M. (eds.) EMCIS 2020. LNBIP, vol. 402, pp. 113–124. Springer, Cham (2020). https://doi.org/10.1007/978-3-030-63396-7_8

5. Detoc, M., Bruel, S., Frappe, P., Tardy, B., Botelho-Nevers, E., Gagneux-Brunon, A.: Intention to participate in a COVID-19 vaccine clinical trial and to get vaccinated against COVID-19 in France during the pandemic, In: Vaccine, vol. 38, pp: 7002–7006 (2020)

6. Kapassa, E., Touloupou, M., Themistocleous, M.: Local electricity and flexibility markets: Swot analysis and recommendations, In: 6th International Conference on Smart and Sustainable Technologies (SpliTech), pp. 1–6, IEEE (2021)

7. Gjorgievski, V.Z., Cundeva, S., Georghiou, G.E.: Social arrangements, technical designs and impacts of energy communities: a review, In Renewable Energy, pp. 1138–1156 (2021)

8. Caramizaru, A., Uihlein, A.: Energy Communities: an Overview of Energy and Social Innovation Luxembourg: Publications Office of the European Union (2020)

9. Parity H2020 – Parity H2020. https://parity-h2020.eu/. Accessed 07 Sept 2022

10. Strepparava, D., Nespoli, L., Kapassa, E., Touloupou, M., Katelaris, L., Medici, V.: Deployment and analysis of a blockchain-based local energy market. Energy Reports **8**, 99–113 (2022)

11. Sukamolson, S.: Fundamentals of quantitative research. In: Language Institute Chulalongkorn University, pp. 1–20 (2007)

12. Energy Market Transformation Survey. https://zenodo.org/record/6420882. Accessed 07 Sept 2022

13. Renewables 2021 - Analysis and forecast to 2026. https://iea.blob.core.windows.net/assets/5ae32253-7409-4f9a-a91d-1493ffb9777a/Renewables2021-Analysisandforecastto2026.pdf. Accessed 07 Sept 2022

14. Short-Term Energy Outlook. https://www.eia.gov/outlooks/steo/. Accessed 07 Sept 2022

15. Gielen, D., et al.: Global energy transformation: a roadmap to 2050 (2019)

16. Spain Renewable Energy Market - Growth, Trends, COVID-19 Impact, and Forecasts (2021–2026). https://www.mordorintelligence.com/industry-reports/spain-renewable-energy-market. Accessed 07 Sept 2022

17. Switzerland Renewable Energy Market - Growth, Trends, COVID-19 Impact, and Forecasts (2022–2027). https://www.mordorintelligence.com/industry-reports/switzerland-renewable-energy-market. Accessed 07 Sept 2022

18. Romande Energie's high-altitude floating solar farm wins renewable energy award in 2021 Watt d'Or competition. https://www.romande-energie.ch/images/files/communiques_archives/210107_communique_en.pdf. Accessed 07 Sept 2022

19. Greek Energy Market Report 2020. https://www.haee.gr/media/1934/haees-greek-energy-market-report-2020-brief-version.pdf. Accessed 07 Sept 2022

20. All you need to know about Greek energy Market, https://greendealflow.com/all-you-need-to-know-about-the-greek-energy-market. Accessed 07 Sept 2022

21. Sweden Renewable Energy Market - Growth, Trends, COVID-19 Impact, and Forecasts (2021–2026). https://www.mordorintelligence.com/industry-reports/sweden-renewable-ene rgy-market. Accessed 07 Sept 2022

22. Innovative Solutions For 100% Renewable Power in Sweden. https://www.irena.org/-/media/Files/IRENA/Agency/Publication/2020/Jan/IRENA_Innovative_power_Sweden_2020_summary.pdf?la=en&hash=9FC47DCAD97F5001B07663FD7D246872DBC0F868. Accessed 07 Sept 2022

23. Solar Photovoltaic (PV) Market size, share and covid-19 impact analysis by technology, by grid type, by installation, by application and regional forecast, 2020–2027. https://www.fortunebusinessinsights.com/industry-reports/solar-pv-market-100263. Accessed 07 Sept 2022

24. Solar Heating and Cooling Technology Roadmap European Technology Platform on Renewable Heating and Cooling. https://www.rhc-platform.org/content/uploads/2019/05/Solar_Thermal_Roadmap.pdf. Accessed 07 Sept 2022

25. Solar Thermal Heating and Cooling - Technology Market Report. https://publications.jrc.ec.europa.eu/repository/bitstream/JRC118312/jrc118312_1.pdf. Accessed 07 Sept 2022

26. Global Wind Report 2022. https://gwec.net/global-wind-report-2022/. Accessed 07 Sept 2022

27. Impact of Russia's invasion of Ukraine on the markets: EU response. https://www.consilium.europa.eu/en/policies/eu-response-ukraine-invasion/impact-of-russia-s-invasion-of-ukraine-on-the-markets-eu-response/. Accessed 07 Sept 2022

28. Directive of the European Parliament and of the Council - amending Directive 2012/27/EU on energy efficiency. https://ec.europa.eu/energy/sites/ener/files/documents/1_en_act_part1_v16.pdf. Accessed 07 Sept 2022

29. CEEE Consumer Guide: Top-Rated Energy Efficient Appliances: Gas and Oil Furnaces. http://www.aceee.org/consumerguide/topfurn.htm. Accessed 07 Sept 2022

30. Renewable energy statistics - European Commission. https://ec.europa.eu/eurostat/statistics-explained/index.php?title=Renewable_energy_statistics. Accessed 07 Sept 2022

31. Reviews of cutting-edge technology and country data on low-carbon industry and transport https://www.irena.org/industrytransport. Accessed 07 Sept 2022

32. The Renewable Route To Sustainable Transport. https://irena.org/-/media/Files/IRENA/Agency/Publication/2016/IRENA_REmap_Transport_working_paper_2016.pdf. Accessed 07 Sept 2022

33. Reducing the carbon footprint of the manufacturing industry through data sharing. https://www.weforum.org/impact/carbon-footprint-manufacturing-industry/. Accessed 07 Sept 2022

34. Zabaleta, K., et al.: Barriers to widespread the adoption of electric flexibility markets: a triangulation approach. In: 2020 5th International Conference on Smart and Sustainable Technologies (SpliTech), IEEE (2020)

35. Themistocleous, M., Stefanou, K., Megapanos, C., Losif, E.: To chain or not to chain? A blockchain case from energy sector. In: Proceedings of the 15th European, Mediterranean, and Middle Eastern Conference, pp. 29–35, Springer Switzerland (2018)

36. Touloupou, M., Kapassa, E., Rizou, S.: Cloud orchestration for optimized computing efficiency: the case of wind resource modelling. In: 2022 IEEE 11th International Conference on Cloud Networking (CloudNet), IEEE (2022)

37. Themistocleous, M., Stefanou, K., Losif, E.: Blockchain in solar energy. Cyprus Rev. **30**(2), 203–212 (2018)

38. Pressmair, G., Kapassa, E., Casado-Mansilla, D., Borges, C.E., Themistocleous, M.: Overcoming barriers for the adoption of local energy and flexibility markets: a user-centric and hybrid model J. Cleaner Prod. **317**, 128323 Elsevier (2021)

39. Strepparava, D., Nespoli, L., Kapassa, E., Touloupou, M., Katelaris, L: Medici V. deployment and analysis of a blockchain-based local energy market. Energ. Rep. 8, 99–113, Elsevier (2022)

40. Kapassa, E., Touloupou, M., Christodoulou, K.: A Blockchain based approach for demand response management in internet of vehicles. In: 7th International Conference on Smart and Sustainable Technologies (SpliTech), pp. 1–6, IEEE (2022)
41. Kapassa, E., Themistocleous, M.: Blockchain technology applied in IoV demand response management: a systematic literature review. Future Internet. vol. 14(5), 135. MDPI (2022)

# IT Governance and Alignment

# Towards a Taxonomy of Strategic Drivers of IT Costs

Constanze Riedinger[1](✉), Melanie Huber[2], Niculin Prinz[1], and Christopher Rentrop[1]

[1] Konstanz University of Applied Sciences, Konstanz, Germany
constanze@htwg-konstanz.de.riedinger, {niculin.prinz,
rentrop}@htwg-konstanz.de
[2] BITCO[3] GmbH, Konstanz, Germany
melanie.huber@bitco3.com

**Abstract.** Nowadays, information technology (IT) is a strategic asset for organizations. As a result, the IT costs are rising and there is a need for transparency about their root causes. Cost drivers as an instrument in IT cost management enable a better transparency and understanding of costs. However, there is a lack of IT cost driver research with a focus on the strategic position of IT within organizations. The goal of this paper is to develop a comprehensive overview of strategic drivers of IT costs. The Delphi study leads to the identification and validation of 17 strategic drivers. Hence, this paper builds a base for cost driver analysis and contributes to a better understanding of the causes of costs. It facilitates future research regarding cost behavior and the business value of IT. Additionally, practitioners gain awareness of levers to influence IT costs and consequences of managerial decisions on their IT spend.

**Keywords:** Delphi Study · Strategic Cost Drivers · IT Cost Drivers · Cost Behavior · Information Technology

## 1 Introduction

A functioning IT Cost Management (ITCM) ensures organizations to make the right decisions when investing in information technology (IT) [1] and to increase the customer value [2]. ITCM also strengthens cost awareness and influences IT costs anticipatorily [1]. These goals require the investigation and a deep understanding of cost and cost behavior [3]. Thus, cost management proposes the identification, classification, and estimation of "factors causing a change in the total cost" [4]. These strategic cost drivers determine cost behavior fundamentally and long-term [1] and therefore enable the establishment of the strategic position of an organization. They describe cost across the entire value chain [5] and impact customer value, revenue, and profitability [6]. Already in 1996 a study by Carr and Tomkins [7] stressed that successful companies focus twice as much on cost driver analysis – regardless of the companies' strategic position - differentiation or cost leadership [8]. Nowadays, IT is seen as a strategic asset [9] and IT costs account for an increasing part of the total cost within organizations

M. Papadaki et al. (Eds.): EMCIS 2022, LNBIP 464, pp. 555–569, 2023.
https://doi.org/10.1007/978-3-031-30694-5_39

[10]. As a first step and to enable cost driver analysis, cost driver identification aims for cost transparency [11]. However, a study of 474 IT managers in 2021 states that 79% of the respondents still face a lack of transparency and understanding of costs for their IT investments [10]. This prevents managers from measuring and highlighting the value proposition of their IT [12]. A comprehensive overview of strategic IT cost drivers would enable the required transparency and the investigation of cost behavior as measure to make the right decisions and to increase business value in the long term. Thus, it is the paper's goal to deliver a structured overview about the strategic drivers of IT costs and assess their influence.

Structured overviews on strategic cost drivers emerged in the 1980s and 1990s as taxonomies presenting a variety of "strategic choices" [13] that drive costs in manufacturing companies [5, 14]. Their main objectives were to better understand incurring costs and related profitability as well as to identify levers of influence for the overall cost of an organization [6]. However, those taxonomies of strategic cost drivers do not include nowadays' IT perspective. Information system (IS) researchers also make use of the analysis and evaluation of cost drivers: they investigate IT cost drivers for the cost estimation in software development [15, 16], software maintenance [17], or for measuring the cost of scrum activities [18]. IT cost driver research focuses on subsections or individual IT solutions [19] and thereby neglects the interaction of systems and information technology as a whole. This impedes comprehensive transparency and understanding across the total IT landscape. Furthermore, they focus on tactical cost drivers rather than giving a strategic perspective. Yet, IT cost driver research lacks a holistic and structured overview of strategic IT drivers. We therefore conduct a Delphi study to get a consistent and verified picture of the strategic drivers and their influence on IT costs. Our study contributes to scientific research by strategically enhancing existing concepts in the context of IT cost drivers and, consequently, by developing and verifying a taxonomy of strategic drivers of IT costs. Additionally, practitioners gain awareness about levers for cost influence and opportunities for strategic discussion on IT costs and the value proposition of IT.

This paper is structured as follows: first, we outline cost driver research. We then introduce the Delphi method as our research method comprising an exploratory study in the preparation phase, the expert selection, and the Delphi survey itself. Finally, we present and discuss our findings, and then draw our conclusions.

## 2   The Importance of Strategic Cost Drivers for IT Management

IT is a strategic asset for most organizations which leads to rising IT budgets [20]. Therefore, the main focus of managers is on the optimal use of IT and thereby to ensure IT's contribution to strategy and business value [12]. Consequently, the transparency and understanding of root causes of IT cost are prerequisites in strategic decision-making [1]. To understand those root causes, cost driver analysis is a strategically important instrument in cost management. The concept of cost drivers describes and analyzes cost behavior [5] to reveal the root causes of costs [21] and to better understand the costs of a whole organization or a certain area [19]. The strategic cost management discipline introduced the term cost driver to describe those factors that cause costs [8] or decision variables that drive costs [6]. We define cost drivers as "factors or decision variables that influence costs". They can be divided into operative, tactical and strategic cost drivers [22]: operative cost drivers, such as production volume, focus on manufacturing on a short-term basis. Tactical cost drivers are identified especially in terms of cost allocation with Activity-Based Costing (ABC) systems and explain mid-term cost dependence. They are determined for specific use cases. Strategic cost drivers are long-term decisions variables that affect tactical and operative cost drivers [23]. They influence the cost position of organizations fundamentally and across periods [22]. For organizations those cost drivers and their relationships are strategically important as they result from strategic management decisions [6]. Therefore, the identification of strategic cost drivers contributes to transparency of IT costs [11] and in a second step their detailed analysis leads organizations to gain understanding about the root causes of their cost position and influence them [5].

The digital transformation of organizations and the resulting increase in IT investments led to growing research interest in IT cost drivers [1]. However, this consideration of cost driver in IT is mainly used in the context of ABC and cost allocation [19], on the level of tactical cost drivers. Raz and Elnathan [24] identify potential cost drivers for typical project activities for their ABC model. Pramono and Suryani [18] consider ABC and corresponding cost drivers to measure benefits of cost usage in scrum-based companies. Ooi et al. [25] present an ABC approach to estimate and recover system development and implementation costs with cost drivers such as project duration, project type, number of features or back-end complexity. Raghu and Chaudhury [26] identify the lines of code and function points as ABC cost drivers in software development. Besides the ABC approach, cost driver models are also proposed for other specific use cases such as cost estimation, IT operations and evaluation of IT solution profitability: Boehm et al. [15] base their software development cost estimation model (COCOMO 2.0) on 23 cost drivers. Factors for a better understanding of software maintenance cost are according to Benaroch [17] divided into system attributes (age, size, complexity, information quality constraints, software volatility), and personnel factors (number of maintainers, location diversity, skill diversity). Herzfeldt [19] presents a cost driver model for IT solutions to measure their profitability over the complete life cycle: the IT cost drivers are divided into three main categories: input, value creation process and uncertainty. A total of 214 cost drivers are assigned to 30 categories in the model.

As pointed out, the existing research presents IT cost drivers on a tactical level for specific use cases. Thereby, there is a lack of the strategic perspective describing those variables that influence the tactical cost drivers in the first place and long-term. Furthermore, presented IT cost driver models describe their influence on individual sub-areas for the development and the evaluation of IT solutions. This does not reflect IT's strategic position within organizations and its increasing importance. Furthermore, it neglects the company-wide evaluation of interrelated IT contributions and their generated business value. To our knowledge, there is no structured overview on strategic IT cost drivers, although, such categories build "a powerful taxonomy for classifying the different types of cost drivers" [27]. Approaches presenting taxonomies on strategic cost drivers [5, 13] emerged mainly in the 1980s and 1990s and set their focus on manufacturing and not

IT. Banker and Johnston [6] give an overview on those cost driver taxonomies. However, those overviews do not yet include todays' position and omnipresence of IT within organizations. These shortcomings of existing approaches lead to the research question of this publication: *What are the strategic drivers of IT costs and what is their influence on IT costs?*

## 3   Research Method

The Delphi method is used to "exploratively examine a complex problem through a group of experts" [28]. In IS research it has been applied for several use cases: validation of definitions and classification [29], development of frameworks [30] or to achieve a consensus on a ranking [31, 32]. The Delphi method is defined by four generic characteristics: anonymity, controlled feedback, an iterative process, and the statistical aggregation of group responses [33]. The objective of our research is to obtain a comprehensive overview of strategic cost drivers and their influence on IT costs. Since strategic cost drivers have a long-term effect, their influence is difficult to measure and is therefore preferably estimated; this estimate can be made by experience or experts [22]. A Delphi study can support this estimate through consolidating expert opinions in an iterative process with several rounds including controlled feedback [34]. Thereby it also omits group dynamics and bias because the experts within a Delphi study are anonymous to each other [33]. We further choose the method as it can be conducted easily gathering multiple expert opinions with flexible and asynchronous processing [35]. Furthermore, the standardized Delphi questionnaire allows us to aggregate and quantify expert opinions statistically [35]. For these reasons, we consider the method to be appropriate to answer our research question.

For the design of our study, we apply the characteristics and specifications of Delphi method variants in IS research according to Strasser [28]: With an objective to generate facts, we design our study to elicit opinions and gain consensus. We choose experts in a wide sense with total anonymity and a not too large size of the expert panel. The predefined set of issues is developed in an exploratory study. This is the basis for the first round of our survey with a quantitative questionnaire including options to provide qualitative explanations [29].

We structure our Delphi study in three steps: we start with an exploratory study, conducted as data triangulation [36]. We follow earlier research for cost driver identification applying data triangulation [21], to make use of multiple sources and develop a predefined set of issues [34]. Second, we define criteria for the expert selection, identify and approach experts and develop the questionnaire for the Delphi survey [36]. Third, we execute the Delphi survey in two rounds with interim analyses to classify the answers and develop a validated overview of strategic drivers of IT costs. The stop criterion for our Delphi study was to reach consensus on a set of strategic drivers. We define consensus for categorical yes/no answers with 75% agreement and for the 5-point Likert scale with the inter-quartile deviation (IQD) $\leq 1$ [37].

# 4   Results

In the following we present the results of our Delphi study following the three steps: exploratory study, expert selection and questionnaire, and execution of Delphi survey.

## 4.1   Exploratory Study

The data triangulation starts with a literature review according to Cooper [38] to ensure a theoretical background of our study. Our search extended to the beginning of 2022. We limited our search to title, abstract, and keywords with the search keywords "IT cost driver", ("cost driver" OR "cost factor") AND "information technology", „influencing factors" AND "information technology" AND "cost". The search was then performed with a focus on research papers published in major conferences or journals (following the IS basket of eight) in four databases: *AISeL, EBSCOHost Business Source Premier, Sciencedirect* and *IEEE.* We removed duplicated and irrelevant publications by applying defined exclusion criteria [39]: publications without focus on IT costs or in other languages than English or German. We conducted both forward and backward search [40] to avoid missing important references. In our literature search we identified 17 publications, of which 11 explicitly specify IT cost drivers. The first author carried out a qualitative coding [41] and identified and categorized these IT cost drivers. They were then reviewed and adjusted in two iterations with the other authors. After two iterations of qualitative coding, we developed a first set of six strategic cost drivers of IT costs with short explanations: *Number of Users, Complexity, Performance Quality, Sourcing Quality, Service Level* as well as the *Size/Service Offer of IT.*

Using the result from the literature review, we applied the research method of expert interviews to obtain a comprehensive picture of the phenomenon in practice [42]. In doing so, we conducted semi-structured and in-depth interviews [42] to have the opportunity to ask more detailed follow-up questions. We interviewed 15 experts from practice, providers, and academia to cover a broad spectrum of people and functions via Microsoft Teams in German language. Finally, we transcribed, and anonymized all interviews. We conducted qualitative coding [41] for the analysis of interview data, using the tool MAXQDA. We used the defined cost drivers from the literature review to assign respective codes to the data but also applied open coding to identify additional strategic drivers and corresponding categories for classification [41]. We thereby expanded our six cost drivers from literature review to a list of 15 after the expert interviews. Those 15 cost drivers are categorized in qualitative coding in the four categories:

- *IT service offer*: IT Scope, Number of IT-based Products
- *Complexity of IT landscape*: Harmonization, Standardization, Technical Complexity
- *Quality*: Performance Quality, Service Level, Sourcing Quality, System Quality, Security Requirements, Compliance Requirements
- *Corporate context*: Number of Users, Organizational Complexity, Business Infrastructure, Corporate Culture

We then conducted an expert workshop following the procedure of Kluge et al. [36]: four researchers revised and specified the results of the exploratory study as a basis to develop the Delphi questionnaire. They examined and formulated detailed descriptions for the strategic cost drivers, and their influence from the codes of the expert interviews and verified the established categories to classify the strategic drivers of IT costs.

## 4.2  Expert Selection

For the selection of experts for our delphi survey, we followed the recommendation of 5–20 panelists by Rowe et al. [33]. When choosing experts, we followed the job title and verified that they have several years of experience in the field and thereby aimed for a heterogeneous panel [31, 34]. We contacted German-speaking experts, what led to a panel of initially 21 experts responding the first round (R1) and 18 in the second round (R2). The panel (Table 1) included practitioners (P) (13 in R1|11 in R2) of a variety of company sizes and sectors, IT consultants (C) (5|4) and academics (A) (3|3).

**Table 1.** Overview of Delphi Study Experts

|    | Function |     | Function |    | Function |
|----|----------|-----|----------|----|----------|
| P1 | Chief Information Officer (CIO) | P8 | IT Portfolio Management | C2 | CEO, IT Service Consultancy |
| P2 | Chief Executive Officer (CEO) | P9 | Head of IT Governance | C3 | ITCM Consultant |
| P3 | Head of IT Governance | P10 | IT Portfolio Management | C4 | IT Management Consultant |
| P4 | CIO | P11 | IT Controlling | C5 | IT Strategy Consultant |
| P5 | Head of IT Governance | P12 | IT Controlling | A1 | Professor, University of Applied Sciences |
| P6 | Chief Financial Officer (CFO) | P13 | IT Portfolio Management | A2 | Academic Consultant ITCM |
| P7 | Head of IT Governance | C1 | CEO, ITCM Tool Supplier | A3 | Academic Consultant IT Management |

For the development of the Delphi questionnaires, we followed the guidelines proposed by Belton et al. [34]: the questionnaire for each round was designed and conducted in the survey tool ZOHO Forms to ensure a fast and efficient procedure. Each draft was then pre-tested to ensure plausibility, comprehensibility, and consistency and refine the questionnaire [34, 36]. Furthermore, each questionnaire was designed for a maximum time exposure of 15 min to limit the fatigue of the participants [34].

## 4.3  Execution of Delphi Survey

Next, we started an online Delphi survey to validate the identified strategic drivers from the exploratory study, complete the list and assess their influence. The execution took place in April and May 2022. It was conducted in iterations of two weeks for each survey round and in between one week of interim analysis.

In the first round of the Delphi survey, we presented the identified cost drivers and their influence deducted from the exploratory study and asked the panel for validation. On a Likert scale with categorical responses (yes/no/don't know), the experts could agree (yes) or disagree (no) on the strategic drivers and on the description of their influence on the level of IT costs. The don't know option was provided to avoid possible misstatements [35]. Furthermore, we included options to provide qualitative remarks on each cost driver. Those remarks allowed us to adjust our descriptions of the drivers and their influence. To complete the list of strategic cost drivers, we provided a qualitative request at the end of the survey for the panel to add missing cost drivers of IT costs. In a third question the expert panel could estimate the intensity of each cost drivers influence on the level of IT costs on a 5-point Likert scale (In your opinion, to what extent can the cost driver influence the level of IT costs? 1 – Very Low,..., 5 – Very High). We then statistically assessed the responses and conducted a qualitative coding of the remarks and additional drivers proposed by the panel [41]. The coding team made various changes on the list of cost drivers, based on the remarks of the panelists [29]; these changes are summarized in Table 2. All in all, the panelists agreed on 9 cost drivers, their influence description, and their influence level (IQD < = 1) in R1.

**Table 2.** Changes made after Round 1

| Cost Drivers | Adaption |
|---|---|
| IT Scope | Adjusted description of cost driver based on remarks |
| Performance Quality | Divided into *personnel quality* and *processes quality* following the remarks |
| Sourcing Quality | Adjusted description of cost driver and extended by a second influence description based on remarks |
| Number of Users | Adjusted description of cost driver and influence based on remarks |
| Organizational Complexity | Adjusted description of cost driver based on remarks |
| IT Setup | Added based on the remarks under category corporate context |
| Strategic Fit | Added based on the remarks under category corporate context |
| Corporate Culture | Removed, as no consensus on a direct influence could be reached and it rather influences diverse strategic cost drivers |

In the second round of the Delphi survey, we provided statistical feedback on the 15 cost drivers from R1. Personal feedback on the qualitative remarks was sent to each panelist privately. We put the adapted cost drivers (Table 2) on vote again in R2 and reached consensus for a total of 17 cost drivers. The panelists provided no further qualitative remarks and we assessed the quantitative responses similar to R1.

Summarizing, our research resulted in 17 drivers clustered in the four categories from the qualitative coding: *service offer of IT, complexity of IT landscape, quality* and *corporate context*. We present the results including the statistics in Table 3. The average influence level (mean ∅) of each factor estimated by the experts in the Delphi survey can range from very low influence on IT costs (1) to a very high influence on IT costs (5). An IQD less or equal to 1 defines consensus on the influence level.

**Table 3.** Strategic Drivers of IT Costs

| Short description (cost driver & cost influence) | Consensus | |
|---|---|---|
| **Service Offer of IT** | **Round** | **Level** |
| **IT Scope** describes the degree of IT coverage within an organization | R2 (100%) | ∅ = 3,68 |
| The higher the degree of IT coverage in an organization, the higher the IT costs | R1 (95%) | IQD: 0,88 |
| **Number of IT-based Products** describes the level of IT in systems or services for internal use or customer products | R1 (100%) | ∅ = 3,6 IQD:0,5 |
| The more systems or customer products are based on IT, the higher the IT costs | R1 (91%) | |
| **Complexity of IT Landscape** | **Round** | **Level** |
| **Technical Complexity** describes the complexity of the landscape in terms of the interaction of the systems in the enterprise architecture as well as the age structure of the landscape | R1 (100%) | ∅ = 4,14 IQD:0,5 |
| The higher the technical complexity of the IT landscape, the higher the IT costs | R1 (95%) | |
| **Standardization** describes the degree of use of standard applications and services in an organization | R1 (95%) | ∅ = 4,05 IQD:0 |
| IT costs decrease until an optimal level of standardization is reached. From this point on, IT costs rise again, because the execution of the processes with standard tools often necessitates corresponding workarounds or additional process steps | R1 (91%) | |
| **Harmonization** describes the degree of consolidation of services, systems and infrastructure components that serve the same purpose | R1 (100%) | ∅ = 3,95 IQD:0,38 |
| IT costs decrease until an optimum level of harmonization is reached. From this point on, IT costs rise again, because excessive harmonization requires a high level of coordination to carry out the processes with unified systems | R1 (100%) | |
| **Quality** | **Round** | **Level** |
| **Personnel Quality** describes the quality and know-how of the personnel from IT and business departments | R2 (100%) | ∅ = 3,11 IQD:0,88 |
| The higher the quality of the personnel, the lower the (additional) IT costs | R2 (86%) | |
| **Process Quality** describes the quality of IT processes. In addition to the classic "plan, build and run" processes, this also applies to management processes and IT service processes | R2 (100%) | ∅ = 3,21 IQD:1,0 |
| The higher the process quality, the lower the (additional) IT costs | R2 (95%) | |
| **Sourcing Quality** describes the quality of IT system purchasing | R2 100%) | ∅ = 3,32 IQD:0,5 |
| 1. The better the quality of sourcing, the lower the (additional) IT costs<br>2. The higher the dependency in sourcing (e.g. negotiating power of suppliers), the higher the IT costs | R1 (95%)<br>R2 (89%) | |
| **System Quality** describes the quality of the individual IT systems | R1 (100%) | ∅ = 3,09 IQD:1,0 |
| The higher the system quality, the lower the IT costs (e.g. for operating and maintaining systems, etc.) | R1 (86%) | |

*(continued)*

**Table 3.** (*continued*)

| Short description (cost driver & cost influence) | Consensus | |
|---|---|---|
| **Quality** | **Round** | **Level** |
| **Service Level** describes the performance level that is defined and required for IT services, including attributes such as availability, flexibility, and speed | R1 (95%) | ⌀ = 3,41 IQD: 0,5 |
| The higher the required service level, the higher the IT costs | R1 (95%) | |
| **Security Requirements** describe the demands for IT security management in the organization and the preventive measures | R1 (100%) | ⌀ = 3,78 IQD:0,5 |
| The higher the security standard and prevention requirements, the higher the IT costs | R1 (100%) | |
| **Compliance Requirements** describe the demand for compliance defined both externally and by the corporate code | R1 (95%) | ⌀ = 2,82 IQD:0,5 |
| The more compliance requirements, the higher the IT costs | R1 (95%) | |
| **Corporate Context** | **Round** | **Level** |
| **Number of Users** describe the number of people using IT systems | R2 (86%) | ⌀ = 3,14 IQD:0,38 |
| The more users, the higher the IT costs (for (software) licenses, hardware equipment, user support etc.) | R2 (100%) | |
| **Organizational Complexity** describes the organizational structure and complexity | R2 (95%) | ⌀ = 2,73 IQD:1,0 |
| The more complex the organizational structure, the higher the IT costs | R1 (91%) | |
| **Business Infrastructure** describes the number of locations of an organization, the global structure and the "global footprint" of an organization | R1 (91%) | ⌀ = 3,45 IQD:0,5 |
| The more complex the business infrastructure, the higher the IT costs | R1 (91%) | |
| **IT Setup** describes the structural setup of IT departments in the organization | R2 (100%) | ⌀ = 3,37 IQD:0,5 |
| The more diverse the IT structure, the higher the (additional) IT costs | R2 (84%) | |
| **Strategic Fit** describes the fit of strategy, structure, and culture | R2 (84%) | ⌀ = 3,0 IQD:1,0 |
| The better the alignment of strategy, structure, and culture in an organization, the lower the (additional) IT costs | R2 (79%) | |

## 5 Discussion

The results from the Delphi study in Table 3 show that apart from strategic decisions that are related to IT at the first sight (*service offer of IT* and the *complexity of the IT landscape*), also strategic decisions regarding the *corporate context* or *quality* have an influence on the level of IT costs.

As a first category of strategic drivers, we identified the *service offer of IT*. In IS cost driver literature, the service offer is mainly related to the size of systems [17], number of features and project duration [24, 25] or the number and size of backlog items [18]. However, for a broad and strategic consideration of IT within an organization, experts proposed the differentiation between *IT scope* and *number of IT-based products*.

The increasing *service offer of IT* often results from management decisions following the industry standard or other external influences: one interviewee explains that IT-based products are especially relevant for a digital industry standard and thereby implies higher

IT costs. A current study stresses this rise of IT costs as they become an increasing part of the total cost within organizations [10]. However, higher investments in IT may also lead to a decrease of overall cost due to automatization and staff savings. This shows that the overall understanding of cost behavior demands to also understand the complex interplay of those cost drivers [13]. Future research should therefore study those interrelations and their impact on the level of cost.

The results show three strategic drivers of IT costs in the category *complexity of the IT landscape*: *technical complexity*, *standardization*, and *harmonization*. The taxonomies on strategic cost drivers of the 1980s consider complexity on the level of a single IT system as a cost driver for manufacturing [5, 13, 43] and equally IS research defines tactical IT cost drivers related to complexity: complexity of the software itself [17], of the type of client [44] or activity [18] as well as related to the number of function points [26]. Our results stress the influence of the overall landscape. The high degree of consensus between the panelists ($IQD \leq 0,5$) highlights this interrelation between the complexity of the IT landscape and the overall IT cost. The differentiation into three strategic drivers stresses that the *complexity of the IT landscape* may have different influence.

*Technical complexity* causes the highest level of influence on IT costs. Thereby, experts mention decisions on the compatibility of legacy systems or new systems within the landscape or the interplay between systems of different technologies. *Standardization* as well as *harmonization* are measures to reduce complexity and thereby are drivers to save IT costs [9]. However, the experts specify that if the level of harmonization or the application of standards exceed the optimum, IT costs increase due to a higher coordination effort, workarounds or even the emergence of shadow IT. A study by Zimmermann and Rentrop [45] confirms this interrelation between the emergence of shadow IT and harmonization. According to one expert, the right balance between individual needs and the standard solution is therefore crucial in these strategic decisions.

The category *quality* includes seven strategic drivers. *Personnel quality* and *process quality* as strategic drivers of IT costs show similarities to the structural factor *experience* proposed by Shank [13], however experience is not specifically related to IT processes or digital skills. The strategic driver *process quality* relates to the definition, execution as well as the continuous improvement of processes. This includes the classical IT processes of "plan, build, run" [9] but also governance or management processes. In IS literature the number of resources, the resource rate [18] and the skill diversity [17] are identified as IT cost drivers. However, our taxonomy explicitly includes the quality of business and IT *personnel*. One expert specifies the skill of an employee to adapt to the situation and use the know-how while another expert stresses the leadership skills of managers. However, organizations struggle to find talents and therefore plan further investments for new talents and skills [20] and consequently the influence of this strategic driver demands detailed further analysis.

Furthermore, we identified the strategic driver *sourcing quality*, adapted from tactical IT cost drivers such as number of suppliers, contracts or the number of requisitions proposed in IS research [24]. In addition to this quantitative focus, our result shows a focus on the cooperation with suppliers and the quality delivered as characteristics of *sourcing quality*. This becomes increasingly relevant, as many organizations move to the cloud and need an elaborated management of vendors [10]. Another strategic driver is the *service level*. IS research proposes tactical drivers for software cost estimation such as required reliability and reusability [15]. However, the strategic driver *service level* also includes attributes such as flexibility or speed, which become increasingly important in the volatile environment. A current study confirms this by highlighting future IT as strategic asset that must be flexible and fast [46]. Earlier IS research does not explicitly include strategic choices on *system quality, security* and *compliance requirements* that drive IT costs. Our result validates those three as strategic drivers and especially strategic choices on the measures and preventions of IT security strongly influence IT costs (3,78). A recent study [20] confirms the increasing *security requirements* which strongly influence organizations investments in IT.

Globalization and digitization require a change in organizational structures and value creation and therefore lead to strategic management decisions for new alignment [9]. Those strategic decisions have an influence on the level of overall IT costs. We identified five strategic drivers and subsumed them in the category *corporate context*. Existing IS research did not propose IT cost drivers with a special focus on the *corporate context*, apart from the number of users [24]. However, the proposed definition of *number of users* incorporates a broader view including internal and external users.

Besides the *business infrastructure*, the expert panel validates *organizational complexity*. Organizational structure can be divisional, functional, following the matrix organization or else, which hampers a clear description of the influence of *organizational complexity* on IT costs (2,73 with IQD of 1). Therefore, the management within those structures is decisive for the description of the influence level. *Organizational complexity* also includes the number of hierarchical levels that may lead to more complicated communication channels. This is in line with difficulties in cooperation between business and IT due to organizational complexity or separation [9]. It also accounts for the structural organization and global footprint of IT departments and IT coordination costs rise with a global dispersion of IT employees. Therefore, the expert panel proposed another strategic driver: *IT setup*.

*Strategic fit* describes the alignment of strategy, structure, and culture and results from the interplay of strategic direction of an organization together with prevailing structure and processes as well as a corporate culture: experts mention collaboration such as DevOps or agile structures to achieve strategic goal and these collaboration models need to be supported by an according mindset or corporate culture. Also, the role of IT department as cost center or service center needs to be aligned to strategy and supported by the corporate culture. However, good alignment may lead to less IT cost for support but to higher total IT costs due to a higher degree of IT usage. These contrasts reflect the difficulties of the panel to reach consensus on the influence and the level estimation (79% and IQD = 1,0).

The developed taxonomy allows the classification of strategic decision variables that drive IT costs. Each of these strategic cost drivers involves a variety of strategic choices and management decisions [13] as well as tactical and operational drivers [23]. Former cost driver research states that hierarchies or taxonomies to classify cost drivers are an important step in the development of new theory [27], to achieve a comprehensive overview to describe cost behavior and to explain cost and competitive position of organizations [13]. In cost driver research the identification of strategic drivers is followed by their analysis [22] with special focus on determinants and consequences of managerial decisions [3]. Managers take a variety of strategic decisions that are not directly IT-related (e.g. *business infrastructure* and *organizational complexity*) however, they may also lead to an influence on IT costs. The drivers therefore are a strategically important instrument of ITCM to enhance cost awareness at the decision locations in management and to foster managers' consciousness for the consequences of their decisions on IT costs. Likewise, the consequences of strategic decisions are generally considered to be long-term and errors in the management of strategic drivers may lead to cost issues [22]. Concerning IT costs, the presented drivers advance strategic ITCM and through this enable benefits such as the identification of saving potentials on a long-term basis.

Furthermore, an important part in IS research is the contribution to business value. Earlier studies already highlighted this relation between cost drivers, customer value, revenue and profitability [6]: they present a model on decision-making based on underlying strategy and ending with profitability and shareholder value in which the drivers represent possible strategic decision variables. The strategic cost drivers presented can therefore also offer a way to show interrelations between the value proposition of IT across the entire value chain and the management decisions to accomplish strategy. Our taxonomy therefore represents a basis for further cost driver analysis and research.

## 6 Conclusion

The identification of IT cost drivers is the first step towards cost driver analysis and thus contributes to a better understanding of the causes of costs as well as the impact of strategic decisions. The goal of this study therefore is to identify strategic drivers of IT costs and to assess their influence. To this end, we conduct an exploratory study including a literature review, expert interviews as well as an expert workshop to identify a first set of strategic drivers of IT costs and their influence. Next, we select experts and prepare a questionnaire for the execution of the Delphi survey to establish a comprehensive and verified overview of those strategic cost drivers. This taxonomy shows that strategic decisions regarding the *service offer of IT*, the *complexity of the IT landscape, quality,* and the *corporate context* drive IT costs on various influence levels. The 17 strategic drivers allow a classification of tactical and operational cost drivers as well as strategic decisions and thereby foster the understanding of cost behavior as well as the explanation of cost and competitive position of IT within organizations.

This paper makes a theoretical contribution by providing a comprehensive and validated overview on strategic drivers of IT costs. Moreover, practitioners gain awareness of levers to influence IT costs and consequences of managerial decisions on their IT spend. In addition, the holistic representation enables the identification of cost reduction points and can be a basis for the evaluation of IT contribution to business value.

However, our study has limitations. The selection of experts is a key stage in Delphi studies [34]. For our study, we chose different experts for the interviews in the exploratory study and the expert panel of the Delphi survey with diverse backgrounds and experience and from different industries and company sizes. However, those, in total, 36 experts from a German-speaking context were not randomly chosen. Thus, the results may not be generalized for all corporate contexts and future research could consider other cultural backgrounds. Furthermore, the study was mainly qualitative. The use of formal mathematical models or quantification in the context of data analyses could provide more detailed information about the impact on the level of IT costs. In addition, the identified cost drivers have a direct impact on IT costs, however, *corporate culture* was excluded due to its indirect influence. The investigation of further indirect influencing factors could expand the picture and contribute to a consideration beyond those examined managerial decisions variables. Nevertheless, the identification of the strategic cost drivers of IT and the given overview are important steps in IT cost driver research and our research serves as a foundation for future research.

The next steps to advance this research are the further analysis of the identified strategic cost drivers of IT and their interrelations as well as the resulting cost behavior. Furthermore, the understanding of managerial decisions that impact the drivers enhances the cost awareness and thereby builds a basis for business IT alignment and strategic cost management. In a third step the relation between strategic cost drivers and business value of IT across the entire value chain should be investigated.

# References

1. Egle, U.: IT-Kostenmanagement. Studie zum Kostenmanagement und zur IT bei Schweizer Unternehmen, 1st edn. GRIN-Verlag, München (2008)
2. Horngren, C.T., Datar, S.M., Rajan, M.V.: Cost accounting. A managerial emphasis, 15th edn. Always learning. Pearson, Boston, MA (2015)
3. Banker, R.D., Byzalov, D., Fang, S., Liang, Y.: Cost Management Research. JMAR, **30**, 187–209 (2018)
4. Bjørnenak, T.: Understanding cost differences in the public sector—a cost drivers approach. Manage. Account. Res. **11**, 193–211 (2000)
5. Porter, M.E.: The Competitive Advantage: Creating and Sustaining Superior Performance. The Free Press, New York, NY (1985)
6. Banker, R.D., Johnston, H.H.: Cost and profit driver research. In: Chapman, C.S., Hopwood, A.G., Shields, M.D. (eds.) Handbooks of Management Accounting Research, vol. 2, pp. 531–556. Elsevier (2007)
7. Carr, C., Tomkins, C.: Strategic investment decisions: the importance of SCM. A comparative analysis of 51 case studies in U.K., U.S. and German companies. Manage. Account. Res. **7**, 199–217 (1996)
8. Lord, B.R.: Strategic management accounting: the emperor's new clothes? Manage. Account. Res. **7**, 347–366 (1996)
9. Urbach, N., Ahlemann, F.: IT Management in the Digital Age. A Roadmap for the IT Department of the Future. Management for Professionals. Springer, Cham (2019)
10. Flexera: State of Tech Spend Report (2021). https://info.flexera.com/SLO-REPORT-State-of-Tech-Spend. Accessed 2 Feb 2022
11. Grytz, R., Krohn-Grimberghe, A.: Service-oriented cost allocation for business intelligence and analytics: who pays for BI&A? In: HICSS 2017 Proceedings, pp. 1043–1052 (2017)
12. Kappelman, L., et al.: IT Trends Study 2021. Issues, Investments, Concerns, & Practices of Organizations and their IT Executives (2020). https://trends.simnet.org/trends-study-archive. Accessed 2 Feb 2022

13. Shank, J.K.: Strategic cost management: new wine, or just new bottles? JMAR **1**, 47–65 (1989)
14. Cooper, R., Kaplan, R.: The Design of Cost Management Systems: Text, Cases, and Readings, 2nd edn. Prentice Hall, Englewood Cliffs, NJ (1998)
15. Boehm, B., Clark, B., Horowitz, E., Westland, C., Madachy, R., Selby, R.: Cost models for future software life cycle processes: COCOMO 2.0. Ann. Soft. Eng. **1**, 57–94 (1995)
16. Singh, C., Sharma, N., Kumar, N.: Analysis of software maintenance cost affecting factors and estimation models. IJSTR **8**, 276–328 (2019)
17. Benaroch, M.: Understanding factors contributing to the escalation of software maintenance costs. In: ICIS 2013 Proceedings, Milan, Italy, pp. 3125–3139 (2013)
18. Pramono, E.A., Suryani, E.: Designing cost measurement system in a small scrum-based software company using activity-based costing model. In: ICOIACT, pp. 943–947 (2019)
19. Herzfeldt, A.: Untersuchung der Profitabilität von IT-Lösungen. Eine Praxisstudie aus Anbietersicht. Springer Gabler, Wiesbaden (2015)
20. State of the CIO. Executive Summary. IDG Inc. (2022). https://foundryco.com/tools-for-mar keters/research-state-of-the-cio/. Accessed 21 May 2022
21. Foster, G., Gupta, M.: Manufacturing overhead cost driver analysis. J. Account. Econ. **12**, 309–337 (1990)
22. Kajüter, P.: Proaktives Kostenmanagement. Konzeption und Realprofile. Gabler, Wiesbaden (2000)
23. Brokemper, A.: Strategieorientiertes Kostenmanagement. Controlling-Praxis. Vahlen, München (1998)
24. Raz, T., Elnathan, D.: Activity based costing for projects. Int. J. Project Manage. **17**, 61–67 (1999)
25. Ooi, G., Soh, C., Lee, P.-M.: An Activity Based Costing Approach to Systems Development and Implementation. In: ICIS 1998 Proceedings, p.35 (1998)
26. Raghu, T., Chaudhury, A.: Modeling IS activities for business process reengineering: a colored petri net approach. In: AMCIS 1995 Proceedings, p.125 (1995)
27. Kaplan, R.S.: Innovation action research: creating new management theory and practice. JMAR **10**, 89–118 (1998)
28. Strasser, A.: Delphi method variants in IS research: a taxonomy proposal. In: PACIS 2016 Proceedings, p.224 (2016)
29. Plessius, H., van Steenbergen, M.: A study into the classification of enterprise architecture benefits. In: MCIS 2019 Proceedings, p.33 (2019)
30. Bacon, C.J., Fitzgerald, B.: A systemic framework for the field of information systems. SIGMIS Database **32**, 46–67 (2001)
31. Hehn, J., Uebernickel, F., Herterich, M.: Design thinking methods for service innovation - a delphi study. In: PACIS 2018 Proceedings, Japan, p.126 (2018)
32. Schmidt, R.C.: Managing Delphi surveys using nonparametric statistical techniques. Decis. Sci. **28**, 763–774 (1997)
33. Rowe, G., Wright, G., Bolger, F.: Delphi: a reevaluation of research and theory. Technol. Forecast. Soc. Chang. **39**, 235–251 (1991)
34. Belton, I., MacDonald, A., Wright, G., Hamlin, I.: Improving the practical application of the Delphi method in group-based judgment: a six-step prescription for a well-founded and defensible process. Technol. Forecast. Soc. Chang. **147**, 72–82 (2019)
35. Häder, M.: Delphi-Befragungen. Ein Arbeitsbuch, 3rd edn. Springer VS, Wiesbaden (2014)
36. Kluge, U., Ringbeck, J., Spinler, S.: Door-to-door travel in 2035 – A Delphi study. Technol. Forecast. Soc. Chang. **157**, 120096 (2020)
37. Diamond, I.R., et al.: Defining consensus: a systematic review recommends methodologic criteria for reporting of Delphi studies. JCE **67**, 401–409 (2014)

38. Cooper, H.M.: Scientific guidelines for conducting integrative research reviews. Rev. Educ. Res. **52**, 291–302 (1982)
39. Cooper, H.M.: Synthesizing Research. A Guide for Literature Reviews, 3rd edn. Sage Publications, Thousand Oaks (1998)
40. Webster, J., Watson, R.: Analyzing the past to prepare for the future: writing a literature review. MIS Q. **26**, xiii–xxiii (2002)
41. Corbin, J.M., Strauss, A.L.: Basics of Qualitative Research: Techniques and Procedures for Developing Grounded Theory, 4th edn. Sage Publications Inc, Thousand Oaks, CA (2015)
42. Saunders, M.: Research Methods for Business Students, 8th edn. Pearson, Harlow (2019)
43. Riley, D.: Competitive cost based investment strategies for industrial companies. Manuf. Issues (1987)
44. Hamel, F., Herz, T.P., Uebernickel, F., Brenner, W.: Towards a better understanding of IT cost drivers of asset management companies. In: Conf-IRM 2012, Vienna, Austria, pp. 587–598 (2012)
45. Zimmermann, S., Rentrop, C.: On the emergence of shadow IT – A transaction cost-based approach. In: ECIS 2014 Proceedings, Tel Aviv, Israel (2014)
46. 2022 CIO Agenda. Create an Action Plan to Master Business Composability. Gartner Inc. (2021)

# The Digital Operational Resilience Act for Financial Services: A Comparative Gap Analysis and Literature Review

Anita Neumannová[1]([✉]) [iD], Edward W. N. Bernroider[1] [iD], and Christoph Elshuber[2]

[1] Institute for Information Management and Control, Vienna University of Economics and Business (WU Vienna), Welthandelsplatz 1/D2/C, 1020 Vienna, Austria
{anita.neumannova,edward.bernroider}@wu.ac.at
[2] NTT DATA Deutschland GmbH, Hans-Döllgast-Straße 26, 80807 Munich, Germany
christoph.elshuber@nttdata.com

**Abstract.** Regulatory bodies, driven by enhanced speed of digital transformations, seek to strengthen the resilience of information and communication technologies (ICT) to ensure their operational integrity. As a result, The Digital Operational Resilience Act (DORA) was recently proposed to unify and enhance ICT risk management of financial institutions by recommending stricter rules. ICT risk management has to date been mainly governed by ISO 27001:2013 standard in the context of information security governance. Based on qualitative content analysis, we firstly mapped ISO 27001:2013 to DORA and identified nine gaps in ISO 27001:2013 in relation to six general DORA requirements. While we find sufficient support in academic literature for six of the nine extensions suggested by DORA, three areas seem less supported: Threat-led penetration testing, major incident management, and ICT third-party risk management. We argue that these topics should serve academic interest to further our understanding of digital operational resilience in theory and practice.

**Keywords:** Digital Operational Resilience · DORA · ISO 27001:2013 · Mapping analysis · IT risk management

## 1 Introduction

To account for the increased number of digital transformations and to ensure the secureness of financial industry, European policy makers have recently introduced the Digital Operational Resilience Act (DORA) for financial services [1]. European policy makers have long been aiming to unify regulations in financial institutions, especially those connected to information and communication technologies (ICT) risks (e.g., digital operations) [2]. ICT risk and security standards are embedded in existing information technology (IT) governance frameworks (e.g., ITIL, COBIT) or certification standards (e.g., ISO) [3, 4 5], which are utilized by numerous financial institutions. Nevertheless, the extant IT governance frameworks or standards have varying focuses [6, 7], which

M. Papadaki et al. (Eds.): EMCIS 2022, LNBIP 464, pp. 570–585, 2023.
https://doi.org/10.1007/978-3-031-30694-5_40

complicates the assessment of DORA's impact on the ICT risk management rules in financial institutions.

According to academic literature, IT governance can either prescribe strategies connected to leadership involvement [7, 8] or can set specific process steps to guard operations [6, 9]. DORA unifies regulations connected to ICT risk management, cybersecurity, third-party risk exposure, and brings to the forefront the need for leadership involvement to portray resilience. Albeit DORA is primarily focused on securing the digital operations of financial institutions, its integration of leadership involvement provides the most comprehensive directive on ICT risks to date.

Financial institutions, to safeguard digital operations, are known to possess the ISO 27001:2013 certification which specifically addresses ICT risk management [10], and which has been used as a baseline for DORA's requirements [11]. Nevertheless, ICT risk management and its specifics have been scarcely addressed as attributes of resilience by the information systems (IS) literature [12]. Consequently, more research is warranted to show how well DORA relates to both, current standards in practices and known attributes of resilience in the IS literature. We seek to uncover DORA's relation with ISO 27001:2013 and examine its requirements from the perspective of prior resilience focused studies. In order to do so, we seek to address the following research questions:

**RQ1:** *How do ICT risk management requirements of DORA compare to ICT security management standards of ISO 27001:2013?*

**RQ2:** *How are (if any) extended DORA requirements supported by IS literature on digital resilience?*

To answer these questions, firstly, we extended a mapping method previously used in a similar context [13, 14, 15] by implementing a comparative qualitative content analysis from medical studies [16] in order to map DORA's requirements to ISO 27001:2013. Secondly, we have aligned DORA's requirements to resilience attributes considered in prior IS literature. Our findings are relevant to both academia and practice in relation to ICT risk management and IT auditing, especially to those trying to understand implications and the rationale of new policy measures to safeguard the ICT of financial institutions.

## 2  Conceptual Background

### 2.1  Digital Operational Resilience Act (DORA)

The need to strengthen the resilience of firms has been a focal point of discussions not only among academia, but also practitioners and policy makers [17, 18]. Policy makers and practitioners mainly attribute resilience to ICT security standards (e.g., ICT risk management principles) [1, 19, 20]. Within the financial industry, the directives partly addressing ICT risk management can be traced back to the 2008 due to the derailing events of 2008 Financial Crisis. The financial crisis underlined the need to strengthen the financial standing of numerous countries which resulted in ongoing regulatory advances aiming to strengthen resilience of financial institutions. Albeit, as became highly visible, these regulations omitted the digitalization advances and only focused on operational resilience – not the specifics of ICTs and risks associated with their implementation [2]. The crisis driven by Covid-19 pandemic shed light onto the risk aspects concerning digital transformations or digital operations in the financial industry, as the switch to fully remote work opened doors to numerous cyber-attacks as well as increased ICT vulnerabilities [21]. This impacted the stability and integrity of the European financial industry, prompting the European Union (EU) to establish detailed and comprehensive framework to maintain operational resilience of ICT – i.e., Digital Operational Resilience Act (DORA) [1].

The general aim of DORA is to ensure consistency in ICT risk management throughout the financial industry and to introduce the concept of digital operational resilience (DOR). As a result, its implications will not only affect large incumbents, but also high-tech growing enterprises (i.e., FinTechs) and their partners [1, 20]. Nevertheless, DORA's diverse requirements do not include the focus on microenterprises, which are, for this purpose, defined as firms employing less than 10 employees, and whose annual turnover or balance sheet does not exceed EUR 2 million [22]. The implementation of DORA is intended to improve and standardize ICT risk management, ICT-related incident reporting, in-depth auditing of ICT systems, and the oversight of critical third-party ICT risks [20]. Additionally, it strives to raise awareness of cyber risks and ICT-related incidents among upper management and supervisory authorities [2].

To ensure digital operational resilience (DOR), DORA stands on six main pillars with each having its own requirements towards financial institution (please see Table 1). Albeit, in its current draft DORA stands on five pillars, we list ICT governance separately to emphasize its importance for resilience [23, 24]. Firstly, DORA explicitly defines the requirements on the management body regarding ICT risks management. The internal governance and control frameworks have to be deployed and the management body shall be accountable for defining, approving and monitoring all arrangements regarding the ICT risk management framework. Moreover, DORA calls for state-of-the art ICT systems, asset inventory and initial audits to ensure the protection and prevention of ICT risks. It further stipulates the need to map the risks both internally and with external service providers. Next, DORA defines processes for ICT incident management. It distinguished two types of incidents (normal and major), and further defines steps on reporting major incidents to government, media, and partners, as well as their supervisory feedback. DORA further underlines the need for regular testing of performance, ICT tools and systems to increase resilience of financial institutions. It states the requirements on threat-led penetration testing and adds advanced testing as an additional part financial institutions have to account for. DORA greatly calls for the assessment of outsourcing agreements and sets their key contractual provisions. In lieu, it provides a framework for critical ICT third-party providers' assessment which is to be defined and implemented by European policy makers. Lastly, it describes information sharing agreements with competitors and partners in regard to cyber intelligence and threats. Altogether these requirements portray DORA's explicit operational demands to ensure ICT risk management of financial institutions.

## 2.2 Information Security Management Systems (ISMS) – ISO 27001:2013

The regulatory directives governing ICT security or risk management (e.g., DORA) are known to draw information from existing certification standards such as the international standards organization (ISO) [11] or the popular IT governance frameworks of ITIL or COBIT [3, 25]. With respect to operation secureness and resilience, ISO offers numerous

Table 1. DORA's main requirements

| Requirement | Aim | DORA Articles |
|---|---|---|
| R1: ICT governance | To set requirements on internal governance and control decisions | Art. 4 |
| R2: ICT risk management | To provide boundary conditions to prevent ICT risk exposure and ensure recovery methods | Art. 5–14 |
| R3: ICT-related incidents management, classification and reporting | To prescribe adequate incident response and communication management | Art. 15–20 |
| R4: Digital operational resilience testing | To set controls for regular audits, performance test and penetration tests | Art. 21–24 |
| R5: Managing of ICT third-party risk | To lessen external risk exposure and assess contractual provisions on ICT third-party providers | Art. 25–39 |
| R6: Information-Sharing Agreements | To prescribe regulations on information exchange regarding cyber threats and intelligence | Art. 40 |

certification choices firms and financial institutions can choose from. Firstly, the choice can range from business capacity management (e.g., ISO 22301:2019), resilience management (e.g., ISO 22313:2020) or even ICT risk management (e.g., ISO 27005). The most comprehensive measure on ICT risk management and operation secureness, however, comprises of ISO 27001:2013 [26]. This certification encompasses every section of information security related measures and has been known to be widely used across numerous financial institutions [27].

Prior to DORA's introduction, the European Banking Authority (EBA) within its final report on ICT risk management, in addition, proposed to tailor ISO 27001:2013 as a baseline on ICT risk management [11]. We, therefore, suggest that the possession of ISO 27001:2013 can be viewed as a standard for ICT risk management of EU financial institutions and shall be contextually comparable to the requirements of DORA. Nevertheless, academic circles argue ISO guidelines are generic in scope and based on universal principles to make them applicable to a wide set of organizations. They are not context-specific and, consequently, cannot account for all aspects and needs of various financial institution [28]. Altogether, it is imperative to understand the support among ISO 27001: 2013 guidelines and DORA's requirements, as DORA is contextually specific to the needs of digitalization advances and their secureness in the financial industry.

## 2.3 Resilience in the IS Literature

Notwithstanding, academics underline resilience as a research concept has been highly fragmented across various research domains, portraying its multidisciplinary aspect [12].

Research on resilience was further influenced by different levels of analysis, consideration of regulatory requirements [29]. Presently, due to the strain on ICT in numerous organizations driven by Covid-19 pandemic and with the introduction of DORA, a new definition emerged – digital operational resilience (DOR). According to policy makers, DOR is defined as "*the ability of (a financial entity) to build, assure and review its operational integrity from a technological perspective*" [1]. Arguably, this new definition brings to the forefront the need to both safeguard and utilize IS to ensure the resilience of the organization as a whole.

According to DORA, in crisis situations IS and their operability need to be safeguarded through ICT risk management and IT governance [19] to portray DOR. Two topics which have rarely been emphasized in prior resilience literature in the IS research [12, 23]. For instance, Sarkar et al. [8, 23] aimed to link IT governance to resilience, however, their focus has been on leadership involvement and commitment as opposed to specific operational secureness practices which are the core of DORA. The focus on securing or recovering operations and their connection to resilience can be, to a certain degree, visible in the research of business continuity plans (BCP) and disaster recovery plans (DRP) [30, 31]. Albeit both Baham et al. [30], Sakurai and Chughtai [31] underline BCP and DRP mainly address the recovery of complex IS as opposed to ensuring their operability. Consequently, this opens questions regarding resilience attributes postulated by academics and their coverage of DORA's requirements.

## 3   Methodology

An effective approach to capture and compare two texts, especially frameworks, is through a mapping method [13, 14]. These methods are derived from the principles of qualitative content analysis to allow for "*rendering the rich meaning*" [32] and enable document observation to make inferences [33]. To ensure reliability of our analysis, we have followed content analysis standards offered by IS research [34, 35], which we enriched by considering comparative content analysis in healthcare research [16, 36]. These additions include the closeness of content and its interpretation among two or more texts [37]. The closeness of content is either manifested (text closeness in wording or interpretation) or latent (describes distance from the text while still rendering close interpretation) [36, 37].

Overall, comparative content analysis should stand on procedural steps which ensure its reliability [38, 39]. Following Nasir [34], firstly, we have chosen the texts to be examined (DORA and ISO 27001:2013). Secondly, we have selected the units of analysis. In DORA, we have identified single articles (4–40) which address ICT security management and in ISO 27001:2013 we have utilized its division into single controls and annexes. Thirdly, we have identified the theme categories. The themes have been determined by the titles of selected DORA's articles. Next, we have pinpointed key aspects of each DORA article and searched for these key aspects within ISO 27001:2013 [35]. Following Barello et al. [16], we have firstly compared the manifested key aspects (i.e., searching for exact wording comparison) before proceeding to search for latent content (i.e., searching for the synonyms of key aspects). This is a step visible in mapping analyses of IT Governance frameworks [14], and which allowed us to render a richer meaning as not every definition, description and interpretation is identical in each of the texts. For example, considering third-party risk exposure, DORA classifies this as *"third-party or external vendors"*, whereas ISO uses the terminology of *"suppliers or supplier services"*. Following, we have categorized each DORA article dependent on the level of support by ISO 27001:2013 controls. When a majority of DORA's key aspects have been supported, we identified the article as largely covered, when part of the DORA's article key aspects could not be mapped, we have classified this article as partly covered, and when no ISO 27001:2013 controls could be mapped to key aspects, we have classified this article as not covered by ISO 27001:2013. Next, we have determined differences among the two texts by inferring which DORA's key aspects were only partly, or not at all supported by ISO 27001:2013 controls. As a last step, the differences were summarized into overall gaps.

To purport and justify our findings in relevance to extant resilience attributes in IS literature, we have surface-searched for peer-reviewed articles, which addressed resilience as well as the identified differences per each gap. Following prior non-exhaustive literature reviews in the IS research [40, 41], we have focused on the following major IS journals and IS conferences: BISE, CAIS, EJIS, ISJ, ISR, JAIS, JIT, JMIS, JSIS, MISQ, MISQe, PAJAIS, AMCIS, ECIS, HICSS, ICIS, PACIS [12, 41]. To assess these outlets, we have queried AIS electronic library and EBSCO Business Premier databases, and searched for the following phrases: "resilient OR resiliency OR resilience" (abstract) AND *"identified difference"* (all text) [12, 41]. We have accumulated number of results for each query as hits and assessed them for final consideration per each gap (please see Table 2). Firstly, we have excluded short-papers, research-in-progress papers, as well as articles stating matters or opinions [41]. Subsequently, we focused on articles on resilience, and which addressed, even if marginally, the differences our prior mapping analysis has identified. Altogether, as some of the final articles addressed more than one gap, we have collected 18 articles spread among 8 outlets out of the 17 given above for our subsequent analysis (please see Table 3).

**Table 2.** Literature search

| Gap | Search Term | Hits | Final |
|-----|-------------|------|-------|
| G1 | Leadership involvement | 33 | 10 |
| | Leadership commitment | 21 | |
| G2 | Risk identification | 27 | 9 |
| | Risk mapping | 29 | |
| G3 | Back-up policy | 13 | 5 |
| G4 | Incident review | 27 | 6 |
| | Incident documentation | 21 | |

*(continued)*

**Table 2.** (*continued*)

| Gap | Search Term | Hits | Final |
|-----|-------------|------|-------|
| G5 | Incident communication | 27 | 5 |
|    | External communication | 48 |   |
| G6 | Major incident | 24 | 0 |
| G7 | Penetration test | 8 | 1 |
| G8 | Third-party risk | 22 | 0 |
|    | Outsourcing risk | 10 |   |
| G9 | Information sharing | 90 | 3 |

**Table 3.** Number of relevant articles per outlet

| Journals | | |
|----------|----|------|
| CAIS | ISJ | MISQ |
| 2 | 1 | 2 |
| MISQe | | PAJAIS |
| 2 | | 1 |
| BISE, EJIS, ISR, JAIS, JIT, JMIS, JSIS | | |
| 0 | | |

| Conferences | | |
|-------------|-------|------|
| AMCIS | HICSS | ICIS |
| 4 | 2 | 4 |
| ECIS, PACIS | | |
| 0 | | |

## 4 Findings

Albeit ISO 27001:2013 is a widely used certification ensuring the ICT security standards of financial institutions, certain DORA's requirements are either partly or not supported by ISO 27001:2013 controls. Overall, we have identified 37 articles (Art. 4–40) in DORA that prescribe its requirements towards ICT risk management of financial institutions. Firstly, out of the 37 Articles, 16 were focused on stating regulations to ESA, EBA, and EU Commission (Articles 14, 18–20, and 28–39). These articles we have omitted from our analysis, as our focus does not comprise of requirements towards policy makers. Secondly, our findings indicate the support of DORA's articles in ISO 27001:2013 is the following: 7 articles were largely mapped (Art. 5–6, 8–10, 21–22), 7 articles were partly mapped (Art. 4, 7, 11–13, 15, 25) and 7 articles could not be mapped (Art. 16–17, 23–24, 26–27, 40) to ISO 27001:2013 controls and annexes. This is mainly driven by abstractedness in the ISO 27001:2013 language when compared to the detailed and descriptive one of DORA. The 14 articles (67% of analyzed articles) that were partly or not mapped to ISO 27001:2013 controls overall culminated in 9 gaps (please see Table 4): Leadership and commitment (Art 4.), Risk mapping and identification (Art. 7), Backup recovery times and policies (Art. 11), ICT incident review and documentation (Art. 12), External incident communication (Art. 13 and Art.15), Major ICT incident (Art. 16–17), Threat-led penetration testing (Art. 23–24), ICT third-party risk assessment (Art. 25–27) and Information-sharing arrangements (Art. 40).

Firstly, the gap on leadership and commitment stems from ISO's 27001:2013 marginal demands to leadership accountability. Even though ISO 27001:2013 controls 5, 9.3, and annex A.7.2.1 require the leadership commitment to a certain degree, majority of other controls prescribe accountability to the organization as opposed to leadership (e.g., controls 6, 7, and 8.1). Next, ISO 27001:2013 controls 4, 6.1.1-.1.2, and 8.2 are not explicit about the timing of internal risk reviews, the mapping of external risk, and the involvement of third-party service providers. This culminates in a gap in risk mapping and identification. The gap on backup recovery times and policies stems from insufficient information on precise incident recovery times, and the handling of third-party service providers in ISO 27001:2013 annexes A.9.1, A.12.2-.3, A.17.2. ICT incident review and documentation discrepancies result from ISO 27001:2013 limited support in controls 9.1, 9.3, 10, and annexes A.12.6.1, A.16.1.6-.7 of post ICT incident reviews, the mapping for future risks, and the specific documentation requirements. Apart from that, ISO 27001:2013 does not ask for communication plans and policies to be aligned both internally and with external partners resulting in a gap on external incident communication. Albeit the external incident communication is partially addressed in ISO 27001:2013 control 7.4 and annex A.16, neither mentions the involvement of governing bodies, media, and external partners. Moreover, ISO 27001:2013 does not define and distinguish major incident in annex A.16. The same applies to threat-led penetration testing which is not supported by a single control or annex. Albeit ISO 27001:2013 marginally addresses ICT third-party risk assessment in controls 4.2, 8.1, and annexes A.14.2.7, A.15, it does not contextually specify the needs or risks associated with "third-party" service providers, does not depict third-party risk assessments or third-party contractual agreements. Lastly, no control or annex in ISO 27001:2013 supports the requirements for information sharing arrangements which culminates in the last observed gap. Our findings overall distinguish DORA's requirements as more complex and explicit when mapped with ISO 27001:2013 controls and annexes.

As a last step, we mapped the uncovered nine gaps of ISO 27001:2013 in relation to DORA to extant resilience attributes within selected IS articles. Our findings indicate that six (G1–5, 9) of the gaps have been addressed by IS research, whereas three (G6–8) were hardly portrayed (please see Table 5): major ICT incidents, threat-led penetration testing, and third-party risk assessment. These three gaps clearly lack academic support and warrant the most attention by IS research according to our analysis. Firstly, albeit Green et al. [42] underlie the need of adequate testing of ICT, they do not stipulate advanced tests or threat-led penetration tests ensure resilience (G7). Secondly, although information on incident management, reporting or communication is addressed within the uncovered articles, neither defines the difference nor distinguishes between major and minor incident resulting in no identified article covering major incidents (G6). The same applies to the gap on third-party risk assessments (G8).

**Table 4.** ISO 27001:2013 support of DORA requirements

| DORA | | ISO 27001 | |
|---|---|---|---|
| Requirement | Key Area | Support | Gaps |
| ICT governance (R1) | Art. 4: Governance and organization | 5.1; 5.2; 5.3; 6.1; 6.2; 7.1; 7.2; 7.3; 7.4; 8.1; 9.2; 9.3; A.5.1; A.6.1; A.7.2.1; A.18.2.2 | Leadership involvement and commitment (G1) |
| ICT risk management (R2) | Art. 7: Identification | 4.1; 4.2; 6.1.1; 6.1.2; 8.2; 9.1; 9.2; A.8.1.1; A.8.2; A.15.1.2; A.15.2.1; A.18.1.1; A.18.2.3 | Risk mapping and identification (G2) |
| | Art. 11: Backup policies and recovery methods | A.9.1; A.12.2; A.12.3; A.17.2 | Back-up recovery times and policies (G3) |
| | Art. 12: Learning and evolving | 9.1; 9.3; 10; A.12.6.1; A.16.1.6; A.16.1.7; A.17.1.3 | ICT incident review and documentation (G4) |
| | Art. 13: Communication | 7.4; A.15.1.3; A.16 | External incident communication (G5) |
| ICT-related incidents (R3) | Art. 15: ICT-related incident management process | A.16 | |
| | Art. 16: Classification of ICT-related incidents | n/a | Major ICT incidents (G6) |
| | Art. 17: Reporting of major ICT-related incidents | n/a | |
| DOR testing (R4) | Art. 23: Advanced testing of ICT tools, systems and processes based on threat led penetration testing | n/a | Threat-led penetration testing (G7) |
| | Art. 24: Requirements for testers | n/a | |
| Managing of ICT third-party risk (R5) | Art. 25: General principles | 4.2; 8.1; A.6.1.3; A.13.1.2; A.13.2.2; A.13.2.4; A.14.2.7; A.15; A.18.1.1 | Third-party risk assessment (G8) |

*(continued)*

**Table 4.** (*continued*)

| DORA | | ISO 27001 | |
|---|---|---|---|
| **Requirement** | **Key Area** | **Support** | **Gaps** |
| | Art. 26: Preliminary assessment of ICT concentration risk and further-outsourcing arrangements | n/a | |
| | Art. 27: Key contractual provisions | n/a | |
| Information-Sharing Agreements **(R6)** | Art. 40: Information-sharing arrangements on cyber threat information and intelligence | n/a | Information-sharing arrangements **(G9)** |

**Table 5.** Mapping of gaps to existing resilience literature in IS

| Gaps | | | | | | | | | |
|---|---|---|---|---|---|---|---|---|---|
| **Articles** | **G1** | **G2** | **G3** | **G4** | **G5** | **G6** | **G7** | **G8** | **G9** |
| [43] | x | x | x | x | | | | | |
| [44] | | x | | | | | | | |
| [45] | | | | | x | | | | |
| [42] | | x | x | x | x | | x | | x |
| [46] | x | | | | | | | | |
| [47] | | x | x | | x | | | | x |
| [48] | x | | | | | | | | |
| [49] | x | | | x | | | | | x |
| [50] | x | | | | | | | | |
| [51] | | x | | x | | | | | |
| [52] | | x | | x | | | | | |
| [53] | | x | | | x | | | | |
| [54] | x | | x | | | | | | |
| [8] | x | x | | | | | | | |
| [23] | x | | | | | | | | |

(*continued*)

**Table 5.** (*continued*)

| Gaps | | | | | | | | | |
|---|---|---|---|---|---|---|---|---|---|
| Articles | G1 | G2 | G3 | G4 | G5 | G6 | G7 | G8 | G9 |
| [55] | x | | | | | | | | |
| [56] | x | x | x | x | | | | | |
| [57] | | | | | x | | | | |

## 5  Discussion, Limitations and Future Research

In this study we have provided a mapping analysis of DORA to ISO 27001:2013, and subsequently to attributes of resilience addressed by the IS literature. Understanding the regulatory extensions made by DORA is important, as its implementation will have a profound impact on numerous financial institutions and their partners. Firstly, our comparative review makes clear that the possession of ISO 27001:2013 certification or the aim to obtain it, cannot be seen as equivalent in terms of complying with DORA as their discrepancies look substantial given the gaps identified. The mapping analysis portrays that 33% of DORA's analyzed articles are partly covered within the ISO 27001:2013, and 33% of DORA's articles cannot be clearly linked to a single ISO 27001:2013 control or annex. Overall, we have identified nine gaps, which should be of interest to practitioners analyzing financial institutions preparedness for DORA. The majority of the identified difference leading to the gaps appear to stem from the lacking presence of a detailed leadership involvement in ISO 27001:2013 controls and annexes (e.g., G1–3, 5, 8–9) [6, 13].

Secondly, we have reflected on the coverage of the identified gaps of ISO 27001:2013 in relation to DORA in resilience focused IS literature. Results show that differences leading to six of the gaps (G1–5, 9) have been at least discussed in some way among IS academics as resilience attributes. Leadership involvement and commitment (G1) is connected to the presence of leadership-focused IT government frameworks [54, 56] or cyber-security cultures [49]. Leadership involvement can further ensure both risk identification and mapping (G2) as well as strengthened back-up policies (G3) [8, 43, 54]. In addition, back-up policy requirements as well as risk identification are implemented through agile principles [44], cybersecurity assessments [42] or BCP/DRP [53], whereas risk mapping steps are attributed to the usage of combined IT governance frameworks [47]. Apart from that, Park et al. [51] state incident tracking software enables future ICT incident reviews, whereas Baham et al. [43] stress the importance of proper ICT incident documentation to portray resilience (G4). IS researchers address external incident communication (G5) through communication protocols [58], DRP [43, 53], situational-crisis communication theory [57], or information secureness [59]. Lastly, Marotta and Pearlson [49] position information sharing (G9) as important trust-building procedure leading to cyber-resilience, and Green et al. [42] addressed both benefits and risk associated with information sharing among competitors, as well as to government. Consequently, this support by prior IS research also justifies the presence of these six extended requirements in DORA to portray DOR (in relation to ISO 27001:2013), and in this sense provides further motivation for practitioners applying DORA respectively.

Notwithstanding, we also contribute by explicating three gaps (G6–8) that were hardly addressed in the considered IS literature related to resilience, which may be contemplated in scoping risk-based control-coverage decisions. Simultaneously, our finding calls for future research to gain more evidence to understand the role of these gaps for achieving DOR. Albeit incident management has been addressed by various academic circles, and has been connected to both firm's absorptive capacity [60, 61], as well as to technical debt [62, 63], the distinction between major and minor incidents (G6) does not appear to be broadly addressed in the IS literature on resilience. This is intriguing as existing IT governance frameworks (i.e., ITIL and COBIT) do distinguish among different incident types [5] indicating relevance for DOR of firms. A promising research area could delve deeper into how specific types of ICT incidents (stemming from DORA, ITIL or COBIT) affect firms and which specific steps firms apply in order to withstand them. Secondly, support in academic literature seems to be lacking in terms of the strengthened demand on the management of third-party risk exposure (G8). Under DORA, external partners, cloud vendors especially, should be heavily guarded [20]. A research area not broadly discussed within our subsequent analysis. Risks associates with third-party involvement are primarily found in research focusing on outsourcing strategies [64], or digital innovations and IS project risks [65]. Future research could perform a more detailed and comprehensive literature review and uncover ingrained stipulations on third-party risk management in connection to DOR. Lastly, threat-led penetration testing (TLPT) or advanced testing (G7) has received more attention by practice-oriented research as opposed to academic research. Even though consulting companies address TLPT as a source of lessened ICT risk exposure [66], our subsequent analysis could not find an IS article specifically connecting TLPT to resilience of firms. Future IS research should include these perspectives and specify under which conditions advanced testing or TLPT is imperative for DOR.

# 6 Conclusion

This paper illustrates differences of what DORA requires in addition to a widely used ICT risk management standard (ISO 27001:2013) through a comparative qualitative content analysis. The comparative analysis identified nine gaps which arose due to ambiguous language of ISO 27001:2013, and the strengthened focus on ICT secureness and third-party risk assessment of DORA. The identified gaps have been subsequently related to prior work in IS with a literature review. Our findings support the view that DORA will demand strengthened controls for financial institutions in terms of ICT risk management based on the applied ISO benchmark. However, some of these demands are barely supported by extant IS literature on resilience, which not only questions the validity of the regulatory extensions in DORA from an academic standpoint, but also calls for future research to better substantiate these extensions. The strengthened demands on threat-led penetration testing (TLPT), third-part risk assessment and the division among major vs. minor incident from DORA should receive further support from academic research. On this basis, we suggest that future IS research should strive to examine the roles of these three gaps in both academic and practical settings. In particular, academics could examine the effectiveness of differentiated approaches for handling different types of ICT incidents, the role of third-party exposure in outsourcing strategies or conduct more empirical research on the role of penetration tests in support of DOR. Further comparative work could include other frameworks such as the CERT-RMM model focusing particularly on ICT operational resilience of diverse enterprises.

# References

1. European Commission: Proposal for a regulation of the European Parliament and of the Council on digital operational resilience for the financial sector and amending Regulations (EC) No 1060/2009, (EU) No 648/2012, (EU) No 600/2014 and (EU) No 909/2014, COM/2020/595 final, 2020/0266 (COD) (2020). https://eur-lex.europa.eu/legal-content/EN/TXT/?uri=CELEX%3A52020PC0595. Accessed 03 Oct 2022
2. Kun, E.: From Operational Risk to Systemic Risk: The EU's Digital Operational Resilience Act for Financial Services (DORA) (2021). https://www.law.kuleuven.be/citip/blog/from-operational-risk-to-systemic-risk/. Accessed 03 Oct 2022
3. Vilarinho, S., da Silva, M.M.: Risk management model in ITIL. Sociotechnical Enterprise Information Systems Design and Integration, pp. 207–214. IGI Global (2013)
4. Bradley, R.V., Byrd, T.A., Pridmore, J.L., Thrasher, E., Pratt, R.M., Mbarika, V.W.: An empirical examination of antecedents and consequences of IT governance in US hospitals. J. Inf. Technol. **27**, 156–177 (2012)
5. Pereira, R., Silva, M.M.D.: A literature review: IT governance guidelines and areas. In: Proceedings of the 6th International Conference on Theory and Practice of Electronic Governance), pp. 320–323. Association for Computing Machinery, Albany, New York, USA (2012)
6. Kumsuprom, S., Corbitt, B., Pittayachawan, S.: ICT risk management in organizations: case studies in Thai business. In: ACIS 2008 Proceedings, 98 (2008)
7. O'Donohue, B., Pye, G., Warren, M.: Improving ICT governance in Australian companies. In: ACIS 2006 Proceedings, 53 (2006)
8. Sarkar, A., Wingreen, S.C., Ascroft, J.: Top management team decision priorities to drive IS resilience: lessons from jade software corporation. In: AMCIS 2016 Proceedings, 10 (2016)
9. Jafarijoo, M., Joshi, K.: IT governance: review, synthesis, and directions for future research. In: AMCIS 2021 Proceedings, 5 (2021)
10. So, I.G., Setiadi, N.J., Papak, B., Aryanto, R.: Action design of information systems security governance for bank using COBIT 4.1 and control standard of ISO 27001. Adv. Mater. Res. **905**, 663–668 (2014)
11. European Banking Authority: Final report on EBA Guidelines on ICT and security risk management, EBA/GL/2019/04 (2019). https://www.eba.europa.eu/regulation-and-policy/internal-governance/guidelines-on-ict-and-security-risk-management. Accessed 03 Oct 2022
12. Weber, M., Hacker, J., vom Brocke, J.: Resilience in information systems research-a literature review from a socio-technical and temporal perspective. In: ICIS 2021 Proceedings, 3 (2021)
13. von Solms, B.: Information Security governance: COBIT or ISO 17799 or both? Comput. Secur. **24**, 99–104 (2005)
14. ITGI, OGC, Information Systems Audit and Control Association: Aligning COBIT, ITIL and ISO 17799 for Business Benefit: Management Summary (2005). http://www.itgi.org. Accessed 03 Oct 2022
15. Sheikhpour, R., Modiri, N.: An approach to map COBIT processes to ISO/IEC 27001 information security management controls. Int. J. Secur. Appl. **6**, 13–28 (2012)

16. Barello, S., Graffigna, G., Vegni, E.: Patient engagement as an emerging challenge for healthcare services: mapping the literature. Nutr. Res. Pract. **2012**, 1–7 (2012)
17. Bhamra, R., Dani, S., Burnard, K.: Resilience: the concept, a literature review and future directions. Int. J. Prod. Res. **49**, 5375–5393 (2011)
18. Hillmann, J., Guenther, E.: Organizational resilience: a valuable construct for management research? Int. J. Manage. Rev. **23**, 7–44 (2021)
19. Leo, M.: Operational resilience disclosures by banks: analysis of annual reports. Risks **8**, 128 (2020)
20. Scott, H.S.: The EU's Digital Operational Resilience Act: Cloud Services & Financial Companies (2021). https://ssrn.com/abstract=3904113
21. Lallie, H.S., et al.: Cyber security in the age of COVID-19: a timeline and analysis of cyber-crime and cyber-attacks during the pandemic. Comput. Secur. **105**, 1–20 (2021)
22. European Commission: Commission recommendation of 6 May 2003 concerning the definition of micro, small and medium-sized enterprises, 2003/361/EC (2003). https://eur-lex.europa.eu/LexUriServ/LexUriServ.do?uri=OJ:L:2003:124:0036:0041:en:PDF. Accessed 03 Oct 2022
23. Sarkar, A., Wingreen, S.C., Ascroft, J.: Towards a practice-based view of information systems resilience using the lens of critical realism. In: Proceedings of the 53rd Hawaii International Conference on System Sciences, pp. 6184–6193 (2020)
24. McManus, S., Seville, E., Brunsden, D., Vargo, J.: Resilience management: a framework for assessing and improving the resilience of organisations. Resilient Organisations Research Report University of Canterbury, Civil and Natural Resources Engineering (2007)
25. Ivanov, M., Stefanov, B.: Approaching risk management in IT by implementing it in ITIL. Electrotech. Electronica **49**(9–10), 2–9 (2014)
26. ISO/IEC 27001:2013. https://www.iso.org/standard/54534.html. Accessed 03 Oct 2022
27. van der Stoop, T.: ISO 27001 in the banking industry: "One standard to rule them all" (2009). https://advisera.com/27001academy/blog/2019/11/25/iso-27001-for-banks-a-game-changing-security-investment/. Accessed 03 Oct 2022
28. Siponen, M., Willison, R.: Information security management standards: problems and solutions. Inf. Manage. **46**, 267–270 (2009)
29. Müller, G., Koslowski, T. G., Accorsi, R.: Resilience - a new research field in business information systems? In: Abramowicz, W. (ed.) BIS 2013. LNBIP, vol. 160, pp. 3–14. Springer, Heidelberg (2013). https://doi.org/10.1007/978-3-642-41687-3_2
30. Baham, C., Hirschheim, R., Calderon, A.A., Kisekka, V.: An agile methodology for the disaster recovery of information systems under catastrophic scenarios. J. Manage. Inf. Syst. **34**, 633–663 (2017)
31. Sakurai, M., Chughtai, H.: Resilience against crises: COVID-19 and lessons from natural disasters. Eur. J. Inf. Syst. **29**, 585–594 (2020)
32. Duriau, V.J., Reger, R.K., Pfarrer, M.D.: A content analysis of the content analysis literature in organization studies: research themes, data sources, and methodological refinements. Organ. Res. Methods **10**, 5–34 (2007)
33. Prasad, B.D.: Content analysis. Res. Methods Soc. Work **5**, 1–20 (2008)
34. Nasir, S.: The development, change, and transformation of management information systems (MIS): a content analysis of articles published in business and marketing journals. Int. J. Inf. Manage. **25**, 442–457 (2005)
35. Gallivan, M.J.: Striking a balance between trust and control in a virtual organization: a content analysis of open source software case studies. Inf. Syst. J. **11**, 277–304 (2001)
36. Graneheim, U.H., Lindgren, B.-M., Lundman, B.: Methodological challenges in qualitative content analysis: a discussion paper. Nurse Educ. Today **56**, 29–34 (2017)
37. Graneheim, U.H., Lundman, B.: Qualitative content analysis in nursing research: concepts, procedures and measures to achieve trustworthiness. Nurse Educ. Today **24**, 105–112 (2004)

38. Insch, G.S., Moore, J.E., Murphy, L.D.: Content analysis in leadership research: examples, procedures, and suggestions for future use. Leadersh. Q. **8**, 1–25 (1997)

39. Krippendorff, K.: Content Analysis: An Introduction to its Methodology. Sage publications, Thousand oaks (2018)

40. vom Brocke, J., Simons, A., Riemer, K., Niehaves, B., Plattfaut, R., Cleven, A.: Standing on the shoulders of giants: challenges and recommendations of literature search in information systems research. Commun. Assoc. Inf. Syst. **37**(1), 205–224 (2015)

41. Kohn, V.: How the coronavirus pandemic affects the digital resilience of employees. In: ICIS 2020 Proceedings, 6 (2020)

42. Green, A.W., Woszczynski, A.B., Dodson, K., Easton, P.: Responding to cybersecurity challenges: Securing vulnerable US emergency alert systems. Commun. Assoc. Inf. Syst. **46**, 187–208 (2020)

43. Baham, C., Calderon, A., Hirschheim, R.: Applying a layered framework to disaster recovery. Commun. Assoc. Inf. Syst. **40**, 277–293 (2017)

44. Baskerville, R., Pries-Heje, J.: Achieving resilience through agility. In: ICIS 2021 Proceedings, 8 (2021)

45. Butler, B.S., Bateman, P.J., Gray, P.H., Diamant, E.I.: An attraction–selection–attrition theory of online community size and resilience. MIS Q. **38**, 699–729 (2014)

46. Heeks, R., Ospina, A.V.: Conceptualising the link between information systems and resilience: a developing country field study. Inf. Syst. J. **29**, 70–96 (2019)

47. Junglas, I., Ives, B.: Recovering IT in a disaster: lessons from hurricane katrina. MIS Q. Exec. **6**, 39–51 (2007)

48. Lacity, M.C., Reynolds, P.: Cloud services practices for small and medium-sized enterprises. MIS Q. Exec. **13**, 31–44 (2014)

49. Marotta, A., Pearlson, K.: A culture of cybersecurity at Banca Popolare di Sondrio. In: AMCIS 2019 Proceedings, 24 (2019)

50. Morisse, M., Prigge, C.: Design of a business resilience model for Industry 4.0 manufacturers. In: AMCIS 2017 Proceedings, 4 (2017)

51. Park, I., Sharman, R., Rao, H.R.: Disaster experience and hospital information systems. MIS Q. **39**, 317–344 (2015)

52. Rehm, S.-V., Georg Schaffner, L., Goel, L.: Framing dialogues on cyber-resilience on boards. In: ICIS 2021 Proceedings, 10 (2021)

53. Sakurai, M., Kokuryo, J.: Design of a resilient information system for disaster response. In: ICIS 2014 Proceedings, 5 (2014)

54. Sarkar, A., Traubinger, T.: IS Resilience decision priorities at german smes: a q-method approach. In: AMCIS 2021 Proceedings, 19 (2021)

55. Sarkar, A., Wingreen, S., Ascroft, J., Sharma, R.: Bouncing back after a crisis: lessons from senior management team to drive IS resilience. In: Proceedings of the 54th Hawaii International Conference on System Sciences, pp. 6712–6721 (2021)

56. Sarkar, A., Wingreen, S.C., Cragg, P.: CEO decision making under crisis: an agency theory perspective. Pac. Asia J. Assoc. Inf. Syst. **9**(2), 1–22 (2017)

57. Syed, R., Dhillon, G.: Dynamics of data breaches in online social networks: Understanding threats to organizational information security reputation. In: ICIS 2015 Proceedings, 14 (2015)

58. Carlo, J.L., Lyytinen, K., Boland, R.J.: Systemic risk, IT artifacts, and high reliability organizations: a case of constructing a radical architecture. All Sprouts Content **4**(2), 57–73 (2008)

59. Conklin, W.: Information sharing and emergency services: an examination using information security principles. In: AMCIS 2008 Proceedings, 12 (2008)

60. Grispos, G., Glisson, W.B., Storer, T.: Rethinking security incident response: the integration of agile principles. In: AMCIS 2014 Proceedings, 9 (2014)

61. Mehrizi, M.H.R., Nicolini, D., Mòdol, J.R.: How do organizations learn from information systems incidents? A synthesis of the past, present and future. MIS Q. **46**, 531–590 (2022)
62. Alves, N.S.R., Mendes, T.S., De Mendonça, M.G., Spínola, R.O., Shull, F., Seaman, C.: Identification and management of technical debt: a systematic mapping study. Inf. Softw. Technol. **70**, 100–121 (2016)
63. Nielsen, M.E., Østergaard Madsen, C., Lungu, M.F.: Technical debt management: a systematic literature review and research agenda for digital government. In: Viale Pereira, G. (ed.) EGOV 2020. LNCS, vol. 12219, pp. 121–137. Springer, Cham (2020). https://doi.org/10.1007/978-3-030-57599-1_10
64. Beck, R., Schott, K., Gregory, R.W.: Mindful management practices in global multivendor ISD outsourcing projects. Scand. J. Inf. Syst. **23**(2), 5–28 (2011)
65. Hoermann, S., Schermann, M., Krcmar, H.: When to manage risks in IS projects: an exploratory analysis of longitudinal risk reports. In: Proceeding of 10th International Conference on Wirtschaftsinformatik, pp. 871–880 (2011)
66. Deloitte: Deloitte's Cyber Risk capabilities, Cyber Strategy, Secure, Vigilant, and Resilient. https://www2.deloitte.com/content/dam/Deloitte/ca/Documents/risk/ca-en-risk-advisory-cyber-brochure.pdf. Accessed 03 Oct 2022

# Management and Organizational Issues
# in Information Systems

# Towards Analytical Business Services as a New Business Model for Shared Service Centers

Jörg H. Mayer[1], Maurin Siegmund[1]([✉]), Markus Esswein[2], Reiner Quick[1], and Dirk Keweloh[3]

[1] Darmstadt University of Technology, Hochschulstrasse 1, 64289 Darmstadt, Germany
maurin.siegmund@gmail.com
[2] Henkel AG & Co. KGaA, Henkelstraße 67, 40589 Düsseldorf, Germany
[3] Deutsche Telekom Services Europe SE, Innere Kanalstraße 98, 50672 Köln, Germany

**Abstract.** Forward-looking companies like Google and Baidu have started to complement their efficiency targets with effectiveness goals such as new or better insights. Based on findings from two reference companies, this article ranks a list of initial analytical business services that combine data across different shared service-center domains. Clustered by the beneficiary functions, that are Human Resources, Supply Chain Management, and Sales, we detail three out of twenty-two initially developed analytical business services and highlight how shared service centers can add value through these activities. Based on discussions with functional experts and practitioners about this finding, we conclude that analytical business services should be an essential lever towards new business models for shared service centers.

**Keywords:** Business Services · Shared Service Center · Digital Transformation · Case Study · Design Science Research in Information Systems

## 1 Introduction

Typically, company processes have to pay off with a strong efficiency target [1]. Translating low-cost strategies into action, shared service centers (SSCs) play an important role. They had their origin in the 1980s, when companies started to outsource finance processes [2]. Supported by service contracts, shared services traditionally embrace repetitive, barely value-creating processes from an autonomous unit to multiple other units within the company [3]. Examples include record-to-report (R2R) processes such as intercompany accounting or consolidation.

However, companies have started to complement this efficiency target with effectiveness goals such as new or better insights [4]. That is a radical departure from the current role of SSCs, but companies such as Google and Baidu demonstrate that combining cross-domain data is fundamental for generating new insights. Attaining competitive advantages, these companies are frontrunners through driving analytical business services across domains – and ideally, customers are willing to pay for it.

M. Papadaki et al. (Eds.): EMCIS 2022, LNBIP 464, pp. 589–602, 2023.
https://doi.org/10.1007/978-3-031-30694-5_41

Analytical business services are end-to-end processes that extend single (finance)-domain analyses. They are directly connected to a business transaction such as procurement or sales, and thus deliver new insights and tangible business value. In doing so, they typically access several data sources [5].

Accordingly, SSCs have to go beyond the finance domain with its core processes of order-to-cash (O2C), purchase-to-pay (P2P), and R2R. Sharing the same infrastructure, such multi-tower SSC additionally handle processes from HR or other domains [6]. Leveraging cross-domain data for business services is rare [7]. The reasons are manifold: (1) Multi-tower SSCs cross board departments, thus inducing ownership conflicts [8]; (2) HR data is subject to special privacy guidelines [9]; (3) there is natural resistance by experts to change their own field of expertise without a compelling reason [10].

Given that the integration of enterprise resource planning (ERP) and business intelligence (BI) solutions is currently top ranked at many companies, it makes sense to examine their impact on analytical business services. A win-win situation is emerging. On the one hand, SSCs with a focus on analytical business services will improve the return-on-investment of ERP/BI integration. On the other hand, an ERP/BI single source of truth including analytical business services can be a focal point for SSC rede-signs. The objective of this article is to present initial analytical business services that leverage data across different SSC domains to generate valuable new insights. We take two companies as our reference case and pose the following research questions (RQ):

1. What constitutes analytical business services and how are they currently supported by digital technology?
2. Do the proposed analytical business services fulfill their intention (validity) and are they useful beyond the environment in which they were developed (utility)?

To create things that serve human purposes [11], finally to build a better world [12], we follow Design Science Research in Information Systems [13, 14], for which the publication scheme from Gregor and Hevner [15] gave us direction. We motivate this present article in terms of challenging current SSC lacking valuable analytical business services and how a new ERP/BI generation can support them (introduction). Highlighting several research gaps, we contextualize our research questions (literature review, Sect. 2). Addressing these gaps, we adopt a dual-case study (method, Sect. 3). Emphasizing a staged process with iterative "build" and "evaluate" activities [16], we capture the lessons learned through the portrayal of the most promising analytical business services (artefact description, related to RQ 1, Sect. 4). Testing their validity and utility, we perform expert interviews within and beyond the reference companies (evaluation, related to RQ 2, Sect. 5). Comparing our results with prior work and examining how they relate back to both the article's objective and research questions, we close with a summary, limitations of our work, and avenues for future research (discussion and conclusion, Sect. 6) [17].

## 2 Literature Review

Following Webster and Watson [18] as well as vom Brocke [19, 20], we identified relevant articles in a four-stage process. (1) We focused on leading IS journals as well as selected business, management, and accounting journals, complemented by proceedings from major IS conferences (outlet search). For a practitioner's perspective, we considered journals such as MIS Quarterly Executive and Harvard Business Review. (2) Accessing the identified outlets, we used ScienceDirect, EBSCOhost, Springer Link, and AIS eLibrary (database search). (3) Then, we searched for articles through their titles, abstracts, and keywords (keyword search).

Combining shared services and digital technology, our research yielded about 1200 hits. We supplemented the term shared services with business model, redesign/transformation, and shared service center as well as digital technology with new insights, analytics, machine learning (ML)/artificial intelligence (AI), and data wrangling/data munging. We then filtered the results with reporting and analysis. This search resulted in fourteen relevant papers. Furthermore, we found 11 practitioner articles.

Finally, we conducted a backward and forward search. Following the citation pearl growing approach, we complemented our search whenever we examined new aspects in the retrieved publications [21]. With references from all relevant publications, we identified another ten articles and ended up with 35 publications in total (Fig. 1).

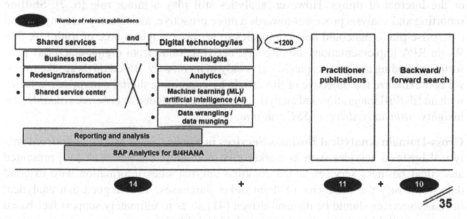

**Fig. 1.** Search strategy

For our gap analysis, we structured the relevant publications into three clusters: (1) We cover the various kinds of motivation behind *SSCs targets*. (2) We divide *digital technologies* threefold, that is automation, analytics, and digital enterprise platform [22, 23]. Articles covering remaining technologies fall into "others." (3) We differentiate *analytical business services* from transactional services.

**Effectiveness as a Complementary Target for SSCs.** Being regularly bench-marked against out-sourcing regarding cost reduction [8], there is a common understanding of efficiency as the main target of SSC designs [3, 24]. In addition, there are complementing targets such as quality improvement or centralized access to expertise [25, 26]. To ensure service quality, service level agreements became a standard [27]. Based on a PWC study [6], the level of standardization is about 61%. Accordingly, gaining efficiency benefits still presents work in progress and is highly relevant.

However, Richter and Brühl [25] argued that SSCs with exclusive efficiency targets often do not provide the expected cost savings and especially fail to leverage big data [28]. Yet, precisely this is currently among the highest priorities for redesigning SSC models – and especially true for multi-tower SSCs [29]. In doing so, Iansiti and Lakhani [30] revealed that leveraging big data for new insights has neither been re-searched nor implemented. This is in line with Deloitte [7] and EY [31], who described the growing interest of companies in generating effectiveness through new insights. Thus, coming to a first finding from our literature review, *we will examine the effectiveness target of SSCs leveraging big data.*

**Supporting Digital Technologies.** Focusing on finance processes, companies have already started to automate their SSC [32]. With their high-volume and standardized processes, robotic process automation (RPA) plays a pivotal role [33, 34], mimicking repetitive human activities such as performing simple bookings [35]. Out of our 35 articles, we identified seven covering the impact of RPA on "reporting and analysis" processes. Cokins [36] and Yaqoob et al. [37] predicted substantial value generation by using analytics, if companies additionally access data from social media listening or the internet of things. However, analytics still play a minor role [6, 7]. Shifting reporting and analysis processes towards a more proactive, anticipatory steering, digital enterprise platforms could help companies to set up a single source of truth [36, 38]. While RPA implementations are ongoing, deriving value from digital enterprise platforms (including analytics capabilities) is only in its infancy. Based on these takeaways, we contextualize the objective of this article by examining digital enterprise platforms with an ERP/BI integration and analytics capabilities included *to generate valuable new insights, ultimately, drive an SSC redesign.*

**Cross-Domain Analytical Business Services in Focus.** Sia et al. [39] considered analytical business services such as market mix modelling. Fischer et al. [40] presented analytical business services which combine internal sales information with external data to classify the fraud risk of flight ticket purchases. They argued that analytical business services should be demand-driven [41] so as to ultimately support fact-based decisions [42]. More generally, Grover et al. [43] defined the requirements for analytical business services threefold: (1) Cross domain data source, (2) analytics portfolio, and (3) employee qualification. However, the body of knowledge on how to develop analytical business services across domains remains insufficient. Cokins [36] went a step further, stating that disruption through digital technologies is not only required in efforts to succeed inside a competitive market, but to survive in the constantly changing digital environment. Concisely, the need for analytical business services is obvious, but "how" to find, examine, and implement them has not yet been researched. Thus, we aim to pave the way for new business models of SSCs *by defining initial analytical business services based on cross-domain big data.*

# 3   Method

Guided by our RQs (Sect. 1) and the findings from our literature review (Sect. 2), we conducted a dual-case study. Given that there is little research, and the practical insights are expected to be significant [44, 45], we opted for an exploratory research approach. *Case studies* are examinations of a contemporary phenomenon within its real-life context, which can be an effective way to bridge the gap between practice and academia [46]. Compared to surveys, they provide more substantial in-depth information and enable researchers to study their artefacts in a natural setting [47]. Furthermore, compared to multiple case studies, which bring in a multitude of results [48], dual-case studies are more suitable when the research topic is complex and, thus, relevant starting points of research are not easy to obtain [49].

We took two reference companies, an integrated tele-communications company, and a leading specialty chemicals company. With about 216,500 employees as well as 235.8 million mobile customers, 27.3 million fixed-networks, and 21.3 million broadband lines, our first reference company generated sales of about 108 billion € in 2021. Focusing on additives, nutrition & care, and performance materials, our second reference company generated sales of 15 billion € in 2021 with more than 33,000 employees.

Aiming to develop initial analytical business services, we commenced our project with desk research at the reference companies [50]. We examined internal documents and decided to start with the finance domain of the reference companies' SSCs, as this is the domain covering the highest volume of data preparation and processing. Complementing the takeaways from our literature, we assessed their main shared services. In parallel, we examined several vendor white papers, handbooks, and attended a series of workshops at the reference companies. Resulting from this, we developed a thorough understanding of the current shared services setup.

After concluding the six weeks of desk research, we organized seven semi-structured interviews with employees from HR, Procurement, Sales, Management Accounting & Finance. The aim of the interviews was to generate ideas for analytics business services based on these future customers' needs. As a starter for the series of individual interviews, we had a joint session introducing the participants to the idea of analytical business services and how the SSC setup can drive implementation and the subsequent delivery of such services. The interviewees then received a template for describing initial analytical business services for discussion during the semi-structured interviews. To analyze the findings, we carried out a qualitative content analysis [51], which enables a systematic procedure by applying a category system to the answers given. With the number of similar answers, we integrated a quantitative step as well (frequency analysis [52]) and, finally, based on the qualitative and quantitative results, we developed the proposed analytical business services – according to the different SSC domains linked with the finance function (RQ 1, Sect. 4.1).

Accompanying the whole process, we discussed our results from Nov 2020 until Apr 2021 with our project expert group: The Senior Vice President Digitalization & Transformation, the Head of Global Financial Services, and an Accountant for Digital Transformation & Change from the first reference company, as well as the Head of Global Financial Services & Processes and Transformation from the second reference company. We revised our findings from the interviews with the project team whenever we identified aspects for improvement and reached closure once our approach was saturated, meaning that adding further aspects led only to incremental improvements.

We then prioritized the ideas for analytical business services considering three dimensions: (1) What *new insight(s)* does the service offer and to what extent does it create value? (2) What kind of *(cross-domain) data* is needed and is it – as of now – possible to access this data? (3) What *type of algorithm* is required to generate the insights for the respective analytical business service and to what extent are the required skills available inside the SSC? Based on these criteria, we ranked the analytical business service for implementation, where each project team member could distribute up to three votes among the various ideas.

Emphasizing a staged process with iterative "build" and "evaluate" research activities, we evaluated the new artefacts [53]: We presented our results in three individual expert interviews of about 45 min, with both heads of SSC from the two reference companies and the Senior Vice President (SVP) Accounting from the second reference company (Sect. 5). During these interviews, we gathered feedback about our artefact's utility and validity (RQ 2, Sect. 1). We considered interviews as an appropriate method, as their purpose is to obtain the interviewee's knowledge as comprehensively as possible, regarding the meaning of the described phenomena [54]. In comparison to surveys, they enable going deeper into a subject matter. In this way, more insights can be collected, which can be useful for a later quantitative evaluation. In particular, we applied semi-structured expert interviews, because they combine both a comparable structure within a series of interviews, whilst still being flexible when interviewees want to share their individual way of thinking and any hidden facets [55].

# 4   Artefact Design

Based on the notes from our content analysis (Sect. 3), we gathered the results of the interviews and summarized our findings for each analytical business service (Fig. 2). We derived twenty-two analytical business services (Fig. 3). In the second step, we clustered ten of them to human resources (HR), six to supply chain management (SCM) and six to sales. The choice of cluster represents the domain which is expected to benefit the most from the respective service, regarding both financial gains as well as new in-sights. In the third step, we prioritized the twenty-two analytical business services. The prioritization is based on the dimensions of novelty and value, required data and availability, as well as underlying algorithm. Based on the project team's ranking, we present the highest ranked analytical business service of each domain in detail (see bold entries in Fig. 3). The remaining ones are presented briefly afterwards.

| Analytical Business Service | # 01: Attrition monitoring |
|---|---|
| Cross-domain influence factors | HR: Age at hiring, hiring type, country, city, manager information, contract type, etc.<br>Finance: Payroll information<br>External: Unemployment rate or similar jobs in the area |
| Comments | Combine information from employee surveys, performance reports and sociodemographic data (e.g., taken from employee CV) to find correlations with "leaves company within the next 12 months". |

**Fig. 2.** Analytical business service # 1 Attrition monitoring

**Extended Procurement Forecast (#11).** With the target of a required amount of raw material [tons] per month, the production of the specialty chemicals company relies on a timely provision of raw materials and semi-finished products, which are ordered by procurement. Like the sales forecasting (#17), the process is based solely on human judgment and simple extrapolation of historical values using Microsoft Excel.

Similar to the first priority, the project team chose this analytical business service, expecting immediate gains in forecast accuracy by using data already available inside the company. Given that sales and procurement are less connected than finance and sales, accessing the required data is slightly more complicated. The forecasting model is based on regression algorithms.

In future, this analytical business service should include actual sales and sales forecast values taken from analytical business service #17. These are currently not reported. Furthermore, to calculate the demand for raw materials, information on the composition of the final products is needed. As of today, this data is not available for procurement, due to the authorization concept in place.

**Attrition Monitoring (# 01).** Understanding fluctuation within the company is relevant for both reference companies, in order to plan hiring and employee training needs. While conversations with employees indicate whether an employee is about to leave, interviewees mentioned that such intentions can frequently be com-pared to a "black box," particularly when considering longer time horizons.

We chose attrition monitoring as the HR-related service with the highest priority, considering that hiring new employees incorporates significant costs and inefficiencies during the employee induction. Accordingly, an early warning mechanism should act as a foundation for preventing employee fluctuation – especially of high performers. The related model is based on classification algorithms.

A future analytical HR business service should forecast earlier, whether a specific employee will quit his or her position. This service should be based on both HR information such as age at hiring, contract type, city of residence and responsible manager, and external data which should include the number of free positions in a similar job in the area of residence or salaries offered from competitors on job portals like LinkedIn. To answer the question of which employee is likely to leave the organization, we considered using classification algorithms such as k-nearest neighbors for our model. Based on the results of our classification, actions can be initiated to retain employees who are classified as "likely to leave," especially in the case of high performers.

**Human Resources.** The majority of the remaining analytical business services are clustered within the HR domain. While several processes such as payroll processing, training management, and offboarding are at the responsibility of the SSCs in our reference companies, two topics were mentioned as potential roadblocks: It is necessary to take into account that processing data at the person level is confronted by compliance and data privacy, making these ideas harder to realize. Furthermore, an additional ethical assessment of HR-related services is necessary.

Both reference companies conducted employee surveys to assess employee satisfaction, but the reasoning behind specific answers is not always evident. #02 Key levers for employee satisfaction is meant to reveal some of the "hidden" levers by combining and processing survey information with employee-specific data, such as recent promotions or the number of trainings taken, to eventually indicate potential correlation. Equally related to analyzing and understanding employee wellbeing, analytical business service #03 Internal customer stress levels is intended to analyze the content, choice of words and length of service tickets, e.g., related to IT or HR services, to determine the stress level of senders and, ultimately, enable targeted improvement actions. To support project business, #04 Demand-driven project staffing combines data from past projects, such as financial success, timeliness and scope, in order to provide project managers with insights which, through project workforce composition, is expected to lead to fruitful results.

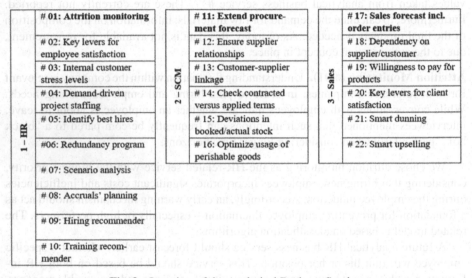

**Fig. 3.** Overview of the Analytical Business Services

To facilitate an optimal workforce composition, #09 Hiring recommender (for recruiters) and #10 Training recommender (for employees) analyze the gap between "as-is" and "to-be" workforce composition, as defined by HR with external information such as LinkedIn data to identify hiring trends or university enrollments to consider future talent supply. Similarly, assessing the available workforce was in the interest of the interviewees. Accordingly, #05 Identify best hires was proposed by interviewees, combining payroll information with recruitment records and performance reports, to highlight employee developments. At the other end of the process, the analytical business service #06 Redundancy program was meant to support identifying the "right" employees to leave, e.g., those that are planning to leave anyway or those of lower value to the company, by processing information such as attrition rate in the respective department, number of recently absolved trainings and compensation compared to the industry average. Both #05 and #06, combined with workforce requirements, could then be used as inputs for #07 Scenario analysis, analyzing the change in FTE on financial figures such as earnings before interest and tax. Analytical business service #08 Accruals' planning was mentioned as a specific part of #07, supporting the planning process of accruals for pensions and settlements.

**Supply Chain Management.** It is essential to use perishable goods before they lose their value. While inventory counts are performed by the supply chain team in the business, all reports are prepared by the SSCs of our reference companies. #16 Optimize usage of perishable goods aims to reduce opportunity cost by combining information such as manufacturing capacity as well as (excess) inventory, in order to recommend preferable manufacturing schedules. Related to inventory is #15 Deviations in booked/ actual stock, aiming to monitor and inform about deviations of inbound and outbound data compared to actual postings. The specialty chemicals company interacts with several suppliers that are customers at the same time. #18 Dependency on supplier/customer is about providing information to sales and procurement employees across all business lines, such as the importance (e.g., by sales volume) of a customer, to facilitate favorable decision making. Also related to supplier relationship management is #12 Ensure supplier relationships, gathering internal and external information to assess supplier solvency. Finally, #14 Check contracted versus applied terms is to check whether newly signed contracts are within the constraints set up by frame agreements.

**Sales.** Representing a connection between sales and procurement, #13 Customer-supplier linkage would provide information on the trade volume between both companies, expected to improve the bargaining power of the specialty chemicals company. Serving a huge number of end-costumers, the telecommunications company would benefit from #21 Smart dunning as well as #22 Smart upselling. In both cases, customer information such as residential area, payment behavior and occupation, as well as macroeconomic factors such as the unemployment rate, would influence the automatized communication with each costumer. Equally, prior sales to a customer, combined with geographical influence factors such as the shop location or the city, constitute central parts of analytical business service #19: Willingness to pay for products. Combined with sales information such as analyzing the complaints using natural language processing, #20 is intended to reveal hidden key levers for client satisfaction.

**Additional Findings.** We asked the interviewees which unit or domain should be the owner of these analytical business services. The answers revealed a common ground: Interviewees mentioned their own domain as a predestined owner. However, the majority was in favor of creating the unit within finance, as this department has interfaces with all other domains. One interviewee mentioned that the process owner should be cross-domain or inter-domain, to ensure that a subordination to the budget of a specific domain does not influence the decision making. Accordingly, the unit responsible for analytical business services should be centralized and consist of process experts from the domains, as well as data scientists. As such, SSCs were identified as being a perfect fit to maintain analytical business services.

# 5  Evaluation

Referring to RQ 2 about evaluating the business services' validity and utility, we conducted three expert interviews of about 45 min with the heads of SSC from the two reference companies and the Senior Vice President (SVP) Accounting from the second reference company. To avoid bias, three measures were taken: (1) The interviewees were not part of the expert group which constituted the design of the analytical business services; (2) two interviewers to avoid misinterpretation; (3) an interview guideline for a semi-structured approach was used to minimize the influence of spontaneous questions. As such, our research applies triangulation of methods and investigators [56].

## 5.1  Validity

All three expert interviews supported the idea of breaking up data silos in SSCs, in order to generate new insights. They further agreed on starting with domains which are less restrictive than HR. Two interviewees pointed out the importance of including most relevant stakeholders from the beginning of implementing analytical business services. Both interviewees agreed that regulating the access to underlying (highly) confidential data must be governed centrally, so as to ensure compliant usage. The SVP Accounting continued, pointing out that solely considering the data access is not enough. She recommended regulating the processing of both the data set used to perform the analysis, as well as the data resulting from the analysis, thus making sure the "data treasure" is safe while handling it in accordance with compliance regulations.

The head of SSC of the specialty chemicals company added that driving a shift towards a "data science mindset" throughout the company is required. Regarding the SSC employees, he considered upskilling as a precondition to benefiting fully from analytical business services, given that some of the algorithms were unknown to the employees. Beyond the workforce, the head of SSC of the telecommunications company outlined that a comprehensive data orchestration plays an important role in implementing analytical business services successfully. He considered three major ways to handle this: (1) Harmonization with the legacy systems; (2) manpower-extensive mapping of new and legacy data structures or (3) the implementation of new tools.

## 5.2  Utility

Regarding the utility of our initial analytical business services, we evaluated to what extent the services themselves, as well as the related lessons learned, are useful even beyond our reference companies and the SAP Data Intelligence environment per se. Especially in publicly listed companies, expanding the access to crucial data is not easy. Thus, having a clear strategy for implementing the authorization processes related to analytical business services was considered relevant by all interviewees, and for external companies as well. However, the interviewees agreed that smaller companies are less affected, given that less internal compliance is needed.

The head of SSC of the telecommunications company pointed out that the transformational potential of analytics business services is comparable among any multi-tower SSC. Based on the fact that the contractual framework of SSCs facilitates the processing of the underlying data, he expects less reluctance and a seamless integration of such services. However, different national standards have to be considered, so that adjustments are required when pursuing implementation in cross-national contexts.

In line with our findings from the idea-generation interviews, the head of SSC from the telecommunications company added that the developed analytical business ser-vices are not equally applicable by every company, e.g., production-related services would add fewer valuable insights to the telecommunications company than to the specialty chemicals company. However, he stated that the concept of combining cross-domain data to generate value is a "must-have" for companies to ensure competitiveness, especially in the upcoming years.

# 6   Discussion and Conclusion

The objective of this paper was to present initial *analytical business services that leverage data across different SSC domains*, in order to generate valuable new in-sights. Based on interviews with future consumers of such services, we developed twenty-two analytical business services. Subsequently, together with the project team, we ranked the analytical business services regarding each service's novelty and value, required data and availability, as well as the underlying algorithm (Sect. 4.1).

For practice, the approach at hand should help to conceptualize the design of analytical business services. Multi-tower SSCs are the preferred choice for implementing analytical business services, but they are valid for the internal departments as well. In doing so, our findings may lead to significant readjustments of current SSCs, not just driven by content, but in a new setup of responsibilities across existing SSC domains. Additionally, the value created by analytical business services could further leverage the investment in state-of-the-art digital platforms, which play a central role in the setup of such services.

For research purposes, we found that value generation by new insights is currently not considered as an integral target of SSCs. We contradicted the widely accepted exclusive target for SSCs of cost reduction (efficiency), as is mentioned by Fielt et al. [3]. In turn, we revealed effectiveness targets such as new insights. Being frequently ignored, investigating the effectiveness dimension offers a great opportunity to further improve the value generation of SSC.

Our research inevitably has certain limitations. Accordingly, there are several avenues for future research. Although case studies offer a broad range of advantages, one critique is their limited generalizability. Firstly, our research should become more multifaceted by initiating complementary use cases. Furthermore, our artefact design itself (Sect. 4) entails limitations. Data access restrictions are one obstacle, especially regarding HR data. The fact that IS infrastructure and authorization concepts differ among companies, may lead to impracticality regarding some of our ideas.

To keep the project tractable, a larger number of algorithms tested is another avenue for future research. However, to fully benefit from analytical business services, it is critical to find ways for companies to enable their employees to work in such a new, data-driven environment.

Another research avenue is that of assessing our results in other setups to expand our findings with new kinds of services and generating further content. Additionally, integrating external data, even streaming data, was mentioned by the interviewees, which we did not consider in our use case. Furthermore, more insights into the organizational structure needed to develop and run analytical business services could potentially increase the speed of adopting such services.

# References

1. Baker, S., Bloom, N., Davis, S., Terry, S.: COVID-induced economic uncertainty (2020). Accessed 03 May 2021
2. Fries, A., Noldus, S.: Shared services trends 2020. Control. Manage. Rev. **60**(3), 46–53 (2016). https://doi.org/10.1007/s12176-016-0083-y
3. Fielt, E., Bandara, W., Miskon, S., Gable, G.: Exploring shared services from an IS perspective: a literature review and research agenda. Commun. Assoc. Inf. Syst. **31**(1), 1–39 (2014)
4. SSON: SSON State of the Global Shared Services Industry Report 2020 (2020). https://www.ssonetwork.com/global-business-services/reports/sson-state-of-the-global-shared-services-industry-report-2020. Accessed 03 May 2021
5. Naous, D., Schwarz, J., Legner, C.: Analytics as a service: cloud computing and the Transformation of business analytics business models and ecosystems. In: ECIS 2017 Proceedings, vol. 25, pp. 487–501 (2017)
6. PWC: Shared Services - Digitalise Your Services (2019). https://www.pwc.de/de/prozessoptimierung/pwc-studie-shared-services.pdf. Accessed 03 May 2021
7. Deloitte: 2019 Global Shared Services Survey Report (2019). https://www2.deloitte.com/content/dam/Deloitte/us/Documents/process-and-operations/2019-global-shared-services-survey-results.pdf. Accessed 03 May 2021
8. Lacity, Mary, Fox, Jim: Creating global shared services: sourcing lessons from reuters. In: The Practice of Outsourcing, pp. 467–487. Palgrave Macmillan UK, London (2009). https://doi.org/10.1057/9780230240841_16
9. European Union: Opinion 03/2013 on purpose limitation (2013). https://ec.europa.eu/info/law/law-topic/data-protection/reform/rules-business-and-organisations/principles-gdpr/purpose-data-processing/can-data-be-processed-any-purpose_en
10. Olsen, T., Welke, R.: Managerial challenges to realizing IT shared services in a public university. TG **13**(1), 76–92 (2019)
11. Simon, H.A.: The Science of the Artificial, 3rd edn. MIT Press, Cambridge, Mass (1996)
12. Walls, J.G., Widmeyer, G.R., El Sawy, O.A.: Building an information system design theory for vigilant EIS. Inf. Syst. Res. **3**(1), 36–59 (1992)
13. Hevner, A.R., March, S.T., Park, J., Ram, S.: Design science in information systems research. MIS Q. **28**(1), 75–105 (2004)
14. vom Brocke, J., Winter, R., Hevner, A., Maedche, A.: Accumulation and evolution of design knowledge in design science research - a journey through time and space. J. Assoc. Inf. Syst. **21**(3), 520–544 (2020)
15. Gregor, S., Hevner, A.R.: Positioning and presenting design science research for maximum impact. MIS Q. **37**(2), 337–355 (2013)
16. Peffers, K., Tuunanen, T., Rothenberger, M.A., Chatterjee, S.: A design science research methodology for information systems research. J. Manag. Inf. Syst. **24**(3), 45–77 (2007)
17. Blinded for review (2022a)
18. Webster, J., Watson, R.T.: Analyzing the past to prepare for the future: writing a literature review. MIS Q. **26**(2), xiii–xxiii (2002)

19. Vom Brocke, J., Simons, A., Niehaves, B., Riemer, K., Plattfaut, R., Cleven, A.: Reconstructing the giant: on the importance of rigor in documenting the literature search process. In: Newell, S, Whitley, E.A., Pouloudi, N. Wareham, J., Mathiassen, L (eds.) Proceedings of the 17th European Conference on Information Systems (ECIS 2009)
20. Vom Brocke, J., Simons, A., Niehaves, B., Riemer, K., Plattfaut, R., Cleven, A.: Standing on the shoulders of giants: challenges and recommendations of literature search in information systems research. Commun. Assoc. Inf. Syst. **37**(1), 205–224 (2015)
21. Rowley, J., Slack, F.: Conducting a literature review. Manag. Res. News **27**(6), 31–39 (2004)
22. Mergel, I., Edelmann, N., Haug, N.: Defining digital transformation: results from expert interviews. Gov. Inf. Q. **36**(4), 1–16 (2019)
23. Mayer, J.H., Esswein, M., Razaqi, T., Quick, R.: Zero-quartile benchmarking – a forward-looking prioritization of digital technologies for a company's transformation. In: ICIS 2018, pp. 1–17 (2018)
24. Schulz, V., Hochstein, A., Uebernickel, F., Brenner, W.: A classification of shared service centers: insights from the IT services industry. In: PACIS 2009 Proceedings, pp. 1–13 (2009)
25. Richter, P.C., Brühl, R.: Shared service center research: A review of the past, present, and future. Eur. Manag. J. **35**(1), 1–13 (2017)
26. Lacity, M.C., Khan, S.A., Yan, A.: Review of the empirical business services sourcing literature: an update and future directions. J. Inf. Technol. **31**(3), 269–328 (2016)
27. Sako, M.: Outsourcing versus shared services. Commun. ACM **53**(7), 27–29 (2010)
28. Lehrer, C., Wieneke, A., vom Brocke, J., Jung, R., Seidel, S.: How big data analytics enables service innovation: materiality, affordance, and the individualization of service. J. Manag. Inf. Syst. **35**(2), 424–460 (2018)
29. Richter, P.C., Brühl, R.: Ahead of the game: Antecedents for the success of shared service centers. Eur. Manag. J. **38**(3), 1–12 (2020)
30. Iansiti, M., Lakhani, K.R.: Digital ubiquity: how connections, sensors, and data are revolutionizing business. Harv. Bus. Rev. **92**(11), 90–99 (2014)
31. EY: How can Global Business Services drive value across continents? (2019). https://www.ey.com/en_gl/consulting/how-global-business-services-can-drive-value-across-continents. Accessed 03 May 2021
32. Accenture: From Bottom Line to Front Line (2018). https://www.accenture.com/_acnmedia/pdf-85/accenture-cfo-research-global.pdf. Accessed 11 Jan 2021
33. Plaschke, F., Seth, I., Whiteman, R.: Bots, algorithms, and the future of the finance function. McKinsey on Finance (65), 18–23 (2018)
34. The Hackett Group: The CFO Agenda: Finances Four Imperatives to Accelerate Business Value (2018). https://www.thehackettgroup.com/key-issues-fin-1801-thankyou/. Accessed 11 Jan 2021
35. Genpact: From robotic process automation to intelligent automation. Six best practices to delivering value throughout the automation journey (2018). https://www.genpact.com/downloadable-content/insight/the-evolution-from-robotic-process-automation-to-intelligent-automation.pdf. Accessed 11 Jan 2021
36. Cokins, G.: Enterprise performance management (EPM) and the digital revolution. Perf. Improv. **56**(4), 14–19 (2017)
37. Yaqoob, I., et al.: Big data: from beginning to future. Int. J. Inf. Manage. **36**(6), 1231–1247 (2016)
38. Lucas, S.: The benefits of the SAP digital enterprise platform (2016). https://blogs.saphana.com/2016/02/03/the-benefits-of-the-sap-digital-enterprise-platform/. Accessed 11 Jan 2021
39. Sia, S.K., Soh, C., Weill, P.: IT governance in global enterprises: managing in Asia. In: Proceedings of the ICIS 2008, pp. 1–10 (2008)

40. Fischer, T.M., et al.: Business analytics in shared service organisationen. In: Fischer, Thomas M., Lueg, Kai-Eberhard. (eds.) Erfolgreiche Digitale Transformation von Shared Services. Z, vol. 74/20, pp. 147–187. Springer, Wiesbaden (2020). https://doi.org/10.1007/978-3-658-30484-3_5
41. Garvin, D.A.: What every CEO should know about creating new businesses. Harv. Bus. Rev. **82**(7/8), 18–21 (2004)
42. Holsapple, C., Lee-Post, A., Pakath, R.: A unified foundation for business analytics. Decis. Support Syst. **64**, 130–141 (2014)
43. Grover, V., Chiang, R.H., Liang, T.-P., Zhang, D.: Creating strategic business value from big data analytics: a research framework. J. Manag. Inf. Syst. **35**(2), 388–423 (2018)
44. Demetriou, D.: Case study. In: The Development of an Integrated Planning and Decision Support System (IPDSS) for Land Consolidation. ST, pp. 133–143. Springer, Cham (2014). https://doi.org/10.1007/978-3-319-02347-2_6
45. Benbasat, I., Goldstein, D.K., Mead, M.: The case research strategy in studies of information systems. MIS Q. **11**(3), 369–385 (1987)
46. Myers, M.D.: Qualitative research in information systems. MIS Q. **21**(2), 241 (1997)
47. Dul, J., Hak, T.: Case Study Methodology in Business Research. Butterworth-Heinemann, Amsterdam (2007)
48. Yin, R.K.: The case study crisis: some answers. Adm. Sci. Q. **26**(1), 58–65 (1981)
49. Gustafsson, J.: Single case studies vs. multiple case studies: a comparative study (2017)
50. Eisenhardt, K.M.: Building theories from case study research. AMR **14**(4), 532–550 (1989)
51. Elo, S., Kyngäs, H.: The qualitative content analysis process. J. Adv. Nurs. **62**(1), 107–115 (2008)
52. Mayring, P.: Qualitative content analysis: demarcation, varieties. Dev. Forum: Qual. Soc. Res. **20**(3), 1–14 (2019)
53. Venable, J., Pries-Heje, J., Baskerville, R.: FEDS: a framework for evaluation in design science research. Eur. J. Inf. Syst. **25**(1), 77–89 (2016)
54. Kvale, S.: Interviews: An Introduction to Qualitative Research Interviewing. SAGE Publications, London (1996)
55. Qu, S.Q., Dumay, J.: The qualitative research interview. Qual. Res. Account. Manag. **8**(3), 238–264 (2011)
56. Carter, N., Bryant-Lukosius, D., DiCenso, A., Blythe, J., Neville, A.J.: The use of triangulation in qualitative research. Oncol. Nurs. Forum **41**(5), 545–547 (2014)

# Remote Working from a Management Perspective

Lea-Christin Hellwig and Matthias Murawski[✉]

FOM University of Applied Sciences, Bismarckstr. 107, 10625 Berlin, Germany
matthias.murawski@fom-net.de

**Abstract.** Due the societal change of experiencing remote working, caused by the COVID-19 pandemic, many corporations are facing the need for an extensive new remote work policy. Thus, corporations and management need to position themselves towards new ways of working and therefore towards the new standard of remote working in order to develop their long-term remote work policy. For this purpose, this study identifies criteria and consequences, as well as advantages and challenges which need to be considered during the policy-making procedure to incorporate a remote working environment. By conducting eleven semi-structured interviews with managers following a maximum variation sampling approach, valuable insights are gained, which deepen the understanding for a management's perspective. Regardless of the industry, this study most notably identifies that managers benefit from similar advantages and are confronted by identical challenges trying to establish remote working policies. Illustrating the different policies and investigating remote working in respect to today's highly relevant impact on the employer's attractiveness, present the key contribution to existing literature. This research provides extensive considerations and recommendations for the management how to deal with remote working.

**Keywords:** Remote Working · Management · Qualitative Research

## 1 Introduction

The transformation of work towards remote working has been discussed for decades by the ideas of *Future of Work* or *New Work* [1, 2]. However, most of the management executives have not seriously considered the trend on the horizon. In 2020, remote work has suddenly become part of almost every corporation due to the COVID-19 pandemic. As a consequence, of the newly implemented short-term policies, the majority of employees and their managers have been separated spatially from each other for the first time [3]. Most managers were unprepared and suddenly confronted with the new situation [4]. Especially for the rather conservatively managed companies, which used to have no established remote work policy, this situation presents a great challenge. However, by now structural barriers have been partly overcome to be able to adapt to new long-term policies. Therefore, corporations are facing the strategic decision of developing a corporate remote work policy which guides beyond the pandemic. This

© The Author(s), under exclusive license to Springer Nature Switzerland AG 2023
M. Papadaki et al. (Eds.): EMCIS 2022, LNBIP 464, pp. 603–615, 2023.
https://doi.org/10.1007/978-3-031-30694-5_42

policy especially concerns knowledge workers, who are preferably able to do their job remotely. Current management approaches and opinions range from returning back to the office full-time, or on strengthening remote working solutions [5].

Strategically paving the way for the future involves many challenges, such as maintaining a corporate culture with corporate values, sharing knowledge and information between coworkers and the essential process of teambuilding. Despite the challenges, remote work has become crucial in regard to the employer's attractiveness. The concept of new ways of working (NWW) shifts the power to the employee and has gained increasing attention during the last decades [6]. Based on the hypothesis that the employees' demand for remote work remains steady throughout the coming years, this paper aims to identify the decisive criteria and consequences of developing a corporate remote work policy which helps corporations to stay attractive for future employees.

The pandemic was able to demonstrate that working in the office during a fixed time frame needs to be rethought [7]. The New York Times stated that *"employers are struggling with how, when and even if they will bring employees back to the office. [...] [CEOs] are struggling to balance rapidly shifting expectations with their own impulse to have the final word on how their companies run. They are eager to appear responsive to employees who are relishing their newfound autonomy, but reluctant to give up too much control"* [8]. This quote illustrates the current situation of corporations and their respective management. In order to lay the pioneering foundation for literature, as this presents an unprecedented societal situation, the demand for structural work change needs to be investigated.

Remote working has been researched in terms of new ways of working by several scholars in the past [9, 10]. Additional research has been conducted in regard to who is able to carry out remote work. Former research has revolved around the advantages and challenges of remote working and has identified the essential factor of work-life balance [11–13]. Furthermore, the impact on motivation, productivity, performance and job satisfaction have been analyzed by different scholars [14]. Due to the key fact, that the employee has not been working in close proximity to the management, the matter of the transforming relationship of managers to the employees has been an additional emerging topic for further scholars [15]. However, by reviewing the literature it is striking that the topic of developing a corporate remote work policy has not been addressed yet. Especially, due to the current and timely relevance of the topic, there is almost no research to be found on strategic corporate remote work policies. The purpose of this paper is to close this research gap with a focus on managerial strategic positioning, concerning remote working. Therefore, this research aims to answer the following research question (RQ):

*What are the criteria and consequences of developing a corporate remote work policy?*

According to the research question, the unexplored subject matter needs to be dealt with exploratively. Fittingly, a qualitative research method through the execution of interviews is applied.

The remainder of the paper is structured as follow: Sect. 2 contains an overview about the current state of research. Section 3 includes the methodology used in this study. Section 4 summarizes the key findings of the study combined with a discussion, based on which a conclusion is provided in Sect. 5.

## 2 Current State of Research

The rise of new ways of working is an ongoing transformation which has started many decades ago [16] but it is still relevant and still in process [17]. One of the crucial elements of the holistic change of work includes the emerging need of remote working [14]. The location where the work is done is key, furthermore time can also be a parameter of new way of working.

In respect to the criteria and consequences of developing a corporate remote work policy, the awareness and openness of transforming the traditional work-settings towards new ways of working is essential. Corporations are required in being able to change their mindset and attitude in order to stay relevant and attractive as an employer.

Remote working is attractive to employees, because of the advantageous possibility to meet one's professional as well as one's private responsibilities. Therefore, the individual work-life balance is improved and corresponds with the idea of new ways of working [12] due to the ability of self-management and autonomy of the employee. Hereby, the terms flexibility and autonomy in respect to employees are the most essentials findings, also in regard to the choice of spatial location of work and possible timely choices [18]. Closely related to the matter of work-life balance is the employee's motivation, productivity, performance and job-satisfaction. These factors influence each other. Thus, they represent advantages to the employee which ultimately lead to beneficial effects on the organizational success [11]. In this respect, the employee's autonomy is discussed as an essential factor, especially for the intrinsic motivation [13]. Due to the gained autonomous and self-determined way of working, the employee feels valued, which leads to higher productivity, performance, job-satisfaction and motivation [19]. The third reappearing subject matter in the examined literature is the relationship between the manager and the employee. The relationship is undergoing a great change, due to the transformation of the traditional idea of work towards new ways of working [14]. The traditional control-based managing structure dissolves and a relation of trust and autonomy arises [20]. Whereby, the managers themself need to act in the sense of new ways of working, in order to successfully establish the changed working conditions [21]. Overcoming the lack of trust presents a beneficial and essential factor for a successful implementation of remote working [22].

By considering the relevant literature it can be concluded, that broad insights and different concepts of remote working can be drawn from the literature review. Since the existing concepts vary in regard to different topics, the research methods are not comparable and there is no value to be added by discussing the individual research methods. Moreover, there are no specific theories to be found in the generated literature and no strongly striking controversies. Further, in regard to key contributors, there are some researchers that have been cited more often, as for example Kurland and Bailey [22]. Among others, they analyzed the challenges and advantages for organizations while also considering the individual perspective [22]. The aspects which are highlighted in Table 1 emerge as well in the literature. Work-life balance presents a decisive factor for a remote work policy, because remote working enables the employee to be flexible, which results in benefits of the private life as well as for work. Further, motivation and performance are a common theme in literature, which present both, an advantage as well as a challenge.

**Table 1.** *Advantages and Challenges of Remote Working*, own illustration based on [22]

| Advantages and Challenges of Remote Working | | |
|---|---|---|
| **Remote Working** | **Advantages** | **Challenges** |
| Perspective of the Organization | • Greater Productivity<br>• Lower Absenteeism<br>• Fewer Interruptions at Office<br>• Lower Turnover<br>• Better Morale<br>• Greater Openness<br>• Reduced Overhead<br>• Wider Talent Pool<br>• Regulation Compliance | • Managerial Control    • Organization Loyalty<br>• Organization Culture    • Interpersonal Skills<br>• Performance Monitoring    • Availability<br>• Performance Measurement    • Schedule Maintenance<br>• Technology    • Work Coordination<br>• Mentoring    • Internal Customers<br>• Jealous Colleagues    • Communication<br>• Synergy    • Guidelines<br>• Informal Interaction    • Performance<br>• Virtual Culture |
| Perspective of the Individual Employee | • Work/Family Balance<br>• More Job Satisfaction<br>• More Autonomy<br>• Less Time Commuting<br>• Less Stress<br>• Schedule Flexibility<br>• Fewer Distractions<br>• Cost Savings<br>• No Need for Relocation<br>• Comfortable Work Environment<br>• Absence of Office Politics<br>• Workplace Fairness | • Organization Culture    • Conducive Home<br>• Work/Family Balance       Environment<br>• Social Isolation    • Focusing on Work<br>• Professional Isolation    • Longer Hours<br>• Reduced Office Influence    • Access to Resources<br>• Informal Interaction    • Technical Savvy |

During the whole literature review only two articles have been detected that reflect upon corporate policies and remote work. Firstly, in Clark's research [23] an economic model for the decision to remote work has been developed by using the model of planned behavior and by representing a strong economic theory approach. Clark's findings concerning corporate policies were *"regardless of what subordinates or superiors think, regardless of resources or personal attitude toward telework, corporate policies may be in place which promote or inhibit telework."* Further, Clark proposed that *"the degree to which these policies endorse telework will influence the degree to which the decision maker feels able to make a decision for telework. The extreme condition in which telework is prohibited allows no (or very little) behavioral control on the part of the individual."* However, if a corporation solely implements remote work, it would have the same consequences [23]. Concluding, Clark considers the aspect of a corporate policy, but this aspect only presents a short paragraph in Clark's paper, which suggest that corporations are required to form a policy. Secondly, in 2001, Daniels et al. addressed remote work policies by presenting *"a framework to help understand the nature of teleworking"* [24]. The framework identified five variables: the location, the amount of ICT usage, the knowledge intensity, and the internal or external organizational work. Nevertheless, their research clearly consists of a different societal situation and aims to achieve the understanding of remote working and not the corporate need to develop a corporate policy.

Due to the current nature of the topic, there is no research to be found on strategic corporate remote work policies. Thus, the purpose of this paper is to close and to investigate this research gap of managerial strategic positions concerning remote working.

## 3  Methodology

This paper follows a qualitative research approach. Due to the literature findings, it becomes clear that the research gap calls for an exploratory investigation. In order to gain broad insights and to generate and gather qualitative data, the semi-structured interview approach is pursed [25].

The interview guideline covers 12 key questions with a few additional flexible questions. Thereby, the questions are prepared in a set order whilst being amendable to flexibility in order to gain sufficient insights from the interviewee [25]. The questions are developed on the theoretical foundation of the literature review and their characteristics are open, neutral and simple [26]. For developing the interview guideline, the following approach is chosen. In the beginning of the interview the interviewees are asked factual questions about their management position and their corporation. Further, the second set of questions relates to their remote work policies before the pandemic and in the aftermath of the pandemic. The third set of questions is strongly focusing on advantages and challenges. The questions regarding advantages and challenges of remote working are firstly asked from the perspective of the individual employee and secondly from the perspective of the corporation or management position. The fourth block of questions describes three different scenarios involving rigorous policy approaches [26]. Based on the hypothetical scenarios, possible consequences for the company need to be identified and thus decisions regarding the remote work policy become more understandable. These types of questions are formulated in a narrative exhilarating way [26].

As a guideline to define potential interviewees, Gläser and Laudel [26] proposed to consider, who has the relevant information and is likely to provide the information. Further, the availability of the informants is decisive. By referencing the RQ, it can be argued that in corporations, management positions define and determine policies and regulations. Therefore, the interviewees are required to have experience in working in a management position. Justified by the exploratory approach of the research design, maximum variation sampling is applied. In order to answer the RQ, it is not needed to limit the interviewees to a certain industry because of the RQ's broad explorative nature, however only the area of knowledge work is considered.

The recruiting of interviewees is needed until the content saturation has been reached. Therefore, this paper tries to recruit as many high-level interviewees as possible, by considering the saturation of statements.

By performing a trial interview, the interview guideline got verified and the questions' comprehensibility and completeness and the duration of the interview were tested beforehand. After conduction of the test interview, the interviewee noted that a few longer pauses for reflection would have been necessary. This feedback helped to restructure some questions for the actual interviews.

**Table 2.** *Maximum Variation Sampling – Interviewees' ID,* own illustration

| Maximum Variation Sampling – Interviewees' ID | | | | | |
|---|---|---|---|---|---|
| Interviewee | Industry | Position | Foundation of the Company | Amount of the Company's Employees | Years of Leadership Position/ Experiences |
| A | Banking | Head of Service and Infrastructure Management | 1958 | 1500 | 13 years |
| B | Media, News and Publishing | Co-managing Director & Head of Product | 1947 | 1200 | 15 years |
| C | Pharmaceutical Industry | Head of Marketing Rx & Business Development | 1945 | 390 | 16 years |
| D | Brand Agency | Team Lead Retail Deployment | 1985 | 850 | 1 year |
| E | Foundation | Deputy Secretary General | 2000 | 14 | 3 years |
| F | Tech and Online Marketing (SaaS) | Vice President Large Customer Sales | 2016 | 255 | 1 year |
| G | Garment Industry | Head of HR | 2017 | 380 | 4 years |
| H | Tech, Software, Developers | COO | 2005 | 146 | 16 years |
| I | Consulting in Banking Industry | External Consultant of Transformation Process | 1846 | 1200 | 25 years |
| J | Educational Industry | Deanery Executive | 1992 | 1890 | 16 years |
| K | Controlling & Finance & Real Estate | Head of Finance & Controlling & Authorized Officer | 1990 | 1000 | 29 years |

Hence, all but one interviews were conducted virtually using digital platforms or software. The interviews took between 40 and 58 min. Ultimately, after conducting eleven interviews (see Table 2 for an overview), the answers had a recognizable pattern and only few new aspects were being reported. A saturation has been reached and the process of conducting interviews was completed. The interviews were recorded in order to use them for the transcription process. For the transcription of the conversations, an online transcription provider was used as support.

In order to analyze the gathered data of the interview transcripts, Mayring's approach of the qualitative content analysis (QCA) is applied [27]. Due to the pioneering research of the previously defined research gap and because of the explorative character of the research question, the summarizing approach (inductive) is applied in order to create a comprehensive overview of the topic.

## 4 Key Findings and Discussion

### 4.1 Findings from the Interviews

Generally, we find that remote working and the development of such a policy is a highly relevant topic, which brings advantages but also great challenges to corporations. Therefore, each interviewee, regardless of their industry and company size, was able to share their personal experiences and each one is dealing with the topic from a different angle on a daily basis. With regards to the RQ, the interviewees shared valuable insights into the criteria and consequences of developing a corporate remote work policy. However, fundamental for the development of a policy is the attitude towards new ways of working and thus towards remote working. At first glance, every interviewee seemed to have a positive attitude towards the topic. They all recognize the advantages and, above all, the urgency to deal with such a policy. However, some comments also reveal skepticism about the issue. Thus, the opinion about remote working reveals to be potentially self-contradictory depending on the manager's attitude. Especially, the managers who work in the office mostly seem to still have doubts about long-term implementation of remote working. However, others fully support remote working and encourage their employees to work remotely and appreciate the advantages for themselves as well. Nevertheless, it needs to be highlighted, that the manager's attitude is key for implementing a remote work policy successfully.

Almost all interviewees face the same problems and challenges of remote working, for which they propose individual solutions. It is common sense that the work-life balance with its various benefits is the decisive advantageous factor of remote working. On the other hand, there is a challenge of remote working for the individual employees, who are on their own and have to complete the tasks independently. The work time gets allocated by the employee individually and the responsibility increases. Further, not every personality is able to work remotely and to stay connected to the team. Compared to the advantages and challenges for the individual employee, the advantages and challenges for a corporation are more predominant in their scope. On the one hand, the advantages are highlighted through the saving of costs due to less office space. However, on the other hand, the decisive advantage is the positive effect on the employees and the applicants. Every interviewee was aware that there is an inevitable need for granting and enabling of remote working in order to stay relevant and attractive in the job market. However, this implementation brings many challenges from the company's point of view. The usual way of managing employees must be rethought so that management also functions virtually and at distance. Moreover, the corporate culture, as well as the way of communicating, must be adapted and changed. Thus, technology and digitization can be helpful here, but there are also challenges of a skillful use of tools and equipment. Thus, in regards to the RQ, fruitful insights have been gathered, which present the criteria and consequences of developing a corporate remote work policy.

In general, it can be argued that the crucial difference between the existing literature and the conducted research is presented by the newly acquired experience of remote working for every manager due to the pandemic since 2020. This radical experience of the sudden requirement to pursue work remotely illustrates a new dimension, especially for companies and managers for whom it would not have been an option to work remotely at all. Further, most of the literature analyzed the phenomenon of remote working as a singularly emerging way of working and not as a widespread transformation of work, with its policies and consequences. Accordingly, there has been detected little overlap within the existing literature. The results present a novel contribution to the existing literature on remote work.

By comparing and adding the advantages and challenges that have been analyzed by the gathered data to Kurland and Bailey's findings (Table 3), it becomes clear that various points made by Kurland and Bailey (Table 1) were also mentioned by the interviewees during the data collection. However, many additional aspects in respect to challenges and advantages were added, due to the broad experiences of each manager. Therefore, an insightful finding to be discussed is that the characteristics of the advantages and challenges of remote working persisted more than 20 years. Consequently, this indicates that the mindset of the management has not changed sufficiently. However, due to the pandemic, managers were forced to gather a wide range of experiences regarding remote working, which calls a rework of the overview by Kurland and Bailey [22].

**Table 3.** *Advantages and Challenges of Remote Working According to Kurland and Bailey and the Gathered Data,* own illustration.

| Advantages and Challenges of Remote Working According to Kurland and Bailey and the Gathered Data | | | | |
|---|---|---|---|---|
| Remote Working | Advantages according to Kurland and Bailey | Advantages according to the gathered data* | Challenges according to Kurland and Bailey | Challenges according to the gathered data* |
| Perspective of the Organization | • Greater Productivity<br>• Lower Absenteeism<br>• Fewer Interruptions at Office<br>• Lower Turnover<br>• Reduced Overhead<br>• Wider Talent Pool<br>• Better Morale<br>• Greater Openness<br>• Regulation Compliance | • Requirement Employer Attractiveness<br>• Ability To Stay Relevant as Employer<br>• Willingness To Do Extra Effort<br>• New Standard To Find and Retain Talent<br>• Flexibility of Number of Employees in Regards to Office Space<br>• Less Office Space Is Needed<br>• Desk Sharing<br>• New Office Concepts<br>• Motivated Employees<br>• Empowering of Employees<br>• Daily Meetings With the Team | • Managerial Control    • Interpersonal Skills<br>• Organization Culture    • Availability<br>• Performance    • Schedule<br>  Monitoring     Maintenance<br>• Performance    • Work Coordination<br>  Measurement    • Internal Customers<br>• Technology    • Communication<br>• Mentoring    • Guidelines<br>• Jealous Colleagues    • Performance<br>• Synergy    • Technology<br>• Informal Interaction<br>• Virtual Culture<br>• Organization Loyalty | • Developing a Remote    • Transparency and Clear<br>  Working Policy     Communication<br>• Changing of Leadership    • Experiencing of Corporate<br>  Style     Culture<br>• Leading by Objectives    • Team Dynamic<br>  (Results & Output)    • Virtual Events<br>• Employee Appraisals    • Social Component<br>• Onboarding    • Technologic Equipment<br>• Leap of Faith    • Training of Using Tools<br>• Measuring of Performance    and Software<br>• Exemplifying of Remote    • Ensuring Remote Setup of<br>  Working, Managers Need    Employees<br>  To "Walk the Talk"    • Digitalisation of Processes<br>• Resistance of Employees To    • Procedures Which Are<br>  Return to the Office    Shaped by Handling of<br>• Spontaneous Exchange/    Paper<br>  Feedback    • Dependency of Customers<br>• Transformation of    and Clients and Their<br>  Traditional Habits and    Working Hours<br>  Culture To Work Virtually |
| Perspective of the Individual Employee | • Work/Family Balance<br>• More Job Satisfaction<br>• More Autonomy<br>• Less Time Commuting<br>• Less Stress<br>• Schedule Flexibility<br>• Fewer Distractions<br>• Cost Savings<br>• No Need for Relocation<br>• Comfortable Work Environment<br>• Absence of Office Politics<br>• Workplace Fairness<br>• More Job Satisfaction | • Ability To Manage Time<br>• Potential for Work Life Balance<br>• Ability To Work<br>• Not Being Bound to the Desk<br>• Ability To Prioritise<br>• Flexibility in Place of Residence<br>• Increase of Life Quality<br>• Being Empowered | • Organization Culture<br>• Work/Family Balance<br>• Social Isolation<br>• Professional Isolation<br>• Reduced Office Influence<br>• Informal Interaction<br>• Conducive Home Environment<br>• Focusing on Work<br>• Longer Hours<br>• Access to Resources<br>• Technical Savvy | • Learning of Handling More Responsibility<br>• Being Able To Manage Time and Task Efficiently<br>• Being Self-Organized<br>• Boundary Management<br>• Extra Hours<br>• Feeling Obligated To Be Available (Easy Access to Work)<br>• Home Environment, Availability of an Adequate Desk Situation<br>• Less Social Interaction |
| *only novelty aspects are added, Kurland and Bailey's aspects remain. | | | | |

By referring to the research question, it can be argued that the advantages and challenges formulated by Kurland and Bailey [22], as well as the new findings present crucial criteria and consequences of developing a corporate remote work policy. In order to be able to develop a solid policy as an organization, the awareness of the impact due to advantages and challenges is essential. It is striking, that the challenges for the corporation are predominant and rarely any comprehensive implementable solutions were formulated by the interviews.

## 4.2 Recommendations for Action

Based on the insights of the individual managers, recommendations for practice can be formulated. Figure 1 visualizes a framework of the criteria and consequences of developing a remote work policy. Kurland and Bailey's remote working research from 1999 [22] presents a way of working that has been partially applied in some companies back then. In respect to the advantages for the company, Kurland and Bailey only list the wider talent pool as a benefit of remote working [22]. Nevertheless, from today's perspective the wider talent pool presents not the key advantage anymore. The impact of the employer's attractiveness presents the decisive factor and the wider talent pool is a further advantage. The necessity to enable employees and future applicants to work remotely is the driving force of today's involvement and consideration of developing a corporate remote work policy. Challenges that affect the entire company have already been included in Kurland and Bailey's work [22], but not to the degree that is crucial today. Additionally, the perspective of the employee has slightly changed, caused by the pandemic situation. Therefore, issues such as an appropriately equipped desk set up or the willingness of gaining more responsibility are newly emerging criteria of remote working.

Remote working itself and its emergence as a new standard of working have a massive impact on employer's attractiveness. Societal change and the expectations of applicants and employees are the driving forces in this respect. Therefore, corporations need to undertake the strategic decision of developing a remote work policy. In the policy-making process, the attitude towards remote working must be defined, since this is crucial for the successful implementation. Hence, managers must be able to exemplify the policy and represent the company's approach accordingly. Further, advantages and challenges need to be considered and a strategy in regards to the flexibility of time and place should be developed. Additionally, the ratio of being present in the office in contrast to working remotely needs to be taken into consideration including the corresponding tasks with the consequences this entails. The policy-making process results in an organizational development, which brings changes and thereby challenging consequences, especially

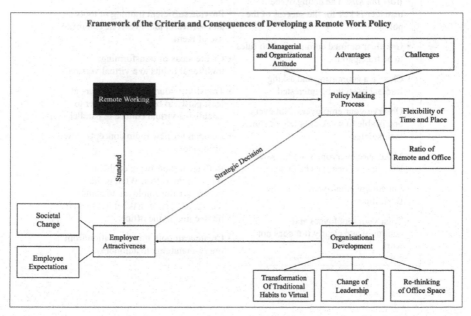

**Fig. 1.** *Framework of Criteria and Consequences of Developing a Remote Work Policy*, own illustration

for the management level. The leadership approach needs to be rethought in a way to also function virtually.

In addition, it is advantageous to rethink the traditional habits that shape the corporate culture in order to establish a virtual culture in the long term. Consequently, to the individual policy, the number of employees who are in the office at the same time changes and, thus, the office also offers an opportunity to improve work in the office through new space concepts. Supplementary, approaches abstracted from the gathered data are formulated in Table 4. These recommendations support the decisions during a policy-making process and help the implementation of a policy in a corporation successfully.

**Table 4.** *Recommendation for the Development of a Corporate Remote Work Policy*, own illustration

| Recommendation for the Development a Corporate Remote Work Policy | |
| --- | --- |
| • Define the corporation's and the management's attitude towards NWW and remote working. | • Present applicants and employees your policy to be an attractive employer. |
| • Develop the policy in a diverse team. | • Ensure your employees have an adequate and comfortable remote setup. |
| • Determine the characteristic of tasks. Which can be done remotely? Which tasks need to be done in the office? | • Ensure that the non-remote workers are treated equally. |
| • *Walk the talk.* The acting of the manager is key for the success of the policy. | • Provide sufficient training of tools and software, that everyone can make use of them. |
| • Use clear defined communication rules to be transparent. | • Create ideas of transforming traditional habits to a virtual version. |
| • Choose a supporting and asking leadership style, be interested. | • Transform your corporate culture at least partly to be virtual in order to establish a virtual culture in parallel. |
| • Observe your employees. Not every one is able to work remotely or some feel isolated. | • Be open for new room concepts, new office ideas. |
| • Make sure everyone is on the same level and no one falls behind. | • Rethink and be aware of the time spend in the office. What is the benefit for the employee to come to the office? Try to make it worth their tie to come to the office. |
| • Encourage employees to use the flexibility. | • Create a smooth transformation from micro-managing to autonomy. |
| • Trust your employees and communicate clearly if it does not work. | |

## 5 Conclusion

This research deepens the understanding of the management decision-making in respect to remote working. The COVID-19 pandemic presents the decisive factor in forcing companies to adopt remote working, and societal change and expectations are the reasons why remote working remains. This study aimed to identify criteria and consequences of developing a corporate remote work policy. While investigating the experiences of managers by applying qualitative research methods, the criteria and consequences, thus the advantages and challenges of developing a remote work policy got revealed.

Even if the challenges outweigh the advantages in terms of quantity, it is the advantages that make the remote work policy urgent. The work-life balance is the decisive criterion for satisfied and motivated employees. Accordingly, the flexibility that the work-life balance enables is demanded by current and future talent, which in turn has a significant impact on employer's attractiveness. Employees or applicants strongly demand the possibility to work remotely and to retain their gained flexibility and autonomy. Therefore, the key consequence of developing a corporate remote work policy is to remain relevant in the labor market and to be able to retain and attract employees. However, due to the policy-making process manifold challenges arise which corporations have to deal with and find ways to overcome them. The qualitative data collection has revealed diverse and detailed insights of the managerial attitude towards the current situation. Beyond, it is remarkable that although the maximum sampling variation method was applied, saturation of the statements occurred. This indicates that the topic has validity regardless of the industry or the size of a company.

The main implications for practice of this study can be derived from the framework of criteria and consequences of developing a remote work policy (Fig. 1) along with the related list of practical recommendations (Table 4). These artefacts enable companies to develop a comprehensive but also individual approach towards remote work. The main implication for theory is that existing studies are still relevant, but require permanent updates, particularly in such a dynamic topic as remote work.

This research opens up a new subject area that has not yet been sufficiently explored by the literature and previous research. However, this study is limited in scope and invites researchers to analyze the topic further to contribute to the literature. To better understand the implications of these results, future studies could address the aspects summarized in Fig. 1 and Table 4. Thereby, the development of the dependencies of the different impacting factors can be researched overtime. Figure 1 presents a first approach in visualizing criteria and consequences of developing a remote work policy, therefore it is necessary to take future considerations into account. For example, further research is required to establish whether the managerial attitude changes towards new ways of working and its impact on the change of leadership. Another approach would be the transformation of the workspace, and the idea of consciously using the time in the office. Additionally, quantitative research could analyze the dependency of impacting aspects in more detail. Since, this research focused on interviewing managers, it would be a suitable approach to also survey employees about the development of a corporate remote work policy in order to learn from their experiences. Moreover, the consideration of individual industries presents an approach, which was not guaranteed by the maximum sampling method. In addition, only one interviewee from a fully remote company was interviewed, whose statements added exciting impulses to the gathered data, but also provided starting points for further research questions. It may add valuable information to the research field to interview other remote first managers to investigate their experiences.

# References

1. Bergmann, F.: New Work, New Culture Work we Want and a Culture that Strengthens Us. Zero Books, Winchester, UK, Washington, USA (2019)
2. Gordon, G.E.: Telecommuting: planning for a new work environment. J. Inf. Syst. Manag. **3**, 37–44 (1986)
3. Larson, B.Z., Vroman, S.R., Makarius, E.E.: A guide to managing your (newly) remote workers. Harvard Bus. Rev. **18**, 27–35 (2020)
4. Thompson, C.: What If Working From Home Goes on … Forever?. https://www.nytimes.com/interactive/2020/06/09/magazine/remote-work-covid.html
5. Seabrook, J.: Has the Pandemic Transformed the Office Forever?. https://www.newyorker.com/magazine/2021/02/01/has-the-pandemic-transformed-the-office-forever
6. Onken-Menke, G., Nüesch, S., Kröll, C.: Are you attracted? Do you remain? Meta-analytic evidence on flexible work practices. Bus. Res. **11**(2), 239–277 (2017). https://doi.org/10.1007/s40685-017-0059-6
7. Gratton, L.: How to do hybrid right. Harvard Bus. Rev. **99**, 65–74 (2021)
8. Gelles, D.: What Bosses Really Think About the Future of the Office. https://www.nytimes.com/2021/11/12/business/corner-office-return.html
9. de Kok, A., Helms, R.W.: Attitude towards the new way of working - a longitudinal study. In: 24th European Conference on Information Systems, ECIS 2016
10. Jemine, G., Dubois, C., Pichault, F.: From a new workplace to a new way of working: legitimizing organizational change. Qual. Res. Organ. Manage.: Int. J. **15**, 257–278 (2020)
11. de Menezes, L.M., Kelliher, C.: Flexible working and performance: a systematic review of the evidence for a business case. Int. J. Manage. Rev. **13**, 452–474 (2011)
12. Felstead, A., Jewson, N., Phizacklea, A., Walters, S.: Opportunities to work at home in the context of work-life balance. Hum. Resour. Manage. J. **12**, 54–76 (2002)
13. Rupietta, K., Beckmann, M.: Working from Home. Schmalenbach Bus. Rev. **70**, 25–55 (2018)
14. Bailey, D.E., Kurland, N.B.: A review of telework research: findings, new directions, and lessons for the study of modern work. J. Organ. Behav. **23**, 383–400 (2002)
15. Cooper, R., Baird, M.: Bringing the "right to request" flexible working arrangements to life: from policies to practices. Empl. Relat. **37**, 568–581 (2015)
16. Olson, M.H.: New information technology and organizational culture. MIS Q. **6**, 71 (1982)
17. Jemine, G., Dubois, C., Pichault, F.: When the Gallic village strikes back: the politics behind 'new ways of working' projects. J. Chang. Manage. **20**, 146–170 (2020)
18. Hill, E., Ferris, M., Märtinson, V.: Does it matter where you work? A comparison of how three work venues (traditional office, virtual office, and home office) influence aspects of work and personal/family life. J. Vocat. Behav. **63**, 220–241 (2003)
19. Chen, Y., Fulmer, I.S.: Fine-tuning what we know about employees' experience with flexible work arrangements and their job attitudes. Hum. Resour. Manage. **57**, 381–395 (2018)
20. Groen, B.A., van Triest, S.P., Coers, M., Wtenweerde, N.: Managing flexible work arrangements: Teleworking and output controls. Eur. Manag. J. **36**, 727–735 (2018)
21. Vroman, S.: Organizational intentions versus leadership impact: the flexible work experience. JVBL **13**, 13 (2020)
22. Kurkland, N.B., Bailey, D.E.: The advantages and challenges of working here, there anywhere, and anytime. Organ. Dyn. **28**, 53–68 (1999)

23. Clark, S.D.: Decision to telework: a synthesized model. In: Proceedings of the Hawaii International Conference on System Sciences, 393–402 (1998)
24. Daniels, K., Lamond, D., Standen, P.: Teleworking: frameworks for organizational research. J. Manage. Stud. **38**, 1151–1185 (2001)
25. Rowley, J.: Conducting research interviews. Manage. Res. Rev. **35**, 260–271 (2012)
26. Gläser, J., Laudel, G.: Experteninterviews und Qualitative Inhaltsanalyse. VS Verlag für Sozialwissenschaften, Wiesbaden (2010)
27. Mayring, P.: Qualitative Content Analysis - Theoretical Foundation, basic Procedures and Software Solution. SSOAR, Klagenfurt (2014)

# Application for Generating Camouflages from Satellite Photographs

Aneta Poniszewska-Marańda$^{(\boxtimes)}$ ⓘ, Krzysztof Stepień, and Michal Suszek

Institute of Information Technology, Lodz University of Technology, Lodz, Poland
aneta.poniszewska-maranda@p.lodz.pl, krzysztof.stepien@dokt.p.lodz.pl

**Abstract.** The camouflages are present on civilian market and in military. First of all they are used by solders but also they can be used by hunters, military enthusiasts, clothing designers and game developers. These people are forced to design their own or choose between already developed camouflages, both military and civilian, so they cannot adapt it to their needs in a fast and easy way.

The paper presents and application as a solution to the problem of generating digital camouflages for civilian and military purposes. The proposed approach uses the satellite images of target areas to analyse colours and apply an algorithm to generate digital camouflages. The implementation of this approach allows users to easily generate digital camouflages to meet their specific needs. Currently, there are few solutions on the market that support this kind of customization and flexibility. The paper also presents the results of camouflage quality analysis. The analysis was carried out on camouflages obtained with the proprietary application, two selected market generators and selected military digital camouflages.

**Keywords:** information systems engineering · digital camouflage · image analysis · generating images · modelling

## 1 Introduction

The use of camouflage has a long history in both nature and human society. In nature, animals use camouflage to avoid detection by predators, while predators use it to avoid detection by their prey. In humans, the use of camouflage originated with hunters who sought to get close to their prey without being detected. The military also began using camouflage during World War I, when it became clear that brightly colored uniforms were not effective for hiding soldiers on the battlefield. Since then, the military has continued to research and develop effective camouflages, particularly in response to advances in technology such as thermal imaging. Camouflage not only helps to hide soldiers and military equipment, it also helps to project an image of professionalism and elite status.

Camouflage is not just used by the military, but also by civilians, such as hunters and clothing designers. However, there are few companies that offer the ability to design custom camouflage, and most of these are focused on hunters. Private camouflage projects on the internet exist, but they have their own disadvantages. There is a need for solutions on the market that allow users to easily and quickly generate custom camouflages based on satellite images.

This paper examines the current state of camouflage generators, including the advantages and disadvantages of their solutions. The paper then presents the development of a camouflage generation algorithm and an accompanying application that allows users to easily generate and customize camouflages. The results of the generator are compared with other commercially available camouflages and those generated by online generators.

The paper is structured as follows: Sect. 2 provides an overview of camouflage and selected online camouflage generators. Section 3 discusses the technical aspects of proposed camouflage generator application, including the algorithm and architecture of the "CammoGenerator" application while Sect. 4 presents an evaluation of camouflage generation application.

## 2   Practical Aspects of Camouflage and Its Generation

The main idea of a camouflage is to hide the selected object such as building, vehicle or person, or change its visible features, e.g. speed or direction. To make that it is necessary to select and use the proper colours, shapes, materials or lighting.

The following types of camouflage can be distinguished depending on the used technique [1,2]: *Decorative* – based on the use of artificial masking structures, which are to show a different state of the object than it is in fact. It can be, for example, an imitation of destroyed building above a combat post, which makes it uninteresting to the enemy, while maintaining its defensive potential. *Imitative* – it consists in making a given stationary object similar to another harmless object of the surroundings, e.g. a camouflage showing the position of a soldier hidden under a trunk, which is a natural object that does not arouse suspicion found in forests. *Colour* or *classic* camouflage – by changing the colours of the object, as well as the appropriate deforming patterns, it blends the object into its surroundings, e.g. the need for at least a minimal similarity of the object's colour to its surroundings. *Natural and vegetal* – it consists in using to mask the natural properties of the terrain, e.g. fog, as well as live or cut vegetation. Live vegetation is used to mask stationary positions, which makes it difficult to see objects from the air and land. The cut vegetation allows masking moving and smaller objects. Such a camouflage is effective in the area from which the vegetation was obtained, and its durability, which translates into effectiveness, decreases with time, e.g. the use of grass to disguise the soldier and his weapons.

The other classification divide the camouflages into three main types: *Mimetic* – usually made up of small spots to make the object resemble its surroundings, which naturally consists of a noise of various colours. These spots are to make the object similar to a given colour noise occurring in a given area. There is a micro and macro pattern in mimetic patterns. *Deformational* – composed of overlapping geometries of soft, natural or sharp geometric shapes. Camouflages of this type are very effective at longer distances, because large shapes do not merge into one colour. *Pixel/Digital* – currently the most popular type of camouflage. Its pattern is built by pixels filled with appropriate colours. The principle of operation is similar to the mimetic camouflage, it also has a micro and macro pattern.

The *micro pattern* is a composition composed of small spots that can be distinguished only when viewed from close range [1–3]. The role of this type of pattern is to hide the object from spatial vision, i.e. the area where the observer does not focus the pattern and thoughts. The *macro pattern* is a large stain designed to break the shapes of camouflaged objects. Its role is mainly to influence central vision, focused on finding shapes. When an object is spotted through spatial vision, the observer will focus on it and try to identify the object by recognizing the shape of the shadows as well as the lines of symmetry.

The colours in camouflage are a very important element that creates it. During the stage of creating the camouflage, it is very important that the colours are as close as possible to those in the real world. Such colours can be obtained from the analysis of photos of areas at different times of the day and year, as well as weather conditions. When creating universal camouflages, the environments of a given area is analysed and the most common colours are chosen.

Nowadays, there are no military camouflage generators available on the market¿. However, the users interested in using the camouflage must use already created commercial or military solutions. There are a few private camouflage generators that can be found in Internet. Two generators were selected for analysis: Camouflage Generator [4] and Digital Camouflage Generator [5].

*Camouflage Generator* [4] is one of the most extensive camouflage generators available. It offers a large selection of types of generated patterns, these are: two types of pixel and four types of deformation with already defined colour palettes. It also provides a limited ability to edit the appearance of the camouflage in terms of: colour, degree of pixelation and size of spots. Thanks to predefined patterns with colours and simple editing, the application allows for intuitive generation of various camouflages. The main disadvantage is the lack of the ability to edit the scale and size of the camouflage, which is a considerable difficulty, especially when there is a need to apply a pattern to, e.g. part of an outfit. Additionally, the user has no support in choosing the colours, so he has to analyse the terrain where the generated camouflage will be used. The generated pattern does not have a macro-pattern, so the pattern loses its effectiveness.

*Digital Cammo Generator* is a simple browser based digital camouflage generator. There is only one type of pattern available in two colour versions: forest and desert. Both colour versions already have a prepared colour palette, without the possibility of changing them. The generated camouflage has a micro-pattern in the form of colour noise, made of rectangles, but it does not have a macro-pattern, which makes it a mimetic type, without deforming properties, which can be a significant difficulty in masking a given object. Likewise, it is not possible to change the scale and size. Digital Cammo Generator is a simple application that allows to generate very simple and not advanced patterns. Its simplicity facilitates the generation process, but it does not allow to adjust the generated pattern to own needs. A predefined colour palette speeds up the process of creating a camouflage, but the inability to change colours means that the given generator creates camouflages that are limited to a very small number of areas.

The use of these both generators does not give the user the possibility to have the tools that allow him to choose the right colours for his own camouflage. The generated patterns do not contain the macro-pattern. It reduces the camouflage properties due to the lack of stains deforming the solids of camouflaged object. Moreover, the size of pattern is always constant, which is the problem when user wants to apply it to a large area. Both generators have predefined colour palettes so that the user can see a sample pattern, but not all of them allow to change these colours.

## 3  Technical Elements of Application for Camouflage Generator

The created camouflage generator, named *CammoGenerator*, was developed to generate the digital camouflages based on satellite images. The user submits the satellite photos that is interested for him and creates a colour palette based on it, which he then uses to compose his own camouflages. The following functional requirements have been formulated to fulfil the needs of its possible users:

- User can add colours to the colour palette in two ways: (a) enter a picture and generate the given number of colours, then add the selected colours to the palette, (b) choose the colour we want from the dialogue box.
- User can remove a colour from the palette.
- User can change the size of generated image and its background colour.
- User has the option to add or remove a colour from the camouflage.
- User has the option to generate a camouflage and save it in a file.
- User has the option to enter the name and format of the file.
- Application checks if the colours in colour palette are not repeated.
- Application stores and displays the colours generated from a photo by user.

The architecture of *CammoGenerator* application was based on a multi-tier pattern that separates the user interface, processing and data storage from each other into separate layers. This approach allows to modify independently each layer separately, without adversely affecting other layers.

The three-tier architecture of *CammoGenerator* application was presented in Fig. 1 by the use of separating three layers. The logic layer consists of two components: *ColorExtractor*, responsible for providing the function of extracting colours from the image, and *CammoGenerator*, which provides a GUI interface for communication with the data layer and a simple interface for generating a camouflage. The presentation layer includes four classes responsible for interaction with the user and presenting data obtained from the logic layer. The particular layers are described in the following subsections.

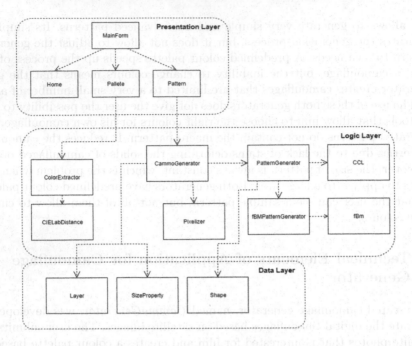

**Fig. 1.** The architecture of *CammoGenerator* application based on multi-tier pattern.

## 3.1   Data Layer

Because there is no the need to store and load data in the created application, the data layer does not communicate with any database or file. This layer consists only of three classes, presented in the class diagram in Fig. 2, storing the data needed for the generator to operate:

1. *SizeProperty* class is responsible for storing information about the size of the generated camouflage. This class is used by the *CammoGenerator* class and is passed on to other classes, which allows to keep the same values of *SizeProperty* class attributes in all classes and edit it globally.
2. *Layer* class stores the information needed to generate a single colour in a camouflage, determined as a layer. Contains private fields and properties needed to gain access. The *color* field contains the colour of the layer, *size* field is an object of *SizeProperty* class and stores the size of layer, *seed* field is an integer containing a number which is a random seed needed to obtain different noise values for each layer. In addition, this class has a *Map* attribute, which is a two-dimensional array of boolean type that holds pattern information for a given layer. The color, size and seed fields are set once by the constructor during object creation and cannot be edited further.
3. *Shape* class contains information about a single spot in a camouflage. Extracting shapes from a layer is a deliberate procedure that allows for a simple and effective implementation of the effect of interpenetrating colour layers, which is described in the next subsection.

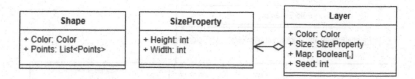

**Fig. 2.** Class diagram presenting the data layer in *CammoGenerator* application.

## 3.2    Logic Layer

Logic layer of *CammoGenerator* application is divided into two components: *ColorExtractor*, which provides an interface for extracting colours from an image, and *CammoGeneratror*, which provides an interface for generating and editing the camouflage, and gives the presentation layer access to the data layer. Such a division allows for the separation of functions into independent components, which can be changed and expanded in the future, depending on users needs.

**ColorExtractor Component.** This component is responsible for providing the function of extracting colours from an images. The class diagram in Fig. 3 shows the relationship between the implemented classes and the classes from external packages. The main class of the component is *ColorExtractor* class with static *GetPallete()* method that takes: a bitmap image, the number of colours the user wants, and BacgroundWorker to monitor and report the progress of colour extraction operation. The *ColorExtractor* class uses the *kMeans* external class from *Accord.MachineLearing* package. It uses the implemented *Distance()* method from *CIELabDistance* class, which implements the *IDistance* interface from *Accord.Math.Distance* package, to calculate the distance between colours. To calculate the distance between colours, the *CIELabDistance* class uses the *CnCComparison* class from the *ColorMine.ColorSpaces* package. The process of selecting the colours from an image is presented in Fig. 4.

The *colour extraction algorithm* is based on the centroid algorithm (*kMeans*), which thanks to the use of *CIELab* colour space allows to group colours according to their similarity, similar to the perception of colour differences in humans. For the grouped colours, the average colour for each of the groups obtained is calculated. The obtained average colours after conversion to the RGB colour space are the final result of the algorithm.

The *Kmeans* class from the *Accord.MachineLearnig* library, contains the implementation of centroid algorithm. This algorithm groups similar points concentrated around the searched centroids, being the mean value of the value of the points assigned to a given group (centroid). When the image is analysed, the algorithm obtains a set of colours that are vectors of the three values. Then, the initial values of centroids are drawn. After they are drawn, the colours are assigned to the centroid to which they are most similar. When all colours are separated, centroids are assigned the average value of the colours in their set. Then, the mean square error is computed. When this error is greater than the

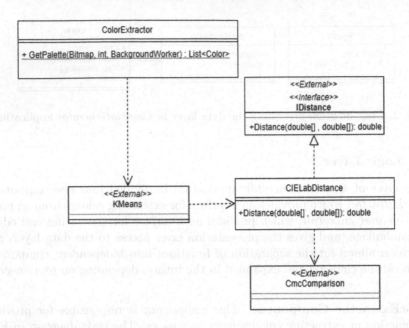

**Fig. 3.** Class diagram for *ColorExtractor* component in logic layer of *CammoGenerator* application.

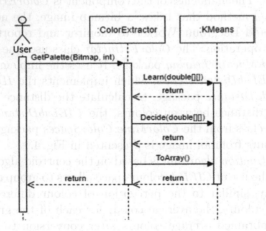

**Fig. 4.** Sequence diagram for the use case of selecting the colours from an image.

assumed tolerance, we repeat all operations, omitting the initial randomization of centroid positions. In the case when calculated error is lower than the assumed tolerance, the algorithm is interrupted.

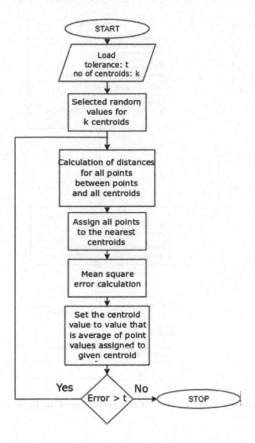

**Fig. 5.** Block diagram of centroid algorithm for *kMeans* class.

**CammoGenerator Component.** This component is based on a facade design pattern that provides a simple interface for generating camouflage and managing its properties. The classes included in this component are shown in the class diagram in Fig. 6. Facade is the *CammoGenerator* class that provides parameters of the generated camouflage such as size, background colour, layers and *Generate()* method which is used to generate the camouflage. The facade needs a *PatternGenerator* class that provides *GetShape()* method to generate a spot list for a given layer, and a *Pixelizer* class that has *Pixelize()* method to convert a bitmap image into a square image of a given size. For the *GenerateMap()* method to work, the *PatternGenerator* class requires the use of the *ConectedComponentLabeling* and *fBMPatternGenerator* classes. The *ConectedComponentLabeling* class is used to convert a two-dimensional list of logical type into a list of spots, while *fBMPatternGenerator* is a class for generating the pattern itself, which is done using *Simplex Noise* [6] and its transformations using the implemented *fractional Brownian motion* [7] The process of generating a camouflage preview from an image is presented in Fig. 7.

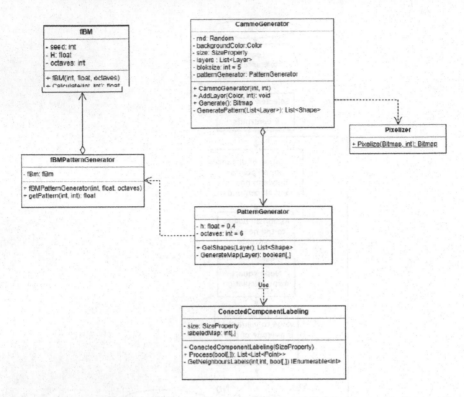

**Fig. 6.** Class diagram for *CammoGenerator* component in logic layer of *CammoGenerator* application.

The schema of proposed camouflage generation algorithm of *Generate()* function is shown in Fig. 8. The following parameters were given to the function: size of created image, background colour and layers needed to generate the camouflage. The generation of camouflage begins with the operation of generating the shapes for each colour, i.e. layer. This operation consists of two stages: generating a map and extracting the shapes. The map generation algorithm consists of noise generation operation *SimplexNoise* (Fig. 9). This noise is then transformed with function *fbm*, which transforms the shapes generated by noise into the camouflage spots: $fbm\left(fbm\left(fbm\left(p + V_1\right) + V_2\right) + V_3\right)$, where: $p$ – pair $(x; y)$, specifying the coordinates on the two-dimensional array of the camouflage pattern, $v_1$; $v_2$; $v_3$ - appropriately selected vectors, transforming the noise into a camouflage pattern, $fbm()$ – *fractional Brownian motion* function [7].

The *fbm()* function adds successive less amplitude noise to the base noise while keeping it dependent on the base noise, resulting in increased detail for the base noise. It allows to get more varied noise values. This effect is shown in Fig. 10. After transforming the noise using *fbm* function, a *modulo 256* operation is performed on the noise values. Depending on whether the value is greater or

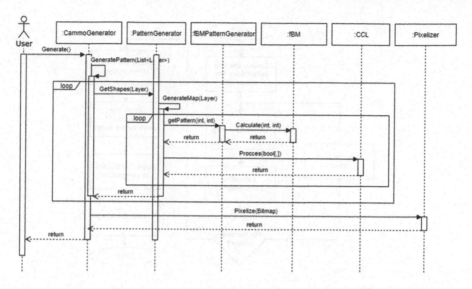

**Fig. 7.** Sequence diagram for the use case of generating a camouflage preview.

less than a predetermined threshold, the pattern table for the layer is set to *true* or *false* at the *p* position.

When the entire pattern table is filled with values, the shapes are created from the *true* values using the algorithm proposed in Fig. 11. It groups points into shapes. It is done by iterating over all *true* values in the pattern table. For a given point in the table, labels of neighbour points are checked. In the case where the neighbours do not have assigned labels, the current value of the label is assigned to the point and the value of the label is increased by one. When a point has at least one neighbour with a label, the smallest neighbour label is assigned to the point and its neighbours' labels, so joining of labels is take place.

After generating the shapes for each layer, a blank image is generated over which the background colour is applied. The shapes are then superimposed on the image in random order. The overlapping effect was easily achieved by mixing the order in which the stains were applied. The final stage of camouflage genera-tion is its deformation, revealing squares of colours, called pixelation. This effect is achieved by dividing the image into squares of the selected size (for Cammo-Generator the square is 5), and then by determining the dominant colour for each square. When selected, the squares are filled with their dominant colour.

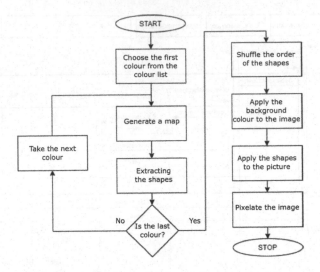

**Fig. 8.** Schema of camouflage generation algorithm.

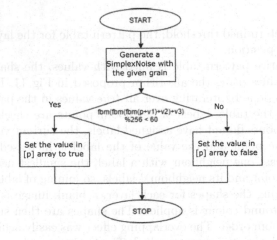

**Fig. 9.** Schema of camouflage map generation algorithm.

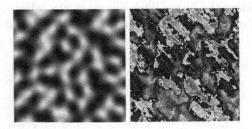

**Fig. 10.** Sample of SimplexNoise noise (on the left) and noise transformed with proposed function *fbm* (on the right).

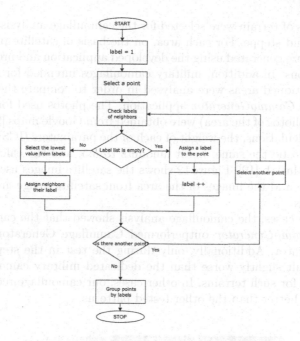

**Fig. 11.** Schema of shape extraction algorithm.

### 3.3   Presentation Layer

The graphic layer of the *CammoGenerator* application was created using the Windows Forms library. It consists of the *FormMain* main view, which contains two objects: a colour palette, which is a list of colours, and an object of the *CammoGenerator* class. The main view has the ability to display three views: Home, Pallete and Pattern. The *Home* view contains information about the application and instructions for generating the camouflage, *Pallete* is the view for creating and editing the colour palette, and *Pattern* is for editing and generating the camouflage. The *FormMain* class gives other views access to the palette and the generator, thanks to which the *Pallete* and *Pattern* views are able to be dynamically replaced without data loss.

## 4   Evaluation of Application for Camouflage Generation

In order to estimate the quality of generated camouflages and compare them with those generated by other camouflage generators, we used the method of estimating the quality of camouflage based on colour difference and size of gradient [8]. The method receives as an input a background slice in which the quality of camouflage is tested and tested camouflage itself (both images must be of the same size). Then the ICSI (Image Colour Similarity Index) [8] and GMSD (Gradient Magnitude Similarity Deviation) [8] values are calculated between the background and the camouflage.

Three types of terrain were selected for the camouflage analysis: desert, temperate forest and steppe. For each area, on the basis of satellite photos [9], the camouflages were generated using the developed application and previously analysed applications. In addition, military camouflages intended for operations in the above-mentioned areas were analysed in order to compare the camouflages created by the *CammoGenerator* application. The photos used for the analysis (satellite and photos of the area) were obtained from Google maps [9]. When performing the calculations, the weight of each of the parameters (ICSI and GMSD) was assumed to be the same, i.e. it amounts to 0:5, and the calculations were performed in Matlab [10]. Figure 12 shows the satellite images used to generate the camouflage and the images of the area from satellite image area needed for quality analysis.

In all three cases, the camouflage analysis showed that the camouflage generated by *CammoGenerator* outperformed Camuflage Generator and Digital Cammo Generator. Additionally, only during the test in the steppe terrain it achieves a result slightly worse than the dedicated military camouflage, which was developed for such terrains. In other cases, our camouflage receives an estimated quality better than the other tested patterns.

**Fig. 12.** Photographs of the terrains (satellite and real) used during the camouflage analysis: forest terrain, steppe terrain and desert terrain.

**Fig. 13.** Camouflage generated by CammoGenerator and the real terrains in forest terrain, steppe terrain and desert terrain.

# 5   Conclusions

The developed algorithm and accompanying CammoGenerator application were designed to streamline and improve the process of selecting colours for camouflage. The solution was implemented using a three-tier architecture, which allows for modifications to be made in each layer without affecting the other layers. Additionally, the logic layer includes separate components for color extraction and camouflage generation, allowing for easy changes to the algorithms used by each component. This modular design allows for flexibility and adaptability in the solution. The performance of the proposed solution was evaluated and compared with other selected generators and military camouflages. The results showed that the approach produces higher-quality camouflages than commercially available generators, and is comparable to or better than military-grade camouflages. This was supported by the analysis results, which showed that the generated camouflages were similar to or better than those used by the military. The results show that the proposed solution is able to easily generate high-quality camouflages based on satellite photos. The CammoGenerator application addresses the shortcomings of existing generators and provides users with a simple and effective tool for generating camouflage. The solution uses a neural network for colour extraction.

# References

1. Pettersson, R.: Visual camouflage. J. Vis. Literacy **37**(3), 181–194 (2018)
2. Newark, T.: The Little Book of Camouflage. Osprey Publishing, Oxford (2013). ISBN: 978 1 78200 831 6
3. Foster, H.-F.: Fm5-20 Camouflage, Basic Principles. War Department, Washington (1944)
4. Astrom, U.: Camouflage generator. https://www.happyponyland.net/camogen.php. Accessed 10 Feb 2021
5. Chandler, A.: Digital Camo Generator. https://cowdd.com/game/canvas/index.php. Accessed 10 Feb 2021
6. Gustavson, S.: Simplex noise demystified (2005)
7. Shevchenko, G.: Fractional Brownian motion in a nutshell. In: International Journal of Modern Physics: Conference Series, Vol. 36, p. 1560002 (2015)
8. Bai, X., Liao, N., Wu, W.: Assessment of camouflage effectiveness based on perceived color difference and gradient magnitude. Sensors **20**(17) (2020). ISSN: 1424–8220. https://www.mdpi.com/1424-8220/20/17/4672
9. Google Maps. https://www.google.com/maps. Accessed 10 Feb 2021
10. Matlab. https://www.mathworks.com/products/matlab.html. Accessed 10 Feb 2021
11. MARPAT Camouflage. https://pl.pinterest.com/pin/802555596073688790/. Accessed 10 Feb 2021
12. EMR Camouflage. https://militaristwear.com/image/cache/data/05/3339298-600x600.jpg. Accessed 10 Feb 2021

# Bibliometric Analysis of Publications on the Evacuation of Hazards Using GIS as a Decision Support Tool from the Scopus Database

Atifa Albalooshi[✉] ⃝, Khalifa Alebri ⃝, and Maria Papadaki

The British University in Dubai, Dubai, UAE
{20196586,20206542}@student.buid.ac.ae,
Maria.papadaki@buid.ac.ae

**Abstract.** The increasing diffusion of geographic information systems (GIS) in the domain of evacuation and disaster management, depicts the contributions and efforts of the countries in risk reduction. It has been witnessed in the last decade the global challenges and impacts caused by natural hazards and climate risk. One of these challenges is to evacuate the exposure community promptly and efficiently, these impacts either are predicted or non-predicted, which shows the lack of understanding of these hazards or mitigation tools and strategies for risk reduction. This study aims to explore the publications in the evacuation domain of multi-hazards using Geographic Information System (GIS) as a decision-making tool, as this domain was found only between 1996 and 2022. The researchers focused only on the Scopus database platform using the VOSviewer to apply the bibliometric analyses to understand the evolution and development of these domains. Different bibliometric analysis such as citation, co-authorship, co-occurrence, and co-citation were performed on different units of analysis. The outcome of the bibliometric analysis will support future researchers to build their road map and develop this domain based on a solid database of the current bibliometric analysis.

**Keywords:** Bibliometrics · Decision Support System · Evacuation · GIS · Multi-Hazards

## 1 Introduction

Emergencies have occurred often in recent years across the world. In addition to traditional natural disasters, and emergencies, such as hurricanes, earthquakes, and floods, new types of events such as terrorist attacks, stampedes, and fires are emerging; these catastrophic events have caused tremendous damage to mankind, human life, and property, and other losses are uncountable. (Galindo and Batta, 2013). Without a doubt, emergency evacuation is a multidisciplinary scientific issue requiring expertise in engineering, operational research, transportation, and social sciences. (Tanimoto et al. 2010).

© The Author(s), under exclusive license to Springer Nature Switzerland AG 2023
M. Papadaki et al. (Eds.): EMCIS 2022, LNBIP 464, pp. 630–648, 2023.
https://doi.org/10.1007/978-3-031-30694-5_44

However, a geographic information system (GIS) is a system of hardware, software, and procedures designed to support the capture, management, manipulation, sophisticated analysis and modelling, and display of spatially referenced data suitable for solving complex planning and management problems. In terms of risk minimization, GIS could be used to detect and model hazards, as well as to begin assessing the impacts of possible emergencies. (Cole, J.W et al. 2005).

A study done by Bonilauri et al. (2021) developed a GIS-based risk analysis tool for emergency evacuation for buildings and coastal zones. The results reveal that 33% of the buildings can be evacuated in 4 min and that a 10-min warning time is necessary for a complete and well-distributed evacuation, with the population evenly dispersed between all evacuation exits to minimize congestion. Another study conducted by Thumerer et al., (2000) shows the creation of GIS-based models for assessing climate change-related coastal flood threats along the east coast of the UK. The findings revealed the significant uncertainty involved with estimates of sea level rise. According to the authors, GIS has the potential to be a significant tool in evaluating not just flood risk, but also a wide variety of interactions within the larger spectrum of services offered by the coastal zone. It also has the potential to be utilized as a tool to aid in the resolution of future resource conflicts.

A bibliometric analysis of a domain is based on information such as publishing patterns, citation patterns, research productivity, and keyword analyses of various entities such as authors, nations, enterprises, sources, and literature. It provides numerous interesting insights and sheds light on various challenges encountered by information scientists in assessing the evolution, progress, and current state of affairs of a topic (Sengupta, 1985). The current research provides a bibliometric analysis that unravels many quantitative features of the multi-hazard, evacuation, and GIS domains. The investigation is based on a collection of 220 Scopus-indexed publications published between 1996 and 2022. This bibliometric research includes several quantitative elements based on domain-specific assessments of publishing trends, citation patterns, and keyword occurrence and co-occurrences. The following are the primary objectives of the current study:

- To explore the publication trends and visualize research collaboration across multiple entities.
- To explore the citations and co-citations trends and show research connections between distinct authors and entities.
- To explore the co-authorship and multiple entities and authors and the connections between them.
- To explore the keyword's co-occurrence patterns to find research directions and frontiers.

The main research questions are as follows:

- RQ1: What is the intellectual structure of knowledge based on hazard evacuation using GIS?
- RQ2: What key concepts have been explored on the topic of hazard evacuation using GIS and how are they related?

● RQ3: What is the nature of the collaboration that is evident in the publications of hazard evacuation using GIS?

The research is divided into four sections to demonstrate the varied significance of the proposed topic. Section 1 introduces the main areas that have been covered in the following sections. Section 2 discussed the research methodology that included the technique used for data collection and processing, as well as the selection of analytical methodologies and tools, for the current research. Section 3 summarized the findings and their analysis. Section 4 explained the study's conclusions and recommendations. These sections of the study are covered in detail below.

## 2   Research Methodology

The bibliometric analysis of this study had been applied through many stages. The first stage was to identify the research objectives and the research questions that need to be investigated and answered to achieve the objectives of the bibliometric analysis of this research. The second stage is determining the publications database source and data collection. There are many database platforms such as Scopus, Web of Science, Dimensions, PubMed and Google Scholar. However, this study focused only on Scopus, which is a popular online multidisciplinary academic database that contains materials from articles, peer-reviewed journals, books, and conference proceedings.

The third stage was the research selection and quality evaluation is linked to the inclusion/exclusion criteria. This was done to ensure that a complete content analysis could be carried out to evaluate how and to what extent the articles contributed to the two research concerns outlined for this study. As a result, the outcomes of the study were confined to peer-reviewed scientific publications and articles from international conference proceedings, excluding book chapters and reviews. Furthermore, this domain of study was found during the period from 1996 to September 2022.

The data were analyzed using the criteria provided in this section in the fourth stage, as the first search was using the search engine in Scopus with the option of searching the keyword within the title, abstract, and keywords. The researchers first used one keyword "evacuation" and found "46,686" documents results, then the results of the document were reduced to "710" once the researchers added the second keyword in the title "GIS". After that, the results were reduced to "226" once added the third keyword was "hazard". Finally, the researchers excluded the book's chapters and review and kept only the Peer-reviewed academic articles and conference papers. Then the final database was exported in CSV format to analyze it in different software. The data was explored and analyzed initially in the Excel software. Then, it was processed and analyzed in both Excel and VOS-viewer software for bibliometric analysis.

# 3 Analysis and Results

Based on the selection criteria outlined in the previous section, 220 articles were selected that met the inclusion and exclusion criteria. As a result, this part provides a summary of these papers as well as answers to the three research questions that have been formulated.

## 3.1 Citation

**Top Countries' Contributions in the Domain of Evacuation, Geographic Information Systems (GIS) for Multi-Hazard management.** The researchers identified the countries with publication documents more than three in Scopus and identified the top ten countries with the highest citation in their publications. Fifty-four countries contributed to the publication's domain of implementing the Geographic Information System (GIS) on the base of disaster management of the evacuation within various fields of hazards. This study focuses only on the contributions of articles, conferences, and review papers. Twenty-two countries out of fifty-four published at least three documents. The United States is the most active country in this domain accounting for (15.63%) of the total publications from 1996 – 2022. As this domain investigated only in the last three decades. The top ten countries with more than three publications are the United States (15.63%), Japan (12.50%), China (12.50%), Indonesia (10.27%), France (5.80%), Philippines (4.91%), Italy (4.46%), India (3.57), South Korea (3.13%) and United Kingdom (3.13%). As shown in the following Table 1, the top ten countries with a higher citation in their publications are Cuba (124), Netherlands (51.33), Bangladesh (44), Australia (39.75), Ireland (35), United States (26.94), India (26), Germany (26), Hungary and Singapore (21). The main clusters with higher collaboration with other countries are the United States, followed by Japan, then China, and Indonesia as shown in (see Fig. 1).

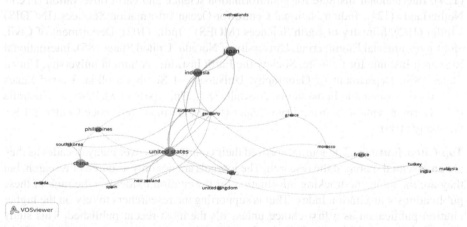

**Fig. 1.** The top four main Clusters of the Countries' collaboration in the study domain.

**Top Cited Organizations.** The nations are developed by involving all the community in identifying the gaps either in academic or practical fields. The contribution commitment

**Table 1.** The top ten countries with a higher average citation in their publications.

| No | Country | Documents | Citations | Av. Citations | Total link strength |
|----|---------|-----------|-----------|---------------|---------------------|
| 1 | Cuba | 1 | 124 | 124.00 | 0 |
| 2 | Netherlands | 3 | 154 | 51.33 | 1 |
| 3 | Bangladesh | 1 | 44 | 44.00 | 5 |
| 4 | Australia | 4 | 159 | 39.75 | 12 |
| 5 | Ireland | 1 | 35 | 35.00 | 1 |
| 6 | United States | 35 | 943 | 26.94 | 43 |
| 7 | India | 8 | 208 | 26.00 | 5 |
| 8 | Germany | 6 | 156 | 26.00 | 17 |
| 9 | Hungary | 1 | 21 | 21.00 | 0 |
| 10 | Singapore | 1 | 21 | 21.00 | 1 |

from organizations has been noticed in different disciplines, as this study focuses on the disaster management domain that is related to the implementation of geographic information systems in developing the evacuation to ensure the safety and mitigation of existing hazards and threats to human lives. As shown in Table 2 the organizations have been selected out of 452 with citations of 100 and above with at least one minimum document. The National Center for Geographic Information and Analysis (NCGIA), and University of California, in United States has accounting the higher average citation (176), followed by Department of geography, university of Utah - United States (174), then Instituto de Geología y Palentología (IGP), vía blanca y carretera central - Cuba (124), International institute for geoinformation science and earth observation (ITC) - Netherlands (124), Indian National Centre for Ocean Information Services (INCOIS) - India (102), Ministry of Earth Sciences (MOES) - India (102), Department of Civil, and Environmental Engineering, University of Nevada, United States (85), International Research Institute for Climate, Society the Earth Institute, columbia university, United States (85), Department of Geography, University of South Carolina, United States (83), Csiro Sustainable Ecosystems, Australia (81), University of Melbourne, Australia (81), German Remote Sensing Data Center (DFD), German Aerospace Center (DLR), Germany (106).

**Top Cited Journals.** The journals proved their contributions over many decades as they are an essential source of the research. The journals are not only a source of research, but they are including the tracking information for the citation to identify the rank of these publications with citation index. That is supporting the researchers to rely on the higher citation publication as a first choice unless it's the most recent published. This study identified the top cited journals in the domain of disaster management which is related to GIS and evacuation to make it easier for future studies to search directly in these journals. As shown in Table 3, the top ten journals out of 147 journals have more than three publications. The Natural Hazards journal is the highest one with 14 publications during the period 1996–2022, followed by Earth and Environmental Science with 13

**Table 2.** Shows the highest cited organizations in the current study domain within 100 citations and above.

| No | Organizations | Documents | Citations | Av. Citations per Document | Total link strength |
|----|--------------|-----------|-----------|---------------------------|---------------------|
| 1 | National Center for Geographic Information and Analysis (NCGIA), University of California, United States | 1 | 176 | 176 | 3 |
| 2 | Department of geography, university of Utah, United States | 1 | 174 | 174 | 2 |
| 3 | Instituto de Geología y Palentología (igp), vía blanca y carretera central, Cuba | 1 | 124 | 124 | 0 |
| 4 | International Institute for geoinformation science and earth observation (ITC), Netherlands | 1 | 124 | 124 | 0 |
| 5 | Indian National Centre for Ocean Information Services (INCOIS), India | 1 | 102 | 102 | 0 |
| 6 | Ministry of Earth Sciences (MOES), India | 1 | 102 | 102 | 0 |

publications. However, the top journal with the highest citations average (89.67) is the international journal of Geographic Information Science, followed by Landslides journal with (53.67) average citations for their publication in the same period (Tables 1 and 2).

**Top Cited Documents.** The researchers identified the top 26 publications in the domain of the study within the same period 1996–2022. These publications have been cited minimally by 30 authors. The top one cited by 176 is about modeling community vulnerability using GIS, followed by the same author (174) however his focus in this study is on microsimulation of neighborhood evacuation in the urban-wildland interface. The

**Table 3.** The top ten journals out of 147 journals published more than three documents.

| No | Source | Documents | Citations | Av. Citations per Document | Total Link Strength |
|---|---|---|---|---|---|
| 1 | Natural Hazards | 14 | 278 | 19.86 | 6 |
| 2 | Iop Conference Series: Earth and Environmental Science | 13 | 14 | 1.08 | 0 |
| 3 | Remote Sensing and Spatial Information Sciences | 6 | 6 | 1.00 | 1 |
| 4 | 37th Asian Conference on Remote Sensing, ACRS 2016 | 5 | 0 | 0.00 | 0 |
| 5 | Journal of Disaster Research | 5 | 103 | 20.60 | 2 |
| 6 | International Journal of Disaster Risk Reduction | 4 | 41 | 10.25 | 4 |
| 7 | International Journal of Disaster Resilience in the Built Environment | 3 | 8 | 2.67 | 0 |
| 8 | International Journal of Geographical Information Science | 3 | 269 | 89.67 | 6 |
| 9 | Journal of Flood Risk Management | 3 | 30 | 10.00 | 1 |
| 10 | Landslides | 3 | 161 | 53.67 | 0 |

third one (124) is about the generation of a landslide risk index map for Cuba using spatial multi-criteria evaluation. The rest of the 26 documents had been cited within a range between 102 to 30 as shown in Table 4 (Tables 5 and 6).

**Table 4.** The top 10 publications in the domain of the study within the same period between 1996–2022

| No | Document | Citations_Scopus | Citations_Google Scholar | Links | Title |
|---|---|---|---|---|---|
| 1 | Cova t.j. (1997) | 176 | 337 | 6 | Modelling community evacuation vulnerability using GIS |
| 2 | Cova t.j. (2002) | 174 | 310 | 4 | Microsimulation of Neighborhood Evacuations in the Urban–Wildland Interface |
| 3 | Abella e.a.c. (2007) | 124 | 239 | 0 | Generation of a landslide risk index map for Cuba using spatial multi-criteria evaluation |
| 4 | Mahendra r.s. (2011) | 102 | 153 | 0 | Assessment and management of coastal multi-hazard vulnerability along the Cuddalore–Villupuram, east coast of India using geospatial techniques |
| 5 | Mosquera-Machado s. (2007) | 85 | 137 | 0 | Flood hazard assessment of Atrato River in Colombia |
| 6 | Kar b. (2008) | 83 | 194 | 3 | A GIS-Based Model to Determine Site Suitability of Emergency Evacuation Shelters |
| 7 | Zerger a. (2004) | 81 | 169 | 0 | Beyond Modelling: Linking Models with GIS for Flood Risk Management |
| 8 | Chen z. (2013) | 63 | 97 | 1 | The temporal hierarchy of shelters: a hierarchical location model for earthquake-shelter planning |
| 9 | Post j. (2009) | 61 | 109 | 4 | Assessment of human immediate response capability related to tsunami threats in Indonesia at a sub-national scale |

*(continued)*

**Table 4.** (*continued*)

| No | Document | Citations_Scopus | Citations_Google Scholar | Links | Title |
|----|----------|------------------|--------------------------|-------|-------|
| 10 | Mas e. (2013) | 56 | 88 | 3 | An Integrated Simulation of Tsunami Hazard and Human Evacuation in La Punta, Peru |

## 3.2 Co-citation

The co-citation analysis of the cited references may rapidly and readily discover the core foundation literature throughout the range of cited literature and portray the inter-relationships between the identified core foundation literature (Zhao et al. 2019). Each node represents a cited reference; the size of the node represents the number of citations received by that reference; the color of the node represents the cluster to which the reference belongs; and the link between reference nodes represents the co-citation link or the relationship between references cited by the same document.

**Cited Authors.** The author co-citation analysis offers a comprehensive picture of the intellectual structure of the domain, identifies the associated influential authors, and aids in the formation of partnerships with the relevant writers for a domain, which aids in the advancement of the domain. The co-citation network of the referenced authors is shown in Fig. 2 using VOS-viewer software. The size of the node shows the number of citations received by that cited author, the color of the node represents the cluster to which the author belongs, and the connection between author nodes represents the co-citation link between the authors that are cited by the same document.

The cluster contains the author nodes that are closely linked. The larger the size of the author node, the more citations the author has gotten. The larger the number of co-citations of the author nodes (on both sides of the connection) by the documents, the stronger the coupling of the author nodes' research interests. The author nodes with the same hue are generally closely linked. The network displayed in Fig. 2 is constructed by examining the top 20 cited authors, each of whom is cited by at least 20 research publications in the domains of GIS, evacuation, and multi-hazards.

The network has highlighted the authors with the strongest co-citation linkages among the 20 authors chosen out of 9773. The authors' co-citation network assists researchers in identifying significantly connected writers who shared research interests. As shown in Table 5 the top 10 co-cited authors, their links, and the total strength of links between them. The table also shows that the author Imamura, f. and Cova, t.j. has the largest number of citations with 57 and 53 each. The total link strength from the list of 20 authors is strongest between Imamura, f and Koshimura, s with 705 and 518. The links and link strength between the author van Westen, c.j., Guzzetti, f., and Ahmad, s. are the weakest.

The bigger circles indicate that these are the authors who have mostly co-cited. In Fig. 2 we can see Imamura.f, Cova, t.j, and Dominey-howes, d. have the largest circles which show their highest number of co-citations as well.

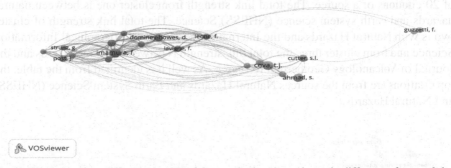

**Fig. 2.** Shows the four different clusters of the co-cited authors shown in different colors and the links between them with a minimum of 20 citations per author.

**Table 5.** Shows the top 10 co-cited authors, their links, and the total strength of links between them.

| No | Authors | Link | Total link strength | Citations |
|----|---------|------|---------------------|-----------|
| 1 | Imamura, f. | 16 | 705 | 57 |
| 2 | Koshimura, s. | 15 | 518 | 33 |
| 3 | Dominey-howes, d | 13 | 405 | 37 |
| 4 | Post, j. | 15 | 373 | 28 |
| 5 | Mas, e. | 15 | 342 | 24 |
| 6 | Suppasri, a. | 14 | 322 | 22 |
| 7 | Baptista, m.a. | 15 | 314 | 24 |
| 8 | Strunz, g. | 15 | 310 | 22 |
| 9 | Wegscheider, s. | 15 | 278 | 21 |
| 10 | Muhari, a. | 15 | 266 | 20 |

**Cited Sources.** Since it is important and helpful to know the co-cited author's network which can assist researchers in identifying connected writers the co-cited sources are useful as well.

Twelve sources met the threshold after cleaning the data as shown in Fig. 3. The sources are clustered into 3 different colors each indicating a relationship between the sources. Cluster 1 has five sources namely Natural Hazards and Earth System Science (NHESS), Science, Geomorphology, Water, and Journal of Volcanology Geothermal Research. Cluster 2 included 6 sources which were Natural hazards, International Journal of Geographical Information Science, International Journal of Disaster Risk Reduction,

Disasters, Natural Hazards Review, and Nature. Cluster 3 included 4 sources which were Natural Hazards, Journal of Volcanology Geothermal Research, Bull Volcanol, and Natural Hazards and Earth System Science (NHESS).

As we can see also from Table 6, the co-citation of the sources with a minimum of 20 citations of a source. The total link strength from cluster one is between natural hazards and earth system science (NHESS) Science. The total link strength of cluster two is with Natural Hazards and the International Journal of Geographical Information Science and from cluster three the total link strength is between Natural hazards and the Journal of Volcanology Geothermal Research. As well it is identified from the table, the top citations are from the sources Natural Hazards and Earth System Science (NHESS) and Natural Hazards.

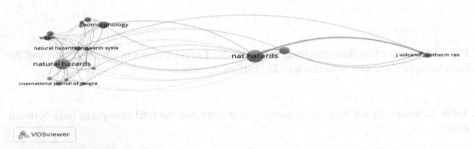

**Fig. 3.** Shows the main Clusters of the co-citations of the cited sources with a minimum of 20 citations of each source.

**Table 6.** Shows the co-citation of the sources with a minimum of 20 citations of a source.

| No | Sources | Cluster | Links | Total link strength | Citations |
|----|---------|---------|-------|---------------------|-----------|
| 1 | Natural Hazards | 3 | 10 | 449 | 75 |
| 2 | Journal of Volcanology Geothermal Research | 3 | 4 | 343 | 30 |
| 3 | Bull Volcanol | 3 | 4 | 312 | 21 |
| 4 | Natural Hazards and Earth System Science (NHESS) | 3 | 9 | 265 | 45 |
| 5 | Natural Hazards | 2 | 13 | 234 | 67 |
| 6 | Natural Hazards and Earth System Science (NHESS) | 1 | 27 | 232 | 78 |
| 7 | Science | 1 | 13 | 115 | 23 |
| 8 | Geomorphology | 1 | 10 | 103 | 43 |
| 9 | International Journal of Geographical Information Science | 2 | 7 | 90 | 25 |

*(continued)*

**Table 6.** (*continued*)

| No | Sources | Cluster | Links | Total link strength | Citations |
|----|---------|---------|-------|---------------------|-----------|
| 10 | International Journal of Disaster Risk Reduction | 2 | 8 | 85 | 22 |
| 11 | Disasters | 2 | 9 | 82 | 21 |
| 12 | Natural Hazards Review | 2 | 8 | 76 | 20 |
| 13 | Nature | 2 | 11 | 72 | 20 |
| 14 | Water | 1 | 6 | 66 | 34 |
| 15 | Journal of Volcanology Geothermal Research | 1 | 3 | 52 | 30 |

### 3.3 Co-occurrence

**All Keywords.** While keywords are connected to the fundamental content of the articles, keyword analysis can assist in the identification of significant study subjects in evacuation, multi-hazards, and GIS studies. After loading the data in the software and after the cleaning process from out of 1987 keywords, 103 had met the threshold. Figure 4 depicts an overview of all the keyword co-occurrence networks built from the core dataset, which included 97 items and 1960 linkages. It shows the five different clusters linked in different nodes and shown in different colors. A node represents a single key phrase extracted from publications. The size of each node is proportional to the frequency of co-occurrence of the associated terms.

From the analysis of all the keywords shown in Fig. 4, cluster one shows the co-occurrence of keywords like disaster management, evacuation, evacuation simulation, climate change, hazard management and assessment, natural hazards, urban planning, safety, tsunami, tsunami events, urban areas, spatial analysis along with GIS and others. Cluster two shows the co-occurrence between a group of keywords like accidence, debris, and debris flow, decision support systems, disaster planning, emergency evacuation, response and traffic control, fires and fire hazards, landslides, real-time, risk analysis, wildfires along with Geographic information systems (GIS) and others. Cluster three; shows co-occurrence between keyword groups namely analytical hierarchy process, coastal zones, decision making, disasters and disaster prevention, earthquakes, emergency services, evacuation planning, plans and routes, hazards, motor transportation, population statistics, tsunami hazards, alongside location, maps, and others. Cluster four; presents the co-occurrence between different keywords such as buildings, early warning system, emergency management, flood control, damage, hazards, risk assessment, and risk management along others. Cluster five; reveals co-occurrences of various other keywords like a disaster, risks, volcanic eruption, hazard, information use, and land use, alongside mapping and remote sensing.

In a nutshell, the clusters reflect tightly linked research between GIS, management, evacuation, and planning alongside various multi-hazards.

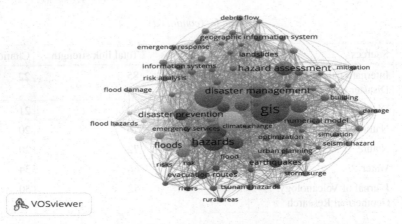

**Fig. 4.** Shows the co-occurrence of all the keywords with a minimum number of 5 occurrences of a keyword.

**Author keywords.** While analyzing the co-occurrence of the author keywords shown in Fig. 5 shows the total strength of the co-occurrence of 17 keywords with other keywords. The clusters are divided into 5 with a link of 53 and a total link strength of 110. we can as well see the links between Geographic information systems (GIS), evacuation, and emergency evacuation alongside various multi-hazards.

Table 7 as well shows the top 13 keywords that have occurred more than 5 times where the total links and link strength of GIS and evacuation are at the top with an occurrence of 92 and 21 followed by other hazards and various keywords like emergency evacuation, risk assessment remote sensing and others.

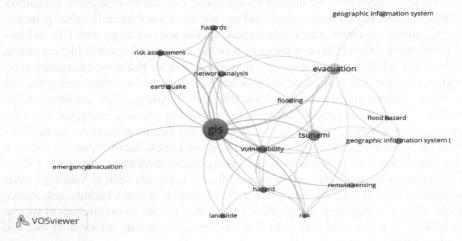

**Fig. 5.** Shows the co-occurrence links between the main five clusters with different strengths.

**Table 7.** Shows the top 13 keywords that occurred more than five times in the study domain.

| No | Author Key Words | Links | Total link strength | Occurrences | Avg. Citations |
|----|------------------|-------|---------------------|-------------|----------------|
| 1 | geographic information system (gis) | 19 | 64 | 92 | 13.03 |
| 2 | evacuation | 12 | 31 | 21 | 11.19 |
| 3 | tsunami | 10 | 19 | 20 | 8 |
| 4 | hazard | 14 | 22 | 16 | 14.23 |
| 5 | vulnerability | 11 | 21 | 12 | 14.83 |
| 6 | flood hazard | 8 | 12 | 11 | 18.02 |
| 7 | network analysis | 7 | 11 | 8 | 3.88 |
| 8 | risk assessment | 5 | 9 | 8 | 14.75 |
| 9 | remote sensing | 6 | 10 | 6 | 4.17 |
| 10 | earthquake | 4 | 6 | 5 | 8 |
| 11 | emergency evacuation | 1 | 2 | 5 | 4.2 |
| 12 | landslide | 3 | 3 | 5 | 15.6 |
| 13 | risk | 6 | 10 | 5 | 7.2 |

## 3.4 Co-authorship

**Authors.**Researchers typically cooperate in academic research, which may increase productivity and access to knowledge while also keeping researchers from becoming isolated. In this research, the minimum number of published publications and author citations were established at one, respectively. The number of citations as well obtained by an author in a domain is used to assess his or her impact in that area. As a result, 739 of the 739 authors satisfied the criteria but not all were linked to each other. Figure 6 below analyzes the citation patterns of the writers to identify the influential authors in the domain of multi-hazard evacuation and GIS. We can see that 22 of the authors are clustered in one cluster with a link of co-authorship of 231 between them (Fig. 6).

**Organizations.** The co-authorship investigation of the top-cited organizations in a domain can assist in the identification of the producer-organizations of the most well-followed research in that subject and their collaborations. Henceforth, the top-cited organizations based on the total received citations for their total publication documents along with their average citations per document, in the domain of hazard evacuation and GIS during the period 1996–2020, are depicted in Table 8. The table also shows the documents by each organization which looks equal in the table with 2 documents each for every organization. The total link strength is also equal for the organizations as well as 25 citations received for the documents of each organization.

Figure 7 as well shows the largest set of connected organizations which consists of the 3 organizations which are Geological Engineering Department, Middle East Technical

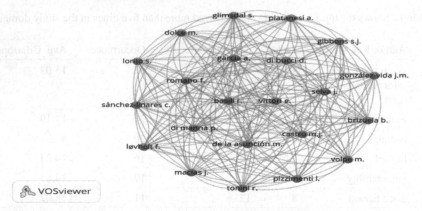

**Fig. 6.** Shows the minimum number of 1 document per author with 22 authors shown in one cluster with the total link strength of co-authorship between them.

University in Turkey, Nizhny Novgorod State Technical University located in the Russian Federation, and the Special Research Bureau for Automation of Marine Research, Russian Academy of Sciences in Russian Federation as well. The data showed 5 other organizations as well, but which had no network connections between them.

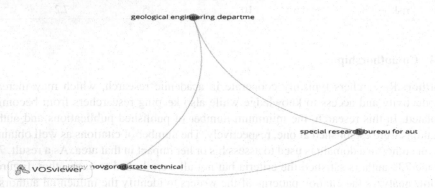

**Fig. 7.** Shows the top cited organizations with a minimum of 2 documents of the organizations and the links between them.

**Countries.** It is critical to understand your research's co-authorship in terms of authors, organizations, and countries as well. Figure 8 shows the 14 countries with a minimum of 5 documents which consist of 5 clusters. Cluster 1 is (China, South Korea, United States), cluster 2 (France, Italy, United Kingdom), Cluster 3 (Germany, Indonesia, and Malaysia), and Cluster 4 (Canada, and Japan). Table 9 shows the total link between the countries, the strength of the link, documents published by the countries, the average publications per year, and the average citations. We can see that Indonesia; Japan and the United States have the strongest link strength. The table also shows the most published documents are from the United States and Japan (35, 31) as well as the highest citations (946, 276). Also seen in Fig. 8 are the clusters of countries with a minimum of 5 documents and

**Table 8.** Shows the top cited organizations with a minimum of 2 documents and their total reviewed documents citations.

| No | Organizations | Cluster | Links | Total link strength | Documents | Citations | Avg. Pub. Year |
|----|---------------|---------|-------|---------------------|-----------|-----------|----------------|
| 1 | Geological engineering department, middle east technical university, Ankara, Turkey | 1 | 2 | 4 | 2 | 25 | 2017 |
| 2 | Nizhny Novgorod state technical university, 24 minin street, Nizhny Novgorod, Russian federation | 1 | 2 | 4 | 2 | 25 | 2017 |
| 3 | special research bureau for automation of marine researches, Russian academy of sciences, yuzhno-sakhalinsk, 693013, Russian Federation | 1 | 2 | 4 | 2 | 25 | 2017 |

the total strength of the co-authorship link between them whereas the United States and Japan have the highest co-authorship link (Fig. 8).

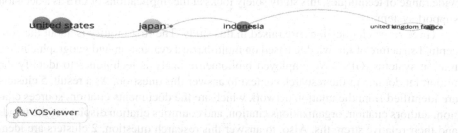

**Fig. 8.** Shows the 4 different clusters of countries with a minimum of 5 documents and the total strength of the co-authorship link between them.

**Table 9.** Shows the countries with a minimum of 5 documents and the total strength of co-authorship links with each other.

| No | Countries | cluster | Links | Total link strength | Documents | Citations | Avg. Citations |
|---|---|---|---|---|---|---|---|
| 1 | united states | 1 | 4 | 7 | 35 | 946 | 27.03 |
| 2 | Germany | 3 | 3 | 4 | 6 | 157 | 26.17 |
| 3 | China | 1 | 1 | 2 | 29 | 328 | 11.31 |
| 4 | Italy | 2 | 2 | 4 | 10 | 109 | 10.90 |
| 5 | Canada | 4 | 1 | 1 | 6 | 61 | 10.17 |
| 6 | France | 2 | 2 | 3 | 13 | 124 | 9.54 |
| 7 | Japan | 4 | 4 | 7 | 31 | 276 | 8.90 |
| 8 | United Kingdom | 2 | 3 | 5 | 7 | 57 | 8.14 |
| 9 | South Korea | 1 | 1 | 2 | 7 | 56 | 8.00 |
| 10 | Indonesia | 3 | 4 | 8 | 23 | 107 | 4.65 |
| 11 | Malaysia | 3 | 1 | 1 | 6 | 24 | 4.00 |

# 4   Recommendation and Conclusion

A bibliometric analysis of the literature was done in this study, in association with multi-hazard, evacuation, and GIS. An initial set of 226 publications was found by bibliometric analysis, which was narrowed down to a concentrated selection of 220 works published between 1996 and September 2022. Although evacuation and multi-hazards involve a wide range of techniques, this study solely looks at the implications of GIS as a decision support system.

Three research questions are raised in this study. The first inquiry is about the conceptual structure of knowledge based on multi-hazard evacuation and geographic information systems (GIS). We employed bibliometric analysis techniques to identify the important domain in the research context to answer this question. As a result, 5 clusters are identified from the citation network which are the documents citation, sources citation, authors citation, organizations citation, and countries citation displaying the trends and their relative strengths. Also, to answer this research question, 2 clusters are identified from the co-citation networks namely cited sources co-citations and cited authors co-citation as well as their trends and relative strengths.

Based on the analysis of the document's citation, it's found that the United States is the most active country in this domain accounting for (15.63%) of the total publications from 1996–2022. While the top two countries with higher citations in their publications are Cuba (124) and the Netherlands (51.33). Moreover, the higher collaboration with other countries the United States, followed by Japan, then China, and Indonesia. While the top two cited organization have been selected out of 452 with higher citations of more than 70 with at least one minimum document is the National Center for Geographic Information and Analysis (NCGIA), and the University of California, in the United States accounting for the higher average citation (176). On the other hand, the top cited journal out of 147 journals having more than three publications is the Natural Hazards journal with 14 publications during the period 1996 – 2022. The top cited documents have been cited by 176 authors; it is about modelling community vulnerability using GIS, while the Co-Citation, top cited authors, are Imamura, f. and Cova, t.j who has the largest number of citations with 57 and 53 each. The total link strength from the list of 20 authors is strongest between Imamura, f and Koshimura, s with 705 and 518.

The author co-citation network assists researchers in identifying significant connected writers who all share the same interests. In the same way, it is also helpful to identify the co-cited sources network. From the co-cited analysis, the total link strength from cluster one was between Natural Hazards and Earth System Science (NHESS) and Science. The total link strength of cluster two was with Natural Hazards and the International Journal of Geographical Information Science and from cluster three the total link strength was between Natural hazards and the Journal of Volcanology Geothermal Research. As well it was identified, the top citations were from the sources Natural Hazards and Earth System Science (NHESS) and Natural Hazards.

The second research question inquires about the key concepts that have been explored on the topic of multi-hazards evaluation and GIS and the relationship between them. The research tries to answer this question with the help of the co-occurrence network by 3 clusters which are the All-keywords co-occurrence and author keywords co-occurrence. The research clarifies the total strength of the co-occurrence links of citations between the keywords. Based on the analysis of all the keywords, the clusters of the keywords reflect the closely linked research between GIS, management, evacuation, and planning alongside various multi-hazards. Otherwise from the author keyword analysis as well we could reflect the GIS linked with evacuation, and emergency evacuation, alongside various multi-hazards keywords. The total link strength was between GIS and evacuation with 64 and 31 each and their occurrences were 92 and 21 each.

Further answering the last and final question of the research which was about the nature of the collaboration that was evident in multi-hazard evacuation and GIS. The research answered this question using the co-authorship network by identifying 3 clusters namely authors, co-authorship, organizations co-authorship, and countries co-authorship. The co-authorship between top authors and organizations and countries can help researchers know top active researchers, organizations, and countries in the domain of their interest and collaborate in case of any collaboration intention. As shown in our analysis the top 22 authors all clustered in one cluster and a total link of 231 between them. Additionally, there in our search domain there is a linkage between concepts and nations such as the United States, Japan, China, Indonesia, and others. These connections represent the nations' research hotspots.

In addition to answering these questions, the added value of this research to the body of knowledge in the multi-hazard's evacuation domain is by giving out a bibliometric analysis that has been discussed in the previous sections. The results of this analysis would be useful for future researchers, practitioners, and officials to be aware of the related subject areas that are associated with GIS, evacuation, and hazards. Furthermore, the gathered publications were acquired solely from the Scopus database, which encourages other researchers to explore other databases platforms such as Web of Science, Dimensions, PubMed and Google Scholar. Moreover, the current study concentrated on the association of GIS and evacuation which is part of disaster response and mitigation strategy using GIS as a decision support tool in the planning and evacuation management of multi-hazards hence more, future research should be conducted to explore the association of GIS with disaster recovery.

# References

Bonilauri, E.M., et al.: Tsunami evacuation times and routes to safe zones: a GIS-based approach to tsunami evacuation planning on the Island of Stromboli, Italy. J. Appl. Volcanol. **10**(1), 1–19 (2021). https://doi.org/10.1186/s13617-021-00104-9

Cole, J.W., et al.: GIS-based emergency and evacuation planning for volcanic hazards in New Zealand. Bull. N. Z. Soc. Earthq. Eng. **38**(3), 149–164 (2005)

Galindo, G., Batta, R.: Review of recent developments in OR/MS research in disaster operations management. Eur. J. Oper. Res. **230**(2), 201–211 (2013)

Sengupta, I.N.: Bibliometrics: a bird's eye view. IASLIC Bull. **30**(4), 167–174 (1985)

Tanimoto, J., Hagishima, A., Tanaka, Y.: Study of bottleneck effect at an emergency evacuation exit using cellular automata model, mean field approximation analysis, and game theory. Phys. A: Stat. Mech. Appl. **389**(24), 5611–5618. Elsevier B.V. (2010)

Thumerer, T., Jones, A.P., Brown, D.: A GIS based coastal management system for climate change associated flood risk assessment on the east coast of England. Int. J. Geogr. Inf. Sci. **14**(3), 265–281 (2000)

Zhao, X., Zuo, J., Wu, G., Huang, C.: A bibliometric review of green building research 2000–2016. Archit. Sci. Rev. **62**(1), 74–88 (2019)

# A Bibliometric Analysis of Digital Transformation for a Resilient Organization

Hind Lootah [ID], Juma Aldhaheri(✉) [ID], and Maria Papadaki

The British University in Dubai, Dubai, UAE
{20198322,20181367}@student.buid.ac.ae

**Abstract.** Digital transformation and resilience are being intensively focused on areas of research in the current period. This is due to the current situation the world is facing because of Covid-19. Many organizations not only failed to grow during this turbulent event but also, to remain functioning in the market which led to their disappearance. On the other hand, organizations that focused their core responsibilities towards being resilient tend to shift from surviving to thriving organizations. Combining both adoptions in literature needs to be more analysed. Therefore, this paper will be focusing on conducting bibliometric analysis on digital transformation and its impact on the resiliency of an organization to conquer the unexpected events such as the pandemic of Covid-19. The bibliometric analysis will be implemented using a software called VOS viewer. It will support in observing the research clusters, emerging topics and leading scholars. From the analysis, information regarding the top countries that contributed to the research, main keywords and number of documents published were defined. Moreover, providing topics for future research was also determined from the analysis.

**Keywords:** Digital transformation · resilience · bibliometric analysis · Scopus · VOS viewer

## 1 Introduction

Nowadays, digital technologies are in rapid development being very critical aspect for organizations to take into their strategic plans for better performance and sustainable development. After the Covid-19 pandemic, many organizations failed to survive in the market. Therefore, the call for a resilient organization is the primary focus of organizations. In this paper, a closer insight will be focused on combing both areas in the sense of digital transformations and organizational resilience to achieve a resilient organization that will not only survive but thrive in the face of any turbulent event in the future. Specifically, this paper will mainly focus on conducting a bibliometric analysis in this research area. The bibliometric analysis discipline is used for a very long time due to its importance in the analysis between journals. The first term of bibliometric analysis emerged to show the connection between journals that were related to information science since 1972 (1972). This paper will be focusing on using this discipline on journals related to organizational resilience and focusing specifically in its impact on organization performance. The bibliometric analysis, computer assisted methodology review, will help

M. Papadaki et al. (Eds.): EMCIS 2022, LNBIP 464, pp. 649–662, 2023.
https://doi.org/10.1007/978-3-031-30694-5_45

in identifying core research of related areas. Also, discover the relationship between authors and their publications. The bibliometric analysis with journals extracted from a database such as Scopus or science web and a software such as VOS viewer which has dimensions that goes beyond the normal literature in order to enable new researchers absorb the extend of the topic and emergent trends in the related field market.

Furthermore, bibliometric analysis is helpful for understanding and outlining the accumulated scientific knowledge and evolutionary opinions of established areas since it has the capacity to rigorously make sense of massive volumes of unstructured data. Therefore, properly conducted bibliometric studies can lay the foundation for novel and important advancements in a field; it enables and empowers scholars to (1) gain a thorough understanding, (2) identify new opportunities, (3) develop original research ideas, and (4) position their intended contributions to the field (Donthu et al. 2021). Many researches indicate that the implementation of digital transformation strategies in businesses contributes to the improvement of organizational resilience. In addition, digital transformation has a beneficial effect on the organizational business resiliency (Zhang et al. 2021).

The objectives of this paper will be deliberated in more details and insights from literature about digital transformation and resilience will be discussed below. Looking into more details, the method of conducting this paper will also be mentioned in the methodology section to show the source of all documents used for the bibliometric analysis and all the software and databases utilised for the data analyses when combining digital transformation and resilience. Finally, the discussion of the analysed data is made with the conclusion for this paper.

## 2  Research Objectives

The main objectives of the current study are to create a bibliometric analysis of digital transformation and its impact on the resilience of the organization through a digital culture. To conduct the bibliometric analysis there are certain steps that need to be implemented. First of all, to create maps using the software VOS viewer based on journals that are published derived from the Scopus network which are related to digital organization, digital culture and organizational resilience. Also, to visualize the network data to be able to investigate the citation trends and links between the published journals. Last but not least, to explore the corresponding geographical distribution of authors, publication sources and the main keywords existence.

## 3  Related Literature

### 3.1  Digital Transformation

Digital transformations are defined as an international fast-moving technology revolution process done by people, organizations, and nations and result from digitization. Expressions such as Mobile applications, the Internet of things, and industry four means have similar meanings. In addition, digital transformations relate to culture span, discipline cross, borderless, and computer-generated data (Collin et al. 2015).

Also, digital transformation is defined as a business stimulating method by technical innovation either to interrupt or to defend a digitalized organization from interruptions (Ali 2019). Similarly, Digital transformation definition according to the US government is "Data and Analytics Innovation, Emerging Opportunities and Challenges" (Gudergan and Mugge 2017).

Moreover, according to (Jakubik and Berazhny 2017) digital transformation is considered a crucial part of the industry four revolution. According to (Bunse et al. 2013) the definitions of Industry 4 are equal to economic change and industrial transition. Industry 4 relates to the 4th industrial age, which is based on technology innovations from physical cybernetic systems to the Internet-Things (IoT) implanted devices. The basis on which innovative solutions are developed is technologies of Data, communication and Information Technology (IT). Also, Industry 4.0 demands technical creativity like companies require a new competitive emphasis, new working competences, and new skills to be established (Berman et al. 2016); (Harvard Business Review Analytic Services 2017).

Moreover, the organization digital strategy is the basis for an effective business transition according to (Wade 2015). The digital strategy essential questions are why changing? what transforming? and how transforming? Different urgings motivate organizations' digitalization, such as digital clients, agile operation, largescale, supervisory modifications, and digital technique. However, new competitors can trigger further incentives for digital change, including expanded options, enhanced engagement mechanisms, and better values (Wade 2015). Furthermore, according to the authors (Schoemaker et al. 2018; Swanson et al. 2017).

Dynamic capabilities address repetitive process problems in which the organization must familiarize regarding resources, processes, products, and services. Also, to enable the development of the organization's dynamic capabilities, organizations must establish adaptable capability and create creativity. Dynamic capabilities modify organization business model to respond to changes and transformation of the market environment. Besides, dynamic capabilities progression can be improved by alignment with flexible strategy (Wasono et al. 2111). Also, Information Management is one of the Digital Transformation prerequisites. It is clear that the informational tools advance Digital transformation and its quality (Mahsud Ali 2019; Williams et al. 2014).

Digital transformations implementations require a profound adjustment which is more considerable than the classical change in organizations (Kwon and Park 2017). Also, according to the authors (Bongiorno et al. 2018) who stated that other enablers alongside the technology platform are vital to transforming a digital company, such as business processes, products, services, business models. Similarly, (Fitzgerald et al. 2013) claimed that practically no company will defend itself from digitalization and aggressive interruption only by embracing modern emerging innovations and organizational business models.

## 3.2 Digital Transformation Technologies

According to the authors (Foerster-Metz et al. 2018) explanations and summaries of the digital transformation technologies (trends) are categorized into 5 categories. Technologies are categorized into five groups which focus on connections and accessibility, data and intelligence, robotics and automation and efficiency, communication and interactions and finally the security. The first group is mainly focusing on mobile applications, internet of things and cloud computing. The second category includes big data and data analysis. Moreover, robotics category comprises of robotics, intelligent automation and artificial intelligence. Generally, the fourth category consists of social media and specifically internet communication channels, engagement and content sharing. The final category is related to new security which focuses on both cybersecurity and physical security.

### 3.3 Digital Transformation Models and Types

On the other hand, according to the report presented by (Brynjolfsson et al. 2016) there are four models of digital organizations named as following; Tactic model focused on opportunity and innovation, Central model focused on setting agenda, a Champion model focused on change and Usual business model focused on Normality. While, the champion model aims at establishing a deep connection between management and staff. Whereas, the common business model aims at business flexibility and agility.

### 3.4 Organizational Resilience

The term "resilience" is originated from the Latin word "resilire" which means to leap or jump back. The concept of resilience has been widely used for a long period of time in many disciplines. Due to this fact, there is no single consistent understanding of resilience construct. Looking into more details, the construct of resilience goes back to the nineteenth century where it was used in engineering as engineering resilience. This term was used to explain the ability of steel to absorb and endure shocks and tension (Holling 1996; Alexander 2013). After that, resilience was used in the psychology field referring to patients of Schizophrenic disorders who are capable of tolerating shocks in the 1950s (Yates and Masten 2004). In the 1970s, the word resilience appeared in the academic field of Ecology by Holling. It emerged to define the development and reorganization of the dynamics of the system and the adaptive capacity. Also, He defined and illustrated the differences between the resilience approaches in engineering and ecology (Holling 1996, p.33). Engineering resilience draws the attention on efficiency, reliability and predictability because it concentrates on handling risks to develop responsive perturbations techniques. While ecological resilience concentrates on changeability and unpredictability to ensure that systems function sustainably in dynamic conditions (Carpenter et al. 2001; Seville et al. 2006).

## 4 Research Methodology

The material utilized for this study is derived from the secondary data to start our research on digital transformation and its impact on an organization to be resilient and be able to face any turbulent event. As well as, reviewed the existing literature on the different topics creating a link between digital transformation and organizational resilience. This study of digital transformation to create a resilient organization is a phenomenon that researchers are taking into consideration as we are still in the Covid-19 era. This is taken into consideration in order to accomplish the objectives of this bibliometric analysis.

Primarily all published journals were extracted from the Scopus network using the search fields terms such as: "digital transformation" and resilience. Initially, a result of 199 documents was derived from digital transformation and resilience only. Therefore, to create a bibliometric analysis from the derived journals, the VOS viewer software was employed for this analysis. This is to support in demonstrating the quality of the studies, scrutinize the key areas of the research and support in the anticipation of the research future direction (Yu et al. 2020). After using the data from Scopus, it turned out that the life span of the research is between the years of 2017 and 2022.

The results of the search in Scopus which included 199 documents included all sort of documents and not limiting the research on specific type of document. Due to the fact that the main purpose of the research to induct a bibliometric analysis on the citation of the selected research area. Therefore, no exclusion of any document type for this research which made it result as 199 documents for this study. For further citation analysis in this paper, not only VOS viewer was utilized but also, did a comparison of citation from Google Scholar for the documents resulted from Scopus. To view data in a clear perspective, tables from excel sheet were created and graphs from VOS viewer and Scopus were extracted.

## 5 Data and Results

### 5.1 Research Areas

Digital transformation and its impact on the resiliency of an organization was introduced in several research fields. The first document published in that specific research area was in 2017 in the engineering field. There are several subject fields focusing on this research area from the total search results of 199 documents. The two main research areas of digital transformation and resilience were in computer science and business, management and accounting. Table 1 demonstrates the top five research areas.

**Table 1.** Top 5 research areas assigned to papers in the sample

| Research areas | Records | % of 443 |
|---|---|---|
| Computer Science | 89 | 20.1% |
| Business, Management and Accounting | 66 | 14.9% |
| Engineering | 58 | 13.1% |
| Social Sciences | 56 | 12.6% |
| Decision Sciences | 44 | 9.9% |
| Total top 5 research areas | 313 | 71% |

As indicated in the table above, computer science field has the highest percentage of publications regarding the topic of digital transformation and organizational resilience.

Lot of research is done in that subject area as derived from the network of Scopus. This can be evident from the computer science percentage which is 20.1% from all other field publications. The second highest subject area is in business, management and accounting resulting in 14.9%. However, engineering and social sciences have very similar percentatges in their publications of the topic showing as 58% and 56% respectiavley. The below pie chart shows the total publication percentages from different subject area and not restricted to the top five subject areas for more information (Fig. 1).

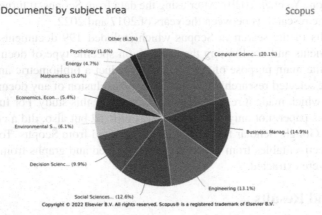

**Fig. 1.** All published documents of the topic from different subject areas, generated with the database Scopus (https://www.scopus.com).

### 5.2 Citation Analysis

**Top Countries Contributed in the Area of Research.** Using the VOS viewer, this paper examined the countries with publications that are more than five documents per country. In a total of 63 countries, 19 of the countries met the threshold of five documents per country. The result ended with a total of 8 clusters for the 19 countries. However, 4 countries in 4 different clusters were not connected to each other. The four countries are Japan, Russian Federation, Saudi Arabia and Spain. Therefore, only the largest sets of connected items were created minimizing the number of clusters into 4 clusters which consist of 15 countries as shown in the figure below (Fig. 2).

The countries that participated in conducting and publishing documents of the topic were minimized to 15 countries that had connections between them. For more details, the first cluster which is in red color included 5 countries which are Australia, Austria, Germany, Romania and Switzerland. The second cluster, the green cluster, included five countries as well which are Brazil, France, Italy, Portugal and United Kingdom. The blue cluster which is the third one included 3 countries which are Greece, India and the United States. Finally, the yellow cluster which included only two countries, Canada and China.

Generally, it is demonstrated from the above graph that the highest document published by a country is 19 documents and the least published documet country wise is 5

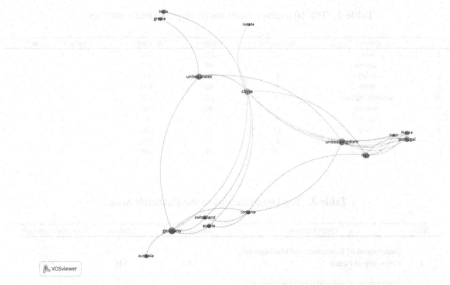

**Fig. 2.** Top countries contributed in the research in four main clusters, generated with VOS viewer (http://www.vosviewer.com).

documents. As indicated Germany and the United Kingdom are the two mian countries with highest published papers conducted in digital transformation and resilience resulting with 19 documents published each and a citation of 77 and 132 citations respectively. On the other hand, China comes in the third place with a total of 17 published documents and a total of 101 citations. Both Spain and the United States have the same number of publications with a total of 16 pubications each. Finally, the least countries among all pulished countries of the research topic are Canada and Brazil with 5 documents each.

Table 2 indicates the 10 highest average citations from the published countries. It is evident that the average citaion is not related with the number of documents published. For instance, the country with the highest average citation is Italy. Yet, the number of published documents is not the highest among all published countries where it is resembled as 14 published documents. On the contrary, Germany which is one of the highest countries in number of oublished documents, has the lowest average citations of its documents with a link strength of 6.

**Top Organizations Contributed in the Area of Research.** All organizations seek for improvements in their strategies and procedures within the organization in order to achieve high perfromance results. This is done throug different business development tools in areas that need improvements. To be more specific, in the current sitaution of Covid-19, many organizations are moving towrads being more resilient which is proven in this study after analysing the fields of digital transformation and resilience. It was noticed from the number of research that was done in the current years as this topic was grabbing the intrest of many countries around the world. In the table below, organizations with the highest citations of their publication will be displayed. From a total of 199 documents in the area of digital transformation and its impact on the resiliency of

**Table 2.** Top 10 average citations from published countries.

|    | Country | Documents | Citation | Avg. Citation | Total Link Strength |
|----|---------|-----------|----------|---------------|---------------------|
| 1  | Italy | 14 | 245 | 17.5 | 6 |
| 2  | Canada | 5 | 68 | 13.6 | 2 |
| 3  | Brazil | 5 | 55 | 11 | 5 |
| 4  | India | 7 | 60 | 8.57 | 3 |
| 5  | United Kingdom | 19 | 132 | 6.95 | 5 |
| 6  | Austria | 7 | 46 | 6.57 | 3 |
| 7  | France | 9 | 54 | 6 | 6 |
| 8  | China | 17 | 101 | 5.94 | 15 |
| 9  | Australia | 8 | 39 | 4.88 | 1 |
| 10 | Germany | 19 | 77 | 4.05 | 6 |

**Table 3.** Top Organizations in the Research Area

| No. | Organization | Documents | Citation | Total Link Strength |
|-----|--------------|-----------|----------|---------------------|
| 1 | Department of Economics and Management, University of Padua | 1 | 148 | 7 |
| 2 | Department of Industrial and Mechanical Engineering, University of Brescia | 1 | 148 | 7 |
| 3 | Department of Industrial Engineering, University of Florence | 1 | 148 | 7 |
| 4 | Department of Management and Engineering, Linkoping University | 1 | 148 | 7 |
| 5 | ICT Industry, United Kingdom | 1 | 67 | 7 |
| 6 | Newcastle University Business School, United Kingdom | 1 | 67 | 10 |
| 7 | Chitkara Nusiness School, India | 1 | 51 | 10 |
| 8 | Department of Engineering, University of Naples Parthenope | 1 | 51 | 2 |
| 9 | Department of Industrial Engineering, University of Naples Federico | 1 | 51 | 2 |
| 10 | Labonfc, Canada | 1 | 51 | 2 |

an organization, 430 organizations contributed in the research area. However, among all organizations, only 8 organizations met the threshold of having atleast 2 documents that are published. Other organizations have done only one publication during the period of 2017 – 2022. The eight organizations that have atleast 2 published documents are Nanfang College in China, Roskilde University in Denmark, Intellectual Capital Association in Portugal, Federal University of Santa Catarina in Brazil, Physikalisch-technische in

**Table 4.** The top 15 Publications in the domain of the study with the same period between 2017 - 2022

| No | Document | Citations_Scopus | Citations_Google Scholar | Links | Title |
|---|---|---|---|---|---|
| 1 | (Rapaccini et al., 2020) | 148 | 298 | 2 | Navigating disruptive crises through service-led growth: The impact of COVID-19 on Italian manufacturing firms |
| 2 | (Papagiannidis et al., 2020) | 67 | 140 | 3 | WHO led the digital transformation of your company? A reflection of IT related challenges during the pandemic |
| 3 | (Shashi et al., 2020) | 51 | 85 | 22 | Agile supply chain management: where did it come from and where will it go in the era of digital transformation? |
| 4 | (Klein & Todesco, 2021) | 49 | 123 | 0 | COVID-19 crisis and SMEs responses: The role of digital transformation |
| 5 | (Korhonen & Halen, 2017) | 48 | 81 | 0 | Enterprise architecture for digital transformation |
| 6 | (Belhadi et al., 2021) | 44 | 67 | 1 | Artificial intelligence-driven innovation for enhancing supply chain resilience and performance under the effect of supply chain dynamism: |
| 7 | (González et al., 2021) | 41 | 57 | 0 | Innovative multi-layered architecture for heterogeneous automation and monitoring systems: Application case of a photovoltaic smart |
| 8 | (Trenerry et al., 2021) | 27 | 68 | 0 | Preparing Workplaces for Digital Transformation: An Integrative Review and Framework of Multi-Level Factors |
| 9 | (Fonseca & Azevedo, 2020) | 27 | 64 | 1 | COVID-19: Outcomes for Global Supply Chains |
| 10 | (Chonsawat & Sopadang, 2020) | 23 | 33 | 0 | Defining smes' 4.0 readiness indicators |
| 11 | (Gürdür Broo et al., 2022) | 22 | 39 | 0 | Rethinking engineering education at the age of industry 5.0 |
| 12 | (Lincaru et al., 2018) | 21 | 33 | 0 | Low-low (LL) high human capital clusters in public administration employment-predictor for digital infrastructure public investment |
| 13 | (Ding et al., 2020) | 19 | 43 | 0 | Building stock market resilience through digital transformation: using Google trends to analyze the impact of COVID-19 pandemic |
| 14 | (Scholz et al., 2020) | 19 | 38 | 2 | Organizational vulnerability of digital threats: A first validation of an assessment method |
| 15 | (Zhang et al., 2020) | 16 | 31 | 4 | How does digital transformation improve organizational resilience?—findings from pls-sem and fsqca |

Germany and The New Club of Paris in France, all organziations with 2 published documents. The only organization with 3 published documents is the Dinamia'cct-iul – iscte-iul in Portugal. The top cited orgniations from all 430 organizations will be illustrated below (Table 3).

**Top Cited Documents in the Area of Research.** The interest in this research area has grown during the short period between 2017 and 2022 among organizations from different parts of the world. Research is still going on since organziations are thriving to be resilient after facing the crisis of Covid-19. As the main network of documents used for this paper analysis which is Scopus, the highest cited documents were derived from Scopus and displayed the number of citations per document. Moreover, a comparison of the citation number was displayed for every document using Google Scholar. It indicates the differences in citation and ease in accessing each network. From all the published documents, the top 15 documents will be the highest citation in Scopus and its comparison with Google Scholar will be illustrated in the below table for the research period (Fig. 3) (Table 4).

## Published Documents from Different Sources

For further details regarding publication, the above graph demonstrates the documemnts that were published annually by source for the time period between 2018 to 2022 derived from Scopus network. In total, five sources for publications were selected to publish the documents that focused on acheving a resilient organization through the implemntation of digital transformation. The sources are ACM International Conference

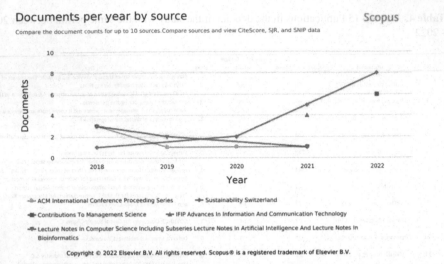

**Documents per year by source**                                           Scopus
Compare the document counts for up to 10 sources.Compare sources and view CiteScore, SJR, and SNIP data

- ACM International Conference Proceeding Series     - Sustainability Switzerland
- Contributions To Management Science     - IFIP Advances In Information And Communication Technology
- Lecture Notes In Computer Science Including Subseries Lecture Notes In Artificial Intelligence And Lecture Notes In Bioinformatics

**Fig. 3.** Publication of all documents on yearly bases from different sources, generated with the database Scopus (https://www.scopus.com).

Proceeding Series, Sustainability Switzerland, Contributions to Management Science, IFIP Advances In Information And Communication Technology and Lecture Notes in subject areas such as, Computer Science with subseries lecture notes in Artificial Intelligence and Bioinformatics. It is evident that all documents were published by different sources in 2018. However, even though all strated published in the same year, only one source was able to thrive to the maxmim of all sources and other declined from their strating point. To be more specific, the Sustainability Switzerland started with the lowest number of published documents in 2018 but started to have a slight increase in the following two years of 2019 and 2020. The turning point for the Sustainability Switzerland source occurred in 2020 in the number of published documents. A significant growth of published documents was clear between the years of 2020 and 2022. It makes this sources the highest source of the published documents among all other sources. On the contrary, the other two sources which are the ACM International Series and the lecture notes started publishing the highest documents among others in 2018. As the time passed, the two sources started having a decline in the number of published documnents. This resulted in being the lowest two sources by the year of 2021.

### 5.3 Co-occurrence Analysis (Keywords)

Figure 4 represents the cloud map of co-citation analysis and keywords used in the search field of "digital transformation" and resilience. In that search the documents were limited more specifically to in the sense of digital transformation in which the final search was resulted to 199 documents rather than 345 documents. In the comparison of the two figures, Figure 4 clearly indicates the link and connection between digital transformation, resilience and COVID 19 being the largest clusters among all others. From the 690 keywords, 52 keywords met the threshold of having 3 as a minimum

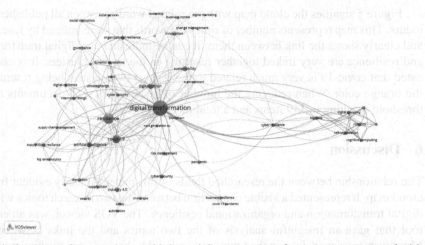

**Fig. 4.** A cloud map of co-occurrence with 3 as minimum number of occurrences, generated with VOS viewer (http://www.vosviewer.com).

number of occurrences of keyword resulting in having 7 clusters with 224 links between them.

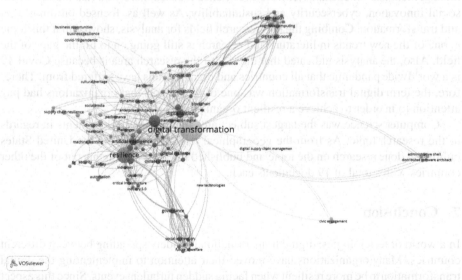

**Fig. 5.** A cloud map of co-occurrence with 2 as minimum number of occurrences, generated with VOS viewer (http://www.vosviewer.com).

Figure 5 signifies the cloud map with the related words between all published documents. This map represents number of times the words that were utilized by researchers and clearly shows the link between them. It can be indicated that digital transformation and resilience are very linked together resulting in the largest clusters. It is also indicated that covid-19 is very much related to the biggest two clusters being resembled in the orange color. When reducing the number of occurrences, 119 documents met the threshold resulting in 119 items and a total of 13 clusters.

## 6  Discussion

The relationship between the researched fields of this paper is clearly evident from the cloud map. It represented a visible connection between the two research topics which are digital transformation and organizational resilience. The VOS viewer, was an essential tool that gave an insightful analysis of the two topics and the links between them. Moreover, gave an insight on the future research fields that are very much related to this paper. The utilization of VOS viewer, indicated a broad image of showing that digital transformation, resilience were the largest two clusters in the map. From those two large fields, other research areas were closely linked.

Looking into more details, the digital transformation showed a strong and multi-connection with other fields such as governance, change management, leadership, risk management and blockchain. However, resilience focused merely on risk management, social innovation, cybersecurity and sustainability. As well as, focused on innovation and transformation. Combing the two research fields for analysis, showed that this topic is one of the new trends in literature and research is still going on to fill the gaps of the field. Also, the analysis indicated that the rise of the research area is because Covid-19 is a worldwide pandemic that all countries and organizations have suffered from. Therefore, the term digital transformation was one of the aspects that organizations had pay attention to in order to achieve a resilient organization.

Computer science was the largest subject area that published documents in regards to the research topic. As from the geographical lens, Germany and the United States published done research on the topic and published the most documents out of the other countries with a total of 19 documents each.

## 7  Conclusion

In a world of technologies, digital transformation is widely spreading between different countries. Many organizations have moved their attention to implementing the digital transformation to be more resilient when facing sudden turbulent events. Since this aspect of research is essential to many countries, it is a call for other countries to consider the topic in future research. As well as, the link between the resulted keywords to address new gap in literature that will have an impact on digital transformation and organizational resilience. The results demonstrated that digital transformation has a positive impact on organizational resilience and by consequence will improve organizational performance. From the analyzed data between digital transformation and resilience, the focus of both terms defined together has gained attention from the past five years only. This indicates the new concentration countries and organizations are moving toward. In literature combining efforts for the digital transformations to have better organizational resilience is one of the business strategies to resist unexpected global market challenges and to stay competitive during and after COVID-19 pandemic.

# References

Alexander, D.E.: Resilience and disaster risk reduction: an etymological journey. Nat. Hazards Earth Syst. Sci. **13**(11), 2707–2716 (2013)

Ali, S.: Digital transformation framework: excellence of things (EoT) for business excellence (BE) (2019)

Belhadi, A., Mani, V., Kamble, S.S., Khan, S.A.R., Verma, S.: Artificial intelligence-driven innovation for enhancing supply chain resilience and performance under the effect of supply chain dynamism: an empirical investigation.Ann. Oper. Res.126 (2021).https://doi.org/10.1007/s10479-021-03956-x

Berman, S.J., Korsten, P.J., Marshall, A.: A four-step blueprint for digital reinvention. Strategy Leadership **44**(4), 18–25 (2016)

Bongiorno, G., Rizzo, D., Vaia, G.: CIOs and the Digital Transformation: A New Leadership Role, pp. 1–9. Springer, Cham (2018)

Broo, D.G., Kaynak, O., Sait, S.M.: Rethinking engineering education at the age of industry 5.0. J. Ind. Inf. Integr. **25**, 100311 (2022)

Brynjolfsson, E., Mcafee, A., Moise-Cheung, R. Sommerfeld, B.: The Second Machine Age 1 The Digitally-Fit Organization (2016)

Bunse, B., Kagermann, H., Wahlster, W.: Industrie 4.0-smart manufacturing for the future. GTIA-Ger. Trade Inv. **40**, 12–23 (2013)

Carpenter, S., Walker, B., Anderies, J.M., Abel, N.: From metaphor to measurement: resilience of what to what? Ecosystems **4**, 765–781 (2001)

Centobelli, P., Cerchione, R., Ertz, M.: Agile supply chain management: where did it come from and where will it go in the era of digital transformation? Ind. Mark. Manage. **90**, 324–345 (2020)

Chonsawat, N., Sopadang, A.: Defining SMEs' 40 readiness indicators. Appl. Sci. **10**(24), 8998 (2020)

Collin, J., Hiekkanen, K., Korhonen, J., Halen, M., Itala, T., Helenius, M.: IT Leadership in Transition, Helsinki (2015)

Ding, D., Guan, C., Chan, C.M.L., Liu, W.: Building stock market resilience through digital transformation: using Google trends to analyze the impact of COVID-19 pandemic. Front. Bus. Res. China **14**(1), 1–21 (2020). https://doi.org/10.1186/s11782-020-00089-z

Donthu, N., Kumar, S., Mukherjee, D., Pandey, N., Lim, W.M.: How to conduct a bibliometric analysis: an overview and guidelines. J. Bus. Res. **133**, 285–296 (2021)

Fitzgerald, M., Kruschwitz, N., Bonnet, D., Welch, M.: Embracing digital technology: a new strategic imperative I capgemini consulting worldwide. MIT Sloan Manag. Rev. **55**(1), 1–13 (2013)

Foerster-Metz, U.S., Marquardt, K., Golowko, N., Kompalla, A., Hell, C.: Digital transformation and its implications on organizational behavior. J. EU Res. Bus. pp. 1–14 (2018)

Fonseca, L., Azevedo, A.L.: COVID-19: outcomes for global supply chains. Manag. Mark. Challenges Knowl. Soc. **15**(s1), 424–438 (2020)

González, I., Calderón, A.J., Portalo, J.M.: Innovative multi-layered architecture for heterogeneous automation and monitoring systems: Application case of a photovoltaic smart microgrid. Sustainability **13**(4), 2234 (2021)

Gudergan, G., Mugge, P.: The gap between the practice and theory of digital transformation. In: Hawaiian International Conference of System Science, (August), pp. 1–15 (2017)

Harvard Business Review Analytic Services. Operationalizing digital transformation: new insights into making digital transformation work (2017)

Holling, C.S.: Engineering resilience versus ecological resilience. Eng. Ecol. Constraints 31(1996), p. 32 (1996)

Jakubik, M., Berazhny, I.: Rethinking leadership and its practices in the digital Era. Venice, 471–483 (2017)

Klein, V.B., Todesco, J.L.: COVID-19 crisis and SMEs responses: the role of digital transformation. Knowl. Process. Manag. 28(2), 117–133 (2021)

Korhonen, J.J., Halén, M.: Enterprise architecture for digital transformation. In: 2017 IEEE 19th Conference on Business Informatics (CBI), vol. 1, pp. 349–358. IEEE (2017)

Kwon, E.H., Park, M.J.: Critical factors on firm's digital transformation capacity: empirical evidence from Korea. Int. J. Appl. Eng. Res. 12(22), 12585–12596 (2017)

Lincaru, C., Pirciog, S., Grigorescu, A., Tudose, G.: Low-Low (LL) high human capital clusters in public administration employment-predictor for digital infrastructure public investment priority-Romania case study. Entrepreneurship Sustain. Issues 6(2), 729 (2018)

Papagiannidis, S., Harris, J., Morton, D.: WHO led the digital transformation of your company? A reflection of IT related challenges during the pandemic. Int. J. Inf. Manage. 55, 102166 (2020)

Rapaccini, M., Saccani, N., Kowalkowski, C., Paiola, M., Adrodegari, F.: Navigating disruptive crises through service-led growth: the impact of COVID-19 on Italian manufacturing firms. Ind. Mark. Manage. 88, 225–237 (2020)

Schoemaker, P.J.H., Heaton, S., Teece, D.: Innovation, dynamic capabilities, and leadership. Calif. Manage. Rev. 61(1), 15–42 (2018)

Scholz, R.W., Czichos, R., Parycek, P., Lampoltshammer, T.J.: Organizational vulnerability of digital threats: a first validation of an assessment method. Eur. J. Oper. Res. 282(2), 627–643 (2020)

Seville, E., Brunsdon, D., Dantas, A., Le Masurier, J., Wilkinson, S., Vargo, J.: Building organisational resilience: a summary of key research findings (2006)

Swanson, D., Jin, Y.H., Fawcett, A.M., Fawcett, S.E.: Collaborative process design: a dynamic capabilities view of mitigating the barriers to working together. Int. J. Logistics Manag. 28(2), 571–599 (2017)

Trenerry, B., et al.: Preparing workplaces for digital transformation: an integrative review and framework of multi-level factors. Front. Psychol. 12, 620766 (2021)

Wade, M.: Digital business transformation. IMD and Cisco, Working Paper, pp. 1–16 (2015)

Wasono, L.W., Furinto, A., Rukmana, R.A.N.: The effect of dynamic, innovation, and alliances capability on sustainable competitive advantage in the digital disruption Era for incumbent telecommunication firm. In: Proceedings of the International Conference on Industrial Engineering and Operations Management, pp. 2111–2121 (2018)

Williams, S.P., Hausmann, V., Hardy, C.A., Schubert, P.: Managing enterprise information: meeting performance and conformance objectives in a changing information environment. Int. J. Inf. Syst. Proj. Manag. 2(4), 5–36 (2014)

Yates, T.M., Masten, A.S.:. Fostering the future: resilience theory and the practice of positive psychology (2004)

Yu, Y.: A bibliometric analysis using VOS viewer of publications on COVID-19. Ann. Transl. Med. 8(13), 816 (2020)

Zhang, J., Long, J., von Schaewen, A.M.E.: How does digital transformation improve organizational resilience?—findings from PLS-SEM and fsQCA. Sustainability 13(20), 11487 (2021)

# Metaverse

# Virtual Immersive Workplaces: The New Norm? – A Qualitative Study on the Impact of VR in the Workplace

Mahdieh Darvish[✉], Laura Keresztyen, and Markus Bick

ESCP Business School Berlin, Heubnerweg 8-10, 14059 Berlin, Germany

{mdarvish,mbick}@escp.eu, laura.keresztyen@edu.escp.eu

**Abstract.** Since the 1970s, telecommuting has generated significant interest among scholars and practitioners alike. However, the topic of flexible working arrangements has never been more relevant than lately, in the face of COVID-19 pandemic. As digital technologies evolve different aspects of today's world, immersive workplaces enabled by technologies such as virtual reality (VR) become more appealing. Therefore, the main objective of this paper is to study the phenomenon of virtual immersive workspaces and the generated impact in the organizations. Conducting semi-structured expert interviews, this study provides insights to better understand the trends driving subsequent immersive technologies, fostering competitive advantages in the web3 era, but also relevant roadblocks organizations face in this context. Our results identify the key benefits, as well as the limitations pertaining to the implementation of VR in the workplace. Moreover, our findings address the importance of purposefully designing a toolset for telecommuters to inspire further debates on the future of work for academia and in practice.

**Keywords:** Telecommuting · Virtual Workspace · Immersiveness · Virtual Immersive Workplace · Virtual Reality · Organizational Behavior

## 1 Introduction

Recently, the COVID-19 pandemic induced the need for social distancing and minimized physical presence at work. Consequently, many organizations were forced to use tools such as videoconferencing, cloud services, and virtual private networks, on an unprecedented level, forming a "new normal". Moreover, throughout the evolution of web, coined as webvolution [1], focus has shifted from access and find (Web 1.0), share, participate, and collaborate (Web 2.0) to immersive collaboration and co-creation (Web 3.0) [2] on a larger scale. Accordingly, new interfaces – defined as means of communication between user and computer or any electronic device [3] – need to be explored in order to establish a sustainable and effective "new normal" for work environments [4].

Considering the alternative workplaces, telecommuting has generated significant interest from scholars and practitioners alike, since its inception in the 70's. Varying terms and conceptualizations, such as remote work, distributed work, virtual work, etc., have been introduced since then [5]. Focusing on the work experience and employee outcomes, existing research on telecommuting has uncovered a number of predictors, mediators and moderators as follows: (1) the characteristics of the work itself, autonomy, schedule control and task interdependence [6], (2) the level of trust, social isolation, influencing knowledge sharing among employees which is critical to the development of social capital, and organizational effectiveness, and (3) information technology induced factors and the ability to transmit social cues [7]. According to the Media Richness Theory [8], different media or forms of communication have different levels of richness in the information that they provide. For instance, tools such as e-mail lack social richness as gestures and emotions are difficult to transmit, hence the success of video tools that can convey some social cues [8]. Indeed, the type of medium used to communicate with the user is just as important as the content presented by the medium itself [9]. However, as the traditional ways of communication have been disrupted due to the rise of new interfaces, the academic research, to this date, lacks the focus on the technological infrastructures within telecommuting, spotlighting "immersiveness". Therefore, this paper combines research on telecommuting and immersiveness through technological solutions of Virtual Reality (VR), and draws on studies from a broad range of fields, with a main focus on management, computer sciences and psychology. More specifically, we explore the adoption of VR in organizations, and study the impact of digital immersiveness in the workplace by answering the following research question: *How do immersive environments impact the work experience in organizations?*

The main objective of our research is to gain a deeper understanding of the perceptions of immersiveness in the workplace. Therefore, we apply a qualitative research approach through an explorative literature review. Semi-structured expert interviews are then conducted to answer the aforementioned RQ within a real-life context. Our study contributes to the body of knowledge on immersive tools, especially VR, by exploring the driving reasons to accelerate the adoption and suggesting approaches to measure the success as well as investigating the significant limitations of this technology in organizations. Moreover, our results indicate that the user experience is pivotal and involves all the key benefits, but also drawbacks, of VR adoption in workplace.

The remainder of this paper is organized as follows: The literature review provides an overview of the academic theoretical framework of telecommuting and a depiction of the benefits and drawbacks identified so far. This is complemented by expanding on the growing demand and macro trends driving the conversation on the future of work. Section 3, describes the methodological approach. Finally, the results and discussion section reveals the major findings and the concluding section addresses the limitations as well as future research.

## 2   Literature Review

Workplace, defined by Jackson and Suomi (2002) as a social entity [10], is not only merely the sum of elements of production, processing and outcomes but also an extensive social environment, where colleagues share interactions affecting each other, and ultimately, the quality of the work produced [11]. Reflecting on Organizational Behavior (OB) – defined as a field of study devoted to understanding, explaining, and ultimately improving the attitudes and behaviors of individuals and groups in organizations [12] – a key component of workplace experiences is communication, through which much of the work in a team is accomplished [13].

In the context of telecommuting, the effectiveness of communication is critical and can be influenced by the competence of the sender and receiver as well as noise, information richness, and network structure [13]. Evidently, virtual teams often have limited communication bandwidth, requiring employees to invest more effort compared to human regular communication patterns, which are transferred through body language and non-verbal cues [14]. Meanwhile, as the economy shifts from a manufacturing to an information economy, the number of telecommuting work possibilities is growing consequently [15]. Therefore, with the rise of new computer interfaces, communication sciences need to be further developed so that organizations can facilitate communication efficiently, through choosing the right media [16].

Considering the multifactorial and complex impact of telecommuting, in this paper, we follow the definition of telecommuting as an alternative work arrangement in which employees can replace or substitute work environments, away from a central workplace, through the use of information communication technology, for at least some portion of their work schedule [5, 17, 18]. Implications of telecommuting for employees have been studied, ranging from work-family issues, attitudes, and work outcomes (including job satisfaction, organizational commitment and identification, stress, performance, wages, withdrawal behaviors, and firm-level metrics) [5]. Examples of benefits are as follows: (1) significantly lower work-role stress and work exhaustion [19], (2) reduced commute times, (3) positively associated job performance and productivity [17] and (4) better-suited workforce as the most qualified individuals can be recruited [20]. However, a number of potential drawbacks have also been identified, such as reduced face-to-face communication [17], increased experience of loneliness [21], social and professional isolation [22], decreased knowledge sharing [23], unclear boundaries between work and family roles [5] or even negative career consequences [24]. In addition to the academic interest, telecommuting has also generated great public debates due to the COVID-19 pandemic-induced need for social distancing and minimizing physical presence at work [25]. Companies with a high level of IT endowment, output-oriented coordination, and experience in providing flexible working hours adapted better to the quarantine and social distancing measures introduced in many countries under the influence of the pandemic [26]. The rapid adaptation of teleworking strategies globally showed how the concept boosted team productivity and creativity with many team managers reporting positive experiences [26]. Evidently, new business practices are needed to establish a sustainable and effective new normal [4]. Indeed, developments in ICTs facilitate more suitable equipment for employees to work outside office spaces [27]. However, higher demands of ICT applications and an reduced rate of face-to-face interactions between team members might impact the subjective work experiences of employees and their motivation to participate in that given team [28]. As a possible solution to these issues, immersive virtual reality technologies, metaverse platforms, and the usage of 5G technologies may enhance teleworking experiences.

The idea of virtual reality arose in the mid-1960s, as a window through which a user perceives the virtual world as if it looked, felt and sounded real and in which the user could act realistically [29]. Varying definitions have been formulated since then, however, all definitions emphasize three common characteristics of virtual reality systems: immersion, presence, and interaction with that environment [30–34] which have a significant impact on the user experience [35, 36]. Today VR is successfully employed for a rather vast range of applications, due to its ability to induce significant improvements and increase effectiveness in various fields such as engineering, medicine [37], design, architecture [38] and construction, education [39], learning and social skills training [40], arts, entertainment, business, communication, marketing, military, and exploration [41].

## 3 Methodology

In order to understand the current state of research on the impact of immersive technologies on the workplace, we adopted an explorative approach. Starting with a brief review of related works, semi-structured expert interviews were conducted to approach the main topics from different viewpoints and expand the data collection by providing time and format for crystallization on practical insider knowledge [42], thus enriching the research [43].

We applied a purposeful sampling strategy to recruit interview partners (Table 1), including the perspectives of technology providers, consultants and researchers who work with VR and experience or study the implications.

**Table 1.** Overview: Interview partners.

| Interviewee | Position | Type of Company |
|---|---|---|
| IP 1 | Global Lead - Go To Market Strategy | Software developing |
| IP 2 | Strategy, XR & Metaverse | Consulting |
| IP 3 | Founder & CEO | Software developing |
| IP 4 | Managing Director: Innovation/AI & Emerging Technology Lead | Consulting |
| IP 5 | Strategic Account Executive | Software developing |
| IP 6 | Founder & CEO | Software developing |
| IP 7 | Head of Sales | Software developing |
| IP 8 | PhD Candidate on User Experience of Virtual Reality | Research |

All participants are currently working on implementing VR in the workplace and were approached through personal contacts or reaching out to industry leaders via LinkedIn. In total, 8 semi-structured interviews were conducted online, in English, lasting 20–56 min. All interviews were recorded, transcribed, and coded accordingly.

### 3.1 Data Analysis

In order to interpret our data, we applied qualitative content analysis [44] and coded each sentence [45]. Thereafter, we used Gioia methodology to categorize codes into 1st-order concepts, distilled into 2nd-order themes, and finally creating aggregate dimensions, turning the full transcript into manageable units [46]. The coding process was supported by ATLAS.ti, which is a qualitative computer software package to manage textual, graphical, audio, and video data. Figure 1 illustrates our data structure as follows.

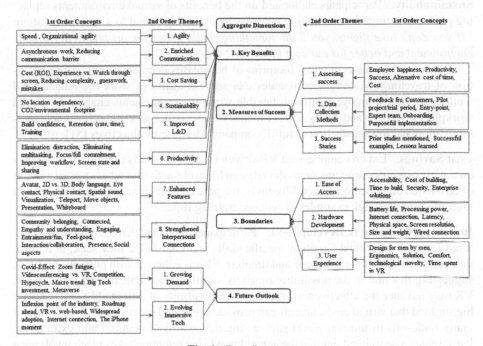

**Fig. 1.** Data Structure.

## 4 Finding and Discussion

Following the Gioia method in our data analysis, we identified four aggregated dimensions regarding virtual immersive workplaces as following: key benefits, measures of success, limitations and future outlook.

### 4.1 Key Benefits

**Agility.** Virtual immersive workplaces provide organizations with the capability to respond and adapt to changing circumstances in a quicker and more resourceful manner. Participants pointed out the opportunity to foster the business agility as a solution for maintaining competitive advantage: *"You are actually faster, then if you work in 2D, you actually arrange information better." (IP 3)*

This becomes increasingly relevant in times of uncertainty and complexity, such as the COVID-19 pandemic [47].

**Enhanced Communication.** Experts highlighted the relevance of having the option of working and communicating both synchronously and asynchronously in teams. The virtual space saves progresses, remembers exactly where team members left the work and the project can carry on 24/7. This finding is consistent with literature that considers virtual environments supporting synchronous and asynchronous collaboration to increase the quality of communication, knowledge sharing and interactions among different stakeholders and multidisciplinary teams [48–50].

**Sustainability.** Participants elaborated on the benefits of virtual environments replacing physical settings for conducting business by not being bound to a specific location: *"If you don't take flights, you'll do something that's better for people, better for the environment and better for our cost base." (IP 4)*

Consequently, the ecological footprint of businesses decreases through the reduction of travelling. Moreover, companies can address issues of work-life-balance and wellbeing of the employees through flexible working arrangements enabled by virtual workplaces. Azeem and Kotey (2021) identify these issues as critical for employee's motivation and job satisfaction to fulfill company's long-term objectives [51].

**Cost Savings.** Experts emphasized VR-driven financial rewards ranging from savings on travel costs, infrastructure, or product design lifecycle-activities where a great amount of cost to the organization incurs. Moreover, the pure cost efficiency has been mentioned multiple times through examples of reduced training time.

**Improved Learning & Development.** Participants explained that trainings can be tailored for each individual targeting specific skills and knowledge they seek, in a much more concise and efficient manner and timeline. This is coherent with previous findings, suggesting that due to the possibility to safely simulate real contexts and experiences, VR may advance the effectiveness, safety and accessibility of training [52–54]. Experts highlighted that virtual environments can provide a repeated practice possibility without many trade-offs in the real world such as injuries, waste of material and extra costs. Participants also pointed out that the actual knowledge retention takes place involving a number of senses leading to deeper formed memories: *"Designed to train people usually in dangerous jobs, jobs that if they fail in the real world, people die. So now they can do it safely." (IP4)*

**Productivity.** All participants underlined productivity, as one of the main benefits stemming from purposefully replacing the real world environment with stimulating VR, which is distraction-minimizing. This finding represents important motivating factors to further tap into the potentials of VR in the workplace, as productivity is directly related to the ability of a system to generate profits [55]. Unlike teleconferencing, VR users benefit from the feeling of co-presence, a sense of being together, a simulation of being in the same room. Thanks to more "psychological connection of minds" [56] this can lead to improved teamwork and productivity as well: *"I've never spent 45 min- 1 h on a resourcing activity like that with my full attention." (IP7)*

*"The number one reason for VR being effective is that you can't be distracted. You got a headset on, you can't be doing your email. You can't be doing a deck, right? You can't be texting your friends or watching YouTube." (IP4)*

**Enhanced Features.** Experts emphasized elements such as handshakes, virtual hugs, gestures, eye contact, and representation through an avatar. Next to body gestures, facially communicated information is centrally important in real life social interactions [57]. Studies have shown that direct stare "evokes" the perceiver's full attention instinctively [58–60]. These observations can explain why the ability to keep eye contact and see the direction of gaze caught special attention to our participants.

**Strengthened Interpersonal Connections.** Strengthened Interpersonal Connections. Results show that in a virtual environment people do feel connected and also understand each other to a greater extent. As unique identities and avatars are visually displayed in 3D virtual environments, a sense of empathy is built among the participants during a virtual meeting [61].

### 4.2 Measures of Success

**Assessing Success.** Experts highlighted, the usefulness of a virtual environment is generally assessed in terms of practical success, such as knowledge transfer and achievement goals, or the measurable return on investment. The reason behind this is the feasibility of measuring these impacts. Furthermore, experts highlighted the transformative impacts, such as generating long-term competitiveness and the power of transformative technology in the long run: *"It would, of course, depend on the reason why it is implemented and the success metrics of the unique company." (IP8)*

**Data Collection Methods.** As stated above, the question of what is a success metric varies according to the business objective of the specific company; hence the form of collecting information varies as well. However, our experts suggested both quantitative and qualitative approaches: (1) retrieving figures from the past that can be compared with current figures to evaluate the impact, such as the cost of an error or cost of travel, (2) measuring the progression over time, for example performance applying a specific skill which is acquired through VR, (3) collecting open and honest feedback through anonymous questionnaires and qualitative interviews investigating how employees perceived the use and usefulness of the technology.

**Success Stories.** Interviewees mentioned many real-world success stories as examples and inspirational ideas to measure the return on VR adoption in the organization. One of the respondents described an example from the Architecture, Engineering and Construction industry (AEC) in which through adding an extra dimension to the digital model reviews, new errors in both construction and model are frequently discovered: *"Instantly within 20 min, they found like 20 errors that they couldn't see when they came onto the screen". (IP1)*

These are errors that occur in large projects, which would otherwise only be discovered later in the process. Extra costs incurs quickly, especially if the error is not detected before the job is done physically on the construction site. The ROI of the implementation of VR arguably pays itself off already after the first use. One of the experts from the consulting world stressed to look at the cost of travel compared to the cost of a headset. As the infrastructure of VR becomes more and more affordable, a virtual conference in

VR can pay off already in the first meeting. Likewise, in the field of learning and development multiple examples of success have been mentioned. In an insurance company, employees have been divided into two distinct groups, training in a traditional setting versus VR. This took place over a 24-month period of time, and the results showed that the people that were trained in VR performed 20% better than those that were trained traditionally. Similar scenarios illustrate that VR training could achieve three times higher productivity and enable senior employees to return to work sooner, hence decreasing revenue loss of the organization.

### 4.3 Boundaries

**Ease of Access.** According to the experts, although the Covid-19 pandemic created an increased demand to adopt VR, it also led to government-imposed lockdowns of manufacturing and other facilities. Furthermore, the workplace usage of VR bears different needs than consumer use cases, as some of the interviewees stated. In addition to the hardware issue, mainstream adoption has been hampered far by the limitation of building software solutions. Despite the fact that many software packages are available off the shelf, many firms tailor software to their specific needs. This process might both take long and become comparably expensive. Moreover, enterprise adoption is challenged due to concerns of security. In specific, there is a limitation towards securing data after they are introduced onto their network environments.

**Hardware Development.** Interviewees noted that in order to achieve large scale adoption, headset development is crucial. Noteworthy recurring variables mentioned were: battery life, tracking, processing power, latency, screen resolution, size, and weight (IP1, IP5, IP6).

*"Hardware has to catch up. Hardware has become more comfortable and has become more powerful. The battery life has to be better". (IP5)*

Beyond that, it has been underlined that the technological advancements in tracking largely contribute to the perceived feeling of users. Tracking allows recording the position and orientation of real objects in physical space and transferring it to VEs, so that there is spatial consistency between real and virtual objects. The more accurate the tracking is, the better the interaction in a VE is [62].

**User Experience.** One of the most mentioned complaints has been the subjective experience of discomfort while wearing the headsets. According to our participants, current adaptors often experience discomfort due to wearing the head mounted display. Moreover, the mental phenomena side of comfort, the so-called "feel-good" factor, has been highlighted as secondary selection criteria of any VR technology. Experts pointed out that although in technology solution selection process organizations primarily evaluate tangible aspects such as features or security, at the final stage of the selection, the feelings associated with the experience play a crucial role (IP3, IP6). The UI/UX paradigms of 3D spatial computing are still new to users and hence the dynamics still to be learned impose another layer of challenge to the users. The majority of employees tend to be novice users, meaning they are inexperienced using the skills and knowledge required to seamlessly operate in VEs. Another factor of limitations imposed by human character is the reluctance to change. In the words of one of our experts: *"people don't like change."* *(IP 1).*

Researchers have been studying the number of reasons why various symptoms including nausea, difficulty concentrating, dizziness, headache, sore/aching eyes, and etc. [63] arise both during and after the VE experience, often referred to as cyber sickness. However, as of today, there are no definitive answers due to human complexity and a number of individual factors such as gender, age, illnesses and position in the simulator that all can influence the usability of the technology [64]. Furthermore, our results pointed out the user experience in terms of user resistance and acceptance involve every key benefits, but also drawbacks, as well as measures of success and future outlook of immersive technologies in the work place. This finding is consistent with the Unified Theory of Acceptance and Use of Technology that holds to factors such as Gender, age, experience, and voluntariness of use to moderate the impact on usage intention and behavior of users towards an information technology [65].

### 4.4  Future Work

**Growing Demand.** According to the experts, demand has been shaped by organic growth and by external shock. Due to technological advancements in recent years, there are already industries such as the simulation industry, where the gains of adoption are so evident that it becomes a necessity to maintain competitive advantage rather than an opt-in option: *"People have invested so much in these remote tools and they've gotten so used to being able to work remotely, and there is now a demand by many employees that they can have the flexibility for where they work. I think it has fundamentally changed how people perceive these types of technologies."* *(IP6)*

Another expert pointed out: *"Everybody has felt the pain and everybody is kind of aware that this new technology isn't so niche gaming anymore."(IP3)*

However, due to the Covid-19 pandemic, the industry observed a sudden surge of organizations looking to solve the lack of social interactions, engagement, and videoconferencing fatigue.

**Evolving Immersive Technology.** The experts agreed that the adoption will be fueled by technological advancements and compared this evolution to the adaptation of computers. However, the adoption of extended reality technologies is still in the early stages, where the actors today can be categorized as early adopters. Nevertheless, the inflexion point is nearing, and once the industry arrives at the so called "iPhone moment" it will take less than 10 years to have full enterprise saturation, similar to the current situation with smartphones (IP4). Experts see new entrants to further drive the market in short-term, increasing the competition, and providing new solutions with possibly faster release cycles. In the long run, participants foresee solutions that can respond to the aforementioned limitations and foster wide scale adoption through mixed reality seamlessly interweaving in our everyday lives.

# 5  Conclusion, Limitations and Future Research

The findings of this research contribute to a greater understanding of virtual immersive tools such as VR in the workplace. As the flexible working arrangements are becoming increasingly relevant, the role of information technologies in organizations evolves further in the context of telecommuting. A brief overview of the academic theoretical framework of telecommuting and a depiction of the benefits and drawbacks are provided with a focus on immersive workplaces enabled by VR. Furthermore, the growing demand and macro trends driving the conversation on the future of work are identified. Moreover, eight representatives with different backgrounds and expertise in the field of VR were interviewed, offering their insights and outlining the adaption of immersive tools in the workplace. At this point, the Gioia data structure and developed aggregate dimensions shed light on the impact of immersive environments on the work experience in the organizations from four aspects of key benefits, measures of success, boundaries and future outlook. In a nutshell, immersive virtual environments enable more engaging communication and collaboration, and provide users with capabilities to achieve higher productivity, in a cost-effective way. However, while the demand for such technology is rising, the provider side must catch up and address the limitations regarding the user experience that are holding back wide-scale adoption. Importantly, the expert interviews indicated that the "VR revolution" in the workplace is only at the early adoption phase and that changes in the years to come will be paramountcy fueled by technological advancements that have long been awaited. Thus, the research sets out those organizations, especially employing knowledge workers, already have a wide range of possibilities to benefit from this technology. Conclusively, organizations are recommended on a case-by-case basis; actively engaging in evaluating the implications of such technology in their field of business and workplace.

This research, like research in general has certain limitations. First, the research was limited to eight interviews with experts in the virtual reality industry. Future empirical evidence from employees, and organizations already using virtual reality technology would allow more generalization. Furthermore, building upon the results provided herein, future research may narrow down the technology and its impacts on the specific industries to develop deeper insights into potentials and challenges. Finally, considering the exploratory nature of this study, in-depth surveys and analyses can be conducted to confirm the findings. Besides, as this study primarily focuses on the technological aspect of telecommuting, the future research could link the impacts observed with different managerial approaches to better capture the complete picture of beneficial potentials of flexible working arrangements rather than conventional office work.

# References

1. Kapp, K.M., O'Driscoll, T.: Learning in 3D: Adding a New Dimension to Enterprise Learning and Collaboration. John Wiley & Sons, Hoboken (2009)
2. Bulu, S.T.: Place presence, social presence, co-presence, and satisfaction in virtual worlds. Comput. Educ. **58**, 154–161 (2012)
3. Chandler, D., Munday, R.: A Dictionary of Media and Communication. OUP Oxford, Oxford (2011)

4. Xiao, Y., Becerik-Gerber, B., Lucas, G., Roll, S.C.: Impacts of working from home during COVID-19 pandemic on physical and mental well-being of office workstation users. J. Occup. Environ. Med. **63**, 181 (2021)
5. Allen, T.D., Golden, T.D., Shockley, K.M.: How effective is telecommuting? Assessing the status of our scientific findings. Psychol. Sci. Public Interest **16**, 40–68 (2015)
6. Golden, T.D., Veiga, J.F.: The impact of extent of telecommuting on job satisfaction: resolving inconsistent findings. J. Manag. **31**, 301–318 (2005)
7. Cascio, W.F., Aguinis, H.: 3 Staffing twenty-first-century organizations. Acad. Manag. Ann. **2**, 133–165 (2008)
8. Daft, R.L., Lengel, R.H.: Organizational information requirements, media richness and structural design. Manage. Sci. **32**, 554–571 (1986)
9. McLuhan, M., Fiore, Q.: The Medium is the Message. New York, vol. 123, pp. 126–128 (1967)
10. Jackson, P.J., Suomi, R.: eBusiness and Workplace Redesign. Routledge, Milton Park (2002)
11. Sackett, P.R., Berry, C.M., Wiemann, S.A., Laczo, R.M.: Citizenship and counterproductive behavior: clarifying relations between the two domains. Hum. Perform. **19**, 441–464 (2006)
12. Colquitt, J., Lepine, J.A., Wesson, M.J.: Organizational Behavior: Improving Performance and Commitment in the Workplace (4e). McGraw-Hill, New York (2014)
13. Langan-Fox, J.: Communication in organizations: speed, diversity, networks, and influence on organizational effectiveness, human health, and relationships. Ind. Work Organ. Psychol. 2188 (2001)
14. Mandal, S.: Brief introduction of virtual reality & its challenges. Int. J. Sci. Eng. Res. **4**, 304–309 (2013)
15. Kizza, J.M.: New frontiers for computer ethics: cyberspace. In: Ethical and Social Issues in the Information Age. TCS, pp. 241–264. Springer, Cham (2017). https://doi.org/10.1007/978-3-319-70712-9_12
16. Cortellazzo, L., Bruni, E., Zampieri, R.: The role of leadership in a digitalized world: a review. Front. Psychol. **10**, 1938 (2019)
17. Gajendran, R.S., Harrison, D.A.: The good, the bad, and the unknown about telecommuting: meta-analysis of psychological mediators and individual consequences. J. Appl. Psychol. **92**, 1524 (2007)
18. Bélanger, F., Watson-Manheim, M.B., Swan, B.R.: A multi-level socio-technical systems telecommuting framework. Behav. Inf. Technol. **32**, 1257–1279 (2013)
19. Sardeshmukh, S.R., Sharma, D., Golden, T.D.: Impact of telework on exhaustion and job engagement: a job demands and job resources model. N. Technol. Work. Employ. **27**, 193–207 (2012)
20. Jarvenpaa, S.L., Shaw, T.R., Staples, D.S.: Toward contextualized theories of trust: the role of trust in global virtual teams. Inf. Syst. Res. **15**, 250–267 (2004)
21. Kaushik, M., Guleria, N.: The impact of pandemic COVID-19 in workplace. Eur. J. Bus. Manag. **12**, 1–10 (2020)
22. Feldman, D.C., Gainey, T.W.: Patterns of telecommuting and their consequences: framing the research agenda. Hum. Resour. Manag. Rev. **7**, 369–388 (1997)
23. Taskin, L., Bridoux, F.: Telework: a challenge to knowledge transfer in organizations. Int. J. Human Res. Manag. **21**, 2503–2520 (2010)
24. Shockley, K.M., Allen, T.D.: When flexibility helps: another look at the availability of flexible work arrangements and work–family conflict. J. Vocat. Behav. **71**, 479–493 (2007)
25. Organisation for Economic Co-operation and Development: Teleworking in the COVID-19 Pandemic: Trends and Prospects. OECD Publishing (2021)
26. Tokarchuk, O., Gabriele, R., Neglia, G.: Teleworking during the COVID-19 crisis in Italy: evidence and tentative interpretations. Sustainability **13**, 2147 (2021)

27. Taser, D., Aydin, E., Torgaloz, A.O., Rofcanin, Y.: An examination of remote e-working and flow experience: The role of technostress and loneliness. Comput. Hum. Behav. **127**, 107020 (2022)
28. Suh, A., Lee, J.: Understanding teleworkers' technostress and its influence on job satisfaction. Internet Res. (2017)
29. Sutherland, I.: The ultimate display (1965)
30. Slater, M., Usoh, M.: Body centred interaction in immersive virtual environments. Artif. life Virtual Reality **1**, 125–148 (1994)
31. Skalski, P., Tamborini, R.: The role of social presence in interactive agent-based persuasion. Media Psychol. **10**, 385–413 (2007)
32. Sundar, S.S., Xu, Q., Bellur, S.: Designing interactivity in media interfaces: a communications perspective. In: Proceedings of the SIGCHI Conference on Human Factors in Computing Systems, pp. 2247–2256
33. Bailenson, J.N., Yee, N., Merget, D., Schroeder, R.: The effect of behavioral realism and form realism of real-time avatar faces on verbal disclosure, nonverbal disclosure, emotion recognition, and copresence in dyadic interaction. Presence Teleoperators Virtual Environ. **15**, 359–372 (2006)
34. Blascovich, J., Loomis, J., Beall, A.C., Swinth, K.R., Hoyt, C.L., Bailenson, J.N.: Immersive virtual environment technology as a methodological tool for social psychology. Psychol. Inq. **13**, 103–124 (2002)
35. Burdea, G.C., Coiffet, P.: Virtual Reality Technology. John Wiley & Sons, Hoboken (2003)
36. Burdea, G., Richard, P., Coiffet, P.: Multimodal virtual reality: input-output devices, system integration, and human factors. Int. J. Human-Comput. Interac. **8**, 5–24 (1996)
37. Gallagher, A.G., et al.: Virtual reality simulation for the operating room: proficiency-based training as a paradigm shift in surgical skills training. Ann. Surg. **241**, 364 (2005)
38. Song, H., Chen, F., Peng, Q., Zhang, J., Gu, P.: Improvement of user experience using virtual reality in open-architecture product design. Proc. Inst. Mech. Eng. Part B: J. Eng. Manuf. **232**, 2264–2275 (2018)
39. Englund, C., Olofsson, A.D., Price, L.: Teaching with technology in higher education: understanding conceptual change and development in practice. High. Educ. Res. Dev. **36**, 73–87 (2017)
40. Schmidt, M., Beck, D., Glaser, N., Schmidt, C.: A prototype immersive, multi-user 3D virtual learning environment for individuals with autism to learn social and life skills: a virtuoso DBR update. In: Beck, D., et al. (eds.) Immersive Learning Research Network. Communications in Computer and Information Science, vol. 725, pp. 185–188. Springer, Cham (2017). https://doi.org/10.1007/978-3-319-60633-0_15
41. Saggio, G., Ferrari, M.: New trends in virtual reality visualization of 3D scenarios. Virtual Reality-Human Computer Interaction (2012)
42. Bogner, A., Littig, B., Menz, W.: Interviewing Experts. Springer, Cham (2009)
43. Mason, J.: Mixing methods in a qualitatively driven way. Qual. Res. **6**, 9–25 (2006)
44. Schreier, M.: Qualitative Content Analysis in Practice. Sage Publications, New York (2012)
45. Mayring, P.: Qualitative Content Analysis. Forum Qual Soc Res 1. 2000 (2014)
46. Gioia, D.A., Corley, K.G., Hamilton, A.L.: Seeking qualitative rigor in inductive research: notes on the Gioia methodology. Organ. Res. Methods **16**, 15–31 (2013)
47. Mathiassen, L., Pries-Heje, J.: Business agility and diffusion of information technology. Eur. J. Inf. Syst. **15**, 116–119 (2006)
48. Narasimha, S., Dixon, E., Bertrand, J.W., Madathil, K.C.: An empirical study to investigate the efficacy of collaborative immersive virtual reality systems for designing information architecture of software systems. Appl. Ergon. **80**, 175–186 (2019)

49. Pedersen, G., Koumaditis, K.: Virtual reality (VR) in the computer supported cooperative work (CSCW) domain: a mapping and a pre-study on functionality and immersion. In: Chen, J.Y.C., Fragomeni, G. (eds.) HCII 2020. LNCS, vol. 12191, pp. 136–153. Springer, Cham (2020). https://doi.org/10.1007/978-3-030-49698-2_10

50. Wolfartsberger, J., Zenisek, J., Sievi, C.: Chances and limitations of a virtual reality-supported tool for decision making in industrial engineering. IFAC-PapersOnLine **51**, 637–642 (2018)

51. Azeem, M.M., Kotey, B.: Innovation in SMEs: the role of flexible work arrangements and market competition. Int. J. Human Res. Manag. 1–36 (2021)

52. Guo, Z., et al.: Applications of virtual reality in maintenance during the industrial product lifecycle: a systematic review. J. Manuf. Syst. **56**, 525–538 (2020)

53. Leyer, M., Aysolmaz, B., Brown, R., Türkay, S., Reijers, H.A.: Process training for industrial organisations using 3D environments: an empirical analysis. Comput. Ind. **124**, 103346 (2021)

54. Wen, J., Gheisari, M.: Using virtual reality to facilitate communication in the AEC domain: a systematic review. Constr. Innov. **20**, 509–542 (2020)

55. Muttaqin, B.I., et al.: VR applications in ergonomic and productive workplace design: a literature review. In: International Seminar on Information and Communication Technologies, pp. 70–82

56. Nowak, K.: Defining and differentiating copresence, social presence and presence as transportation. In: Presence 2001 Conference, Philadelphia, PA, pp. 686–710. Citeseer

57. Syrjämäki, A.H., Isokoski, P., Surakka, V., Pasanen, T.P., Hietanen, J.K.: Eye contact in virtual reality–a psychophysiological study. Comput. Hum. Behav. **112**, 106454 (2020)

58. Conty, L., George, N., Hietanen, J.K.: Watching Eyes effects: When others meet the self. Conscious. Cogn. **45**, 184–197 (2016)

59. Lyyra, P., Astikainen, P., Hietanen, J.K.: Look at them and they will notice you: distractor-independent attentional capture by direct gaze in change blindness. Vis. Cogn. **26**, 25–36 (2018)

60. von Grünau, M., Anston, C.: The detection of gaze direction: a stare-in-the-crowd effect. Perception **24**, 1297–1313 (1995)

61. Lanier, J.: Dawn of the New Everything: Encounters with Reality and Virtual Reality. Henry Holt and Company, New York (2017)

62. Rolland, J.P., Davis, L.D., Baillot, Y.: A survey of tracking technologies for virtual environments. In: Fundamentals of Wearable Computers and Augmented Reality, pp. 83–128. CRC Press (2001)

63. Yu, M., Zhou, R., Wang, H., Zhao, W.: An evaluation for VR glasses system user experience: the influence factors of interactive operation and motion sickness. Appl. Ergon. **74**, 206–213 (2019)

64. LaViola, J.J., Jr.: A discussion of cybersickness in virtual environments. ACM Sigchi Bull. **32**, 47–56 (2000)

65. Venkatesh, V., Morris, M.G., Davis, G.B., Davis, F.D.: User acceptance of information technology: toward a unified view. MIS Q. 425–478 (2003)

# An Educational Metaverse Experiment: The First On-Chain and In-Metaverse Academic Course

Marinos Themistocleous[✉], Klitos Christodoulou, and Leonidas Katelaris

University of Nicosia, Nicosia, Cyprus
{themistocleous.m,christodoulou.kl,katelaris.l}@unic.ac.cy

**Abstract.** The Metaverse is not a new concept; nor its use for educational purpose. Despite its considerable attention and recent popularity, largely due to Facebook's strategic bet, educators have been experimenting with such an environment for a while. Still though, educational activities that are delivered over a Metaverse environment require further exploration to fully understand their full potential, opportunities, and challenges, since such activities differ from traditional on-campus or e-learning education. Nowadays, the area of on-chain education in the Metaverse remains unexplored. In this paper, we are presenting the findings from our attempt to deliver the first of its kind on-chain and in the Metaverse course delivered by the University of Nicosia (UNIC). The course was offered in Fall 2022 as a Massive Open Online Course (MOOC) and was attended by 22,500 students. The paper reports interesting observations from this experiment highlighting lessons learned and open issues for further research. The findings reveal that a Metaverse-like environment that is structured for offering educational experiences to potential learners is likely to disrupt existing educational paradigms and academic practices. We are at a stage where significant amount of research and development is required to further understand and explore the opportunities for the application of a "meta" environment for education.

**Keywords:** Metaverse · on-chain · education · MOOC · experiment

## 1 Introduction

The term Metaverse was coined in 1992, by Neal Stephenson, in his science-fiction novel "Snow Crash" (Stephenson 1992). Since then, the term is used to refer to a virtual reality-based successor of the Internet. However, as discussed by Tlili et al. (2022), various authors have since proposed variations with different foci. Interestingly, some conceptualizations "*imply that the Metaverse does not simply combine the physical and virtual worlds; it is instead a continuity of the physical world in the virtual world to create an ecosystem that merges both worlds*" (Tlili et al. 2022). This view is in stark contrast with fears that it may "*lead to the neglect of their physical surroundings*" or "*lose pre-existing relationships and the ability to form new ones*" (Harris Poll 2022).

© The Author(s), under exclusive license to Springer Nature Switzerland AG 2023
M. Papadaki et al. (Eds.): EMCIS 2022, LNBIP 464, pp. 678–690, 2023.
https://doi.org/10.1007/978-3-031-30694-5_47

Further, Hwang and Chien (2022) highlight the fact that a Metaverse space should be shared, persistent, and decentralized. In brief, the authors suggest that such a shared virtual environment should be publicly accessible via the Internet layer and that it should not be experienced alone (it is eminently social). Regarding the decentralization aspect, the authors refer to the use of blockchain technology to ensure ownership of digital objects, safety of economic activities, and traceability of the information. Although such concerns are valid and have been discussed in the literature, and blockchain can address them, we argue that Metaverse *decentralization* should have a more ample meaning, aligned with Knox (2022), who emphasizes the fact that such a shared virtual environment should not be dominated by a single company or a powerful state entity or some government. On another note, a paper authored by Christodoulou et al. (2022) mentions that a blockchain network, can be leveraged to enable a ubiquitous and a persistent architectural layer for the Metaverse that could eventually led to the Spatial Web vision.

A group of futurists, technology architects, academics, and entrepreneurs set out a 10-year roadmap for the Metaverse (Smart et al., 2007). Given the complexity and uncertainty of the subject, they have used a scenario approach to explore possible its possible directions. In doing so, they selected two key continua likely to influence the unfolding of the Metaverse: *"the spectrum of technologies and applications ranging from augmentation to simulation; and the spectrum ranging from intimate (identity-focused) to external (world-focused)"* (Smart et al. 2006). These two orthogonal continua define four quadrants that configure four different types of Metaverse, as shown in Fig. 1.

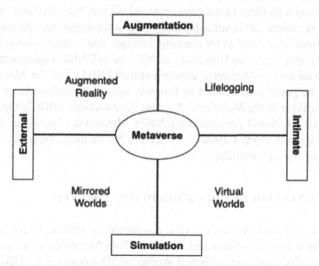

**Fig. 1.** Four Metaverse types (adapted from Smart et al. 2006)

According to Smart et al. (2006):

- **Augmentation** refers to *"technologies that add new capabilities to existing real systems; in the Metaverse context, this means technologies that layer new control systems and information onto our perception of the physical environment."*
- **Simulation** refers to *"technologies that model reality (or parallel realities), offering wholly new environments; in the Metaverse context, this means technologies that provide simulated worlds as the locus for interaction."*
- **Intimate** technologies are focused inwardly *"on the identity and actions of the individual or object; in the Metaverse context, this means technologies where the user (or semi-intelligent object) has agency in the environment, either through the use of an avatar/digital profile or through direct appearance as an actor in the system."*
- **External** technologies are focused outwardly *"towards the world at large; in the Metaverse context, this means technologies that provide information about and control of the world around the user."*

Despite the educational experiments conducted in the Metaverse, the findings in the literature point to a lack of exploitation of the technology to a higher complexity and call for higher levels of sophistication (Tlili 2022). Additionally, since the Metaverse roadmap has been published, another landmark technology has been introduced – Blockchain (Nakamoto 2008) – whose inherent decentralization matches well with the idea of having a trust layer for the Metaverse, while its characteristics (e.g., decentralization, immutability, transparency) open new possibilities for building an open Metaverse.

Since Facebook rebranded to "Meta", Metaverse has attracted a lot of attention and a race for building a plethora of use case applications had been initiated. An interesting use-case for a Metaverse environment is how to use it for delivering educational courses, share course material, as well as for creating a unique and immersive experience for the students. This paper describes University of Nicosia's (UNIC) experiment for building the first on-chain and in-Metaverse course entitled "NFTs and the Metaverse"[1]. The remainder of the paper is structured as follows: Sect. 2 provides a brief overview for delivering education in the Metaverse. Section 3, introduces UNIC's Open Metaverse Initiative (OMI). Section 4 presents the UNIC's Metaverse experiment followed by a reflection and lessons learned discussion. Section 5 concludes the paper and discusses open challenges and opportunities.

## 2   A Brief Overview of Education in the Metaverse

According to a systematic literature review conducted by Tlili et al. (2022), educational applications are not evenly distributed across the four Metaverse types as illustrated in Fig. 1. The majority make use of Virtual Worlds in 3D environments (Kye et al. 2021). These have proven useful for teaching and delivering simulation experiments that would otherwise be expensive or risky to conduct (Kanematsu et al. 2014; Siyaev and Jo 2021a, Siyaev and Jo 2021b). However, as these environments become more commonplace, such as in game environments, they have also been used in less stringent scenarios.

---

[1] https://www.unic.ac.cy/openmetaverse/mooc-nfts-metaverse/.

Conversely, Augmented Reality, Lifelogging, and Mirror Worlds Metaverse types are considerably less used, according to the literature.

An example of building Metaverse education in Virtual Worlds type is reported by Rapanotti (2010), where the authors describe the *"experience of developing a learning virtual space based on Second Life as part of an innovation project at The Open University, UK."* Second Life is a platform created by Linden Lab[2] where users create their own avatars and interact with other players and content over an on-line virtual world. The platform can also be used to support an immersive learning environment. Other experiments with similar technologies e.g., OpenSim[3] have been leveraged to create Massive Open Online Courses (MOOC) for distance education of vocational courses in Brazil (Wagner et al. 2013).

Virtual worlds have been used in various educational levels, for example such environments are mainly used (62.9%) in higher education (Tlili et al. 2022). Also, 53% of the uses of Metaverse in education are in natural science, mathematics, and engineering, motivated by the technical background and support for the disciplines. An example of STEM education is the case reported by Kanematsu et al. (2014), where a blend of real-world and Second Life-supported Metaverse are used. Furthermore, it is reported that Metaverse education can enable a higher interactive and learning experience especially when this experience is enhanced with gamification versus the more static traditional classroom setup (Estudante and Dietrich 2020). Another positive impact refers to the promotion of multilingual communication among students and better learning experience (Kanematsu et al. 2010) where the interactive 3D learning environment promotes an interesting and cooperative learning environment for the students (Barry et al. 2015).

Finally, it also worth mentioning that Metaverses can help blur the boundaries between virtual and physical learning environments, as discussed by Wang et al. (2022), who aimed to *"transform the traditional physical classroom into virtual-physical cyberspace as a new social network of learners and educators connected at an unprecedented scale."*

In summary, the literature reveals the following findings in relation to the area of applying a Metaverse environment for delivering educational activities:

- **Literature Finding 1:** Most of the existing works in this space focuses on applications related to virtual worlds that emphasize on the use of simulation technologies and have a focus on identity and interaction.
- **Literature Finding 2:** Less or limited attention is paid on the other three types of Metaverses (i.e., mirror worlds, augmented reality and lifelogging).
- **Literature Finding 3:** There are no cases reported for on-chain education in Metaverse or on-chain education in an open Metaverse space.

Based on these findings we observed that there is a gap in using decentralized technologies like blockchain to support the enrollment, distribution of educational material, and delivery of courses in Metaverse environments. As a result, we are proposing the following Research Questions (RQ):

---

[2] https://www.lindenlab.com.
[3] https://opensim.stanford.edu.

- **RQ 1:** How can we offer education on-chain and in an open Metaverse?
- **RQ 2:** How the development of an on-chain Metaverse educational environment relates to the four types of Metaverse applications?

Due to the considerable scope involved in answering these research questions, we decided to focus on RQ 1 and leave RQ 2 for future research.

Regarding the selection of a research methodology, we decided to employ a qualitive approach that focuses on action research (MacIsaac, 1995, Susman 1983). Action research is appropriate for real situations or pilot studies where the researchers are part of the team. For the investigation of RQ1 we selected the case of the University of Nicosia (UNIC) as it conducts research on this topic. In addition, the authors are part of the team that works on the development of the Metaverse education pilot study.

## 3   The University of Nicosia Open Metaverse Initiative (OMI)

The University of Nicosia is a pioneer in Blockchain, cryptocurrencies, NFTs and Metaverse education and research. Since 2013, UNIC has significantly contributed to these areas by dedicating time, human and financial resources to provide high quality research and education (Themistocleous et al. 2020, Themistocleous 2023, Themistocleous et al. 2023). As a result, it is reported as the first university in the world to:

- accept cryptocurrency as tuition fees (2013);
- launch a full accredited MSc in Blockchain and digital currency (2014);
- issue blockchain based academic certificates (2015);
- issue all University diplomas on the Blockchain since 2017;
- issue certificates in the forms of Non-Fungible Tokens (NFTs) (2021) and SoulBound Tokens (SBT) NFTs (2022);
- run the first course on-chain and in the Metaverse (2022);
- airdrop course material in the form of NFTs (2022);
- launch a full MSc in Metaverse (Fall 2023 - expected).

In Fall 2021, the University launched the UNIC Open Metaverse Initiative (OMI), a new comprehensive initiative that focuses on academic, research and policy issues relating to the Metaverse, with a particular emphasis on open public systems and standards. The UNIC OMI focuses on increasing and accelerating Metaverse adoption. To meet this goal, attention should be placed on education, sourcing, and regulation.

- **Education** as well as training, awareness, and certification, significantly increases adoption. Those who study Metaverse can learn how to participate and provide support in various stages of the lifecycle of a Metaverse product/service. In doing so, they can be part of business or technical teams spotting opportunities, conceptualise products, design, implement, offer, extend, support and gain value from its use. Education and awareness can further improve understanding in terms of the potential and the opportunities that Metaverse provides to enterprises, and society.

- **Sourcing** is vital for the take up of Metaverse. Currently there is a high demand for talents with appropriate skill set to assist organizations building their Metaverse solutions and products. This demand is expected to rise in the next years as Metaverse economy is growing fast and its market cap is estimated to reach 1 trillion US dollars by 2030.
- **Regulation** is also a key parameter that influences Metaverse adoption and development. Governments and decision-making bodies need to fully understand the nature of Metaverse and assess its potential and impact. In doing so, a critical investigation of ethical, legal, social, financial and compliance issues is required before introducing Metaverse related measurements.

UNIC OMI builds on UNIC's eight-year track record as the leading university in the field of blockchain and digital currency. Specifically, OMI focuses on the following areas/pillars which support education, sourcing, and regulation:

- academic and professional training programmes;
- research and policy;
- innovation and entrepreneurialism;
- NFTs on campus.

The aim of the first pillar is twofold: (a) to introduce *a Massive Open Online Course* entitled "NFTs and the Metaverse" and (b) to develop a new Masters programme in Metaverse Systems. The former was launched on October 3, 2022, and has attracted a lot of interest. Overall, 22,500 participants (from more than 120 countries) registered for the first iteration of the course. This is the first-of-its-kind course globally that is delivered in the Metaverse, and it is on-chain. A series of smart contracts is handling the delivery of the material for the course as non-fungible tokens that exist on-chain. The action to fulfil the second aim of the this pillar is the development of a complete MSc program in Metaverse. For this an application is submitted to the Cyprus Agency of Quality Assurance and Accreditation in Higher Education (CYQAA) for accreditation. At the time, the programme is under evaluation, and it is expected to be launched in Fall 2023.

The next section analyses the UNIC Metaverse experiment that mainly focuses on the delivery of the MOOC that took place from October 2022 to January 2023.

## 4 The UNIC Metaverse Experiment

Internally, UNIC allocated human resources (e.g., faculty, and researchers) to work in parallel to: (a) design and submit for accreditation a new MSc in Metaverse, (b) systematically develop UNIC Metaverse campus (UNIC meta-campus or Meta-U), and (c) prepare the delivery of the MOOC to speed up and spread the adoption of Metaverse. UNIC decided to offer the MOOC using a Metaverse space to test and support the development of its virtual meta campus.

The UNIC Metaverse experiment is based on the following design principles:

- **In the Metaverse:** UNIC decided to conduct a research experiment and provide novel designs for Metaverse based education. UNIC is persuaded all educational and academic activities will be disrupted by the Metaverse. Thus, it is decided that the new MOOC should leverage a 3D virtual space in the Metaverse focusing on offering a new social and interactive learning experience to all participants.
- **Decentalized, on-chain and transparent** are fundamental design principles of this experimentation. For this research, UNIC decided to run the course without using its centralized information systems. This principle introduces a series of innovative practices such as the use of NFTs. More specifically, NFTs are used as part of the registration/enrollment processes and for enabling token gated access to the learning environment. For instance, anyone could register to the course using a digital wallet (e.g., Metamask on Ethereum Mainnet). Participants can mint the course NFT on-chain and use it as a ticket to access the virtual environment, and for accessing course material. The course material is also represented on-chain as NFTs. Students can lazy mint the material as NFTs (if they wish to) by issuing a transaction on-chain.
- **Open and scalable:** The last two design principles refer to scalability and openness. UNIC initially evaluated various Metaverse environments and it finally selected the Open Metaverse platform (OM) Oncyber[4] to implement its Metaverse university campus (meta-campus). Apart from openness, the OM environment offers scalability as it allows both UNIC and its participants to: (a) scale in terms of functionality (e.g., participants per class, instances used, functions developed etc.) and (b) co-develop the virtual campus (e.g., students can create their own spaces (e.g., dormitory rooms, exhibition spaces) within UNIC's meta-campus.

**The Technological Environment of UNIC Meta-campus:** The UNIC meta-campus is an ongoing project and at the time, this is the first university campus in the world that has been built, operated, and offered on-chain education. The UNIC meta-campus resembles a real-life university campus and consists of auditoriums, teaching classes, laboratories, special learning facilities, exhibition spaces, administrative offices, faculty offices, common rooms, library, outdoor meeting areas, students' dormitories, social meeting rooms, museum etc. Figure 2 depicts the current state of the virtual meta-campus facilities. Screens 1 and 2 depict part of the UNIC meta-campus where Screen 3 shows a live session that took place at the outdoor lecture theatre and Screen 4 shows an NFT exhibition gallery. Currently, a new implementation is carried out regarding the outdoor lecture theatre that will resemble a digital twin of the real outdoor amphitheater at UNIC's main campus.

From a technological perspective, UNIC's meta-campus has been developed using Oncyber. Regarding our design principles, Oncyber is an open and scalable environment that allows us to operate in a decentralized and transparent manner and supports on-chain activities. In addition to these, Oncyber was selected based on various other criteria such as functionality, operation, customization, user friendliness. UNIC's meta-campus uses rendering techniques which make it highly scalable since the rendering happens at the client side. This rendering feature allows UNIC meta-campus to run in a decentralized

---

[4] www.oncyber.io.

Screen 1: UNIC meta-campus          Screen 2: UNIC meta-campus

Screen 3: UNIC meta-campus– outdoor lecture theatre     Screen 4: UNIC meta-campus – NFT exhibition room

**Fig. 2.** UNIC's meta-campus

way; in doing so, its performance remains the same as participants' web browser uploads the UNIC Metaverse environment. In that way, the system can host thousands of users without performance issues.

Participants enrolment: In August 2022, UNIC designed a unique NFT (Fig. 3) to be used as a free ticket to access the UNIC Metaverse campus and enrol to the course. The NFT was issued on-chain using the Ethereum ERC721 standard. Participants enrolled to the course using their own self-custodial Ethereum wallet (e.g., Metamask) and by minting the course NFT. This registration process proposes a new decentralized way to register for a course which opens new research avenues. With the decentralized enrollment, the university does not maintain any details about its participants or use any internal university system to enroll them; registration process is pseudonymous, all information is available on-chain and easily auditable with the use of a blockchain explorer (e.g., Etherscan[5]). Such practice is in line with the design principles of this experiment about decentralization, transparency and on-chain registration and delivery.

The first iteration of the course was offered in Fall 2022 and has been of great success as it attracted 22.500 students from more than 120 different countries. The data analysis reveals that only 19,440 of the participants were unique wallet addresses. This indicates that some students minted more than one NFTs as they are expecting to gain value from these tokens in the future. However, from a design point of view, the valuation of the course NFT and its future price are outside the scope of this experiment.

Learning and teaching on-chain and in the Metaverse: For this experiment the delivery of the classes was restricted to an outdoor virtual auditoriums since alternative

---

[5] https://etherscan.io/.

**Fig. 3.** NFT Access token for META511

teaching areas are under construction For experimentation purposes, users can join the UNIC meta-campus using a multiuser view or a single user view option[6].

Once students enter the virtual meta-campus environment, they are directed to the area where the live lecture takes place. To better accommodate the large number of participants during the live lectures several instances of the environment have been created utilized. We started our experimentation from maximum 300 participants per instance and extend it to 600 participants. To provide support to participants teaching assistants (TAs) were assigned to each instance. Although students' microphones were muted during the live sessions, they had the option to request access to voice, contribute to the class discussions, and ask questions. Students could also interact with their fellow students and the TAs within their instance using the live-chat service. Twitter was also used as an alternative medium for asking questions about the course logistics and material. Students could write their questions in twitter and answers are given by the UNIC staff. All Q&A sessions were open to the public domain and responses have been exposed to the public's critique enabling a transparent discussion cycle. The MOOC delivers multiple live-sessions over a period of 12 weeks, and the plan is to repeat it twice a year. All live sessions delivered in the Metaverse and have been recorded and transcribed. Apart from the instructors more than 50 world class collectors, artists, investors, thinkers, and entrepreneurs participated to the course as guest speakers. The course material is recorded and those who cannot attend the live sessions can access it on demand. All course material, like presentations and video recordings are available in the form of an NFT and students can claim it to their wallets. Additionally, the material is published in a publicly available github repository, also stored in using a peer-to-peer (P2P) file storage networks like Arweave. In a similar fashion, assignments and final examinations are represented on-chain with the use of NFT tokens. The actual data that relate to the assignments and the final examination are stored as encrypted objects in a p2p storage network (e.g., IPFS or Arweave) and various techniques are used for their distribution.

Students use their own avatars and are free to navigate in the virtual meta-campus not only during the lecture times but at any time. This allows students to meet in common

---

[6] https://oncyber.io/ommetacampus.

areas or in meeting places socialize and interact with each other similarly to a real-life campus. In doing so, they can collaborate for their assignments discuss various topics, etc. Although a comprehensive analysis for measuring the presence pf participants and their interactions with specific metrics is pending we observed that there was a high interest from participants to spend additional time either before or after the lectures to communicate with their fellow students in the meta-campus. Although promising, this observation requires further investigation and evaluation.

**Assessment:** UNIC meta campus is continuously assessed and tested to improve its quality and enhance user experience. Table 1 presents our findings from our initial evaluation and suggestions made to overcome the identified issues. These issues are classified into three categories namely: interactivity, visual, and persistence. The development team noted the issues reported (summarized in Table 1) and will soon address them all.

**Table 1.** Initial evaluation findings and suggestions

| Category | Issue | Suggestion |
|---|---|---|
| Interactivity | Single user view does not support user-to-user interactivity | Single user view should only be used for testing and experimentation. Students should not have access to it |
| Interactivity | Breakout sessions can not be implemented in the current version | Design and implement rooms for the breakout sessions. Employ teleporting mechanisms to transfer users to these designated rooms |
| Interactivity | No support for in-class polls | Further research is required. Develop software that support in-class polls |
| Interactivity | Chat: For this experimentation the classes were too big. It is therefore difficult to attend the chat discussion during the lecture as 300–600 share their views in the chat | Use smaller classes. In a regular class size this will not be a problem as there will be small number of users (e.g., maximum 30 students per class) |
| Visual | Avatars bounce during class | Control this function during class (if required by the instructor(s)) |
| Visual | Avatars are moving and disturb the class | Explore and assess various scenarios like (a) restrict users' movement; (b) offer the option to exit the area without disturbing the class; (c) offer a pre-seat option |
| Visual | Avatars shape | Replace them with a human body figure |
| Persistence | Refresh option moves students in a different instance (class) and thus cannot follow the discussion in the chat | Students to join the same instance after refresh |

# 5  Reflection and Lessons Learned

On-chain education disrupts the existing educational models: We argue that on-chain education is a disruptive innovation concept that brings new ideas and practices for implementation. We will soon observe significant developments in this area, and we predict that this on-chain educational paradigm will open new avenues for education. As a new concept, it requires time and space to evolve and flourish.

Decentralization will affect academic practices and processes: The development of Soulbound tokens (SBTs) can lead to a new generation of wallets. We envision that in future, students' digital wallet will include all relevant documents in the form of SBTs that will provide enhanced transparency. Soulbound tokens have unique characteristics as they do not have monetary value, they cannot be sold or bought and are not transferable; SBTs can be given upon the completion of an achievement (e.g., university diploma, driving license, birth certificate etc.). As such they are used to representing a person's identity and reputation for life.

Rendering has a positive impact on scalability and performance: In this experiment we focus on testing the performance of the Metaverse environment using large classes of students. To maintain a reasonable performance, the platform uses a rendering mechanism that allows the UNIC meta-campus environment to run in a decentralized way. This practice allows the system to run smoothly and with no delays during the students' interactions within the meta-campus. After login, any delay caused is attributed to the number of tasks that are executed on the end-user computer and not to the number of users that are on meta-campus that moment. Thus, delays on ones computer do not affect the overall meta-campus performance or other users. This practice allows us to provide enhanced scalability. We can easily increase the number of the participants of a course from ten students to thousands with no performance issues.

Instances are important for the management of large and too large classes, as we can split them into smaller sub-classes and assign Teaching Assistants (TAs) to support their sub-class. For this experiment we assigned several dedicated TAs to support each class. It was observed that it was easier to manage these too large classes using this practice compared to other practices. TAs coordinated the chat discussion and collected all questions. TAs answer to students' questions through chat discussion. They only ask the instructor to those questions that they could not answer themselves.

Open issues: Our experimentation, has uncovered a number of open-issues for further research including but not limited to: (a) the on-chain distribution of assignments and final examinations, (b) the restriction of public from revealing details about examination or assignments, (c) the automatic assessment and validation of students responses for exams or assignments, and (d) the specification of a method that applies generative art for the issuance of customizable and unique academic certificates based on specific attributes. Further research, experimentation, and validation is required before disclosing any results from our preliminary research on the above.

# 6  Conclusion

This paper explores the case of using a Metaverse for educational purposes. Metaverse and its applications in education are not new with the normative literature demonstrating that Metaverse education and its evolution is affected by technological innovations. An example of this, is the utilization of the computer game "Second life" and the impact it had in Metaverse education. The literature also demonstrates that the majority of the Metaverse cases focuses on virtual worlds with limited attention to the other three types of Metaverse, namely mirror worlds, augmented reality and lifelogging (see Fig. 1). Moreover, on-chain and in Metaverse education has not yet been fully explored or realized. To this end, we argue that Blockchains have the ability to provide a trusted representation layer for educational material, as well as, to support the decentralization and economy layers for Metaverse education.

Furthermore, we explored the research question: "how can we offer education on-chain and in the open Metaverse?" To investigate this RQ we adopted action research and we studied UNIC's on-chain metaverse experiment. The experiment was conducted in Fall 2022 and concentrated on the delivery of a MOOC. The experiment took place on-chain, in UNIC Metaverse virtual campus and the course was attended by 22,500 participants. The design principles of the experiment deal with decentralization, transparency, on-chain delivery in metaverse, scalability and openness. Although this is just the first case of its kind, several interesting and ground-breaking ideas were implemented and tested including: (a) students enrollment using their digital wallet and without accessing any centralized infrastructure, (b) NFT gated access to the course, (c) provision of course material in the form of non-fungible tokens that are claimable by the students, and (d) course material airdrops. Our research highlighted several issues that need to be addressed as these are presented in Table 1 and are classified into persistence, visual and user interactivity categories. All reported issues have been taken into consideration by the development team.

The case presented herein also presents a number of open issues for further research. Issues related to the exams and assignments like their distribution, privacy issues, or automatic assessment and validation are highlighted for future research. Another interesting topic is the design and implementation of a method that creates unique academic certificates and their issuance as NFTs using generative art.

# References

Barry, D.M., et al.: Evaluation for students' learning manner using eye blinking system in Metaverse. Procedia Comput. Sci. **60**, 1195–1204 (2015). https://doi.org/10.1016/j.procs.2015. 08.181

Christodoulou, C., Katelaris, L., Themistocleous, M., Christoudoulou, P., Iosif, E.: NFTs and the metaverse revolution: research perspectives and open challenges. In: Lacity, M., Treiblmaier, H. (eds.) Blockchains and the Token Economy: Theory and Practice, pp. 139–178. Palgrave Macmillan, Cham (2022)

Estudante, A., Dietrich, N.: Using augmented reality to stimulate students and diffuse escape game activities to larger audiences. J. Chem. Educ. **97**(5), 1368–1374 (2020). https://doi.org/ 10.1021/acs.jchemed.9b00933

Hwang, G., Chien, S.: Definition, roles, and potential research issues of the Metaverse in education: an artificial intelligence perspective. Comput. Educ. Artif. Intell. **3**,100082 (2022). https://www. sciencedirect.com/science/article/pii/S2666920X22000376?via%3Dihub

Kanematsu, H., Fukumura, Y., Barry, D.M., Sohn, S.Y., Taguchi, R.: Multilingual discussion in metaverse among students from the USA, Korea and Japan. In: Setchi, R., Jordanov, I., Howlett,

R.J., Jain, L.C. (eds.) KES 2010. LNCS (LNAI), vol. 6279, pp. 200–209. Springer, Heidelberg (2010). https://doi.org/10.1007/978-3-642-15384-6_22

Kanematsu, H., Kobayashi, T., Barry, D.M., Fukumura, Y., Dharmawansa, A., Ogawa, N.: Virtual STEM class for nuclear safety education in Metaverse. Procedia Comput. Sci. **35**, 1255–1261 (2014). https://doi.org/10.1016/j.procs.2014.08.224

Knox, J.: The metaverse, or the serious business of tech frontiers. Postdigital Sci. Educ. **4**(2), 207–215 (2022). https://doi.org/10.1007/s42438-022-00300-9

Kye, B., Han, N., Kim, E., Park, Y., Jo, S.: Educational applications of Metaverse: possibilities and limitations. J. Educ. Eval. Health Professionals (2021). https://www.jeehp.org/journal/view.php?doi=10.3352/jeehp.2021.18.32

Nakamoto, S.: Bitcoin: A peer-to-peer electronic cash system (2008)

MacIsaac, D.: An Introduction to action research (1995). http://www.phy.nau.edu/~danmac/act ionrsch.html

Poll, H.: (2022). https://theharrispoll.com/briefs/covid-19-tracker-wave-106/

Rapanotti, L., Hall, J.: Lessons learned in developing a second life educational environment (2010). https://www.scitepress.org/Link.aspx?doi=10.5220/0002777200330038

Smart, J., Cascio, J., Paffendorf, J.: Metaverse roadmap overview: pathways to the 3D web (2007). https://www.Metaverseroadmap.org/overview/

Siyaev, A., Jo, G.S.: Neuro-symbolic speech understanding in aircraft maintenance Metaverse. IEEE Access **9**, 154484–154499 (2021a). https://doi.org/10.1109/access.2021.3128616

Siyaev, A., Jo, G.S.: Towards aircraft maintenance Metaverse using speech interactions with virtual objects in mixed reality. Sensors **21**(6), 2066 (2021b). https://doi.org/10.3390/s21062066

Stephenson, N.: Snow Crash. Bantam Spectra Books USA (1992). ISBN: 9780553351927

Susman, G.: Action research: a sociotechnical systems perspective. In: Morgan, G. (ed.). Sage Publications, London (1983)

Themistocleous, M., Christodoulou, K., Iosif, E., Louca, S., Tseas, D.: Blockchain in academia: where do we stand and where do we go? In: The Proceedings of the 53rd Annual Hawaii International Conference on System Sciences, (HICSS 53), 7–10 January 2020, Maui, Hawaii, USA. IEEE Computer Society, Los Alamitos (2020). https://scholarspace.manoa.hawaii.edu/handle/10125/64398

Themistocleous, M.: Teaching blockchain: the case of the MSc in blockchain and digital currency of the university of Nicosia. In: Paliwoda-Pękosz, G., Soja, P., (eds.) Transforming Higher Education with Blockchain. Routlege Publishing, UK (2023)

Themistocleous, M., Christodoulou, K., Iosif, E.: Academic certificates issued on blockchain: the case of the University of Nicosia and Block.co. In: Paliwoda-Pękosz, G., Soja, P., (eds.) Transforming Higher Education with Blockchain. Routlege Publishing, UK (2023)

Tlili, A., et al.: Is Metaverse in education a blessing or a curse: a combined content and bibliometric analysis. Smart Learn. Environ. **9**, 24 (2022). https://doi.org/10.1186/s40561-022-00205-x

Wang, Y., Lee, L., Braud, T., Hui, P.: Re-shaping post-COVID-19 teaching and learning: a blueprint of virtual-physical blended classrooms in the metaverse era (2022)

Wagner, R., Piovesan, S.D., Passerino, L.M., de Lima, J.: Using 3D virtual learning environments in new perspective of education. In 2013 12th International Conference on Information Technology Based Higher Education and Training (ITHET), pp. 1–6. IEEE (2013). https://doi.org/10.1109/ithet.2013.6671019

https://about.fb.com/news/2021/10/facebook-company-is-now-meta/

# Author Index

M. Papadaki et al. (Eds.): EMCIS 2022, LNBIP 464, pp. 691–692, 2023.
https://doi.org/10.1007/978-3-031-30694-5

Printed in the United States
by Baker & Taylor Publisher Services